Cisco®:
The Complete Reference

About the Author

Brian Hill, CCNP, CCNA, MCSE+I, MCSE, MCT, INet+, Net+, and A+, currently holds the position of Lead Technology Architect and Senior Technical Trainer for TechTrain, a fast-growing training company based in Charlotte, NC. Brian has been in the computer industry since 1995, and has been an avid home-computing enthusiast since he was eight years old. In previous positions, he has shouldered responsibilities ranging from PC Technician to Senior Technical Director. Currently, he is responsible for all network design duties and all technical interviews for new staff members, and he is the technical trailblazer for the entire company. Brian also holds the distinction of being one of the first 2,000 people in the world to achieve an MCSE in Windows 2000.

Brian's Cisco background consists of over four years of in-depth, hands-on experience with various models of routers and switches, as well as teaching accelerated Cisco classes. Brian designed TechTrain's internal network, consisting of Cisco 3500 switches, Cisco 2600 and 3600 routers, Cisco PIX firewalls, Cisco CE-505 cache engines, Cisco 2948G layer 3 switches, and several HP Procurve access switches. In addition, he designed TechTrain's expansive CCNA, CCNP, and CCIE router racks, consisting of Catalyst 5500, 1900, and 3500 switches, Cisco 2600 and 3600 routers, Cisco 2511 RJ access servers, and an Adtran Atlas for POTS and ISDN emulation. He is currently pursuing his Cisco CCIE certification.

You can reach Brian through his web site at http://www.alfageek.com.

Cisco®:
The Complete Reference

Brian Hill

McGraw-Hill/Osborne

New York   Chicago   San Francisco
Lisbon   London   Madrid   Mexico City
Milan   New Delhi   San Juan
Seoul   Singapore   Sydney   Toronto

McGraw-Hill/Osborne
2600 Tenth Street
Berkeley, California 94710
U.S.A.

To arrange bulk purchase discounts for sales promotions, premiums, or fund-raisers, please contact **McGraw-Hill**/Osborne at the above address. For information on translations or book distributors outside the U.S.A., please see the International Contact Information page immediately following the index of this book.

Cisco®: The Complete Reference

1234567890 DOC DOC 0198765432

ISBN 0-07-219280-1

Publisher
  Brandon A. Nordin

Vice President & Associate Publisher
  Scott Rogers

Editorial Director
  Tracy Dunkelberger

Acquisitions Editor
  Steven Elliot

Project Editor
  Laura Stone

Acquisitions Coordinator
  Alexander Corona

Technical Editors
  Henry Benjamin, Tom Graham

Copy Editor
  Chrisa Hotchkiss

Proofreader
  Linda Medoff

Indexer
  Valerie Perry

Computer Designers
  Carie Abrew, Elizabeth Jang,
  Melinda Moore Lytle, Lauren McCarthy

Illustrator
  Jackie Sieben

Series Design
  Peter F. Hancik

This book was composed with Corel VENTURA™ Publisher.

To my wife and children: Beth, Chris, Jesse, and Tylor.

Yes, daddy is done with the book, and yes, he will leave
the computer room soon (maybe).

Contents at a Glance

Part III Cisco LAN Switching

Part IV Cisco Routing

Contents

Part I

Networking Basics

Part II

Cisco Technology Overview

Part III

Cisco LAN Switching

Part IV

Cisco Routing

Acknowledgments

Whew! It's been a long eight months, but the book is finally complete and in your hands. Although it was a lot of fun to write, I have to admit that I am glad that the writing is over (at least for a little while). I would like to take the time to thank all of the people who have helped shape this book.

To Henry Benjamin, for his insight and words of encouragement, as well as his ability to see what I was trying to accomplish and provide creative input.

To all of the people at McGraw-Hill/Osborne for help along the way and understanding as the deadlines crept up. A heck of a lot of work went into this thing, and you guys deserve a lot of the credit.

To Cisco, for cramming onto their web site nearly anything you could need to know about anything they make. Although the explanations are sometimes lacking, the quantity of information makes up for it.

To Google, for providing a great tool for searching the Cisco web site in the form of the Google toolbar. I have retained most of my hair because of them.

To Paul Piciocchi, Brad Baer, and everyone else at TechTrain, for the understanding and support along the way.

To Jeremy Beyerlein, for looking over the first few chapters and giving me his honest opinion.

To my many students, who have helped me understand how to teach people and get a point across.

To my mother, for telling me as a child that I could do anything I set my mind to.

To my wife, for believing in me when believing in me was unpopular.

To everyone who didn't believe in me. You gave me a reason to succeed.

And to everyone who chose this book out of the multitude of titles lining the bookshelves. Thank you all, and I hope this book will be a useful reference for years to come.

Introduction

C*isco: The Complete Reference* is a lofty title for a book, and one that you could take in a multitude of different directions. Some think that a book with this title should be the "end all and be all" of Cisco books, including every possible Cisco technology and the most obscure details of those technologies. Unfortunately, that book would consist of over 50,000 pages, and it would be obsolete by the time you got it. (Cisco has been trying for years to write that book; it's called the Cisco web site, and it's still not complete.)

Rather, the tactic I chose for this book was to cover the most commonly used technologies in most networks, with detailed explanations and a focus on practical understanding and use. In most cases, although obscure details are somewhat interesting, they really don't help much unless you are a contestant on "Cisco Jeopardy." Therefore, I wrote a book that I feel people have the most need for: a book designed to explain Cisco technology to the average network administrator or junior network engineer who may need to understand and configure Cisco devices. The goal of this book is not to help you pass tests (although it may do that) and not to be the final word on any subject. Rather, the goal is to give you a complete understanding of Cisco technologies commonly used in mainstream networks, so that you can configure, design, and troubleshoot on a wide variety of networks using Cisco products.

The book starts out innocently enough, beginning with Part I: Networking Basics, to give you a refresher course on LAN and WAN protocols and general-purpose protocol suites. In many cases, I also provide links to web sites to help you locate additional reading materials. I suggest that you examine Part I in detail, especially

Chapter 6 on advanced IP, even if you feel you already know the subjects covered. Without a solid understanding of the fundamentals, the advanced concepts are much harder to grasp.

Part II, Cisco Technology Overview, provides an overview of Cisco networking technologies, including references to most of the currently available Cisco networking products. In this section, I provide reference charts with product capabilities and port densities to help you quickly find the Cisco product you need to support your requirements, which I hope will save you from hours of looking up datasheets on Cisco's web site. Part II culminates with a look at common IOS commands for both standard IOS and CatOS devices.

Part III, Cisco LAN Switching, covers Cisco LAN-switching technologies. Layers 2 through 4 are covered, including VLAN configuration, STP, MLS, queuing techniques, and SLB switching. Like all chapters throughout the rest of the book, these chapters focus first on understanding the basic technology, and second on understanding that technology as it applies to Cisco devices.

Part IV, Cisco Routing, covers routing on Cisco devices. It begins with a chapter explaining the benefits and operation of static routing, and progresses through more and more complex routing scenarios before ending with a chapter on securing Cisco routers with access lists. All major interior routing protocols are covered, including RIP, EIGRP, and OSPF.

The appendix contains a complete index of all 540 commands covered in the book, complete with syntax, descriptions, mode of operation, and page numbers. This appendix is designed to be your quick reference to IOS commands on nearly any Cisco device.

Finally, many enhanced diagrams and errata can be found on my personal web site, at http://www.alfageek.com.

Thanks again, and enjoy!

The Complete Reference

Cisco

Part I

Networking Basics

In this, the first of four parts, we will go over the basics of networking. This is one of the few sections in the book where we will not be focusing solely on Cisco gear. This information is important, however, because to understand the advanced topics we will be exploring, you must have a solid grasp of the basic principles. Included in this part of the book is a section on network models, a section on LAN technologies, and a section on WAN technologies. This information will be invaluable later in the book when we look at how Cisco devices use these principles. In addition, these sections will help you understand all network environments, not just those dominated by Cisco devices. That being said, I invite you to sit back, relax, and breathe in the technology behind networking.

Chapter 1

The OSI Model

The OSI (Open Systems Interconnection) model is a bit of an enigma. Originally designed to allow vendor-independent protocols and to eliminate monolithic protocol suites, the OSI model is actually rarely used for these purposes today. However, it still has one very important use: it is one of the best tools available today to describe and catalog the complex series of interactions that occur in networking. Because most of the protocol suites in use now (such as TCP/IP) were designed using a different model, many of the protocols in these suites don't match exactly to the OSI model, which causes a great deal of confusion. For instance, some books claim that Routing Information Protocol (RIP) resides at the network layer, while others claim it resides at the application layer. The truth is, it doesn't lie solely in either layer. The protocol, like many others, has functions in *both* layers. The bottom line is, look at the OSI model for what it is: a tool to teach and describe how network operations take place.

For this book, the main purpose of knowing the OSI model is so that you can understand which functions occur in a given device simply by being told in which layer the device resides. For instance, if I tell you that physical (Media Access Control—MAC) addressing takes place at layer 2 and logical (IP) addressing takes place at layer 3, then you will instantly recognize that an Ethernet switch responsible for filtering MAC (physical) addresses is primarily a layer 2 device. In addition, if I were to tell you that a router performs path determination at layer 3, then you already have a good idea of what a router does.

This is why we will spend some time on the OSI model here. This is also why you should continue to read this chapter, even if you feel you know the OSI model. You will need to fully understand it for the upcoming topics.

What Is a Packet?

The terms *packet, datagram, frame, message,* and *segment* all have essentially the same meaning—they just exist at different layers of the OSI model. You can think of a packet as a piece of mail. To send a piece of snail mail, you need a number of components (see Figure 1-1):

- **Payload**   This component is the letter you are sending, say, a picture of your newborn son for Uncle Joe.

- **Source address**   This component is the return address on a standard piece of mail. This indicates that the message came from you, just in case there is a problem delivering the letter.

- **Destination address**   This component is the address for Uncle Joe, so that the letter can be delivered to the correct party.

- **A verification system**   This component is the stamp. It verifies that you have gone through all of the proper channels and the letter is valid according to United States Postal Service standards.

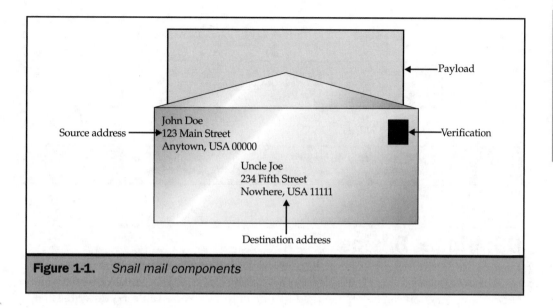

Figure 1-1.   *Snail mail components*

A packet is really no different. Let's use an e-mail message as an example—see Figure 1-2. The same information (plus a few other pieces, which we will cover later in the chapter) is required:

■ **Payload**   This component is the data you are sending, say, an e-mail to Uncle Joe announcing your newborn son.

■ **Source address**   This component is the return address on your e-mail. It indicates that the message came from you, just in case there is a problem delivering the e-mail.

■ **Destination address**   This component is the e-mail address for Uncle Joe, so that the e-mail can be delivered correctly.

■ **Verification system**   In the context of a packet, this component is some type of error-checking system. In this case, we will use the frame check sequence (FCS). The FCS is little more than a mathematical formula describing the makeup of a packet. If the FCS computes correctly at the endpoint (Uncle Joe), then the data within is expected to be valid and will be accepted. If it doesn't compute correctly, the message is discarded.

The following sections use the concept of a packet to illustrate how data travels down the OSI model, across the wire, and back up the OSI model to arrive as a new message in Uncle Joe's inbox.

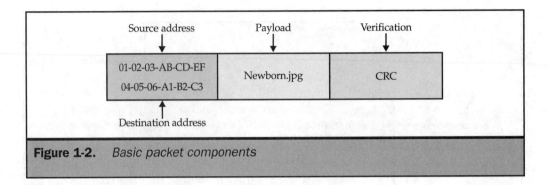

Figure 1-2. *Basic packet components*

OSI Model Basics

The OSI model is a layered approach to networking. Some of the layers may not even be used in a given protocol implementation, but the OSI model is broken up so that any networking function can be represented by one of the 7 layers. Table 1-1 describes the layers, beginning with layer 7 and ending with layer 1. I am describing them in this order because, in most cases, people tend to understand the model better if introduced in this order.

Layer	Function
Application (layer 7)	This layer is responsible for communicating directly with the application itself. This layer allows an application to be written with very little networking code. Instead, the application tells the application-layer protocol what it needs, and it is the application layer's responsibility to translate this request into something the protocol suite can understand.
Presentation (layer 6)	This layer is responsible for anything involved with formatting of a packet: compression, encryption, decoding, and character mapping. If you receive an e-mail, for instance, and the text is gobbledygook, you have a presentation-layer problem.

Table 1-1. *The Layers of the OSI Model*

Layer	Function
Session (layer 5)	This layer is responsible for establishing connections, or sessions, between two endpoints (usually applications). It makes sure that the application on the other end has the correct parameters set up to establish bidirectional communication with the source application.
Transport (layer 4)	This layer provides communication between one application program and another. Depending on the protocol, it may be responsible for error detection and recovery, transport-layer session establishment and termination, multiplexing, fragmentation, and flow control.
Network (layer 3)	This layer is primarily responsible for logical addressing and path determination, or routing, between logical address groupings.
Datalink (layer 2)	This layer is responsible for physical addressing and network interface card (NIC) control. Depending on the protocol, this layer may perform flow control as well. This layer also adds the FCS, giving it some ability to detect errors.
Physical (layer 1)	The simplest of all layers, this layer merely deals with physical characteristics of a network connection: cabling, connectors, and anything else purely physical. This layer is also responsible for the conversion of bits and bytes (1's and 0's) to a physical representation (electrical impulses, waves, or optical signals) and back to bits on the receiving side.

Table 1-1. *The Layers of the OSI Model* (continued)

When data is sent from one host to another on a network, it passes from the application; down through the model; across the media (generally copper cable) as an electrical or optical signal, representing individual 0's and 1's; and then up through the model at the other side. As this happens, each layer that has an applicable protocol adds a *header* to the packet, which identifies how that specific protocol should process the packet on the other side. This process is called *encapsulation*. See Figure 1-3 for a diagram (note that AH stands for application header, PH stands for presentation

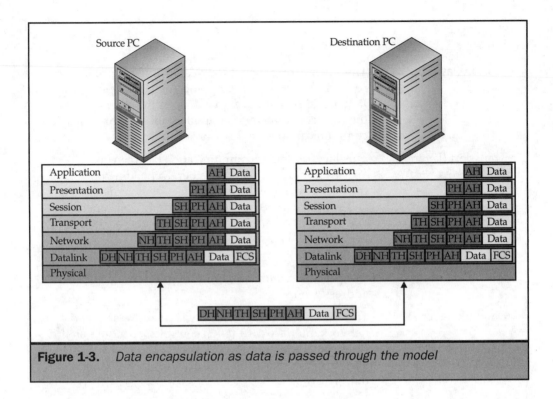

Figure 1-3.   *Data encapsulation as data is passed through the model*

header, and so on). Upon arriving at the destination, the packet will be passed back up the model, with the protocol headers being removed along the way. By the time the packet reaches the application, all that remains is the data, or *payload*.

Now we will delve into the specifics of each layer and the additional processes for which each layer is responsible.

Layer 7: The Application Layer

The application layer is responsible for interacting with your actual user application. Note that it is not (generally) the user application *itself*, but, rather, the network applications used by the user application. For instance, in web browsing, your user application is your browser software, such as Microsoft Internet Explorer. However, the network application being used in this case is HTTP, which is also used by a number of other user applications (such as Netscape Navigator). Generally, I tell my students that the application layer is responsible for the initial packet creation; so if a protocol seems to create packets out of thin air, it is generally an application- layer protocol. While this is not always the case (some protocols that exist in other layers create their own packets), it's not bad as a general guideline. Some common application-layer protocols are HTTP, FTP, Telnet, TFTP, SMTP, POP3, SQL, and IMAP. See Chapter 5 for more details about HTTP, FTP, SMTP, and POP3.

Layer 6: The Presentation Layer

The presentation layer is one of the easiest layers to understand because you can easily see its effects. The presentation layer modifies the format of the data. For instance, I might send you an e-mail message including an attached image. Simple Mail Transport Protocol (SMTP) cannot support anything beyond plain text (7-bit ASCII characters). To support the use of this image, your application needs a presentation-layer protocol to convert the image to plain text (in this case, Multi-purpose Internet Mail Extensions, or MIME). This protocol will also be responsible for converting the text back into an image at the final destination. If it did not, the body of your message would appear like this:

BCNHS ^%CNE (37NC UHD^Y 3cNDI U&">{ } | __D Iwifd YYYTY TBVBC

This is definitely not a picture, and is obviously a problem, proving my point that a presentation-layer problem is generally easy to recognize. The presentation layer is also responsible for compression and encryption, and pretty much anything else (such as terminal emulation) that modifies the formatting of the data. Some common presentation-layer data formats include ASCII, JPEG, MPEG, and GIF.

Layer 5: The Session Layer

Conversely, the session layer is one of the most difficult layers to understand. It is responsible for establishing, maintaining, and terminating sessions. This is a bit of a broad and ambiguous description, however, because several layers actually perform the function of establishing, maintaining, and terminating sessions on some level. The best way to think of the session layer is that it performs this function between two applications. However, as we will see in Chapter 5, in TCP/IP, the transport layer generally performs this function, so this isn't always the case. Some common session-layer protocols are RPC, LDAP, and NetBIOS Session Service.

Layer 4: The Transport Layer

The transport layer performs a number of functions, the most important of which are error checking, error recovery, and flow control. The transport layer is responsible for reliable internetwork data transport services that are transparent to upper-layer programs. The first step in understanding transport-layer error checking and recovery functions is to understand the difference between connection-based and connectionless communication.

Connection-Based and Connectionless Communication

Connection-based communication is so named because it involves establishing a connection between two hosts *before* any user data is sent. This ensures that bidirectional communication can occur. In other words, the transport-layer protocol sends packets to the destination specifically to let the other end know that data is coming. The destination then sends a packet back to the source specifically to let the source know that it received the "notification" message. In this way, both sides are assured that communication can occur.

In most cases, connection-based communication also means guaranteed delivery. In other words, if you send a packet to a remote host and an error occurs, then either the transport layer will resend the packet, or the sender will be notified of the packet's failed delivery.

Connectionless communication, on the other hand, is exactly the opposite: no initial connection is established. In most cases (although not all), no error recovery exists. An application, or a protocol above or below the transport layer, must fend for itself for error recovery. I generally like to call connectionless communication "fire and forget." Basically, the transport layer fires out the packet and forgets about it.

In most cases, the difference between connection-based and connectionless protocols is very simple. You can think of it like the difference between standard mail and certified mail. With standard mail, you send off your message and hope it gets there. You have no way of knowing whether the message was received. This is connectionless communication. With certified mail, on the other hand, your message is either delivered correctly and you get a receipt, or your message is attempted to be delivered many times before it times out and the postal service gives up—and you *still* get a receipt. Either way, you are guaranteed to be notified of what happened so that you can take appropriate measures. This is typical connection-based communication.

Flow Control

In it's simplest form, *flow control* is a method of making sure that an excessive amount of data doesn't overrun the end station. For example, imagine that PC A is running at 100 Mbps and PC B is running at 10 Mbps. If PC A sends something to PC B at full speed, 90 percent of the information will be lost because PC B cannot accept the information at 100 Mbps. This is the reason for flow control.

Currently, flow control comes in three standard flavors, as described in the following sections.

Buffering   Commonly used in conjunction with other methods of flow control, buffering is probably the simplest method. Think of a *buffer* as a sink. Imagine you have a faucet that flows four gallons of water a minute, and you have a drain that accepts only three gallons of water a minute. Assuming that the drain is on a flat countertop, what happens to all of the excess water? That's right, it spills onto the floor. This is the same thing that happens with the bits from PC A in our first example. The answer, as with plumbing, is to add a "sink," or buffer. However, this solution obviously leads to its own problems. First, buffers aren't infinite. While they work well for bursts of traffic, if you have a continuous stream of excessive traffic, your sink space will eventually run out. At this point, you are left with the same problem—bits falling on the floor.

Congestion Notification   *Congestion notification* is slightly more complex than buffering, and it is typically used in conjunction with buffering to eliminate its major problems. With congestion notification, when a device's buffers begin to fill (or it notices excessive congestion through some other method), it sends a message to the originating station basically saying "Slow down, pal!" When the buffers are in better shape, it then relays another message stating that transmission can begin again. The obvious problem with this situation is that in a string of intermediate devices (such as routers), congestion notification just prolongs the agony by filling the buffers on every router along the path.

For example, imagine Router A is sending packets to Router C through Router B (as in Figure 1-4). As Router C's buffer begins to fill, it sends a congestion notification to Router B. This causes Router B's buffer to fill up. Router B then sends a congestion notification to Router A. This causes Router A's buffer to fill, eventually leading to a "spill" (unless, of course, the originating client understands congestion notifications and stops the flow entirely). Eventually, Router C sends a restart message to Router B, but by that time, packets will have already been lost.

Windowing   The most complex and flexible form of flow control, windowing, is perhaps the most commonly used form of flow control today. In *windowing*, an agreed-upon number of packets are allowed to be transferred before an acknowledgment from

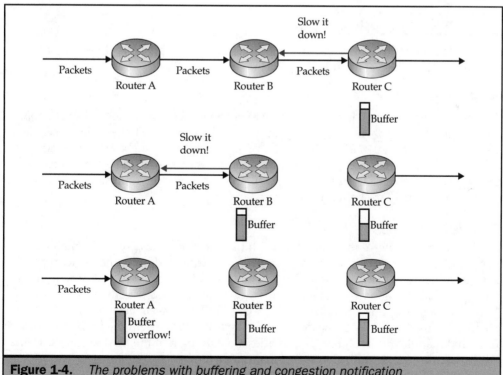

Figure 1-4.   *The problems with buffering and congestion notification*

the receiver is required. This means that one station should not be able to easily overload another station: it must wait on the remote station to respond before sending more data. In addition to flow control, windowing is also used for error recovery, as we will see in Chapter 5.

Some common transport-layer protocols are TCP, UDP, and SPX, which will be covered in more detail in Chapters 5 and 7.

Layer 3: The Network Layer

The network layer deals with logical addressing and path determination (routing). While the methods used for logical addressing vary with the protocol suite used, the basic principles remain the same. Network-layer addresses are used primarily for locating a host geographically. This task is generally performed by splitting the address into two parts: the group field and the host field. These fields together describe which host you are, but within the context of the group you are in. This division allows each host to concern itself only with other hosts in its group; and the division allows specialized devices, called *routers*, to deal with getting packets from one group to another.

Some common network-layer protocols are IP and IPX, which are covered in Chapters 5 through 7.

Layer 2: The Datalink Layer

The datalink layer deals with arbitration, physical addressing, error detection, and framing, as described in the following sections.

Arbitration

Arbitration simply means determining how to negotiate access to a single data channel when multiple hosts are attempting to use it at the same time. In half-duplex baseband transmissions, arbitration is required because only one device can be actively sending an electrical signal at a time. If two devices attempt to access the medium at the same instant, then the signals from each device will interfere, causing a collision. This phenomenon is perhaps better demonstrated in Figure 1-5.

Physical Addressing

All devices must have a physical address. In LAN technologies, this is normally a MAC address. The *physical address* is designed to uniquely identify the device globally. A MAC address (also known as an Ethernet address, LAN address, physical address, hardware address, and many other names) is a 48-bit address usually written as 12 hexadecimal digits, such as 01-02-03-AB-CD-EF. The first six hexadecimal digits identify the manufacturer of the device, and the last six represent the individual device from that manufacturer. Figure 1-6 provides a breakdown of the MAC address. These addresses were historically "burnt in," making them permanent. However, in rare cases, a MAC address is duplicated. Therefore, a great many network devices today have configurable MAC addresses. One way or another, however, a physical address of some type is a required component of a packet.

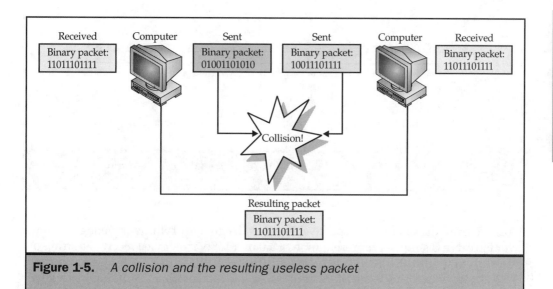

Figure 1-5.   *A collision and the resulting useless packet*

Error Detection

Another datalink-layer function, *error detection*, determines whether problems with a packet were introduced during transmission. It does this by introducing a trailer, the FCS, before it sends the packet to the remote machine. This FCS uses a Cyclic Redundancy Check (CRC) to generate a mathematical value and places this value in the trailer of the packet. When the packet arrives at its destination, the FCS is examined and the reverse of the original algorithm that created the FCS is applied. If the frame was modified in any way, the FCS will not compute, and the frame will be discarded.

Note   *The FCS does not provide error recovery, just error detection. Error recovery is the responsibility of a higher layer, generally the transport layer.*

Framing

Framing is a term used to describe the organization of the elements in a packet (or, in this case, a frame). To understand why this task is so important, we need to

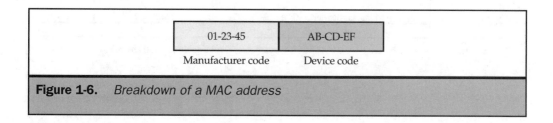

Figure 1-6.   *Breakdown of a MAC address*

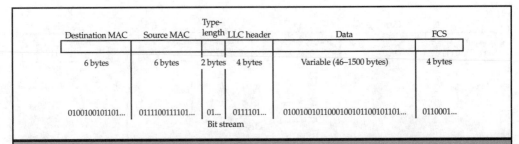

Figure 1-7. An Ethernet 802.3 framing key being applied to the bit stream, breaking it into sections

look at it from the device's perspective. First, realize that everything traveling over the cable is simply a representation of a 0 or a 1. So, if a device receives a string of bits, such as 01101010001010111101011111010101010010100010101010111, and so on, how is it to know which part is the MAC address, or the data, or the FCS? It requires a key. This is demonstrated in Figure 1-7.

Also, because different frame types exist, the datalink layers of both machines must be using the same frame types to be able to tell what the packet actually contains. Figure 1-8 shows an example of this.

Notice that the fields do not line up. This means that if one machine sends a packet in the 802.3 format, but the other accepts only the Sub-Network Access Point (SNAP)

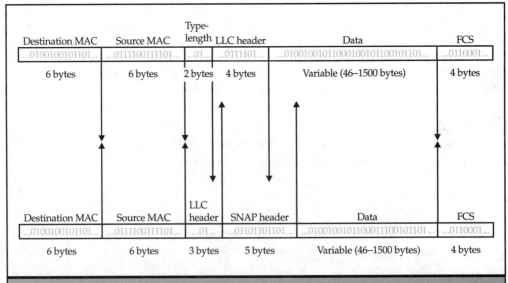

Figure 1-8. Misaligned fields due to incorrect frame type

format, they will not be able to understand each other because they are looking for different components in different bytes of the packet.

Some common datalink-layer protocols are the following: virtually all of the 802 protocols (802.2, 802.3, 802.5, and so on), LAPB, LAPD, and LLC.

Layer 1: The Physical Layer

The physical layer is responsible for the most substantial of all functions. All connectors, cabling, frequency specifications, distances, propagation-delay requirements, voltages—in short, all things physical—reside at the physical layer.

Some common physical-layer protocols are EIA/TIA 568A and B, RS 232, 10BaseT, 10Base2, 10Base5, 100BaseT, and USB.

Peer Communication

Peer communication is the process in networking whereby each layer communicates with its corresponding layer on the destination machine. Note that the layers do not communicate directly, but the process is the same as if they were communicating directly. A packet is sent from one host to another with all headers attached; but, as the packet passes up through the model on the other side, each layer is solely responsible for the information in its own header. It views everything else as data. This process is shown in Figure 1-9.

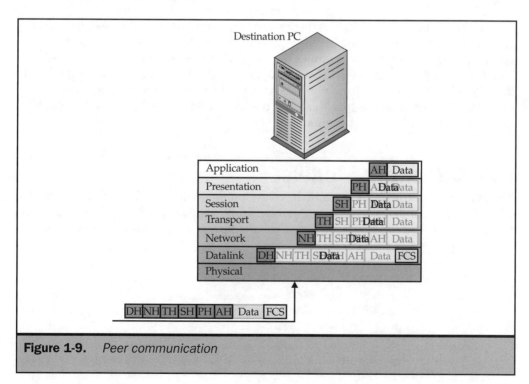

Figure 1-9. *Peer communication*

Note that a layer is concerned only with the header from the exact same layer on the other device. It treats everything else as data (even though it isn't). Therefore, one layer can, in a sense, communicate with its twin layer on the other device.

Bringing It All Together

Finally, I have included a sample network communication between two devices, broken down by layer (see Figure 1-10). Note that this sample is *not* technically accurate. I have included it only for illustrative purposes because it shows how each layer performs a specific function, even if that function isn't performed in exactly the same manner in real life. The major technical problem with this diagram lies at the network layer, in the "Intermediate Destination Address" field. There is no Intermediate Address field in reality, but because we have not discussed how routing really works yet, this example illustrates the point well enough for now.

In this example, we are sending an e-mail using TCP/IP. As we transmit the message, it begins at layer 7 by adding a Mail Application Programming Interface (MAPI) header. Then it passes to the presentation layer, which adds a MIME header to explain the message format to the other side. At the session layer, name resolution is performed, resolving techtrain.com to 209.130.62.55. At the transport layer, the 256KB message is segmented into four 64KB chunks, and a TCP session is established, using windowing for flow control. At the network layer, routing is performed, and the path is sent to the nearest router (represented here by the Intermediate Destination Address).

Also note that the IP addresses (logical) are resolved to MAC addresses (physical) so that they can be understood by the next layer. At the datalink layer, the packet is segmented again, this time into frames that conform to the Maximum Transmission Unit (MTU) of the media. At the physical layer, the data is sent as electrical signals. At the other side, the communication passes back up through the model, performing the opposite of the sending machine's calculations to rebuild the packet into one 256KB chunk of raw data for the application.

Other Network Models

The DOD model is important because it is the foundation for TCP/IP, not the OSI model. While the DOD model matches the OSI model fairly well, the fact that it is the foundation for TCP/IP can lead to some confusion when attempting to learn the OSI model. The

Application (E-mail Program)	Application (E-mail Program)
Application Layer Data Format: Mail to (via MAPI): Brian Hill \<BEGIN MESSAGE\>	**Application Layer** Data Format: Received from (via MAPI): Brian Hill \<MESSAGE RECEIVED\>
Presentation Layer Data Format: Send to: \<SMTP via TCP/IP\> Bhill@techtrain.com \<Mime-Version: 1.0; Content-Type: text/plain; charset=US-ASCII; Compression: No\>	**Presentation Layer** Data Format: Received from: \<SMTP via TCP/IP\> Bhill@techtrain.com \<Begin Decoding; Mime-Version: 1.0; Content-Type: text/plain; charset=US-ASCII; Compression: No\>
Session Layer Data Format: Send to: \<TCP/IP via DNS\> bhill@209.130.62.55	**Session Layer** Data Format: Received from: \<TCP/IP via DNS\> bhill@209.130.62.55
Transport Layer Data Format: Send to: bhill@209.130.62.55 \<Establish TCP Session; Window Size: 32KB; Segment Format: 64KB; Total Segments: 4; Total Bytes: 256KB\>	**Transport Layer** Data Format: Received from: bhill@209.130.62.55 \<Begin Packet Reception; Begin Segment Creation; Acknowledge Receipt; Packet Format: 64KB; Total Packets: 4; Total Bytes: 256KB\>
Network Layer Data Format: Send to: \<via MAC Address\> 2d6c9e446a32 \<Source Address: 165.200.2.23 = 6587da32b5d6; Final Destination Address: 209.130.62.55 = 2d6c9e446a32; Intermediate Destination Address: 209.215.192.10 =4adc500ad23a; Packet Size: 64KB; Total Packets: 4; Total Bytes: 256KB\>	**Network Layer** Data Format: Received from: \<via MAC Address\> 6587da32b5d6 \<Source Address: 165.200.2.23 = 6587da32b5d6; Final Destination Address: 209.130.62.55 = 2d6c9e446a32; Begin Packet Creation; Packet Size: 64KB; Total Packets: 4; Total Bytes: 256KB\>
Datalink Layer \<Establish Link: Transceiver Number: 0; Frame Size: 1500 bytes; Total Frames: 175; Total Bytes: 256KB; Begin Frame Numbering; Begin Transmission\>	**Datalink Layer** \<Link Established: Transceiver Number: 0; Frame Size: 1500 bytes; Total Frames: 175; Total Bytes: 256KB; Begin Sequence Check; Begin Frame Recreation\>
Physical Layer Data Format: 01000101011101000101110 1 0111010101000010111010101011101011...	**Physical Layer** Data Format: 01000101011101000101110 1 0111010101000010111010101011101011...

Data flow — Data flow — Data flow

Figure 1-10. *Processes performed by each layer of the model*

upper layers of the DOD model don't match the upper layers of the OSI model, which can lead to different books listing protocols in different places within the OSI model. The key here is to understand that unless you are studying for a test, it doesn't really matter too much where you place a given protocol in the OSI model, as long as you understand the functionality of each layer of the model. Figure 1-11 depicts how the OSI and DOD models match up.

Whereas the OSI and DOD models present a model of how network-based communication occurs, Cisco's hierarchical internetworking model is a layered approach to the topological design of an internetwork. It is designed to help improve performance, while at the same time allowing optimum fault tolerance. When you use this model, you simplify the network design by assigning various roles to the layers of the network design. The obvious drawback of using this model in a small- to medium-sized network is cost; however, if you require a high-performance, scalable, redundant internetwork, using this approach is one of the best ways to design for it.

The hierarchical internetworking model consists of three layers:

■ **Core layer** This layer is the network backbone. As such, the main issue here is that any major problem will likely be felt by everyone in the internetwork. Also, because speed is very important here (due to the sheer volume of traffic that will be entering the backbone), few activities that consume significant routing or switching resources should be applied in this layer. In other words, routing, access lists, compression, encryption, and other resource-consuming activities should be done *before* the packet arrives at the core.

■ **Distribution layer** This layer is the middle ground between the core and access layers. Clients will not be directly connected to this layer, but most of their packet processing will be performed at this layer. This is the layer where most supporting functions take place. Routing, Quality of Service (QoS), access

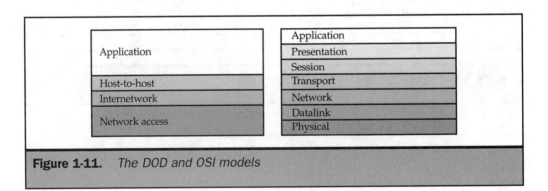

Figure 1-11.   *The DOD and OSI models*

lists, encryption, compression, and network address translation (NAT) services are performed at this layer.

■ **Access layer**   This layer provides user access to local segments. The access layer is characterized by LAN links, usually in a small-scale environment (like a single building). Put simply, this layer is where the clients plug in. Ethernet switching and other basic functions are generally performed here.

Figure 1-12 provides an example of the model in action.

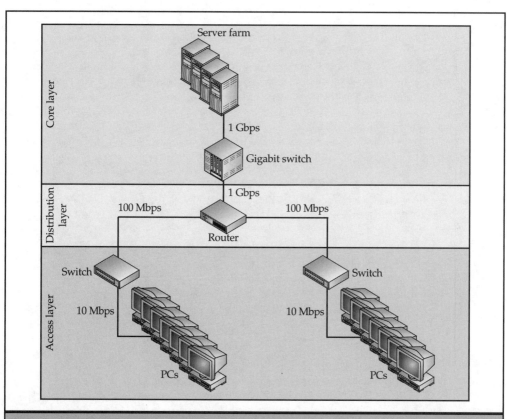

Figure 1-12.   *The Cisco hierarchical internetworking model*

Summary

In this chapter, we have reviewed the most popular networking models, including the OSI, Cisco, and DOD models. This information will help us understand references to the layered networking approach examined throughout the book, and will serve as a guide to understanding the place of routing and switching in any environment.

The Complete Reference

Cisco

Chapter 2

Ethernet and Wireless LANs

Because up to 80 percent of your Internet traffic takes place over local area network (LAN) technologies, it makes sense to spend a bit of time discussing the intricacies of LANs. While a great many LAN technologies exist, we will focus solely on Ethernet and 802.11b wireless LANs (WLANs). What about Token Ring, Fiber Distributed Data Interface (FDDI), LocalTalk, and Attached Resource Computer Network (ARCnet), you may ask? I chose not to include these technologies for a number of reasons. While I do recognize that they exist, in most cases, they have been replaced with some form of Ethernet in the real world. This means that over 90 percent of the time, you will never deal with any of these technologies. For this reason, I feel that our time will best be spent going over technologies that you *will* encounter in nearly every network.

Ethernet Basics

Ethernet is a fairly simple and cheap technology. It is also the most prevalent technology for local area networking, which is why we are going to spend most of this chapter covering the intricacies of Ethernet.

Topology

Ethernet operates on a *bus* or *star-bus* topology model. What does this mean? Well, ThickNet (also known as 10Base5) and ThinNet (also known as 10Base2) operate like a series of bus stops. No stop may be skipped, even if there are no passengers at those stops. If you send a packet to PC C from PC B, that packet will also be transmitted to PCs A and D, as shown in Figure 2-1.

Your first problem with a bus is that, even though the packet is not destined for the other PCs, due to the electrical properties of a bus architecture, it must travel to *all* PCs on the bus. This is because in a true *physical* bus topology, all PCs share the same cable. While there may be multiple physical segments of cable, they are all coupled to one

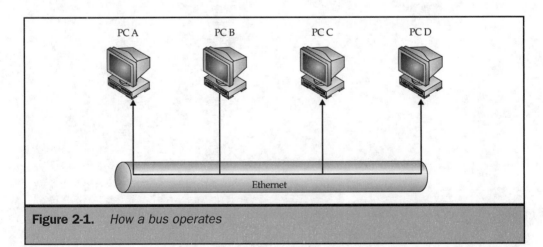

Figure 2-1. *How a bus operates*

another and share the same electrical signal. Also, the electrical signal travels down all paths in the bus at exactly the same rate. For instance, in the previous example, the electrical signal from PC B will reach PC A and PC C at exactly the same time, assuming all cable lengths are exactly the same.

Another problem with the physical bus topology is that if a cable breaks, all PCs are affected. This is because a physical bus must have a resistor, called a *terminator*, placed at both ends of the bus. A missing terminator causes a change in the cable's impedance, leading to problems with the electrical signal. Any break in the cable—at any point— effectively creates two segments, neither of which will be adequately terminated.

10Base5 and 10Base2 are good examples of both logical and physical buses. They are physical buses because all PCs share a common cable segment. They are logical buses because every PC in the network receives exactly the same data, and the bandwidth is shared.

The star-bus architecture, which is much more prevalent today than a standard bus architecture, operates slightly differently than the bus architecture. The star bus is a physical star but a logical bus. The physical star means that all of the devices connected to the network have their own physical cable segment. These segments are generally joined in a central location by a device known as a *multiport repeater*, or *hub*. This device has absolutely no intelligence. Its entire purpose is to amplify and repeat the signals heard on any port out to all other ports. This function creates the logical bus required by Ethernet. The advantage of a physical star is that if any one cable segment breaks, no other devices are affected by the break. That port on the hub simply becomes inoperative until the cable is repaired or replaced.

As for the logical bus in star-bus Ethernet, it operates exactly as in the physical bus architecture. If a signal is sent to PC B from PC C, it still must be transmitted to PCs A and D. This is demonstrated in Figure 2-2.

Bandwidth

Ethernet operates in a variety of speeds, from 10 Mbps to 1 Gbps. The most common speed currently is 100 Mbps, although you must take this speed with a grain of salt. First, you need to understand that in standard Ethernet, you are using one logical cable segment (logical bus) and running at *half-duplex*—only one machine may send at any time. This leads to one major problem: *shared bandwidth*. For example, if you have 100 clients connected to your 100 Mbps Ethernet, each client has an average maximum transfer speed of about 1 Mbps. Then you must remember that this is mega*bits* per second, not mega*bytes* per second. This means you have an average peak transfer rate of around 125 KBps.

So, doing a little math, you figure you should be able to transfer a 1MB file in around eight seconds. Not so. Without even counting packet overhead (which can be significant, depending on the protocol suite used), you must take another factor into account— *collisions*. Because it is a shared, half-duplex connection, you will begin experiencing collisions at around 20 percent utilization. You will most likely never see a sustained

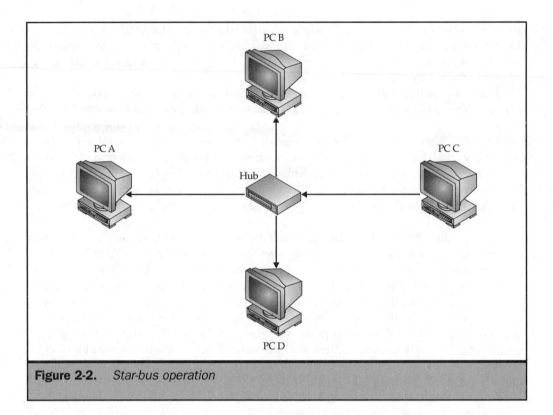

Figure 2-2.   *Star-bus operation*

utilization of over 50 percent without excessive collisions. So, what is the best average *sustained* data rate attainable on this network? Theoretically, around 60 KBps would be a good bet. 100 Mbps Ethernet doesn't sound so fast anymore, huh?

Duplex

Ethernet can operate at either half- or full-duplex. At half-duplex, you can either send or receive at any given instant, but you can't do both. This means that if another device is currently sending data (which you are receiving), you must wait until that device completes its transmission before you can send. This also means that collisions are not only possible, but likely. Imagine half-duplex Ethernet as a single lane road. Then imagine packets as city buses screaming down the road. If one is coming toward you and your bus is screaming toward it, there will be a collision. This results in the loss of both buses—or packets.

Full-duplex Ethernet solves this little dilemma, however, by allowing a two-lane road. With full-duplex, you can send *and* receive at the same time. This feat is accomplished by using separate physical wires for transmitting and receiving. However, it requires the hub to support full-duplex because it must do a bit of a juggling act. In full-duplex, whenever something comes into the hub from a transmit pair, the hub must know to send that signal out of all receive pairs. Today's hubs generally also have a small buffer (512KB to 1MB) in case the hub is overloaded with traffic.

Full-duplex communication can also be accomplished between two devices by using a *crossover cable*: a cable in which the send and receive wires are crossed, or flipped, so that the send on one end goes into the receive on the other. This is similar to a lap link cable. In this scenario, no hub is needed.

 Full-duplex communication cannot be accomplished on a physical bus topology (such as with 10Base2 or 10Base5).

Attenuation

Although not an issue solely with Ethernet, attenuation is a major concern in Ethernet. *Attenuation* is the degradation of a signal over time or distance. This degradation occurs because the cable itself provides some resistance to the signal flow, causing a reduction in the electrical signal as the signal travels down the cable. You can think of it like a car traveling down a road. If you accelerated to 60 mph, the minimum speed was 40 mph, and you had just enough gas to accelerate but not enough to maintain your speed, your car would fall below the minimum speed in only a short distance. This analogy can be compared to Ethernet. The signal is sent at a certain voltage and amperage, but over distance, resistance in the cable causes the signal to degrade. If the signal degrades too much, the *signal-to-noise ratio* (how much signal is present compared to noise in the wire) will drop below minimum acceptable levels, and the end device will not be able to determine what is signal and what is noise.

You can deal with attenuation in a number of ways, but two of the most common solutions are setting maximum cabling lengths and repeating, or amplifying, the signal. Setting maximum cabling lengths helps alleviate the problem by establishing maximum limits on the total resistance based on cable type. Repeating the signal helps alleviate the problem (to a degree) by amplifying the signal when it gets too low. However, this strategy will work only a few times because a repeater does not rebuild a signal; it simply amplifies whatever is present on the wire. Therefore, it amplifies not only the signal, but the noise as well. As such, it does not improve the signal-to-noise ratio, it just ensures that the signal "volume" remains at an acceptable level.

Chromatic Dispersion

Attenuation occurs only on copper-based media. With fiber-optics, light is used, and light does not suffer from the same problem. It does, however, suffer from a different problem: chromatic dispersion. *Chromatic dispersion* occurs when an "impure" wavelength of light is passed down the cable. This light is not composed of a single wavelength, but, rather, many differing wavelengths. If you remember physics from high school, you might already recognize the problem. The wavelengths of light are most commonly called colors in everyday life. If you send an impure light signal down the cable, over time, it will begin to break apart into its founding components, similar to a prism. Because wavelengths are just a fancy way of describing speed, this means that some rays of light in a single bit transmission may reach the other end sooner, causing false bits to be detected.

Chromatic dispersion is a problem with all fiber-optic transmissions, but it can be greatly reduced in a number of ways. The first method is by reducing the frequency of signals sent down the cable. You will be sending less data, but you can send it for longer distances. The basic rule is that every time you halve the cabling distance, you can double the speed. Unfortunately, this strategy doesn't work with Ethernet due to its inherent timing rules, but it's good information to know, anyway.

The second method involves using high-quality components, such as single-mode fiber and actual laser transmitters. Normally, multimode fiber-optics are used, which leads to greater dispersion due to the physical size of the fiber and the impurities present in the glass. Also, light-emitting diodes (LEDs) are typically used instead of lasers. LEDs release a more complex signal than a laser, leading to a greater propensity to disperse. Still, dispersion is not a major issue with fiber-optics unless you are running cable for many miles, and it is generally not an issue at all with Ethernet because the maximum total distance for Ethernet is only 2,500 meters (again, due to timing constraints).

Electromagnetic Interference

Electromagnetic interference (EMI) is another problem that is not specific only to Ethernet. EMI occurs with all copper cabling to some degree. However, it is usually more pronounced with Ethernet due to the predominant use of unshielded twisted-pair (UTP) cabling. Electronic devices of all types generate EMI to some degree. This is due to the fact that any time you send an electrical pulse down a wire, a magnetic field is created. (This is the same principle that makes electromagnets and electric motors work, and the opposite of the effect that makes generators work, which move a magnetic field over a cable.) This effect leads to magnetic fields created in other devices exerting themselves on copper cabling by creating electric pulses, and electric pulses in copper cabling creating magnetic fields that exert themselves on other devices. So, your two real problems with EMI are that your cabling can interfere with other cabling, and that other electronic devices (such as high-powered lights) can interfere with your cabling.

EMI is also the reason for a common cabling problem called crosstalk. *Crosstalk* occurs when two cables near each other generate "phantom" electrical pulses in each other, corrupting the original signal. This problem can be reduced considerably in UTP cabling by twisting the cables around each other. (In fact, the number of twists per foot is one of the major differences between Category 3 and Category 5 cabling.)

All in all, EMI shouldn't be a major concern in most environments; but if you have an environment in which high-powered, nonshielded electrical devices are used, you might want to consider keeping data cables as far away as possible.

Ethernet Addressing

Ethernet uses Media Access Control (MAC) addresses for its addressing structure. Whenever you send an Ethernet frame, the first two fields are the destination MAC address and the source MAC address. These fields must be filled in. If the protocol stack does not yet know what the MAC address of the intended recipient is, or if the message is to be delivered to all hosts on the network, then a special MAC address, called a *broadcast address*, is used. The broadcast address is all *F*s, or FF-FF-FF-FF-FF-FF. This address is important because in Ethernet, although every client normally receives every frame (due to the logical bus topology), the host will not process any frame whose destination MAC address field does not equal its own.

Note	*Multicast frames may also be passed up to higher layers of the protocol stack, as multicast MAC addresses can be matched to the multicast IP address.*

For example, if my PC's MAC address is 01-02-03-AA-BB-CC, and my PC receives a frame with a destination MAC address of 55-55-55-EE-EE-EE, my network interface card (NIC) will discard the frame at the datalink layer, and the rest of my protocol suite will never process it. There are two exceptions to this rule, however. The first is a broadcast, and the second is a promiscuous mode.

Broadcasts are handled differently than normal frames. If my PC receives a frame with the destination MAC address of all *F*s, the PC will send the frame up the protocol stack because occasionally a message needs to be sent to all PCs. In general, only one PC actually needs the message, but my PC has no idea which PC that is. So my PC sends the frame to all PCs, and the PC that needs to receive the frame responds. The rest discard the frame as soon as they realize that they are not the "wanted" device. The down side to broadcasts is that they require processing on every machine in the broadcast domain (wasting processor cycles on a large scale); furthermore, in a switched network, they waste bandwidth.

Promiscuous mode is the other exception to the rule. When a NIC is placed in promiscuous mode, it keeps all frames, regardless of the intended destination for those frames. Usually, a NIC is placed in promiscuous mode so that you can analyze each individual packet by using a device known as a *network analyzer* (more commonly called a sniffer). A *sniffer* is an extremely useful piece of software. If you understand the structure of a frame and the logic and layout of the protocol data units (PDUs—a fancy way of saying packets), you can troubleshoot and quickly solve many seemingly unsolvable network problems. However, sniffers have a rather bad reputation because they can also be used to view and analyze data that you aren't supposed to see (such as unencrypted passwords).

Ethernet Framing

Ethernet has the ability to frame packets in a variety of ways. The standard Ethernet framing types are shown in Figure 2-3, and in the following paragraphs.

The first frame type listed in Figure 2-3 is known as Ethernet 2, or DIX (Digital Intel Xerox—named for the companies that wrote the specification). This frame type is the most commonly used type today. The other frame types are not used in most networks, although Cisco devices support all of them.

The second and third frame types shown in Figure 2-3 are IPX/SPX specific and used primarily by Novell. Although both frame types technically contain a "type" field, it is only used to specify the total length of the packet, not the type of protocol used,

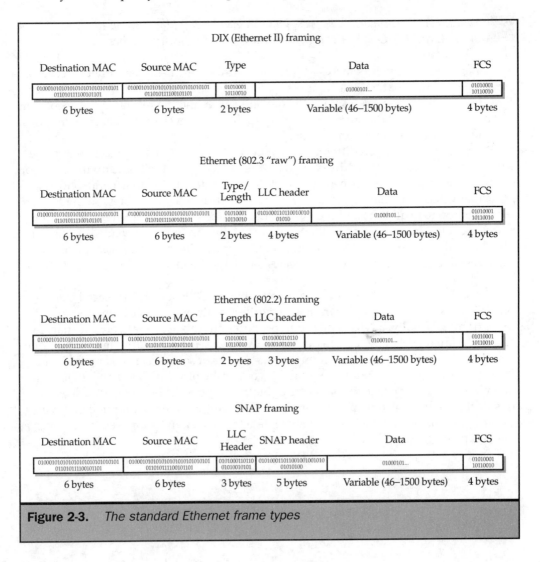

Figure 2-3. *The standard Ethernet frame types*

making both frame types suitable only for IPX/SPX. The second frame type is generically known as 802.3 "raw." Novell terms this frame type 802.3, whereas Cisco calls it Novell, making the generic name even more confusing. The 802.3 "raw" frame type is primarily used in Netware 3.11 and earlier for IPX/SPX communications. The third frame type is known generically as the IEEE 802.3 frame type, or 802.2/802.3, by Novell as 802.2, and by Cisco as LLC. This is the frame type created and blessed by the IEEE, but to date its only use has been in Novell 3.12 and later and in the OSI protocol suite.

Finally, 802.3 SNAP (Sub-Network Access Protocol—known in Novell as Ethernet_ SNAP and in Cisco as SNAP) was created to alleviate the problem of an Ethernet frame being able to support only two bytes of protocol types. For this reason, a SNAP header was added to permit three bytes for an "Organizational ID," allowing different vendors of different protocols to further differentiate themselves. Unfortunately, this header was hardly ever used (except by AppleTalk), and the end result is that almost no one uses the SNAP specification.

Note   *For a further discussion of the various frame types used in Ethernet, including the political pressures that caused their creation, visit ftp://ftp.ed.ac.uk/pub/ EdLAN/provan-enet.*

Typically (in TCP/IP), the DIX or Ethernet II frame type is used. However, this depends somewhat on the OS because some operating systems (such as Advanced Interactive executive, IBM's UNIX OS) can use multiple frame types. Remember, the main point to keep in mind is that for two hosts to communicate directly, they must understand the same frame types.

Now we will look at the different fields contained in these frame types. First, note that there are two additional fields added by the physical layer that are not shown. Every frame begins with a *preamble,* which is 62 bytes of alternating 0's and 1's. This lets the machines on the network know that a new frame is being transmitted. Then a *start frame delimiter (SFD)* is sent, which is just a binary code of 10101011 that lets all stations know that the actual frame itself is beginning. From there, we get into the datalink sections of the frame, listed and analyzed here:

- **Destination address**   This field lists the destination MAC address.
- **Source address**   This field lists the source MAC address.
- **Type**   This field is used to designate the type of layer 3 protocol in the payload of the frame. For instance, a type designation of 0800 hex indicates that an IP header follows in the payload field. This field allows multiple layer 3 protocols to run over the same layer 2 protocol.
- **Length**   This field is used to designate how long the frame is so that the receiving machine knows when the frame is completed. It really isn't needed in most cases, however, because Ethernet adds a delay between frames to perform the same function.

- **DSAP**   The destination service access point (DSAP) field is used to tell the receiving station which upper-layer protocol to send this frame to (similar to the type field). It is part of the LLC header.

- **SSAP**   The source service access point (SSAP) field is used to tell which upper-layer protocol the frame came from. It is part of the LLC header.

- **Control**   This field is used for administrative functions in some upper-layer protocols. It is part of the LLC header.

- **OUI**   The organizationally unique ID (OUI) field is used only on SNAP frames. It tells the other side which vendor created the upper-layer protocol used.

- **Frame Check Sequence (FCS)**   This field is a complex mathematical algorithm used to determine whether the frame was damaged in transit.

Arbitration

Arbitration is the method of controlling how multiple hosts will access the wire—and, specifically, what to do if multiple hosts attempt to send data at the same instant. Ethernet uses Carrier Sense Multiple Access with Collision Detection (CSMA/CD) as an arbitration method. To better understand this concept, let's look at its components.

Carrier sense means that before sending data, Ethernet will "listen" to the wire to determine whether other data is already in transit. Think of this as if you have multiple phones hooked to the same phone line. If you pick up one phone and hear someone else talking, you must wait until that person finishes. If no one else is using the line, you may use it. This is what Ethernet does.

Multiple access means multiple machines have the ability to use the network at the same time, leading to collisions—which brings up the need for the last component.

Collision detection means that Ethernet can sense when a collision has occurred and resends the data. How Ethernet actually accomplishes this is really unimportant, but when sending a frame, Ethernet will listen to its own frame for the first 64 bytes just to make sure that it doesn't collide with another frame. The reason it doesn't listen to the entire transmission is because by the time the sixty-fourth byte is transmitted, every other machine on the network should have heard at least the first byte of the frame, and, therefore, will not send. The time this process takes is known as *propagation delay*. The propagation delay on standard 10 Mbps Ethernet is around 51.2 microseconds. On 100 Mbps Ethernet, this delay drops to 5.12 microseconds, which is where distance limitations involved with timing begin to crop up in Ethernet. If your total cable length on a logical Ethernet segment (also known as a collision domain) is near or greater than the recommended maximum length, you could have a significant number of late collisions.

Late collisions occur after the "slot," or 64-byte time in Ethernet, and therefore are undetectable by the sending NICs. This means that the NIC will not retransmit the data, and the client will not receive it. In any case, assuming a collision occurs that is detectable by the NIC, the NIC will wait a random interval (so that it does not collide with the other station yet again) and then resend the data. It will do this up to a maximum of 16 times, at which point it gives up on the frame and discards it.

While the entire Ethernet specification uses CSMA/CD as an arbitration method, it is actually needed only for half-duplex operation. In full-duplex environments there are no collisions, so the function is disabled.

Basic Ethernet Switching

Layer 2 Ethernet switching (also known as *transparent bridging*) is a fairly simple concept. Unlike a hub, an Ethernet switch processes and keeps a record of the MAC address used on a network and builds a table (called a content-addressable memory, or CAM table) linking these MAC addresses with ports. It then forwards an incoming frame out only from the port specifically associated with the destination MAC address in the frame. The Ethernet switch builds its CAM table by listening to every transmission occurring on the network and by noting which port each source MAC address enters through. Until the CAM table is built, it forwards the frame out from all ports except the originating port because it doesn't yet know which port to send the frame to. This concept is known as *flooding*. To demonstrate these ideas, Figures 2-4 through 2-10 show how a switch works.

First, the switch starts with a blank CAM table. Figure 2-4 shows the initial switch configuration.

Next, PC A sends a frame destined for the server to its hub, as shown in Figure 2-5.

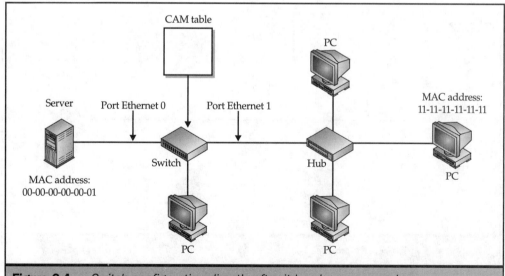

Figure 2-4.   *Switch configuration directly after it has been powered on*

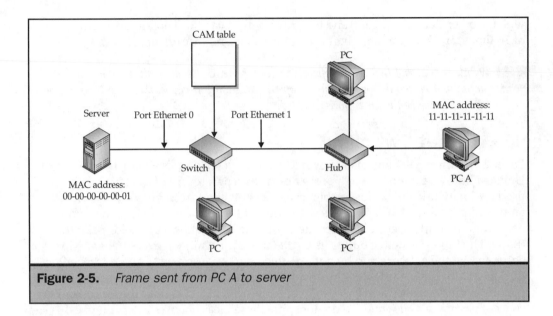

Figure 2-5. *Frame sent from PC A to server*

Then the hub repeats the frame out from all ports, as shown in Figure 2-6.
The switch adds the source address to its CAM table, as shown in Figure 2-7.

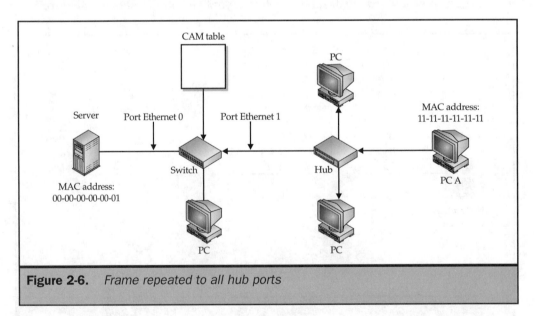

Figure 2-6. *Frame repeated to all hub ports*

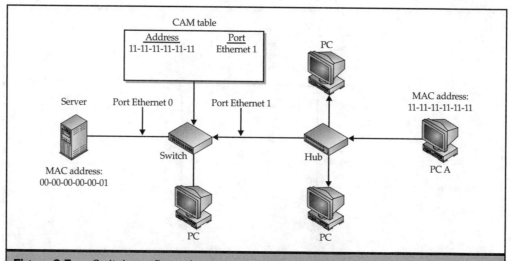

Figure 2-7. *Switch configuration upon receiving first frame*

At this point, the switch cannot find the destination address in the CAM table, so it forwards out from all ports except for the originating port, as shown in Figure 2-8.

The server responds, so the switch adds its MAC address to the CAM table, as shown in Figure 2-9.

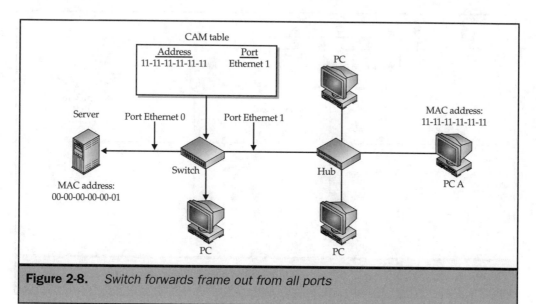

Figure 2-8. *Switch forwards frame out from all ports*

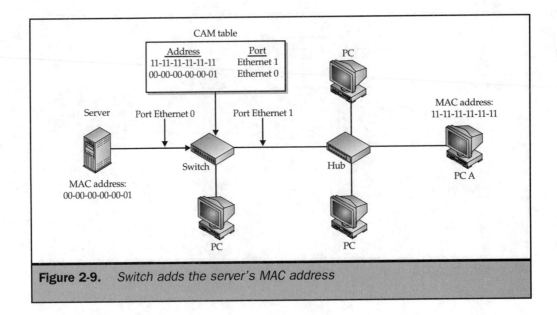

Figure 2-9.   *Switch adds the server's MAC address*

Then, the switch looks up the MAC address and forwards the packet out from port E1, where it reaches the workgroup hub that repeats the frame out from all ports. See Figure 2-10.

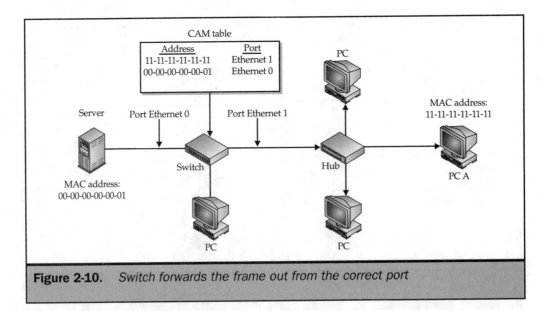

Figure 2-10.   *Switch forwards the frame out from the correct port*

The major benefit of switching is that it separates, or logically segments, your network into collision domains. A *collision domain* is an area in which collisions can occur. If you are in a given collision domain, the only devices that your frames can collide with are devices on the same collision domain. Using the previous example network, Figure 2-11 demonstrates how a network is broken into collision domains.

Segmenting a network into collision domains provides two main benefits. The first is obviously a reduction in the number of collisions. In Figure 2-11, it is very unlikely that a collision will ever occur on collision domains 1 or 2 because only one device resides on these domains. (You will be able to collide only with the host you are directly transmitting to at any given time.) This benefit should cut collisions on this network by 40 percent because now only three PCs (all on collision domain 3) are likely to have a collision. The second benefit is simply that segmenting the network with a switch also increases available bandwidth.

In Figure 2-11, if we replace the central switch with a hub, returning the network to one collision domain, we will have only 1.2 Mbps of bandwidth available to each device (assuming 10 Mbps Ethernet links). This is because all devices have to share the same bandwidth. If one device is sending, all the rest must wait before sending. Therefore, to determine the available bandwidth in one collision domain, we must take the maximum speed (only around 6 Mbps is possible with 10 Mbps Ethernet, due to collisions, frame size, and gaps between the frames) and divide it by the number of hosts (five). However, with our segmentation, each collision domain has a full 6 Mbps of bandwidth. So the server in collision domain 1 and the PC in collision domain 2 both have a full 6 Mbps. The three PCs in collision domain 3, however, have only 2 Mbps.

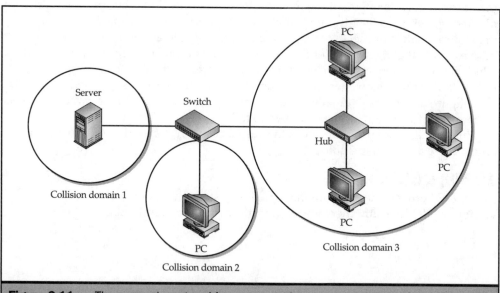

Figure 2-11.   *The example network's segmentation*

Ethernet Technologies

This section discusses some of the issues specific to each major Ethernet technology. Much of this section is devoted to tables detailing the specifications of each technology for your reference. Let's begin by looking at some key points valid to all Ethernet specifications.

First, note that all common Ethernet technologies use the *baseband* signaling method, which means that only one frequency can be transmitted over the medium at a time. Contrast baseband signaling to *broadband* signaling, in which multiple frequencies (usually called *channels*) are transmitted at the same time. To give you an example of the difference, cable TV uses broadband transmissions. This is why you can switch channels instantaneously. The cable receiver simply tunes itself to a different frequency. If cable TV were baseband, your cable receiver would have to make a request to receive a separate data stream each time you wanted to change channels, leading to a delay.

> **Note** *Ethernet specifications for broadband transmission, such as 10Broad36, do exist, but are very, very rare. For that reason, they have been omitted from our discussion of Ethernet.*

Second, note that all common Ethernet specifications conform to the following naming scheme: speed/base/cable type. Speed is the speed in Mbps. Base stands for baseband, the signaling technology; and the cable type is represented with a variety of terms, such as T (for twisted pair) and 2 (for ThinNet coaxial, a .25-inch cable). So 10BaseT means 10 Mbps baseband signaling, using twisted-pair cable.

Third, All Ethernet specifications conform to what is known as the 5/4/3 rule. This rule states that you can have up to five segments and four repeaters, with no more than three segments populated with users. In the star-bus Ethernets, this works slightly differently, but the rule of thumb is that no two hosts should be separated by more than four repeaters or five times the maximum segment length in meters of cable.

Next, note that most types of Ethernet on Cisco equipment support *auto negotiation*, which allows the hub or switch to automatically configure the link to be 10 or 100 Mbps, full- or half-duplex. However, you should also be aware that the suggested practice is to configure the duplex manually, because autonegotiation (especially between products from different vendors) has been known to fail.

Finally, a little information about the various cable types used in Ethernet is in order. These are listed in Table 2-1.

10 Mbps Ethernet

10 Mbps Ethernet comes in many flavors, the most common of which is 10BaseT. All versions of 10 Mbps Ethernet, however, conform to the same standards. The main

Cable Type	Maximum Speed	Duplex	Topology	Benefits	Drawbacks
Unshielded twisted pair (UTP)	1 Gbps (Category 5e)	Both	Star bus	Easy to install, cheap, plentiful, wide support	Moderately susceptible to EMI
ThickNet coaxial (RG-8)	10 Mbps	Half	Bus	Long distances, decent EMI resistance	Difficult to install, expensive, hard to find
ThinNet coaxial (RG-58 A/U)	10 Mbps	Half	Bus	Good for fast temporary connections, decent EMI resistance	Prone to problems
Fiber-optic	1 Gbps (10 Gbps in process)	Full	Star bus	Very fast, extremely long distances, fast, immune to EMI, fast (Did I mention it was fast?)	Somewhat difficult to install, and can be expensive

Table 2-1.   *Ethernet Cabling Specifics*

variations in these types are due to the cabling system used. Table 2-2 details the specifics of 10 Mbps Ethernet.

Cable Type	10Base5	10Base2	10BaseT	10BaseFL
	RG-8	RG-58 A/U	Category 3, 4, or 5 UTP	Fiber-optic (single or multimode)

Table 2-2.   *10 Mbps Ethernet Technologies*

Cable Type	10Base5	10Base2	10BaseT	10BaseFL
Maximum Segment Length	500 meters	185 meters	100 meters	2000 meters
Maximum Overall Length	2500 meters	925 meters	500 meters	2500 meters (due to timing constraints)
Maximum Nodes per Segment	100	30	2	2
Topology	Bus	Bus	Star bus	Star bus
Capable of Full-Duplex?	No	No	Yes	Yes

Table 2-2.   *10 Mbps Ethernet Technologies* (continued)

100 Mbps Ethernet (Fast Ethernet)

Table 2-3 describes the most common 100 Mbps Ethernet technologies, both over copper and fiber cabling.

	100BaseTX	100BaseT4	100BaseFX
Cable Type	Category 5 UTP	Category 3, 4, or 5 UTP, STP	Fiber-optic (single or multimode)
Maximum Segment Length	100 meters	100 meters	2000 meters
Maximum Overall Length	500 meters (full-duplex), ~250 meters (half-duplex)	~250 meters	2500 meters
Maximum Nodes per Segment	2	2	2

Table 2-3.   *100 Mbps Ethernet Technologies*

	100BaseTX	100BaseT4	100BaseFX
Topology	Star bus	Star bus	Star bus
Capable of Full-Duplex?	Yes	No	Yes

Table 2-3.   *100 Mbps Ethernet Technologies* (continued)

Note that 100BaseT4 uses all four pairs of cable, whereas 100BaseTX uses only two of the four pairs. Also, the distance limitations of half-duplex Fast Ethernet are due to propagation delay and may vary according to the hubs and switches used.

1000 Mbps Ethernet (Gigabit Ethernet)

Table 2-4 describes the most common 1000 Mbps Ethernet technologies, both over copper and fiber cabling.

	1000 BaseCX	1000 BaseT	1000 BaseSX	1000 BaseLX	1000 BaseLH
Cable Type	STP	Category 5e UTP	Fiber-optic (multimode)	Fiber-optic (single or multimode)	Fiber-optic (single or multimode)
Maximum Segment Length	25 meters	100 meters	550 meters	550 meters (multimode), 5000 meters (single mode)	100,000 meters
Maximum Overall Length	25 meters	200 meters	2750 meters	2750 meters (multimode), 20,000 meters (single mode)	Varies

Table 2-4.   *1 Gbps Ethernet Technologies*

	1000 BaseCX	1000 BaseT	1000 BaseSX	1000 BaseLX	1000 BaseLH
Maximum Nodes per Segment	2	2	2	2	2
Topology	Star bus	Star bus	Star bus	Star bus	Star bus
Capable of Full-Duplex?	Yes	Yes	Yes	Yes	Yes

Table 2-4.   *1 Gbps Ethernet Technologies* (continued)

Note that as of this writing, no Cisco products support 1000BaseT, and very few products from any vendor have been designed for 1000BaseCX. Also note that 1000BaseLH (LH for *long haul*) is drastically different from the other types relative to timing constraints.

10 Gigabit Ethernet

10 Gigabit Ethernet is an emerging standard due for release sometime in 2002. It makes some changes to the Ethernet standard to support long-haul cabling distances for wide area network (WAN) and metropolitan area network (MAN) links. Currently, 10 Gbps Ethernet is being proposed only over fiber-optic cabling. You can find out more about 10 Gigabit Ethernet at http://www.10gea.org.

WLANs

Wireless LAN technologies are a bit different from everything we have discussed so far. First, wireless LAN technologies must contend with many more obstacles than their wired counterparts. Obstacles like radio interference, walls and doors, metal girders in the floor, and even microwave ovens can cause signal loss.

Currently, two major technologies are in use for short-range wireless communication. The first is *infrared (IR)*, a line-of-sight technology, meaning that in order to communicate, a clear path must exist between the sending station and the receiving station. Even window glass can obstruct the signal. Because of this, IR is typically used only for very short-range communications (1 meter or less) at a limited bandwidth (around 1 Mbps), such as data exchange between two laptops or other handheld devices. For this reason (because the standard is more like a lap-link standard than a networking standard), we will focus most of our energy on the other technology—radio.

Radio-based wireless LANs (WLANs) are defined in the IEEE 802.11 specification. This specification defines a spread-spectrum radio technology over the unlicensed 2.4-GHz frequency channel, running at 1 to 2 Mbps, with a range between 30 and 300 meters. The specification also allows the use of the IR technology, as well as two differing types of spread-spectrum technologies. However, direct sequence spread spectrum (DSSS) is the only one of these three technologies that has made it to high speed. The other two technologies reach only 1 to 2 Mbps, making them somewhat unattractive for common use. For this reason, all of the issues covered in this chapter assume the use of DSSS technology.

The 802.11 technology specification was updated later to the 802.11b standard, which defines a speed improvement from 2 Mbps to 11 Mbps. This IEEE 802.11b standard, which uses DSSS, is quickly becoming widely adopted, and this is the technology we will focus our attention on.

How IEEE 802.11b Works

Wireless communication over the 802.11b specification specifies only two devices: an access point (AP) and a station (STA). The AP acts as a bridge device, with a port for the wired LAN (usually Ethernet) and a radio transceiver for the wireless LAN. STAs are basically network interfaces used to connect to the AP. In addition, the 802.11b standard allows for two modes of operation: infrastructure mode and ad hoc mode.

With *infrastructure mode,* all STAs connect to the AP to gain access to the wired network, as well as to each other. This mode is the most commonly used for obvious reasons. The AP controls all access to the wired network, including security, which we will discuss in more detail in the Security section of this chapter. In this environment, you would typically place the AP in a centralized location inside the room in which you wish to allow wireless connectivity (such as a cubical "farm"). In addition, multiple APs may be set up within the building to allow *roaming*—extending the range of your wireless network.

With *ad hoc mode,* no AP is required. Rather, the STAs connect to each other in a peer-to-peer fashion. While this mode allows no connectivity to the wired network, it can be extremely useful in situations in which several PCs need to be connected to each other within a given range.

Radio Communication

Initially, radio communication for network traffic may seem to be a fairly simple proposition. It sounds like you just plug a NIC into a walkie-talkie and go cruising. Unfortunately, it's not quite that simple. With a fixed frequency (like that of two-way radios), you run into a number of problems, the biggest of which are lack of security and interference. In the network environment, you need a method for reducing the ability of someone to tap your communications, and, in the case of interference, some way to recover and resend the data. This is why wireless networking uses a technology called spread spectrum.

Spread spectrum operation in 802.11b works as follows. The 802.11b standard specifies the use of the unlicensed radio range from 2.4465 GHz to 2.4835 GHz. In this range, the frequencies are split up into 14 22-MHz channels that are usable by wireless networking devices. Without going into all of the engineering details of how this works, suffice it to say that wireless LAN devices hop frequencies within a given channel at regular intervals. Because of this, if another device causes interference in one frequency, the wireless LAN simply hops to another frequency. This technique also helps the LAN be more secure—someone scanning one particular frequency will have only limited success in intercepting the transmissions.

Arbitration

Wireless LANs lend themselves to much larger hurdles relative to collisions. In a wireless LAN, you cannot "hear" a collision like you can in Ethernet. This is because the signal you are transmitting drowns out all of the other signals that could be colliding with it. The signal is strongest at the source; so even though another signal sounds equally strong at the AP, from your STA, all you can hear is your signal. Therefore, 802.11b does not use the CSMA/CD arbitration method. Instead, it has the ability to use two different methods of arbitration: CSMA/CA and request to send/clear to send (RTS/CTS).

Normally, CSMA/CA works as follows: Before sending a data packet, the device sends a frame to notify all hosts that it is about to transmit data. Then, all other devices wait until they hear the transmitted data before beginning the process to send their data. In this environment, the only time a collision occurs is during the initial notification frame. However, in the 802.11 specification, this process works a little differently (because STAs cannot detect a collision—period). Instead, the STA sends the packet and then waits for an acknowledgment frame (called an *ACK*) from the AP (in infrastructure mode) or from the receiving STA (in ad hoc mode). If it does not receive an ACK within a specified time, it assumes a collision has occurred and resends the data. Note that an STA does not send data if it notices activity on the channel (because that would lead to a guaranteed collision).

RTS/CTS works similarly to a modem. Before sending data, the STA sends a request to send (RTS) frame to the destination. If there is no activity on the channel, the destination sends a clear to send (CTS) frame back to the host. This is done to "warn" other STAs that may be outside the range of the originating STA that data is being transferred, thus preventing needless collisions. Figure 2-12 shows an example of this problem. In general, RTS/CTS is used only for very large packets, when resending the data can be a serious bandwidth problem.

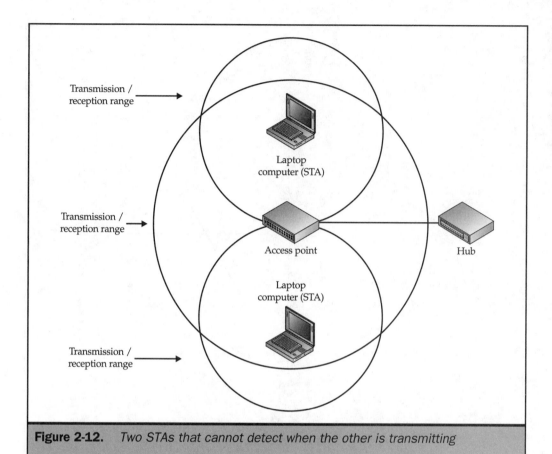

Transmission / reception range

Laptop computer (STA)

Transmission / reception range

Access point

Hub

Laptop computer (STA)

Transmission / reception range

Figure 2-12. *Two STAs that cannot detect when the other is transmitting*

These two standards are not mutually exclusive. In some implementations, both are used in a sequential fashion to further reduce collisions. Figures 2-13 through 2-16 demonstrate this. In Figure 2-13, Station A sends an RTS packet, including the length of the data to be transmitted and the intended destination (Station B), to the rest of the WLAN. The length field is included so that all STAs will have an estimated duration for the conversation and can queue packets accordingly.

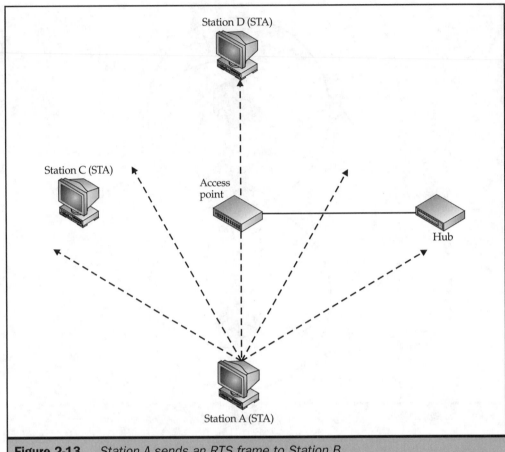

Figure 2-13.   *Station A sends an RTS frame to Station B.*

Station B then listens for traffic and, if none is present, releases a CTS packet to Station A, as shown in Figure 2-14.

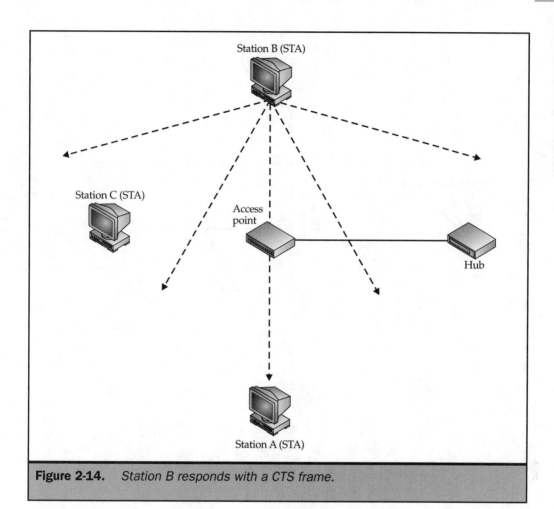

Figure 2-14. *Station B responds with a CTS frame.*

Upon receiving the CTS, Station A sends the data packet to Station B, as shown in Figure 2-15.

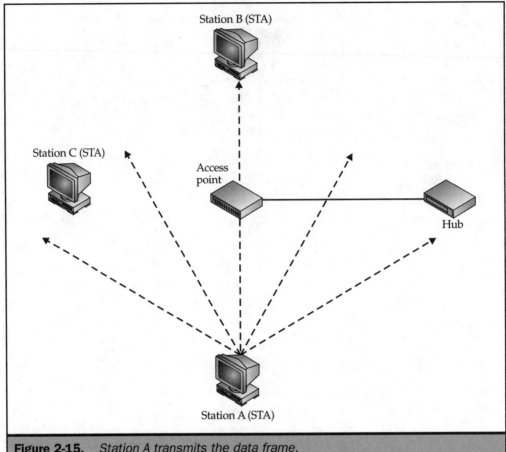

Figure 2-15. *Station A transmits the data frame.*

Finally, Station B, in turn, sends an ACK back to Station A, letting Station A know that the transmission was completed successfully, as shown in Figure 2-16.

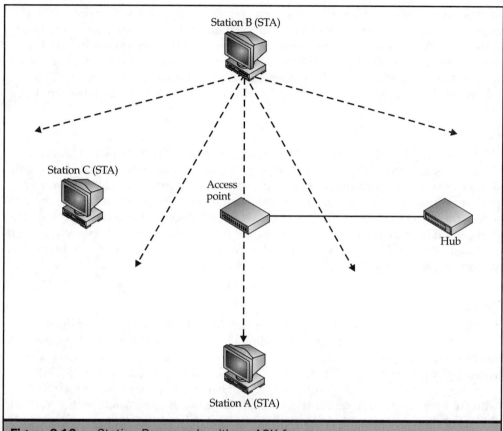

Figure 2-16. *Station B responds with an ACK frame.*

Fragmentation

The datalink layer of the 802.11b specification also allows frame *fragmentation* to assist in error recovery. This means that if a frame is consistently receiving errors (collisions), an STA or AP can choose to fragment a message, chopping it into smaller pieces to try to avoid collisions. For example, if a station is transmitting a frame that consistently experiences a collision on the 800th byte, the station could break the packet into two 750-byte packets. This way, the offending station would (hopefully) send its packet during the time period between the two 750-byte transmissions. One way or another, if a collision is experienced with the fragmented packet, chances are that it will be experienced during one transmission, requiring that only 750 bytes be retransmitted.

Cells

WLANs generally operate in a *cellular* topology, with access points placed around the building to allow users to roam from place to place. When an STA is first initialized, it listens to the signals around it and chooses an AP. This process is called *association*. It performs this process by comparing signal strength, error rates, and other factors until it finds the best choice. Periodically, it rescans to see whether a better choice has presented itself, and it may choose to switch channels (or APs) based on what it sees. With a well-designed cellular infrastructure, you could build your WLAN such that no matter where you are in the building, you will have network access.

Security

By nature, WLANs are security nightmares because you eliminate most of the physical security inherent in your network by sending the data over radio waves. However, you can use three techniques to help eliminate most of the security problems in WLANs: wired equivalent privacy (WEP), data encryption, and access control lists (ACLs).

Wired equivalent privacy is used to prevent hackers from associating their STA with your private AP. It requires that each STA that wants to associate with the AP use the AP's preprogrammed extended service set ID (ESSID). In this manner, only clients who have been configured with the ESSID can associate with the AP.

With *data encryption*, an AP can be told to encrypt all data with a 40-bit shared key algorithm. Also, to associate with the AP, you are required to pass a challenge issued by the AP, which also requires that your STA possess this key. Unfortunately, a 40-bit key is not incredibly difficult to break. Luckily, WLANs also support all standard LAN encryption technologies, so IPSec, for instance, could be used to further encrypt communications.

Finally, access control lists can be used to secure communications by requiring that every STA that wishes to associate with the AP be listed in the access control entry. However, these entries are MAC addresses, so upkeep can be difficult in a large environment.

Bandwidth and Range

The current specification for WLANs allows a maximum speed of 11 Mbps, with the provision to fall back to 5.5 Mbps, 2 Mbps, and 1 Mbps if conditions do not exist to support 11 Mbps.

The range for modern WLANs depends on the environment in which the WLAN is deployed. Some environments can support distances of 300 feet, while some struggle at 40 feet. The speed and distance attainable with radio technology is limited by a number of factors. The amount of metal in the structure, the number of walls, the amount of electrical interference, and many other things contribute to lower the attainable bandwidth and range. In addition, even though amplifying the power of the signal overcomes most of these issues, because the 2.4-GHz range is an unlicensed range, the Federal Communications Commission (FCC) limits power output for devices in this range to 1 watt.

Summary

This chapter explained how Ethernet, the most prevalent LAN technology, works. Knowing the intricacies of Ethernet will help you immeasurably in almost any environment. You also learned about the benefits of wireless networking and how WLANs operate. While WLANs are not the end all and be all of wireless networking, they are currently the fastest, most secure option available for cheap wireless networks. This advantage alone is leading to hefty adoption of the standard and even further research into increasing its speed, interference rejection, and security. Armed with this understanding of Ethernet and WLANs, you are prepared to face the complexities of most LAN environments and ready to pursue the next level: WAN technologies.

The Complete Reference

Cisco

Chapter 3

Frame Relay

F rame Relay is one of the most widely used WAN protocols today. Part X.25 (an older, highly reliable packet switched network), part ISDN, *Frame Relay* is a high-speed packet-switching WAN networking technology with minimal overhead and provisions for advanced link-management features. In a nutshell, Frame Relay is the technology of choice for moderately high speed (up to around 45 Mbps, or T3 speed) packet-switched WANs.

Frame Relay is really not a complicated technology, but it does have some new terms and concepts that can be a bit alien if you come from a LAN background. The first step in understanding Frame Relay is understanding how the technology works.

How Frame Relay Works: Core Concepts

Frame Relay is a technology designed to work with multivendor networks that may cross international or administrative boundaries. As such, it has a number of concepts that, while fairly common in WAN environments, are rarely seen in LANs. Frame Relay came about because of an increase in the reliability of the links used in WAN connections that made the significant overhead and error correction involved in X.25 unnecessary and wasteful. A protocol was needed that could provide increased speed on these higher-reliability links. Frame Relay leveraged the technologies that existed at the time, but cut out all of the additional overhead that was not needed on these new high-speed links.

Virtual Circuits

Frame Relay uses virtual circuits (VCs) to establish connections. *Virtual circuits* are like "pretend" wires. They don't really exist physically; rather, they exist logically. But, like a real wire, they connect your device to another device. VCs are used because in a large ISP, telephone company (telco), or even a moderate-sized company, running multiple physical wires would be wasteful and cost-prohibitive. This is because, in general, higher port density on a router leads to higher costs. For example, take a network with five remote offices linked to a central home office. Without VCs, at the home office, we would need a router with five WAN ports and at least one Ethernet port. We can't do this type of configuration with a low-end router, like a 2600 series, so we would need to step up to either two 2600 routers or one 3620 router. This configuration is shown in Figure 3-1.

However, with VCs, we could use a single 2600 router with five VCs and one physical WAN port, as shown in Figure 3-2. In just this (rather small) example, the cost difference would be around $12,000. In other words, going with multiple physical links is going to cost us *five times* what a single link with VCs would cost. By consolidating these links, we are putting the responsibility for high port density on our provider, who probably has more than enough ports anyway.

How this process is actually accomplished is a bit more complicated. Frame Relay uses a process known as *multiplexing* to support these VCs. Multiplexing can be done

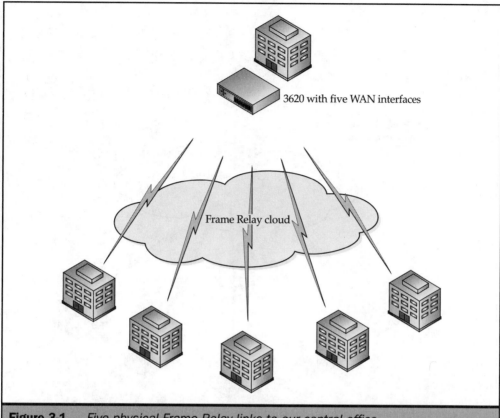

3620 with five WAN interfaces

Frame Relay cloud

Figure 3-1. *Five physical Frame Relay links to our central office*

in one of two ways, depending on whether the media is baseband or broadband. Because Frame Relay is baseband, we will concentrate on how VCs are multiplexed in baseband media.

In baseband technologies, data is typically multiplexed by using what is known as *time-division multiplexing (TDM)*. In TDM, packets on different channels are sent in different "time slots." This flow is similar to traffic on a single-lane road. Just because it is a single lane does not necessarily mean that only one car can travel down the road. The cars just need to travel in a straight line. This principle is the same in TDM. More than one data stream can be sent or received on the line. They just have to do so in "single file." This concept is illustrated in Figure 3-3.

Frame Relay uses *statistical multiplexing*, which differs from TDM in that it uses VCs, which are variable slots, rather than fixed slots like channels. This technique allows Frame Relay to better divide bandwidth among differing applications. Instead of just handing out a fixed amount of bandwidth to a given connection, statistical multiplexing alters

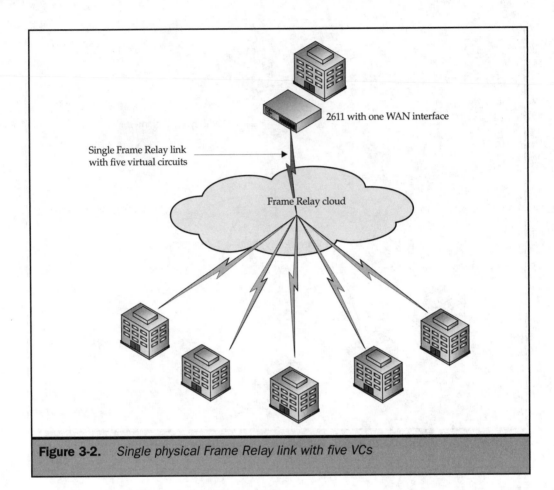

Single Frame Relay link
with five virtual circuits

2611 with one WAN interface

Frame Relay cloud

Figure 3-2.   *Single physical Frame Relay link with five VCs*

the bandwidth offered based on what the application requires. It also allows Frame Relay to be less wasteful with "bursty" traffic. Rather than allocating a fixed amount of bandwidth to the connection, even when no data is being transmitted, Frame Relay can allocate only what each connection needs at any given time.

In addition, VCs are divided into two types: a permanent virtual circuit (PVC) and a switched virtual circuit (SVC). Currently, SVCs are used rarely or not at all, so we will concentrate mostly on PVCs.

In a *permanent virtual circuit*, the connection is always up, always on, and always available. You can think of this like the bat phone. When Batman picks up the bat phone, he always gets Commissioner Gordon. He never has to dial. The connection is always there. This is how a PVC works. Once established, it's like having a direct piece of cable connecting two locations, which makes it ideal for replacing dedicated links.

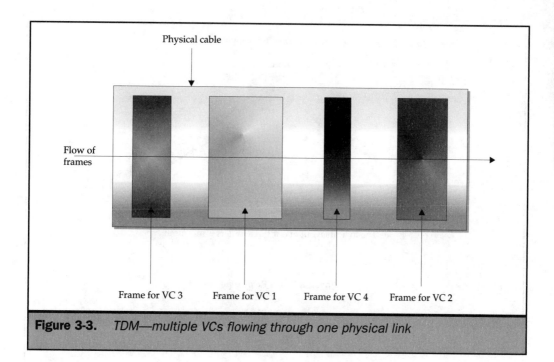

Figure 3-3. *TDM—multiple VCs flowing through one physical link*

A *switched virtual circuit* works similarly to a standard phone. You need to dial a number to establish a connection. Once established, you can communicate. When you have finished, you terminate the connection, and you are free to contact a different party. SVC standards exist, but they are currently rare and used mostly when running Frame Relay in a LAN such as a Lab environment.

Addressing

Frame Relay is a technology that covers both the physical and datalink layers of the OSI model. As such, it has its own physical addressing structure, which is quite different from the Media Access Control (MAC) addressing structure.

A Frame Relay address is called a *Data Link Connection Identifier (DLCI)*. Unlike MAC addresses, DLCIs do not specify the physical port; rather, they specify the logical link between two systems (virtual circuit). In this manner, each physical Frame Relay port can have multiple DLCIs because multiple VCs may be associated with that port. For example, Figure 3-4 shows both the logical and physical layout of a Frame Relay implementation.

In all of these Frame Relay implementations, you may be wondering what the "cloud" is. Put simply, the Frame Relay cloud is your telco. It is drawn as a cloud because it is "voodoo." As long as the packet enters where we want it to and leaves where we want

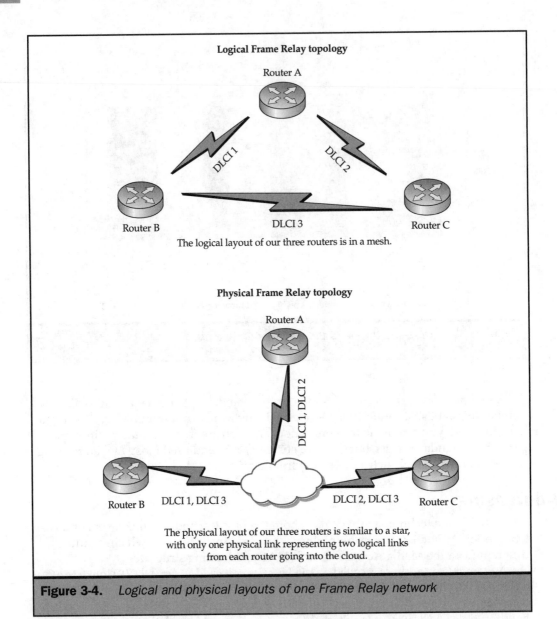

Figure 3-4.   *Logical and physical layouts of one Frame Relay network*

it to, we don't care how it works. Most likely, if you call your telco and have them try to explain it to you, they will say "Uguh Buguh Alacazam" and hang up. OK, that last part was a joke—but honestly, we don't really care from an implementation perspective. However, I will shed some light on this voodoo for you just so you fully understand the technology.

The Frame Relay cloud is really just a huge bank of Frame Relay switches. In Frame Relay terminology, two types of devices exist: the *Data Communications Equipment* or *Data Circuit-Switching Equipment (DCE)*, and the *Data Terminal Equipment (DTE)*. The DCEs are the frame switches in the cloud. The DTEs are the routers. The DCEs operate on the same basic principles as the frame routers—they are just built to handle a *lot* of VCs simultaneously. In addition, the DCEs provide what's called a *clocking signal* to the DTEs. The clocking signal is needed because Frame Relay is a *synchronous* protocol: the frames are synchronized with the clocking signal, so no start bit and stop bit are needed. As a result, Frame Relay becomes a little more efficient and, subsequently, faster.

To put these concepts into perspective, the cloud represents a fragmented path to the final destination. In other words, you do not have a dedicated piece of cable supporting each of your PVCs. Instead, you have a VC that may be switched through ten, hundreds, or even thousands of smaller lines on its way to its final destination. This keeps the telco from having to build a new trunk for every client they add. Rather, they are just reselling their current infrastructure. Figure 3-5 shows an example of a simple frame cloud and how these VCs are switched.

However, this depiction is a bit oversimplified. In truth, another aspect of Frame Relay addressing comes into play: *local significance*. In reality, very few Frame Relay DLCIs are globally significant, as shown in Figure 3-5. Instead, most of them are locally significant. Therefore, it doesn't matter if the DLCI changes throughout the cloud, as long as your DTE and its corresponding DCE both use the same DLCI number. What your DCE uses as the DLCI for your connection with another DCE in the cloud doesn't matter to your DTE. All your DTE is concerned with is what the DLCI means to it. For example, let's take a simple Frame Relay implementation and show DLCI-switching in the cloud.

In Figure 3-6, two routers (our DTEs) are communicating with each other through a Frame Relay cloud composed of three switches. At Router A, DLCI 50 identifies the VC to Router B. At Router B, DLCI 240 identifies the connection to Router A. Throughout the frame cloud, mappings take place to associate these DLCIs. This example is mapping DLCIs (layer 2 addresses) to other DLCIs; but, in reality, the DLCIs are mapped to layer 3 (such as IP) addresses, and routing takes place to send the packet down the correct path. However, because we have yet to discuss IP addressing, this example suffices. One way or another, the principle is the same.

The telco gives you your DLCIs, which will most likely not be the same on either side of the connection, and the telco sorts out how the DLCI-switching occurs. Note that the reason locally significant DLCIs are needed in the first place is because no structure exists to assure uniqueness of DLCIs. In fact, global uniqueness of DLCIs would bring the whole frame infrastructure to its knees because most providers use a frame structure that provides for only 10-bit DLCI numbers. This means that only 1,024 DLCIs could exist at any given time (because only 1,024 values can be represented with ten bits), and there are a *lot* more frame VCs than 1,024. With the current locally significant structure, *each* DCE can have up to 1,024 VCs (minus a few reserved DLCIs), and these VCs can be different from the VCs present on every other DCE, which makes

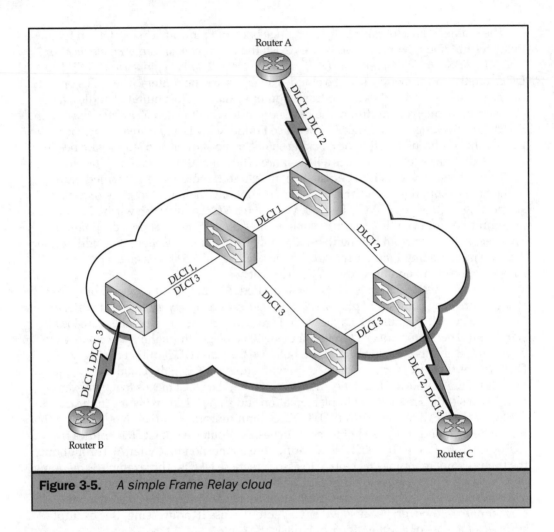

Figure 3-5.   *A simple Frame Relay cloud*

a locally significant DLCI structure much more scalable than a globally significant DLCI structure.

LMI

Frame Relay uses an interesting concept for link management called *local management interface (LMI)*. The LMI is an extension of Frame Relay that provides a number of benefits, as described in the following sections.

Globally Significant DLCIs

Your telco can assign certain links DLCIs of global significance. This strategy allows you to use Frame Relay just like a big LAN, with each individual PVC having its own static address. Unfortunately, it also limits the scalability of the frame cloud.

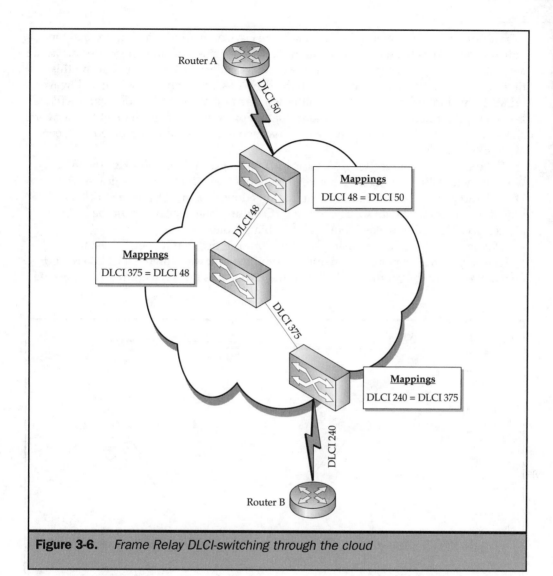

Figure 3-6. *Frame Relay DLCI-switching through the cloud*

Status Inquiries

Possibly the most useful benefit of LMI is that status messages can inform a DTE of a
PVC going down inside the cloud, preventing the DTE from forwarding frames into
a black hole. Status inquiries work like this: Once every so often, the DTE contacts its
directly connected DCE and asks for an update on its PVCs. How often the LMI does
this is called a *heartbeat*. After a certain number of heartbeats have passed, the LMI also
requests a full update on all PVCs that terminate at the LMI. This process allows a DTE
to remove any PVCs for which the status inquiry comes back negative.

This function is extremely useful in a frame environment in which more than one path or route exists to any given destination. Without this function, the router may take a significant amount of time to determine that its primary link is down. During this time, it continues to send frames through the dead PVC, resulting in lost data. However, if LMI status inquiries are used, the router will know that the PVC is dead and will instead send packets down the secondary path. For example, take a look at the network shown in Figure 3-7. Let's see what would happen if a PVC failure occurred between Networks 1 and 2.

If the DTE was not receiving periodic LMI messages, as shown in Figure 3-8, it wouldn't yet realize that the link between Network 1 and Network 2 is down. The DTE would proceed as normal, sending packets down DLCI 200, but the packets would never reach Network 2 due to the PVC failure. The packets would be lost for all eternity. This is known as "routing into a black hole."

However, if the DTE was using LMI, it would notice that its periodic status messages were not arriving, and would assume the link was down. DLCI 200 would be removed from the routing table due to the LMI status message failure. An alternate

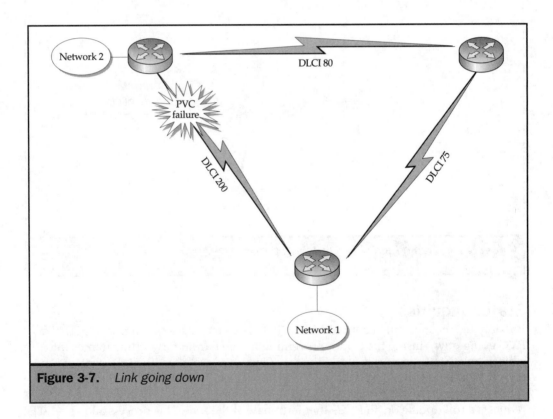

Figure 3-7. *Link going down*

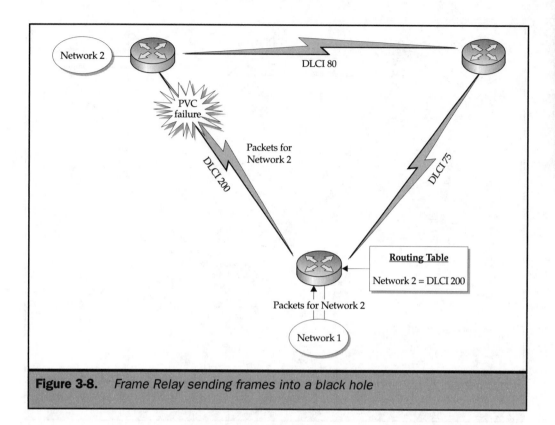

Figure 3-8.   *Frame Relay sending frames into a black hole*

path, via DLCI 75, would be entered in the routing table, allowing the DTE to route packets down DLCI 75, as shown in Figure 3-9.

DLCI Autoconfiguration

This benefit is useful for automatically configuring Frame Relay DLCIs in a small frame environment. The telco simply sets the DLCIs on a DCE, and the directly connected DTEs automatically initialize the correct DLCI numbers.

Multicasting

LMI also allows multicasting over Frame Relay by allowing multicast groups to be established and multicast status inquiries to be sent over the PVCs. Multicasting is discussed in more detail in Chapter 6.

Now that you know what the LMI does, we can look at the different LMI types. Unfortunately, as with most aspects of networking, no one could seem to agree on one single standard. As a result, we now have three distinct, separate LMI types.

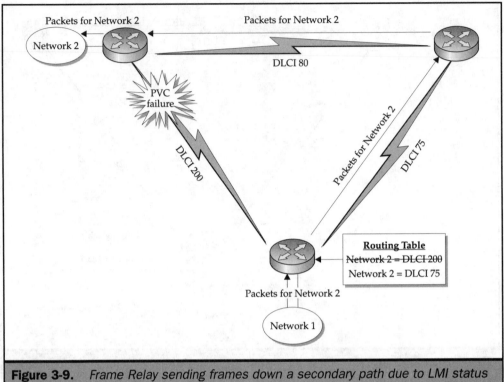

Figure 3-9.   *Frame Relay sending frames down a secondary path due to LMI status messages*

They all perform basically the same tasks, but they just accomplish them differently. And, of course, they are not compatible with one another. The three LMI types are Cisco (sometimes called Gang of Four after the four companies who created it), ITU Q.933-A (sometimes called Annex-A), and ANSI T1.617-D (sometimes called Annex-D).

Basically, the point to remember about LMI types is that you do not have to have the same LMI type on two DTEs in a given VC, but they need to be the same from the DCE to the DTE on each side. Figure 3-10 illustrates this point. Luckily, this limitation isn't much of an issue today because Cisco routers running IOS 11.2 or later can automatically detect and configure the LMI type (and most providers use Annex-D, anyway).

Framing

Frame Relay has a number of frame formats, depending on the layer and sublayer in question. For our purposes, only two are of real significance: Cisco and the Internet Engineering Task Force (IETF).

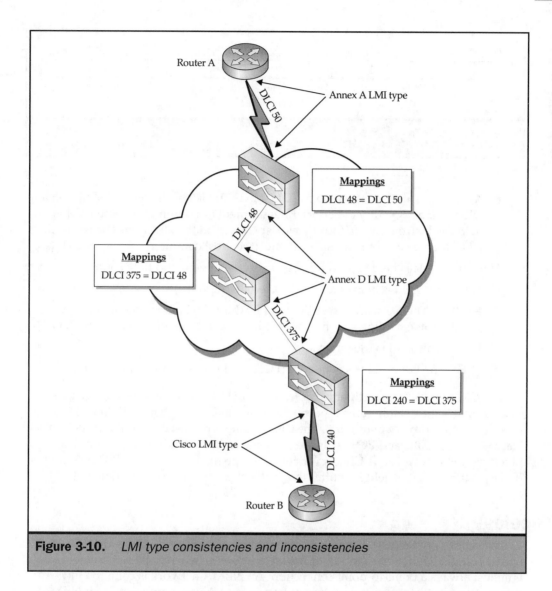

Figure 3-10.   *LMI type consistencies and inconsistencies*

The IETF encapsulation was defined in 1992 in RFC 1294 and then replaced in 1993 by RFC 1490. This encapsulation is generally supported by all Frame Relay equipment, including Cisco devices. The IETF encapsulation is shown in Figure 3-11.

The IETF frame consists of a number of sections:

■ **Flags**   The flags at the beginning and end of the frame are simply the pattern of 01111110 to delimit the beginning and end of the frame.

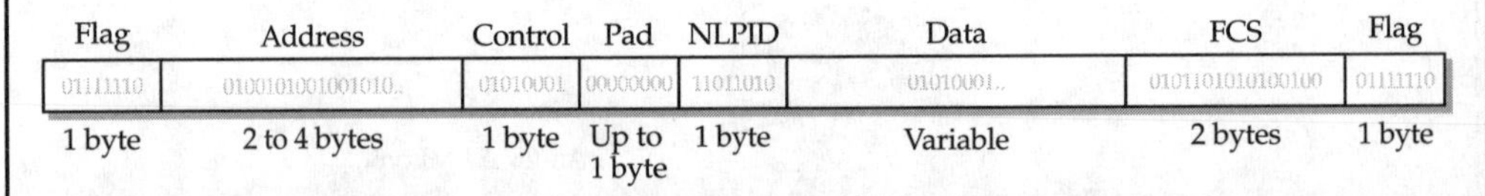

Flag	Address	Control	Pad	NLPID	Data	FCS	Flag
01111110	010010100100101 0..	01010001	00000000	11011010	01010001..	010110101010100100	01111110
1 byte	2 to 4 bytes	1 byte	Up to 1 byte	1 byte	Variable	2 bytes	1 byte

Figure 3-11. *IETF framing*

- **Address**   This field is where the DLCI lives. It also hosts the FECN, BECN, and DE bits. (See "Quality of Service Enhancements," later in this chapter.) This field can be from two to four bytes, depending on the DLCI sizes being used. DLCIs can be one of three sizes: 10 bits (the most common), 16 bits, or 23 bits.

- **Control**   This field is used for session establishment and flow control.

- **Pad**   This field is blank and can be used for alignment.

- **NLPID**   The Network Layer Protocol ID (NLPID) field is used to identify which protocol follows in the data field. You can think of it as the type field.

- **Data**   This field is the payload.

- **FCS**   The Frame Check Sequence (FCS) field is used for error detection.

The Cisco frame format differs a bit from the IETF format. On Cisco devices, the Cisco format is the default frame type, but be aware that you will need to run whichever frame type your provider supports. The Cisco frame type is shown in Figure 3-12. The Flags, Address, Data, and FCS fields are the same as in the IETF specification. The Ethertype field, however, is Cisco's version of a type field. Like the NLPID field in the IETF specification, this field identifies the protocol encapsulated in the data field.

Topology

You can configure Frame Relay in a number of logical topologies, depending on your needs. Because it is a nonbroadcast, multiaccess (NBMA) network, its physical topology is almost always a point-to-point connection. An NBMA network is common in WAN technologies, but it is extremely different from most LAN technologies. In an NBMA environment, broadcasts are not allowed. Every data packet must have a specific destination, which can lead to a number of issues. (NBMA issues are discussed more thoroughly in Chapters 22–26.) The main issue here is that any device that is not participating in the specific VC does not receive any data from the VC—which leads to a physical point-to-point topology. Logically, Frame Relay can be configured with two types of connections: point to point and point to multipoint.

Flag	Address	Ethertype	Data	FCS	Flag
01111110	010010100100101010...	0100010100101 100	01010001...	010110101010100100	01111110
1 byte	2 to 4 bytes	2 bytes	Variable	2 bytes	1 byte

Figure 3-12. *Cisco framing*

A *point-to-point connection* is easy to understand. Basically, every device is linked to another device using a single logical network connection. Generally, this means that only two devices share the same layer 3 network addressing structure. All other communications must be routed between the VCs. Logically, this is probably the simplest way to set up your frame topology. Figure 3-13 provides an example of a point-to-point topology. In this configuration, if we send a packet to Network 1, it will travel down only VC1. All other VCs will not receive the packet.

In a *point-to-multipoint connection*, multiple VCs share the same layer 3 addressing structure, making this similar (logically, anyway) to a bus topology. When a packet is sent to one VC on the network, it is sent to all VCs that are a part of the network. Note that the reason the packet travels to all VCs isn't because the packet is "repeated" over the logical ports; it's because the packet is routed to the network address that is shared by the VCs in question. Figure 3-14 provides an example of this topology. Notice how,

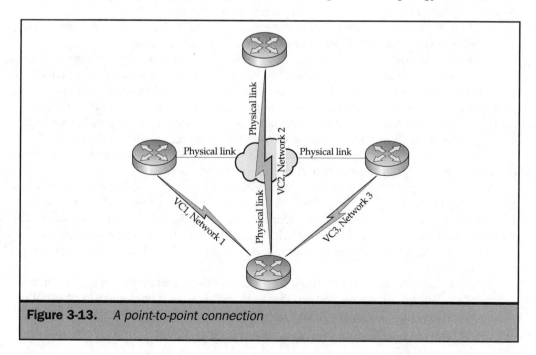

Figure 3-13. *A point-to-point connection*

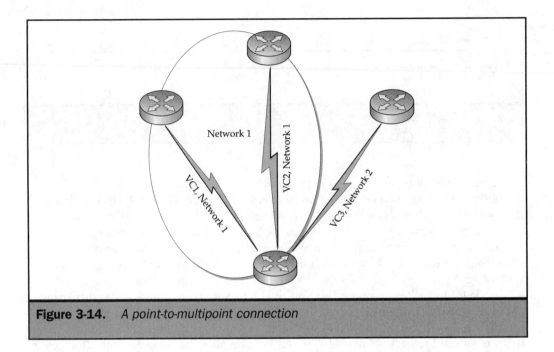

Network 1

VC2, Network 1

VC1, Network 1

VC3, Network 2

Figure 3-14. *A point-to-multipoint connection*

in this example, the packet we send to Network 1 is sent down all VCs participating in Network 1, but not down the VC that is participating in Network 2.

In addition, you can use several different topology types with each type of connection. These fall into two basic categories: hub-and-spoke (or star) and mesh.

The hub-and-spoke topology is probably the simplest and generally the easiest to set up. In a *hub-and-spoke topology*, every device is connected to a central hub device, which performs the lion's share of the work. This setup is best for environments in which everyone needs to access centralized resources from the hub location but seldom, if ever, need to access other remote locations. This setup is typical for a company that has a large number of branch offices that need to access network resources at the headquarters. The hub-and-spoke topology is typically implemented by using point-to-point connections. The example in Figure 3-13 is representative of this topology.

A *mesh topology* is a type of setup in which at least two paths exist from a point in the network to another point in the network. Meshes are further divided according to how meshed they are. A *hybrid mesh* (sometimes called a partial mesh) simply has multiple paths available somewhere in the network. Multiple paths may not be available to all devices, but at least one redundant path must exist. Figure 3-15 shows an example of a hybrid mesh.

A *full mesh*, on the other hand, has a direct connection from every device to every other device. This type of connection can lead to increased costs if implemented physically, but it isn't quite as bad (at least in small networks) when done over Frame Relay VCs.

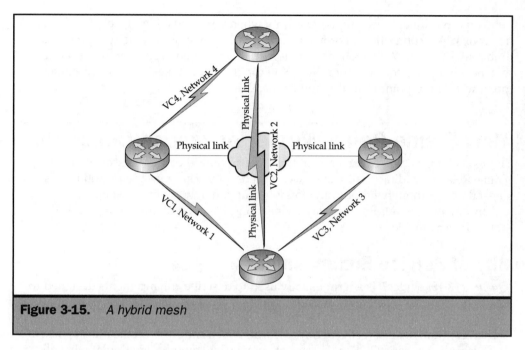

Figure 3-15. *A hybrid mesh*

With meshed VCs, the most significant cost is usually in terms of time—a large mesh can take forever (well, months, anyway) to configure. Figure 3-16 shows a full mesh.

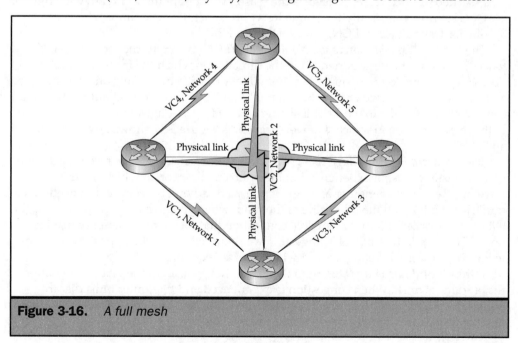

Figure 3-16. *A full mesh*

You typically use mesh topologies only when you need a high level of redundancy for your WAN connections. For instance, in a hub-and-spoke topology, if the central hub router fails, all WAN connectivity is lost. In a mesh environment, however, there will be other paths for the connection. You can implement mesh topologies either with point-to-point or point-to-multipoint connections.

How Frame Relay Works: Advanced Concepts

In this section, we will cover some of the more advanced concepts associated with Frame Relay, including quality of service, speeds, physical connectivity, and multiservice capabilities. Although it is entirely possible that you may never use some of these enhancements on your Frame Relay connections, it's good to know that these advanced options are available.

Quality of Service Enhancements

Quality of service (QOS) is a term usually used to describe enhancements designed for improving or guaranteeing throughput. Most QOS enhancements are like complex flow control. They try to keep traffic from backing up—but if it does, they drop packets selectively (unlike most forms of flow control, which simply drop the oldest packets). For example, by using QOS in the case of an overflow, you could stipulate that all business traffic takes precedence and that other traffic (such as from playing Quake) be discarded (or vice versa—Quake is the reason we bought those T3s anyway, right?).

QOS services specifically for Frame Relay are generally very simple and consist of four major technologies: FECN, BECN, DE, and FRTS.

The *Forward Explicit Congestion Notification (FECN)* field tells the receiving station that congestion is being experienced in the direction in which the frame is traveling. To give you an analogy, if you were driving to work, hit a traffic jam, and got stuck for an hour, when you arrived at work, you might tell your coworkers about the horrible congestion on the way to work. This would be an FECN. Congestion was experienced in the same direction in which you were traveling. Figure 3-17 shows an example of an FECN.

The *Backward Explicit Congestion Notification (BECN)* field tells the receiving station that congestion is being experienced in the opposite direction in which the frame is traveling. To give you another analogy, if you were driving to work and noticed that a traffic jam built up on the other side of the road, when you arrived at work, you might tell your coworkers about the horrible congestion that they may experience driving home. This would be a BECN. Congestion was experienced in the opposite direction in which you were traveling. Figure 3-18 shows an example of a BECN.

If the *Discard Eligibility (DE)* bit in the Frame Relay header is set, it basically means that a frame is trash. When congestion is experienced and the router must discard frames, the first frames to go are those with the DE bit set. The DE bit tells the router that the frames aren't really needed, so it trashes them.

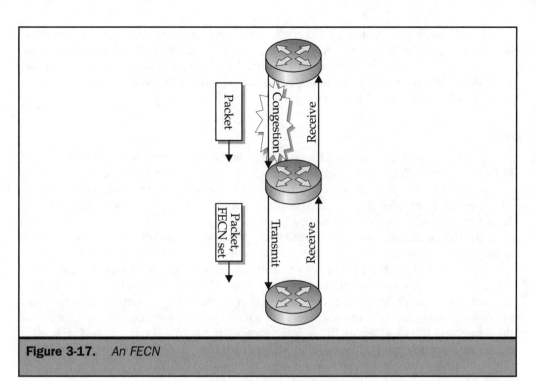

Figure 3-17. *An FECN*

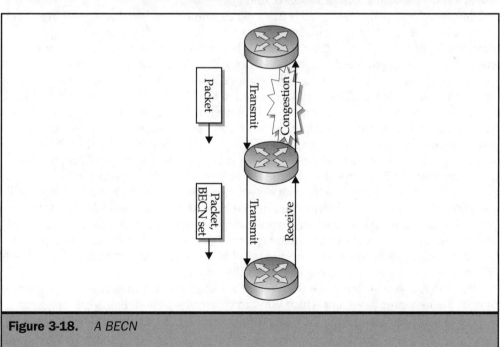

Figure 3-18. *A BECN*

Frame Relay Traffic Shaping (FRTS) is used to dynamically control the amount of data sent over each VC at any time. FRTS uses the CIR, Bc, and Be of a VC (described in the next section) to determine when traffic needs to be modified to adhere to the purchased bandwidth for a given VC. It can also be set to respond to BECNs and DEs to dynamically modify traffic flow.

Speeds

Frame Relay can run at a variety of speeds, from less than 1 Mbps to well over 40 Mbps. In fact, the Frame Relay technology has no real speed boundary because it has no real physical-layer specifications. It's just generally not used past T3 speed because Asynchronous Transfer Mode (ATM) and Synchronous Optical Network (SONET) are the technologies of choice at higher speeds.

In fact, the entire discussion of Frame Relay speed is somewhat ambiguous. Because of the issues involved with billing for a medium that is extremely burst intensive by nature, Frame Relay has a number of methods for measuring speed. Frame Relay—and truly, any synchronous protocol—is required to be burst-dependent because there are no start or stop bits anywhere in the transmission. If data is to be sent, it must be sent in one lump chunk. You cannot send half of a frame, wait a few seconds, and then send the other half. It would throw the whole scheme off. So, to bill fairly based on your bandwidth requirements, Frame Relay bandwidth is sold a bit differently than say, xDSL or Cable. You are charged for the amount of bandwidth you use through the use of a few new terms: CIR, line speed, and burst rate.

Committed information rate (CIR) is the most common term thrown around in a Frame Relay sales meeting. CIR is basically the amount of bandwidth the provider agrees to have available for you at any time. Your CIR is your real bandwidth, which is what you can count on getting out of the line—everything else is gravy.

The burst rate and line speed are intertwined. The *line speed* is the maximum speed of the wire based on the clock signal provided at the DCE, while the *burst rate* is the maximum speed you can attain from your connection based on the number of channels you are using. For instance, you may have a fractional T1 coming into your building, which means that you are not using the full potential of the wire. Perhaps you are using only 12 of the 24 channels, giving you a burst rate of around 768 Kbps, but the line is capable of 1.544 Mbps. In this case, 1.544 Mbps is the line speed.

Unfortunately, talking to a telco about your speed is not that simple. The telco will probably use two additional terms, committed burst (Bc) and excess burst (Be). The *committed burst* is what the provider agrees to let you burst to under most circumstances. In other words, this is the maximum *unregulated* burst speed. The unregulated part is the important piece because it means the provider will allow your frames to travel unhindered as long as you do not break this threshold.

However, if you do break the threshold, the excess burst comes into play. The *excess burst* is the maximum speed at which you can transmit—period. For everything over the committed burst, the provider generally sets the DE bit on your frames, making those frames eligible for discard, and absolutely will not allow anything over the excess burst rate. Therefore, you should be able to transmit at the CIR at all times. You will be able to burst over the CIR up to the Bc pretty much unhindered; but you can't rely on being able to do this, and you will not be able to continuously transmit at this rate. In addition, for short amounts of time, you will be able to burst up to the Be; but if you attempt to sustain that speed, the DE bit will be set, and you will have packet loss.

In addition, you need to realize that these speeds are averaged over time. For instance, say your CIR is 256 Kbps, your Bc is 512 Kbps, and your Be is 768 Kbps. Now let's say the sampling period is ten seconds. For the first two seconds, you send 768 Kbps, the next four you send 128 Kbps, and the final four you send 256 Kbps. Your total rate over the sampling period is then 307 Kbps—slightly over your CIR, but well within your Bc.

Finally, while Frame Relay speeds are set per VC, the combined speed of all VCs cannot be above the line speed. So, for instance, if you had a full T1 (1.544 Mbps), and you bought four VCs, each with a CIR of 512 Kbps, a problem will occur somewhere down the road when you oversubscribe the line. If you had three of the four VCs running at full CIR during a certain time, for instance, then the total bandwidth they are using would be 1536 Kbps, or, basically, the entire bandwidth of the T1. If data needs to be transmitted over the fourth VC, then the other VCs have to slow down to allow transmission. Basically, when you made the line and VC purchases, you were banking on the improbability of all four VCs needing to transmit at full CIR at the same time.

Table 3-1 provides the speeds at which the physical links for Frame Relay are currently implemented, along with their names and channelization. Note that these lines are not specific to Frame Relay. They are simply the lines you will most likely be offered when purchasing a Frame Relay connection. Frame Relay itself does not define what devices are implemented at the physical layer.

Name	Speed	Channels	Country
DS0	64 Kbps	1	All
DS1 (NA, also called T1)	1.544 Mbps	24	North America, Japan
DS1 (Europe, also called E1)	2.048 Mbps	30+2D	Europe
DS1C	3.152 Mbps	48	North America
DS2 (NA, also called T2)	6.312 Mbps	96	North America, Japan

Table 3-1.   *Standard Telco Line Speeds and Channels*

Name	Speed	Channels	Country
DS2 (Europe, also called E2)	8.448 Mbps	120+2D	Europe
DS3 (NA, also called T3)	44.736 Mbps	672	North America
DS3 (Japan)	32.064 Mbps	480	Japan
DS3 (Europe, also called E3)	34.368	480+2D	Europe

Table 3-1. *Standard Telco Line Speeds and Channels* (continued)

Note that the European versions use something called a *data channel* (*D channel*)—it is referred to as a *data* channel even though no user data is sent down it. The D channel is used by the line for control information (such as busy signals, connection establishment, and basic signaling processes). In North America, a D channel is not used, but the same control information must be sent by using one of two methods.

Bit robbing was the original way of dealing with this problem, and it consists of "stealing" a certain number of bits out of so many frames to use for control information. Originally, a single bit was stolen out of every six frames. This strategy worked fine for voice transmissions because a single bit loss out of every thousandth of a second or so makes little or no difference in line quality. However, for data connections, this strategy is unacceptable and will be noticed. So, for the data connections, every eighth bit was stolen automatically, reducing each channel to 56 Kbps of user bandwidth. *Reduction of channels* is another technique—one or more channels are simply used for signaling, while the rest carry user data at the full 64 Kbps data rate.

In addition, the observant and mathematically gifted (or calculator-owning) reader might notice that if you multiply 64 Kbps by 24, you end up with 1.536 Mbps, not 1.544 Mbps, as is listed under the North American DS1 in the table. This is because an extra framing bit is included for every 192 data bits, bringing up the total number of bits transmitted per second to 1.544 Mbps. So, truthfully, how much data do you get to transmit? Well, that is a rather loaded question because the amount actually depends on the upper-layer protocols in use. In most cases, a base T1 is probably capable of only 1.344 Mbps, once framing and robbed bits are counted, and you will likely not get more than 1.25 Mbps of actual data transmitted once you consider upper-layer packet overhead.

Error Recovery

Frame Relay is actually a connection-based protocol, if you use the technical definition of connection-based communication (a connection must be established before data is transmitted; this is taken care of by configuring VCs in Frame Relay). However, Frame

Relay is usually implemented as an unreliable protocol. Although standards within the Frame Relay protocol do exist to allow reliable communication, if you use these standards, you fall back to the same shortcomings X.25 was plagued with (most notably, low performance and high overhead). However, Frame Relay still includes an FCS field, so you can still check data integrity by using this field. Unfortunately, FCS checking is still error detection, not error correction, so the packet is just thrown away if this field does not compute properly. As such, Frame Relay leaves the job of error correction and recovery to the upper-layer protocols, like TCP (further explained in Chapter 5).

Physical Connections

Frame Relay itself does not stipulate any specific physical medium. However, Frame Relay is normally implemented with standard unshielded twisted-pair (UTP) cabling from the provider, which then enters a device known as a *channel services unit/data services unit (CSU/DSU)*, and is then connected to the customer's router with a high-speed serial (v.35) connection. The CSU/DSU has a number of responsibilities. Generally, it takes the original signal from the provider and converts its electrical properties into the serial format for your router. The CSU/DSU also sometimes acts as a channel bank, or *multiplexer*, breaking the channelized connection apart so that the voice and data can be separated. (The next section delves deeper into this topic.) In most modular Cisco routers, you can get a *line card* (similar to an expansion card in a PC) that puts the CSU/DSU right into your router, allowing you to just plug the provider's cabling directly into your router. Figure 3-19 shows a typical Frame Relay cabling scenario.

Multiservice and Channelization

One of Frame Relay's biggest advantages is the ability to carry both voice and data. When Frame Relay was originally built, it was designed as a flexible medium able to transmit both voice and data traffic. By using Voice over IP (VoIP) or Voice over Frame Relay (VoFR), you can designate VCs to carry voice traffic to other Frame Relay voice devices. This method, however, is a fairly complicated way of transmitting voice traffic that requires specialized devices on each end of the connection. Typically, Frame Relay itself is not the method for transmitting voice traffic on your channelized (i.e., T1) circuit. More commonly in a small to mid-sized business, certain channels are split off for voice and others are split off for data. While splitting these channels has the disadvantage of requiring that you have a fixed amount of bandwidth available for voice and data, it has the advantage of reduced complexity—and, in a small environment, reduced costs. Therefore, we will go over the channelization of typical circuits used in most environments, even though it is not specific to our discussion of Frame Relay.

When you transmit a digital voice conversation, you are essentially transforming an analog signal into a digital data signal, so the switch from voice into data is simple. With data communications, the DS circuits are *channelized*, or broken into distinct time slots using TDM, to carry more than one voice conversation over a single physical cable. The channels are set at 64 Kbps each because, to provide a quality telephone conversation, the human voice needs to be sampled at 8-bit 8 KHz. 8-bit, 8 KHz means an 8-bit recording

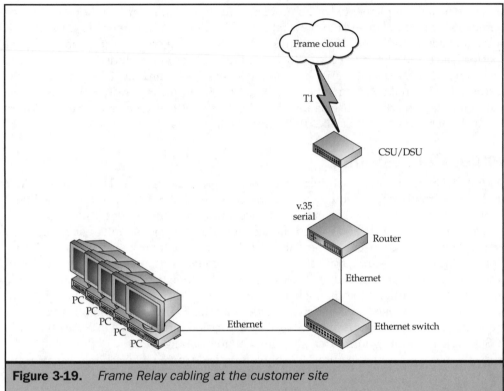

Figure 3-19.   *Frame Relay cabling at the customer site*

needs to be made of your voice 8,000 times per second. Well, doing the math, 8 (bits) times 8,000 equals 64,000, so the 64 Kbps "channel" was born.

By using TDM, a cable with the ability to transfer more data than 64 Kbps could be split into separate channels to carry more that one phone conversation at a time. A T1 connection, for instance, can carry up to 24 separate phone conversations over the same physical cable. Eventually, someone got the bright idea to send data over these channels as well. The logic is valid: if the voice communications are already being converted into data, why not send regular data down the line as well? So now we have the ability to send data (using Frame Relay) and voice communications across the same connections.

The main advantage of the ability to send both voice and data (also known as multiservice communication) is that the cost is significantly less than having a separate physical leased line for data and multiple phone lines. Note that a *leased line* is a connection from your location directly to the location you want to connect to (such as an ISP). In this way, it is your own cable. You share the bandwidth on the cable with no one, which leads to greater data rates and (usually) higher reliability, but leased lines tend to cost quite a bit more than Frame Relay.

For example, a 768 Kbps leased-line data connection to your ISP will run you somewhere around $1,000 a month, depending on your distance from your ISP's point of presence (POP). If you need 12 phone lines as well, these will run you around $2,400 a month for 12 individual lines. This brings your total cost to $3,500 a month for voice and data. However, if you went with Frame Relay, you could buy a single T1 connection and a single PVC, and split 12 channels off for data (768 Kbps) and 12 channels off for voice. Your total cost, in this case, would be around $1,200 per month. After looking at the pricing, it's easy to see why Frame Relay is used so often.

Summary

A solid understanding of Frame Relay will help you immeasurably when you deal with WAN connectivity in any environment. You have seen how this diverse and robust technology operates, both within your infrastructure and through your provider's cloud. You have also seen how Frame Relay, with its ability to support both voice and data communications, can help you reduce costs.

You can find additional information about Frame Relay at the following web sites:

- http://www.cisco.com/univercd/cc/td/doc/cisintwk/ito_doc/frame.htm
- http://www.alliancedatacom.com/Frame Relay-success-tutorial.htm
- http://www.alliancedatacom.com/Frame Relay-white-papers.asp
- http://www.trillium.com/whats-new/wp_frmrly.html

The Complete Reference

Chapter 4

ATM and ISDN

This chapter covers two widely adopted WAN technologies: Asynchronous Transfer Mode (ATM) and Integrated Services Digital Network (ISDN). ATM covers the need for incredibly high-speed data, voice, and video transfers, whereas ISDN covers the need for moderate-speed data and voice connections in areas where no other low-cost possibilities (such as Digital Subscriber Line [DSL] or cable) exist, as well as providing demand-dial connections that are significantly faster than Plain Old Telephone Service (POTS).

What Is Asynchronous Transfer Mode?

Asynchronous Transfer Mode (ATM) is a relatively new technology (compared to Frame Relay, anyway) that is becoming the technology of choice for high-speed multiservice connections. ATM came about because of a growing need for a technology that could support many different types of traffic over varying distances, across administrative and national boundaries, at a wide variety of speeds, and with full quality of service (QOS) support. As a result, ATM is one of the most complex and versatile protocols in use today. Luckily, most network administrators and engineers don't need to understand all of the complexity behind ATM. Therefore, this chapter simply covers the basics of ATM and how it works in most environments.

How ATM Works: Core Concepts

ATM is a technology that doesn't map very well to the OSI model because it was built using the broadband ISDN (BISDN) model. However, it maps primarily to the datalink layer, with some specifications that could be placed in the physical and network layers. Central to ATM is the concept of multiservice transmissions. The requirement that ATM be able to adequately handle many types of transmissions has led it to become quite a bit different from the other protocols that we have discussed. However, it is also quite similar to Frame Relay, including the concept of virtual circuits (VCs).

Virtual Circuits

Like frame relay, ATM uses virtual circuits—both permanent virtual circuits (PVCs) and switched virtual circuits (SVCs)—to make and address connections to other ATM devices. Unlike Frame Relay, however, ATM defines one additional type of virtual circuit: the soft permanent virtual circuit (SPVC).

An SPVC is basically a PVC that is initiated on demand in the ATM switches. From each endpoint's perspective, the SPVC appears to be a standard PVC; but to the ATM switches in the cloud, the SPVC differs in one significant way. With a PVC, a VC is created statically throughout the cloud and is always up. With an SPVC, however, the connection is static only from the endpoint (DTE) to the first ATM switch (DCE). From

DCE to DCE within the cloud, the connection can be built and rebuilt on demand. The connection is still static once built, unless a link problem causes the VC within the cloud to come down.

In this case, without the manual intervention required on a PVC, the SPVC automatically attempts to reestablish the VC by using another route. With a PVC, the provider needs to manually rebuild the PVC in the cloud. This feature of SPVCs leads to higher reliability of the link, with no additional configuration required on the DTE end.

In addition, while SVCs in Frame Relay are not very well defined, the call establishment and termination features that an SVC requires are pretty well defined in ATM. Consequently, you can actually get SVC service from many providers for your ATM WAN connections. While SVC service results in dynamically available connections based on current needs and a possible reduction in costs, it is considerably more difficult to configure. In addition, SVCs are usually *tariffed*, or priced based on connect time, class of service, and bandwidth used, which can actually lead to an increase in cost if connections stay active for a long time. For these reasons, we will still concentrate on PVCs.

ATM adds a few more pieces to the VC puzzle as well, assaulting you with an "acronym soup" of different classifications of VCs. In addition to the differentiation between SVCs, PVCs, and SPVCs, you now have to differentiate between User to Network Interface (UNI) and Network to Network Interface (NNI) connections. On top of this, UNI and NNI connections are further differentiated between public and private connections. (Sometimes, you have to wonder if the unstated goal for designing a new technology is to confuse as many people as humanly possible.) Let's tackle the UNI versus NNI differentiation first.

User to Network Interface (UNI) is the connection from a user device (DTE) to a network device (DCE). In a simple ATM scenario in which your company is connected to the provider using a router, and you are connected to the router through an Ethernet, the connection from the router to the provider would be a UNI connection. The connection from you to the router would *not* be a UNI connection, because to be an ATM UNI connection, it must first be an ATM connection.

Network to Network Interface (NNI) is the connection from ATM switch (DCE) to ATM switch (DCE) within the cloud. You will not need to deal with this type of connection in most cases, unless you work for a provider or have your own large-scale ATM backbone in-house.

As for the public versus private designation, this simply defines whether the connection is over a public infrastructure (such as a provider's cabling) or a private infrastructure (such as your own ATM backbone). The main difference is in the cabling and distance requirements for the two; typically, determining whether your link is public or private is not very difficult. Figure 4-1 illustrates the UNI/NNI and public/private designations.

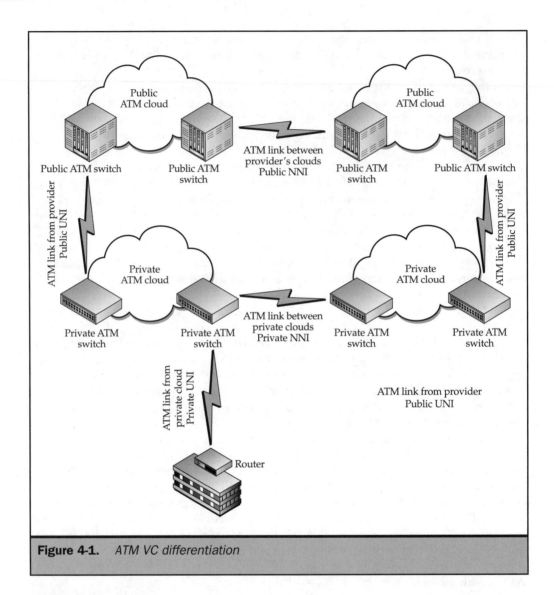

Figure 4-1. *ATM VC differentiation*

ATM Addressing

ATM addressing is an interesting topic because ATM defines addresses on two separate levels. First, ATM addresses its VCs, similar to frame relay DLCIs. Second, ATM addresses the ATM device itself, similar to Ethernet's Media Access Control (MAC) address. We will discuss the VC addressing first.

ATM addresses VCs with virtual path identifier (VPI) and virtual circuit identifier (VCI) pairs. The VPI sets parameters like the bandwidth for the connection and QOS options. VCIs have the same QOS requirements as the VPI and fit within the bandwidth allocation of the VPI, and are then run inside the VPI to actually make the connection. The basic premise of VPI/VCI pairings is that it is easier for the provider to manage than DLCIs. For example, rather than having 15 DLCIs for your company at odd intervals such as DLCI 534, 182, 97, 381, and so on, and having to look at a chart to figure it all out, the provider can have one VPI for your company with 15 VCIs inside it. All the engineer needs to know is which VPI you have.

VPI/VCI pairings are also more scalable than DLCIs; for example, rather than having 10 bits to uniquely identify the connection, you have 24 bits (8 VPI plus 16 VCI) to identify a UNI VC. In addition, VPI/VCI pairings allows most of the call setup overhead to be completed on the VPI, with the VCI simply accepting the VPI's parameters. Figure 4-2 shows an example of VPI/VCI pairings.

As for the ATM device addressing, it is a bit more complicated. ATM addresses are 20 bytes long and have a considerably more complex addressing structure than most other address structures. Luckily, you generally won't have to deal with ATM device addressing. Cisco ATM devices come with an ATM address already set, and as long as you are not building a large ATM infrastructure, the preset address should suffice. If you have public ATM links, your provider can often give you an ATM address. If you are building a large public infrastructure (in other words, if your company becomes an ATM provider), you will have to apply for a block of ATM addresses to be assigned to you by the sanctioning body in your country.

However, if you are building a large private ATM infrastructure, you will need to bone up on ATM addressing and a lot of other topics. Because you are unlikely to find yourself in such a situation, ATM addressing is beyond the scope of this book. However, ATM's addressing structure, shown in Figure 4-3, can be briefly explained as follows:

- **ESI** End System Identifier. Used to identify your individual device (and is typically a MAC address).

- **Selector Byte** Used for local multiplexing at the end station. Insignificant to other devices.

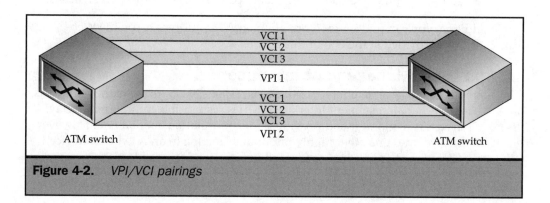

Figure 4-2. *VPI/VCI pairings*

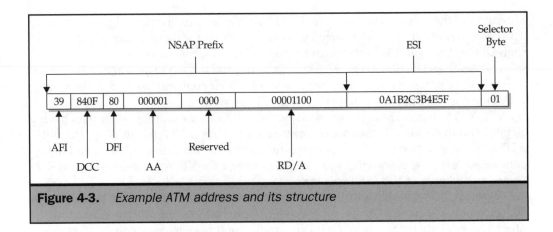

Figure 4-3. *Example ATM address and its structure*

■ **NSAP Prefix** Network Service Access Point. Your ATM network. The NSAP Prefix is also further broken down. There are actually three different formats for the NSAP portion: DCC, ICD, and E.164. Figure 4-3 uses the DCC format, which is broken down as follows:

■ **AFI** Address Format Indicator. Used to determine the layout of the NSAP (DCC, ICD, or E.164).

■ **DCC** Data Country Code. Used to determine which country you are in (840F is U.S.).

■ **DFI** Domain Format Identifier. Used to identify who the registering body for your address is (ANSI is 80).

■ **AA** Administrative Authority. Used to identify the unique number for the registering company (your company).

■ **Reserved** Reserved for future use.

■ **RD/A** Routing Domain and Area. Used to address your ATM switches.

In addition, you can find more information on ATM addressing on the Web by using the links provided at the end of this chapter.

Interim Local Management Interface

The Interim Local Management Interface (ILMI) is responsible for many of the same tasks in ATM as the Local Management Interface (LMI) is responsible for in Frame Relay. Its main purpose is to monitor and provide status information about a link. However, the ILMI also provides one other highly important function in an ATM UNI device: address registration. ATM addresses for UNI devices are designed to be automatically configured. As soon as an ATM UNI device boots, it sends out a message over the ILMI channel (VPI 0, VCI 16, by default) announcing its MAC address to the ATM switch.

The ATM switch then attaches the NSAP to the MAC address and sends it back to the client device as the device's new ATM address. In this manner, the client device (DTE) never needs to know the ATM addressing structure of the network, and it doesn't need to be manually reconfigured if the structure changes.

Framing

ATM framing is a bit different than other protocols. As mentioned at the beginning of this chapter, ATM was designed to handle multiple types of data over the same link. Initially, the primary types of data that needed to be considered fell into three categories: voice, data, and video. The requirement for ATM to handle multiple data types almost ensured that ATM would be unlike most LAN or WAN protocols of the time. By taking a close look at the three types of traffic ATM was initially designed to support, we can see why:

- **Voice traffic**   Voice traffic has some simple requirements. It requires a constant data rate, but it can be fairly slow (64 Kbps). It can also accept the loss of small amounts of data without significantly affecting the communication. However, latency and sequencing are very important. In other words, data must arrive in the correct order (or you might end up sounding like a bad imitation of Yoda: "Go home we will, yes"). And the data must have a constant delay; otherwise, artificial pauses may be introduced.

- **Video traffic**   Video traffic is a bit more complicated. It can be sent over a constant or variable data rate (if compression is used), and it generally needs fairly high speeds (256 Kbps+). It can accept the loss of small amounts of data as well. Latency and sequencing are still extremely important, however.

- **Data traffic**   Data traffic requirements are completely different from voice and video requirements. Data traffic rarely runs at a steady bit rate. More commonly, it is very bursty, with little or no traffic for extended periods of time, and then large quantities of traffic sent as fast as possible. In addition, data traffic generally cannot accept the loss of any data because even a single bit error can corrupt the entire transmission. However, latency is generally unimportant. Sequencing is still important, but in a very different way. It no longer matters whether the frames *arrive* in the correct order, as long as they can be *reassembled* in the correct order at the destination. This contrast is stark compared to the other types of traffic, in which the chunks of data must arrive in the correct order.

As you can see, the standards committees had their hands full when trying to build a technology that not only could fill the immediate need of supporting all of these traffic types, but also could be adapted to future types of traffic that we cannot even imagine today. The committees chose two main methods of dealing with these problems: fixed-length frames and adaptation layers.

ATM frames are called *cells*. (Therefore, ATM is sometimes referred to as *cell relay*.) They are *fixed length*: unlike Ethernet or frame-relay frames, they have a set size and can never be any bigger or smaller. Any additional space in the cell that isn't occupied by data must be filled with padding. The cell must always equal 53 bytes. (The voice group wanted a 32-byte payload and the data group wanted a 64-byte payload, so they added them together and divided by 2 to reach a compromise of 48 bytes for the payload.)

Compared to most other technologies (such as Ethernet with its 1,500-byte Maximum Transmission Unit), this number seems small and inefficient. However, having a fixed-size cell provides one important benefit: low latency. In other technologies, the switches and endpoints have to look at the length field in the frame and determine where the frame is supposed to end. ATM switches eliminate this step entirely. The cell will always be 53 bytes long. This leads to a fixed amount of latency per frame, which is important for voice and video traffic. Figure 4-4 shows a basic ATM cell.

In addition, ATM cells come in two basic types: UNI cells and NNI cells (detailed in Figures 4-5 and 4-6). The reason for the different cell types is that a UNI cell is assumed to need a smaller number of VPIs than an NNI cell. This assumption is reasonable because the NNI cell will presumably be the provider and may be carrying a huge number of ATM VCs. In the UNI cell, an additional field takes up the four extra bits used for VPI addressing in the NNI cell. This field is called Generic Flow Control (GFC). The idea behind this field was that UNI devices could use it for some future purpose. However, it has not been formally defined, so it is typically unused.

After the VPI and VCI fields, there is a Payload Type field. This field is used to tell the ATM devices whether the cell is a data cell, an Operations and Management (OAM) cell, or an idle cell, and whether congestion is being experienced. An OAM cell is used for functions similar to those undertaken by the ILMI. In addition, the OAM cell can carry information for loopback and other link-monitoring purposes.

An *idle* cell is a cell that is inserted into the stream because no user data is being sent at that time. Contrary to its name, ATM actually runs over synchronous links. It is called *asynchronous* because ATM does not require that *data* be sent in a synchronous fashion. ATM has idle cells to deal with times when no data needs to be transmitted. However, ATM does require a synchronous physical layer protocol (like SONET) to send data because ATM has no start or stop signals to identify frames. ATM needs a clocking source, just like frame relay, to synchronize the transmissions.

However, it's how ATM detects the clocking signal that makes idle cells important. Because ATM does not require a clocking *signal*, just a clocking *source*, it needs a

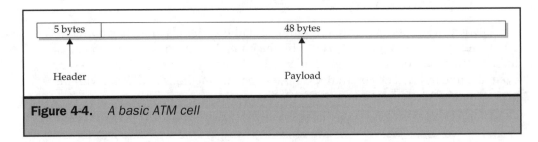

Figure 4-4.   *A basic ATM cell*

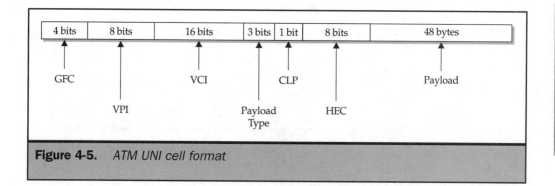

Figure 4-5. *ATM UNI cell format*

way to determine what the clock is set at. It does this by looking for the Header Error Control (HEC) field in the frame. Once it finds this field, it knows (based on the delay values between HEC fields) the clocking of the connection. Determining the clocking rate based on the HEC field requires that a constant stream of cells be transmitted so that the HEC fields can be checked, whether data needs to be transmitted or not.

After the Payload Type field, the Cell Loss Priority (CLP) field is listed. The CLP field is similar to the DE bit in Frame Relay. It tells which cells are eligible for discard.

Finally, the HEC field (mentioned previously) comes into play. This field is a checksum used primarily to detect errors in the ATM cell header. If an error is found, the HEC field contains enough parity information to fix any single bit error. This field is also used to synchronize ATM to the clock source.

After using fixed-length frames, the second step in dealing with multiple traffic types involves using a layered approach to cell preparation. Because of the differing needs of voice, video, and data traffic, various techniques are used to package each type of traffic. ATM accomplishes this (frame repackaging based on traffic type) with an adaptation layer. The ATM adaptation layer (AAL) is responsible for packaging the payload of the cells based on the type of traffic that is to be sent. The ATM adaptation layer encapsulates the data again before placing it in the payload area of the ATM cell, as shown in Figure 4-7. ATM formally defines five different adaptation layers, AAL 1 through 5, but layers 3 and 4 are combined into AAL 3/4, so there are really only four layers.

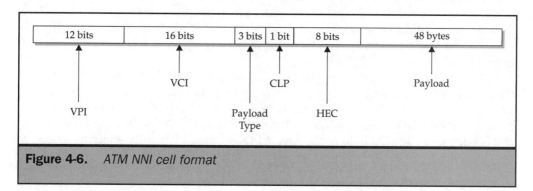

Figure 4-6. *ATM NNI cell format*

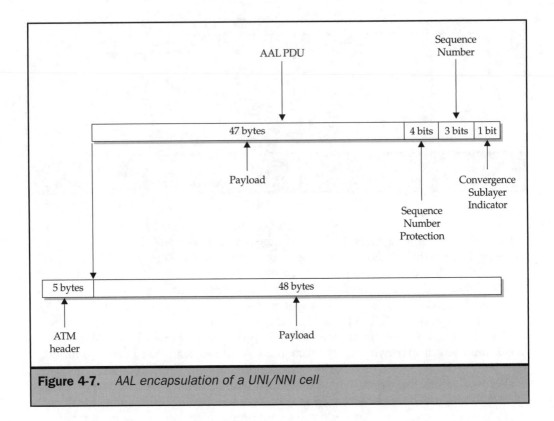

Figure 4-7.   *AAL encapsulation of a UNI/NNI cell*

AAL 1 is designed for voice traffic. Voice traffic is highly tolerant of errors, so the AAL 1 PDU does not have an FCS or CRC field. Voice traffic does have to be in sequential order, however, so sequence number and sequence number protection fields are inserted. Figure 4-8 shows an AAL 1 PDU.

AAL 2 was designed for video transmissions. Video transmissions are somewhat tolerant of errors, but much less so than voice transmissions. Therefore, a CRC field (similar to the FCS in Ethernet) is required in the AAL 2 PDU. In addition, each frame of a video transmission will most likely span more than one 53-byte cell, so a Length field, an Information Type field, and a Sequence Number field tell ATM how many cells this transmission spans and which part of the transmission any particular cell is. Figure 4-9 shows an AAL 2 PDU.

AAL 3 was designed for connection-based transmissions. AAL 4 was designed for connectionless transmissions. They are very similar in structure, and are therefore generally combined as AAL 3/4. The only new fields in these PDUs are the RES and Multiplex ID (MID) fields. The RES field in the AAL 3 PDU is reserved for future use and is not used. The MID field in the AAL 4 PDU is used to identify which *datagram*

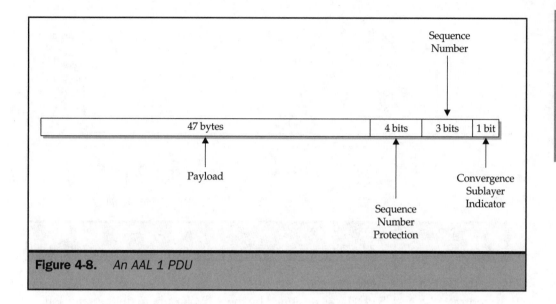

Figure 4-8. *An AAL 1 PDU*

(large packet made out of many smaller cells) this cell is a member of, because in connectionless communication, there is no initial communication to establish this piece of information. Figures 4-10 and 4-11 show AAL 3 and 4 PDUs. The addition of the RES and MID headers leads to a fairly inefficient PDU with a data payload of only 44 bytes per frame.

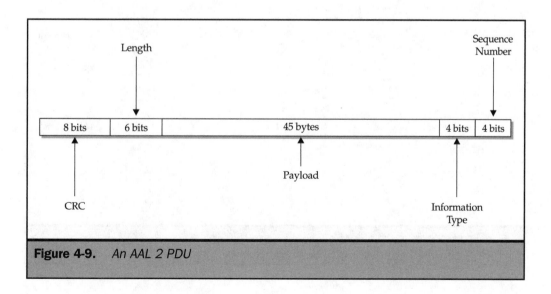

Figure 4-9. *An AAL 2 PDU*

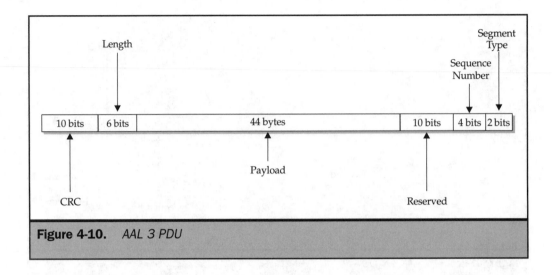

Figure 4-10.   *AAL 3 PDU*

In contrast, the AAL 5 PDU was designed by the ATM Forum to do the same job as the AAL 3/4 PDU, but with less overhead. The AAL 5 PDU relies on upper-layer protocols (like TCP) to perform the error recovery, and it expects that the cabling used for ATM will be fairly error free. Because both of these assumptions are typically correct, AAL 5 can eliminate all of the additional information included in every cell. The AAL 5 PDU simply adds all of this information into the larger PDU (usually a TCP segment), and then it chops this segment into cells.

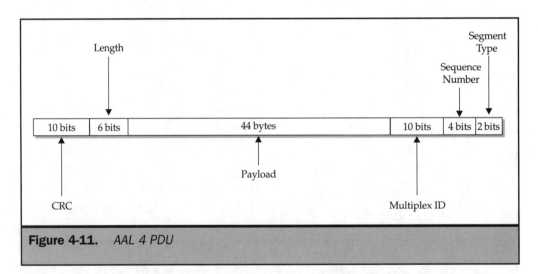

Figure 4-11.   *AAL 4 PDU*

Upon reassembly, all of the needed information is there; it just wasn't duplicated on every cell. This lack of duplication allows an AAL 5 PDU to squeeze in four more bytes of data per cell, for a full 48 bytes of data per cell. (I know it doesn't sound like much, but it is nearly a 10 percent increase over the AAL3/4 PDU.) The kicker is that for every segment transmitted, a large part of one cell's payload for that segment is taken up by overhead. However, if you are using the maximum TCP segment size of 65,536 bytes, you lose only one cell out of every 1,365 or so to overhead: not a bad trade-off. Therefore, the AAL 5 PDU is probably the most-used PDU for data.

Topology

ATM is built using the same logical topology models as frame relay. ATM uses both point-to-point and point-to-multipoint connections and can be configured in either a hub-and-spoke or mesh (full or hybrid) topology.

How ATM Works: Advanced Concepts

Advanced ATM is a subject that can span several volumes of books. In this section, we briefly overview some of the more advanced features of ATM, including QOS, Error Recovery mechanisms, and physical connectivity.

Quality of Service Enhancements

Basic QOS features in ATM are similar to those in Frame Relay. ATM uses the Protocol Type field in the cell to indicate per-VC link congestion (similar to Forward Explicit Congestion Notification [FECN] fields in Frame Relay), and it uses the Cell Loss Priority (CLP) field in the same manner as the DE bit in Frame Relay. However, the primary ATM QOS mechanism consists of three parts: traffic contracts, traffic shaping, and traffic policing. All of them work on the same principles as Frame Relay Traffic Shaping, using the committed information rate (CIR) (called sustainable cell rate [SCR] in ATM), committed burst (Bc), and excess burst (Be) to determine bandwidth on a per-VC basis.

However, in ATM, these parameters are established through all switches in the cloud at VC creation. In other words, the UNI device (DTE) makes a request to the first hop NNI device (DTE), saying "I want an SCR of 256 Kbps, a Bc of 512 Kbps, and a Be of 756 Kbps on VPI 1 VCI 1 that should connect me to that host over there." The NNI device then finds a route through the cloud and, hop by hop, proceeds to verify that every switch along the way agrees to these parameters.

If a switch rejects these parameters (perhaps the other VCs on the switch are taking up too much bandwidth and the switch has only 128 Kbps available), the other switches try to locate another path and begin the process along that path. If no suitable path is found, the switch rejects the connection. In this way, if a connection is established, bandwidth is practically guaranteed throughout the entire cloud.

Let's take a look at the individual pieces of this QOS puzzle.

Traffic Contracts   *Traffic contracts* are the QOS parameters of the connection that are guaranteed at connection establishment. Basically, just like the name implies, these are contracts between the ATM devices. These contracts are the rules by which both sides agree to play. I say I won't send any more sustained traffic than 256 Kbps, and the network agrees to provide me with this amount of data throughput.

Traffic Shaping   *Traffic shaping* is a tool to help the UNI devices keep their end of the bargain. For instance, in most networks, you will have one or more Ethernet segments connected to a router (your UNI device or DTE), which, in turn, provides the connection to the ATM provider's switch. The problem with this situation is that Ethernet has absolutely no idea about ATM QOS functions. Ethernet sends as much as possible, as quickly as possible. So the Ethernet client sends data as fast as it can to the router to go to the ATM network.

Just for kicks, let's say the Ethernet client sends a sustained data stream at 20 Mbps. Your ATM SCR is 256 Kbps. Something is obviously wrong. The router will not be able to buffer all of the information: some will get lost, while other information will get sent, all in a fairly randomized manner. Traffic shaping helps solve this problem by selectively modifying the queues when the traffic begins to exceed the capacity of the VC. Traffic shaping begins to drop packets as soon as it notices the Ethernet client breaking the rules, rather than waiting until the router or switch's buffers are overflowing.

This solution may sound just as bad as the original problem right now, but in Chapter 5, you will see how this is actually helpful. If the client is using TCP, TCP will automatically adjust fairly rapidly and realize that some of its packets are getting lost. It will then assume that an error condition has occurred and reduce the rate of transmission while also retransmitting the lost frames.

Traffic Policing   *Traffic policing* is similar to traffic shaping, except that it occurs on your provider's switches. Traffic policing is the bouncer of ATM QOS. If you set a contract, yet your router refuses to abide by the contract and sends data for too long at or above the Bc, traffic policing steps in and sets the CLP bit on the errant cells, causing them to be "bounced" into the bit bucket.

Speeds

As far as speeds are concerned, ATM is similar in some respects to Frame Relay. ATM does not provide for a physical-layer specification. Rather, like Frame Relay, it relies on the physical-layer specifications of other protocols—namely, Synchronous Optical Network (SONET), Synchronous Digital Hierarchy (SDH), and Plesiochronous Digital Hierarchy (PDH). Generally, SONET is assumed, but it is not necessarily required. PDH is used for slower bandwidth connections, and it is available in speeds from DS-0 (64 Kbps) to DS-4 (274.167 Mbps), but ATM is not typically used at speeds below DS-1. SONET/SDH is available in the following speeds: OC1 (54.84 Mbps), OC3 (155.52 Mbps), OC12 (622.08 Mbps), OC48 (2.488 Gbps), OC96 (4.976 Gbps), OC192 (9.953 Gbps), and OC768 (35.813 Gbps).

Obviously, ATM Internet connections are not for the faint of heart and wallet. ATM also shares its per-VC bandwidth classifications with Frame Relay (SCR, Be, Bc, line speed), although the naming conventions may be a bit different. Also, ATM PVCs are priced individually in addition to the line charge, like Frame Relay. ATM SVCs, however, are usually tariffed as well, meaning that you are also charged for connect time and data transmitted. All in all, ATM is very high speed and can scale to just about any speed required—assuming, of course, that your pockets are deep enough.

Error Recovery

Like Frame Relay, even though ATM is technically a connection-based protocol (because a connection, or VC, must be established before data is sent), ATM does not include end-to-end error correction mechanisms, making ATM cell transmissions best effort or unreliable at the datalink layer. Remember, however, that this does not mean that all ATM transmissions are unreliable. Just as with Frame Relay, ATM relies on the upper-layer protocols (such as TCP at the transport layer) to ensure reliable data transfer. ATM does, however, use the HEC field in the ATM header to reconstruct any single bit error in the ATM header. Also, depending on the AAL used, it performs error checking (but not correction) on the payload of the ATM frame.

Physical Connections

Similar to Frame Relay, ATM physical connections can be made by using an ATM DSU (ADSU) over a High Speed Serial Interface (HSSI) connection. ADSUs are used for devices that do not inherently support direct ATM connections, but that can support the ATM Data Exchange Interface (DXI). You can also implement ATM in certain devices as a native connection. Native ATM support requires that the switch or router have the appropriate connection type and port adapter installed as well (such as the correct speed and type of fiberoptic SONET interface). Cisco's ATM-switching products include the Catalyst 5500, Lightstream 1010, and BPX switches. Individual switch models are discussed further in Chapters 9 and 10.

What Is Integrated Services Digital Network?

Integrated Services Digital Network (ISDN) is a specification that defines high-speed digital data transfer over standard phone lines. ISDN was designed to replace POTS connections for customers who required higher data transfer speeds than POTS could provide. ISDN provides the same features as a POTS connection (namely, on-demand connections, voice and data transfer capability, and caller ID), and it also adds a few of its own features into the mix.

In most environments today, you will see ISDN in one of two roles: as a backup link for a router in case the router's primary link (usually Frame Relay) goes down, or as a data communications link for the home/office user who cannot get cable or asymmetric digital subscriber line (ADSL) services. Figure 4-12 shows examples of these uses.

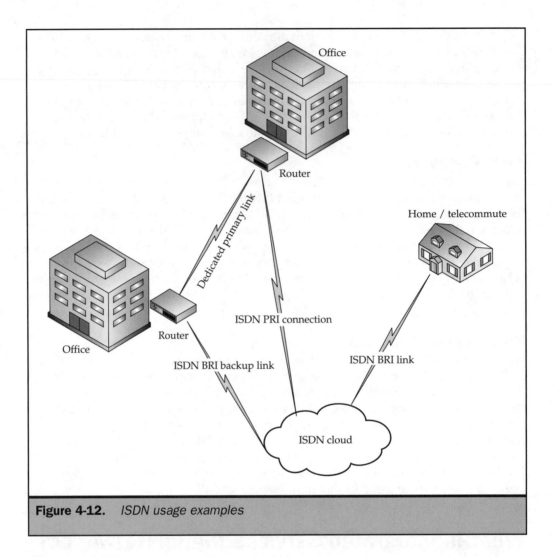

Figure 4-12.   *ISDN usage examples*

Unfortunately for ISDN, cable, and xDSL (any type of DSL connection, including SDSL and ADSL), connections are faster and cheaper in most cases, forcing ISDN out of the marketplace. However, ISDN still has its uses, and it is still present in many environments. Therefore, the rest of this chapter is dedicated to the functions of ISDN.

How ISDN Works: Core Concepts

ISDN is a different type of technology than most of what we have seen thus far. Unlike Frame Relay or ATM, ISDN does not use VCs. ISDN is channelized, like a T1 line, using time-division multiplexing (TDM). The first step in understanding ISDN operation is to understand its channelization.

ISDN Basic Rate Interface and Primary Rate Interface

ISDN comes in two basic flavors: Basic Rate Interface (BRI) and Primary Rate Interface (PRI). BRI connections consist of two bearer (B) channels and one delta (D) channel (sometimes called a *data channel*). For this reason, BRI is sometimes referred to as 2B+D. ISDN BRI connections run over standard copper telephone wiring, so rewiring of the customer's premises is not usually required. Each B channel in ISDN can support up to 64 Kbps of data transfer, or one standard voice conversation. In BRI, the D channel is a 16-Kbps channel used for signaling (call setup, teardown, and so on). This gives ISDN BRI a maximum uncompressed data throughput of 128 Kbps.

While significantly faster than a standard phone connection, this data rate is a bit slow for most businesses. ISDN PRI is designed to address this problem. ISDN PRI in North America contains 23 B channels and 1 D channel. (In Europe, it consists of 30 B channels and 1 D channel.) The D channel in PRI is 64 Kbps; but other than this (and different physical cabling, which we will discuss further in the "Physical Connections" section, later in this chapter), the PRI just provides higher bandwidth and better response times than a BRI.

ISDN Addressing

ISDN has its own addressing structure, consisting of a service access point identifier (SAPI) and a terminal endpoint identifier (TEI). In basic terms, the SAPI is the address of the ISDN link, and the TEI is the address of your specific device on that ISDN link. This address structure allows you to have multiple devices (up to eight) sharing a single ISDN link because each device has its own TEI. However, in most cases, knowledge of ISDN addressing is not required because, from the customer's perspective, ISDN is typically addressed with telephone numbers. In some cases, both B channels may use the same telephone number; whereas in other cases, each B channel may get its own number. When you want to connect to another location using ISDN, you have your ISDN "modem" (it's really a terminal adapter, not a modem) dial the phone number for the other ISDN device and connect, the same as with a standard analog modem.

In addition, you may be assigned a service provider ID (SPID) from your provider. The SPID is usually used only in North America, and it typically contains your ISDN phone number plus a few additional digits. The SPID is not universally defined, and each ISDN provider may have a different SPID specification in use, but the purpose of the SPID is to let the provider's ISDN switches know how to route the calls from your ISDN device and to reveal what the device is capable of (voice traffic, data traffic, or both). One way or the other, if your provider assigns you a SPID, you will have to set it correctly on your ISDN device to get service.

In addition to the SPID, several different types of ISDN-switching equipment exist. These devices all perform tasks slightly differently, making it imperative that you specify the correct type of switch for your ISDN connection. Most Cisco equipment supports four types of U.S. ISDN switches: 4ESS, 5ESS, DMS100, and NI-1. In addition, several international switches are supported. Although the differences between these switch types are not usually important to know (unless, of course, you are an ISDN provider), you will need to know which switch type you are connected to in order to configure your ISDN equipment properly. Your provider should relay this information to you.

Framing

ISDN uses a different framing specification for B channels and D channels. For D channels, Link Access Procedure over the D channel (LAPD) and International Telecommunications Union (ITU) specification Q.931 are typically used to perform signaling and multiplexing. In B channels, Point-to-Point Protocol (PPP), the most common protocol for TCP/IP internetworking over modems, is typically used. PPP is defined in RFC 1662, at http://www.rfc-editor.org/rfc/rfc1662.txt, and is covered more thoroughly in Chapter 5.

How ISDN Works: Advanced Concepts

The ISDN concepts discussed in this part of the chapter, while still technically "core" to ISDN functionality, are a bit more complicated than the previous concepts. In this section, we will examine ISDN speeds, error-recovery mechanisms, and (the part that confuses most people) physical connectivity.

Speeds

ISDN can run at a variety of speeds by using a technique known as bonding. *Bonding* occurs when multiple data connections are combined to provide a single channel of higher throughput. For instance, if you have a BRI connection with two B channels, then you could have two simultaneous connections: either voice or data. So you could be talking to a buddy while surfing the Internet. In this case, one B channel is allocated to data traffic and the other is allocated to voice traffic. However, when you've finished your voice conversation, your second B channel is now inactive and unused. By using Multilink PPP and Bandwidth Allocation Protocol (BAP), however, you could have the second B channel bonded to the first B channel upon the completion of your voice conversation, giving you the full 128 Kbps of bandwidth. In addition, you do not lose any voice functionality because ISDN uses the D channel for signaling.

In other words, if your buddy calls back while you are using both B channels, your ISDN device can remove one of your B channels from the data connection and reallocate it to the voice connection. This is because ISDN uses the separate D channel to perform the initial call connection. So, even though you were using both B channels, ISDN can still detect an incoming call because the initial call establishment message travels over the D channel. With bonding, you can dynamically assign bandwidth where it is needed, up to the B channel limit, without sacrificing other services (such as voice).

ISDN physical speeds are limited to 128 Kbps for ISDN BRI and T1/E1 for ISDN PRI.

Error Recovery

For LAPD on the D channel, both acknowledged and unacknowledged services are defined. This means that, in the acknowledged transfer, error detection and recovery is supported using windowing. In the unacknowledged transfer service, error detection is still supported using a CRC, but error recovery is not supported. For the B channels, using PPP provides error detection, but error recovery is the responsibility of upper-layer protocols (such as TCP).

Physical Connections

Probably the most complicated specification in ISDN is the physical connection. ISDN physical devices are separated into two main groupings: function groups and reference points.

ISDN *function groups* are most easily described as hardware devices. Terminal adapters (ISDN modems), native ISDN equipment, and channel services unit/data services unit (CSU/DSU) (in PRI circuits) are all defined as function groups. You need to be aware of five different function groups. (Several other function groups exist, but they are for the provider's side of the connection.)

The first type, terminal adapters (TAs), are responsible for converting an ISDN signal into something that a non-ISDN-compatible piece of equipment (TE2) can understand. A good example of this is an ISDN modem that is plugged into your PC's USB port.

Next, the network terminators (NTs) are responsible for powering the circuit, performing multiplexing, and monitoring activity on the circuit. They are the points at which your network meets the provider's network. There are two types of NTs: NT1 and NT2. NT1 devices are the most common. These devices simply convert the U reference point (or two-wire cabling) into the standard S/T reference point (four-wire cabling). NT1 devices are included in some models of ISDN modems, making them both TAs and NT1s. NT2 devices are typically complex devices like Private Branch Exchange (PBX, an internal telecom switching system), and they may perform other functions with the ISDN link, such as converting the PRI circuit into several BRI circuits. NT2 devices are optional.

Finally, terminal equipment (TE) also comes in two varieties: TE1 and TE2. The TE1 is a device that natively understands ISDN. This device has no need for an adapter, and it can plug directly into the four-wire S/T cable. An example is a digital phone. TE2 devices, on the other hand, do not natively understand ISDN, and they require a converter, or TA, to communicate over the ISDN network. They use a variety of interfaces to plug into the TA, from a PCI bus on a PC to an RS-232 serial connection. Regardless of the connection used, however, it is always called an R reference point.

ISDN *reference points* are most easily described as cables and connectors. We must concern ourselves with four main reference points: U, T, S, and R.

- **U**   A two-wire circuit directly from your provider. Typically, this reference point is your ISDN phone line. This cable plugs into an NT1 device to be converted into the next reference point, T (or S/T).

- **T**   Used to connect to an NT2 device. In an environment in which no NT2 device exists, the T reference point becomes an S/T reference point for TE1 and TA function groups.

- **S**   Used to connect to TE1 devices or TAs.

- **R**   Used to connect from TAs to TE2 devices. An example of this is the USB cable that connects your ISDN modem to your PC.

Figures 4-13 through 4-17 show examples of the ISDN function groups and reference points. Figure 4-13 depicts a home setup with two devices: one for the TA role and one for the NT1 role, with a S/T reference point (a UTP cable) connecting the two.

Figure 4-14 depicts a home setup with combined devices; the ISDN "modem" performs both the TA and NT1 roles, and the S/T reference point is internal on the circuit boards within the modem.

In Figure 4-15, you can see a simple example of sharing a single ISDN line among multiple devices, which is similar to using "splitters" for POTS lines.

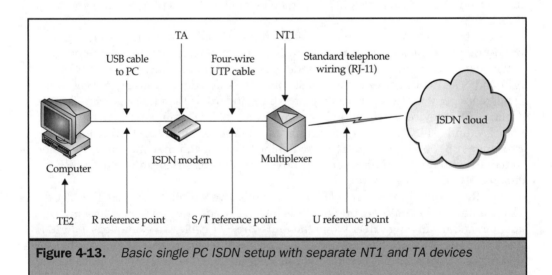

Figure 4-13.   *Basic single PC ISDN setup with separate NT1 and TA devices*

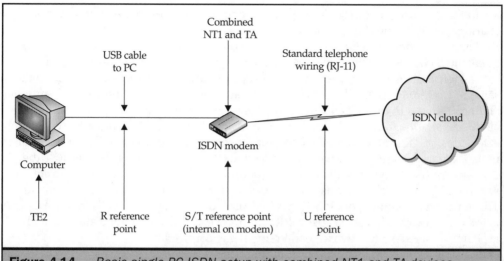

Figure 4-14. *Basic single PC ISDN setup with combined NT1 and TA devices*

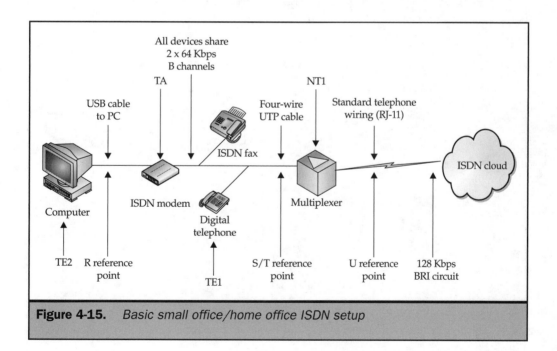

Figure 4-15. *Basic small office/home office ISDN setup*

In Figure 4-16, you can see a slightly more complicated example, with the PCs sharing the ISDN line using a router.

Figure 4-17 depicts a more-complex, large office ISDN setup, including a PBX to break the B channels of the ISDN PRI into voice or data lines for individual users.

In Figures 4-16 and 4-17, you may have noticed multiple devices connected to the same S/T reference point. This is possible because of the Terminal Endpoint Identifier (TEI) mentioned in the "ISDN Addressing" section earlier in this chapter. Although all devices sharing the same S/T reference point have the same SAPI, they all have unique extensions, or TEIs, to distinguish them from each other. Also notice in Figure 4-17 that there is a variable amount of bandwidth for data traffic. This is possible because the PBX system (NT2) dynamically combines (a process known as *inverse multiplexing*) or separates bonded B channels as needed for data and voice traffic, based on demand. This function is common for an NT2 device. Also notice that a CSU/DSU in this example is the NT1 device. This is because a PRI circuit is typically run over a standard T1 UTP link, just like some lower-speed Frame Relay connections.

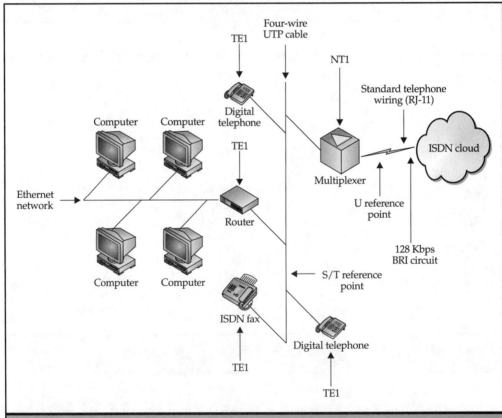

Figure 4-16.   *Small branch office ISDN setup*

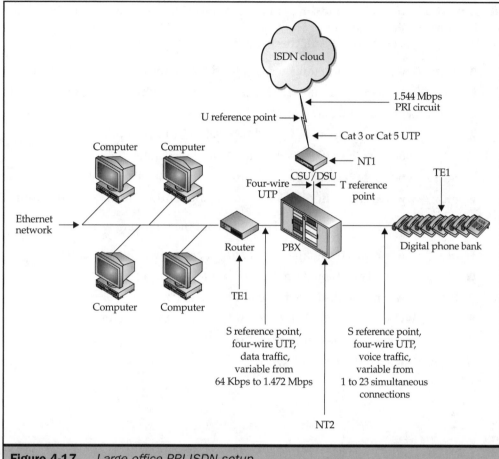

Figure 4-17. *Large office PRI ISDN setup*

Summary

In this first part of the chapter, you learned why ATM was created and how its unique features can help you support a wide variety of traffic types. You also investigated the similarities and differences between ATM and Frame Relay. While ATM is still in its infancy (as far as networking standards go), the rapid growth of this technology is already felt, and its future impact on our networks as the high-speed WAN technology of choice is almost assured. Armed with an understanding of how ATM works, now you are better prepared to configure your ATM devices when the need arises.

In the final section of the chapter, you learned that ISDN can provide higher-speed connections for small environments, as well as backup connections for larger offices. In addition, you learned about the different types of ISDN connections and how they are used in common environments.

For more information on ATM, check out the following web sites:

- http://www.cisco.com/univercd/cc/td/doc/cisintwk/idg4/nd2008.htm
- http://www.cisco.com/univercd/cc/td/doc/cisintwk/ito_doc/atm.htm
- http://www.tticom.com/atmglosy/atmlex.htm
- http://www.employees.org/~ginsburg/0201877015ch1-3.pdf
- http://media.it.kth.se/SONAH/RandD/atmindex.htm
- http://www.scan-technologies.com/tutorials/ATM%20Tutorial.htm
- http://www.sci.pfu.edu.ru/telesys/studwork/telfut96/Isdn/atm/tute.html

For additional information about ISDN, check out the following web sites:

- http://www.cisco.com/univercd/cc/td/doc/cisintwk/idg4/nd2011.htm
- http://w10.lanl.gov/lanp/atm.tutorial.html
- http://www.nationalisdncouncil.com/isdnassistance/
- http://www.scan-technologies.com/tutorials/ISDN%20Tutorial.htm
- http://www.protocols.com/pbook/isdn.htm
- http://www.ralphb.net/ISDN/

The Complete Reference

Cisco

Chapter 5

TCP/IP Fundamentals

First implemented in the early 1970s, TCP/IP has grown to become the dominant protocol suite. Because it was initially designed for the Department of Defense's (DOD) ARPANET—the "mother" of the Internet—TCP/IP is one of the few protocol suites in use today that is completely vendor independent. Amazingly, its growth has skyrocketed to the point that nearly every OS in existence now supports it natively.

TCP/IP History

Transmission Control Protocol/Internet Protocol (TCP/IP) is a protocol surrounded by a lot of history. On an interesting side note, TCP/IP was nearly killed in 1988 when the U.S. government decreed that all federal network devices would run the OSI protocol suite, and only the OSI protocol suite, by August of 1990. Luckily, no one listened and TCP/IP has flourished, with current reports estimating that over 100 million TCP/IP devices are connected to the Internet.

TCP/IP is also one of the few protocol suites with an open forum for change. Originally, ARPANET was maintained and improved by a small group of researchers and consultants. These individuals used a system of notes (the proverbial "network on a napkin"), experiments, and discussions to create the protocols we now use every day. Eventually, standards bodies, such as the Internet Engineering Task Force (IETF), were formed and brought order out of the chaos, but the spirit of open communication and unobstructed change still exists. Today, documents known as Request for Comments (RFCs) define the protocols used in the TCP/IP suite. RFCs are not limited. You do not have to be anyone special to submit one or for someone else to use your proposed change. In the words of David Clark (the original chairman of the Internet Architecture Board (IAB)), the IETF credo states

> We reject kings, presidents, and voting.
> We believe in rough consensus and running code.

However, just because you submit a protocol does not mean it will be used. Protocols proposed in RFCs are tested by a wide variety of people. If the protocol proves to be useful and stable enough, it may eventually become an Internet standard. However, most protocols never reach this level, and, eventually, a better solution makes them obsolete.

Luckily, because of this open forum, a protocol exists for nearly every possible use of network resources, with more being added every day. Also, because of the open forum, nearly all RFCs are available for perusal for no fee whatsoever, making it very easy to learn about any protocol—that is, if you have the fortitude to read through them. (Most people find that RFCs have a peculiar effect on the human body; the act of reading them tends to induce a comatose state almost immediately.) You can find RFCs at this web site: http://www.rfc-editor.org/.

On that note, we will enter the world of TCP/IP with a look at a few of the most commonly used protocols.

Individual Protocols

TCP/IP is not a suite with a set number of used protocols. On the contrary, literally thousands of protocols are defined for use in TCP/IP. No single platform supports all proposed protocols, but several protocols are used almost universally. We will concentrate our efforts on the most significant of these.

Serial Line Internet Protocol

Serial Line Internet Protocol (SLIP) defines encapsulation of data over serial connections. SLIP is a de facto standard that most closely resembles a datalink-layer protocol. It is designed to encapsulate upper-layer protocols for transmission over serial connections (that is, modems). SLIP is a simple protocol, designed when modem connections were excruciatingly slow and standards for serial communications clumsy. For this reason, it does not support a number of features that are supported by Point-to-Point Protocol (PPP): compression, error detection, dynamic addressing, or a type field to identify the layer 3 protocol used (SLIP is IP only). SLIP is still supported by a number of operating systems, but for the most part, it has been replaced by the more robust (and complicated) PPP.

SLIP is defined in RFC 1055, Standard 47, and it is available at this address: ftp://ftp.isi.edu/in-notes/std/std47.txt.

Point-to-Point Protocol

Point-to-Point Protocol (PPP) is another data-encapsulation protocol that can be used with a wide variety of links. PPP is a highly robust and configurable protocol, and it is currently the major standard for modem-based serial communications. PPP is actually a suite of protocols that most closely resembles a datalink-layer protocol. Some sub-protocols of PPP are the following:

- **Link Control Protocol (LCP)**   Used to negotiate encapsulation options, detect configuration errors, perform peer authentication, and terminate connections.

- **Password Authentication Protocol (PAP)**   Used to perform user authentication across serial links. Cleartext authentication (unencrypted).

- **Challenge Handshake Authentication Protocol (CHA)**   Also used to perform user authentication across serial links. Encrypted authentication.

- **Network Control Protocol (NCP)**   Used to perform functions specific to each type of upper-layer network protocol supported.

PPP supports the following features:

- IP header compression (RFC 2509)

- Data compression using a variety of algorithms, including Microsoft Point-to-Point Compression (MPPC), Predictor, and STAC

- Multiple upper-layer protocols using various NCPs

- Multiplexing of multiple upper-layer protocols over a single physical connection (in other words, running Internetwork Packet Exchange/Sequenced Packet Exchange (IPX/SPX) and TCP/IP over the same dial-up connection)

- Dynamic address allocation (using the appropriate NCP)

- Error detection with a frame check sequence (FCS)

- Tunneling using Level 2 Tunneling Protocol (L2TP) (RFC 2661) or Point-to-Point Tunneling Protocol (PPTP) (RFC 2637)

- A wide variety of physical- and datalink-layer technologies including X.25, serial, Frame Relay, and Asynchronous Transfer Mode (ATM) using AAL5

- Bonding of connections with Multilink PPP (MP) (RFC 1990)

- Dynamic allocation of bandwidth using Bandwidth Allocation Protocol (BAP) (RFC 2125)

- Many other less common enhancements

PPP is formally defined by RFCs 1661 and 1662, Standard 51, and is available at ftp://ftp.isi.edu/in-notes/std/std51.txt.

Internet Protocol

Internet Protocol (IP) has a fairly simple job. IP defines network-layer addressing in the TCP/IP suite and how, when, and where to route packets. Version 4 is the current version of IP (defined in RFC 791, Standard 5), and version 6 is the latest version (defined in RFC 1884). We will concentrate our efforts on version 4 because version 6 is not yet widely used and is not expected to be for quite some time.

IP is responsible for two major jobs: addressing and fragmentation. Fragmentation occurs when the layer below IP (datalink) cannot accept the packet size IP is attempting to send it. In this case, IP chops the packets into smaller pieces for the datalink layer. It also sets the More Fragments (MF), Fragment Offset (FO), and Fragment Identification fields to aid the endpoint in reassembly.

IP addresses (in version 4) are 32-bit addresses generally represented with four octets in dotted decimal notation. For example, a common IP address is 192.168.1.1. Each section, or *octet*, consists of eight bits, which means the maximum value for any given octet is 255 and the minimum is 0. IP specifies an additional part of the address known as a *subnet mask*, which masks, or hides, part of the address to split an IP address into its two basic components: Network and Host. (We will examine IP addressing in more detail in the next chapter.)

Finally, IP specifies four additional fields in the IP header of a packet that require further explanation: Type of Service (TOS), Time to Live (TTL), an Options field for additional customization, and a Header Checksum field. Figure 5-1 provides a breakdown of an IP packet, along with a basic description of each of the fields.

- The TOS field in an IP packet tells network devices along the path what priority, or precedence, the packet has, and how much delay, throughput, and reliability it requires. The TOS field is generally used by protocols at other layers (such as ATM) to define quality of service (QOS) parameters and modify queues along the path of a connection.

- The TTL field in an IP packet is used to discard packets that have been en route for too long. The TTL is measured in seconds and is an 8-bit field, making the maximum time a packet could be in transit 255 seconds. In reality, a packet will never reach this limit because each device (mainly routers) operating at layer 3 or above along the path is required to decrement the TTL by at least one, even if the device has the packet for less than one second. When the TTL field reaches zero, the packet is discarded. This field exists so that, in case a packet is caught in a routing loop, it will not circle the network endlessly.

- The Options field is of variable length and is rarely used (it would be the last field in an IP header, but is not shown in Figure 5-1). A few standard options have been defined, which consist mainly of fields for source routing (where the source device determines the route for a packet) and security descriptions for the DOD (secret, top secret, and so on).

- The Header Checksum field is a simple computation that can adequately detect bit errors in the IP header. If bit errors are detected in the header, the packet will be discarded.

Finally, the IP version 4 specification, as defined in RFC 791, is available at the following address: ftp://ftp.isi.edu/in-notes/rfc791.txt.

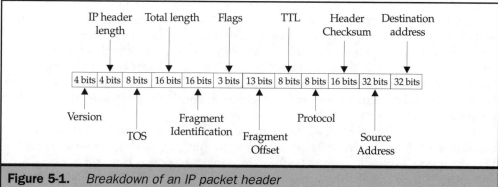

Figure 5-1.   *Breakdown of an IP packet header*

Internet Control Messaging Protocol

Internet Control Messaging Protocol (ICMP) is actually considered a subprotocol of IP. This is why it is defined in the same Internet Standard Document (Standard 5) as IP. ICMP, however, acts more like an upper-layer protocol than a subprotocol, using IP in the same manner as an upper-layer protocol (such as TCP) by letting IP encapsulate its data. As such, ICMP still uses IP addressing and fragmentation mechanisms. The reason it is considered a subprotocol is because its role is considered to be so pivotal to the successful operation of IP that it is a required component of any IP implementation. So what does this obviously important protocol do? Put simply, it reports errors.

ICMP's purpose as a protocol is to provide information to IP devices about the status of their IP transmissions. Typically, it is most often used to inform hosts that an error condition has occurred. For example, let's say we are attempting to connect to a remote host using Telnet. We type in the appropriate host name in the Telnet application, and a few seconds later, we are greeted with—Destination Unreachable. How did the Telnet application know that the destination was unreachable? If the router had just dropped the packet, we would have no way of knowing.

This is where ICMP comes in. Rather than your Telnet application waiting forever for the connection to be established (or at least for four minutes or so until the maximum TTL has expired), it got an ICMP message from the router indicating that the destination was unreachable. The Destination Unreachable message type is only one of the many types that ICMP defines. Table 5-1 provides the full list of ICMP message types, as defined in RFC 792.

Let's take a further look at the purpose of each message type. Note that while all IP implementations are technically required to *recognize* each message type, not all of them actually *use* all of the message types. In other words, depending on the OS in use, your mileage may vary.

ICMP Echo Request and Echo Reply

The ICMP Echo messages are used for diagnostic purposes in one of the most commonly used network applications to date: Ping. If you are unfamiliar with Ping, you can think of it as a violent way of checking vital signs. For instance, if I am teaching a class and someone is slumped over in her chair, I may want to determine whether she is alive and sleeping or dead. In this case, I might take a baseball (doesn't every instructor bring one to the classroom?) and "ping" this individual on the head with it. Now, if she is just sleeping, I am sure to get a reaction of some type. If she's not sleeping, well, then I don't have to run away.

By using the Ping network application, we perform a similar task. We want to determine whether a network device is up and running. To do this, we send an ICMP Echo Request message, hoping to jar the device into a response. If we hit our target, it will respond with an Echo Reply message, letting us know that it exists and is accepting communications.

ICMP Code	ICMP Type
0	Echo Reply
3	Destination Unreachable
4	Source Quench
5	Redirect
8	Echo Request
11	Time Exceeded
12	Parameter Problem
13	Timestamp Request
14	Timestamp Reply
15	Information Request
16	Information Reply

Table 5-1.   *ICMP Codes and Message Types*

Note   *The common syntax for the* ping *command in most operating systems is PING <IP Address or DNS Name>.*

Destination Unreachable

This message is used to indicate to a host that the destination for his or her IP packet cannot be found. This problem could be caused by a number of things, depending on the subcode within the Destination Unreachable message. The Destination Unreachable message defines six codes: Network Unreachable, Host Unreachable, Protocol Unreachable, Port Unreachable, Unable to Fragment, and Source Route Failed:

■ **The Network Unreachable code**   Indicates that the routers are saying that they can't get there from here. In other words, the network may exist, but the routers don't know where it is or how to get there, so they dropped the packet and sent you a message.

■ **The Host Unreachable code**   States that the network in question is up and running, the router can get it to the network just fine, but the destination host is not responding for some reason. The destination host sends the Protocol Unreachable and Port Unreachable messages.

- **The Protocol Unreachable code**   States that the host is up and running, but that the protocol you have decided to use is not supported.
- **The Port Unreachable code**   States that the requested protocol is running, but that the requested port for that protocol is unavailable. (We will delve more into ports later in this chapter.)
- **The Unable to Fragment code**   States that a router along the way needed to fragment the packet to adhere to the Maximum Transmission Unit (MTU) for the datalink-layer protocol that it is running, but it could not because the Do not Fragment (DF) bit was set on the packet.
- **The Source Route Failed code**   Lets the host know that the route the host specified to the destination is not valid. Typically, this message is not used because source routing is not usually done in IP.

Source Quench

The ICMP Source Quench message is a typical example of congestion notification. When a device's buffers begin to fill, it can send an ICMP Source Quench message to the next hop device, telling that device to slow down its transmission—similar to the congestion notification scheme we examined in Chapter 1. The only down side is that a device typically will not send a Source Quench message unless it has already dropped that device's packet. Source Quench messages can be generated by any device along the path, whether they are routers or endpoints (hosts). Any device along the path (routers or hosts) may also respond to a Source Quench message—although, again, your mileage may vary.

Redirect

The ICMP Redirect message informs a device that it is using the wrong gateway (router) to get to a given address. If a router receives a packet from a host to go to a specific remote network, it performs a lookup in its routing table for the destination network and forwards the packet to the next router in the path to that network. If it happens to notice that the next hop, or router, is on the same network as the client, it forwards the packet normally but then issues a Redirect message to the client, informing the client that the faster way to get to the network in question is to talk directly to the other router. The idea is that the redirect message causes the client to enter the destination network and alternate router as a new entry in its local routing table, shortening the path to get to the remote network.

Time Exceeded

To inform the host that its message took too long to reach the intended destination, the Time Exceeded message is sent when the TTL on an IP packet expires. A common use of the TTL field is in a program called trace route (*trace* in Cisco and *tracert* in Windows). The purpose of trace route is to locate every hop along the path to a given destination. Knowing this information informs you of the exact path to the destination—extremely useful information to have when troubleshooting.

For instance, to get to Microsoft's web site, my usual path is from my home to Raleigh, North Carolina; then to Washington, D.C.; and then to Seattle, Washington. However, if I notice that Microsoft's web site is responding particularly slowly one day, I may wonder whether the problem is on my end, my ISP's end, or just an issue with my normal backbone. If I performed a trace route and noticed that I was taking a path through Florida to get to Seattle, I would have the answer. Most likely, the backbone I normally take is clogged or down; therefore, I am taking an alternate route, which is a bit slower.

Trace route accomplishes this feat very simply. Remember earlier, when you learned that upon expiration of the TTL, a packet is dropped and an ICMP Time Exceeded message is sent to the host? Well, trace route makes use of this little setting. By setting the TTL of the first packet to a destination to 1, trace route receives a Time Exceeded message back from the first router on the path (unless, of course, that router is configured not to send Time Exceeded messages). The reason the Time Exceeded is sent is simple: every router along the path must reduce the TTL by *at least* 1. Well, if the TTL was 1 and the router reduces it by 1, it will be 0, forcing the packet to be discarded and a Time Exceeded message to be sent.

Trace route simply performs this process many, many times, increasing the TTL by 1 every time. So, on the first message, the TTL is 1, which gives us the first hop. On the second message, it is 2, which gives us the second hop, and so on.

Parameter Problem

The Parameter Problem message is sent to a device whenever a receiving device cannot understand a setting in the IP header and must discard the packet. Typically, this message is sent only when some options are set in the packet that the receiving device does not understand.

Timestamp Request and Timestamp Reply

The ICMP Timestamp messages mark a packet with the time—in milliseconds past midnight—that the sender sent the message, the receiver first received the message, and the receiver sent the message back. The most obvious use for this message is delay computation (determining how much latency a device inserts into the transmission), but it has also been used for clock synchronization.

Information Request and Information Reply

The ICMP Information message type is used by a device to determine which network it is on. A device sends an IP packet with the source and destination address fields set to 0.0.0.0. A different device (usually a router or server) then replies with the network address of the network, letting the sender know which network it is a part of.

You can find further information on ICMP in RFC 792, Standard 5, at the following address: ftp://ftp.isi.edu/in-notes/rfc792.txt.

Address Resolution Protocol

Address Resolution Protocol (ARP) is used by a device to resolve a MAC address from a known IP address. ARP is required because, to send a packet to a host, you need a physical address to send it to. In Ethernet (and most other LAN technologies), ARP is used.

ARP works like this: When you want to communicate with a host on your network, ARP sends a broadcast to the all *F*'s MAC address (FF-FF-FF-FF-FF-FF); but in the IP header of the packet, the destination IP address is listed normally. Also, the source device's MAC and IP addresses are listed normally in the packet. When all local network devices receive the frame, they add the source device's MAC address and corresponding IP address to an ARP table for future reference and then pass the packet up to IP for processing. IP checks the destination IP address and, if it matches its IP address, it responds with an ARP reply directly to the original host (unicast) containing both its IP address and MAC address.

The host then adds the IP and MAC address to its ARP table and begins communications. The ARP table is used to keep the devices from having to broadcast for the MAC address of the destination device every time they send a packet. Rather, they just look in the ARP table first, and if the IP address in question is listed, they use the entry from the table.

ARP is defined in RFC 826, Standard 37, and can be found at ftp://ftp.isi.edu/in-notes/rfc826.txt.

Inverse ARP

Inverse ARP (InARP) is basically ARP for Non-Broadcast Multi Access (NBMA) networks. Most WAN links are NBMA environments, so broadcasts are not possible. Instead, another mechanism must be used to resolve physical addresses to logical addresses. However, in an NBMA environment, the problem isn't usually that you know the IP and need the MAC (or other datalink-layer address); it is exactly the opposite. For instance, in Frame Relay, we already know the physical address (DLCI) that we use to communicate with our neighbors because we had to use that to establish the connection in the first place. However, we do not know our neighbors' IP addresses.

InARP solves this problem as follows. Upon activation (when the interface first comes up), the sending device (we'll call it Router A) sends an InARP message down all virtual circuits (VCs) on the interface to the routers or switches on the other side of the connection (we'll call this one Router B). This message contains the source hardware and protocol address (from Router A) and the destination hardware address (for Router B). The destination protocol address is left blank, so that the receiving device (Router B) knows that the sending device (Router A) is requesting its hardware address. The receiving station (Router B) enters the IP and hardware address of the sender (Router A) into its ARP cache and sends a response back to the sender containing its IP address. Router A then adds Router B's IP address and hardware address into its ARP table, and the cycle is complete.

Inverse ARP is defined in RFC 2390 and can be found at ftp://ftp.isi.edu/in-notes/rfc2390.txt.

User Datagram Protocol

User Datagram Protocol (UDP) is the connectionless transport protocol in the TCP/IP suite. UDP performs the service of multiplexing between applications that require a minimum of overhead. Unlike TCP, UDP has no mechanisms for error recovery, sequencing, or flow control. Its sole error-related encapsulation feature is the inclusion of a checksum that can detect errors in any part of the UDP datagram, including the data portion. Because of these features (or the lack thereof), however, UDP is an extremely efficient protocol, requiring much less overhead than TCP. Whereas all of the additional fields in TCP make the TCP header upward of 200 bits, the UDP header is only 64 bits. (A UDP datagram is shown in Figure 5-2.)

In addition, the simplicity of UDP means that the client and server devices have to spend fewer resources to send data with UDP. However, because it is unreliable, only a few applications use UDP; most notably, Trivial File Transfer Protocol (TFTP) and Domain Naming System (DNS) both use UDP.

UDP is defined in RFC 768, Standard 6, and is available at ftp://ftp.isi.edu/in-notes/std/std6.txt.

Transmission Control Protocol

Transmission Control Protocol (TCP) is the connection-based transport protocol in the TCP/IP suite. TCP includes a lot of functionality that is not present in UDP. One of TCP's primary features is a method of flow control and error correction called windowing.

Windowing is an efficient and graceful form of flow control. As explained in Chapter 1, windowing allows a host to send only a set amount of bytes before expecting a response, or *acknowledgment* (*ACK*), from the receiver. Windowing, like all forms of flow control, keeps one device from overrunning the other in a network communication. TCP uses a

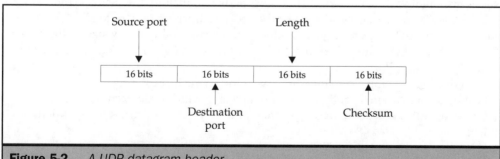

Figure 5-2.   *A UDP datagram header*

type of windowing based on *forward acknowledgments* (*FACKs*), and its limits are based on bytes. With a FACK, if I sent 50 bytes to you and you received all 50, you would send back an ACK of 51. The ACK of 51 tells me that all 50 bytes were received and that byte 51 is expected next. I would not send byte 51 until I received an ACK for the previous bytes. If I never receive an ACK for any bytes, I will retransmit all 50 bytes after a timer expires (known as the *Retransmit Timer*). If I receive an ACK for some but not all of the bytes, I will retransmit only those that you did not receive. For instance, if you send me an ACK with a value of 39, I will send you all bytes after byte 38. This example is a bit simplified, however, because another concept comes into play: sliding windows.

Sliding windows describes how TCP "moves" its window across the data stream in response to ACKs. This concept is probably best shown in an example. In Figure 5-3, the PC named Bart is sending a message to the PC named Homer. The *window size*, or the amount of bytes that are allowed to be transmitted before an acknowledgment is expected, is 500 bytes. Bart sends 500 bytes, starts his timer, and waits for either Homer to respond or for the timer to expire. Homer receives all 500 bytes and sends an ACK segment back to Bart numbered 501 to let Bart know that byte 501 is expected next. Bart then "slides" his window down to the next 500 bytes and sends bytes 501 through 1000 to Homer.

Windowing in TCP is dynamic, meaning that the window size may (and nearly always will) change throughout the session in response to errors (or the lack thereof). During the initial session establishment, the base window size will be set. The initial Sequence Number (SEQ), declaring the beginning byte number, will also be sent. For every TCP segment sent, a Window Size field in the TCP header tells that side which window size the other side has set. If an error occurs and some data is lost (causing TCP to retransmit the data), the host may reduce the window size. On the other hand, if massive amounts of data have been sent with no errors, the host may increase the window size.

For instance, in Figure 5-4, Chandler is sending data to Rachael. Chandler's initial window size is 100 bytes. He sends 100 bytes to Rachael, who receives all 100, sends an ACK back for byte 101, and increases the window size to 200 bytes. Chandler then slides his window down to byte 101 and expands it to include the next 200 bytes. He then sends 200 more bytes to Rachael, who receives through only byte 250. Rachael then sends an ACK back for byte 251, along with a reduced window size of 100 bytes. Chandler slides the window up to byte 251 and reduces the window to include only bytes 251 through 350.

In addition to being a dynamic process, windowing in TCP is also full-duplex, which means that there is both a send and a receive window on each host. Each window operates independent of the other, and may grow or shrink independent of the other.

Another major feature of TCP is the ability to multiplex logical sessions between remote hosts. This feature is found in both TCP and UDP, and allows you to have multiple connections open at the same time by providing the transport protocol a

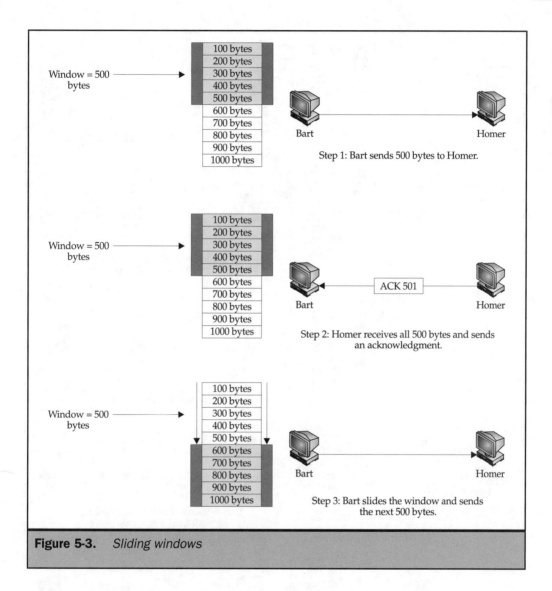

Figure 5-3. *Sliding windows*

method of distinguishing between the connections. This subject is covered further in Chapter 6.

Finally, we must discuss the part that makes all of the features of TCP work: the session itself. The TCP session establishment has three main goals:

- To verify connectivity between the two hosts involved in the session
- To inform each side of the starting sequence number to be used
- To inform each side of the starting window size to be used

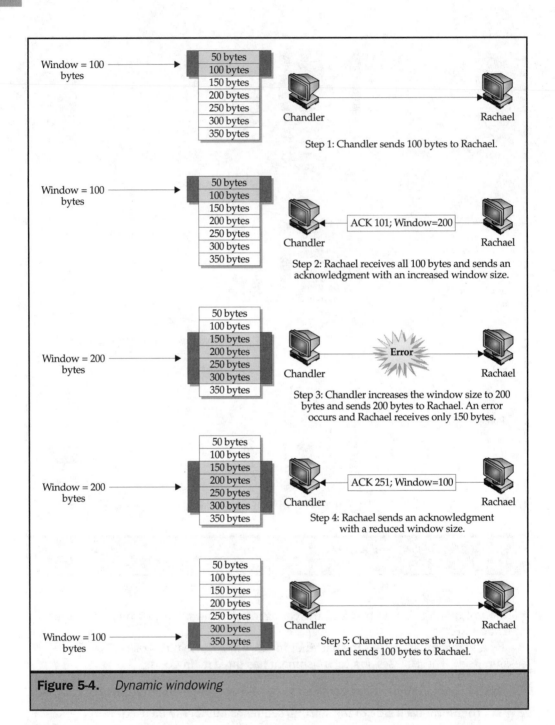

Figure 5-4. *Dynamic windowing*

The TCP session establishment process (also known as the *three-way handshake*) goes as follows. The client initiates the session by sending a segment with the Synchronization (SYN) bit set. This segment includes the client's window size and its current sequence number. The server responds with an ACK to the client's SYN request and also includes its own SYN bit, window size, and initial sequence number. Finally, the client responds with an ACK to the server's SYN. Figure 5-5 shows this process.

The TCP session teardown process is similar to the establishment process. First, the side that wants to close the session (we will assume it is the client for this example) initiates the termination by setting the Finish (FIN) bit. The server responds by ACKing the client's FIN. Then the server sends its own segment with the FIN bit set. The client then ACKs the server's FIN, and the session is closed. This process is shown in Figure 5-6.

Finally, Figure 5-7 shows the TCP protocol data unit (PDU).

TCP is defined by RFC 793, Standard 7, and is available at the following address: ftp://ftp.isi.edu/in-notes/std/std7.txt.

Dynamic Host Configuration Protocol

Dynamic Host Configuration Protocol (DHCP) is a subprotocol of Boot Protocol (BOOTP) that, true to its name, is used to dynamically assign IP addresses to hosts. How DHCP works is fairly simple. First, you must assign to the DHCP server a scope, or range, of addresses to be used for your clients. Then you must enable DHCP on the clients. Upon boot, the clients broadcast for an IP address. The DHCP server, upon hearing the cry for help, looks for a suitable address in its scope. If one is available, it sends out a packet offering this address to the client, along with a lease time.

If the client accepts the address (it accepts the first offer it gets and ignores the rest), it sends a packet to the DHCP server acknowledging that it is using that address. The DHCP server then removes the address from the usable range until the lease expires

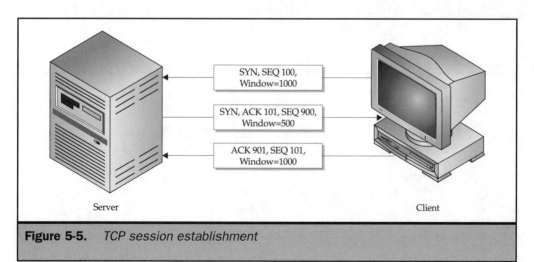

Figure 5-5. *TCP session establishment*

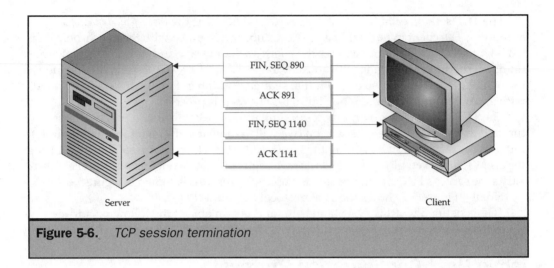

Figure 5-6. *TCP session termination*

(or is released by the client). This removal keeps multiple clients from using the same IP address (well, from the same server, anyway). At regular intervals (generally every four days or so), the client renews its IP address lease with the DHCP server, ensuring its right to continue using the same address.

Pretty simple, right? In most cases, this process works like a charm, but a few snags can trip you up if you're not careful:

■ *DHCP is a broadcast-based protocol.* This means that it will not pass through most routers. Luckily, you can configure Cisco routers to forward DHCP broadcasts, if needed. Also, a specialized DHCP client called a Proxy Agent can be set up to forward the DHCP requests, using unicast, through a router. One way or the other, however, you will need to keep this issue in mind.

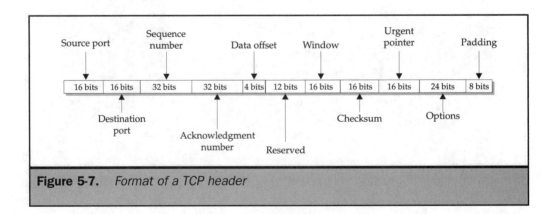

Figure 5-7. *Format of a TCP header*

■ *DHCP servers have no way of contacting each other and sharing information.* This leads to a very common problem whereby two or more DHCP servers with the same scope are set up on the same subnet. Remember, DHCP clients just accept the first address they are offered, regardless of whether that address conflicts with another host. If two or more DHCP servers are using the same scope, this means that they will eventually offer the same IP address to two different clients, causing an address conflict.

Overall, though, DHCP is a reliable and easy way to assign addresses to your clients without a whole lot of hassle.

Basic DHCP functionality is defined in RFC 2131 and is available at ftp:// ftp.isi. edu/in-notes/rfc2131.txt.

Domain Naming System

Domain Naming System (DNS) is defined by no less than 40 RFCs. The pivotal RFCs for DNS are considered to be 1034 and 1035, but the addition of many other specifications in later RFCs truly makes DNS a workable system. The purpose of DNS is to resolve fully qualified domain names (FQDNs) to IP addresses and, in some cases, vice versa. *Fully qualified domain names* are the primary naming convention of the Internet. They follow a hierarchical convention, with the most specific part of the name on the left and the least specific on the right. This configuration is shown in Figure 5-8.

DNS is absolutely pivotal in the functioning of today's global Internet. Without it, you would have to keep a list of all of the IP addresses for every web site you wanted to get to, rather than being able to just type in **www.cisco.com**. How DNS works is a complicated discussion and could easily comprise an entire book itself. For our purposes, we will take a look at the basics of how DNS operates by beginning with a few definitions:

■ **Namespace**   A domain.

■ **Zone**   An area of DNS authority. This can be a single namespace (such as corp.com) or a collection of contiguous namespaces (for example, corp.com, sales.corp.com, and support.corp.com). If a DNS server contains a given DNS zone, it is said to be *authoritative* for that zone. That is, it may answer name queries for that zone with firsthand knowledge, rather than having to ask another DNS server about the zone.

■ **Zone file**   The DNS database for a given zone.

■ **Zone transfer**   The process of copying a DNS zone from one server to another.

■ **Primary zone**   The writable copy of a DNS zone. Only one server can contain the primary zone.

■ **Secondary zone**   A read-only copy of the DNS zone. Many servers can contain a secondary copy of the zone to spread out, or load balance, the DNS queries across multiple servers.

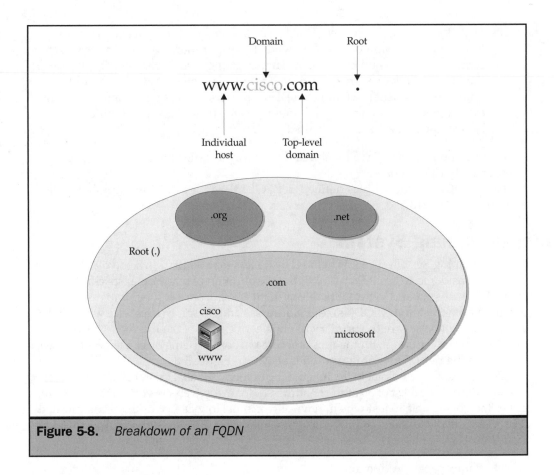

Figure 5-8. *Breakdown of an FQDN*

■ **Name cache**   A copy of all lookups the DNS server has performed during a given time, generally held in RAM. The name cache is used to increase response time for DNS queries by not requiring the server to perform a lookup for the name resolution request. Instead, if the same name has been requested before, the server can just pull it out of the cache.

■ **Root servers**   The servers that contain listings of all of the top-level domain servers. These are contacted during recursive queries. Represented by a single dot (.).

■ **Resource record**   An individual entry in the DNS database. Resource records are divided into several types to further define their uses. For instance, an A record is called a host record and is used to list name-to-address mappings (for example, PC1 = 192.168.1.1). A PTR record is a pointer record and does the opposite of an A record (for example, 192.168.1.1 = PC1).

- **Forward lookup zone**   A zone whose primary job is to resolve names to addresses. Contains mostly A, or host, records.

- **Reverse lookup zone**   A zone whose primary job is to resolve addresses to names. Contains mostly PTR, or pointer, records.

- **Iterative query**   A query that is resolved on the first DNS server asked. This type of response is known as authoritative because the DNS server has authority over the zone by having a copy of the zone file for that zone.

- **Recursive query**   A query in which the DNS server must find the server authoritative for the zone in order to answer the query. This type of response is known as nonauthoritative because the DNS server does not have authority over the zone in question.

Note   *Although this explanation of the differences between iterative and recursive queries usually serves fine, the actual definition of an iterative query is a query that can only be answered with a "Yes, I found the lookup" or a "No, I didn't find the lookup." There may not be a referral to another DNS server. Your client (known as the resolver) typically issues iterative queries. Recursive queries, on the other hand, may be responded to with a referral to another DNS server. Servers typically issue recursive queries.*

Now that we have the definitions out of the way, let's look at how DNS operates. In Figure 5-9, the namespaces we will use in this example are defined, and all of the servers are listed. Assume I am PC1.corp.com, and I am attempting to get to server1.corp.com. I send a query to my DNS server (DNS.corp.com), and it checks to see whether it is authoritative for the corp.com zone. Finding that it is, it looks up the resource record for server1 and lets me know that the IP address for server1 is 192.168.1.100.

I then contact server1 by using that IP address. This query is iterative because my DNS server could resolve the query without asking for help. Let's say a little later, I decide to visit www.realcheapstuff.com. I send a query to my DNS server. My DNS server realizes it is not authoritative for realcheapstuff, so it goes out to the root (.) server and says, "Do you know where I can find www.realcheapstuff.com?" The root server replies, "I don't have a clue, but try the .com server, it might know."

So my DNS server asks the .com server the same question. The .com server replies, "I don't have a clue, but try the realcheapstuff server, it might know." My DNS server asks the realcheapstuff server, and bingo, gets the address. My DNS server then enters this address into its cache and hands me the answer to my query. Now if PC2 wants to get to www.realsheapstuff.com, the DNS server will just give PC2 the information out of its name cache, rather than having to perform the entire process again. This process is known as a *recursive query*. These two processes (recursive and iterative queries) form the basis of how DNS works.

Of course, DNS is not used in every case. Sometimes, it is easier just to manually add names and their corresponding addresses into a *host table*—a tiny version of DNS used for only one device. A good example of the use of a host table is in a Cisco router.

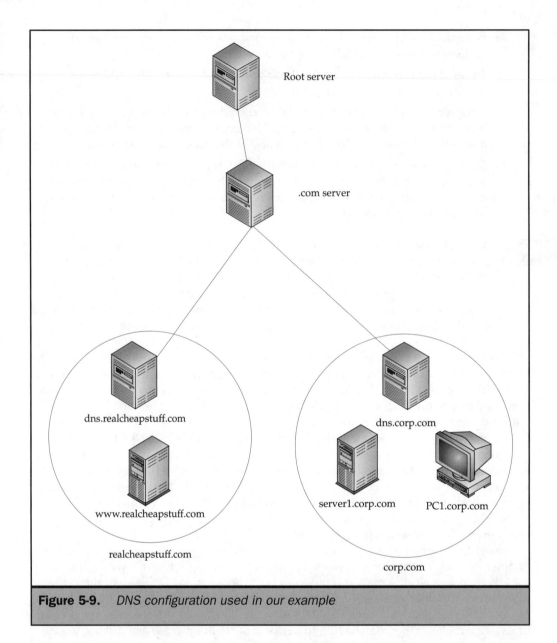

Figure 5-9. *DNS configuration used in our example*

Basically, because you rarely need to resolve more than a few names from inside the Internetwork Operating System (IOS), most routers just have those few that are needed manually entered, and DNS resolution through the corporate DNS server is turned off. Disabling DNS resolution is done mostly to keep the router from trying to resolve every mistyped command into an IP address. Chapters 16 and 17 discuss how to configure your host table inside IOS.

Trivial File Transfer Protocol

Trivial File Transfer Protocol (TFTP) is a very light, simple protocol used to copy files to and from TCP/IP devices. TFTP is one of the more important protocols, however, when dealing with Cisco gear, because it is the primary method of transferring configuration files and IOS images to and from routers and switches.

TFTP uses UDP as its transport protocol, making TFTP unreliable at the transport layer. However, TFTP uses its own acknowledgment system to make it reliable. When you want to transfer a file using TFTP, that file is broken into and sent in 512-byte blocks. The other device must send an ACK after each 512-byte block. No windowing is involved; rather, the host device just keeps track of the block for which it has not received an ACK, and it will not send any more data until it receives an ACK for that block. This functionality leads to a simple and easy-to-implement error-recovery mechanism.

TFTP does not support many of the major features of FTP, which is one of the reasons it is such a small, lightweight protocol. User authentication (user name and password) is not supported in TFTP. Directory listings and browsing for files are also not supported. When you use TFTP, it is assumed that the file requester knows what the exact name of the file is, and the sender knows where to find the file. If either of these assumptions proves incorrect, an error occurs and the transfer is terminated. Overall, TFTP is a simple, easy-to-use protocol, and it is one that we will be dealing with a lot in our interactions with Cisco devices.

TFTP is defined in RFC 1350, Standard 33, and can be found at ftp://ftp.isi.edu/in-notes/std/std33.txt.

File Transfer Protocol

File Transfer Protocol (FTP) is the more robust of the two basic file transfer protocols discussed in this chapter. FTP supports several features to facilitate advanced data transfers, such as the following: user authentication (using plaintext); TCP for flow control and error recovery; the ability to browse directory structures and files; the ability to create new directory structures or delete directories within FTP; and compression of data transfers.

Because of these additional features, FTP is considerably more complex than TFTP, and it requires that two TCP sessions be established: one for commands and responses to those commands, and one for the data transfer. We will look at this in more detail in Chapter 6. FTP is one of the most commonly used protocols for transferring files over the Internet. (In fact, because most web browsers natively support FTP, in many cases, you use FTP without ever knowing it.)

FTP is defined in RFC 959, Standard 9, and is available at ftp://ftp.isi.edu/in-notes/std/std9.txt.

Telnet

Telnet is the primary terminal protocol in the TCP/IP suite. The idea of Telnet is to allow any TCP/IP host to emulate a dumb terminal and get full command-line access to another device. A good example of how Telnet is used in the context of Cisco equipment is when you use Telnet to connect to a router or switch for remote management. It provides you with an interface that is practically identical to what you would see if you were plugged directly into the router with a physical terminal. Telnet's only major drawback is that with the basic specification, it sends all communication in plaintext, making it a security hole. However, you can encrypt Telnet traffic (as explained in optional RFCs, such as RFC 2946), which can alleviate this problem.

Telnet's basic specification is defined in RFC 854, Standard 8, and it is available at ftp://ftp.isi.edu/in-notes/std/std8.txt.

Hypertext Transfer Protocol

Nearly everyone is aware of Hypertext Transfer Protocol (HTTP). HTTP forms the basis for the World Wide Web: it defines how clients and servers communicate to fulfill web requests. HTTP uses TCP as its transport protocol and leaves error correction and flow control to it.

The intricacies of HTTP are beyond the scope of this book, but we will spend a little time learning how HTTP works on a basic level. First, if you want to connect to an HTTP resource (such as a web page, Java applet, image file, and so on), you need to enter the URL for the resource. HTTP uses Universal Resource Locators (URLs) to define not only which items it needs, but also which protocol will be used to retrieve the item. The URL is structured as follows: Service://FQDN/Path, as in http://www.cisco.com/public/pubsearch.html. This information tells HTTP everything it needs to know to make the connection to the server.

Once a TCP session has been established, HTTP formally requests the resource (pubsearch.html). Once it gets the initial resource, it then requests any other resources specified by the initial resource (such as .gif or .jpg images). Each one of these resources is requested separately, but (in HTTP 1.1) under the same TCP session. This is HTTP's primary job: getting resources.

HTTP 1.1 is the most current version, and it is formally defined in RFC 2616. It can be found at ftp://ftp.isi.edu/in-notes/rfc2616.txt.

Simple Mail Transport Protocol

Simple Mail Transport Protocol (SMTP) is probably the most widely used service on the Internet. Put simply, it is SMTP's job to get a message from your computer to the mail server for your final destination. It deals with routing mail messages (using MX, or mail exchanger, and A, or host records in DNS), formatting mail messages, and establishing sessions between mail clients and mail servers. SMTP typically uses TCP as its transport protocol, but it can use others, as defined in RFC 821.

SMTP does not have any responsibility for mail reception. That is, SMTP does not define how mailboxes for individual users are set up, nor does it define any of the other issues (such as authentication) that mail reception requires. It simply defines mail transfers from the source to the destination.

SMTP does define mail format and the character sets used in e-mail messages, however. Originally, SMTP defined only the use of 7-bit ASCII characters. However, the advent of Multipurpose Internet Mail Extensions (MIME) and the expansion of the Internet into a truly worldwide entity has prompted other character sets to be included in optional specifications. For this reason, an e-mail may now be sent in nearly any language, and with nearly any type of data encoded into it, such as pictures or executable files. All of these extensions have made SMTP more complex but have also made it much more flexible.

SMTP and its extensions are defined in several RFCs, but the most recent specification for basic SMTP is in RFC 2821, and it can be found at ftp://ftp.isi.edu/in-notes/rfc2821.txt.

Post Office Protocol Version 3

Post Office Protocol version 3 (POP3) is the other end of the e-mail equation for most mail servers. POP3 is used to deliver mail from mailboxes to end users. Once SMTP transfers the mail to the appropriate server, POP3 can then insert the mail into the appropriate mailbox for you to pick up later. Of course, other standards also exist for performing this function (such as IMAP4); but of the ones officially recognized as part of the TCP/IP suite, POP3 is probably the most common. POP3 supports authentication of the user in both encrypted and unencrypted form, but unencrypted is probably the most common.

POP3 is defined in RFC 1939, Standard 53, and can be found at ftp://ftp.isi.edu/in-notes/std/std53.txt.

Simple Network Management Protocol

Simple Network Management Protocol (SNMP) is a standard that has literally changed the way networks are analyzed. The initial purpose of SNMP was to provide a quick fix to the problem of networks getting so large that it was difficult to monitor every device for errors and failures. SNMP was born out of a need for a standardized system to detect—and, in some cases, repair—these error conditions. The initial SNMP specification (SNMP version 1), although considerably better than any other industry-standard management suite at the time, was still lacking in many areas. SNMP version 2 improved on version 1 by adding a number of features (such as additional information-gathering commands), but it was still somewhat lacking. SNMP version 3 improves on possibly the most significant problem with SNMP version 1 and version 2: security.

With SNMP versions 1 and 2, the only security implemented is the use of community names and, optionally, the requirement that the Management Station have a specific IP address. With community names, each device participating in the SNMP communication must use the same community name, which is almost like a password. The problem with this setup is that it is not very secure because many networks just use the default community names (typically, public or private). With SNMP version 3, additional authentication features, including optional encryption, are used to prevent security problems.

How SNMP works, like DNS, is the subject of entire books. Unfortunately, we don't have the ability to go into quite that much detail here, but we will go through SNMP on a basic level. SNMP consists of three major components:

- **SNMP Management Station**   The software that collects information from multiple agents and allows that information to be analyzed. Many vendors produce a variety of management stations, each with their own proprietary enhancements. Some common management solutions are IBM Tivoli, HP Openview, and CA Unicenter.

- **SNMP Agent**   The software that enters data from the monitored device into the Management Information Base (MIB). Its job is to collect data from the device so that the Management Station can collect that data from it.

- **MIB**   The monitored data in a hierarchical database. Several different MIBs exist, most with a highly specialized purpose and set of trackable objects. The Agent updates the local MIB, while the Management Station builds the global MIB with the data gained from the Agents.

Basically, the process works like this. The Agents are set to record certain events (such as average throughput) into the MIB. The Management Station then contacts the Agents and collects the data. The Management Station then builds a global MIB with all of the data collected from all of the Agents and, depending on the Management Station, may perform different functions with that data (like tracking trends and predicting failures). With version 3, the lines between Management Stations and Agents become a bit blurred because version 3 is based on a modular architecture, which allows the functions of both to be implemented in the same device, but the basic process is the same.

In addition to the ability to manually collect information from the Agents, the Agent can also dynamically inform the Management Station of problems with *traps*: special messages sent from the Agents to the Management Station based on "problem" conditions. They can be sent when certain events happen, such as disk space getting low or an attempted security breach.

Finally, the Management Station can sometimes change the configuration of the device with a type of message known as a *set* message. This message allows the Management Station to actively modify the Agent and, in some cases, the remote device itself.

SNMP, in all its versions, is by far the most common management protocol in use today. While SNMP version 3 is the most recent and complete version of the specification, SNMP version 2 and even SNMP version 1 are still used in many networks. And, of course, all versions of SNMP are supported on Cisco routers. (SNMP version 2 requires IOS 11.2(6)F or later, whereas SNMP version 3 requires IOS 12.0(3)T or later.)

SNMP is defined by literally hundreds of RFCs, all of which can be found at http://www.rfc-editor.org/cgi-bin/rfcsearch.pl.

Bringing It All Together

To solidify some of the concepts this chapter introduced, let's take a look at an example of how it all works in a common setting: connecting to www.cisco.com with a web browser. To help with this scenario, Figure 5-10 provides a diagram of the example.

After typing **www.cisco.com** in the address bar, my PC must first resolve this name to an IP address. To perform this task, it needs to communicate with my DNS server. To perform this task, it needs to resolve the IP address of my DNS server into a MAC address. ARP is used to formulate a request in the form of a broadcast. The DNS server, noticing its IP address in the ARP broadcast, replies with its MAC address. My PC adds the DNS server's MAC and IP addresses into the ARP table, and it sends out a DNS request for the IP address for www.cisco.com.

My DNS server looks in its zone files and determines that it is not authoritative for the cisco.com domain. It then begins the recursive lookup process by asking the root servers where the cisco.com DNS server can be found. The root servers respond, and

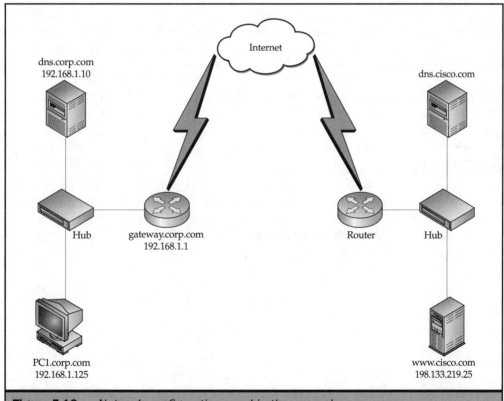

Figure 5-10.   *Network configuration used in the example*

my DNS server eventually locates the authoritative server. My DNS server then asks what the IP address for the host www in the domain cisco.com is. The Cisco DNS server responds, and my DNS server finally formulates an answer to my initial question: "What is the IP address for www.cisco.com?"

At this point, my PC recognizes that the IP address for www.cisco.com is not in the same network of which it is a part (by using a process called ANDing, which is described in Chapter 6). So, rather than formulating an ARP request (because ARP is broadcast and it will not pass through the router), my PC looks in its routing table for the best path to 198.133.219.25. The best path, in this case, is the default gateway (explained further in Chapter 6), so my PC sends the packet to my router (192.168.1.1).

This packet is the first part of the TCP three-way handshake that has to take place before my PC can communicate with the web server using HTTP (because HTTP relies on TCP). This packet contains the SYN flag to let the web server know that this is the start of a new session. The web server responds with a packet with its own SYN bit set, as well as the ACK bit set to acknowledge my SYN transmission. This completes step two of the handshake.

To complete the handshake, my PC sends a packet with an ACK for the web server's SYN. At this point, my PC formulates the HTTP request, and the web server responds by retrieving, fragmenting, and sending the web page to my PC. If any errors occur, the machines reduce the window size appropriately and retransmit. If no errors occur, the window size grows incrementally until it reaches its maximum size.

At the end of this process, my PC begins the teardown routine by sending a packet with the FIN bit set. The server responds by ACKing my FIN, and then it sends its own FIN packet. My PC responds with an ACK, and the session is closed.

Summary

In this chapter, we have learned about some of the more common protocols in the TCP/IP suite and gotten a taste for how they work together in a standard network communication. A solid, basic understanding of these topics will help us immeasurably in the next chapter, where we will address more advanced TCP/IP topics such as subnetting, NAT, and private addressing.

The
Complete
Reference

Cisco

Chapter 6

Advanced IP

IP addressing is a topic that tends to confuse people. However, it's really not all that complicated; it's just a bit alien. The key to truly understanding IP addressing is to understand how the TCP/IP device actually perceives IP addressing. To do this, you have to get into the binary math. There is no other way to fully grasp the concept. For this reason, we will spend some time understanding IP addressing the hard way (pure binary); and then, once you fully understand the concepts, I'll show you some shortcuts to allow you to do it the easy way (decimal). One thing is for sure: learning IP addressing is not a spectator sport. You have to get your hands dirty to understand it. Therefore, we will spend a good bit of time going over example problems and solving them in this chapter.

IP Addressing Basics

In this section, we will cover the fundamental components of IP addressing—namely, binary, ANDing, and address classes. These are the absolute *minimum* components you should understand to be able to understand subnetting, so make sure you have a solid grasp of these core concepts before continuing on to subnetting.

Basic Binary Math

Because computers don't natively understand the decimal system (or base 10 numbering), the first topic we need to cover is basic binary math. Computers natively understand the binary system (or base 2 numbering). To use decimal numbering, they need to do a conversion. For this reason, IP addresses are based on binary numbering.

Acquiring a basic understanding of binary numbering is easy. First, let me show you the significance of placement. Let's take a decimal number, say 762. Is this number two hundred sixty-seven, seven hundred sixty-two, or fifteen (7 + 6 + 2)? Obviously, the answer is seven hundred sixty-two, but how did you know that? By the placement of the numbers. The seven was in the hundreds place, the six was in the tens place, and the two was in the ones place. This is the same way binary works; the places just have different values. In decimal, each place to the left is a power of ten greater than the one directly to the right of it (in other words, 10 is ten times as much as 01); but in binary, each place to the left is a power of two greater. This is probably best shown in Figure 6-1.

Binary also uses only two numbers for each place: one and zero. This is because, in a numbering system based on powers of two, there can be only two values for each digit; so an example of a binary number is 100010100. Now, because most of us were probably taught to understand numbering using the decimal system, to understand binary, we will need to convert it to decimal. This, again, isn't that difficult.

First, let's examine how to establish what the decimal representation of a number actually is. In other words, how do we know 762 is seven hundred sixty-two? This is fairly basic. We know that anything in the hundreds place is that number multiplied by 100. This gives us 700. And anything in the tens place is multiplied by 10. This gives us 60.

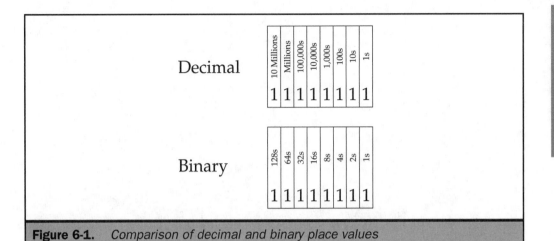

Figure 6-1. *Comparison of decimal and binary place values*

Finally, anything in the ones place is multiplied by 1. This gives us 2. If we add them all together (700 + 60 + 2), we have the full number, 762.

Binary is the same way. If we have the binary number 1100, we know, based on the place values, that anything in the eights place is multiplied by 8 ($1 \times 8 = 8$), anything in the fours place is multiplied by 4 ($1 \times 4 = 4$), anything in the twos place is multiplied by 2 ($0 \times 2 = 0$), and anything in the ones place is multiplied by 1 ($0 \times 1 = 0$). Now we add them all together ($8 + 4 + 0 + 0$), and we get 12. So 1100 in binary equals 12 in decimal.

```
Binary place value ─────────▶    8     4     2     1

Binary number ──────────▶ ✕     1     1     0     0
                                ─────────────────────
                                 8  +  4  +  0  +  0    =    12
```

Result: 1100 in binary = 12 in decimal

Here's another important point to keep in mind: any numbering scheme is infinite because numbers never stop. Therefore, we need to know how to figure out the place values for any number. This is also pretty easy. In binary, every time you go one place to the left, you simply multiply by 2. This is shown in Figure 6-2.

Finally, let me show you some shortcuts to binary conversions. First, if you are trying to figure out how many values (numbers) are in a given number of **binary digits** (bits), you can simply multiply the value of the most significant bit (MSB)—the last one to the left—by 2 (LSB stands for least significant bit, or "littlest one"). This tells you

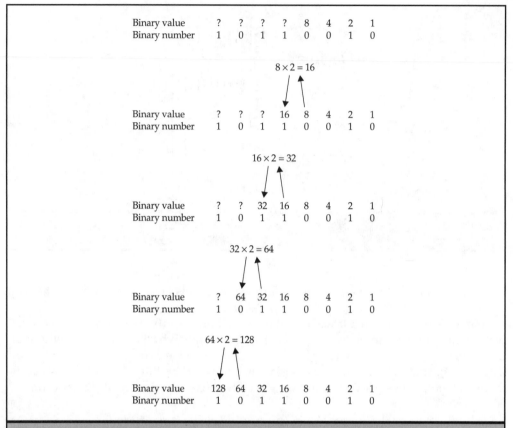

Figure 6-2.   *Example of multiplying by 2 to find unknown place values*

how many individual numbers are represented by that number of bits. An example is shown in Figure 6-3.

Remember, unlike grade-school math, where they taught you that zero was nothing, in computers, it represents an individual state and is counted. So from 0 through 6 is actually seven individual numbers, not six.

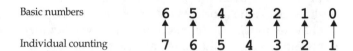

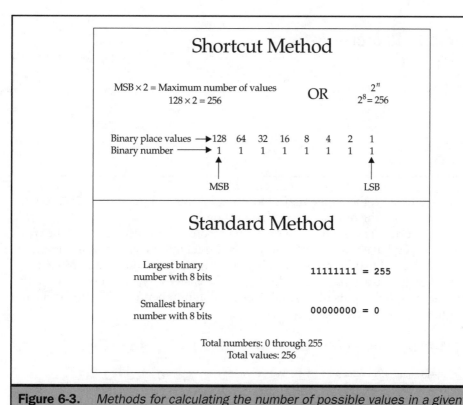

Figure 6-3. *Methods for calculating the number of possible values in a given number of bits*

This is because you count the number you start from. So, with eight bits (which we will be working with most), there are actually 256 possible values, ranging from 0

Exercise 6-1: Binary-to-Decimal Conversions

Convert the following binary numbers to decimal numbers. Answers are provided in the sidebar on the next page. Do *not* use a calculator (it will only defeat the purpose of this exercise).

1. 10001101

2. 011011

3. 110

4. 111100001010

5. 1001001001

Answers to Exercise 6-1

1. 141

2. 27

3. 6

4. 3,850

5. 585

---

through 255. By the way, the mathematical formula for this computation is 2^n (2 to the power of n), where n is the number of bits in question.

The second shortcut is that when doing a binary conversion, always count the type of digits (0 or 1) you have the least of, and then subtract or add as necessary. For instance, if you wanted to determine the value of 11110111, you could add 128 + 64 + 32 + 16 + 0 + 4 + 2 + 1, as we have been doing; but the easier way would be to count the place with the 0 in it (the eights place) and subtract that number from 255 (the value the number would be if all of the bits were a 1).

Binary place values ⟶	128	64	32	16	8	4	2	1
Binary number ⟶	1	1	1	1	0	1	1	1

255 − 8 = 247

Now that you have an understanding of binary, let's take a look at the IP address structure.

IP Structure

As mentioned previously, IP addresses are composed of two primary units: the network portion and the host portion.

Network | Host

192.168.1.200

In addition, every IP connection requires at least one *unique* IP address. However, two additional pieces are commonly required to achieve IP connectivity: a subnet mask and a default gateway.

The *subnet mask* is required in all IP implementations. It defines which part of the address is the host and which part is the network by using a Boolean process known as *ANDing*. ANDing is very simple in binary. It is basically multiplication. For example,

Figure 6-4 shows an IP address and a subnet mask. To arrive at the *network address* (the portion that defines which network you are on), you simply multiply the IP address (in binary) by the subnet mask (in binary). The result is a binary number that you can convert back to decimal to learn the network address. This is where routing comes into play. Routing is the act of forwarding a packet from one logical network or subnetwork to another. A router is simply the device that performs this process.

When a host wants to communicate with another host, it takes that host's IP address and its own IP address, and performs the ANDing process with both addresses. If the result for both addresses is the same, the host is assumed to be on the same logical network as the other host, and it should be able to communicate directly. If the result is different, the hosts are assumed to be on different logical networks (that is, separated by a router), and they must use a router to reach each other. (Figure 6-5 shows an example.)

The router's job is fairly simple on the surface. First, it needs to be aware of any networks that its directly connected hosts may wish to reach. Second, it needs to forward packets from hosts to those networks. Finally, it limits (or eliminates) broadcasts. This last function is actually why the first and second functions are needed.

Think about it like this: to reach a host, any host, you need a MAC address (or other layer 2 address). On a *flat* network ("flat" is a term typically used to describe a network that is not segmented with a router), this wouldn't be an issue. You would just use an ARP broadcast to find the MAC address. However, flat networks have serious scalability problems—when a broadcast is sent, everyone receives and must process that broadcast.

For example, imagine the Internet, with 100 million hosts broadcasting for an address every five seconds or so. Approximately 20 million packets would be delivered to every host every second. If you suspect that this number would be difficult for most hosts to process (which are connected with low-speed lines like 56K modem connections), you are right. Assuming 60-byte packet sizes, this would eat around one gigabyte per second (that's giga*byte*, not giga*bit*). Therefore, you would need an OC 192 (9.953 Gbps)

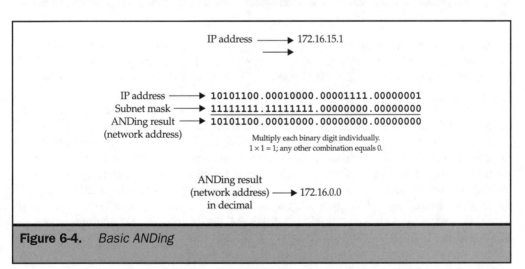

Figure 6-4.   *Basic ANDing*

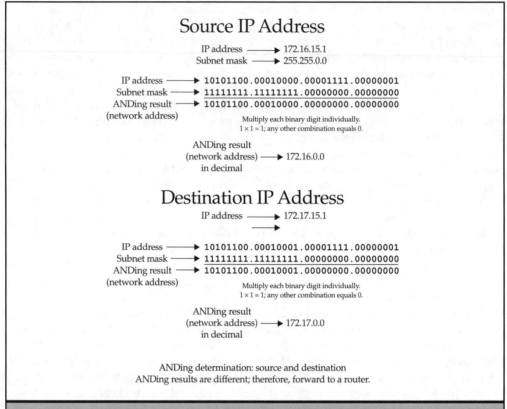

Figure 6-5. *ANDing for network determination with multiple hosts*

or faster connection to be able to transmit *any* data—and this is without even considering the possibility of a *broadcast storm,* which occurs when a device begins spitting out broadcast packets as fast as it can. This malfunction can saturate even LAN connections, so you can imagine what it would do to most WAN connections.

For this reason, routers are used to segment flat networks into hierarchical networks consisting of multiple broadcast domains. A *broadcast domain* is the area in which a broadcast is contained. (Figure 6-6 shows an example.) Routers segment networks into broadcast domains to keep broadcasts from overrunning every device on the network.

So the job of Boolean ANDing is to determine whether the host can just broadcast to reach the other host, or send the packet to a router instead. Remember, however, that a MAC address is still required. If the host can't broadcast for the MAC address of a remote host, how does it get its MAC address? Easy. It doesn't. It uses the router's MAC address instead.

NETWORKING BASICS

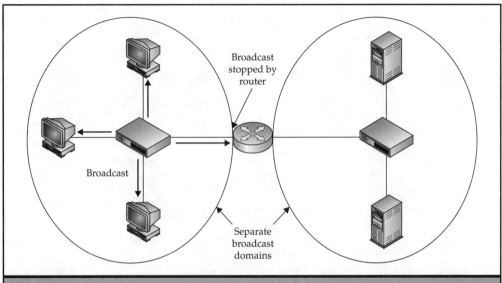

Figure 6-6. *Segmentation into broadcast domains*

If a host determines that the host it is trying to reach is on a different network, it sends the packet to a router that is likely to reach the remote network. Typically, this router is the host's default gateway. (This router could also be one listed in the host's routing table, but we will discuss that further in Chapter 22.) A *default gateway* is the IP device to which a host sends a packet when it doesn't know where else to send the packet. When it sends the packet to the default gateway, it uses the default gateway's MAC address as the destination MAC address rather than the remote host's, but it still uses the remote host's IP address. The router then determines where to send the packet and inserts the new next hop MAC address, as required. Figure 6-7 illustrates this process.

In this example, host Einstein is trying to send a packet to host Darwin. Two routers, Beethoven and Mozart, block their path. Einstein determines that Darwin is a remote host, so he inserts Beethoven's MAC address into the packet and transmits. Beethoven, upon receiving the packet, realizes that he must go through Mozart to get the packet to Darwin. He therefore inserts Mozart's MAC address into the packet and transmits. Mozart knows that Darwin should be reachable directly from his Ethernet interface, but he doesn't yet know Darwin's MAC address, so he ARPs for it. Upon receiving Darwin's MAC address, Mozart inserts it into the packet and transmits the packet directly to Darwin. When Darwin wishes to reply back to Einstein, the process is repeated in reverse.

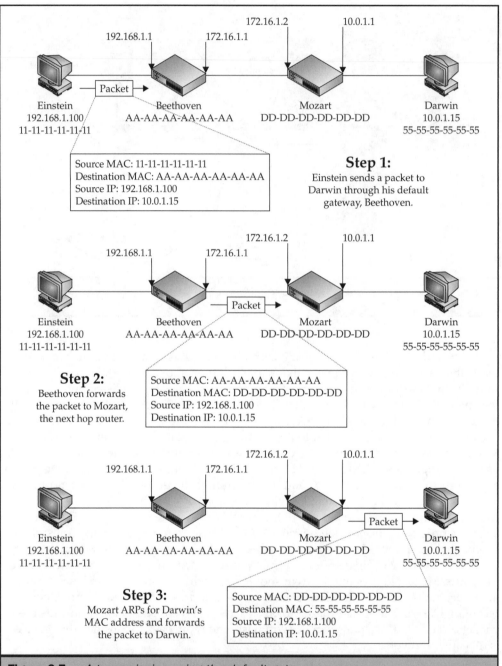

192.168.1.1   172.16.1.1   172.16.1.2   10.0.1.1

Einstein
192.168.1.100
11-11-11-11-11-11

Beethoven
AA-AA-AA-AA-AA-AA

Mozart
DD-DD-DD-DD-DD-DD

Darwin
10.0.1.15
55-55-55-55-55-55

Packet

Source MAC: 11-11-11-11-11-11
Destination MAC: AA-AA-AA-AA-AA-AA
Source IP: 192.168.1.100
Destination IP: 10.0.1.15

Step 1:
Einstein sends a packet to
Darwin through his default
gateway, Beethoven.

Step 2:
Beethoven forwards
the packet to Mozart,
the next hop router.

Source MAC: AA-AA-AA-AA-AA-AA
Destination MAC: DD-DD-DD-DD-DD-DD
Source IP: 192.168.1.100
Destination IP: 10.0.1.15

Packet

Step 3:
Mozart ARPs for Darwin's
MAC address and forwards
the packet to Darwin.

Source MAC: DD-DD-DD-DD-DD-DD
Destination MAC: 55-55-55-55-55-55
Source IP: 192.168.1.100
Destination IP: 10.0.1.15

Packet

Figure 6-7. *A transmission using the default gateway*

NETWORKING BASICS

Now that you have seen this entire process, let's try our hand at determining whether two hosts are on the same network. In Figure 6-8, two hosts, Michelangelo and Renoir, are attempting to communicate. Michelangelo has an IP address of 172.16.1.1, with a subnet mask of 255.255.0.0. Renoir has the IP address of 172.16.5.202, with a subnet mask of 255.255.255.0. Can they communicate? Take a second to work out the ANDing on your own before continuing.

The answer is no. Why? Take a look at Figure 6-8 again. If you perform the ANDing process, it looks as if Michelangelo will attempt to send the packet to a router because his network address is 172.168.0.0 and Renoir's is 172.16.5.0. Actually, the reverse is true. Michelangelo has no way of knowing that Renoir is using a different subnet mask than he is. He will assume that, because Renoir is using the same address block, he must be using the same subnet mask. Therefore, when Michelangelo performs his ANDing, the result will match, and he will attempt to send it directly. The unusual part is that Michelangelo's send operation will actually succeed because they really *are* on the same logical network. However, when Renoir attempts to respond, he will fail because he will perform his own ANDing operation. The result will be that they are on different networks. Renoir will attempt to send the packet to a router, but the packet will never be delivered. This process is shown in Figure 6-9.

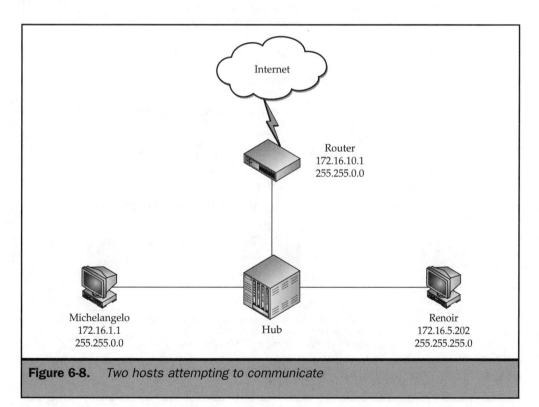

Figure 6-8. *Two hosts attempting to communicate*

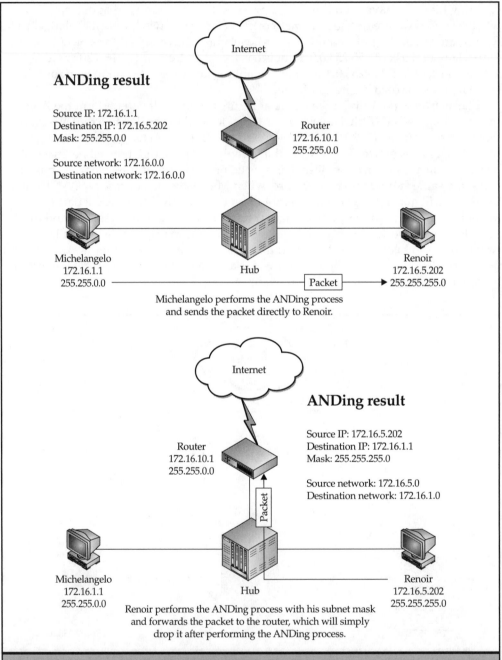

Figure 6-9. *Visual explanation of the failure*

The purpose of this final ANDing example is for you to realize that not only IP addresses but also subnet masks must match if hosts are on the same logical network. If they do not match, the ANDing will be performed incorrectly, and the communication will fail.

IP Classes and Rules

Originally, IP addresses were divided into classes to allow companies of different sizes to use different numbers of addresses. Table 6-1 lists the IP address classes.

For our purposes, we are concerned only with class A, B, and C addresses. (Class D is discussed in the "Multicast Addressing" section, later in this chapter.)

Class A addresses support over 16 million hosts per network. This class is obviously for very large networks (generally top-level ISPs). There are 126 valid class A networks, all of which were registered long ago. Public IP addresses must be registered with the Internet Assigned Numbers Authority (IANA), the sanctioning body controlling the use of public or Internet-accessible IP addresses.

With a class A address, the MSB value in the first octet is always 0. This means that the smallest number you could have is 00000000 (0), and the largest is 01111111 (127). However, a few rules come into play in this case. First, the class A network address of 0 is reserved. This address is used to mean "this network," or the network the sending host is actually physically connected to. Second, the class A network address of 127 is used for a *loopback*: when the TCP/IP suite simply performs a test on itself. By sending packets to the loopback destination address, it doesn't actually send them out on the network. It simply loops them back into itself to make sure that the TCP/IP stack is not corrupting the data. (A common troubleshooting step is to ping the loopback address.) So, any time you send a packet to the all 0's network, you are sending to local hosts.

Class	First Octet Range	Total Networks	Total Hosts Per Network	Default Mask	Use
A	1–126	126	16.7 million	255.0.0.0	Standard class
B	128–191	16,384	65,534	255.255.0.0	Standard class
C	192–223	2 million	254	255.255.255.0	Standard class
D	224–239	N/A	N/A	224.0.0.0	IP multicast
E	240–255	N/A	N/A	N/A	Experimental

Table 6-1.   *IP Address Classes Summary*

Any time you send something to the 127 network, you are performing a loopback. Because of loopback and the reserved network 0, useable class A addresses fall in the range of 1 through 126 in the first octet.

Class A addresses use a 255.0.0.0 subnet mask, also known as an 8-bit subnet mask because the mask has eight ones in a row, and then all zeros (11111111.00000000. 00000000.00000000). This means that in a basic class A network, the first octet is for the network address and the last three octets are for the host address. This goes back to ANDing. If you were to take a class A address of 10.1.1.1 with the standard subnet mask of 255.0.0.0 and AND them, you would end up with 10.0.0.0 for the network address. The rest is for the host. The following diagram shows the breakdown of a class A address when using the default subnet mask.

Network | Host

$$10 \,.\, 0 \,.\, 0 \,.\, 1$$

Class B addresses support 65,534 hosts per network. Class B addresses are designed for smaller (but still pretty large) networks. A little over 16,000 class B networks exist, all of which are already registered.

Class B addresses always start with the binary 10 (as in *10*101100.00010000.00000001. 00000001, or 172.16.1.1), meaning that the first octet must be in the range of 128 (10000000) through 191 (10111111). There are no unusable class B networks.

Class B networks have a 16-bit default mask (255.255.0.0), meaning that the first 16 bits signify the network address and the last 16 bits signify the host address. Here's the breakdown of a class B IP address:

Network | Host

$$172 \,.\, 16 \,.\, 0 \,.\, 1$$

Class C networks can support only 254 hosts per network. Class C addresses are designed for small networks. Over two million class C networks exist, most of which are already registered.

Class C networks must begin with the binary 110 (as in *11*000000.10101000. 00000001.00000001, or 192.168.1.1). Again, there are no unusable class C networks.

Class C networks have a 24-bit default mask, meaning that 24 bits are used for the network portion and 8 bits are used for the host portion. Here's the breakdown of a class C IP address:

Network | Host

$$192 \,.\, 168 \,.\, 1 \,.\, 200$$

At this point, you may have noticed that a few (specifically, two) IP addresses are missing in each network. For example, a class C IP address has only 254 hosts available per network, when there should be 256 hosts (using the 2^n formula, or 2^8). If you caught this "error," I applaud you; but it's not an error. In each network, two addresses are reserved: specifically, the highest address (all ones) and the lowest address (all zeros). The all ones host address signifies a broadcast, and the all zeros address signifies "this network." These two addresses are unusable as host addresses. This is another TCP/IP rule.

The all zeros address is the easiest to understand, so we will cover it first. If you perform an ANDing operation with an IP address/subnet mask pair, the host portion will be all zeros. For instance, if we AND the IP address 192.168.1.1 with the default subnet mask of 255.255.255.0, we will end up with 192.168.1.0, the network address. So the address 192.168.1.0 is the network address and cannot be used as a host address.

The all ones address is also pretty easy to understand. This is reserved for layer 3 broadcasts. For instance, in the IP address 10.255.255.255, the host address is all ones (00001010.*11111111.11111111.11111111*). This means all hosts on this network: a broadcast.

Note that *all* of the bits in the host or network portions have to be all zeros or all ones to be an invalid host address, not just a portion. For instance, the IP address 172.16.0.255 is valid (10101100.00010000.*00000000.11111111*) because the *entire* host address is neither all zeros nor all ones. However, the address 192.168.1.255 is invalid because the entire host portion *is* all ones (11000000.10101000.00000001.*11111111*).

For this reason, when you are calculating the number of hosts per network, the formula is $2^n - 2$, not just 2^n. For instance, if you knew you had ten host bits, you would calculate 2^n ($2 \times 2 \times 2 \times 2 \times 2 \times 2 \times 2 \times 2 \times 2 \times 2 = 1024$) and then subtract 2 ($1024 - 2 = 1022$).

To help solidify these concepts, let's practice in Exercise 6-2.

Exercise 6-2: ANDing, Network Class Determination, and Validity Determination

In the following problems, find the network address, which class of address it is, and whether it is a valid host IP address. Do not use a calculator, but feel free to look back over the material if you get stuck. The answers are in a sidebar on the next page.

1. Address: 222.10.17.1   Mask: default subnet mask

2. Address: 127.12.1.98   Mask: 255.0.0.0

3. Address: 189.17.255.0   Mask: 255.255.0.0

4. Address: 97.1.255.255   Mask: default subnet mask

5. Address: 197.17.0.255   Mask: 255.255.255.0

6. Address: 0.12.252.1   Mask: 255.0.0.0

7. Address: 10.0.1.0   Mask: default subnet mask

8. Address: 220.0.0.254   Mask: default subnet mask

Answers to Exercise 6-2

Question	Network Address	Class	Valid?
1	222.10.17.0	C	Yes
2	127.0.0.0	A	No; loopback address
3	189.17.0.0	B	Yes
4	97.0.0.0	A	Yes
5	197.17.0.0	C	No; host all ones
6	0.0.0.0	A	No; reserved class A network
7	10.0.0.0	A	Yes
8	220.0.0.0	C	Yes

Now that we have covered the basics of IP addressing, let's move on to the real prize: subnetting.

Simple Subnetting

Subnetting is the act of breaking a larger network into smaller, more manageable, components. Before we get into the math of subnetting, you need to know why it is used. As you may have noticed, between class C and class B IP addresses, there is a huge difference in number of hosts per network. The same divide exists between class B and class A addresses. What if you needed 1,000 hosts per network? You would need to get a class B address, even though you would waste over 64,000 addresses. Also, what about the number of hosts you can reasonably support on one flat, nonrouted network? I mean, let's get real here. Can you squeeze 65,000 PCs in one network without using any WAN links? Think about it. Do you think you could squeeze 65,000 PCs into one building within a 500-meter circle using Ethernet? Because of this problem, there needed to be a way to separate the larger network into multiple subnetworks. This is the principle behind subnetting.

Subnetting works because of the binary math behind it. By using the subnet mask, we can "steal" bits from the host portion of the IP address and add to the network portion. This will give us more networks, but fewer hosts per network. Figure 6-10 provides an example.

By using subnetting, we can split the larger network into smaller networks that are more appropriate for the number of hosts we can reasonably squeeze into one broadcast domain. Say, for example, that our company has the registered class B network of

NETWORKING BASICS

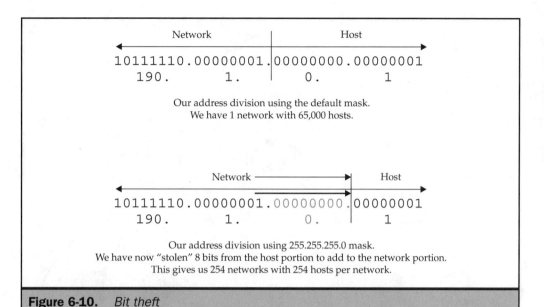

Network | Host

10111110.00000001.00000000.00000001
190.   1.   0.   1

Our address division using the default mask.
We have 1 network with 65,000 hosts.

Network ——————————▶ | Host

10111110.00000001.00000000.00000001
190.   1.   0.   1

Our address division using 255.255.255.0 mask.
We have now "stolen" 8 bits from the host portion to add to the network portion.
This gives us 254 networks with 254 hosts per network.

Figure 6-10.   *Bit theft*

190.1.0.0. With the default mask (16-bit, or 255.255.0.0), we have only one network with 65,000 or so hosts. Well, let's assume our company has 200 or so locations, each with less than 200 hosts. What we could really use is a few hundred class C addresses.

Luckily, by using subnetting, we can split our class B address into 254 class C–sized networks. We do this by giving the network a 24-bit mask (255.255.255.0) instead of a 16-bit mask. With a 24-bit mask, when you perform the ANDing operation, an address like 190.1.1.1 shows the network address as being 190.1.1.0 instead of 190.1.0.0, as it would with a 16-bit mask. This ends up creating a network similar to the one shown in Figure 6-11.

For a more realistic example, let's say that our company splits the broadcast domains of its network by floors. The main building has 48 floors and can squeeze around 600 users, plus around 150 assorted IP network devices (printers, managed switches, IP phones, and so on) onto each floor. Therefore, we need at least 48 subnets with at least 750 hosts per subnet. To accomplish this feat, we could subnet a class B IP address with a 22-bit mask (255.255.252.0). This allows six bits for the subnet address (62 possible subnets) and ten bits for the host address (1,022 possible hosts). The math behind this is simpler than it initially appears.

To determine how many subnets and hosts a certain IP address/subnet mask combination gives us, we need to first break the address and the default subnet mask into binary. Then we need to draw a line, called the *network line,* after the network portion of the address (signified by the last 1 in the *default* subnet mask). Next, we determine how many bits are needed to give us our required subnets. In this case, we

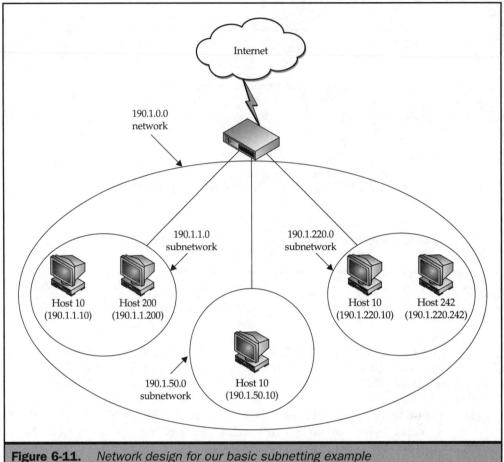

Figure 6-11. *Network design for our basic subnetting example*

need six bits; five bits will give us only 30 ($2^5 - 2$). We then add six bits (six additional 1's) to the subnet mask and draw our second line, called the *subnet line*. Everything to the left of this line (up to the network line) signifies the subnet address. Everything to the right of this line signifies the host address. Figure 6-12 illustrates this process.

Note that when selecting a subnet mask, you can use a mask only with consecutive 1's. In other words, a subnet mask of 01010011 (83) is not valid. However, a subnet mask of 11110000 (248) is. Because of this, there are only nine valid binary combinations in each octet: 00000000 (0), 10000000 (128), 11000000 (192), 11100000 (224), 11110000 (240), 11111000 (248), 11111100 (252), 11111110 (254), and 11111111 (255).

To solidify this concept, let's walk through an example. In this example, our company has a class C address block (200.10.1.0), 6 buildings, and less than 30 hosts per building.

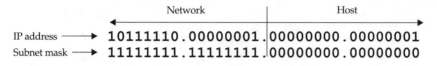

First, we draw our network line after
the last 1 in the *default* subnet mask.

$$2^6 - 2 = 62$$
$$(2^6 = 64; 64 - 2 = 62)$$

Next, we determine how many bits are needed
to make the required number of subnets.

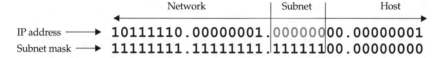

Finally, we add the required number of bits to the mask
and draw our subnet line after the last 1 in our custom mask.
Everything between the network line and the subnet line
is the subnet address.

Figure 6-12.   *The mask determination process*

We need to determine the appropriate subnet mask for the company. (Figure 6-13 illustrates the process.)

First, we turn the IP address and subnet mask into binary. We then draw the first line, the network line, after the last 1 in the mask. Now we need to determine how many bits are needed to give us the required number of subnets and hosts. We have eight bits to work with (the host portion of the IP address), and we need to determine how many of those are needed to create six subnets. One bit would give us two subnets (2^n) minus two subnets (the all zeros and all ones subnets), leaving us with no possible subnets. That obviously won't work. Two bits would give us four subnets (2^n) minus two subnets (the all zeros and all ones subnets), leaving us with two possible subnets. Closer, but still no cigar. Three bits would give us eight subnets (2^n) minus two subnets (the all zeros and all ones subnets), leaving us with six possible subnets. Bingo.

Now we need to make sure that we have enough hosts per subnet. Because we're using three bits for the subnet, we have five bits left for the host (8 − 3 = 5). Five bits gives us 32 hosts, minus 2 hosts, leaves us with 30 hosts per subnet. Perfect. So we will use a 27-bit mask (255.255.255.224) for this network.

Network | Host

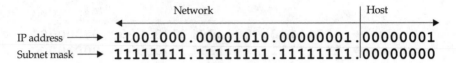

IP address ⟶ 11001000.00001010.00000001.00000001
Subnet mask ⟶ 11111111.11111111.11111111.00000000

First, we draw our network line after
the last 1 in the *default* subnet mask.

$$2^1 - 2 = 0$$
$$2^2 - 2 = 2$$
$$2^3 - 2 = 6$$

Next, we determine how many bits are needed
to make the required number of subnets.

Network | Subnet / Host

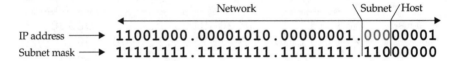

IP address ⟶ 11001000.00001010.00000001.00000001
Subnet mask ⟶ 11111111.11111111.11111111.11000000

Finally, we add the required number of bits to the mask
and draw our subnet line after the last 1 in our custom mask.
Everything between the network line and the subnet line
is the subnet address.

Mask: 27-bit (255.255.255.224)
Subnet bits: 3
Host bits: 5
Subnets: 6 ($2^3 - 2$)
Hosts per subnet: 30 ($2^5 - 2$)

Figure 6-13.   *Walkthrough of the mask determination process*

To recap, you can look at the mask determination as a five-step process:

1. Convert the IP address and default mask into binary.

2. Draw a line after the last 1 in the subnet mask. Everything to the left of this line is your base network address.

3. Determine how many bits are required to achieve the desired number of subnets. Extend the 1's in your subnet mask by this many bits.

4. Draw a second line after the last 1 in the new subnet mask. Everything from the left of this line until the first line is the subnet portion. Everything to the right of this line is the host portion.

5. Make sure the remaining bits in the mask provide enough hosts per subnet.

That's it. To make sure you can perform this task before going forward, let's do a few practice problems. Take a look at Exercise 6-3. As usual, the answers are provided on the next page.

Once you have determined which mask you need to use, the next step is to determine the range of IP addresses that belong to each subnet. —This is required because you need to know into which subnet a given IP address falls. With a simple subnet mask (like 255.255.255.0), it is easy to determine the range. If you had the IP address 172.16.1.1 with a 255.255.255.0 mask, you know, without doing the binary, that the range of hosts is 172.16.1.1 through 172.16.1.254. You can also easily recognize that 172.16.200.1 and 172.16.50.1 are on different subnets. However, it is not as easy to recognize that with a 255.255.224.0 mask, 172.16.34.1 and 172.16.73.1 are on different subnets, while 172.16.130.1 and 172.16.150.1 are on the same subnet. To determine the subnet ranges, we need to do some binary. Let's take a look at the process (this process is diagrammed in Figures 6-14 through 6-16).

First, we need to take the IP address and mask, convert them into binary, and draw two lines: the network line (right after the last 1 in the default subnet mask, which is determined by the class of IP address), and the subnet line (right after the last 1 in the custom subnet mask). Figure 6-14 details this first step.

Exercise 6-3: Subnet Mask Determination

In the following problems, find the appropriate subnet mask to meet the requirements. Do not use a calculator, and remember to start with the default subnet mask for the particular class of address.

	Address	Hosts Required	Subnets Required
1.	192.168.1.0	60	2
2.	172.16.0.0	2,000	28
3.	10.0.0.0	30,000	As many as possible
4.	192.168.1.0	8	As many as possible
5.	172.16.0.0	As many as possible	16

Answers to Exercise 6-3

1. Mask: 255.255.255.192 (26 bits)   Explanation: This is the only mask that meets the criteria.

2. Mask: 255.255.248.0 (21 bits)   Explanation: This is the only mask that meets the criteria.

3. Mask: 255.255.128.0 (17 bits)   Explanation: Any mask with less than 18 bits would work, but a 17-bit mask gives us the most subnets (510).

4. Mask: 255.255.255.240 (28 bits)   Explanation: Any mask with less than 29 bits would work, but a 28-bit mask gives us the most subnets (14).

5. Mask: 255.255.248.0 (21 bits)   Explanation: Any mask with more than 21 bits would work, but a 21-bit mask gives us the most hosts (2,046).

> **Caution**   *Anything to the left of the network line cannot be changed. If you were to change something to the left of this line, you would be on a different network altogether, not just on a different subnet.*

Next, we need to determine all of the different binary combinations possible with the bits in the subnet portion of the address. For instance, if we had four bits in the subnet portion, the possible combinations are the following: 0000 (0), 0001 (1), 0010 (2), 0011 (3), 0100 (4), 0101 (5), 0110 (6), 0111 (7), 1000 (8), 1001 (9), 1010 (10), 1011 (11), 1100 (12), 1101 (13), 1110 (14), and 1111 (15). These are our 16 subnets. However, the all zeros and the all ones subnets are invalid, so 0000 (0) and 1111 (15) have to be thrown away. The reasoning behind this is that the all zeros subnet is the base network address, and the all ones is an "all subnets" broadcast. In truth, this is just a recommendation, not a rule, but we will discuss that later in this chapter in the sections about VLSM and CIDR. The removal of these two subnets leaves us 14 valid subnets, as shown in Figure 6-15.

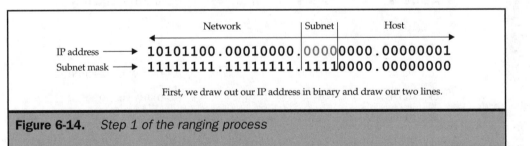

First, we draw out our IP address in binary and draw our two lines.

Figure 6-14.   *Step 1 of the ranging process*

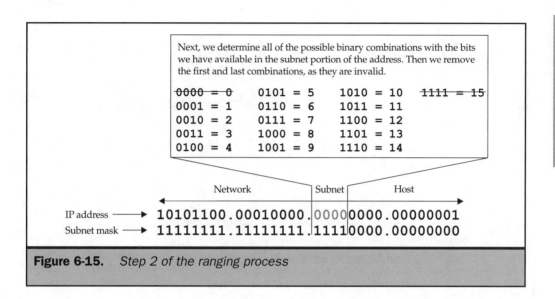

Figure 6-15.   *Step 2 of the ranging process*

Now that we have the subnet binary combinations, we have to put them in the context of the rest of the IP address, as shown in Figure 6-16. For instance, the binary combination for the first valid subnet is 0001. In the context of the rest of the address, this becomes *0001*0000. This makes the base address of the subnet 16 (0 + 0 + 0 + 16 + 0 + 0 + 0 + 0 = 16). Now, without changing the network or subnet portion, we have to determine the highest and lowest numbers in the range. This task is fairly easy: the lowest number is the base subnet address, and the highest is the subnet's binary combination with all ones in the host portion. In this case, this number is *0001*1111, or 31. Finally, we have to remove the first and last IP addresses in the range. Remember, the first address means "this network," and the last is the broadcast address.

Unfortunately, the ranging process is one of the most difficult IP addressing concepts to grasp, so we will go through several examples to see the process in action.

Ranging Example 1 (Class A)

In this example, we will take the class A network address of 10.0.0.0 with a 255.240.0.0 mask and determine the ranges of IP addresses for each subnet. The first step is to take the IP address and mask, convert them into binary, and draw lines. This is shown in Figure 6-17.

Next, we compute all of the possible binary combinations with the number of bits in the subnet portion (in this case, four). This gives us 0000 (0), 0001 (1), 0010 (2), 0011 (3), 0100 (4), 0101 (5), 0110 (6), 0111 (7), 1000 (8), 1001 (9), 1010 (10), 1011 (11), 1100 (12),

First, we need to take our binary combinations and put them into context in our IP address. We do this by inserting the combination into the subnet portion of the address. Then we change the host portion of the address to figure out what the lowest and highest IP addresses in the range are.

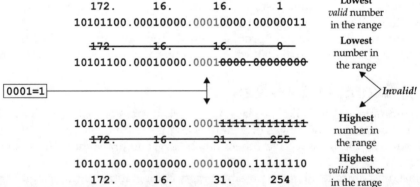

172.	16.	16.	0	Lowest number in the range
10101100.	00010000.	00010000.	00000000	

0001=1

10101100.	00010000.	00011111.	11111111	Highest number in the range
172.	16.	31.	255	

172.	16.	32.	0	Lowest number in the range
10101100.	00010000.	00100000.	00000000	

0010=2

10101100.	00010000.	00101111.	11111111	Highest number in the range
172.	16.	47.	255	

172.	16.	48.	0	Lowest number in the range
10101100.	00010000.	00110000.	00000000	

0011=3

10101100.	00010000.	00111111.	11111111	Highest number in the range
172.	16.	63.	255	

Finally, we have to take each range and remove the first and last IP addresses, as both are invalid for host addresses.

172.	16.	16.	1	Lowest *valid* number in the range
10101100.	00010000.	00010000.	00000011	

~~172.~~	~~16.~~	~~16.~~	~~0~~	Lowest number in the range
10101100.	00010000.	0001~~0000.~~	~~00000000~~	

0001=1   Invalid!

10101100.	00010000.	0001~~1111.~~	~~11111111~~	Highest number in the range
~~172.~~	~~16.~~	~~31.~~	~~255~~	

10101100.	00010000.	00010000.	11111110	Highest *valid* number in the range
172.	16.	31.	254	

Figure 6-16.   *Step 3 of the ranging process*

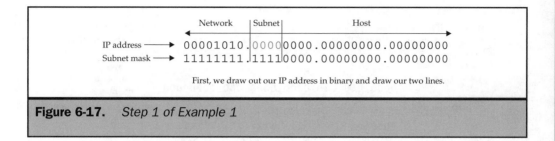

First, we draw out our IP address in binary and draw our two lines.

Figure 6-17.   *Step 1 of Example 1*

1101 (13), 1110 (14), and 1111 (15). This is shown in Figure 6-18. We then remove the all zeros and all ones subnets (0000 and 1111).

Finally, we figure out what those combinations give us in terms of IP address ranges. To do this, we take each subnet combination individually and put all zeros and all ones in the host portion. This is shown in Figure 6-19.

For instance, the first valid subnet is 0001, so we take the second, third, and fourth octets and put all zeros in the host portion for *0001*0000.00000000.00000000 (16.0.0). Then we put all ones in the same section for *0001*1111.11111111.11111111 (31.255.255). This makes the base IP address range 10.16.0.0 through 10.31.255.255. We then remove the first and last IP address from each subnet (10.16.0.0 and 10.31.255.255), because the first one is the all zeros address (network address), and the last is the all ones address

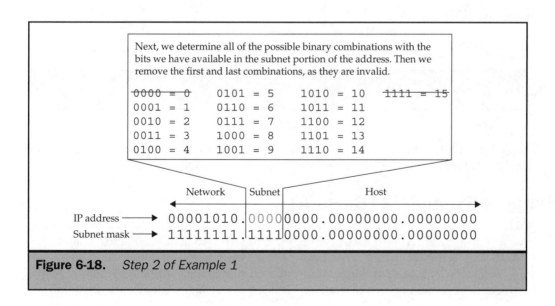

Figure 6-18.   *Step 2 of Example 1*

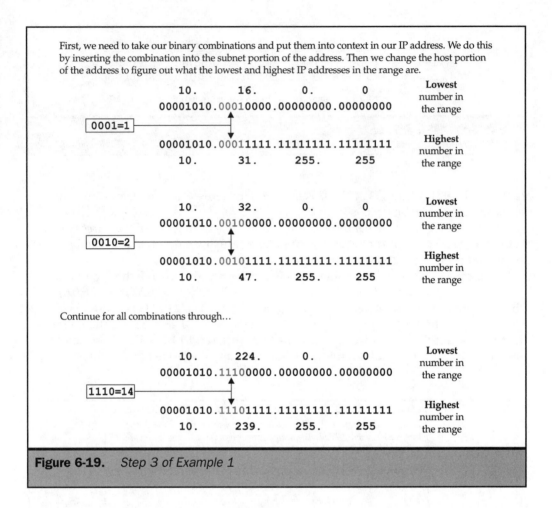

First, we need to take our binary combinations and put them into context in our IP address. We do this by inserting the combination into the subnet portion of the address. Then we change the host portion of the address to figure out what the lowest and highest IP addresses in the range are.

Figure 6-19. *Step 3 of Example 1*

(broadcast address for this subnet). This leaves us with 10.16.0.1 through 10.31.255.254 as the valid address range for this subnet. This is shown in Figure 6-20.

Ranging Example 2 (Class A)

In this example, we will take the same class A network address of 10.0.0.0, but this time we will give it a 255.224.0.0 mask. Again, the first step is to take the IP address and mask, convert them into binary, and draw lines. This is shown in Figure 6-21.

Next, we compute all of the possible binary combinations with the number of bits in the subnet portion (in this case, three). This gives us 000 (0), 001 (1), 010 (2), 011 (3),

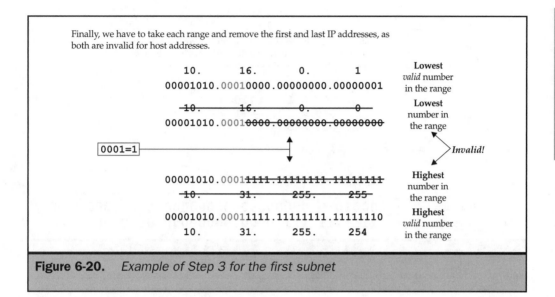

Finally, we have to take each range and remove the first and last IP addresses, as both are invalid for host addresses.

Figure 6-20.   *Example of Step 3 for the first subnet*

100 (4), 101 (5), 110 (6), and 111 (7). This is shown in Figure 6-22. We then remove the all zeros and all ones subnets (000 and 111).

Finally, we figure out what those combinations give us in terms of IP address ranges. To do this, we take each subnet combination individually and put all zeros and all ones in the host portion. This is shown in Figure 6-23.

Again, we will show the process with the first valid subnet (001). We take the second, third, and fourth octets and put all zeros in the host portion for *001*00000.00000000. 00000000 (32.0.0). Then we put all ones in the same section for *001*11111.11111111. 11111111 (63.255.255). This makes the base IP address range 10.32.0.0 through 10.63. 255.255. We then remove the first and last IP address from each subnet (10.32.0.0 and 10.63.255.255) because the first one is the all zeros address (network address), and the last is the all ones address (broadcast address for this subnet). This leaves us with 10.32.0.1 through 10.64.255.254 as the valid address range for this subnet. This is shown in Figure 6-24.

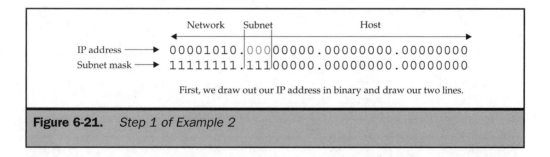

First, we draw out our IP address in binary and draw our two lines.

Figure 6-21.   *Step 1 of Example 2*

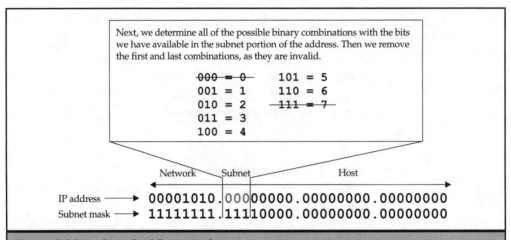

Figure 6-22. *Step 2 of Example 2*

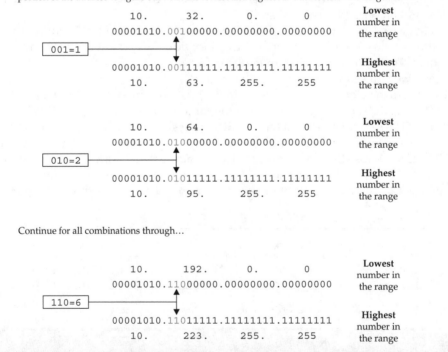

Figure 6-23. *Step 3 of Example 2*

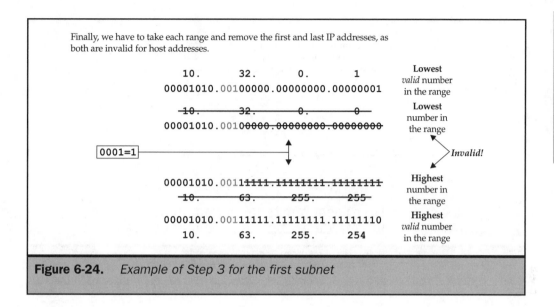

Finally, we have to take each range and remove the first and last IP addresses, as both are invalid for host addresses.

10.	32.	0.	1	**Lowest** *valid* number in the range

```
00001010.00100000.00000000.00000001
```

~~10.   32.   0.   0~~   **Lowest** number in the range
```
00001010.001
```
~~00000.00000000.00000000~~

0001=1   Invalid!

```
00001010.0011
```
~~1111.11111111.11111111~~   **Highest** number in the range

~~10.   63.   255.   255~~

```
00001010.00111111.11111111.11111110
```
```
10.        63.        255.        254
```
Highest *valid* number in the range

Figure 6-24. *Example of Step 3 for the first subnet*

Ranging Example 3 (Class B)

In this example, we will take a class B network address of 172.16.0.0 with a 255.255.192.0 mask. Again, the first step is to take the IP address and mask, convert them into binary, and draw lines. This is shown in Figure 6-25.

Next, we compute all of the possible binary combinations with the number of bits in the subnet portion (in this case, two). This gives us 00 (0), 01 (1), 10 (2), and 11 (3). This is shown in Figure 6-26. We then remove the all zeros and all ones subnets (00 and 11).

Finally, we figure out what those combinations give us in terms of IP address ranges. To do this, we take each subnet combination individually and put all zeros and all ones in the host portion. This is shown in Figure 6-27.

Finally, we will once again show the process with the first valid subnet (01). This time, we take the third and fourth octets and put all zeros in the host portion for

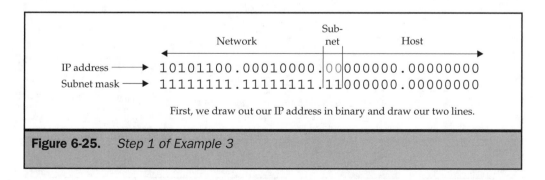

	Network	Sub-net	Host

IP address ──────▶ `10101100.00010000.00000000.00000000`
Subnet mask ─────▶ `11111111.11111111.11000000.00000000`

First, we draw out our IP address in binary and draw our two lines.

Figure 6-25. *Step 1 of Example 3*

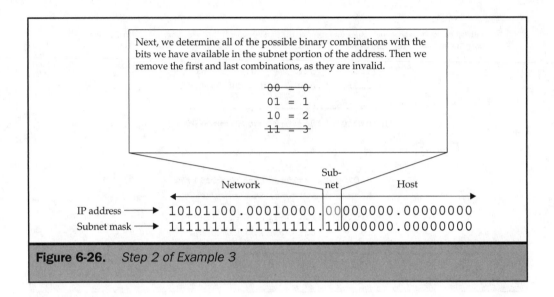

Figure 6-26. *Step 2 of Example 3*

*01*000000.00000000 (64.0). Then we put all ones in the same section for *01*111111.
11111111 (127.255). This makes the base IP address range 172.16.64.0 through 172.16.
127.255. We then remove the first and last IP address from each subnet (172.16.64.0
and 172.16.127.255). This leaves us with 172.16.64.1 through 172.16.127.254 as the
valid address range for this subnet. See Figure 6-28.

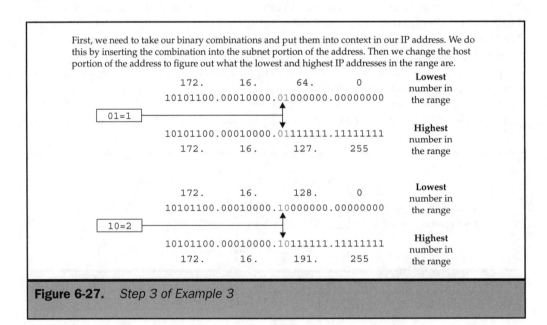

Figure 6-27. *Step 3 of Example 3*

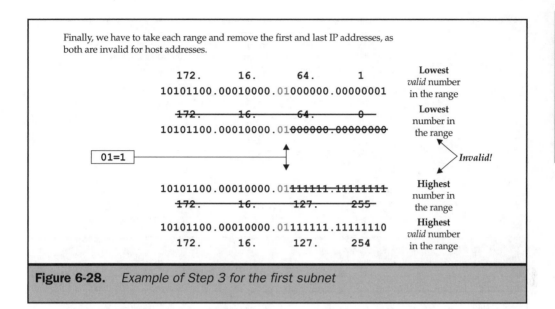

Figure 6-28. *Example of Step 3 for the first subnet*

Ranging Example 4 (Class B)

In this example, we will take the same class B network address of 172.16.0.0, but with a 255.255.240.0 mask. Again, the first step is to take the IP address and mask, convert them into binary, and draw lines. This is shown in Figure 6-29.

Next, we compute all of the possible binary combinations with the number of bits in the subnet portion (in this case, four). This gives us 0000 (0), 0001 (1), 0010 (2), 0011 (3), 0100 (4), 0101 (5), 0110 (6), 0111 (7), 1000 (8), 1001 (9), 1010 (10), 1011 (11), 1100 (12), 1101 (13), 1110 (14), and 1111 (15). (See Figure 6-30.) We then remove the all zeros and all ones subnets (0000 and 1111).

Finally, we figure out what those combinations give us in terms of IP address ranges. To do this, we take each subnet combination individually and put all zeros and all ones in the host portion. This is shown in Figure 6-31.

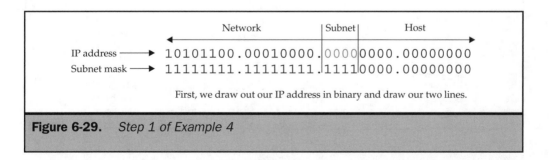

Figure 6-29. *Step 1 of Example 4*

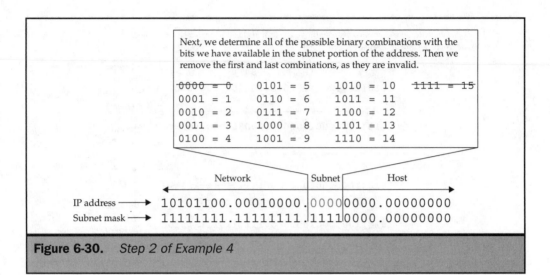

Next, we determine all of the possible binary combinations with the bits we have available in the subnet portion of the address. Then we remove the first and last combinations, as they are invalid.

~~0000 = 0~~	0101 = 5	1010 = 10	~~1111 = 15~~
0001 = 1	0110 = 6	1011 = 11	
0010 = 2	0111 = 7	1100 = 12	
0011 = 3	1000 = 8	1101 = 13	
0100 = 4	1001 = 9	1110 = 14	

Network   Subnet   Host

IP address ⟶ 10101100.00010000.00000000.00000000
Subnet mask ⟶ 11111111.11111111.11110000.00000000

Figure 6-30. *Step 2 of Example 4*

First, we need to take our binary combinations and put them into context in our IP address. We do this by inserting the combination into the subnet portion of the address. Then we change the host portion of the address to figure out what the lowest and highest IP addresses in the range are.

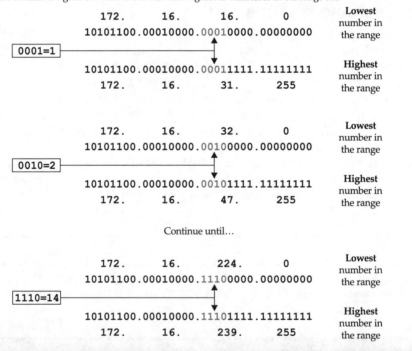

	172.	16.	16.	0	**Lowest** number in the range
	10101100.00010000.00010000.00000000				
0001=1					
	10101100.00010000.00011111.11111111				
	172.	16.	31.	255	**Highest** number in the range

	172.	16.	32.	0	**Lowest** number in the range
	10101100.00010000.00100000.00000000				
0010=2					
	10101100.00010000.00101111.11111111				
	172.	16.	47.	255	**Highest** number in the range

Continue until...

	172.	16.	224.	0	**Lowest** number in the range
	10101100.00010000.11100000.00000000				
1110=14					
	10101100.00010000.11101111.11111111				
	172.	16.	239.	255	**Highest** number in the range

Figure 6-31. *Step 3 of Example 4*

Finally, we will once again show the process with the first valid subnet (0001). This time, we take the third and fourth octets and put all zeros in the host portion for *0001*0000.00000000 (16.0). Then we put all ones in the same section for *0001*1111. 11111111 (31.255). This makes the base IP address range 172.16.16.0 through 172.16. 31.255. We then remove the first and last IP address from each subnet (172.16.16.0 and 172.16.31.255). This leaves us with 172.16.16.1 through 172.16.31.254 as the valid address range for this subnet. This is shown in Figure 6-32.

Ranging Example 5 (Class C)

In this example, we will subnet a class C network address of 192.168.1.0 with a 255.255.255.192 mask. Again, the first step is to take the IP address and mask, convert them into binary, and draw lines. This is shown in Figure 6-33.

Next, we compute all of the possible binary combinations with the number of bits in the subnet portion (in this case, two). This gives us 00 (0), 01 (1), 10 (2), and 11 (3). This is shown in Figure 6-34. We then remove the all zeros and all ones subnets (00 and 11).

Next, we figure out what those combinations give us in terms of IP address ranges. To do this, we take each subnet combination individually and put all zeros and all ones in the host portion. This is shown in Figure 6-35.

Finally, we will once again show the process with the first valid subnet (01). This time, we take the last octet and put all zeros in the host portion for *01*000000 (64). Then we put all ones in the same section for *01*111111 (127). This makes the base IP address range 192.168.1.64 through 192.168.1.127. We then remove the first and last IP addresses

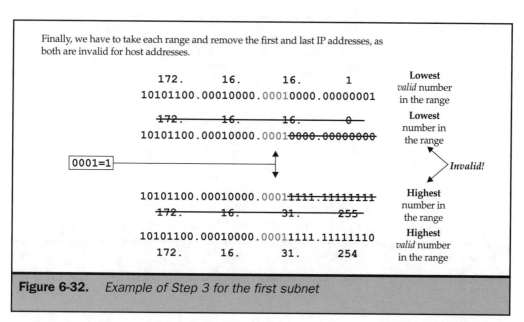

Figure 6-32.   *Example of Step 3 for the first subnet*

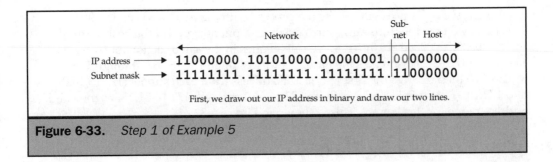

First, we draw out our IP address in binary and draw our two lines.

Figure 6-33. *Step 1 of Example 5*

from each subnet (192.168.1.64 and 192.168.1.127). This leaves us with 192.168.1.65 through 192.168.1.126 as the valid address range for this subnet. This is shown in Figure 6-36.

Ranging Example 6

In this example, we will subnet the same class C network address of 192.168.1.0, but this time, we will use a 255.255.255.240 mask. Again, the first step is to take the IP address and mask, convert them into binary, and draw lines. This is shown in Figure 6-37.

Next, we compute all of the possible binary combinations with the number of bits in the subnet portion (in this case, four). This gives us 0000 (0), 0001 (1), 0010 (2), 0011 (3), 0100 (4), 0101 (5), 0110 (6), 0111 (7), 1000 (8), 1001 (9), 1010 (10), 1011 (11), 1100 (12),

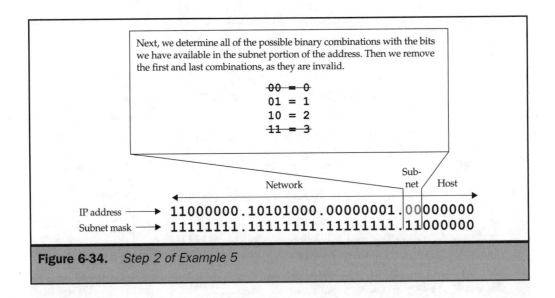

Figure 6-34. *Step 2 of Example 5*

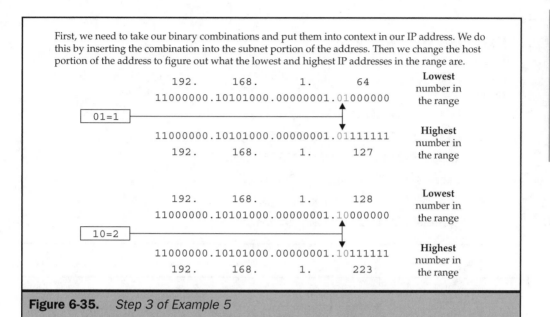

Figure 6-35. *Step 3 of Example 5*

1101 (13), 1110 (14), and 1111 (15). This is shown in Figure 6-38. We then remove the all zeros and all ones subnets (00 and 11).

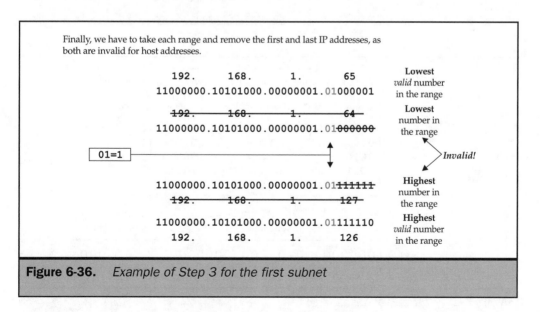

Figure 6-36. *Example of Step 3 for the first subnet*

Exercise 6-4: Ranging

In this exercise, find the range of valid IP addresses for the *first two* valid subnets of each problem. Do not use a calculator. The answers are provided in a sidebar on the next page.

1. Address: 10.0.0.0   Mask: 255.240.0.0

2. Address: 172.16.0.0   Mask: 255.255.224.0

3. Address: 192.168.1.0   Mask: 255.255.255.224

4. Address: 10.0.0.0   Mask: 255.252.0.0

5. Address: 172.16.0.0   Mask: 255.255.254.0

6. Address: 192.168.1.0   Mask: 255.255.255.248

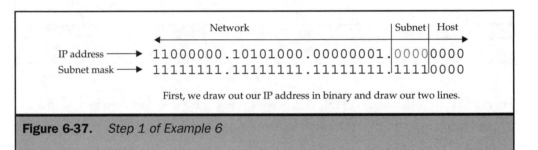

	Network			Subnet	Host

IP address ⟶   11000000.10101000.00000001.0000 0000
Subnet mask ⟶   11111111.11111111.11111111.1111 0000

First, we draw out our IP address in binary and draw our two lines.

Figure 6-37.   *Step 1 of Example 6*

Next, we determine all of the possible binary combinations with the bits we have available in the subnet portion of the address. Then we remove the first and last combinations, as they are invalid.

~~0000 = 0~~	0101 = 5	1010 = 10	~~1111 = 15~~
0001 = 1	0110 = 6	1011 = 11	
0010 = 2	0111 = 7	1100 = 12	
0011 = 3	1000 = 8	1101 = 13	
0100 = 4	1001 = 9	1110 = 14	

IP address ⟶   11000000.10101000.00000001.0000 0000
Subnet mask ⟶   11111111.11111111.11111111.1111 0000

Figure 6-38.   *Step 2 of Example 6*

Next, we figure out what those combinations give us in terms of IP address ranges. To do this, we take each subnet combination individually and put all zeros and all ones in the host portion. This is shown in Figure 6-39.

Finally, we will once again show the process with the first valid subnet (0001). This time, we take the last octet and put all zeros in the host portion for *00010000* (16). Then we put all ones in the same section for *00011111* (31). This makes the base IP address range 192.168.1.16 through 192.168.1.31. We then remove the first and last IP addresses from each subnet (192.168.1.16 and 192.168.1.31). This leaves us with 192.168.1.17 through 192.168.1.30 as the valid address range for this subnet. This is shown in Figure 6-40.

Shortcuts (aka "The Easy Way")

After learning the hard way, most of my students want to strangle me when I show them this method. The reason we go through the binary first, however, is simple: it

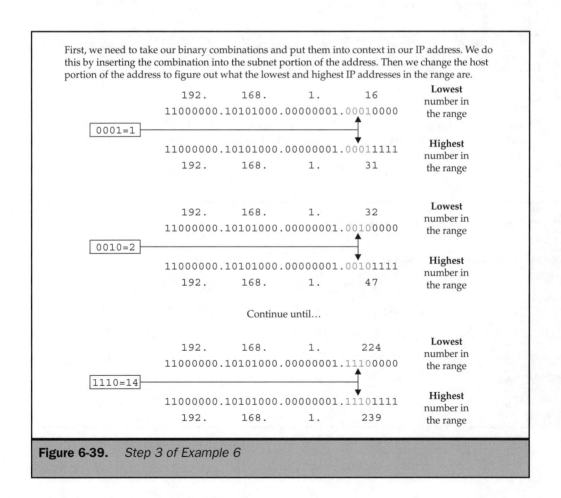

Figure 6-39.   *Step 3 of Example 6*

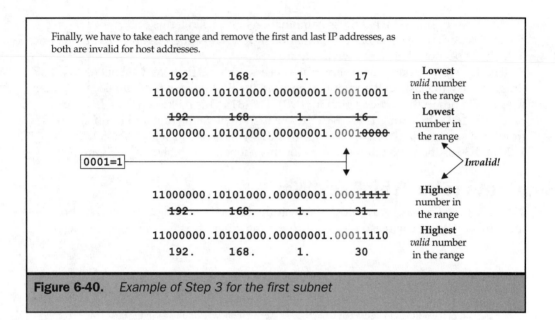

Finally, we have to take each range and remove the first and last IP addresses, as both are invalid for host addresses.

192.	168.	1.	17	**Lowest** *valid* number in the range
11000000.	10101000.	00000001.	00010001	
~~192.~~	~~168.~~	~~1.~~	~~16~~	**Lowest** number in the range
11000000.	10101000.	00000001.	0001~~0000~~	

0001=1 ⟶ Invalid!

11000000.	10101000.	00000001.	0001~~1111~~	**Highest** number in the range
~~192.~~	~~168.~~	~~1.~~	~~31~~	
11000000.	10101000.	00000001.	00011110	**Highest** *valid* number in the range
192.	168.	1.	30	

Figure 6-40. *Example of Step 3 for the first subnet*

provides the framework that allows you to fully understand subnetting. Without understanding the binary, you may know *how* to subnet, but you will have no idea *why* it works.

Answers to Exercise 6-4

Question	Subnet 1	Subnet 2
1	10.16.0.1–10.31.255.254	10.32.0.1–10.63.255.254
2	172.16.32.1–172.16.63.254	172.16.64.1–172.16.91.254
3	192.168.1.33–192.168.1.62	192.168.1.65–192.168.1.91
4	10.4.0.1–10.7.255.254	10.8.0.1–10.11.255.254
5	172.16.2.1–172.16.3.254	172.16.4.1–172.16.5.254
6	192.168.1.9–192.168.1.14	192.168.1.17–192.168.1.22

For the mask selection portion, unfortunately, there is no real shortcut. However, for ranging, the shortcut is very easy. It follows a four-step process, outlined here:

1. Find the "interesting" octet—the one that has a mask that is not either 0 or 255. So in the subnet mask 255.255.192.0, the interesting octet is the third octet (192).

2. Find the range by subtracting the interesting octet from 256. For instance, 256 − 192 = 64.

3. Begin to calculate the range for each subnet by starting with zero in the interesting octet and beginning the next subnet at a multiple of the range. For instance, if the base network address is 172.16.0.0 with a mask of 255.255.192.0, the range is 64, and the interesting octet is the third octet. So the first subnet is from 172.16.0.0 through 172.16.63.255, the second is from 172.16.64.0 through 172.16.127.255, and so on.

4. Finally, remove the first and last subnets and the first and last IP addresses for each subnet.

To show you how this works, let's take an example and perform the ranging process on it. (See Figure 6-41 for a visual representation of the example.)

Let's take 192.168.1.0 with a 255.255.255.248 mask. First, we find the interesting octet, the last octet. Then we subtract from 256 the number of the subnet mask in the

First, we find the interesting octet.

```
IP address      192.168. 1 . 0
Subnet mask     255.255.255.248
```

Then, we subtract the subnet mask in the interesting octet from 256. This gives us our range.

```
    256
 −  248
      8
```

Finally, we simply increment the number in the interesting octet of the IP address by our range, starting at 0. This gives us our base address range for each subnet.

```
192.168.1.0
192.168.1.8
192.168.1.16
192.168.1.24
192.168.1.32
```
and so on...

Figure 6-41.   *Illustration of the example*

interesting octet: 256 – 248 = 8. So eight is our range. Now we calculate the subnets by using the range, starting with 192.168.1.0, as listed here:

1. 192.168.1.0–192.168.1.7
2. 192.168.1.8–192.168.1.15
3. 192.168.1.16–192.168.1.23
4. 192.168.1.24–192.168.1.31
5. 192.168.1.32–192.168.1.39
6. 192.168.1.40–192.168.1.47
7. 192.168.1.48–192.168.1.55
8. 192.168.1.56–192.168.1.63
9. 192.168.1.64–192.168.1.71
10. 192.168.1.72–192.168.1.79
11. 192.168.1.80–192.168.1.87
12. 192.168.1.88–192.168.1.95
13. 192.168.1.96–192.168.1.103
14. 192.168.1.104–192.168.1.111
15. 192.168.1.112–192.168.1.119
16. 192.168.1.120–192.168.1.127
17. 192.168.1.128–192.168.1.135
18. 192.168.1.136–192.168.1.143
19. 192.168.1.144–192.168.1.151
20. 192.168.1.152–192.168.1.159
21. 192.168.1.160–192.168.1.167
22. 192.168.1.168–192.168.1.175
23. 192.168.1.176–192.168.1.183
24. 192.168.1.184–192.168.1.191
25. 192.168.1.192–192.168.1.199
26. 192.168.1.200–192.168.1.207
27. 192.168.1.208–192.168.1.215
28. 192.168.1.216–192.168.1.223
29. 192.168.1.224–192.168.1.231
30. 192.168.1.232–192.168.1.239
31. 192.168.1.240–192.168.1.247
32. 192.168.1.248–192.168.1.255

Finally, we have to remove the two invalid subnets, as well as the first and last IP addresses in each subnet. This leaves us with a revised list, as follows:

1. ~~192.168.1.0–192.168.1.7~~
2. 192.168.1.9–192.168.1.14
3. 192.168.1.17–192.168.1.22
4. 192.168.1.25–192.168.1.30

   …24 subnets later…

28. 192.168.1.217–192.168.1.222
29. 192.168.1.225–192.168.1.230
30. 192.168.1.233–192.168.1.238
31. 192.168.1.240–192.168.1.246
32. ~~192.168.1.248–192.168.1.255~~

That's it. Pretty simple, huh? Now let's get into something a little more complicated: variable length subnet masking (VLSM) and classless interdomain routing (CIDR).

Complex Class-Based Subnetting and VLSM

Before we go into VLSM, you need to get used to a few new concepts. First, you need to understand the "slash" style of writing subnet masks. This style involves putting a slash and then the number of bits (consecutive ones) in the mask after an IP address instead of writing the mask out in decimal. For instance, if you wanted to represent 172.16.1.0 255.255.224.0, you would write it as 172.16.1.0/19. They both mean the same thing; the slash style is just easier to write.

Second, we need to go into some complex subnetting. Complex subnetting involves subnetting more than one octet. Up to this point, we have subnetted only a single octet. Subnetting multiple octets really isn't any harder—just a bit weirder. For instance, let's take an example with a 10.0.0.0 network address. Say we needed 2,000 subnets with 8,000 hosts per subnet. The subnet requirement forces us to use at least 11 bits in the subnet portion ($2^{11} - 2 = 2046$). This gives us a mask of 255.255.224.0, or 10.0.0.0/21. This means that all of the second and part of the third octets are in the subnet section of the address. Determining the ranges might seem a little more complicated for this address, but it really isn't.

	Network	Subnet	Host
IP address ⟶	00001010.	00000000.00000	000.00000000
Subnet mask ⟶	11111111.	11111111.11111	000.00000000

Using the shortcut, we perform the same process. First, we determine which octet is the interesting octet. The third octet fits the bill, so we start there. We take 224 and subtract it from 256. This leaves the range of 32. However, in this case, we must remember that we have a whole octet in front of this one before we begin the next step. Because of this, when we begin the next step, we must take the second octet (the noninteresting one) into account. We do this by beginning at zero on the second octet, increasing the third octet by the range up until we reach the last range, and then adding one to the second octet and starting over. For instance:

1. 10.0.0.0–10.0.31.255
2. 10.0.32.0–10.0.63.255
3. 10.0.64.0–10.0.95.255
4. 10.0.96.0–10.0.127.255
5. 10.0.128.0–10.0.159.255
6. 10.0.160.0–10.0.191.255
7. 10.0.192.0–10.0.223.255
8. 10.0.224.0–10.0.255.255
9. 10.1.0.0–10.1.31.255
10. 10.1.32.0–10.1.63.255
11. 10.1.64.0–10.1.95.255
12. 10.1.96.0–10.1.127.255
13. And so on.

Then we go back and remove the two invalid subnets (first and last) and the two invalid addresses in each subnet (first and last). This leaves us with something similar to this:

1. ~~10.0.0.0–10.0.31.255~~
2. 10.0.32.1–10.0.63.254
3. 10.0.64.1–10.0.95.254
4. 10.0.96.1–10.0.127.254
5. 10.0.128.1–10.0.159.254
6. 10.0.160.1–10.0.191.254
7. 10.0.192.1–10.0.223.254
8. 10.0.224.1–10.0.255.254
9. 10.1.0.1–10.1.31.254
10. 10.1.32.1–10.1.63.254

Then, about 2,000 subnets later…

2042. 10.255.96.1–10.255.127.254

2043. 10.255.128.1–10.255.159.254

2044. 10.255.160.1–10.255.191.254

2045. 10.255.192.1–10.255.223.254

2046. ~~10.255.224.1–10.255.255.254~~

That's all there is to subnetting multiple octets. You just have to remember to keep increasing the number in the noninteresting octet every time you complete a full set of ranges in the interesting octet.

Now we will move on to variable length subnet masking (VLSM), which is used to take a class-based address and make it a bit more scalable and less wasteful. The problem with class-based addresses is that they are typically either too big or too small to be of use in most situations. For instance, assume that we have the network layout pictured in Figure 6-42. With the class B address subnetted using a 20-bit (255.255.240.0) mask, we have 14 subnets and 4,094 hosts per subnet. This is what we need in Building 1 and Building 5 because they both have nearly 3,000 hosts. However, the rest of the locations have significantly fewer, and they are wasting addresses. Out of the 12 additional locations, none of them are using more than 500 IP addresses each, but they all have the /20 mask. This means we are wasting well over 40,000 IP addresses.

All told, VLSM isn't that complicated. It basically consists of subnetting a class-based address space and then subnetting the subnets until you reach the desired number of hosts for a given network.

With VLSM, however, a couple of new rules significantly reduce this waste. First, we do not have to remove the all zeros and all ones subnets. We are allowed to use these to contain hosts. (We still must remove the first and last IP addresses from each subnet, however.) Second, we are allowed to have different masks applied to different sections of the network. This allows us to divide the network up into smaller and smaller pieces as needed (as shown in Figure 6-43). The only trick is to make sure that no address ranges overlap each other.

The trick to making sure no overlap happens is to perform the computations in binary. First, we determine how many hosts are required for the largest networks. In this case, at least 3,000 hosts are required on the two largest networks, so we will start from there. Supporting those hosts requires the 20-bit mask, which gives us 16 subnetworks (remember, we don't have to throw away the first and last subnets with VLSM), with 4,094 hosts each (because we still do have to throw away the first and last IP addresses in each subnet). We use two of these networks for Buildings 1 and 5. The rest of the hosts require only around 6,000 IP addresses, so we need two of the larger 4,094 subnets to support these subnets.

We take the first one (172.16.32.0) and divide it among the eight networks with 450 hosts, each using a 23-bit mask. The three bits added to the subnet mask (making a subsubnet portion) allow us eight subnets with 510 hosts each. Looking at the binary in Figure 6-44, you can see that none of the ranges overlap.

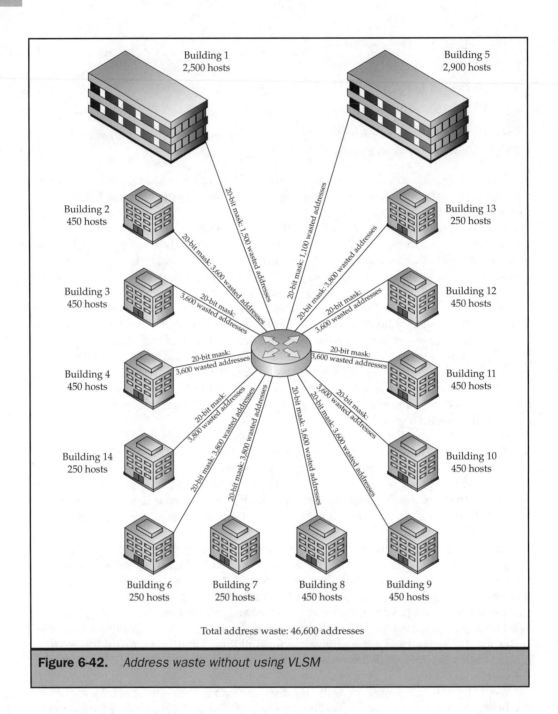

Figure 6-42. *Address waste without using VLSM*

Finally, the last four networks all need less than 254 hosts. This requires a 24-bit mask, so we take one of the 20-bit subnets and subdivide it using this mask. This gives

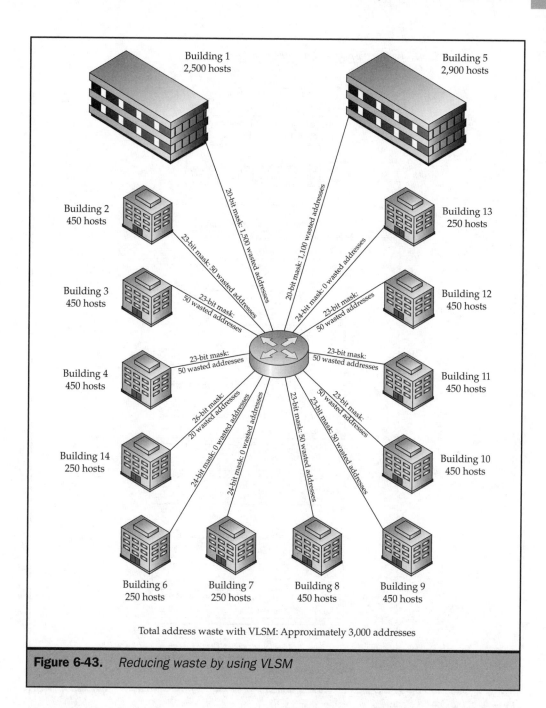

Building 1
2,500 hosts

Building 5
2,900 hosts

Building 2
450 hosts

Building 13
250 hosts

Building 3
450 hosts

Building 12
450 hosts

Building 4
450 hosts

Building 11
450 hosts

Building 14
250 hosts

Building 10
450 hosts

Building 6
250 hosts

Building 7
250 hosts

Building 8
450 hosts

Building 9
450 hosts

20-bit mask: 1,500 wasted addresses

20-bit mask: 1,100 wasted addresses

23-bit mask: 50 wasted addresses

24-bit mask: 0 wasted addresses

23-bit mask: 50 wasted addresses

23-bit mask: 50 wasted addresses

23-bit mask: 50 wasted addresses

23-bit mask: 50 wasted addresses

26-bit mask: 20 wasted addresses

24-bit mask: 0 wasted addresses

24-bit mask: 0 wasted addresses

23-bit mask: 50 wasted addresses

23-bit mask: 50 wasted addresses

Total address waste with VLSM: Approximately 3,000 addresses

Figure 6-43.   *Reducing waste by using VLSM*

us 16 subnetworks within the single 172.16.48.0/20 network, each with 254 hosts. We use four of these for the four subnets, leaving a grand total of 12 subnets with 254 hosts

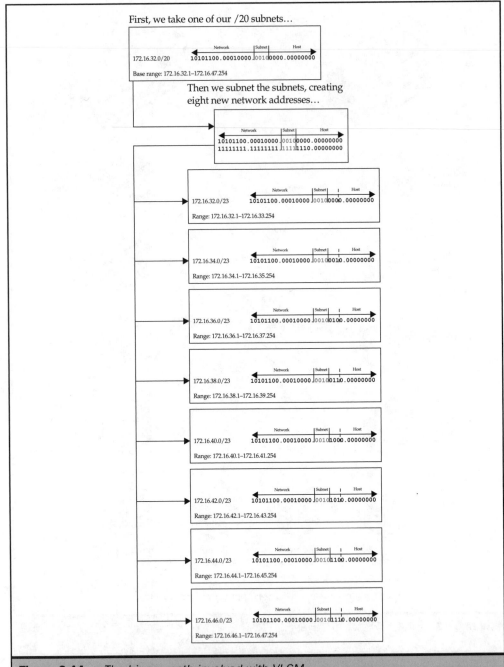

Figure 6-44. *The binary math involved with VLSM*

and 12 subnets with 4,094 hosts left for future expansion. See Figure 6-45 for a logical breakdown of the subdivision.

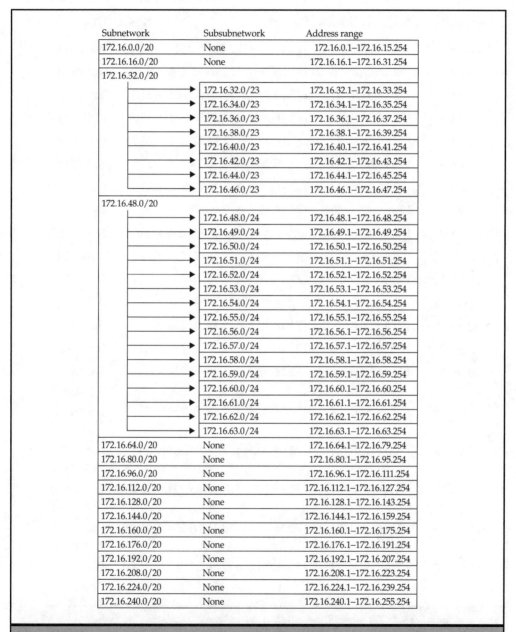

Subnetwork	Subsubnetwork	Address range
172.16.0.0/20	None	172.16.0.1–172.16.15.254
172.16.16.0/20	None	172.16.16.1–172.16.31.254
172.16.32.0/20		
	172.16.32.0/23	172.16.32.1–172.16.33.254
	172.16.34.0/23	172.16.34.1–172.16.35.254
	172.16.36.0/23	172.16.36.1–172.16.37.254
	172.16.38.0/23	172.16.38.1–172.16.39.254
	172.16.40.0/23	172.16.40.1–172.16.41.254
	172.16.42.0/23	172.16.42.1–172.16.43.254
	172.16.44.0/23	172.16.44.1–172.16.45.254
	172.16.46.0/23	172.16.46.1–172.16.47.254
172.16.48.0/20		
	172.16.48.0/24	172.16.48.1–172.16.48.254
	172.16.49.0/24	172.16.49.1–172.16.49.254
	172.16.50.0/24	172.16.50.1–172.16.50.254
	172.16.51.0/24	172.16.51.1–172.16.51.254
	172.16.52.0/24	172.16.52.1–172.16.52.254
	172.16.53.0/24	172.16.53.1–172.16.53.254
	172.16.54.0/24	172.16.54.1–172.16.54.254
	172.16.55.0/24	172.16.55.1–172.16.55.254
	172.16.56.0/24	172.16.56.1–172.16.56.254
	172.16.57.0/24	172.16.57.1–172.16.57.254
	172.16.58.0/24	172.16.58.1–172.16.58.254
	172.16.59.0/24	172.16.59.1–172.16.59.254
	172.16.60.0/24	172.16.60.1–172.16.60.254
	172.16.61.0/24	172.16.61.1–172.16.61.254
	172.16.62.0/24	172.16.62.1–172.16.62.254
	172.16.63.0/24	172.16.63.1–172.16.63.254
172.16.64.0/20	None	172.16.64.1–172.16.79.254
172.16.80.0/20	None	172.16.80.1–172.16.95.254
172.16.96.0/20	None	172.16.96.1–172.16.111.254
172.16.112.0/20	None	172.16.112.1–172.16.127.254
172.16.128.0/20	None	172.16.128.1–172.16.143.254
172.16.144.0/20	None	172.16.144.1–172.16.159.254
172.16.160.0/20	None	172.16.160.1–172.16.175.254
172.16.176.0/20	None	172.16.176.1–172.16.191.254
172.16.192.0/20	None	172.16.192.1–172.16.207.254
172.16.208.0/20	None	172.16.208.1–172.16.223.254
172.16.224.0/20	None	172.16.224.1–172.16.239.254
172.16.240.0/20	None	172.16.240.1–172.16.255.254

Figure 6-45.   *The subdivision of one of our /20 networks*

Table 6-2 lists the final ranges of IP addresses.

Network	Address/Mask	Range	Hosts	Use
Subnet 1	172.16.0.0/20	172.16.0.1–172.16.15.254	4,094	Building 1
Subnet 2	172.16.16.0/20	172.16.16.1–172.16.31.254	4,094	Building 5
Subnet 3	172.16.32.0/23	172.16.32.1–172.16.33.254	510	Building 2
Subnet 4	172.16.34.0/23	172.16.34.1–172.16.35.254	510	Building 3
Subnet 5	172.16.36.0/23	172.16.36.1–172.16.37.254	510	Building 4
Subnet 6	172.16.38.0/23	172.16.38.1–172.16.39.254	510	Building 6
Subnet 7	172.16.40.0/23	172.16.40.1–172.16.41.254	510	Building 7
Subnet 8	172.16.42.0/23	172.16.42.1–172.16.43.254	510	Building 8
Subnet 9	172.16.44.0/23	172.16.44.1–172.16.45.254	510	Building 9
Subnet 10	172.16.46.0/23	172.16.46.1–172.16.47.254	510	Building 10
Subnet 11	172.16.48.0/24	172.16.48.1–172.16.48.254	254	Building 11
Subnet 12	172.16.49.0/24	172.16.49.1–172.16.49.254	254	Building 12
Subnet 13	172.16.50.0/24	172.16.50.1–172.16.50.254	254	Building 13
Subnet 14	172.16.51.0/24	172.16.51.1–172.16.51.254	254	Building 14
Subnet 15	172.16.52.0/24	172.16.52.1–172.16.52.254	254	Future expansion
Subnet 16	172.16.53.0/24	172.16.53.1–172.16.53.254	254	Future expansion
Subnet 17	172.16.54.0/24	172.16.54.1–172.16.54.254	254	Future expansion
Subnet 18	172.16.55.0/24	172.16.55.1–172.16.55.254	254	Future expansion
Subnet 19	172.16.56.0/24	172.16.56.1–172.16.56.254	254	Future expansion
Subnet 20	172.16.57.0/24	172.16.57.1–172.16.57.254	254	Future expansion

Table 6-2.   *Subnet Ranges in the VLSM Example*

Network	Address/Mask	Range	Hosts	Use
Subnet 21	172.16.58.0/24	172.16.58.1–172.16.58.254	254	Future expansion
Subnet 22	172.16.59.0/24	172.16.59.1–172.16.59.254	254	Future expansion
Subnet 23	172.16.60.0/24	172.16.60.1–172.16.60.254	254	Future expansion
Subnet 24	172.16.61.0/24	172.16.61.1–172.16.61.254	254	Future expansion
Subnet 25	172.16.62.0/24	172.16.62.1–172.16.62.254	254	Future expansion
Subnet 26	172.16.63.0/24	172.16.63.1–172.16.63.254	254	Future expansion
Subnet 27	172.16.64.0/20	172.16.64.1–172.16.79.254	4,094	Future expansion
Subnet 28	172.16.80.0/20	172.16.80.1–172.16.95.254	4,094	Future expansion
Subnet 29	172.16.96.0/20	172.16.96.1–172.16.111.254	4,094	Future expansion
Subnet 30	172.16.112.0/20	172.16.112.1–172.16.127.254	4,094	Future expansion
Subnet 31	172.16.128.0/20	172.16.128.1–172.16.143.254	4,094	Future expansion
Subnet 32	172.16.144.0/20	172.16.144.1–172.16.159.254	4,094	Future expansion
Subnet 33	172.16.160.0/20	172.16.160.1–172.16.175.254	4,094	Future expansion
Subnet 34	172.16.176.0/20	172.16.176.1–172.16.191.254	4,094	Future expansion
Subnet 35	172.16.192.0/20	172.16.192.1–172.16.207.254	4,094	Future expansion

Table 6-2.   *Subnet Ranges in the VLSM Example* (continued)

Network	Address/Mask	Range	Hosts	Use
Subnet 36	172.16.208.0/20	172.16.208.1–172.16.223.254	4,094	Future expansion
Subnet 37	172.16.224.0/20	172.16.224.1–172.16.239.254	4,094	Future expansion
Subnet 38	172.16.240.0/20	172.16.240.1–172.16.255.254	4,094	Future expansion

Table 6-2.   *Subnet Ranges in the VLSM Example* (continued)

Now let's take a slightly more complicated example: four groups of buildings connected together with WAN links, each with multiple buildings and different numbers of hosts per building, as shown in Figure 6-46.

In this situation, we need to do a little more complicated address grouping. The first group of buildings, Campus Area Network (CAN) 1, has four buildings. Building 1 needs 16,000 hosts, Building 2 needs 8,000 hosts, Building 3 needs 6,000 hosts, and Building 4 needs 2,000 hosts, for a grand total of 32,000 hosts. CAN 2 has two buildings. Building 5 needs 4,000 hosts, and Building 6 needs 12,000 hosts, for a grand total of 16,000 hosts. CAN 3 has three buildings. Building 7 needs 4,000 hosts, Building 8 needs 3,000 hosts, and Building 9 needs 1,000 hosts, for a grand total of 8,000 hosts. Finally, CAN 4 has five buildings. Building 10 needs 1,000 hosts, Building 11 needs 500 hosts, Building 12 needs 250 hosts, and Buildings 13 and 14 need 100 hosts each, for a grand total of 2,000 hosts.

The first step in dividing the IP address space here is to take the default class-based network address (a class B address of 172.16.0.0, in this example) and split it into a small number of pieces (two, if possible). In this case, CAN 1 needs 32,000 addresses, while CANs 2 through 4 need 26,000 addresses. The easiest solution here is to initially split the networks off at this point with a 17-bit mask, giving the first 32,000 addresses to CAN 1, and splitting the other 32,000 addresses among CANs 2, 3, and 4. This is shown in Figure 6-47.

Next, we will subdivide the 172.16.0.0/17 network for CAN 1 into another group of two, with Building 1 in one group (it needs exactly half of the addresses) and the other three buildings in the second group. We do this by adding another bit to the mask,

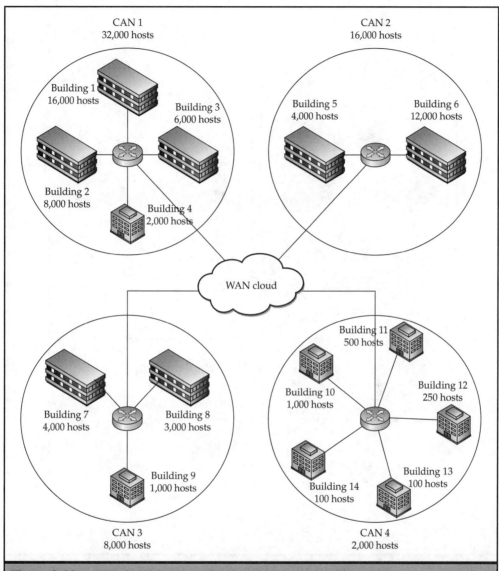

Figure 6-46.   *Initial requirements for the complicated VLSM example*

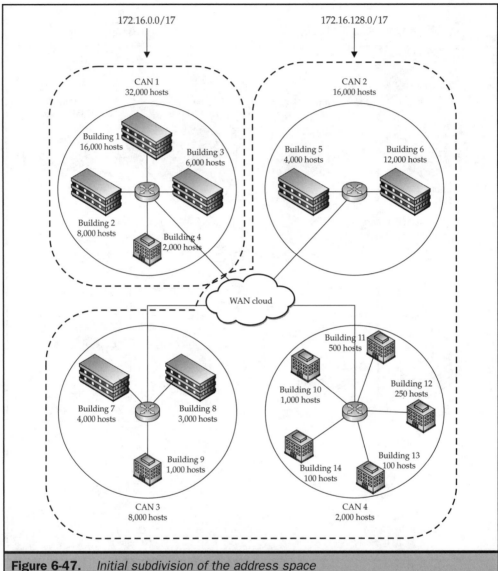

172.16.0.0/17

172.16.128.0/17

CAN 1
32,000 hosts

CAN 2
16,000 hosts

Building 1
16,000 hosts

Building 3
6,000 hosts

Building 2
8,000 hosts

Building 4
2,000 hosts

Building 5
4,000 hosts

Building 6
12,000 hosts

WAN cloud

Building 11
500 hosts

Building 10
1,000 hosts

Building 12
250 hosts

Building 7
4,000 hosts

Building 8
3,000 hosts

Building 9
1,000 hosts

Building 13
100 hosts

Building 14
100 hosts

CAN 3
8,000 hosts

CAN 4
2,000 hosts

Figure 6-47.   *Initial subdivision of the address space*

making the address for the first group (Building 1) 172.16.0.0/18, and the base address for the second group 172.16.64.0/18. This is shown in Figure 6-48.

Now we will subdivide the second address group in CAN 1 (172.16.64.0/18) into two more groups, using a 19-bit mask. The first group will contain only Building 2 because it needs 8,000 hosts, and the second group will contain Buildings 3 and 4.

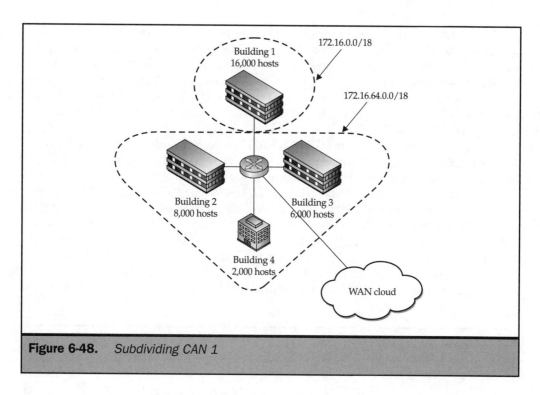

Figure 6-48. *Subdividing CAN 1*

Building 2's IP address space will therefore be 172.16.64.0/19, and the group containing Buildings 3 and 4 will have a base address of 172.16.96.0/19. This is shown in Figure 6-49.

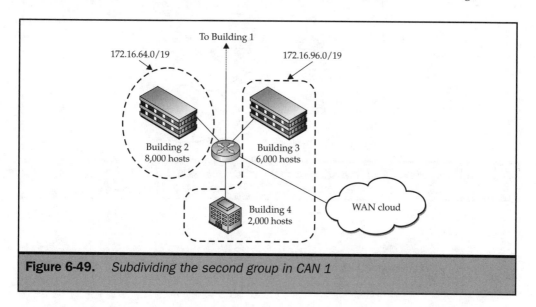

Figure 6-49. *Subdividing the second group in CAN 1*

Now we need to subdivide the group that Buildings 3 and 4 are included in, but we can't do it quite as simply as we have been. Building 3 requires 6,000 hosts, while Building 4 requires 2,000. Obviously, just dividing the address space in half will not work. However, dividing the address space by 4 will. To do this, we add two bits to the 19-bit mask to make a 21-bit mask, giving us four networks with 2,000 hosts each. We then assign three of them to Building 3 (172.16.96.0/21, 172.16.104.0/21, and 172.16.112.0/21) and one of them to Building 4 (172.16.120.0/21). This is shown in Figure 6-50.

Now for the second /17 group (172.16.128.0/17), which included CANs 2 through 4. Once again, the easiest way to proceed here is to divide the address space right down the middle by adding a single bit to the mask. (See Figure 6-51.)

This creates 172.16.128.0/18 and 172.16.192.0/18. We will assign 172.16.128.0/18 to CAN 2 because it requires 16,000 total hosts. We will then subdivide this network into four smaller networks of 4,000 hosts, each with a /20 mask, and assign three of these (172.16.128.0/20, 172.16.144.0/20, and 172.16.160.0/20) to Building 6 and assign one (172.16.176.0/20) to Building 5. This is shown in Figure 6-52.

We will split the 172.16.192.0/18 group down the middle with a 19-bit mask and assign 172.16.192.0/19 to CAN 3; and we will assign 172.16.224.0/19 to another group,

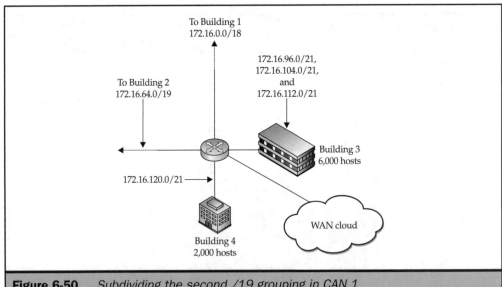

Figure 6-50. *Subdividing the second /19 grouping in CAN 1*

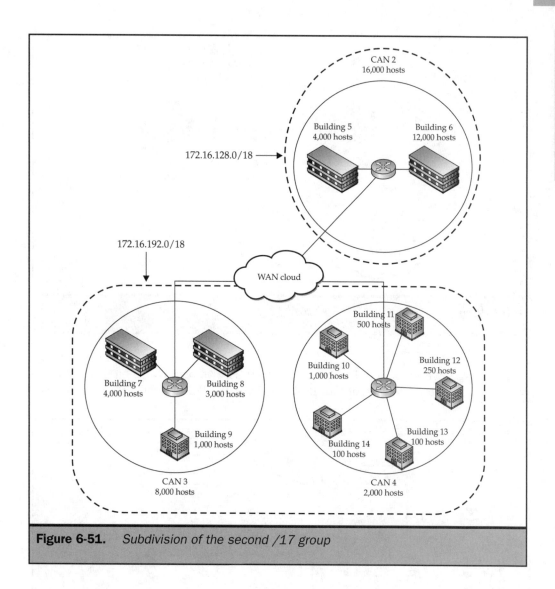

Figure 6-51.   *Subdivision of the second /17 group*

including CAN 4 and three other CANs (that do not currently exist) for future expansion (as shown in Figure 6-53).

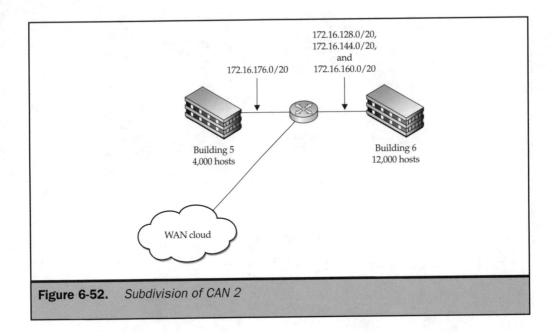

Figure 6-52. *Subdivision of CAN 2*

For CAN 3, we will subdivide the 172.16.192.0/19 group with a 20-bit mask. We will use the 172.16.192.0/20 space for Building 7 and the 172.16.208.0/20 space for the

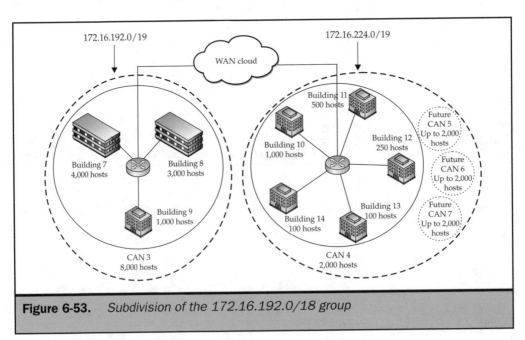

Figure 6-53. *Subdivision of the 172.16.192.0/18 group*

group of buildings including Buildings 8 and 9. We will then subdivide the 172.16.208.0/ 20 space with a 22-bit mask, creating four subnets with 1,000 hosts each. We will assign three of these (172.16.208.0/22, 172.16.212.0/22, and 172.16.216.0/22) to Building 8, and one of these (172.16.220.0/22) to Building 9. This is shown in Figure 6-54.

Finally, we will take the final address space, 172.16.224.0/19, and subdivide it into four groups of 2,000 hosts by using a /21 mask. One of these groups (172.16.224.0/21) will be used for CAN 4, and the rest (172.16.232.0/21, 172.16.240.0/21, and 172.16.248.0/21) will be used for future needs (perhaps another small CAN at a later date).

As for CAN 4, we will take the 172.16.224.0/21 space and subdivide it with a 22-bit mask, giving us two networks with 1,000 hosts per network. The first network (172.16. 224.0/22) we will assign to Building 10, and the other network (172.16.228.0/22) we will assign to the group of buildings including Buildings 11 through 14. This is shown in Figure 6-55.

We will then take the 172.16.228.0/22 network and subdivide it into two networks with 510 hosts each using a 23-bit mask. This creates subnet 172.16.228.0/23, which we will assign to Building 11, and 172.16.230.0/23, which we will assign to the group of buildings containing Buildings 12 through 14. This is shown in Figure 6-56.

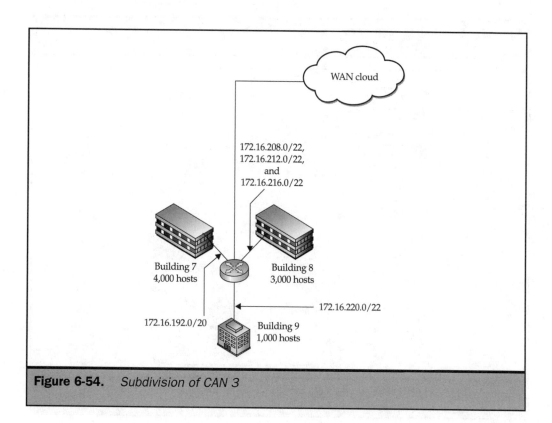

Figure 6-54.   *Subdivision of CAN 3*

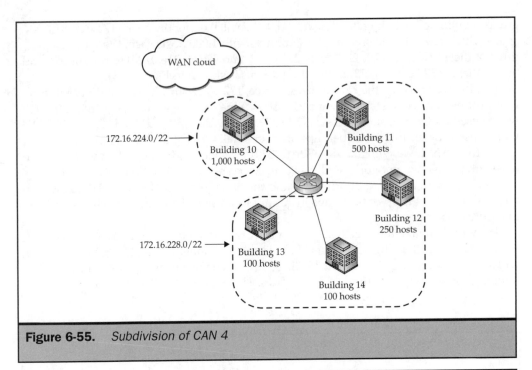

Figure 6-55. *Subdivision of CAN 4*

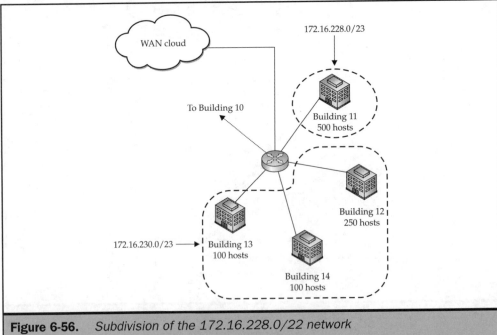

Figure 6-56. *Subdivision of the 172.16.228.0/22 network*

Finally, we will take the 172.16.230.0/23 space and divide it with a 24-bit mask, creating two networks with 254 hosts each. We will then assign 172.16.230.0/24 to Building 12, and we will assign 172.16.231.0/24 to the group of buildings containing Buildings 13 and 14. We will then split the 172.16.231.0/24 network into two separate networks using a 25-bit mask, creating two networks with 126 hosts per network. We will assign the 172.16.231.0/25 network to Building 13 and the 172.16.231.128/25 network to Building 14. This last subdivision is shown in Figure 6-57.

That's all there is to VLSM. It's not very complicated; it just requires a good understanding of the binary behind TCP/IP.

Other TCP/IP Enhancements

In this section, we will cover other enhancements to the TCP/IP suite, such as classless addressing, private addressing, ports, and network address translation. With the sole exception of CIDR, these topics are used daily in nearly all TCP/IP environments (Cisco or not), and will be required in order to fully understand some of the topics in later chapters.

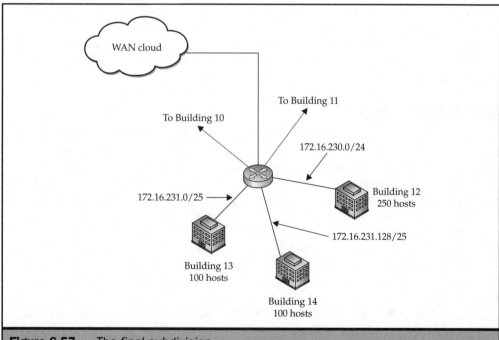

Figure 6-57.   *The final subdivision*

CIDR

Classless interdomain routing (CIDR) is pretty similar to VLSM in concept. CIDR is used for public addresses now because of the shortage of available public IP addresses. CIDR practically eliminates the idea of class-based networks in favor of address spaces defined only by the mask. The initial purpose of CIDR was to allow ISPs to hand out smaller or larger chunks, or blocks, of IP addresses than a given class. For instance, if you get a T1 Internet connection, your ISP will probably hand you a block of public IP addresses to go along with it. They may hand you as few as 2 or as many as 64, depending on your needs. You can also request (and usually pay a fee for) additional public addresses if needed, and they will proportion you out a block of their address space as necessary.

The way this works is fairly simple. If you told your ISP that you needed 25 public IP addresses, they would look at their address space and may find that the range from 64.90.1.32 to 64.90.1.63 is available. They would then issue you the network address 64.90.1.32/27. With this address block, you now have 30 valid IP addresses, from 64.90.1.33 to 64.90.1.62. You don't have the full class A. Your ISP owns that. You just have a piece of their address space to use until such time as you switch ISPs.

The basic idea of CIDR is to forget about the class. Just realize that if your ISP hands you a 22-bit mask, you cannot make that mask any smaller (use less bits), even if the IP address you are given is a class A. You don't own the entire class; you are just being allowed to use the specific section of the address space that has been given to you. You can make the mask larger (use more bits) and subnet your section of the address space to your heart's content; however, just make sure you use only the addresses that have been assigned to you.

Layer 4 Addressing

This section discusses TCP and UDP ports. While most books will not call this layer 4 addressing because technically they aren't addresses, I feel the term fits. Ports are identifiers, just like addresses; but rather than identifying a host, like an address, they identify an application on that host. For example, have you ever been surfing the Web and had a couple of browser windows open at once? Did you ever wonder how the PC kept track of which web page was supposed to open in which browser? I mean, think about it like this: if you have the same web site open in two different browser windows and you click the Search button in one of them, when the PC receives the packets to bring up the search page, how does it know whether the page in question should appear in browser window #1 or #2, or in both? It uses port numbers to identify each instance of the application.

To give you an analogy, imagine a house. Now imagine that 500 people are living in the house but only one mailing address is used. Now imagine that no one put names on envelopes, just addresses, because it is assumed that every individual person has a separate mailing address. If you had 500 people living with you in this house, when you went out to check your mail every day, what would you have to do? That's right, you would not only have to sort through all of the mail, but because there are no names on the envelopes, you would have to open every piece of mail and read it to find out whom it was destined for!

This is like a PC without ports. The PC would have to take messages destined for 500 different applications all arriving at the same address, and read a bit of each of them to determine whether it should send them to Outlook, Ping, or Internet Explorer. With ports, the PC does what you would probably do in the house analogy. Whereas you would probably put up separate mailboxes, each with the same address but different apartment numbers, the PC puts up different ports, each with the same address, but with different port numbers. Therefore, if a PC is a web server and a Telnet server, it sets port 23 to be the incoming port for Telnet, and port 80 to be the incoming port for HTTP. So, if something comes in on port 80, the PC assumes the data is for the web server.

TCP and UDP each have 65,536 ports (0–65,535). Ports 0 through 1023 are considered well-known ports. These numbers are managed by the IANA and should not be changed. Your most common server-side network applications use specific ports in this range by default. Table 6-3 provides a brief list of the most common ports.

TCP Port Number	Protocol
20	FTP Data
21	FTP Control
23	Telnet
25	SMTP
53	DNS
67	BOOTP Server
68	BOOTP Client
79	Finger
80	HTTP
110	POP3
119	NNTP (Network News Transfer Protocol)
137	NBNS (NetBIOS Name Service)
143	IMAP (Internet Message Access Protocol)
161	SNMP

Table 6-3. *Some Common TCP Ports*

TCP Port Number	Protocol
162	SNMP Trap
194	IRC (Internet Relay Chat)
443	HTTPS (HTTP over SSL)

Table 6-3. *Some Common TCP Ports* (continued)

Ports 1,024 through 49,151 are called *registered ports.* They are also typically used by server-side network applications, although they are not as well defined. Any port over 49,151 is considered a *dynamic port* and may be used by applications as needed.

Ports are used on both sides of a connection. For instance, imagine that you are communicating with the same web server from two separate browser windows. When you send a request using one of these windows (say, browser window #1), a source port field and a destination port field are included in the TCP header. The destination port (because the destination is a web server) will most likely be set to 80, but the source port will be set to a random number over 49,151 (like 54,200). Your PC will keep track of that number, remembering that it is to be used specifically for this particular communication from browser window #1.

If you send a request using browser window #2, the PC will put a different number in the source-port portion of the TCP header (like 62,900). When you get the packet back, the source port is now 80 (the packet is coming from the web server), but the destination port to which it is sent is 54,200. This way, your PC can recognize that the message is destined for browser window #2, not #1. This process is shown in Figure 6-58.

NAT and Private Addressing

Private addressing and network address translation (NAT) are two intertwined topics that have helped enormously in reducing the number of public IP addresses required. The basic idea behind them is simple: Give the hosts on the private network an address that is not for use on the public network (the Internet), and when the hosts want to communicate on the public network, they contact a gateway device that translates the private IP address into a public IP address. This reduces the number of public IP addresses required because, instead of needing a public IP address for every host that needs to communicate on the Internet, you simply assign one (or several) to the NAT device and let it handle Internet access.

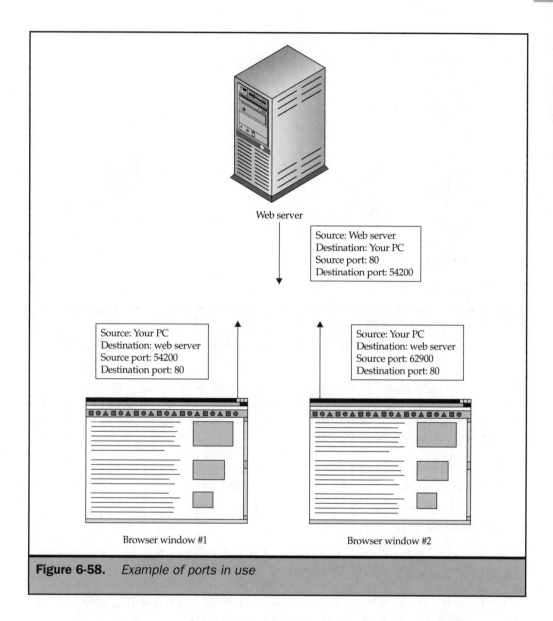

Web server

Source: Web server
Destination: Your PC
Source port: 80
Destination port: 54200

Source: Your PC
Destination: web server
Source port: 54200
Destination port: 80

Source: Your PC
Destination: web server
Source port: 62900
Destination port: 80

Browser window #1

Browser window #2

Figure 6-58. *Example of ports in use*

As usual, however, the actual workings of NAT are a bit more complicated than this. To understand the interactions that occur, let's take a look under the hood with an example.

In Figure 6-59, we are attempting to communicate with www.cisco.com from within a private network. We have a private address, so we need to go through a NAT

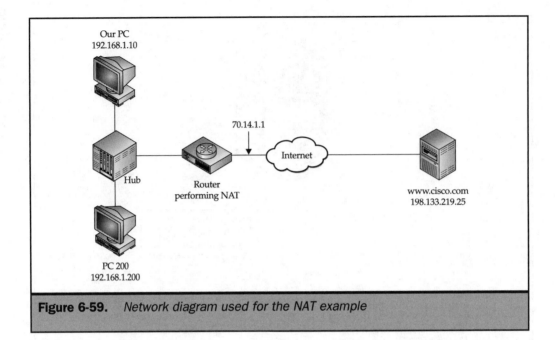

Figure 6-59.   *Network diagram used for the NAT example*

device to reach the Internet. In this case, the router is performing NAT. The PC simply performs the ANDing process, realizes that the packet is destined for another network, and sends it to the default gateway (the NAT-enabled router). Upon reaching the router, the packet has the following properties:

- Source address: 192.168.1.10 (us)
- Source port: 44,000 (our port for this communication)
- Destination address: 198.133.219.25 (www.cisco.com)
- Destination port: 80 (HTTP)

At this point, the NAT server takes the packet and translates it to its public IP address (in this example, 70.14.1.1), adding its own port number so it can tell the difference between a packet sent back to us (192.168.1.10) and a packet sent back to 192.168.1.200. It catalogs the conversions it has performed in a translation table and sends the packet out. At this point, the packet looks like this:

- Source address: 70.14.1.1 (the router's public IP)
- Source port: 60,000 (the router's port for our communication)
- Destination address: 198.133.219.25 (www.cisco.com)
- Destination port: 80 (HTTP)

Let's assume that while all of this was going on, another host (192.168.1.200) was also attempting to communicate with www.cisco.com. The same process occurs for him. His initial packet (before translation) appears like this:

- Source address: 192.168.1.200 (PC 200's IP)

- Source port: 55,555 (PC 200's port for this communication)

- Destination address: 198.133.219.25 (www.cisco.com)

- Destination port: 80 (HTTP)

PC 200's IP address is also translated, and, after translation, it appears like this:

- Source address: 70.14.1.1 (the router's public IP)

- Source port: 50,600 (the router's port for PC 200's communication)

- Destination address: 198.133.219.25 (www.cisco.com)

- Destination port: 80 (HTTP)

After all of this, the NAT device has built up a translation table that looks similar to the one in Table 6-4.

When the packet arrives at the router from www.cisco.com destined for IP 70.14.1.1, port 60,000, the NAT device looks in its translation table and finds that the packet is destined for 192.168.1.1 on port 44,000. It then performs the translation and sends out the packet. Our PC receives the packet completely oblivious to the entire process.

NAT can be done on a router; but in many cases, it is performed on a specialized device commonly called a proxy. A *proxy* performs basic NAT, and it usually performs a number of other processes, including the following: content caching (storing files, typically web pages, that are commonly accessed on a hard disk so that when a host asks for that file, it does not need to download it from the Internet); converting other protocols (such as IPX/SPX) into IP; and filtering packets (accepting or rejecting packets based on criteria, such as the source IP address or port number). No matter what performs the NAT process, NAT operates more efficiently with an IP address pool rather than a single address because, with more than one IP address, the NAT device does not have to translate ports as often.

Source IP	Source Port	Translated IP	Translated Port
192.168.1.10	44,000	70.14.1.1	60,000
192.168.1.200	55,555	70.14.1.1	50,600

Table 6-4.   *NAT Translation Table*

Although NAT can be performed with any IP address structure (including public addresses), it is generally best to use an address space that is reserved for private use. The reason for this is simple: if you are using a public address, then the public resource that that address signifies (and everything specified by the mask you use) will not be accessible from your private network. For example, if you used the 207.46.230.0/24 address internally, then you would never be able to get to any machine on the Internet that resided on the 207.46.230.0 network (one of which is Microsoft) because your machine would think it was a local address and would not send it to the NAT server.

For this reason, a block of addresses for every class of network (A–C) has been defined as "private" and acceptable for use with NAT, as listed here:

- 10.0.0.0
- 172.16.0.0 through 172.31.0.0 (16 networks)
- 192.168.0.0 through 192.168.255.0 (256 networks)

Private addressing is defined in RFC 1918, and NAT is defined in RFC 3022.

Multicasting

IP multicasting is an interesting idea. The point of *multicasting* is to save bandwidth. Unlike *unicasts*, which travel to a single host, or *broadcasts*, which travel to all hosts, *multicasts* travel to a select group of hosts. This is extremely useful in a situation in which many hosts are all requesting the same data stream (such as streaming video). With a broadcast, the packet needs to be processed by all hosts, and it is flooded down all segments, wasting both bandwidth and processor time. In addition, broadcasts are limited to local networks. With a unicast, the packet is processed only by the specified hosts; but multiple packets need to be sent, wasting bandwidth. With a multicast, only one packet is sent, and only the group that needs it receives and processes it. This is shown in Figure 6-60.

In this example, if we are streaming a video feed across the WAN links using unicasts, we have to send one feed for every host, which is a bit wasteful. In this case, it requires 128 Kbps for each stream, for a total of 12 Mbps for 100 hosts. To show you why this is wasteful, think about it this way: if we are all watching the CEO give a speech live, is the version of the video I am getting any different from the version you are getting? Hopefully, the answer is no, and multicasting can save us bandwidth.

With multicasting, instead of sending a stream for each host, we just send a single stream to the multicast group. The multicast-enabled routers along the path take this stream and split it off as necessary to make sure that the single packet gets to all necessary networks, and multicast-enabled layer 2 switches send the packet down only those segments that require the feed, rather than flooding the stream down all segments. The end result is that we use only 128 Kbps of WAN bandwidth, and we use no more LAN bandwidth than is absolutely necessary.

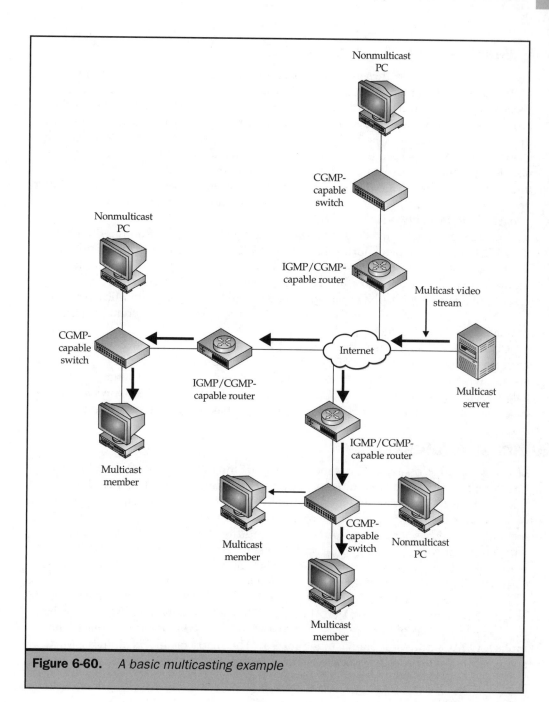

Figure 6-60.   *A basic multicasting example*

How Multicasting Works

The basic principle of multicasting is fairly simple. Hosts join one or more multicast groups using Internet Group Management Protocol (IGMP). These groups specify a specific session of multicast traffic (such as a live video feed). Any host that joins the group begins to use the group's multicast address and receives any traffic destined for the group. Groups can be joined or left dynamically by hosts, so membership is constantly changing. The routers keep track of this membership and attempt to form a loop-free path to the multicast members, with the primary concern being that no packets are unnecessarily repeated. IGMP also makes sure that when no hosts for a given multicast group exist on a given network, the multicast packets are no longer sent to—or are *pruned*—from that network.

Multicast routing protocols, such as Distance Vector Multicast Routing Protocol (DVMRP), Multicast Open Shortest Path First (MOSPF), and Protocol Independent Multicast (PIM), are used to inform multiple routers along the path of the multicast groups in use and the optimal paths to reach these groups. Once the packet has reached the destination LAN, it may be flooded to all hosts by a nonmulticast-enabled layer 2 switch, or forwarded only to those hosts participating in the group by a CGMP-enabled Cisco layer 2 switch.

Cisco Group Management Protocol (CGMP) is used by Cisco layer 2 switches to keep from flooding multicast traffic to all segments. Because IGMP is a layer 3 protocol, layer 2 switches cannot use IGMP to find out about multicast hosts. Therefore, routers tell the switches about multicast group membership using CGMP. This allows multicast-enabled switches to forward multicasts only to hosts that are participating in the group.

Multicast Addressing

Multicasting relies on a number of inclusions to IP and datalink-layer technologies to function properly. For LAN multicasts, an additional function must be included in Ethernet to allow multicasts to work. Normally, Ethernet accepts a frame only if the MAC address is either its MAC or the broadcast MAC. For multicasting, a new address, the multicast address, must be supported. This address begins with 01-00-5E and ends with a number corresponding to the last 23 bits of a multicast group's IP address. Note that, with multicasts, host addresses are unimportant. The hosts take an address corresponding to the multicast group and use that address.

IP addresses in multicasts are class D addresses. Several class D addresses are reserved; however, the range from 226.0.0.0 to 231.255.255.255 is not formally reserved and may be used. The last 23 bits of the multicast IP address are mapped to the last 24 bits of the multicast MAC address, as shown in Figure 6-61. This mapping allows multicast MAC and IP addresses to be easily associated with each other.

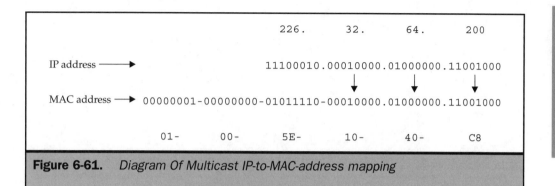

Figure 6-61. *Diagram Of Multicast IP-to-MAC-address mapping*

Summary

In this chapter, you have learned about advanced functionality in the TCP/IP suite, including IP addressing, port mapping, and NAT. These topics (and IP addressing in particular) will be invaluable to us in later chapters, as we begin to look at the core functionality of Cisco networking devices.

For additional information on TCP/IP, check out the following web sites:

- http://www.learntosubnet.com
- http://www.cisco.com/univercd/cc/td/doc/product/iaabu/centri4/user/scf4ap1.htm
- http://www.cisco.com/univercd/cc/td/doc/cisintwk/ito_doc/ip.htm
- http://www.cisco.com/univercd/cc/td/doc/product/aggr/vpn5000/5000sw/conce60x/50028con/ipaddr.htm
- http://www.3com.com/solutions/en_US/ncs/501302.html
- http://mdaugherty.home.mindspring.com/tcpip.html
- http://www.webzone.net/machadospage/TCPIP.html
- ftp://ftp.netlab.ohio-state.edu/pub/jain/courses/cis788-97/ip_multicast/index.htm
- http://www.ipmulticast.com/techcent.htm
- http://www.vtac.org/Tech/TNtcpip.htm
- http://www.private.org.il/tcpip_rl.html
- http://minnie.cs.adfa.edu.au/PhD/th/2_Existing_Congestion_Control_Schemes.html
- http://www.rfc-editor.org

The
Complete
Reference

Cisco

Chapter 7

The IPX/SPX Suite

This chapter covers the other major protocol suite in use, Internetwork Packet Exchange (IPX)/Sequenced Packet Exchange (SPX). Novell created IPX/SPX for their NetWare line of software (although many components of the suite are based on Xerox's XNS). At one time, IPX/SPX was one of the most common protocols in use; however, the decline of NetWare and the rise of the Internet have led to it being used less frequently over the years. In fact, current Novell servers support TCP/IP natively, further reducing the use of the IPX/SPX protocol suite. However, it is often still used, most commonly in an environment that contains NetWare 3.*x* servers. Therefore, a good understanding of IPX/SPX is still useful today.

The IPX/SPX Suite

The IPX/SPX suite is in many ways less varied than the TCP/IP suite, with fewer protocols, each supporting a specialized function. Most of the same functions supported in TCP/IP are supported in IPX/SPX. However, rather than having to choose between 20 different protocols and sometimes many different revisions of the same protocol, IPX/SPX tends to be more simple—one advantage of having a proprietary protocol suite. Unfortunately, this simplicity also leads to less variety in implementation, which leads to less flexibility in the suite. However, IPX/SPX is still considered by many to be the most efficient suite for LAN environments, with TCP/IP being superior in the WAN arena.

Let's take a look at the protocols that make up this suite.

IPX

IPX is the protocol responsible for addressing in IPX/SPX, as well as being the connectionless transport protocol of the suite. In this way, it is like User Datagram Protocol (UDP) and Internet Protocol (IP) combined into one. IPX addresses, like IP addresses, consist of a network and node portion. Unlike IP addresses, however, IPX addresses do not use a mask. Instead, the division between the network and node is of a fixed length. This "limitation" is not really a problem because IPX addresses are 80 bits (10 bytes), whereas IP addresses are 32 bits (4 bytes). This leads to an incredibly large address space. So large, in fact, that address waste is not a concern.

IPX addresses are typically written in hexadecimal (base 16, like MAC addresses), and they consist of 20 hexadecimal digits. A typical IPX address looks like this: 0123.4567.89AB.CDEF.0F1E. The first 32 bits of an IPX address (the first eight hex digits, or 0123.4567, in the example) represent the network, and the last 48 bits represent the node. This provides a total of over 4 billion networks with over 262 trillion hosts per network, which should be plenty! Figure 7-1 shows a breakdown of an IPX address.

The *host* portion of an IPX address is typically the MAC address of the device's network interface. The *network* portion can be whatever you choose because, unlike in TCP/IP, no central authority controls IPX network numbers. Therefore, it is sometimes easier to identify the network/node division in an IPX address by counting the last 12 hex digits (from right to left), rather than counting the first 8 hex digits. For instance,

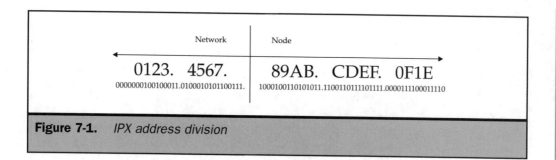

Figure 7-1.   *IPX address division*

if your network address were 0000.0001 and your node address were 1234.5678.ABCD, your address would actually be written as 1.1234.5678.ABCD, with all of the leading zeros dropped. In this case, it is easier to count 12 digits from the right to find the divider.

Just like in TCP/IP, in IPX, the network number determines whether a host is local or remote. If the number is the same for both hosts, they are local. If not, a router must be used to connect.

One area that can be a bit complicated is internal network addresses used on some IPX servers. An *internal network address* is a logical (in other words, *pretend*) network used by Novell servers to identify the operating system and network services. The internal network address must, of course, be unique, and includes only one host: the server on which it resides. The host address for the server on the internal network is 0000.0000.0001. Figure 7-2 shows an example of an internal network address.

IPX also has provisions for sending broadcasts. In IPX, the broadcast address is all binary 1's, just like in IP, except because IPX addresses are in hex, this is represented by the all Fs (FFFF.FFFF.FFFF.FFFF.FFFF) address. The all 0's address is also used in IPX to represent "this network," just like IP.

IPX also includes a specification for multiplexing between transmissions using sockets. IPX sockets are 2-byte numbers that are similar to UDP ports, and they are used for the same purpose. (You may also hear the term *sockets* used in TCP/IP; in TCP/IP, a socket is the full address, including the IP address, transport protocol, and port.)

All in all, IPX addressing is much simpler than IP addressing because there is no need (or functionality) for subnetting.

Finally, Figure 7-3 depicts an IPX packet. A short description of each of the fields is provided here.

- **Checksum**   This field is unused and always set to FFFFH.
- **Length**   This field is the datagram length in octets.
- **Transport Control**   This field is used by IPX routers.
- **Packet Type**   This field specifies the type of the protocol header after the IPX header (for example, SPX).
- **Destination Network Number, Node Number, and Socket Number**   These fields are self-explanatory.

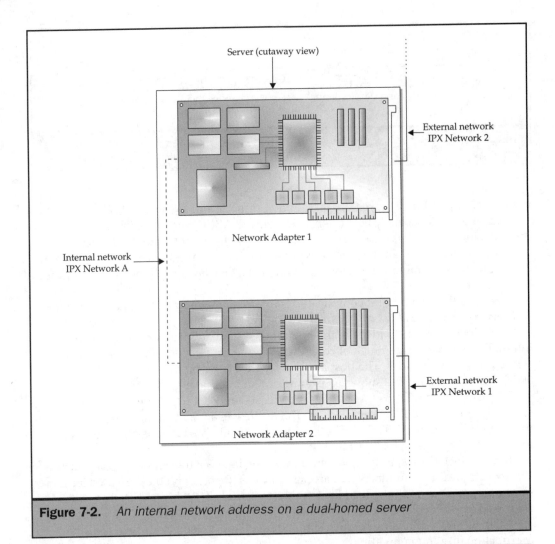

Figure 7-2. *An internal network address on a dual-homed server*

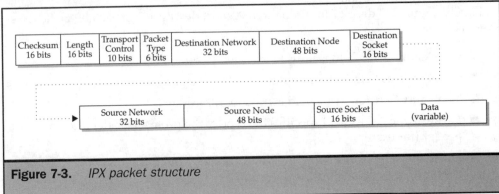

Figure 7-3. *IPX packet structure*

- **Source Network Number, Node Number, and Socket Number** These fields are self-explanatory.
- **Data (variable)** This field includes the data. If SPX is used, the SPX header will also be encapsulated here.

SPX

SPX is the transport-layer protocol that performs connection-based services in IPX/SPX. It is very similar to TCP: it establishes and terminates sessions, as well as performs flow control using windowing, thereby providing for reliable packet delivery. One difference in SPX and TCP, however, is that SPX uses a field called the Connection ID to discriminate between sessions. Another difference is in how SPX performs windowing. In SPX, windowing is not a dynamic process. Window sizes are constant and will not change throughout the session. This feature makes SPX a bit less flexible, but also less complex, than TCP.

Finally, Figure 7-4 shows the structure of an SPX packet. An explanation of the fields follows.

- **Connection Control** Splits up into sections, and signals functions like acknowledgements and message termination
- **Datastream Type** Specifies the type of data contained within the packet
- **Source Connection ID and Destination Connection ID** Identifies the session
- **Sequence Number** Identifies the number of packets transmitted
- **Acknowledgment Number** Identifies the next expected packet (FACK)
- **Allocation Number** Identifies the number of packets unacknowledged
- **Data (Variable)** This is the actual payload of the message. If any upper-layer protocol headers (Application, Presentation, and so on) are used, they will be encapsulated here as well.

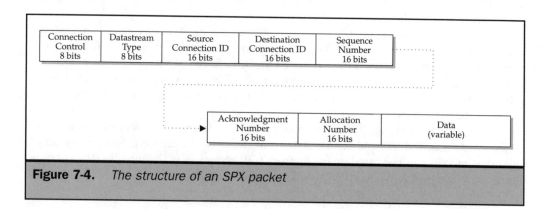

Figure 7-4. *The structure of an SPX packet*

DIAG

DIAG is a protocol used to provide functions similar to some of those in ICMP. Like ICMP, DIAG is used to check connectivity. However, rather than just giving you a success or failure, DIAG can also inform you about which services are running on the remote nodes and which networks the remote nodes are aware of, and it can mass test a group of hosts. Overall, DIAG is an extremely useful protocol for testing your IPX network.

SAP and GNS

The Service Advertising Protocol (SAP) has an extremely important use in an IPX network: it locates servers and services provided by those servers. It works like this: Servers send out an SAP broadcast at a regular interval (every 60 seconds, by default), advertising the services it has available to all directly connected hosts. Other servers receive these broadcasts and update a table, called the SAP table, with the information. They then propagate this information in their next SAP broadcast. Routers can also play a part in this exchange by keeping, updating, and advertising their own SAP tables, including any server they are aware of.

At any time, if there is a change to a server's SAP table, the server can (if so configured) broadcast a new update, known as a *triggered update*. The combination of triggered and timed updates allows all servers on the network to acquire and maintain a SAP table with every available server on the entire network, even those spanning multiple subnets. This is because, as mentioned previously, the routers build an SAP table and then broadcast that table to other subnets. This process allows clients to find any server in the entire IPX network by using another protocol: Get Nearest Server (GNS).

When a client wishes to find a server (to log in, for instance), it will broadcast a GNS request. GNS finds the closest server offering the service that the client requires. The GNS reply is sent back by a server on the same subnet as the client, letting the client know how to reach its closest server. From this process, another very important function occurs: the client builds its IPX address. When the client receives a GNS reply, it looks at the network portion of the IPX address for the server and uses this address for its IPX network address. It then adds its MAC address for the node portion, and it has a complete IPX address.

SAP updates are used primarily by bindery-based (3.1 and 4x with bindery emulation) Netware servers to locate resources. The *bindery* is the accounts database in earlier versions of Netware. In networks based on Netware Directory Service (NDS)—a distributed account database used in Netware 4x and later—SAP broadcasts are used by clients only at bootup to find the nearest server and obtain an IPX address, so SAP-related traffic on the LAN is much lower.

Through this process, an IPX client can locate all needed network services, as well as get a correct and valid IPX address for its specific needs. Unfortunately, SAP broadcasts ultimately lead to another problem (aside from traffic): lack of scalability. SAP broadcasts can cause major problems over WAN links, where bandwidth is limited.

For instance, in Figure 7-5, a router is broadcasting SAP updates every 60 seconds to five permanent virtual circuit (PVCs) over a single T1 (1.544 Mbps) connection. Remember from Chapter 3 that most WAN links are NBMA networks, and they cannot use broadcasts. Consequently, for every PVC, an individual group of packets containing the SAP update must be sent each time. So, assuming your SAP updates take 500KB every 60 seconds, then you are using 2500KB of your T1 simply for SAP updates every 60 seconds.

But wait, it gets worse. The other routers on the far side of the PVCs are also broadcasting the same 500KB SAP table every 60 seconds. This makes the total used bandwidth 5000KB every 60 seconds. Converting this to kilobits, 40,000Kb are being used every 60 seconds—for every minute of time, a full 26 seconds of bandwidth is eaten up by SAP updates. Admittedly, this example is a bit extreme, but it illustrates the problem.

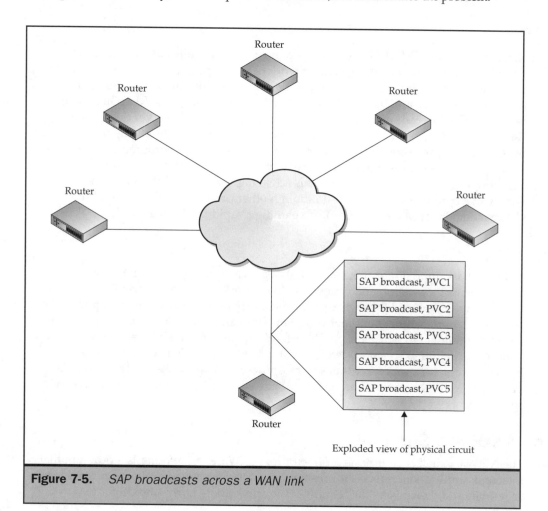

Figure 7-5. *SAP broadcasts across a WAN link*

Fortunately, some solutions exist. The first solution is to use TCP/IP. (OK, that was a joke.) Seriously, Cisco routers can filter SAP updates across WAN links. Therefore, you can choose not to send certain updates over the WAN link, thus reducing the bandwidth. You do this by using access list statements (covered in more detail in Chapter 27). The trade-off for this solution, however, is the possibility of being unable to locate the required resources.

Another solution is to disable triggered updates and increase the time between SAP broadcasts to a more reasonable level. The trade-off for this solution, however, is slower convergence. In other words, it will take the servers much longer to build a complete list of resources or to recognize that a resource has failed. Can't have your cake and eat it, too.

Overall, however, despite their problems, SAP and GNS are integral parts of IPX/SPX and perform some of the most important functions in the suite.

RPC

Remote Procedure Call (RPC) is a function most protocol suites use in one form or another. RPC allows one device on a network to use a program or service on another device without needing to know the specifics involved with the application (such as directory structures and path statements) on the other device. The RPC client simply uses variables (procedure calls) in an RPC message to the RPC server describing what it wants to do. The server performs the requested action and submits the result back to the client. This process allows a lot of modularity and minimal programming effort for developers. They simply make use of the network functions that RPC provides rather than rewriting the book themselves. This advantage has made RPC one of the most widely used protocols in most OSs (although other protocols, such as Distributed Component Object Model (DCOM), are gaining a rapid foothold as well).

NetBIOS

Network Basic Input Output System (NetBIOS) is a system used primarily in Microsoft environments for locating and connecting to network resources. NetBIOS is supported in Novell environments through an *emulator*: a program that attempts to replicate the environment produced by another program. This capability allows applications that were programmed to work in a NetBIOS (and probably Microsoft) environment to still perform the required functions (well, usually, anyway) in a NetWare environment. The NetBIOS emulator simply translates Microsoft-ese into Netware-ese. This capability makes the NetBIOS emulator an important part of the IPX/SPX suite in multivendor environments.

The NetWare Shell

The NetWare shell is your basic redirector for NetWare. A *redirector* is a program that intercepts application requests for resources, determines whether the resource is local or remote, and then passes the request to the appropriate party (the OS for local requests and the network for remote requests). The redirector enables a developer who doesn't know the difference between a mapped network drive and a local drive to write an application. If a drive is mapped as H:, and an application needs to write to the H: drive, the redirector intercepts the request for H: and passes it to the appropriate service or protocol, keeping the application in the dark. This functionality makes the NetWare shell one of the most important pieces of any NetWare network.

NCP

Netware Core Protocol (NCP) is used primarily for file and printer sharing on a Novell network. NCP is used by other protocols, such as the NetWare shell, to provide access to remote resources.

NCP uses two types of messages: requests and replies. Requests are sent by clients for access to resources, and the servers reply to these requests. Every individual resource that a client wishes to use must be requested and replied to individually. Multiple replies cannot be sent based on a single request. As a result, the term *ping-pong protocol* is applied to NCP, and this problem is the major cause of NCP's inefficiencies. NCP, however, is still a very important protocol in the suite, and it performs many valuable functions in Novell networks.

IPX Framing

IPX is unusual in that it has used a wide variety of frame types over the years. In Ethernet-based networks, IPX can be used with the following: Ethernet 2, IEEE 802.3 (the default frame type in Novell 4x and later, and called 802.2 by Novell), 802.3 "raw" (the default frame type in Novell 3.11 and earlier, and called 802.3 by Novell), and 802.3 SNAP. Several other frame types can be used as well if Fiber Distributed Data Interface (FDDI) or Token Ring are being used.

The main point to remember about IPX framing was in Chapter 2: for two hosts to communicate, they must be using the same frame type. However, this restriction can be a bit difficult in a complex Novell network, because several hosts and servers may be on the same logical segment (broadcast domain) that use different frame types. In this situation, you simply configure a router to use both frame types for the same physical interface and have different network numbers assigned to each of the different frame types. The router takes any frames from a host that uses a different frame type than the server they are trying to communicate with and routes them to the server.

For instance, in Figure 7-6, a Novell client (Jason) is attempting to communicate with a Novell server (Freddy). Jason is using the 802.3 "raw" encapsulation, and Freddy is using the IEEE 802.3 encapsulation. Notice that Jason is on logical network A, and Freddy is on logical network B, but they are connected to the same hub, so they are on the same physical network. When Jason sends a request to Freddy, Freddy can actually hear the request; he just can't understand it because Jason is using a different frame type. So, Jason actually sends the packet to the router, Chucky, who routes it back to

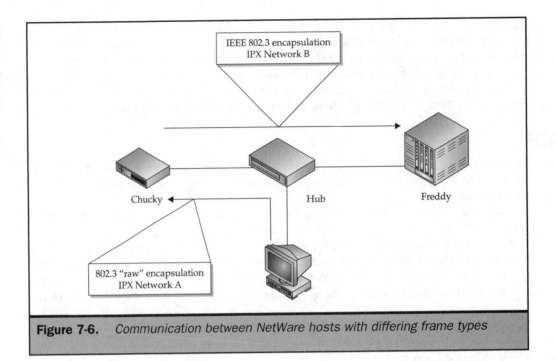

Figure 7-6.   *Communication between NetWare hosts with differing frame types*

Freddy, stripping the 802.3 "raw" encapsulation off the packet and reencapsulating the data with the IEEE 802.3 frame type in the process. When Freddy replies back to Jason, the same process happens, just in reverse.

Bringing It All Together

To illustrate these concepts, let's take a look at the diagram of the IPX network in Figure 7-7. In this example, a client (Wang) is attempting to communicate with a Novell server (Jack) through a router (Lo Pan).

First, Jack issues a SAP broadcast at 60-second intervals, informing all routers and servers of which services he is offering. Lo Pan hears these broadcasts and fills his SAP table with the entries from Jack. At the same time, Gracie is also broadcasting SAP updates, which Lo Pan also records in his SAP table. Lo Pan then broadcasts his SAP table to both Gracie and Jack to ensure that all servers have a complete copy of the table. Wang finally boots and issues a GNS broadcast for the nearest server.

The only server on Wang's subnet is Gracie, who immediately responds. Wang examines the network portion of the IPX address for Gracie and assigns himself the same network address (using his MAC address for the host portion). Wang then queries Gracie to find his preferred server (Jack). Gracie replies with Jack's address, and Wang realizes that Jack lives on a separate network.

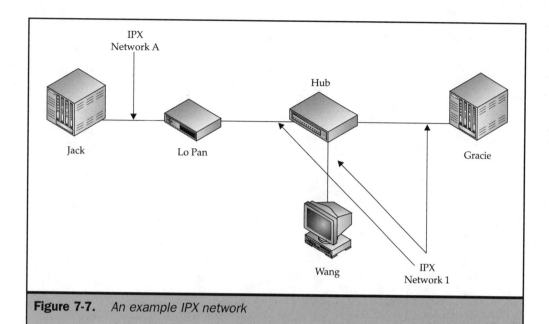

Figure 7-7.   *An example IPX network*

During this time, Wang has been listening to the Routing Information Protocol (RIP) broadcasts on the network, which contain information about all of the networks Lo Pan is aware of, and realizes that Lo Pan knows how to reach Jack's network. He therefore sends his packets to Lo Pan in order to log on to Jack. (RIP is explained further in Chapter 23.)

Summary

In this chapter, you learned how many of the protocols associated with the IPX/SPX suite function, including IPX addressing and SAP. A number of networks still use the IPX suite and will continue to do so for a long time to come. Now when you encounter them, you will not be caught off guard.

Check out these web sites for additional information on IPX:

- http://www.odyssea.com/whats_new/tcpipnet/tcpipnet.html
- http://www.cisco.com/pcgi-bin/Support/PSP/psp_view.pl?p=Internetworking:IPX
- http://www.protocols.com/pbook/novel.htm
- http://www.novell.com

The
Complete
Reference

Cisco

Part II

Cisco Technology Overview

Part II of this book examines Cisco technologies—from switching products to remote access products. Chapter 8 covers the basics of standard Cisco hardware. Chapter 9 analyzes terms and technologies specific to Cisco switching, classifying each model of Cisco switch. Chapter 10 examines each switching line in detail, pointing out uses and target environments for each line. Chapter 11 introduces terminology specific to routing and classifies each model of router, while Chapter 12 details each line. Chapter 13 examines the rest of the Cisco product line. Finally, Chapters 14, 15, and 16 delve into the Cisco IOS, including command references for both the standard and set-based IOS.

Chapter 8

Cisco Hardware Layout

This chapter explores Cisco's hardware layout in general terms. You will learn about hardware configurations, router and switch components, modularity, mounting, and cabling. These topics will help you understand Cisco devices from a purely physical perspective, as well as from a configuration standpoint.

Hardware Form Factors

Cisco devices come in two basic hardware configurations: rack and pedestal. Some actually ship as a pedestal mount but include rack-mount hardware.

Pedestal-Mount Devices

A *pedestal-mount* device is meant to sit on a shelf or table. Typically, the device has small rubber "feet" on the bottom to support it. Most Cisco devices can be used as pedestal-mount devices, which are good for situations when a rack is not available and space is tight.

If you are going to use a pedestal-mount device, be sure to check where the fans on the device are placed. On some Cisco devices, they will be in the back; on others they may be on the bottom; and on rare occasions, they may even be in the front. This is important, as you will need to make sure to place the device in such a way that the fans are not blocked so that the device receives adequate cooling.

Rack-Mount Devices

A *rack-mount* device, on the other hand, is mounted in a 19-inch communications rack with other devices. This configuration provides a secure, grounded mounting point for your equipment, and it is definitely the better way to go, if possible. Rack-mount devices have a standard width (19 inches), but the depth and height of a device may vary. Most Cisco devices have a very shallow depth, typically using only half of the available depth in a 19-inch rack. Some devices, such as the 5500 series switches, may take the full depth of the rack. Luckily, the Cisco web site provides full dimensioning information for all the current products.

As for height, rack-mount devices are typically listed as using so many rack units (RU). Each RU is approximately 1.75 inches. Racks are typically listed by their RU as well as by their size. In a 72-inch relay rack, approximately 39 RU should be available. Cisco products do not typically take any less than 1 RU, but each device can take anywhere from half an RU to several RU.

Most of the Cisco products you will deal with are designed to be rack-mounted. Rack-mounting equipment is great, but it leads to its own special set of problems. One of the most serious of these problems is heat dissipation. When rack-mounting equipment, you are by definition placing the equipment as close together as possible. This setup makes it rather difficult to dissipate heat from the chassis of the devices because they all tend to share the warmth (similar to several people huddled in one sleeping bag). Add to this problem the increasing trend in the computer industry to cram as much as possible into as small a space as possible, and you can end up with a large, expensive heater.

To help alleviate this problem, you have a couple of options. First, if possible, mount your high-density racks in a room (such as a server room) with its own high-power AC system. If you are lucky enough to have such a system, crank the AC down to 65 degrees Fahrenheit or so to help keep the rack cool.

Second, pay special attention to where the fans are located on your Cisco devices. Most devices have the fans either in the front or back of the chassis, which is fine; but like most companies, Cisco isn't consistent with this practice. In the 2500 series routers, for example, the fan is on the bottom. Having a fan on the bottom is not a big concern for pedestal-mount equipment; but for rack-mount equipment, the bottom is one of the worst places for it. In a pedestal-mount situation, there is about ¼ inch of clearance between the bottom of the device and the surface of the shelf due to the rubber feet.

However, in a rack, you typically mount units back to back, with little or no space between them. This setup effectively chokes the fan on the 2500s and can lead to heat-related problems (such as router failure). The solution to this problem is to leave a small bit of space between devices with bottom-mounted fans. Just skip a hole between each of these devices in the rack. You'll lose about ¼ inch of space per unit, but your devices will run noticeably cooler.

Internal and External Components

The internal components of Cisco devices vary a bit depending on a device's function, power requirements, form factor, and modularity (explained later in this chapter), but a few basic components are nearly always included. In many cases, a router or switch is similar to a specialized PC, with similar components used for similar purposes.

However, nearly everything in a Cisco device is specially designed not just for Cisco products, but, in many cases, for a specific model of Cisco product. For this reason, you must purchase hardware components such as memory upgrades specifically for your particular Cisco device, often paying a premium for it. For instance, a 256MB dual inline memory module (DIMM) for a 3660 router can cost you anywhere from $3,000 to $11,000.

In addition to internal components, Cisco devices' external components also vary, again depending on the model of device in question. The following sections describe these internal and external components in detail.

Internal Components

The common internal components include RAM modules, flash memory, ROM, CPU, backplane, and nonvolatile RAM (NVRAM).

RAM Modules

Cisco devices use *dynamic RAM (DRAM)* just like a PC—for working memory. The RAM in your Cisco devices holds your current configuration, called the running config (covered in more detail in Chapter 14), and the run-time version of the Internetworking Operating System (IOS)—the operating system for most Cisco products.

Luckily, in most cases, you won't need as much memory for your Cisco devices as you may require in a regular PC because the IOS is considerably smaller than most PC operating systems today. A typical amount of RAM for most routers is 16MB. If some of the higher-end features—such as Border Gateway Patrol (BGP)—are needed in your router, you may have to increase the RAM to support them, however.

Cisco devices typically use single inline memory modules (SIMMs) and dual inline memory modules (DIMMs) for RAM, just like a PC; but these are not usually standard DIMMs or SIMMs that you can buy (or find in the bottom of Cracker Jack boxes, as cheap as they are now) for your PC. Rather, they are specially made for your Cisco device (usually the slit in the module is in a different spot), so expect to pay a good bit more.

Flash Memory

Flash memory is used on a Cisco device similar to the way a hard disk is used on a PC. Flash contains the stored IOS, and it is used for more permanent storage than RAM because it does not lose information when the router is powered down. Flash memory comes in two basic types on Cisco devices: SIMMs or DIMMs that can be inserted like standard RAM, or flash memory PCMCIA cards (or PC Cards, if your prefer).

Once again, no matter how the flash is inserted, it is a required part of your Cisco device. Also, just because it contains an industry standard PCMCIA slot doesn't mean that you can insert any old flash memory card into it. Again, it typically needs to be a custom card for your particular model of device.

ROM

ROM on a Cisco device is typically used to provide a basic backup version of the IOS for use when you cannot boot the device in any other fashion. ROM contains the ROM Monitor code, which is used if the IOS in flash memory is corrupt and unbootable, or for low-level diagnostics and reconfiguration (like when someone changes the password and locks you out of the router). Because ROM is read-only by definition, to upgrade to a new version of ROM, you must usually take an integrated circuit (IC) puller, extract the old ROM chip from its socket, and then insert the new ROM chip. Luckily, this operation is not typical, and you should only rarely need to do it.

CPU

The *CPU* is used for the same purpose as the CPU on a PC: it is the "brain" of the device. Most Cisco devices perform many software computations, which are carried out by the CPU. Cisco uses many different models of CPUs in various devices, depending on the intended use of the device. In a router, the CPU is especially important because most of a router's functions are carried out in software, and the CPU has a dramatic impact on performance. In switches, the CPU is generally less important because most of a switch's calculations are carried out in specialized pieces of hardware known as *application-specific integrated circuits (ASICs)*. One way or another, knowing where the CPU resides on the mainboard is not important because you cannot upgrade it.

Backplane

The "backplane width" is a term that's thrown around a lot in conversations about network devices (usually switches). The *backplane* is like the bus that all network communication travels over inside the network device. The backplane is of primary importance in switches and other high-port-density devices. This is just basic math. If you have a Cisco Catalyst 3548 switch, the device ships with an 8 Gbps backplane, 48 100BaseT ports, and 2 gigabit slots. If all slots and ports are filled and every device is transmitting at full throttle (which should be an exceedingly rare occurrence), the total data bandwidth used (based solely on wire speed) would be 13.6 Gbps.

However, if you start counting *interframe gaps*—the delay between the time one packet is transmitted and another can transmit—that number drops down to a more realistic 8–10 Gbps. This number is still higher than the backplane speed; so in this situation, the bottom line is that the switch will be a bottleneck. But this situation should happen so rarely that this issue is almost nonexistent. However, imagine that the backplane is only 2 Gbps, and you start to see how important the backplane speed is to a switch.

For a standard low-port-density router, the backplane speed is of minimal importance. The packets per second (PPS) rate on a router has much less to do with the backplane speed and more to do with the processor speed. Typically, a router never operates at wire speed anyway. On switches and other high-port-density devices, however, the backplane is one of the most important considerations.

NVRAM

Nonvolatile RAM (NVRAM) is a type of RAM that does not lose information upon power loss. The amount of NVRAM in most Cisco devices is very small (typically between 32 and 256KB), and it is used to store the configuration used upon boot, called the startup config (discussed further in Chapter 14).

External Components

Some of the most common external devices include the console port, the auxiliary (AUX) port, Ethernet ports, serial ports, and PCMCIA slots. Figure 8-1 shows most of these devices.

Console Port

The *console port* is the primary interface to most Cisco devices (except for Cisco Aironet Basestations). You use this port to get into the IOS for initial configuration, and it

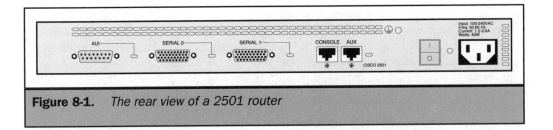

Figure 8-1.   *The rear view of a 2501 router*

consists of a single RJ-45 connector. The console port is actually a low-speed asynchronous serial port (like the serial ports on PCs) that has a special pinout and cable type that must be used to connect (discussed in more detail in the "Cabling" section). By default, the console port also echoes all notifications that occur, making it a great port to use while troubleshooting.

Auxiliary Port

Labeled "AUX" in Figure 8-1, the *auxiliary port* is another low-speed, asynchronous serial port, typically used to connect a modem to the Cisco device to allow remote administration. Most Cisco devices include an AUX port.

Ethernet Ports

10BaseT *Ethernet ports* may or may not be provided on your Cisco device (depending on the model), although most devices do include at least one. The 10BaseT ports generally come with the standard RJ-45–style connector, although you may also have an attachment unit interface (AUI) port. The AUI port is a DB-15 connector (the same connector as a joystick connector on a PC) that requires the use of a transceiver (short for **trans**mitter/re**ceiver**) to make a connection to the network. This allows for the use of other Ethernet cabling types (such as coaxial cable).

Serial Ports

Several models of Cisco devices include both high- and low-speed serial interfaces. Routers typically use a high-speed synchronous serial interface for communication with a WAN channel services unit/data services unit (CSU/DSU), whereas access servers typically have multiple low-speed asynchronous *serial ports* for communication with modems. Unfortunately, Cisco has a variety of serial connectors, including the DB-60 connector shown in Figure 8-1 (all of which are detailed in the "Cabling" section), making cabling a bit of a nightmare.

PCMCIA slots

Some Cisco devices (like the 3660 router) include *PCMCIA slots* (commonly called PC Cards) so that you can easily add flash memory. If your routers are so equipped, upgrading the IOS on multiple routers of the same model will be extremely easy. Rather than having to use TFTP to transfer the IOS image to every router, you simply load it onto a PC Card and transfer it to each router from the card.

Modularity

Cisco devices come in one of two basic configurations: fixed or modular. *Fixed devices* cannot be expanded. They simply ship with a set number and type of interfaces (like the 2501 router) and cannot be upgraded with more interfaces later. *Modular devices* come with module slots, which do allow expansion later.

Modular devices come in one of four basic styles. In the simplest style, the device simply accepts line cards (similar to expansion cards for PCs), called *WAN interface cards (WICs)*, to add functionality. The device typically comes with a set number of card slots for expandability. If you wish to add an additional serial port and a card slot is open, simply purchase the correct card type (a WIC-1T, in this case), insert the card into the chassis, and configure the connection. Figure 8-2 shows an example of this type of modular device.

The second type of device uses a larger card slot, called a *network module,* which typically includes one or more LAN or WAN interfaces, and may also have slots within the network module for other WICs. An example of this type of device, the 3640 router, is shown in Figure 8-3.

The third type of device uses a larger component (called a *blade* in geek-slang), which Cisco may call by different names depending on the model of the device in question. (On a 5500 switch, they are called *modules;* on a 7500 router, they are called *network interface processors;* and so on.) These blades typically support one type of network interface each, but they may support a large number of ports per blade. Figure 8-4 shows an example of this type of device.

Finally, the last type of device has a combination of WIC slots and network module slots. This setup allows you a lot of freedom in choosing your features. A good example of this is the 2611 router (shown in Figure 8-5). It comes standard with two Ethernet ports, two WIC slots, and a single network module slot. With this much modularity, you could have a 2611 routing between two separate Ethernet networks and purchase

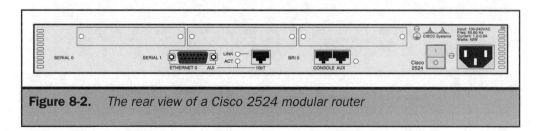

Figure 8-2.   *The rear view of a Cisco 2524 modular router*

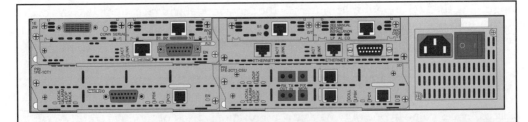

Figure 8-3. *The rear view of a Cisco 3640 modular router*

two dual-port synchronous serial WICs, giving you the ability to route over four total Frame Relay links, and then add a DS3 ATM network module in your free network module slot! And this is on a router that Cisco describes as being a fairly simple, branch-office router.

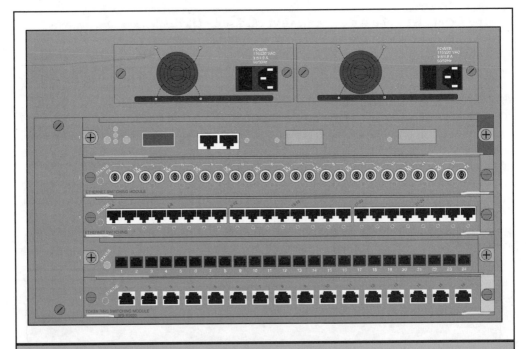

Figure 8-4. *The front view of a Cisco Catalyst 5500 switch*

Figure 8-5.   *The rear view of a Cisco 2611 modular router*

Cabling

Determining the correct cable for a given interface on a Cisco device can be extremely confusing. Cisco devices use a number of different types of connections to support essentially the same features. This point is probably best illustrated by the sheer enormity of different serial options. Let's take a look at some of the most common cables and determine which type of cable is appropriate for a given connector and interface.

Note   *A lot more connector types exist than the ones presented here, but none of them are nearly as common. Following the basic tips described in the following sections should get you through most of your connector nightmares.*

Console Connections

Console connections are fairly simple, and they include the correct type of cable and adapter with the device. *Console connections* are essentially asynchronous serial connections with a strange pinout. They use a cable known as a rollover cable, and an adapter, called a DCE adapter, to convert the RJ-45 connector into a DB-9 or DB-25 for connection to your PC's serial port.

The rollover cable that ships with the device is a bit short (around 10 feet). You may need to crimp your own cable if the one provided isn't long enough. This task is fairly simple. Just acquire a length of standard Cat 5 UTP, a few RJ-45 connectors, and a crimping tool. Strip the cable and, on one end, line the eight individual cables up in any order. Then, on the other end, perform the same task, but reverse the order. Crimp them down, and you are done.

Note   *Another neat trick for extending your console connections is to use standard Ethernet cabling (not crossover) from the patch panel in the NOC to a wall jack. Then, insert the rollover cable into the wall jack and blamo!—instant rollover extension.*

AUX Port

The *AUX port* is similar to the console port as far as physical connections are concerned. The real difference in the two is how they are used. The AUX port can be used as a terminal connection, although typically, it isn't. If you want to use it for this purpose, you will need to attach an RJ-45 rollover cable to the AUX port and use a DTE DB-25 or DB-9 adapter (typically included with the router, but can be purchased separately from Cisco).

The more common use of an AUX port is to connect to a modem. For this, you will also need a rollover cable; but this time, you use a DCE adapter. Note that two types of DCE adapters exist: modem and nonmodem. The modem adapter is the one you want; the nonmodem adapter is obsolete.

RJ-45 Ethernet Connection

An *RJ-45 Ethernet connection* simply uses standard UTP patch cables to make the connection. Cisco switches do not include a crossover button; so to connect a Cisco switch to another Cisco switch (or any switch that does not include a crossover switch, commonly labeled MDI/MDX), you will need a special type of cable—known as a *crossover cable*—that simply crosses the transmit and receive pairs in the cable to allow the transmit from one switch to enter the receive on the other. These can also be made fairly simply by swapping the wires at one end of the cable.

AUI Port   Typically, all you need to connect a UTP cable to the *AUI port* is a decent transceiver. You simply plug the transceiver into the AUI port and plug your UTP cable into the RJ-45 jack on the transceiver. Alternatively, you can use an AUI cable to connect to a detached transceiver, but the procedure is essentially the same. If you need to connect to a coaxial Ethernet network, the procedure is a bit more complicated. However, you still just connect the transceiver to the media like any other device, and then connect the transceiver to the AUI port on the router.

Single Serial Connections

With *single serial connections,* you are immediately faced with the most complex variety of cabling choices you are ever likely to see. There are six common specifications (EIA/TIA-232, X.21, V.35, EIA/TIA-449, EIA-530, and HSSI), and to make it worse, synchronous serial connections require that you choose between DCE and DTE cabling.

Luckily, the choices of single serial interfaces on the backside of your router are fairly standardized, so you will probably have to deal with only two types of connectors there: DB-60 and smart serial. The DB-60 connector is physically the larger of the two, and it is typically used when space on the router or WIC is not at a premium—such as in the 2500 series routers and the WIC-1T single-serial interface cards. You will typically see the smart-serial connection in cases where two or more serial connections need to be squeezed into a relatively small area (such as in the WIC-2T, a dual-port serial interface card).

Both types of serial connectors are what Cisco calls *5-in-1* connectors. This means that they can be used to connect to EIA/TIA-232, X.21, V.35, EIA/TIA-449, and EIA-530 devices; each device just requires a different connector on the far end of the cable.

To choose which cable to use to connect your router's serial port to the required device (usually a CSU/DSU for synchronous serial connections, and a modem for asynchronous connections), you need to follow a couple of simple steps.

First, determine which type of connector you need on the far end (again, either EIA/TIA-232, X.21, V.35, EIA/TIA-449, or EIA-530).

Then choose whether you need a DCE or DTE cable (for synchronous serial connections). This decision is usually fairly easy to determine because, in most cases, your router will be a DTE, so you should buy the DTE cable. Typically, you can just ask the provider of your WAN link and they can tell you. If you are actually running your own WAN links (such as if you were setting up a home lab with full Frame Relay emulation), you would connect the DCE cable to the WAN-switching device and the DTE cable to the site router.

Finally, examine whether you need male or female connectors (which is one difference I will *not* attempt to explain).

That's really all there is to it. Just follow those three steps and you will know what to look for.

Octal Serial Connections

Octal (for 8 into 1) cables are used for specialized routers that require a high number of asynchronous serial ports available in one chassis. An *octal cable* begins at the router with a DB-68 connector, but it splits off into either eight RJ-45 connectors (in the CAB-OCTAL-ASYNC) or eight DB-25 connectors (in the CAB-OCTAL-MODEM).

The RJ-45 style requires a DCE adapter, just like for a console port, to connect to standard serial ports. The RJ-45 style is quite useful, however, for connecting into the console ports of other routers to allow for remote management. Because octal cables are already "rolled," you simply plug them into the console ports of the target routers, and you can Telnet into the access server and then reverse Telnet out of the access server into each of the connected routers—just as if you were sitting directly at a terminal connected to the console port.

The octal modem cable is more useful for connecting to a bank of modems, perhaps to set up a small dial-in point of presence (POP) "on the cheap" using standard external modems. One way or another, if you need a lot of asynchronous serial connections from one router, the octal-style serial connection is the way to go.

Summary

This chapter described which functions are used for the basic hardware components of nearly all Cisco devices. You learned how to cable most of these devices, how modularity in some Cisco devices provides an incredible range of connection options, and even how to use some basic tips to help you mount Cisco gear. All of these topics will come in handy as we continue to explore Cisco products in the coming chapters.

For additional information, check out the following web sites:

- http://www.cisco.com/warp/public/534/index.shtml
- http://www.cisco.com/univercd/cc/td/doc/pcat/se____c1.htm

The
Complete
Reference

Chapter 9

Basic Switch Terminology

This chapter explores Cisco's line of switching hardware. This hardware includes a variety of equipment, from fairly simple, fixed-configuration, layer 2 Ethernet switches, to the modular layer 3 and layer 4 LAN/WAN switches. The goal in this chapter is not to make you an expert on Cisco switching terminology, but rather to give you a basic understanding of the terminology so that you can choose which features you require in your environment and which ones you don't.

Cisco Switching Fundamentals

We'll look more closely at the details for each individual model of Cisco switch later in this chapter. However, let's examine some fundamentals of switching hardware to start out.

Cisco Switch Types

Cisco tends to define devices by their proposed role in the Cisco Hierarchical Internetworking Model (which is explained in Chapter 1) and typically describe a switch by calling it a *core* switch or a *distribution* switch. To help you understand what these distinctions identify, let's examine what each layer is responsible for from a switching standpoint.

The *core layer* is the backbone of the network. All traffic on the core layer has been concatenated from the other two layers and typically exists in large quantities, like a lot of tiny streams joining a great river. The sole objective of the core layer is to switch these large volumes of traffic at wire speed, or as close to that as possible. Therefore, very few processor-intensive activities (such as compression, software-based routing, access lists, and so on) should occur here because these activities detract from the speed with which packets are switched. Typically, at the core layer, only layer 2 switching is performed, and processes like the Spanning Tree Algorithm (STA) should be cut off if at all possible. (STA is discussed further in Chapter 19.)

The *distribution layer* performs the bulk of the processing in packets. This layer is where the following should occur: routing, access lists (discussed further in Chapter 27), firewalls and network address translation (NAT), address aggregation, and media translations (like going from Ethernet to ATM). The distribution layer provides the links to the access layer, which provides the links to the clients. As such, the distribution layer concatenates the client bandwidth to the enterprise services, and then all of the distribution-layer devices are concatenated into the core. Distribution-layer switching is the piece that separates local traffic from remote traffic, and it prioritizes traffic based on bandwidth requirements.

The *access layer* is responsible for providing links to the end user. The access layer provides individual access links for desktops, network printers, and other network devices. Access-layer switching involves basic layer 2 switching and virtual LAN (VLAN) membership (discussed further in Chapter 19); it can also include simple access lists and port security as required for individual hosts.

Typically, Cisco core switches are designed to provide extremely high-speed switching (100,000,000+ *packets per second,* or *PPS*) with a high density of gigabit or better ports. Cisco switches classified as core switches include the Lightstream 1010 ATM switch, the 6500 series, and the 8500 series switches. Nearly all of Cisco's core switches also have the capability to perform as distribution-layer switches, which is extremely useful in *collapsed core environments*: the core functions and distribution functions are combined into one device, a common occurrence for smaller businesses.

Cisco distribution switches are designed to provide moderately high-speed to high-speed switching with a plethora of features. Most distribution switches allow for both medium-speed (100 Mbps) and high-speed (1 Gbps+) interfaces on the same chassis, allowing for concatenation of access links while also allowing for high-speed uplinks to the core. Cisco switches classified as distribution switches include the 6000 and 6500 series, the 5000 and 5500 series, the 4003 switch, the 4840G switch, and the 4200 series switches. Again, some of these switches are designed to perform the functions of two layers: in some cases, distribution and core, and in other cases, distribution and access.

Finally, Cisco access-layer switches are designed to provide reasonably high port density, medium-speed connections to the client, and an ample supply of additional features to provide for the needs of an individual client. Most access-layer switches have at least one higher-speed (either 100 Mbps or 1 Gbps) uplink to the distribution layer, used to concatenate client bandwidth back to the enterprise services. Access-layer devices are typically focused on providing a lower cost per port because more connections— and, therefore, more ports—are used at the access layer than at any other layer. Cisco access switches include the 6000 series, the 5000 and 5500 series, the 4000 series, the 3500 series, the 2900 and 2900XL series, the 2820 switch, and the 1900 series switches.

Switch Modularity

Switches can be either fixed configuration or modular. Typically, the lower-end (access) switches are fixed configuration, and the higher-end (distribution and core) switches are modular. Modular switches (like the 5500 series) use *blades,* usually called *modules,* to add features or ports to the base switch. These modules can perform a number of functions. The most basic modules simply add ports to the switch, but other modules can be added (depending on the switch) to provide additional features like a redundant processing system or layer 3 switching capability. Most of the modular chassis also include provisions for a redundant power supply.

Note that modular switches typically come with no modules installed. At least one module must be purchased, the master module known as a *supervisor module,* to get any functionality from the switch. Supervisor modules vary in functionally, and different versions are offered for different needs. Also, supervisor modules may include their own slots for additional cards, like the Route Switch Feature Card (RSFC) available for certain supervisor modules that adds routing capabilities to a Catalyst 5000 switch.

Also, most gigabit modules (and even some gigabit interfaces in fixed configuration switches like the 3548) use an additional module for each gigabit interface, known as a *GigaBit Interface Converter (GBIC)*: a tiny transceiver that is used to eliminate the need for changing an entire module (or having to buy an entirely new switch in fixed-configuration models) if you need a different type of connector. Because fiberoptic links have several specifications, the use of GBICs reduces some of the headache associated with choosing the correct gigabit module. However, it also adds a little to the total cost of the solution.

Finally, you need to be aware of the interface numbering scheme in modular switches. (This concept will become even more important when you begin to study the IOS in detail.) Modular switches number their ports a bit differently than standard, fixed-configuration switches. Most fixed-configuration switches number their ports very simply. First, you specify the type of interface (such as Ethernet), and then you specify the port number, starting with 1. For example, if you want the first Ethernet port, the port is called Ethernet 1.

However, modular switches (and even some fixed-configuration switches, like the 1900 series) number their ports in one of three ways: module/port, module/slot/port, or (in the case of the 8500) card/subcard/port. The module/port scheme is fairly easy to understand. If you had a 24-port Ethernet module and it was module 5, and you needed to configure port 16, the number would be 5/16.

The module/slot/port is a little more complicated. In some modules, you may have multiple slots for further expansion, similar to the network modules discussed in Chapter 8 that have WIC slots. In this case, you must first identify the module, then the slot within that module, and then the port within that slot. In most cases, no slots exist in the module, so you simply specify slot 1 in the command. For instance, if you had a single module that was module number 4 with 24 Ethernet ports but no additional slots, and you needed to configure port 5, you would specify 4/1/5.

The final type, card/subcard/port, works the same way as the module/slot/port. It just has a different name in the documentation.

Switch Hardware Terms

To understand the technical documentation used to describe switches (Cisco or otherwise), you must learn some new terms. The first term, *backplane,* you are already familiar with from Chapter 8. However, a second term, switch fabric, also needs to be explained.

Switch fabric is a collective term used to describe all of the hardware and software specifically involved with the switching process. In other words, switch fabric does not include the cabling to the client or the client itself, but it does include all of the application-specific integrated circuits (ASICs) within the switch, the algorithms used to determine the switching logic, the backplane, and the switching process inside the switch. Numerous Cisco technical documents use the term "switch fabric" to describe the performance of a given switch, as well as when combining or stacking switches together, so it is a useful term to know, even if it is a bit overgeneralized.

Switch IOS

Cisco's lineup includes a wide variety of switches, all of which are slightly different from one another in some minor, but still significant, way. Unfortunately, the software revisions loaded in switches are, if anything, more complex than the other differences in the switches. Basically, two main types of operating system software exist in switches: IOS and Custom.

IOS-based switches are generally the easiest to get accustomed to because the command line interface (CLI) in the switches matches up pretty well to the CLI used in Cisco routers. However, two types of IOS are used in switches. Both are very similar to the C shell CLI in UNIX, but they do have some striking differences.

The first type, known as the *standard IOS*, is the one that is similar to the CLI used in routers. This IOS uses a relatively large number of commands to accomplish all tasks on the device, from monitoring to configuration. Chapter 15 covers this type of IOS.

The second type, called the *set-based IOS*, is used only in Catalyst 4000, 5000, and 6000 series switches. This type of IOS makes use of three primary commands (*set, show,* and *clear*) and a number of secondary commands to perform all tasks, and it is extremely different from the IOS used in routers. Chapter 16 covers this type of IOS.

Some switches, however, don't use an IOS at all. These switches have what is known as a *custom* interface that is applicable only to that specific switching line. Some custom interface switches, such as the 1900 and 2820 switches, are similar to the standard IOS. Others, such as the 3100, 3200, and 3900 series, are drastically different from the standard IOS, and they use a GUI for interfacing with the switch. You need to examine each custom IOS switch individually because the process for configuring the devices may be significantly different from what this book outlines.

In addition to the CLI for the individual switches, most switches ship with a GUI interface based on HTTP. In other words, you can configure rudimentary settings in most switches from a browser. Some other switches (such as the 1900 series) also provide a menu-driven interface, in addition to a CLI and browser-based interface. The menu-driven interface is somewhat more versatile than the GUI interface; but if you become comfortable with the CLI, you'll find that it takes far less time to configure a switch with the CLI than by using any other method.

Switch Classification Chart

This section presents three reference charts (Tables 9-1 through 9-3) detailing each model of Cisco switch, complete with port types and densities, interface numbering style, IOS type, and features. But first, you need to know the definitions of a few terms and abbreviations used in the tables.

- **Series**   Denotes the line of switch. This is the broader term you will hear used a lot. Most switches in a given series are based on the same hardware and software and have similar features.

- **Switch model**   Denotes the individual models of switches.

- **Classification**   Denotes the switch's classification in the Cisco Hierarchical Internetworking Model.

- **Configuration**   Tells whether the switch is fixed configuration or modular configuration.

- **IOS type**   Tells whether the switch uses the set-based IOS, a custom interface, or the standard IOS.

- **Interface numbering**   Tells the style of interface numbering in this switch, from the following major types:

  - **Type p**   First you specify the type of interface (like Ethernet), and then the port.

  - **Type m/p**   First you specify the type of interface (like Ethernet), then the module followed by a slash (/), and then the port.

  - **Card/subcard/port**   First you identify the card, and then the subcard (or 1, if no subcards exist), and then the port.

- **Port types (Ethernet, Fast Ethernet, and so on)**   Lists the *maximum* ports available in this model of switch. Note that in modular switches with multiple port types available, you cannot have all available port types at the listed capacity. For instance, if you want 528 Ethernet ports in a 5500 switch, you use this configuration at the expense of all other ports. Also, the modular switchers have an asterisk (*) beside all optional features, including ports.

- **MAC address limit**   Denotes the maximum number of MAC addresses that can be in the MAC address table at any given time.

- **Switching methods**   Lists the layer 2 and layer 3 (if applicable) switching methods available in this model. The available switching methods are

  - **SF**   Store-And-Forward

  - **CT**   Cut-Through

  - **FF**   Fragment Free

  - **Netflow**   Netflow layer 3 switching

  - **CEF**   Cisco Express Forwarding layer 3 switching

- **Backplane speed**   Denotes the listed backplane speed in this model.

- **PPS**   Denotes the PPS in this model, as reported by Cisco. Note that most switches do not achieve the quoted rate in real-world situations. This rate is nearly always based on 64-byte packets, which are the minority in most data communications. Larger packet sizes invariably lead to lower PPS scores in most cases. This guide is still a decent indication of a switch's raw speed, however.

The following list describes some of the most common switch features. Note that *all* features in any given switch are not listed. Chapter 10 provides more information on each switch line. The following guide simply presents the most common features of each model. Also note that higher-end switches generally provide a more thorough implementation of a given feature.

- **VLANs** Indicates that VLANs (covered in Chapter 19) are supported in this model. Note that higher-end models support more VLANs than lower-end models.

- **Congestion control** Indicates that some type of congestion control is supported in this model. This control could range from broadcast throttling in the lower-end models to full-blown quality of service (QOS) functionality with advance queuing techniques in the higher-end models

- **Port security** Indicates that this switch has the ability to block or allow access to certain devices based on the port you are attached to and/or your MAC address.

- **DHCP** Indicates that the switch supports the dynamic allocation of its IP address using DHCP.

- **DNS** Indicates that the switch supports the resolution of names from within the interface using DNS.

- **ISL** Indicates that the switch supports interswitch link (ISL) and/or dynamic interswitch link (DISL) for tagging frames with VLAN information (covered in detail in Chapter 19). In higher-end switches, 802.1q is also typically supported. Note that for frame tagging (also known as *trunking*) to be supported, at least one 100 Mbps or faster interface must be available.

- **CGMP** Indicates that the switch supports multicasting, or at least has Cisco Group Management Protocol (CGMP) support. Higher-end switches typically have full multicasting support as well (IGMP, CGMP, PIM, MOSPF, and so on).

- **RMON** Indicates that the switch supports *remote monitoring* (*RMON*), a network management specification, and, typically, it supports SNMP as well.

- **SPAN** Indicates that the switch supports Switched Port Analyzer (SPAN) and, in higher-end switches, Remote Switched Port Analyzer (RSPAN). SPAN and RSPAN allow an administrator to selectively analyze network traffic destined for any given host on the network, using a protocol analyzer.

- **Routing** Indicates that the switch can be enabled to support routing. Note that routing and layer 3 switching are basically the same thing, but routing is performed in software (making it slower), and switching is performed in ASICs, or hardware. If a switch has this designation, it means that a routing module, not a layer 3 switching module, is available for it.

- **L3-IP**   Indicates that layer 3 switching over IP (covered further in Chapter 20) is supported in this switch.

- **L4-IP**   Indicates that layer 4 switching over IP (covered further in Chapter 21) is supported in this switch.

- **L3-IPX**   Indicates that layer 3 switching over IPX is supported in this switch.

- **L4-IPX**   Indicates that layer 3 switching over IPX is supported in this switch.

- **Access lists**   Indicates that access lists are supported in this switch to some degree. Chapter 27 covers access lists in detail.

- **Multiservice**   Indicates that this switch has extended capabilities for supporting voice or video traffic.

- **SLB**   Indicates that this switch has server load balancing (SLB) capabilities, allowing it to provide intelligent load balancing to a bank of servers based on traffic parameters.

- **ATM switching**   Indicates that this switch can perform as an ATM switch, including full support for ATM QOS functionality.

- **TACACS+**   Indicates that this switch supports robust authentication, authorization, and accounting features, including Terminal Access Controller Access Control System+ (TACACS+), a centralized method of controlling access to Cisco devices.

Series	Switch Model	Classification	Configuration	IOS Type	Interface Numbering
1900	1912-A	Access	Fixed	Custom	Type m/p
	1912C-A	Access	Fixed	Custom	Type m/p
	1912-EN	Access	Fixed	Custom	Type m/p
	1912C-EN	Access	Fixed	Custom	Type m/p
	1924-A	Access	Fixed	Custom	Type m/p
	1924C-A	Access	Fixed	Custom	Type m/p
	1924-EN	Access	Fixed	Custom	Type m/p
	1924C-EN	Access	Fixed	Custom	Type m/p
	1924F-A	Access	Fixed	Custom	Type m/p
	1924F-EN	Access	Fixed	Custom	Type m/p

Table 9-1.   *Switch Classification Chart—Basic Information*

Series	Switch Model	Classification	Configuration	IOS Type	Interface Numbering
2820	2828-A	Access	Modular	Custom	Type m/p
	2828-EN	Access	Modular	Custom	Type m/p
2900	2926T	Access	Fixed	Set based	Type m/p
	2926F	Access	Fixed	Set based	Type m/p
	2926GS	Access	Fixed	Set based	Type m/p
	2926GL	Access	Fixed	Set based	Type m/p
	2980G	Access	Fixed	Set based	Type m/p
	2980G-A	Access	Fixed	Set based	Type m/p
	2948G	Access/distribution	Fixed	Set based	Type m/p
	2948G-L3	Access/distribution	Fixed	Standard	Type m/p
2900 XL	2912-XL-EN	Access	Fixed	Standard	Type p
	2912MF-XL	Access/distribution	Modular	Standard	Type p
	2924-XL-EN	Access	Fixed	Standard	Type p
	2924C-XL-EN	Access	Fixed	Standard	Type p
	2924M-XL-EN	Access/distribution	Modular	Standard	Type p
3000	3016B	Access	Modular	Custom	Non IOS
3100	3100B	Access	Modular	Custom	Non IOS
3200	3200B	Access	Modular	Custom	Non IOS
3500 XL	3508G-XL-EN	Access/distribution	Fixed	Standard	Type p
	3512-XL-EN	Access/distribution	Fixed	Standard	Type p
	3524-XL-EN	Access/distribution	Fixed	Standard	Type p
	3548-XL-EN	Access/distribution	Fixed	Standard	Type p
3900	3900	Access	Modular	Custom	Non IOS
	3920	Access	Fixed	Custom	Non IOS

Table 9-1.   *Switch Classification Chart—Basic Information* (continued)

CISCO TECHNOLOGY OVERVIEW

Series	Switch Model	Classification	Configuration	IOS Type	Interface Numbering
4000	4003-S1	Access/ distribution	Modular	Set based	Type m/p
	4006-S2	Access/ distribution	Modular	Set based	Type m/p
	4908G-L3	Access/ distribution	Fixed	Set based	Type m/p
	4912G	Access/ distribution	Fixed	Set based	Type m/p
4840	4840G	Access/ distribution	Fixed	Standard	Type p
5000	5000	Access/ distribution	Modular	Set based	Type m/p
	5002	Access/ distribution	Modular	Set based	Type m/p
5500	5500	Access/ distribution	Modular	Set based	Type m/p
	5505	Access/ distribution	Modular	Set based	Type m/p
	5509	Access/ distribution	Modular	Set based	Type m/p
6000	6006	Access/ distribution	Modular	Set based	Type m/p
	6009	Access/ distribution	Modular	Set based	Type m/p
6500	6506	Distribution/ core	Modular	Set based	Type m/p
	6509	Distribution/ core	Modular	Set based	Type m/p
8500	8510CSR	Core	Modular	Standard	Card/ subcard/port
	8540CSR	Core	Modular	Standard	Card/ subcard/port
	8510MSR	Core	Modular	Standard	Card/ subcard/port
	8540MSR	Core	Modular	Standard	Card/ subcard/port

Table 9-1. *Switch Classification Chart—Basic Information* (continued)

Series	Switch Model	Ethernet Ports	Fast Ethernet (100 Mbps or 10/100) Ports	Gigabit Ethernet Ports	Token Ring Ports	FDDI Ports	ATM Ports	Other Ports
1900	1912-A	12	2	0	0	0	0	0
	1912C-A	12	2	0	0	0	0	0
	1912-EN	12	2	0	0	0	0	0
	1912C-EN	12	2	0	0	0	0	0
	1924-A	24	2	0	0	0	0	0
	1924C-A	24	2	0	0	0	0	0
	1924-EN	24	2	0	0	0	0	0
	1924C-EN	24	2	0	0	0	0	0
	1924F-A	24	2	0	0	0	0	0
	1924F-EN	24	2	0	0	0	0	0
2820	2828-A	24	16*	0	0	0	0	0
	2828-EN	24	16*	0	0	0	0	0
2900	2926T	0	26	0	0	0	0	0
	2926F	0	26	0	0	0	0	0
	2926GS	0	24	2	0	0	0	0
	2926GL	0	24	2	0	0	0	0
	2980G	0	80	2	0	0	0	0
	2980G-A	0	80	2	0	0	0	0
	2948G	0	48	2	0	0	0	0
	2948G-L3	0	48	2	0	0	0	0
2900 XL	2912-XL-EN	0	12	0	0	0	0	0
	2912MF-XL	0	12	2*	0	0	2*	0
	2924-XL-EN	0	24	0	0	0	0	0
	2924C-XL-EN	0	24	0	0	0	0	0
	2924M-XL-EN	0	24	2*	0	0	2*	0

Table 9-2.   *Switch Classification Chart—Port Densities*

Series	Switch Model	Ethernet Ports	Fast Ethernet (100 Mbps or 10/100) Ports	Gigabit Ethernet Ports	Token Ring Ports	FDDI Ports	ATM Ports	Other Ports
3000	3016B	24*	4*	0	0	0	2*	0
3100	3100B	28*	4*	0	0	0	2*	1 WAN*
3200	3200B	28*	14*	0	0	0	7*	1 WAN*
3500 XL	3508G-XL-EN	0	0	8	0	0	0	0
	3512-XL-EN	0	12	2	0	0	0	0
	3524-XL-EN	0	24	2	0	0	0	0
	3548-XL-EN	0	48	2	0	0	0	0
3900	3900	0	2*	0	24*	0	1*	0
	3920	0	0	0	24	0	0	0
4000	4003-S1	0	96*	36*	0	0	0	0
	4006-S2	0	240*	142*	0	0	0	0
	4908G-L3	0	0	8	0	0	0	0
	4912G	0	0	12	0	0	0	0
4840	4840G	0	40	2	0	0	0	0
5000	5000	192*	96*	36*	64*	4*	4*	0
	5002	96*	48*	18*	32*	2*	2*	0
5500	5500	528*	264*	38*	172*	11*	96*	0
	5505	192*	96*	36*	64*	4*	48*	0
	5509	384*	192*	36*	128*	8*	96*	0

Table 9-2. *Switch Classification Chart—Port Densities* (continued)

Series	Switch Model	Ethernet Ports	Fast Ethernet (100 Mbps or 10/100) Ports	Gigabit Ethernet Ports	Token Ring Ports	FDDI Ports	ATM Ports	Other Ports
6000	6006	240*	192*	80*	0	0	5*	10 WAN*, 120 FXS analog phone*, 40 T1 voice circuits*
	6009	384*	192*	130*	0	0	8*	16 WAN*, 192 FXS analog phone*, 72 T1 voice circuits*
6500	6506	240*	192*	80*	0	0	0	8 WAN*, 96 FXS analog phone*, 32 T1 voice circuits*
	6509	336*	168*	114*	0	0	0	14 WAN*, 168 FXS analog phone*, 64 T1 voice circuits*
8500	8510CSR	0	32*	4*	0	0	0	0
	8540CSR	0	128*	64*	0	0	1 Uplink*	0
	8510MSR	0	32*	4*	0	0	32*	0
	8540MSR	0	128*	64*	0	0	128*	0

Table 9-2. *Switch Classification Chart—Port Densities* (continued)

Series	Switch Model	MAC Address Limit	Switching Methods	Backplane Speed	PPS	Capabilities
1900	1912-A	1024	CT, FF, SF	1 Gbps	550,000	Congestion control, port security
	1912C-A	1024	CT, FF, SF	1 Gbps	550,000	Congestion control, port security
	1912-EN	1024	CT, FF, SF	1 Gbps	550,000	VLANs, congestion control, port security, ISL, CGMP, RMON
	1912C-EN	1024	CT, FF, SF	1 Gbps	550,000	VLANs, congestion control, port security, ISL, CGMP, RMON
	1924-A	1024	CT, FF, SF	1 Gbps	550,000	Congestion control, port security
	1924C-A	1024	CT, FF, SF	1 Gbps	550,000	Congestion control, port security
	1924-EN	1024	CT, FF, SF	1 Gbps	550,000	VLANs, congestion control, port security, ISL, CGMP, RMON
	1924C-EN	1024	CT, FF, SF	1 Gbps	550,000	VLANs, congestion control, port security, ISL, CGMP, RMON
	1924F-A	1024	CT, FF, SF	1 Gbps	550,000	Congestion control, port security
	1924F-EN	1024	CT, FF, SF	1 Gbps	550,000	VLANs, congestion control, port security, ISL, CGMP, RMON
2820	2828-A	8192	CT, FF, SF	1 Gbps	550,000	Congestion control, port security, DHCP, DNS
	2828-EN	8192	CT, SF	1 Gbps	550,000	VLANs, congestion control, port security, DHCP, DNS, ISL, CGMP, RMON, TACACS+

Table 9-3.   *Switch Classification Chart—Capabilities*

Series	Switch Model	MAC Address Limit	Switching Methods	Backplane Speed	PPS	Capabilities
2900	2926T	16383	CT, SF	1.2 Gbps	1,000,000	VLANs, congestion control, port security, DHCP, ISL, CGMP, RMON, SPAN
	2926F	16383	CT, SF	1.2 Gbps	1,000,000	VLANs, congestion control, port security, DHCP, ISL, CGMP, RMON, SPAN
	2926GS	16383	CT, SF	1.2 Gbps	1,000,000	VLANs, congestion control, port security, DHCP, ISL, CGMP, RMON, SPAN
	2926GL	16383	CT, SF	1.2 Gbps	1,000,000	VLANs, congestion control, port security, DHCP, ISL, CGMP, RMON, SPAN
	2980G	16383	SF	24 Gbps	18,000,000	VLANs, congestion control, port security, DHCP, ISL, CGMP, RMON, SPAN
	2980G-A	16383	SF	24 Gbps	18,000,000	VLANs, congestion control, port security, DHCP, ISL, CGMP, RMON, SPAN
	2948G	16383	SF	24 Gbps	18,000,000	VLANs, congestion control, port security, DHCP, RMON, SPAN, access lists
	2948G-L3	16383	CT, SF	24 Gbps	18,000,000	VLANs, congestion control, port security, DHCP, DNS ISL, CGMP, RMON, SPAN, L3-IP, L3-IPX, access lists

Table 9-3.   *Switch Classification Chart—Capabilities* (continued)

Series	Switch Model	MAC Address Limit	Switching Methods	Backplane Speed	PPS	Capabilities
2900 XL	2912-XL-EN	2048	SF	3.2 Gbps	3,000,000	VLANs, congestion control, port security, DHCP, DNS, ISL, CGMP, RMON, TACACS+
	2912MF-XL	8192	SF	3.2 Gbps	3,000,000	VLANs, congestion control, port security, DHCP, DNS, ISL, CGMP, RMON, TACACS+
	2924-XL-EN	2048	SF	3.2 Gbps	3,000,000	VLANs, congestion control, port security, DHCP, DNS, ISL, CGMP, RMON, TACACS+
	2924C-XL-EN	2048	SF	3.2 Gbps	3,000,000	VLANs, congestion control, port security, DHCP, DNS, ISL, CGMP, RMON, TACACS+
	2924M-XL-EN	8192	SF	3.2 Gbps	3,000,000	VLANs, congestion control, port security, DHCP, DNS, ISL, CGMP, RMON, TACACS+
3000	3016B	10000	CT, FF, SF	480 Mbps	714,000	VLANs, SPAN, ISL, RMON
3100	3100B	10000	CT, FF, SF	480 Mbps	714,000	VLANs, SPAN, ISL, RMON, routing*
3200	3200B	10000	CT, FF, SF	480 Mbps	714,000	VLANs, SPAN, ISL, RMON, routing*

Table 9-3. *Switch Classification Chart—Capabilities* (continued)

Series	Switch Model	MAC Address Limit	Switching Methods	Backplane Speed	PPS	Capabilities
3500 XL	3508G-XL-EN	8192	SF	10.8 Gbps	7,500,000	VLANs, congestion control, port security, DNS, SPAN, ISL, CGMP, RMON, TACACS+
	3512-XL-EN	8192	SF	10.8 Gbps	4,800,000	VLANs, congestion control, port security, DNS, SPAN, ISL, CGMP, RMON, TACACS+
	3524-XL-EN	8192	SF	10.8 Gbps	6,500,000	VLANs, congestion control, port security, DNS, SPAN, ISL, CGMP, RMON, TACACS+
	3548-XL-EN	8192	SF	10.8 Gbps	8,000,000	VLANs, congestion control, port security, DNS, SPAN, ISL, CGMP, RMON, TACACS+
3900	3900	10000	CT (SRB, SRT, SRS)	520 Mbps	N/A	VLANs, SPAN, ISL, RMON
	3920	10000	CT (SRB, SRT, SRS)	520 Mbps	N/A	VLANs, SPAN, ISL, RMON

Table 9-3. *Switch Classification Chart—Capabilities* (continued)

Note   *Asterisks (*) indicate that the ports in question rely on module slots, and the full listed port densities are only available if all slots are devoted to the port type in question.*

CISCO TECHNOLOGY OVERVIEW

Series	Switch Model	MAC Address Limit	Switching Methods	Backplane Speed	PPS	Capabilities
4000	4003-S1	16383	SF	24 Gbps	18,000,000	VLANs, congestion control, port security, DHCP, DNS ISL, CGMP, RMON, SPAN, L3-IP*, L3-IPX*, access lists
	4006-S2	16383	SF	24 Gbps	18,000,000	VLANs, congestion control, port security, DHCP, DNS ISL, CGMP, RMON, SPAN, L3-IP*, L3-IPX*, access lists
	4908G-L3	16383	SF	24 Gbps	11,000,000	VLANs, congestion control, port security, DHCP, DNS ISL, CGMP, RMON, SPAN, L3-IP, L3-IPX, access lists
	4912G	16383	SF	24 Gbps	18,000,000	VLANs, congestion control, port security, DHCP, DNS ISL, CGMP, RMON, SPAN
4840	4840G	16383	SF	24 Gbps	9,000,000	VLANs, congestion control, port security, DHCP, DNS ISL, CGMP, RMON, SPAN, L3-IP, access lists, SLB, TACACS+

Table 9-3. *Switch Classification Chart—Capabilities* (continued)

Series	Switch Model	MAC Address Limit	Switching Methods	Backplane Speed	PPS	Capabilities
5000	5000	16383	SF, Netflow*	1.2 Gbps	1,000,000	VLANs, congestion control, port security, DHCP, DNS ISL, CGMP, RMON, SPAN, L3-IP*, L4-IP*, L3-IPX*, L4-IPX*, access lists, SLB, TACACS+
	5002	16383	SF, Netflow*	1.2 Gbps	1,000,000	VLANs, congestion control, port security, DHCP, DNS ISL, CGMP, RMON, SPAN, L3-IP*, L4-IP*, L3-IPX*, L4-IPX*, access lists, SLB, TACACS+
5500	5500	16383	SF, Netflow*	3.6 Gbps	56,000,000	VLANs, congestion control, port security, DHCP, DNS ISL, CGMP, RMON, SPAN, L3-IP*, L4-IP*, L3-IPX*, L4-IPX*, access lists, SLB, TACACS+
	5505	16383	SF, Netflow*	3.6 Gbps	56,000,000	VLANs, congestion control, port security, DHCP, DNS ISL, CGMP, RMON, SPAN, L3-IP*, L4-IP*, L3-IPX*, L4-IPX*, access lists, SLB, TACACS+
	5509	16383	SF, Netflow*	3.6 Gbps	56,000,000	VLANs, congestion control, port security, DHCP, DNS ISL, CGMP, RMON, SPAN, L3-IP*, L4-IP*, L3-IPX*, L4-IPX*, access lists, SLB, TACACS+

Table 9-3. *Switch Classification Chart—Capabilities* (continued)

Series	Switch Model	MAC Address Limit	Switching Methods	Backplane Speed	PPS	Capabilities
6000	6006	32536	SF, CEF*	32 Gbps	15,000,000	VLANs, congestion control, port security, DHCP, DNS ISL, CGMP, RMON, SPAN, L3-IP*, L4-IP*, L3-IPX*, L4-IPX*, access lists, SLB, multiservice, TACACS+
	6009	32536	SF, CEF*	32 Gbps	15,000,000	VLANs, congestion control, port security, DHCP, DNS ISL, CGMP, RMON, SPAN, L3-IP*, L4-IP*, L3-IPX*, L4-IPX*, access lists, SLB, multiservice, TACACS+
6500	6506	32536	SF, CEF*	256 Gbps	170,000,000	VLANs, congestion control, port security, DHCP, DNS ISL, CGMP, RMON, SPAN, L3-IP*, L4-IP*, L3-IPX*, L4-IPX*, access lists, SLB, multiservice, TACACS+
	6509	32536	SF, CEF*	256 Gbps	170,000,000	VLANs, congestion control, port security, DHCP, DNS ISL, CGMP, RMON, SPAN, L3-IP*, L4-IP*, L3-IPX*, L4-IPX*, access lists, SLB, multiservice, TACACS+

Table 9-3. *Switch Classification Chart—Capabilities* (continued)

Series	Switch Model	MAC Address Limit	Switching Methods	Backplane Speed	PPS	Capabilities
8500	8510CSR	32536	SF, CEF	10 Gbps	6,000,000	VLANs, congestion control, port security, DHCP, DNS ISL, CGMP, RMON, SPAN, L3-IP, L4-IP, L3-IPX, L4-IPX, access lists, SLB, TACACS+
	8540CSR	32536	SF, CEF	40 Gbps	24,000,000	VLANs, congestion control, port security, DHCP, DNS ISL, CGMP, RMON, SPAN, L3-IP, L4-IP, L3-IPX, L4-IPX, access lists, SLB, TACACS+
	8510MSR	32536	SF, CEF	10 Gbps	6,000,000	VLANs, congestion control, port security, DHCP, DNS ISL, CGMP, RMON, SPAN, L3-IP, L4-IP, L3-IPX, L4-IPX, access lists, SLB, ATM switching, TACACS+
	8540MSR	32536	SF, CEF	40 Gbps	24,000,000	VLANs, congestion control, port security, DHCP, DNS ISL, CGMP, RMON, SPAN, L3-IP, L4-IP, L3-IPX, L4-IPX, access lists, SLB, ATM switching, TACACS+

Table 9-3.   *Switch Classification Chart—Capabilities* (continued)

CISCO TECHNOLOGY OVERVIEW

Summary

This chapter examined some basic terminology related to Cisco switches and familiarized you with the different lines and models of Cisco switching equipment. The next chapter examines the individual switching lines more thoroughly and begins to piece together which switch works best in nearly any situation.

For additional information on Cisco's line of switching products, check out http://www.cisco.com/warp/public/44/jump/switches.shtml.

Chapter 10

Individual Switch Series Reference

This chapter takes a closer look at the individual switch lines available from Cisco. The goal is to familiarize you with the individual lines of switches so that you can clearly see which switch is right for a given need. This chapter covers only the switch lines that you are most likely to see in any standard environment, as well as those switches that are useful in any environment.

The 1548 Series Microswitches

Cisco's 1548 series is a low-cost line of switches designed to appeal to the small office that needs 100-Mbps access, but that doesn't need all the features (or cost) of Cisco's higher-end 100-Mbps switching lines. There are two switches in the 1548 line: the 1548U and the 1548M. (Figure 10-1 depicts the 1548 series.)

The 1548U is a basic Fast Ethernet switch. It includes eight auto-sensing, auto-negotiating 10/100 ports for client access. No management features are provided on the 1548U, making it ideal for organizations that just need to plug in and go. The 1548U supports only 2048 MAC addresses; but in its typical environment (a small office), this shouldn't be a concern. The 1548U also does not support VLANs (Virtual LANs, discussed further in Chapter 19); but if this is a concern, you should step up to the 1548M or a higher-end switch.

The 1548M is a managed version of the 1548U. It supports more MAC addresses than the 1548U—up to 4048. It also supports SNMP and RMON, and it has a CLI as well as a web-based management console. In addition, the 1548M supports Cisco Discovery Protocol (CDP) for automatically detecting directly connected Cisco devices. (See Chapter 14 for more information on CDP.) The 1548M supports SPAN, as well as VLANs, to a maximum of four. However, the 1548M does not support ISL (Inter Switch Link, discussed further in Chapter 19) tagging, 802.1q, and VLAN Trunking Protocol (VTP). (For more on VTP, see Chapter 19.) If you require these features, then a higher-end switch (like the 1900 series or the 3500 series) is required.

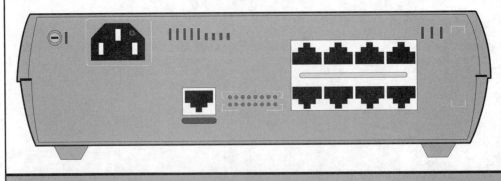

Figure 10-1.   *The rear view of a 1548 series switch*

Both of the 1548 series switches are rather slow; however, in a small office environment, speed is not the primary concern. In this switch's target environment, cost and ease of use are of primary importance. You can find more information about the 1548 line of switches at the following addresses:

- http://www.cisco.com/warp/public/cc/pd/si/1548/prodlit/index.shtml
- http://www.cisco.com/warp/public/cc/pd/si/1548/prodlit/1548_ds.htm

The 1900 and 2820 Series

The 1900 and 2820 series of switches are based around the same hardware and software platforms, with basically the same capabilities but differing port densities. They are also designed for somewhat different uses, as we will see in a moment. For now, let's concentrate on what they have in common.

Both switch lines come in two versions: the standard and enterprise editions. The difference in the two editions is simply in software. The hardware is the same, allowing upgrades from standard to enterprise edition with a simple software upgrade.

The standard edition is designed for networks that require only a relatively basic feature set. This software supports a web-based interface for configuring the switch. As for switching features, the standard edition supports backpressure on half-duplex 10-Mbps links.

Note Backpressure *is a method for controlling congestion whereby, when the switch gets overloaded, it "jams" the port that is overloading its buffers to reduce frame loss.* Congestion control *is a scheme that notifies the device that is overloading the switch, telling it to pause until such time as the switch's buffers have cleared.*

The standard edition supports IEEE 802.3x congestion control on full-duplex links. Also, it supports broadcast storm control by shutting down a port that is experiencing excessive broadcasts.

Finally, the standard edition also supports cut-through, fragment-free, and store-and-forward switching methods; auto-negotiated or manually configured full-duplex communication on all ports; and no per-port MAC address limitations (although there is still a limit to the number of MAC addresses on the entire system, depending on the model). The standard edition also supports CDP and port security.

The enterprise-edition switches support all of the features of the standard edition. In addition, the enterprise edition supports VLAN (including ISL frame tagging), CGMP for multicasting, and SNMP and RMON for management. It adds a CLI (custom, but very similar to the standard IOS), as well as support for Fast Etherchannel.

Note Fast Etherchannel *is a method of "bonding" multiple Ethernet links together to create a larger pipe (discussed in more detail in Chapter 19).*

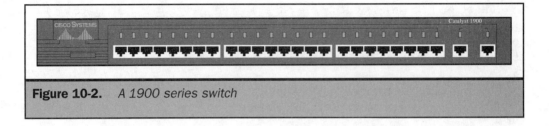

Figure 10-2. *A 1900 series switch*

The 1900 and 2820 series also share most of their hardware. Both lines come primarily with 10-Mbps Ethernet ports (12 or 24 on the 1900 and 24 on the 2820), and both lines include a 10-Mbps AUI port for connection to legacy or fiberoptic Ethernet networks. They include the same performance characteristics and, for the most part, the same feature set.

For the individual lines, the basic difference in the two lies in the environment targeted by each switch line. The 1900 series is designed to be Cisco's entry-level, full-featured switching product, and it offers 12 or 24 10-Mbps Ethernet ports and 2 Fast Ethernet uplink ports. It is designed for companies that need a full-featured, 10-Mbps switching solution at a relatively low cost per port (compared to other Cisco switches, anyway). Figure 10-2 illustrates a 1900 series switch.

The 2820 series is designed for the company that needs a fairly low cost per port, but that also requires modularity and more 100-Mbps connections than the 1900 is capable of. The 2820 supports up to 16 100-Mbps connections with UTP and 8 100-Mbps connections with fiberoptic media. This allows the 2820 to not only aggregate bandwidth to the distribution layer, but also provide for local servers that may require the additional bandwidth offered by Fast Ethernet connections. The 2820 series also provides a few additional features, such as TACACS+, DNS, and DHCP support, to make it a slightly more manageable switch. See Figure 10-3 for an illustration of a 2820 series switch.

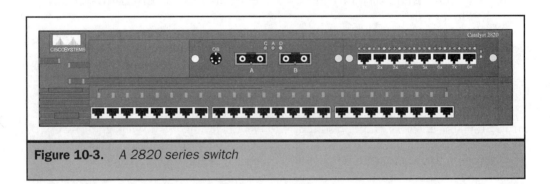

Figure 10-3. *A 2820 series switch*

All things considered, the 1900 and 2820 lines of switches are full-featured switches that, although showing their age, still have a lot to offer. You can find more information about the 1900 and 2820 lines at the following addresses:

- http://www.cisco.com/univercd/cc/td/doc/pcat/ca2820.htm
- http://www.cisco.com/univercd/cc/td/doc/pcat/ca1900.htm
- http://www.cisco.com/warp/public/cc/pd/si/casi/ca1900/prodlit/index.shtml

The 2900 and 2900XL Series

The 2900 series of switches is less of a single line and more of a combination of three individual lines of switches, all with differing benefits, focuses, and capabilities. For this reason, this section is split into four parts: the 2926 line, the 2980 line, the 2948 line, and the 2900XL line of switches are covered individually.

The 2926 Line

As of this writing, the 2926 line has been end-of-lifed (EOL'd)—it is no longer sold—but you may still see it in use. The 2926 line is based on the Catalyst 5000 switching engine, and therefore includes many of the 5000's features. 2926 switches use the set-based IOS, have a 16,000 MAC address/MAC table capacity, use a 1.2-Gbps backplane, and have a PPS rate greater than 1,000,000. Standard features include the ability to set port priority levels to provide enhanced performance to specific devices, Fast Etherchannel support, congestion control, and port security.

The 2926 line also includes SNMP, RMON, CDP, SPAN, DHCP, and VTP management features. Up to 1000 VLANs are supported, and ISL, DISL, and 802.1q frame-tagging support is included. All of the 2926 models include 24 10/100 auto-sensing, auto-negotiating Fast Ethernet ports, and two uplink ports for aggregation back to distribution-layer devices. On the 2926T model, the two uplink ports are 100Base-TX ports; and on the 2926 model, the uplinks are 100Base-FX ports. On the 2926GS and 2926GL switches, the two uplink ports are Gigabit Ethernet, with 1000Base-SX on the GS and 1000Base-LX/LH on the GL. Redundant power supplies were supported on all models.

The 2926 line was designed for networks that needed some of the power inherent in the 5000's feature set, but that didn't require the port density offered by the Catalyst 5000 or that needed this feature set in a more compact chassis. It fits the bill well when used as a feature-rich, midrange access-layer switching solution.

For more information on the 2926 switching line, visit http://www.cisco.com/univercd/cc/td/doc/pcat/ca2926.htm.

CISCO TECHNOLOGY OVERVIEW

The 2980 Line

The 2980 line is based on an extremely high-performance chassis, including a 24-Gbps backplane and an amazing 18,000,000 PPS forwarding rate. This switch line is designed for environments that require no-compromise performance and high port density at the access layer, and gigabit uplinks to the distribution layer—and it does all of this without skimping on the feature set, which supports the entire feature set of the 2926 line (including the set-based IOS).

The 2980 line includes 80 10/100 Fast Ethernet ports and 2 GBIC slots for Gigabit Ethernet uplinks. For the GBIC slots in the 2980 line, customers can choose between these interfaces: 1000Base-SX (for distances up to 550 meters), 1000Base-LX/LH (for distances up to 10 kilometers), or 1000Base-ZX (for distances up to 100 kilometers).

The 2980 line includes two switches: the 2980G and the 2980G-A. The only additional feature included with the G-A model is its ability to support a redundant power supply.

For more information on the 2980 switching line, visit http://www.cisco.com/warp/public/cc/pd/si/casi/ca2900/prodlit/2980g_ds.htm.

The 2948 Line

The 2948 line of switches actually consists of two switches that are very different from one another. The 2948G switch is an access-layer switch that includes 48 10/100 ports and two GBIC slots for gigabit uplinks. It uses the same set-based IOS that the 2980 line uses; it has the same backplane and forwarding speed; and it supports the same feature set, with the exception of ISL/DISL frame tagging (it supports only 802.1q).

The 2948G-L3, on the other hand, supports all of the features of 2948G, but it uses a standard IOS and includes ISL support and layer 3 switching for IP, IPX, and IP multicast. This means that the 2948G-L3 can be used as both a high-speed layer 2 switch and a high-speed router. The 2948G-L3 supports all common routing protocols (including EIGRP, IGRP, BGP, OSPF, and RIP, all, with the exception of BGP, discussed in Part IV of this book); enhanced routing features, including Hot Standby Router Protocol (HSRP); and full IP multicast routing support.

HSRP is similar to clustering for routers: it allows two routers to share the same "virtual" IP and MAC address; and if the primary router fails, the secondary router will take its place.

In addition, the 2948G-L3 supports extensive QoS queuing features and access lists for basic security and firewall duties. All of this leads to a layer 2/3 switching system with incredible power and speed that can easily serve as both an access-layer and distribution-layer device in moderate-sized offices. Figure 10-4 shows a 2948 switch.

Figure 10-4. *A 2948 series switch*

For more information on the 2948 line of switches, visit the following addresses:

- http://www.cisco.com/warp/public/cc/pd/si/casi/ca2900/prodlit/2948g_ds.htm
- http://www.cisco.com/warp/public/cc/pd/si/casi/ca2900/prodlit/29gl3_ds.htm

The 2900XL Line

The 2900 XL line is ideally suited for access-layer duties in a 10/100 switched network. The line comes in two basic layouts: a 12-port version (the 2912XL and the 2912MF-XL) and the 24-port version (the 2924XL, 2924C-XL, and 2924M-XL). Both versions offer a moderate-speed backplane (3 Gbps) and PPS rate (3,000,000), and they support the same feature set: VLAN; congestion control; port security; DHCP, DNS, ISL, and 802.1q frame tagging; and CGMP, RMON, and TACACS+ support. In addition, all 2900XL switches support *stacking*, a technique used to centrally manage and configure a group of switches using a single IP address.

The 2912MF-XL and 2924M-XL also include two module slots for expansion; and they support up to four 100-Mbps Ethernet ports, one 1-Gbps Ethernet port, or one ATM 155 port per module, allowing these switches to perform distribution-layer switch-aggregation duties in a small environment. Figure 10-5 shows a 2900XL switch.

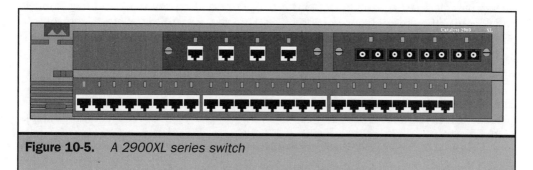

Figure 10-5. *A 2900XL series switch*

For more information on Cisco's 2900XL line of switches, visit the following address: http://www.cisco.com/warp/public/cc/pd/si/casi/ca2900xl/prodlit/index.shtml.

The 3000, 3100, and 3200 Series

All of the switches in the Catalyst 3000 switching line have been officially EOL'd as of the end of 1999, but you may still occasionally see the products in use. The 3000 series switches (including the 3016, 3100, and 3200) all had similar features and architectures. They were modular access-layer switches with stacking capability and a relatively low-speed bus. They ran a custom GUI-based IOS, and they supported 64 VLANs, ISL frame tagging, SPAN, and RMON.

The 3016 and 3100 included two module slots plus a single stack slot, into which a special stack adapter could be inserted to combine up to eight 3000 series switches into one logical switch. The two module slots could be fitted with 10-Mbps Ethernet, 100-Mbps Ethernet, or ATM 155 modules. In the 3100 and 3200, one of these modules could also be used for a 3011 WAN router module, which was essentially a router on a card and brought routing capabilities to the switch. The 3016 had 16 fixed 10Base-T ports, whereas the 3100 had 24. The 3200 had no fixed ports, but included 7 module slots.

For more information about the 3000 series switches, check out http://www.cisco.com/warp/public/cc/pd/si/casi/ca3000/prodlit/index.shtml.

The 3500XL Series

The 3500 series is a high-performance suite of switches well suited to either demanding access-layer duties or moderate distribution-layer duties. The four switches in the line—the 3508, 3512, 3524, and 3548—all share the same high-performance 10-Gbps backplane and a standard IOS CLI. They also share a common feature set, which includes multiple queues per interface (to provide for QoS needs); per-port QoS priorities; Fast Etherchannel and Gigabit Etherchannel; per-port broadcast, unicast, and multicast storm control; clustering support; CGMP, ISL, and 802.1q frame tagging; VTP support; CDP, SNMP, and RMON support; DNS support; TACACS+ support; port security; and optional redundant power supplies.

The 3508 switch is probably the most likely candidate for distribution-layer duties. The 3508 comes with eight GBIC slots, each of which can support 1000Base-FX; 100Base-LX/LH; and 1000Base-ZX, Cisco Gigastack, and 1000Base-T gigabit interfaces. (Although I have yet to see any 1000Base-T GBICs, they are listed on Cisco's web site and should be out by the time you read this.) This makes the 3508 nearly perfect for gigabit link consolidation from access-layer switches with gigabit uplinks. The 3508 can also serve a role in the access layer, however, by providing gigabit links to servers in a high-performance environment. An illustration of a 3508 is shown in Figure 10-6.

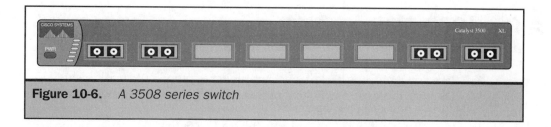

Figure 10-6. *A 3508 series switch*

The 3512, 3524, and 3548 all are more suited for midrange to high-end, access-layer duties than distribution-layer duties. All of these models provide a fixed number of Fast Ethernet ports (12, 24, or 48) for client connections, and two GBIC interfaces for server connections or uplinks to the distribution layer. They support the same GBICs as the 3508, making purchasing GBICs less of a headache.

For more information on the 3500XL series, visit http://www.cisco.com/warp/public/cc/pd/si/casi/ca3500xl/prodlit/index.shtml.

The 3900 Series

The 3900 series is Cisco's primary line of stand-alone Token Ring switches. Because Token Ring is a topic beyond the scope of this book, we will not spend any significant time detailing this switch line. Important features of the 3900 series include modularity (in the 3900, but not in the 3920), which allows it to include ATM or Ethernet interfaces and act as a Token Ring to Ethernet or Token Ring to ATM bridge; adaptive cut-through switching; support for Source Route Bridged (SRB), Source Route Switched (SRS), or Source Route Transparently Bridged (SRT) frames; MAC address and protocol filtering; support for 4, 16, and 32-Mbps operation; priority queuing; stacking; RMON; VLANs; and ISL and SNMP. Figure 10-7 shows a 3900 series switch.

You can find more information about the 3900 series of switches at http://www. cisco.com/warp/public/cc/pd/si/casi/ca3900/prodlit/index.shtml.

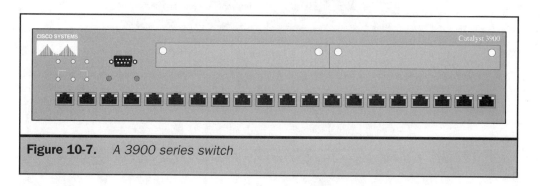

Figure 10-7. *A 3900 series switch*

The 4000 Series

The 4000 series of switches perform admirably in either an access-layer capacity or a distribution-layer capacity due to a wide variety of features, port options, and raw speed. Based on a 24-Gbps backplane, these switches pump out an astonishing 18,000,000 PPS. With the layer 3 switching option card, they are able to switch (route) at layer 3 at an amazing 6,000,000 PPS.

Powered by a set-based IOS CLI and a web-based GUI, the 4000 series supports all of the major features of the 3500 series, plus the following additional features in the 4003 and 4006 modular switches: modular architecture to support future upgrades, including an expanded 64-Gbps backplane; voice features; and upper-layer (4–7) switching and high-density line cards with up to 48 Fast Ethernet or 18 Gigabit Ethernet ports per card. An example of a 4000 series switch is shown in Figure 10-8.

The 4900 switches, which technically belong to the 4000 series, are a bit different than the 4003 and 4006, and deserve their own special section. The 4912G is similar to the 4003 and 4006, but it is a fixed-configuration unit designed solely for Gigabit Ethernet layer 2 switching duties. It is used for the same purposes as the 3508 switch (Gigabit Ethernet aggregation and high-speed server connectivity). The major difference is that it has a 50 percent higher port capacity; a much larger 24-Gbps backplane; and, of course, the correspondingly higher PPS rate.

The 4008G-L3, on the other hand, is a much different animal. It is a fixed-architecture switch with eight GBIC slots, but it has an extremely high-speed layer 3 switching architecture already built in. This architecture allows it to perform layer 3 switching of IP, IPX, and IP multicast at an incredible 11,900,000 PPS, while performing standard layer 2 switching at the same 18,000,000 PPS rate as the other 4000 series switches. On top of this, the 4008G-L3 has additional QoS features to make it a standout winner for

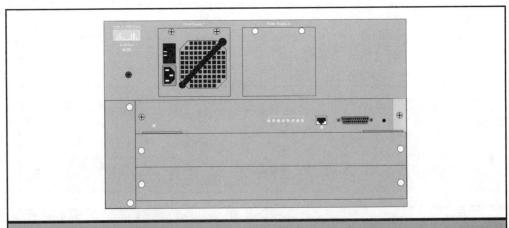

Figure 10-8.   *A 4000 series switch*

midsized distribution-layer duties. In an access-layer capacity, this switch could provide high-speed switched connectivity to local servers while providing routed connectivity to distribution-layer devices over one or more gigabit connections.

For more information about the 4000 series switches, refer to http://www.cisco.com/warp/public/cc/pd/si/casi/ca4000/prodlit/index.shtml.

The 4840 Series

The 4840 switch is technically a member of the 4000 series of switches, but its positioning and use is so drastically different from the other 4000 series switches that it warrants its own section.

The 4840 switch includes most of the features of the 4000 series, including the same feature set and a 24-Gbps backplane. However, it uses the standard IOS rather than the set-based IOS, and it supports layer 3 switching of IP (at 9,000,000 PPS), TACACS+, and a few new features—including Server Load Balancing (SLB, discussed in more detail in Chapter 21). SLB is a technique used by Catalyst 6000 series switches and 7200 series routers to perform load balancing among a group, or cluster, of servers.

The algorithms used for SLB in the 4840 are very flexible and allow you to modify the operation of the load balancing to favor faster servers or reduce the load on slower servers. The SLB algorithm can also take into account which servers are the most and least busy, and allocate the connections appropriately. Although this feature may not be needed in all but the largest private networks, this product has one very important niche market: ISPs. With high-volume web sites, multiple servers may be required in order to respond to HTTP requests within a reasonable time frame. SLB helps enable this with extremely high-speed, dynamic load balancing.

For more information on the 4840 switch, please visit http://www.cisco.com/warp/public/cc/pd/si/casi/ca4800/prodlit/4840_ds.htm.

The 5000 Series

The 5000 series is a switching line designed to be highly modular and flexible, allowing it to be used equally well in both access-layer and distribution-layer scenarios. The 5000 series includes both a 5-slot (5000) and a 2-slot (5002) version. Both versions use the same line cards, and they also share line cards with the 5500 series. The 5000 series supports Ethernet, ATM, FDDI, and Token Ring interfaces, making it an extremely versatile access-layer switching system.

The 5000 series can also be used for a low- to midlevel, distribution-layer switch through the inclusion of the Route Switch Module (RSM) or Route Switch Feature Card (RSFC), both of which allow the 5000 to perform layer 3 switching. Unfortunately, due to the relatively low speed (1.2-Gbps backplane and 1,000,000 PPS) of the 5000, this is useful only in smaller environments.

As a hardware platform, the 5000 series comes in an extremely versatile rack-mount configuration that can be mounted in either a 19-inch or 23-inch rack, and it includes redundant power supplies for high availability. The line cards for the 5000/5500 series come in a wide variety of port densities, with up to 48 ports per card. Figure 10-9 illustrates a 5000 series switch.

The software features depend primarily on the version of the supervisor module that is used in the switch. All current supervisor modules support the following feature sets: 16,000 MAC addresses, 1024 VLANs, switch status reporting, SNMP, RMON, redundant supervisor support, VLANs, congestion control, port security, DHCP, DNS, ISL, CGMP, SPAN, access lists, SLB, and TACACS+ support. In addition, all current supervisor modules (Supervisor III, Supervisor IIIG, and Supervisor IIG) include the NetFlow Feature Card II (NFFC II), which provides high-speed (millions of PPS) layers 3 and 4 switching, advanced QoS features, and the ability to filter broadcasts based on layer 3 protocol information (for instance, not sending an IPX broadcast to a port with only IP hosts).

All supervisor modules also include a console port, an uplink port (modular in the Supervisor IIG and Supervisor III, and can be fitted with a four-port 100-Mbps module or a two-port GBIC module in those models), and an internal card slot for adding additional features in the future.

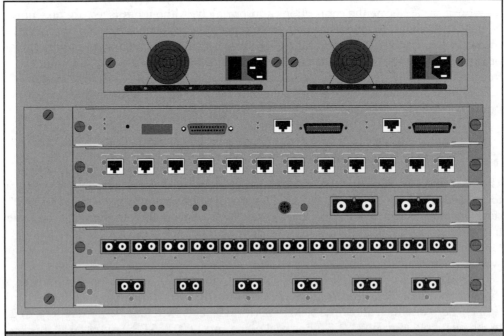

Figure 10-9.   *A 5000 series switch*

As for the differences in the supervisor engines, the Supervisor IIG is the baseline module. Although the IIIG loses the modular uplink (in favor of fixed GBIC uplinks), it gains a slightly faster processor and the ability to triple the backplane speed to 3.6 Gbps in the 5500 models. The Supervisor III is the top-of-the-line model, and it gains a much faster processor, an additional 32MB of DRAM, PC Card slots, and an enhanced switching architecture.

Although not as powerful as its sister line, the 5500 series, the 5000 series of switches are still well suited for environments that need high port density and a comprehensive feature set. For more information on the 5000 series, please visit http://www.cisco.com/warp/public/cc/pd/si/casi/ca5000/prodlit/index.shtml.

The 5500 Series

The 5500 series is a substantial update to the 5000 series. While retaining all of the features that made the 5000 such a success, the 5500 includes a number of improvements that move it firmly into the realm of high-end access and distribution-layer switching.

First and foremost, the 5500 series is built around an updated 3.6-Gbps backplane. This, combined with specialized ASICs on the modules themselves, allows the Catalyst 5500 series to reach an astounding peak forwarding rate of 56,000,000 PPS. Along with this, the number of line-card slots in the chassis (in the 5500 model) has been increased to 13. There are also two smaller models, the 5505 and 5509, with five and nine slots, respectively. Note, however, that on the 5500 model (but not the 5505 or 5509), two of these slots—the first one and the last one—are reserved. The topmost slot (slot 1) may be used only for the supervisor engine, and the bottommost slot (slot 13) may be used only for the LS1010 ASP or the 8510 SRP (discussed further in the next section). An illustration of a 5500 series switch is shown in Figure 10-10.

The 5500 series also breaks new ground by adding support for line cards from the Catalyst 8500 series, Lightstream 1010 ATM switches, and even 7200 series routers, making the versatility of the 5500s truly formidable. When the 5500 model is fitted with the LS1010 ATM Switch Processor (ASP) or the Catalyst 8510 Switch Route Processor (SRP), the 5500 also has an additional 10-Gbps switching fabric specifically for WAN or ATM interfaces.

The LS1010 ASP also gives the 5500 full ATM switching capability. (Note that neither the ASP nor the SRP are available in the 5505 or 5509 models.) Due to this switching capability, the high speed of the switching engine, and the native support of ATM and WAN connections that these additional models can bring, the 5500 series can serve not only as a premier high-density access-layer switch, but it can also double as a high-performance distribution-layer switch in most environments.

The interface modules in the 5500 series can also be taken directly from any 5000 series switch. In addition, the 5500 series uses the same supervisor modules as the 5000 series. Both of these features make it considerably less of a headache to keep spares on the shelves.

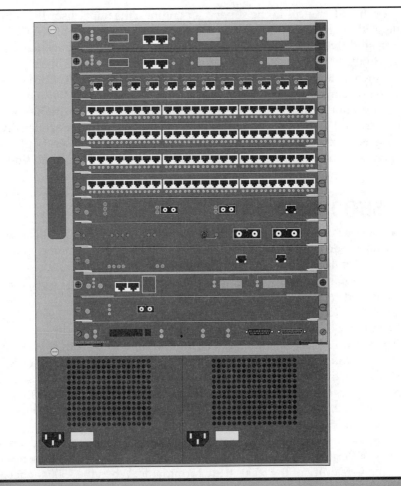

Figure 10-10.   *A 5500 series switch*

All things considered, the 5500 series is a line that any IT manager or closet gearhead would be proud to have in either the NOC or the wiring closet. For more information on the 5500 series, please visit http://www.cisco.com/warp/public/cc/pd/si/casi/ca5000/prodlit/index.shtml.

The 6000 Series

The Catalyst 6000 series of switches is the line of choice for environments that require more functionality than the 5500 series can provide, but that don't necessarily need the raw speed of the 6500 series. The 6000 series can perform as both an ultra-high-end, access-layer switch—providing Gigabit Ethernet or OC-12 ATM connections to high-performance servers, and a midlevel distribution-layer switch that aggregates large quantities of Gigabit Ethernet (130), ATM, or even WAN connections.

The 6000 series comes in two versions: a six-slot version known as the 6006, and a nine-slot version, the 6009. Both versions support the same line cards and supervisor modules, and they can even share their line cards and supervisor modules with their big-brother line, the 6500 series. In fact, using the FlexWAN module, the 6000 series can even support single interface cards from 7200 and 7500 series routers. Both models of the 6000 series switches also support high levels of redundancy, with redundant power supplies, system clocks, supervisor engines, and uplinks all supported. Figure 10-11 shows a 6000 series switch.

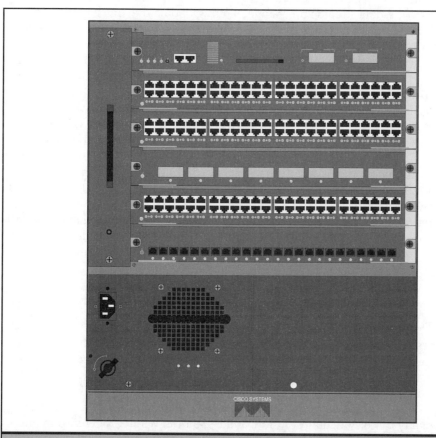

Figure 10-11.   *A 6000 series switch*

As for performance, the 6000 series, although not as powerful as the 6500 series, still offers impressive performance. It uses a 32-Gbps backplane and can switch at 30,000,000 PPS for layer 2 switching, or 15,000,000 PPS for layer 3 switching.

Like with the 5000 series, the features that the 6000 series supports largely depend on the supervisor module installed. The Supervisor engine 1 is designed to provide high-performance wiring closet (access-layer) duties, and includes VLAN support; ISL and 802.1q frame tagging; Fast Etherchannel and Gigabit Etherchannel; protocol filtering; broadcast suppression; load balancing; full multicasting support; TACACS+ and RADIUS authentication; port security; access lists; RMON and SNMP support; DNS and DHCP support; and CDP and SPAN support. All supervisor modules have two GBIC ports for gigabit uplinks.

The Supervisor engine 1A is designed for distribution-layer duties, and it includes all of the features of the Supervisor 1, while also including the Policy Feature Card (PFC) and Multilayer Switch Feature Card (MSFC). The PFC allows the switch to apply traffic policies to the data stream and change queuing of data dynamically based on either layer 2 or layer 3 information in the data stream. This allows you to apply some fairly complex QoS rules to prioritize traffic on an individual basis.

The MSFC, true to its name, allows you to use multilayer switching in your 6000 series switch. The MSFC supports IP, IPX, and IP multicast, and it can also support AppleTalk, DECnet, and Banyan Vines. The MSFC also supports all common routing protocols, and it adds additional layer 3 and 4 QoS features beyond what the PFC provides.

Finally, the Supervisor 2 includes all of the features of the Supervisor 1A, plus the addition of the Cisco Express Forwarding (CEF) switching architecture, the inclusion of the PFC2, and support for the Switch Fabric Module (SFM) in the 6500 series. (SFM is an additional module for increasing the backplane speed.) With these enhancements, the Supervisor 2 provides even higher layer 3 switching support at 30,000,000 PPS and improved QoS capabilities.

Overall, the 6000 series is a high-speed, feature-rich series of switches that are perfect for high-performance access-layer or distribution-layer duties. For more information about the Catalyst 6000 series of switches, please visit http://www.cisco.com/warp/public/cc/pd/si/casi/ca6000/prodlit/index.shtml.

The 6500 Series

The 6500 series is the big brother of the 6000 series and supports much higher switching speeds. It is designed for high-performance, core- and distribution-layer duties.

The 6500 series supports Cisco's CEF switching architecture and, with the SFM, supports a whopping 256-Gbps backplane—allowing the switch to reach incredible forwarding rates of 170,000,000 PPS. This makes the 6500 series one of the highest performers of any Cisco switching solution, and a standout winner for environments that can make use of this raw power. An illustration of a 6500 series switch is shown in Figure 10-12.

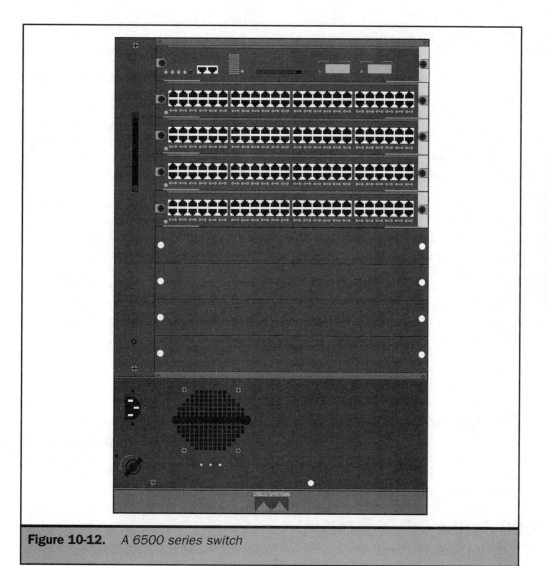

Figure 10-12.   *A 6500 series switch*

The 6500 series uses the same supervisor modules, line cards, and port densities as the 6000 series, and consequently it has the same feature set. The primary difference between the 6000 series and the 6500 series is raw speed. For more information about the Catalyst 6500 series of switches, please visit http://www.cisco.com/warp/public/cc/pd/si/casi/ca6000/prodlit/index.shtml.

The 8500 Series

8500 series switches come in two separate versions: the Multiservice Switch Router (MSR) and the Campus Switch Router (CSR). Both lines are designed for high-performance core-layer duties, and they provide in excess of 6,000,000 PPS for the 8510 and 24,000,000 PPS for the 8540.

The 8500 CSR line is designed to be the core switch in CANs and provides extremely high-speed CAN, MAN, and WAN uplinks. By supporting ATM (up to OC-48c), Gigabit Ethernet, and Fast Ethernet in the same chassis, it can provide a concentration point for distribution-layer devices, a connection point to other CANs in the enterprise, and a direct link to servers within the core. It supports all of the features of the 6500 series, as well as advanced layer 3 switching functionality and QoS features.

The MSR line is designed for environments that require native ATM switching capability combined with layers 2 and 3 Ethernet and WAN switching capability. The primary benefit of this is in the versatility of this switch. Rather than supporting separate core ATM and Ethernet switching platforms, the 8500 MSR can combine them both and reduce costs. The MRS line supports the same features the CSR line supports. The primary difference in the two lines is that the MSR includes advanced ATM switching functionality.

For more information on the 8500 line, check out http://www.cisco.com/warp/ public/cc/pd/si/casi/ca8500/prodlit/index.shtml.

Specialized Switching Solutions

In addition to the switching lines we have covered, some additional switching lines are used for various specialized duties, such as providing DSL connectivity. Because these are very specialized lines, and therefore less likely to be used in most environments, we will only briefly cover the features of each line.

6000 Series DSL Switches

The 6000 series of Digital Subscriber Line (DSL) switches (6015, 6130, 6160, and 6260) are designed to provide ISPs and telcos with a scalable switching architecture for central offices (COs) that provide DSL services to clients. They support several different types of DSL (including synchronous and asynchronous) and ATM in the same platform. This allows the switching line to provide access links for clients, as well as use a high-speed ATM uplink to the provider's cloud. More information on the 6000 DSL series can be found at http://www.cisco.com/warp/public/cc/pd/si/6000/prodlit/index.shtml.

The 6400 Series Concentrator

The 6400 line is alternatively called a "broadband aggregator" or an "access concentrator" by Cisco. The basic purpose of the 6400 series is to take individual sessions from access-layer telco switches (like the 6000 DSL series) and route, apply QoS, and consolidate the individual ATM links from the access layer into larger ATM links. As such, it includes features for both ATM switching and fully functional routing. For more information on the 6400 series, check out http://www.cisco.com/warp/public/cc/pd/as/6400/prodlit/index.shtml.

MGX 8200 Gateways

The MGX 8200 series is a line of concentration switches for service providers. The line includes support for ATM, Frame Relay, and ISDN in one chassis, as well as the ability to bridge between the differing network types. Its primary market is service providers who need to provide a wide variety of access links to clients. More information on the MGX 8200 series can be found at http://www.cisco.com/warp/public/cc/pd/si/mg8200/index.shtml.

The IGX 8400 Series

The IGX 8400 series is a multiservice concentrator for enterprise MAN and WAN applications. It supports Frame Relay and ATM links, and it includes additional functionality for voice, data, and video traffic to provide the most efficient switching possible. It can also be used as a WAN consolidation switch, concentrating separate and varied WAN links from remote offices into one connection at the central office. You can find more information on the IGX 8400 series at http://www.cisco.com/warp/public/cc/pd/si/ig8400/prodlit/index.shtml.

The BPX 8600 Series

The BPX 8600 series is a multiservice switch with X.25, Frame Relay, and ATM switching capabilities. It provides advanced QoS and traffic management features for service providers or enterprise cores that require multiservice switching in a single platform. Its primary use is in service providers who need both ATM and Frame Relay switching capabilities. More information about the BPX 8600 series can be found at http://www.cisco.com/warp/public/cc/pd/si/bp8600/prodlit/index.shtml.

The MGX 8850 Series

The MGX 8850 is a highly versatile WAN switch that can be used for either link aggregation or as an access switch. The features the MGX 8850 supports are diverse, and include VPN connectivity, ATM, Frame Relay, and voice. The MGX 8850 can also be used in a very large enterprise environment for dedicated link aggregation or ATM core switching duties. More information on the MGX 8850 can be found at http://www.cisco.com/warp/public/cc/pd/si/mg8800/prodlit/index.shtml.

The CSS 11000 Series

The Cisco CSS 11000 series of switches are upper-layer (layers 5–7) switches for web-based traffic. They are used to provide high-speed switching and load balancing (up to 11 billion hits per day) to web-based data using upper-layer (URL, cookies, and so on) information. They support additional features for ISPs or high-end web server farms as well, such as denial of service (DOS) prevention and proxy capabilities. There are three models in the series: CSS 11050, CSS 11150, and CSS 11800. The CSS 11050 model is the lower-end solution, and the CSS 11800 is at the upper end. For more information on the CSS 11000 series, go to http://www.cisco.com/warp/public/cc/pd/si/11000/prodlit/index.shtml.

The LightStream LS 1010

Although, technically, the LightStream is a line of its own, the only switch in the line as of this writing is the LS 1010. The LightStream is Cisco's line of premier dedicated ATM switches. Support for speeds up to OC-12 is available in the LS 1010. In addition, full ATM switching—including support for SVCs, as well as full ATM QoS functionality—is included in the feature set of the LS 1010. In addition, the switch processor and port modules from the LS 1010 can be used in the 5500 series of switches, enabling these switches to perform as fully featured ATM switches as well. More information on the LS 1010 can be found at http://www.cisco.com/warp/public/cc/pd/si/lsatsi/ls1010/prodlit/index.shtml.

Summary

This chapter examined each major switching line in detail, providing you with a good understanding of which switch works best in a given situation. This information will help you enormously as we begin to look into configuring and supporting these switches.

For additional information on Cisco's switching line, check out http://www.cisco.com/warp/public/44/jump/switches.shtml.

Chapter 11

Basic Router Terminology

This chapter explores Cisco's line of routers. Like Cisco switching hardware, there is a wide variety of equipment in this group, from simple DSL and cable router/firewalls to high-end enterprise routers with several advanced features. Once again, the goal in this and the following chapter is not to make you an expert on Cisco routers, but rather to teach you about what differentiates each line so that you can make a well-advised decision when you choose routers for your environment.

Cisco Router Types

Unlike switches, Cisco routers are not typically defined by their role in the Cisco internetworking model. This is because routing almost always takes place at the distribution layer, making routers almost universally a distribution-layer device (although on rare occasions, you may see a router placed in the access layer as well). However, Cisco does classify their routers by their intended environment, divided into four categories: Small Office/Home Office (SOHO), branch office, midsize office, and enterprise.

The SOHO designation is for routers that are designed to be used in very small environments, typically with fewer than 50 users. Good examples of this environment are an extremely small office or a home that needs to provide Internet access to multiple PCs from a single DSL line. These routers typically include features such as NAT, basic firewall (packet filtering) functionality, and a DHCP server feature. They may also include a simple GUI or web-based interface and multiple Ethernet ports (allowing them to be used without an additional hub). In general, however, they are fairly limited in performance and don't scale well past 50 users or so as a result. They also do not typically have the ability to accept higher-speed (T1 or higher) Frame Relay interfaces.

The branch office designation is used for routers that are typically used in a smaller branch office, with somewhere around 200 hosts. These routers are mainly used as a gateway into the rest of the organization. As such, their primary function is simply to perform basic routing and to provide a LAN-to-WAN media translation. These routers are typically not overloaded with features, but they usually contain a standard IOS interface and come with a wide variety of interface options, from low-speed asynchronous serial to ATM. Again, although some of these routers support higher-end features, you must be careful because the processor and memory subsystems installed in most of these routers is a bit limited. If too many advanced services are running, poor performance can result.

The midsize office routers bridge the gap between the branch office routers and the enterprise routers. They are typically modular, have a wide variety of possible interfaces, and support most commonly used features. Their typical role is in large branch offices (500 or more users) and smaller central offices that require higher performance and extended features, but which do not require the performance and feature sets that an enterprise network or service provider requires.

The enterprise routers are designed for huge private networks and service providers (typically ISPs). They support extremely advanced features run on application-specific integrated circuits (ASICs) rather than in software. Therefore, they are more correctly termed *switches* than routers, but the function they perform is the same.

Router Modularity

Most of Cisco's router products are modular. The numbering conventions used on routers are similar to those used in switches, with the type port (for example, Ethernet 0) and type module/port (for example, Ethernet 0/0) designations being the most prevalent convention. In routers, numbering generally begins at zero, so the first interface of any given type is the zero interface.

Physical interfaces on a modular router follow the same conventions discussed in Chapter 8. The smaller cards are interface cards (like the cards used in the 2600 series). These are sometimes used in larger cards, called *modules,* like in the 3600 series. Finally, the higher-end routers (like the 7200 series) have large *blades,* variously called port adapters, line cards, or modules.

Router Hardware Terms

In a router, much of the hardware is typically less important than in a switch. Backplane speed in a router is much less important because, in most routers, the backplane is not the bottleneck. Routing is typically done in software (which is one of the key differentiators between a router and a layer 3 switch), so the processor and memory subsystems are usually the limiting performance factor. For this reason, the main hardware concerns for routers are similar to those for a PC server. We are concerned primarily with four aspects: how fast the processor is; how much DRAM, flash RAM, and NVRAM is supported; how expandable and versatile the processor is (for a modular router); and how much hardware redundancy (power supplies, supervisor engines, and so on) the processor supports.

Router IOS

Because routers perform nearly all processing in software instead of hardware, the IOS version, not the router line, is the primary concern when choosing a feature set. Cisco maintains many different versions of IOS with different feature sets for different needs. The advantage of this diversity in a router is that to get a new feature, you don't typically need to buy new hardware. Rather, you can simply upgrade the IOS and use the new feature. The disadvantage is that because the processing is done in software, the PPS rate is considerably slower than in a switch (typically thousands of PPS versus millions of PPS).

The following list details the basic lines of IOS available on nearly all types of routers. (The features included in these descriptions are based on the 2600 series.) Note that some routers may have even more versions of IOS to choose from, or slightly different features for each line of IOS, depending on the series of router in question. If you are not sure which line of IOS you need, consult the product datasheets for your router series.

> **Tip**   *For a great reference to the feature sets supported in the different lines of Cisco IOS version 12.1, check out http://www.cisco.com/warp/public/cc/pd/iosw/iore/iomjre121/ prodlit/1070_pp.htm.*

- **IP-only IOS**   This IOS supports only the TCP/IP protocol, with no provisions for IPX/SPX or other protocols. In many cases, this IOS is considered the standard IOS, and it includes the features required in most IP-only situations, including support for the following: IP routing protocols; basic multicast; basic Frame Relay and ISDN; and access lists. The IP-only IOS does not include advanced functionality, however, like advanced firewall features or Voice over Frame Relay (VoFR).

- **IP Plus IPSec IOS**   This line of IOS includes most of the features (although not necessarily all) from the IP-only line, but it adds additional QOS and VPN functionality like support for Layer 2 Tunneling Protocol (L2TP).

- **IP FW/IDS IOS**   This line of IOS is a reduction of the IP-only IOS feature set, but it adds advanced firewall features.

- **IP FW/IDS Plus IPSec IOS**   This line of IOS supports the advanced firewall feature set and adds additional VPN functionality.

- **IP/IPX/AT/DEC IOS**   This line's feature set is nearly identical to the IP-only IOS, but it adds support for the IPX/SPX, AppleTalk, and DECnet protocol suites.

- **IP/IPX/AT/DEC Plus IOS**   This line's feature set is nearly identical to the IP/IPX/AT/DEC IOS, but it adds support for advanced features like L2TP, advanced QOS, and inverse multiplexing over ATM.

- **IP/IPX/AT/DEC/FW/IDS Plus IOS**   This line's feature set is similar to the IP/IPX/AT/DEC Plus IOS, but it adds support for advanced firewall functionality.

- **Enterprise Plus IOS**   This line of IOS supports nearly all available features except advanced firewall functionality, Systems Network Architecture (SNA) switching, and voice features.

- **Enterprise Plus IPSec IOS**   This line's feature set is nearly identical to the Enterprise Plus IOS, with the addition of IPSec support.

- **Enterprise FW/IDS Plus IPSec IOS**   This line of IOS is similar to the IP/IPX/AT/DEC/FW/IDS Plus IOS, but with additional features from the Enterprise IOS.

- **Enterprise SNASW Plus**   This line of IOS includes features enabling SNA switching.

Router Classification Chart

This section presents 12 quick-reference charts (Tables 11-1 through 11-12) detailing each model of Cisco router, complete with port types and densities, interface numbering style, and features. But first, you need to learn a few definitions and abbreviations used in the chart.

Series   This denotes the line of router. "Series" is the broader term you will hear thrown around a lot. Most routers in a given series are based on the same hardware and software, and they have similar features.

Router Model   This denotes the individual models of routers in the series. All router models available as of this writing are included (even EOL products, in some cases).

Classification   This denotes the router's classification, either SOHO, branch, midsize, or enterprise. Note that some midsize and nearly all enterprise routers are actually feature-rich switching platforms.

Configuration   This denotes whether the router is fixed configuration or modular configuration. Fixed configuration routers are typically not expandable, though exceptions do exist.

Interface Numbering   This denotes the style of interface numbering on this router, from the following major types:

- **Type port**   First, you specify the type of interface (such as Ethernet), and then the port.

- **Type module/port (or type slot/port)**   First, you specify the type of interface (such as Ethernet), then the module or slot followed by a slash (/), and then the port.

- **Type card/subcard/port**   First, you identify the type of interface, then the card, then the subcard, and then the port.

Port Types (Ethernet, Fast Ethernet, and so on)   This lists the *maximum* ports available on this model of router. Note that on modular routers with multiple port types available, you cannot have all available port types at the listed capacity.

In the tables, the modular routers have an asterisk (*) beside all optional features, including ports.

Routing Protocols   This denotes the major routing protocols supported by the router. These protocols will be covered in depth later in Chapters 22 through 26. Note that the *other protocols* designation for some of the routers means that the router supports other routing protocols (such as IS-IS) that are not covered in this reference.

CPU Model   This lists the model of CPU used as the primary CPU in this router. Note that on higher-end routers, there is not only a primary CPU, but there may also be several secondary CPUs, further increasing the router's processing power.

Standard and Maximum DRAM   This lists the standard and maximum route processor DRAM available for this model. Note that in the lower-end routers, this DRAM is shared between processing duties and buffering of incoming frames. In the midrange routers, these duties are divided into two separate memory architectures. (Only the route processor's DRAM is detailed in these tables.) Finally, in the enterprise models, the memory architecture may be separated even further, with each individual line card having its own DRAM.

Standard and Maximum Flash   This details the base and maximum flash RAM usable in this system. Again, this just refers to main system RAM in high-end routers.

PPS   This denotes the packets per second on this model, as reported by Cisco. Note that most routers will not achieve this rate in real-world situations. In fact, this rating is even less reliable than the rating for switches because routers need to perform considerably more processing per packet, leading to higher delay variability. PPS is still a decent guide to a router's raw speed, however.

Capabilities   These detail some of the more common features of the router. Note that all of the features in any given router *are not* listed. More information on each router line will be provided in the next chapter. This chapter simply provides a guide to the most common features of each model. Also note that higher-end routers generally provide a more thorough implementation of a given feature. The following list provides a key to the feature listing:

- **DHCP server**   Indicates that the router has the capability to dynamically assign IP addresses to hosts using DHCP. This is an important feature of the SOHO routers, but it becomes less important in the larger environments (where a dedicated server typically performs this duty).

- **DHCP relay**   Indicates that the router can be configured to forward DHCP broadcasts as a unicast directly to a configured DHCP server, allowing clients on remote subnets that do not have a DHCP server to be dynamically assigned an IP address.

- **NAT/PAT**   Indicates that the router is capable of performing network or port address translation for clients.

- **SNMP**   Indicates that the router supports Simple Network Management Protocol.

- **RMON**   Indicates that the router supports remote monitoring.

- **Advanced queuing**   Indicates that the router supports queuing packets differently, based on lower-layer information such as IP precedence.

- **Access lists**   Indicates that access lists are supported on this router to some degree. (Access lists are covered in detail in Chapter 27.)

- **QOS**   Indicates that the router supports advanced quality of service features, such as traffic shaping.

- **Multiservice**   Indicates that this router has extended capabilities for supporting voice or video traffic.

- **X.25/ATM/Frame Relay switching**   Indicates that this router can perform as an X.25, ATM, or Frame Relay switch.

- **TACACS+**   Indicates that this switch supports robust authentication, authorization, and accounting features, including Terminal Access Controller Access Control System.

- **RADIUS**   Indicates that the router supports Remote Authentication Dial-In User Service, a method of distributing and controlling remote access authentication and accounting used by many ISPs.

■ **AAA**   Indicates that the router supports authentication, authorization, and accounting services. These services allow the router to fully authenticate a connection, allow that connection certain privileges based on the authentication, and then keep a record of activities performed by that session.

■ **Compression**   Indicates that the router is capable of performing hardware or software compression on data transfers, increasing the efficiency of WAN links.

■ **VPN**   Indicates that the router is capable of establishing and maintaining virtual private network connections.

■ **Tag switching**   Indicates that this router can "tag" packets with additional information to allow for more efficient routing and QOS settings to be applied along the path. This feature also indicates that this router can respond to tag-switched packets from other devices.

■ **CEF**   Indicates that this router can use Cisco Express Forwarding, a method of switching packets using ASICs and a compact routing database rather than routing packets using software and information solely from the routing tables. The end result is faster, more efficient routing.

■ **NetFlow**   Indicates that the router can perform NetFlow switching, a method of layer 3 switching whereby the first packet in a data stream (or flow) is routed and the rest of the packets in the flow are switched. This allows for higher-speed routing because a "decision" has to be made only on the first packet. The rest simply follow the leader. NetFlow switching is available only for certain interfaces (mostly LAN), however.

■ **PXF**   Allows the 10000 and 12000 series of routers to reach extremely high channel densities. Parallel eXpress Forwarding is a hardware and software architecture that allows for extremely high aggregation of links, with full processing performed on each individual link. This is done by distributing the processing among many individual processors using symmetric multiprocessing techniques.

■ **IPSec**   Indicates that this router supports IP Security, a popular method of key-based packet-level data encryption and authentication.

■ **Firewall**   Indicates that this model supports Cisco's firewall feature set.

■ **IPX, SNA, AppleTalk, DECnet, OSI, VINES, XNS**   Indicates that the router supports the listed desktop protocol suites.

■ **Dual-router architecture**   Indicates that the router is actually two completely separate routers combined into one chassis. The benefit of this is increased processing power and redundancy. This is available only in the 7576 series.

CISCO TECHNOLOGY
OVERVIEW

Series	Router Model	Classification	Configuration	Interface Numbering
600	673	SOHO	Fixed	Type port
	675	SOHO	Fixed	Type port
	677	SOHO	Fixed	Type port
700	761	SOHO	Fixed	N/A
	775	SOHO	Fixed	N/A
	776	SOHO	Fixed	N/A
800	801	SOHO	Fixed	Type port
	802	SOHO	Fixed	Type port
	803	SOHO	Fixed	Type port
	804	SOHO	Fixed	Type port
	805	SOHO	Fixed	Type port
	826	SOHO	Fixed	Type port
	827	SOHO	Fixed	Type port
	827-4V	SOHO	Fixed	Type port
1000	1003	SOHO	Fixed	Type port
	1004	SOHO	Fixed	Type port
	1005	SOHO	Fixed	Type port
1400	1401	SOHO	Fixed	Type port
	1417	SOHO	Fixed	Type port
1600	1601	SOHO	Modular	Type port
	1602	SOHO	Modular	Type port
	1603	SOHO	Modular	Type port
	1604	SOHO	Modular	Type port
	1605	SOHO	Modular	Type port

Table 11-1.   *SOHO Router Classifications and Configurations*

Series	Router Model	Ethernet Ports	Fast Ethernet (100 Mbps or 10/100) Ports	Gigabit Ethernet Ports	Token Ring Ports	FDDI Ports	ATM Ports	Synchronous Serial Ports	Synchronous Serial Speed	Asynchronous Serial Ports	Other Ports
600	673	0	1	0	0	0	0	0	N/A	0	1 SDSL
	675	0	1	0	0	0	0	0	N/A	0	1 ADSL
	677	0	1	0	0	0	0	0	N/A	0	1 ADSL
700	761	0	1	0	0	0	0	0	N/A	0	1 ISDN BRI
	775	0	4	0	0	0	0	0	N/A	0	1 ISDN BRI, 2 POTS
	776	0	4	0	0	0	0	0	N/A	0	1 ISDN BRI, 2 POTS
800	801	1	0	0	0	0	0	0	N/A	0	1 ISDN BRI
	802	1	0	0	0	0	0	0	N/A	0	1 ISDN BRI/IDSL
	803	4	0	0	0	0	0	0	N/A	0	1 ISDN BRI, 2 POTS
	804	4	0	0	0	0	0	0	N/A	0	1 ISDN BRI/IDSL, 2 POTS
	805	1	0	0	0	0	0	1	512Kbps	0	None
	826	1	0	0	0	0	0	0	N/A	0	1 ISDN BRI/IDSL
	827	1	0	0	0	0	0	0	N/A	0	1 ADSL
	827-4V	1	0	0	0	0	0	0	N/A	0	1 ADSL, 4 POTS
1000	1003	1	0	0	0	0	0	0	N/A	0	1 ISDN BRI
	1004	1	0	0	0	0	0	0	N/A	0	1 ISDN BRI
	1005	1	0	0	0	0	0	1	2Mbps	0	None
1400	1401	1	0	0	0	0	0	0	N/A	0	1 ADSL
	1417	1	0	0	0	0	0	0	N/A	0	1 ADSL DMT Issue 2

Table 11-2. *SOHO Router Port Options*

Series	Router Model	Ethernet Ports	Fast Ethernet (100 Mbps or 10/100) Ports	Gigabit Ethernet Ports	Token Ring Ports	FDDI Ports	ATM Ports	Synchronous Serial Ports	Synchronous Serial Speed	Asynchronous Serial Ports	Other Ports
1600	1601	2 (RJ-45 and AUI)	0	0	0	0	0	2*	2 Mbps	2*	1 T1*, 1 56Kbps*, 1 ISDN*
	1602	2 (RJ-45 and AUI)	0	0	0	0	0	1*	2 Mbps	1*	1 T1*, 2 56Kbps*, 1 ISDN*
	1603	2 (RJ-45 and AUI)	0	0	0	0	0	1*	2 Mbps	1*	1 T1*, 1 56Kbps*, 1 ISDN BRI S/T
	1604	2 (RJ-45 and AUI)	0	0	0	0	0	1*	2 Mbps	1*	1 T1*, 1 56Kbps*, 1 ISDN BRI U
	1605	3 (2 RJ-45 and 1 AUI)	0	0	0	0	0	1*	2 Mbps	1*	1 T1*, 1 56Kbps*, 1 ISDN*
		0	2*	0	0	0	0	4*	2 Mbps	4*	2 T1*, 2 56Kbps*, 2 ISDN*

Table 11-2. *SOHO Router Port Options* (continued)

Series	Router Model	Routing Protocols	CPU Model	Standard DRAM	Maximum DRAM	Standard Flash RAM	Maximum Flash RAM	PPS	Capabilities
600	673	RIP	I960	4MB	4MB	2MB	2MB	<10,000	DHCP server, DHCP relay, RADIUS, NAT/PAT, SNMP
	675	RIP	I960	4MB	4MB	2MB	2MB	<10,000	DHCP server, DHCP relay, RADIUS, NAT/PAT, SNMP
	677	RIP	I960	4MB	4MB	2MB	2MB	<10,000	DHCP server, DHCP relay, RADIUS, NAT/PAT, SNMP
700	761	RIP	i386	1.5MB	1.5MB	1MB	1MB	<10,000	DHCP server, DHCP relay, RADIUS, NAT/PAT, SNMP
	775	RIP	i386	1.5MB	1.5MB	1MB	1MB	<10,000	DHCP server, DHCP relay, RADIUS, NAT/PAT, SNMP
	776	RIP	i386	1.5MB	1.5MB	1MB	1MB	<10,000	DHCP server, DHCP relay, RADIUS, NAT/PAT, SNMP
800	801	RIP, EIGRP*	MPC 850	4MB	12MB	8MB	12MB	<10,000	DHCP server, DHCP relay, NAT/PAT, SNMP, access lists, advanced queueing, TACACS+, compression, VPN*, IPSEC*, firewall*, IPX*
	802	RIP, EIGRP*	MPC 850	4MB	12MB	8MB	12MB	<10,000	DHCP server, DHCP relay, NAT/PAT, SNMP, access lists, advanced queueing, TACACS+, compression, VPN*, IPSEC*, firewall*, IPX*
	803	RIP, EIGRP*	MPC 850	4MB	12MB	8MB	12MB	<10,000	DHCP server, DHCP relay, NAT/PAT, SNMP, access lists, advanced queueing, TACACS+, compression, VPN*, IPSEC*, firewall*, IPX*
	804	RIP, EIGRP*	MPC 850	4MB	12MB	8MB	12MB	<10,000	DHCP server, DHCP relay, NAT/PAT, SNMP, access lists, advanced queueing, TACACS+, compression, VPN*, IPSEC*, firewall*, IPX*
	805	RIP, EIGRP*	MPC 850	4MB	12MB	8MB	12MB	<10,000	DHCP server, DHCP relay, NAT/PAT, SNMP, access lists, advanced queueing, TACACS+, compression, VPN*, IPSEC*, firewall*, IPX*

Table 11-3. SOHO Router Capabilities

Series	Router Model	Routing Protocols	CPU Model	Standard DRAM	Maximum DRAM	Standard Flash RAM	Maximum Flash RAM	PPS	Capabilities
	826	RIP, EIGRP*	MMC 855T	16MB	32MB	8MB	20MB	< 10,000	DHCP server, DHCP relay, NAT/PAT, SNMP, access lists, advanced queueing, QoS, TACACS+, compression, VPN*, IPSEC*, firewall*, IPX*
	827	RIP, EIGRP*	MPC 850	16MB	32MB	8MB	20MB	< 10,000	DHCP server, DHCP relay, NAT/PAT, SNMP, access lists, advanced queueing, QoS, TACACS+, compression, VPN*, IPSEC*, firewall*, IPX*
	827-4V	RIP, EIGRP*	MPC 850	24MB	32MB	8MB	20MB	< 10,000	DHCP server, DHCP relay, NAT/PAT, SNMP, access lists, advanced queueing, QoS, TACACS+, compression, voice/PBX features, VPN*, IPSEC*, firewall*, IPX*
1000	1003	RIP	Motorola 69360	4MB	20MB	0MB	2MB	< 10,000	DHCP server, DHCP relay, NAT/PAT, SNMP, access lists, compression, IPX, AppleTalk
	1004	RIP	Motorola 69360	4MB	20MB	0MB	2MB	< 10,000	DHCP server, DHCP relay, NAT/PAT, SNMP, access lists, compression, IPX, AppleTalk
	1005	RIP	Motorola 69360	4MB	20MB	0MB	2MB	< 10,000	DHCP server, DHCP relay, NAT/PAT, SNMP, access lists, compression, IPX, AppleTalk
1400	1401	RIP, EIGRP*	Motorola QUICC 68360	16MB	24MB	4MB	16MB	< 10,000	DHCP server, DHCP relay, NAT/PAT, SNMP, access lists, advanced queueing, QoS, TACACS+, compression, VPN*, IPSEC*, firewall*, IPX*
	1417	RIP, EIGRP*	Motorola QUICC 68360	16MB	24MB	4MB	16MB	< 10,000	DHCP server, DHCP relay, NAT/PAT, SNMP, access lists, advanced queueing, QoS, TACACS+, compression, VPN*, IPSEC*, firewall*, IPX*

Table 11-3. SOHO Router Capabilities (continued)

Series	Router Model	Routing Protocols	CPU Model	Standard DRAM	Maximum DRAM	Standard Flash RAM	Maximum Flash RAM	PPS	Capabilities
1600	1601	RIP, EIGRP*	Motorola QUICC 68360	8MB	24MB	4MB	16MB	<10,000	DHCP server, DHCP relay, NAT/PAT, SNMP, access lists, advanced queueing, QoS, TACACS+, RADIUS, compression, VPN*, IPSEC*, firewall*, IPX*, SNA*, AppleTalk*
	1602	RIP, EIGRP*	Motorola QUICC 68360	8MB	24MB	4MB	16MB	<10,000	DHCP server, DHCP relay, NAT/PAT, SNMP, access lists, advanced queueing, QoS, TACACS+, RADIUS, compression, VPN*, IPSEC*, firewall*, IPX*, SNA*, AppleTalk*
	1603	RIP, EIGRP*	Motorola QUICC 68360	8MB	24MB	4MB	16MB	<10,000	DHCP server, DHCP relay, NAT/PAT, SNMP, access lists, advanced queueing, QoS, TACACS+, RADIUS, compression, VPN*, IPSEC*, firewall*, IPX*, SNA*, AppleTalk*
	1604	RIP, EIGRP*	Motorola QUICC 68360	8MB	24MB	4MB	16MB	<10,000	DHCP server, DHCP relay, NAT/PAT, SNMP, access lists, advanced queueing, QoS, TACACS+, RADIUS, compression, VPN*, IPSEC*, firewall*, IPX*, SNA*, AppleTalk*
	1605	RIP, EIGRP*	Motorola QUICC 68360	8MB	24MB	4MB	16MB	<10,000	DHCP server, DHCP relay, NAT/PAT, SNMP, access lists, advanced queueing, QoS, TACACS+, RADIUS, compression, VPN*, IPSEC*, firewall*, IPX*, SNA*, AppleTalk*

Table 11-3. *SOHO Router Capabilities (continued)*

Series	Router Model	Classification	Configuration	Interface Numbering
1700	1720	Branch office	Modular	Type port
	1750	Branch office	Modular	Type port
	1750-2V	Branch office	Modular	Type port
	1750-4V	Branch office	Modular	Type port
2500	2501	Branch office	Fixed	Type port
	2502	Branch office	Fixed	Type port
	2503	Branch office	Fixed	Type port
	2504	Branch office	Fixed	Type port
	2505	Branch office	Fixed	Type port
	2507	Branch office	Fixed	Type port
	2509	Branch office	Fixed	Type port
	AS2509-RJ	Branch office	Fixed	Type port
	2510	Branch office	Fixed	Type port
	2511	Branch office	Fixed	Type port
	AS2511-RJ	Branch office	Fixed	Type port
	2512	Branch office	Fixed	Type port
	2513	Branch office	Fixed	Type port

Table 11-4. Branch Router Classifications and Configurations

Series	Router Model	Classification	Configuration	Interface Numbering
	2514	Branch office	Fixed	Type port
	2515	Branch office	Fixed	Type port
	2516	Branch office	Fixed	Type port
	2520	Branch office	Fixed	Type port
	2521	Branch office	Fixed	Type port
	2522	Branch office	Fixed	Type port
	2523	Branch office	Fixed	Type port
	2524	Branch office	Modular	Type port
	2525	Branch office	Modular	Type port
2600	2610	Branch office	Modular	Type module/port
	2611	Branch office	Modular	Type module/port
	2612	Branch office	Modular	Type module/port
	2613	Branch office	Modular	Type module/port
	2620	Branch office	Modular	Type module/port
	2621	Branch office	Modular	Type module/port
	2650	Branch office	Modular	Type module/port
	2651	Branch office	Modular	Type module/port

Table 11-4. *Branch Router Classifications and Configurations (continued)*

Series	Router Model	Ethernet Ports	Fast Ethernet (100Mbps or 10/100) Ports	Gigabit Ethernet Ports	Token Ring Ports	FDDI Ports	ATM Ports	Synchronous Serial Ports	Synchronous Serial Speed	Asynchronous Serial Ports	Other Ports
1700	1720	1*	1	0	0	0	0	4*	2 Mbps	4*	2 T1*, 2 56 Kbps*, 2 ISDN*
	1750	1*	1	0	0	0	0	4*	2 Mbps	4*	2 T1*, 2 56 Kbps*, 2 ISDN*, voice upgradable*
	1750-2V	1*	1	0	0	0	0	4*	2 Mbps	4*	2 T1*, 2 56 Kbps*, 2 ISDN*, 2 voice*
	1750-4V	1*	1	0	0	0	0	4*	2 Mbps	4*	2 T1*, 2 56 Kbps*, 2 ISDN*, 4 voice*

Table 11-5. *Branch Router Port Options*

Series	Router Model	Ethernet Ports	Fast Ethernet (100Mbps or 10/100) Ports	Gigabit Ethernet Ports	Token Ring Ports	FDDI Ports	ATM Ports	Synchronous Serial Ports	Synchronous Serial Speed	Asynchronous Serial Ports	Other Ports
2500	2501	1	0	0	0	0	0	2	2 Mbps	0	None
	2502	0	0	0	1	0	0	2	2 Mbps	0	None
	2503	1	0	0	0	0	0	2	2 Mbps	0	1 ISDN BRI
	2504	0	0	0	1	0	0	2	2 Mbps	0	1 ISDN BRI
	2505	8	0	0	0	0	0	2	2 Mbps	0	None
	2507	16	0	0	0	0	0	2	2 Mbps	0	None
	2509	1	0	0	0	0	0	2	2 Mbps	8	None
	AS2509-RJ	1	0	0	0	0	0	1	2 Mbps	8	None
	2510	0	0	0	1	0	0	2	2 Mbps	8	None
	2511	1	0	0	0	0	0	2	2 Mbps	16	None
	AS2511-RJ	1	0	0	0	0	0	1	2 Mbps	16	None
	2512	0	0	0	1	0	0	2	2 Mbps	16	None
	2513	1	0	0	1	0	0	2	2 Mbps	0	None
	2514	2	0	0	0	0	0	2	2 Mbps	0	None
	2515	0	0	0	2	0	0	2	2 Mbps	0	None
	2516	14	0	0	0	0	0	2	2 Mbps	0	1 ISDN BRI
	2520	1	0	0	0	0	0	4	2 Mbps (2 low-speed)	2	1 ISDN BRI

Table 11-5. Branch Router Port Options (continued)

Series	Router Model	Ethernet Ports	Fast Ethernet (100Mbps or 10/100) Ports	Gigabit Ethernet Ports	Token Ring Ports	FDDI Ports	ATM Ports	Synchronous Serial Ports	Synchronous Serial Speed	Asynchronous Serial Ports	Other Ports
	2521	0	0	0	1	0	0	4	2 Mbps (2 low-speed)	2	1 ISDN BRI
	2522	1	0	0	0	0	0	10	2 Mbps (8 low-speed)	8	1 ISDN BRI
	2523	0	0	0	1	0	0	10	2 Mbps (8 low-speed)	8	1 ISDN BRI
	2524	1	0	0	0	0	0	2*	2 Mbps	0	2 T1*, 2 56 Kbps*, 1 ISDN*
	2525	0	0	0	1	0	0	2*	2 Mbps	0	2 T1*, 2 56 Kbps*, 1 ISDN*

Table 11-5.   *Branch Router Port Options* (continued)

Series	Router Model	Ethernet Ports	Fast Ethernet (100Mbps or 10/100) Ports	Gigabit Ethernet Ports	Token Ring Ports	FDDI Ports	ATM Ports	Synchronous Serial Ports	Synchronous Serial Speed	Asynchronous Serial Ports	Other Ports
2600	2610	5*	0	0	0	0	1*	16*	8 Mbps*	32*	2 T1*, 2 56 Kbps*, 4 ISDN*, 16 modem*, 2 ADSL*, 2 voice*
	2611	6*	0	0	0	0	1*	16*	8 Mbps*	32*	2 T1*, 2 56 Kbps*, 4 ISDN*, 16 modem*, 2 ADSL*, 2 voice*
	2612	5*	0	0	1	0	1*	16*	8 Mbps*	32*	2 T1*, 2 56 Kbps*, 4 ISDN*, 16 modem*, 2 ADSL*, 2 voice*
	2613	4*	0	0	1	0	1*	16*	8 Mbps*	32*	2 T1*, 2 56 Kbps*, 4 ISDN*, 16 modem*, 2 ADSL*, 2 voice*
	2620	4*	1	0	0	0	1*	16*	8 Mbps*	32*	2 T1*, 2 56 Kbps*, 4 ISDN*, 16 modem*, 2 ADSL*, 2 voice*
	2621	4*	2	0	0	0	1*	16*	8 Mbps*	32*	2 T1*, 2 56 Kbps*, 4 ISDN*, 16 modem*, 2 ADSL*, 2 voice*
	2650	4*	1	0	0	0	1*	16*	8 Mbps*	32*	2 T1*, 2 56 Kbps*, 4 ISDN*, 16 modem*, 2 ADSL*, 2 voice*
	2651	4*	2	0	0	0	1*	16*	8 Mbps*	32*	2 T1*, 2 56 Kbps*, 4 ISDN*, 16 modem*, 2 ADSL*, 2 voice*

Table 11-5. *Branch Router Port Options* (continued)

Series	Router Model	Routing Protocols	CPU Model	Standard DRAM	Maximum DRAM	Standard Flash RAM	Maximum Flash RAM	PPS	Capabilities
1700	1720	RIP, IGRP, EIGRP, OSPF, BGP*	Motorola QUICC MPC860T	32MB	48MB	8MB	16MB	< 10,000	DHCP server, DHCP relay, NAT/PAT, SNMP, access lists, advanced queueing, QoS, TACACS+, RADIUS, compression, VPN*, IPSEC*, firewall*, IPX*, SNA*, AppleTalk*
	1750	RIP, IGRP, EIGRP, OSPF, BGP*	Motorola QUICC MPC860T	16MB	48MB	8MB	16MB	< 10,000	DHCP server, DHCP relay, NAT/PAT, SNMP, access lists, advanced queueing, QoS, TACACS+, RADIUS, compression, VPN*, IPSEC*, firewall*, IPX*, SNA*, AppleTalk*, multiservice*
	1750-2V	RIP, IGRP, EIGRP, OSPF, BGP*	Motorola QUICC MPC860T	32MB	48MB	8MB	16MB	< 10,000	DHCP server, DHCP relay, NAT/PAT, SNMP, access lists, advanced queueing, QoS, TACACS+, RADIUS, compression, VPN*, IPSEC*, firewall*, IPX*, SNA*, AppleTalk*, multiservice
	1750-4V	RIP, IGRP, EIGRP, OSPF, BGP*	Motorola QUICC MPC860T	32MB	48MB	8MB	16MB	< 10,000	DHCP server, DHCP relay, NAT/PAT, SNMP, access lists, advanced queueing, QoS, TACACS+, RADIUS, compression, VPN*, IPSEC*, firewall*, IPX*, SNA*, AppleTalk*, multiservice

Table 11-6.   Branch Router Capabilities

Series	Router Model	Routing Protocols	CPU Model	Standard DRAM	Maximum DRAM	Standard Flash RAM	Maximum Flash RAM	PPS	Capabilities
2500	2501	RIP, IGRP, EIGRP, OSPF, BGP, others*	68030	2MB*	18MB	8MB	16MB	< 10,000	DHCP server, DHCP relay, NAT/PAT, SNMP, access lists, advanced queueing, QoS, TACACS+, RADIUS, compression, VPN*, IPSEC*, firewall*, IPX*, SNA*, AppleTalk*, DECNet*, OSI*, VINES*, XNS*
	2502	RIP, IGRP, EIGRP, OSPF, BGP, others*	68030	2MB*	18MB	8MB	16MB	< 10,000	DHCP server, DHCP relay, NAT/PAT, SNMP, access lists, advanced queueing, QoS, TACACS+, RADIUS, compression, VPN*, IPSEC*, firewall*, IPX*, SNA*, AppleTalk*, DECNet*, OSI*, VINES*, XNS*
	2503	RIP, IGRP, EIGRP, OSPF, BGP, others*	68030	2MB*	18MB	8MB	16MB	< 10,000	DHCP server, DHCP relay, NAT/PAT, SNMP, access lists, advanced queueing, QoS, TACACS+, RADIUS, compression, VPN*, IPSEC*, firewall*, IPX*, SNA*, AppleTalk*, DECNet*, OSI*, VINES*, XNS*
	2504	RIP, IGRP, EIGRP, OSPF, BGP, others*	68030	2MB*	18MB	8MB	16MB	< 10,000	DHCP server, DHCP relay, NAT/PAT, SNMP, access lists, advanced queueing, QoS, TACACS+, RADIUS, compression, VPN*, IPSEC*, firewall*, IPX*, SNA*, AppleTalk*, DECNet*, OSI*, VINES*, XNS*

Table 11-6. *Branch Router Capabilities (continued)*

Series	Router Model	Routing Protocols	CPU Model	Standard DRAM	Maximum DRAM	Standard Flash RAM	Maximum Flash RAM	PPS	Capabilities
	2505	RIP, IGRP, EIGRP, OSPF, BGP, others*	68030	2MB*	18MB	8MB	16MB	< 10,000	DHCP server, DHCP relay, NAT/PAT, SNMP, access lists, advanced queueing, QoS, TACACS+, RADIUS, compression, VPN*, IPSEC*, firewall* IPX*, SNA*, AppleTalk*, DECNet*, OSI*, VINES*, XNS*
	2507	RIP, IGRP, EIGRP, OSPF, BGP, others*	68030	2MB*	18MB	8MB	16MB	< 10,000	DHCP server, DHCP relay, NAT/PAT, SNMP, access lists, advanced queueing, QoS, TACACS+, RADIUS, compression, VPN*, IPSEC*, firewall* IPX*, SNA*, AppleTalk*, DECNet*, OSI*, VINES*, XNS*
	2509	RIP, IGRP, EIGRP, OSPF, BGP, others*	68030	2MB*	18MB	8MB	16MB	< 10,000	DHCP server, DHCP relay, NAT/PAT, SNMP, access lists, advanced queueing, QoS, TACACS+, RADIUS, compression, VPN*, IPSEC*, firewall* IPX*, SNA*, AppleTalk*, DECNet*, OSI*, VINES*, XNS*
	AS2509-RJ	RIP, IGRP, EIGRP, OSPF, BGP, others*	68030	2MB*	18MB	8MB	16MB	< 10,000	DHCP server, DHCP relay, NAT/PAT, SNMP, access lists, advanced queueing, QoS, TACACS+, RADIUS, compression, VPN*, IPSEC*, firewall*, IPX*, SNA*, AppleTalk*, DECNet*, OSI*, VINES*, XNS*

Table 11-6. Branch Router Capabilities (continued)

Series	Router Model	Routing Protocols	CPU Model	Standard DRAM	Maximum DRAM	Standard Flash RAM	Maximum Flash RAM	PPS	Capabilities
	2510	RIP, IGRP, EIGRP, OSPF, BGP, others*	68030	2MB*	18MB	8MB	16MB	<10,000	DHCP server, DHCP relay, NAT/PAT, SNMP, access lists, advanced queueing, QoS, TACACS+, RADIUS, compression, VPN*, IPSEC*, firewall*, IPX*, SNA*, AppleTalk*, DECNet*, OSI*, VINES*, XNS*
	2511	RIP, IGRP, EIGRP, OSPF, BGP, others*	68030	2MB*	18MB	8MB	16MB	<10,000	DHCP server, DHCP relay, NAT/PAT, SNMP, access lists, advanced queueing, QoS, TACACS+, RADIUS, compression, VPN*, IPSEC*, firewall*, IPX*, SNA*, AppleTalk*, DECNet*, OSI*, VINES*, XNS*
	AS2511-RJ	RIP, IGRP, EIGRP, OSPF, BGP, others*	68030	2MB*	18MB	8MB	16MB	<10,000	DHCP server, DHCP relay, NAT/PAT, SNMP, access lists, advanced queueing, QoS, TACACS+, RADIUS, compression, VPN*, IPSEC*, firewall*, IPX*, SNA*, AppleTalk*, DECNet*, OSI*, VINES*, XNS*
	2512	RIP, IGRP, EIGRP, OSPF, BGP, others*	68030	2MB*	18MB	8MB	16MB	<10,000	DHCP server, DHCP relay, NAT/PAT, SNMP, access lists, advanced queueing, QoS, TACACS+, RADIUS, compression, VPN*, IPSEC*, firewall*, IPX*, SNA*, AppleTalk*, DECNet*, OSI*, VINES*, XNS*

Table 11-6. *Branch Router Capabilities (continued)*

Series	Router Model	Routing Protocols	CPU Model	Standard DRAM	Maximum DRAM	Standard Flash RAM	Maximum Flash RAM	PPS	Capabilities
	2513	RIP, IGRP, EIGRP, OSPF, BGP, others*	68030	2MB*	18MB	8MB	16MB	<10,000	DHCP server, DHCP relay, NAT/PAT, SNMP, access lists, advanced queueing, QoS, TACACS+, RADIUS, compression, VPN*, IPSEC*, firewall*, IPX*, SNA*, AppleTalk*, DECNet*, OSI*, VINES*, XNS*
	2514	RIP, IGRP, EIGRP, OSPF, BGP, others*	68030	2MB*	18MB	8MB	16MB	<10,000	DHCP server, DHCP relay, NAT/PAT, SNMP, access lists, advanced queueing, QoS, TACACS+, RADIUS, compression, VPN*, IPSEC*, firewall*, IPX*, SNA*, AppleTalk*, DECNet*, OSI*, VINES*, XNS*
	2515	RIP, IGRP, EIGRP, OSPF, BGP, others*	68030	2MB*	18MB	8MB	16MB	<10,000	DHCP server, DHCP relay, NAT/PAT, SNMP, access lists, advanced queueing, QoS, TACACS+, RADIUS, compression, VPN*, IPSEC*, firewall*, IPX*, SNA*, AppleTalk*, DECNet*, OSI*, VINES*, XNS*
	2516	RIP, IGRP, EIGRP, OSPF, BGP, others*	68030	2MB*	18MB	8MB	16MB	<10,000	DHCP server, DHCP relay, NAT/PAT, SNMP, access lists, advanced queueing, QoS, TACACS+, RADIUS, compression, VPN*, IPSEC*, firewall*, IPX*, SNA*, AppleTalk*, DECNet*, OSI*, VINES*, XNS*

Table 11-6. *Branch Router Capabilities* (continued)

Series	Router Model	Routing Protocols	CPU Model	Standard DRAM	Maximum DRAM	Standard Flash RAM	Maximum Flash RAM	PPS	Capabilities
	2520	RIP, IGRP, EIGRP, OSPF, BGP, others*	68030	2MB*	18MB	8MB	16MB	< 10,000	DHCP server, DHCP relay, NAT/PAT, SNMP, access lists, advanced queueing, QoS, TACACS+, RADIUS, compression, VPN*, IPSEC*, firewall*, IPX*, SNA*, AppleTalk*, DECNet*, OSI*, VINES*, XNS*
	2521	RIP, IGRP, EIGRP, OSPF, BGP, others*	68030	2MB*	18MB	8MB	16MB	< 10,000	DHCP server, DHCP relay, NAT/PAT, SNMP, access lists, advanced queueing, QoS, TACACS+, RADIUS, compression, VPN*, IPSEC*, firewall*, IPX*, SNA*, AppleTalk*, DECNet*, OSI*, VINES*, XNS*
	2522	RIP, IGRP, EIGRP, OSPF, BGP, others*	68030	2MB*	18MB	8MB	16MB	< 10,000	DHCP server, DHCP relay, NAT/PAT, SNMP, access lists, advanced queueing, QoS, TACACS+, RADIUS, compression, VPN*, IPSEC*, firewall*, IPX*, SNA*, AppleTalk*, DECNet*, OSI*, VINES*, XNS*

Table 11-6. Branch Router Capabilities (continued)

Series	Router Model	Routing Protocols	CPU Model	Standard DRAM	Maximum DRAM	Standard Flash RAM	Maximum Flash RAM	PPS	Capabilities
	2523	RIP, IGRP, EIGRP, OSPF, BGP, others*	68030	2MB*	18MB	8MB	16MB	< 10,000	DHCP server, DHCP relay, NAT/PAT, SNMP, access lists, advanced queueing, QoS, TACACS+, RADIUS, compression, VPN*, IPSEC*, firewall*, IPX*, SNA*, AppleTalk*, DECNet*, OSI*, VINES*, XNS*
	2524	RIP, IGRP, EIGRP, OSPF, BGP, others*	68030	2MB*	18MB	8MB	16MB	< 10,000	DHCP server, DHCP relay, NAT/PAT, SNMP, access lists, advanced queueing, QoS, TACACS+, RADIUS, compression, VPN*, IPSEC*, firewall*, IPX*, SNA*, AppleTalk*, DECNet*, OSI*, VINES*, XNS*
	2525	RIP, IGRP, EIGRP, OSPF, BGP, others*	68030	2MB*	18MB	8MB	16MB	< 10,000	DHCP server, DHCP relay, NAT/PAT, SNMP, access lists, advanced queueing, QoS, TACACS+, RADIUS, compression, VPN*, IPSEC*, firewall*, IPX*, SNA*, AppleTalk*, DECNet*, OSI*, VINES*, XNS*

Table 11-6. Branch Router Capabilities (continued)

Series	Router Model	Routing Protocols	CPU Model	Standard DRAM	Maximum DRAM	Standard Flash RAM	Maximum Flash RAM	PPS	Capabilities
2600	2610	RIP, IGRP, EIGRP, OSPF, BGP, others*	Motorola QUICC MPC860T (40MHz)	32MB	64MB	8MB	16MB	15,000	DHCP server, DHCP relay, NAT/PAT, SNMP, RMON, access lists, advanced queueing, QoS, TACACS+, RADIUS, compression, VPN*, IPSEC*, firewall*, IPX*, SNA*, AppleTalk*, DECNet*, OSI*, VINES*, XNS*, multiservice*
	2611	RIP, IGRP, EIGRP, OSPF, BGP, others*	Motorola QUICC MPC860T (40MHz)	32MB	64MB	8MB	16MB	15,000	DHCP server, DHCP relay, NAT/PAT, SNMP, RMON, access lists, advanced queueing, QoS, TACACS+, RADIUS, compression, VPN*, IPSEC*, firewall*, IPX*, SNA*, AppleTalk*, DECNet*, OSI*, VINES*, XNS*, multiservice*
	2612	RIP, IGRP, EIGRP, OSPF, BGP, others*	Motorola QUICC MPC860T (40MHz)	32MB	64MB	8MB	16MB	15,000	DHCP server, DHCP relay, NAT/PAT, SNMP, RMON, access lists, advanced queueing, QoS, TACACS+, RADIUS, compression, VPN*, IPSEC*, firewall*, IPX*, SNA*, AppleTalk*, DECNet*, OSI*, VINES*, XNS*, multiservice*
	2613	RIP, IGRP, EIGRP, OSPF, BGP, others*	Motorola QUICC MPC860T (40MHz)	32MB	64MB	8MB	16MB	15,000	DHCP server, DHCP relay, NAT/PAT, SNMP, RMON, access lists, advanced queueing, QoS, TACACS+, RADIUS, compression, VPN*, IPSEC*, firewall*, IPX*, SNA*, AppleTalk*, DECNet*, OSI*, VINES*, XNS*, multiservice*

Table 11-6. Branch Router Capabilities (continued)

Series	Router Model	Routing Protocols	CPU Model	Standard DRAM	Maximum DRAM	Standard Flash RAM	Maximum Flash RAM	PPS	Capabilities
	2620	RIP, IGRP, EIGRP, OSPF, BGP, others*	Motorola QUICC MPC860T (50MHz)	32MB	64MB	8MB	16MB	25,000	DHCP server, DHCP relay, NAT/PAT, SNMP, RMON, access lists, advanced queueing, QoS, TACACS+, RADIUS, compression, VPN*, IPSEC*, firewall*, IPX*, SNA*, AppleTalk*, DECNet*, OSI*, VINES*, XNS*, multiservice*
	2621	RIP, IGRP, EIGRP, OSPF, BGP, others*	Motorola QUICC MPC860T (50MHz)	32MB	64MB	8MB	16MB	25,000	DHCP server, DHCP relay, NAT/PAT, SNMP, RMON, access lists, advanced queueing, QoS, TACACS+, RADIUS, compression, VPN*, IPSEC*, firewall*, IPX*, SNA*, AppleTalk*, DECNet*, OSI*, VINES*, XNS*, multiservice*
	2650	RIP, IGRP, EIGRP, OSPF, BGP, others*	Motorola QUICC MPC860T (80MHz)	32MB	128MB	8MB	32MB	37,000	DHCP server, DHCP relay, NAT/PAT, SNMP, RMON, access lists, advanced queueing, QoS, TACACS+, RADIUS, compression, VPN*, IPSEC*, firewall*, IPX*, SNA*, AppleTalk*, DECNet*, OSI*, VINES*, XNS*, multiservice*
	2651	RIP, IGRP, EIGRP, OSPF, BGP, others*	Motorola QUICC MPC860T (80MHz)	32MB	128MB	8MB	32MB	37,000	DHCP server, DHCP relay, NAT/PAT, SNMP, RMON, access lists, advanced queueing, QoS, TACACS+, RADIUS, compression, VPN*, IPSEC*, firewall*, IPX*, SNA*, AppleTalk*, DECNet*, OSI*, VINES*, XNS*, multiservice*

Table 11-6. Branch Router Capabilities (continued)

Series	Router Model	Classification	Configuration	Interface Numbering
3600	3620	Midsize	Modular	Type module/port
	3640	Midsize	Modular	Type module/port
	3660	Midsize	Modular	Type module/port
7100	7120-4T1	Midsize/enterprise	Modular	Type module/port
	7120-T3	Midsize/enterprise	Modular	Type module/port
	7120-E3	Midsize/enterprise	Modular	Type module/port
	7120-AT3	Midsize/enterprise	Modular	Type module/port
	7120-AE3	Midsize/enterprise	Modular	Type module/port
	7120-SMI3	Midsize/enterprise	Modular	Type module/port
	7140-2FE	Midsize/enterprise	Modular	Type module/port
	7140-8T	Midsize/enterprise	Modular	Type module/port
	7140-2T3	Midsize/enterprise	Modular	Type module/port
	7140-2E3	Midsize/enterprise	Modular	Type module/port
	7140-2AT3	Midsize/enterprise	Modular	Type module/port
	7140-2AE3	Midsize/enterprise	Modular	Type module/port
	7140-2MM3	Midsize/enterprise	Modular	Type module/port
7200	7204VXR	Midsize/enterprise	Modular	Type module/port
	7206	Midsize/enterprise	Modular	Type module/port
	7206VXR	Midsize/enterprise	Modular	Type module/port

Table 11-7. *Midsize Router Classifications and Configurations*

Series	Router Model	Ethernet Ports	Fast Ethernet (100 Mbps or 10/100) Ports	Gigabit Ethernet Ports	Token Ring Ports	FDDI Ports	ATM Ports	Synchronous Serial Ports	Synchronous Serial Speed	Asynchronous Serial Ports	Other Ports
3600	3620	8*	8*	0	2*	0	16*	16*	8 Mbps*	64*	16 T1*, 16 56 Kbps*, 8 ISDN*, 32 analog modem*, 60 digital modem*, 2 HSSI*, 8 voice*
	3640	16*	16*	0	4*	0	32*	32*	8 Mbps*	128*	32 T1*, 32 56 Kbps*, 16 ISDN*, 64 analog modem*, 120 digital modem*, 4 HSSI*, 16 voice*
	3660	24*	24*	0	6*	0	48*	48*	8 Mbps*	192*	48 T1*, 48 56 Kbps*, 24 ISDN*, 96 analog modem*, 180 digital modem*, 6 HSSI*, 24 voice*

Table 11-8.   *Midsize Router Port Options*

Series	Router Model	Ethernet Ports	Fast Ethernet (100 Mbps or 10/100) Ports	Gigabit Ethernet Ports	Token Ring Ports	FDDI Ports	ATM Ports	Synchronous Serial Ports	Synchronous Serial Speed	Asynchronous Serial Ports	Other Ports
7100	7120-4T1	8*	2*	1*	4*	0	1*	12*	2 Mbps*	8*	2 HSSI*, 2 T3*, 12 T1*, 8 ISDN*
	7120-T3	8*	2*	1*	4*	0	1*	8*	2 Mbps*	8*	2 HSSI*, 3 T3*, 8 T1*, 8 ISDN*
	7120-E3	8*	2*	1*	4*	0	1*	8*	2 Mbps*	8*	2 HSSI*, 3 E3*, 8 E1*, 8 ISDN*
	7120-AT3	8*	2*	1*	4*	0	2*	8*	2 Mbps*	8*	2 HSSI*, 3 T3*, 8 T1*, 8 ISDN*
	7120-AE3	8*	2*	1*	4*	0	2*	8*	2 Mbps*	8*	2 HSSI*, 3 E3*, 8 E1*, 8 ISDN*
	7120-SMI3	8*	2*	1*	4*	0	2*	8*	2 Mbps*	8*	2 HSSI*, 2 T3*, 8 T1*, 8 ISDN*
	7140-2FE	8*	3*	1*	4*	0	1*	8*	2 Mbps*	8*	3 HSSI*, 2 T3*, 8 T1*, 8 ISDN*
	7140-8T	8*	3*	1*	4*	0	1*	16*	2 Mbps*	8*	2 HSSI*, 2 T3*, 16 T1*, 8 ISDN*
	7140-2T3	8*	3*	1*	4*	0	1*	8*	2 Mbps*	8*	2 HSSI*, 4 T3*, 8 T1*, 8 ISDN*
	7140-2E3	8*	3*	1*	4*	0	1*	8*	2 Mbps*	8*	2 HSSI*, 4 E3*, 8 E1*, 8 ISDN*
	7140-2AT3	8*	3*	1*	4*	0	3*	8*	2 Mbps*	8*	2 HSSI*, 4 T3*, 8 T1*, 8 ISDN*
	7140-2AE3	8*	3*	1*	4*	0	3*	8*	2 Mbps*	8*	2 HSSI*, 4 E3*, 8 E1*, 8 ISDN*
	7140-2MM3	8*	3*	1*	4*	0	3*	8*	2 Mbps*	8*	2 HSSI*, 2 T3*, 8 T1*, 8 ISDN*

Table 11-8. *Midsize Router Port Options (continued)*

Series	Router Model	Ethernet Ports	Fast Ethernet (100 Mbps or 10/100) Ports	Gigabit Ethernet Ports	Token Ring Ports	FDDI Ports	ATM Ports	Synchronous Serial Ports	Synchronous Serial Speed	Asynchronous Serial Ports	Other Ports
7200	7204VXR	32*	4*	2*	16*	0	4*	32*	8 Mbps*	32*	8 HSSI*, 8 T3*, 32 T1*, 32 ISDN*, 6 IBM mainframe*, 8 voice T1*
	7206	48*	3*	2*	24*	3*	3*	48*	8 Mbps*	48*	6 HSSI*, 6 T3*, 48 T1*, 48 ISDN*, 3 IBM mainframe*
	7206VXR	48*	6*	2*	24*	0	6*	48*	8 Mbps*	48*	12 HSSI*, 12 T3*, 48 T1*, 48 ISDN*, 6 IBM mainframe*, 12 voice T1*

Table 11-8.   *Midsize Router Port Options (continued)*

Series	Router Model	Routing Protocols	CPU Model	Standard DRAM	Maximum DRAM	Standard Flash RAM	Maximum Flash RAM	PPS	Capabilities
3600	3620	RIP, IGRP, EIGRP, OSPF, BGP, others*	80 MHz IDT R4700	32MB	64MB	8MB	32MB	50,000	DHCP server, DHCP relay, NAT/PAT, SNMP, RMON, access lists, advanced queueing, QoS, TACACS+, RADIUS, AAA, compression, VPN*, IPSEC*, firewall*, IPX*, SNA*, AppleTalk*, DECNet*, OSI*, VINES*, XNS*, multiservice*
	3640	RIP, IGRP, EIGRP, OSPF, BGP, others*	100 MHz IDT R4700	32MB	128MB	8MB	32MB	60,000	DHCP server, DHCP relay, NAT/PAT, SNMP, RMON, access lists, advanced queueing, QoS, TACACS+, RADIUS, AAA, compression, VPN*, IPSEC*, firewall*, IPX*, SNA*, AppleTalk*, DECNet*, OSI*, VINES*, XNS*, multiservice*
	3660	RIP, IGRP, EIGRP, OSPF, BGP, others*	225 MHz RISC QED RM5271	32MB	256MB	8MB	64MB	70,000	DHCP server, DHCP relay, NAT/PAT, SNMP, RMON, access lists, advanced queueing, QoS, TACACS+, RADIUS, AAA, compression, VPN*, IPSEC*, firewall*, IPX*, SNA*, AppleTalk*, DECNet*, OSI*, VINES*, XNS*, multiservice*

Table 11-9. Midsize Router Capabilities

Series	Router Model	Routing Protocols	CPU Model	Standard DRAM	Maximum DRAM	Standard Flash RAM	Maximum Flash RAM	PPS	Capabilities
7100	7120-4T1	RIP, IGRP, EIGRP, OSPF, BGP, others*	MIPS 5000	64MB	256MB	48MB	256MB	70,000	DHCP server, DHCP relay, NAT/PAT, SNMP, RMON, access lists, advanced queueing, QoS, TACACS+, RADIUS, AAA, compression, VPN, IPSEC*, firewall*, IPX*, SNA*, AppleTalk*, DECNet*, OSI*, VINES*, XNS*
	7120-T3	RIP, IGRP, EIGRP, OSPF, BGP, others*	MIPS 5000	64MB	256MB	48MB	256MB	70,000	DHCP server, DHCP relay, NAT/PAT, SNMP, RMON, access lists, advanced queueing, QoS, TACACS+, RADIUS, AAA, compression, VPN, IPSEC*, firewall*, IPX*, SNA*, AppleTalk*, DECNet*, OSI*, VINES*, XNS*
	7120-E3	RIP, IGRP, EIGRP, OSPF, BGP, others*	MIPS 5000	64MB	256MB	48MB	256MB	70,000	DHCP server, DHCP relay, NAT/PAT, SNMP, RMON, access lists, advanced queueing, QoS, TACACS+, RADIUS, AAA, compression, VPN, IPSEC*, r*, IPX*, SNA*, AppleTalk*, DECNet*, OSI*, VINES*, XNS*

Table 11-9. *Midsize Router Capabilities (continued)*

Series	Router Model	Routing Protocols	CPU Model	Standard DRAM	Maximum DRAM	Standard Flash RAM	Maximum Flash RAM	PPS	Capabilities
	7120-AT3	RIP, IGRP, EIGRP, OSPF, BGP, others*	MIPS 5000	64MB	256MB	48MB	256MB	70,000	DHCP server, DHCP relay, NAT/PAT, SNMP, RMON, access lists, advanced queueing, QoS, TACACS+, RADIUS, AAA, compression, VPN, IPSEC*, firewall*, IPX*, SNA*, AppleTalk*, DECNet*, OSI*, VINES*, XNS*
	7120-AE3	RIP, IGRP, EIGRP, OSPF, BGP, others*	MIPS 5000	64MB	256MB	48MB	256MB	70,000	DHCP server, DHCP relay, NAT/PAT, SNMP, RMON, access lists, advanced queueing, QoS, TACACS+, RADIUS, AAA, compression, VPN, IPSEC*, firewall*, IPX*, SNA*, AppleTalk*, DECNet*, OSI*, VINES*, XNS*
	7120-SMI3	RIP, IGRP, EIGRP, OSPF, BGP, others*	MIPS 5000	64MB	256MB	48MB	256MB	70,000	DHCP server, DHCP relay, NAT/PAT, SNMP, RMON, access lists, advanced queueing, QoS, TACACS+, RADIUS, AAA, compression, VPN, IPSEC*, firewall*, IPX*, SNA*, AppleTalk*, DECNet*, OSI*, VINES*, XNS*

Table 11-9. *Midsize Router Capabilities (continued)*

Series	Router Model	Routing Protocols	CPU Model	Standard DRAM	Maximum DRAM	Standard Flash RAM	Maximum Flash RAM	PPS	Capabilities
	7140-2FE	RIP, IGRP, EIGRP, OSPF, BGP, others*	MIPS 7000	64MB	256MB	48MB	256MB	150,000	DHCP server, DHCP relay, NAT/PAT, SNMP, RMON, access lists, advanced queueing, QoS, TACACS+, RADIUS, AAA, compression, VPN, IPSEC*, firewall*, IPX*, SNA*, AppleTalk*, DECNet*, OSI*, VINES*, XNS*
	7140-8T	RIP, IGRP, EIGRP, OSPF, BGP, others*	MIPS 7000	64MB	256MB	48MB	256MB	150,000	DHCP server, DHCP relay, NAT/PAT, SNMP, RMON, access lists, advanced queueing, QoS, TACACS+, RADIUS, AAA, compression, VPN, IPSEC*, firewall*, IPX*, SNA*, AppleTalk*, DECNet*, OSI*, VINES*, XNS*
	7140-2T3	RIP, IGRP, EIGRP, OSPF, BGP, others*	MIPS 7000	64MB	256MB	48MB	256MB	150,000	DHCP server, DHCP relay, NAT/PAT, SNMP, RMON, access lists, advanced queueing, QoS, TACACS+, RADIUS, AAA, compression, VPN, IPSEC*, firewall*, IPX*, SNA*, AppleTalk*, DECNet*, OSI*, VINES*, XNS*

Table 11-9.   *Midsize Router Capabilities* (continued)

Series	Router Model	Routing Protocols	CPU Model	Standard DRAM	Maximum DRAM	Standard Flash RAM	Maximum Flash RAM	PPS	Capabilities
	7140-2E3	RIP, IGRP, EIGRP, OSPF, BGP, others*	MIPS 7000	64MB	256MB	48MB	256MB	150,000	DHCP server, DHCP relay, NAT/PAT, SNMP, RMON, access lists, advanced queueing, QoS, TACACS+, RADIUS, AAA, compression, VPN, IPSEC*, firewall*, IPX*, SNA*, AppleTalk*, DECNet*, OSI*, VINES*, XNS*
	7140-2AT3	RIP, IGRP, EIGRP, OSPF, BGP, others*	MIPS 7000	64MB	256MB	48MB	256MB	150,000	DHCP server, DHCP relay, NAT/PAT, SNMP, RMON, access lists, advanced queueing, QoS, TACACS+, RADIUS, AAA, compression, VPN, IPSEC*, firewall*, IPX*, SNA*, AppleTalk*, DECNet*, OSI*, VINES*, XNS*

Table 11-9. Midsize Router Capabilities (continued)

Series	Router Model	Routing Protocols	CPU Model	Standard DRAM	Maximum DRAM	Standard Flash RAM	Maximum Flash RAM	PPS	Capabilities
	7140-2AE3	RIP, IGRP, EIGRP, OSPF, BGP, others*	MIPS 7000	64MB	256MB	48MB	256MB	150,000	DHCP server, DHCP relay, NAT/PAT, SNMP, RMON, access lists, advanced queueing, QoS, TACACS+, RADIUS, AAA, compression, VPN, IPSEC*, firewall*, IPX*, SNA*, AppleTalk*, DECNet*, OSI*, VINES*, XNS*
	7140-2MM3	RIP, IGRP, EIGRP, OSPF, BGP, others*	MIPS 7000	64MB	256MB	48MB	256MB	150,000	DHCP server, DHCP relay, NAT/PAT, SNMP, RMON, access lists, advanced queueing, QoS, TACACS+, RADIUS, AAA, compression, VPN, IPSEC*, firewall*, IPX*, SNA*, AppleTalk*, DECNet*, OSI*, VINES*, XNS*

Table 11-9.   *Midsize Router Capabilities (continued)*

Series	Router Model	Routing Protocols	CPU Model	Standard DRAM	Maximum DRAM	Standard Flash RAM	Maximum Flash RAM	PPS	Capabilities
7200	7204VXR	RIP, IGRP, EIGRP, OSPF, BGP, others*	MIPS (Dependent on Network Processor Model)	128MB	512MB	48MB	128MB	400,000*	DHCP server, DHCP relay, NAT/PAT, SNMP, RMON, access lists, advanced queueing, QoS, TACACS+, RADIUS, AAA, compression, VPN, X.25 switching, ATM switching, Frame Relay switching, tag switching, multiservice, IPSEC*, firewall* IPX*, SNA*, AppleTalk*, DECNet*, OSI*, VINES*, XNS*
	7206	RIP, IGRP, EIGRP, OSPF, BGP, others*	MIPS (Dependent on Network Processor Model)	128MB	512MB	48MB	128MB	400,000*	DHCP server, DHCP relay, NAT/PAT, SNMP, RMON, access lists, advanced queueing, QoS, TACACS+, RADIUS, AAA, compression, VPN, X.25 switching, ATM switching, Frame Relay switching, tag switching, multiservice, IPSEC*, firewall* IPX*, SNA*, AppleTalk*, DECNet*, OSI*, VINES*, XNS*
	7206VXR	RIP, IGRP, EIGRP, OSPF, BGP, others*	MIPS (Dependent on Network Processor Model)	128MB	512MB	48MB	128MB	400,000*	DHCP server, DHCP relay, NAT/PAT, SNMP, RMON, access lists, advanced queueing, QoS, TACACS+, RADIUS, AAA, compression, VPN, X.25 switching, ATM switching, Frame Relay switching, tag switching, multiservice, IPSEC*, firewall* IPX*, SNA*, AppleTalk*, DECNet*, OSI*, VINES*, XNS*

Table 11-9.   Midsize Router Capabilities (continued)

Series	Router Model	Classification	Configuration	Interface Numbering
7500	7505	Enterprise	Modular	Type module/port
	7507	Enterprise	Modular	Type module/port
	7513	Enterprise	Modular	Type module/port
	7576	Enterprise	Modular	Type module/port
10000	10005 ESR	Enterprise	Modular	Type card/subcard/port
	10008 ESR	Enterprise	Modular	Type card/subcard/port
12000	12008	Enterprise	Modular	Type slot/port
	12012	Enterprise	Modular	Type slot/port
	12016	Enterprise	Modular	Type slot/port
	12410	Enterprise	Modular	Type slot/port
	12416	Enterprise	Modular	Type slot/port

Table 11-10.   *Enterprise Router Classifications and Configurations*

Series	Router Model	Ethernet Ports	Fast Ethernet (100 Mbps or 10/100) Ports	Gigabit Ethernet Ports	Token Ring Ports	FDDI Ports	ATM Ports	Synchronous Serial Ports	Synchronous Serial Speed	Asynchronous Serial Ports	Other Ports
7500	7505	64*	8*	4*	16*	8*	4*	64*	8 Mbps*	64*	16 HSSI*, 8 T3*, 240 T1*, 4 IBM mainframe*
	7507	80*	10*	6*	20*	10*	5*	80*	8 Mbps*	80*	20 HSSI*, 12 T3*, 256 T1*, 5 IBM mainframe*
	7513	175*	22*	11*	44*	22*	11*	176*	8 Mbps*	176*	44 HSSI*, 22 T3*, 256 T1*, 11 IBM mainframe*
	7576	175*	22*	11*	44*	22*	11*	176*	8 Mbps*	176*	44 HSSI*, 22 T3*, 256 T1*, 11 IBM mainframe*
10000	10005 ESR	0	1	5*	0	0	5*	0	N/A	0	1008 T1*, 36 T3*
	10008 ESR	0	1	8*	0	0	8*	0	N/A	0	2000 T1*, 72 T3*
12000	12008	0	56*	21*	0	0	28*	0	N/A	0	1176 T1*, 42 T3*
	12012	0	88*	33*	0	0	44*	0	N/A	0	1848 T1*, 66 T3*
	12016	0	120*	45*	0	0	60*	0	N/A	0	2520 T1*, 90 T3*
	12410	0	72*	27*	0	0	36*	0	N/A	0	1512 T1*, 54 T3*
	12416	0	120*	45*	0	0	60*	0	N/A	0	2520 T1*, 90 T3*

Table 11-11.   Enterprise Router Port Options

Series	Router Model	Routing Protocols	CPU Model	Standard DRAM	Maximum DRAM	Standard Flash RAM	Maximum Flash RAM	PPS	Capabilities
7500	7505	RIP, IGRP, EIGRP, OSPF, BGP, others*	MIPS (dependent on RSM model)	32MB*	256MB*	16MB*	110MB*	350,000	DHCP server, DHCP relay, NAT/PAT, SNMP, RMON, access lists, advanced queueing, QoS, TACACS+, RADIUS, AAA, compression, VPN, X.25 switching, ATM switching, Frame Relay switching, tag switching, CEF, NetFlow, multiservice, IPSEC*, firewall*, IPX*, SNA*, AppleTalk*, DECNet*, OSI*, VINES*, XNS*
	7507	RIP, IGRP, EIGRP, OSPF, BGP, others*	MIPS (dependent on RSM model)	32MB*	256MB*	16MB*	110MB*	500,000	DHCP server, DHCP relay, NAT/PAT, SNMP, RMON, access lists, advanced queueing, QoS, TACACS+, RADIUS, AAA, compression, VPN, X.25 switching, ATM switching, Frame Relay switching, tag switching, CEF, NetFlow, multiservice, IPSEC*, firewall*, IPX*, SNA*, AppleTalk*, DECNet*, OSI*, VINES*, XNS*

Table 11-12. *Enterprise Router Capabilities*

Series	Router Model	Routing Protocols	CPU Model	Standard DRAM	Maximum DRAM	Standard Flash RAM	Maximum Flash RAM	PPS	Capabilities
	7513	RIP, IGRP, EIGRP, OSPF, BGP, others*	MIPS (dependent on RSM model)	32MB*	256MB*	16MB*	110MB*	1,000,000	DHCP server, DHCP relay, NAT/PAT, SNMP, RMON, access lists, advanced queueing, QoS, TACACS+, RADIUS, AAA, compression, VPN, X.25 switching, ATM switching, Frame Relay switching, tag switching, CEF, NetFlow, multiservice, IPSEC*, firewall*, IPX*, SNA*, AppleTalk*, DECNet*, OSI*, VINES*, XNS*
	7576	RIP, IGRP, EIGRP, OSPF, BGP, others*	MIPS (dependent on RSM model)	32MB*	256MB*	16MB*	110MB*	1,000,000	Dual-router architecture, DHCP server, DHCP relay, NAT/PAT, SNMP, RMON, access lists, advanced queueing, QoS, TACACS+, RADIUS, AAA, compression, VPN, X.25 switching, ATM switching, Frame Relay switching, tag switching, CEF, NetFlow, multiservice, IPSEC*, firewall*, IPX*, SNA*, AppleTalk*, DECNet*, OSI*, VINES*, XNS*

Table 11-12.   Enterprise Router Capabilities (continued)

Series	Router Model	Routing Protocols	CPU Model	Standard DRAM	Maximum DRAM	Standard Flash RAM	Maximum Flash RAM	PPS	Capabilities
10000	10005 ESR	RIP, IGRP, EIGRP, OSPF, BGP, others*	16 RM7000 MIPS processors	512MB*	512MB*	32MB*	256MB*	2,600,000	DHCP server, DHCP relay, NAT/PAT, SNMP, RMON, access lists, advanced queueing, QoS, TACACS+, RADIUS, AAA, compression, VPN, X.25 switching, ATM switching, Frame Relay switching, tag switching, CEF, NetFlow, PXF, IPSEC*, firewall*, IPX, SNA, AppleTalk, DECNet, OSI, VINES, XNS
	10008 ESR	RIP, IGRP, EIGRP, OSPF, BGP, others*	16 RM7000 MIPS processors	512MB*	512MB*	32MB*	256MB*	2,600,000	DHCP server, DHCP relay, NAT/PAT, SNMP, RMON, access lists, advanced queueing, QoS, TACACS+, RADIUS, AAA, compression, VPN, X.25 switching, ATM switching, Frame Relay switching, tag switching, CEF, NetFlow, PXF, IPSEC*, firewall*, IPX, SNA, AppleTalk, DECNet, OSI, VINES, XNS

Table 11-12.   *Enterprise Router Capabilities (continued)*

Series	Router Model	Routing Protocols	CPU Model	Standard DRAM	Maximum DRAM	Standard Flash RAM	Maximum Flash RAM	PPS	Capabilities
12000	12008	RIP, IGRP, EIGRP, OSPF, BGP, others*	RM5000 MIPS + interface processors	128MB*	256MB*	20MB*	80MB*	28,000,000	DHCP server, DHCP relay, NAT/PAT, SNMP, RMON, access lists, advanced queueing, QoS, TACACS+, RADIUS, AAA, compression, VPN, X.25 switching, ATM switching, Frame Relay switching, tag switching, CEF, NetFlow, PXF, IPSEC*, firewall*
	12012	RIP, IGRP, EIGRP, OSPF, BGP, others*	RM5000 MIPS + interface processors	128MB*	256MB*	20MB*	80MB*	44,000,000	DHCP server, DHCP relay, NAT/PAT, SNMP, RMON, access lists, advanced queueing, QoS, TACACS+, RADIUS, AAA, compression, VPN, X.25 switching, ATM switching, Frame Relay switching, tag switching, CEF, NetFlow, PXF, IPSEC*, firewall*
	12016	RIP, IGRP, EIGRP, OSPF, BGP, others*	RM5000 MIPS + interface processors	128MB*	256MB*	20MB*	80MB*	60,000,000	DHCP server, DHCP relay, NAT/PAT, SNMP, RMON, access lists, advanced queueing, QoS, TACACS+, RADIUS, AAA, compression, VPN, X.25 switching, ATM switching, Frame Relay switching, tag switching, CEF, NetFlow, PXF, IPSEC*, firewall*

Table 11-12. Enterprise Router Capabilities (continued)

Series	Router Model	Routing Protocols	CPU Model	Standard DRAM	Maximum DRAM	Standard Flash RAM	Maximum Flash RAM	PPS	Capabilities
	12410	RIP, IGRP, EIGRP, OSPF, BGP, others*	RM5000 MIPS + interface processors	128MB*	256MB*	20MB*	80MB*	225,000,000	DHCP server, DHCP relay, NAT/PAT, SNMP, RMON, access lists, advanced queueing, QoS, TACACS+, RADIUS, AAA, compression, VPN, X.25 switching, ATM switching, Frame Relay switching, tag switching, CEF, NetFlow, PXF, IPSEC*, firewall*
	12416	RIP, IGRP, EIGRP, OSPF, BGP, others*	RM5000 MIPS + interface processors	128MB*	256MB*	20MB*	80MB*	375,000,000	DHCP server, DHCP relay, NAT/PAT, SNMP, RMON, access lists, advanced queueing, QoS, TACACS+, RADIUS, AAA, compression, VPN, X.25 switching, ATM switching, Frame Relay switching, tag switching, CEF, NetFlow, PXF, IPSEC*, firewall*

Table 11-12.   Enterprise Router Capabilities (continued)

Summary

This chapter examined some basic terminology related to Cisco routers and provided an overview of the different lines and models of Cisco routing equipment. The next chapter examines the individual routing lines more thoroughly so that you can begin to determine which line is best for your specific needs. For additional information on Cisco's routing lines, check out http://www.cisco.com/warp/public/44/jump/routers.shtml.

Chapter 12

Individual Router Series Reference

This chapter concentrates on each line of Cisco routing hardware. As in Chapter 10, the goal is not to become an expert on each line of router but, rather, to understand where each one fits both in Cisco's technology lineup and in your own environments.

The 600 Series

The 600 series of routers are small, simple DSL routers for SOHO environments. The primary use of these devices is for sharing a DSL Internet connection in a SOHO environment. They include features such as DHCP and NAT to allow for easy configuration of client machines in environments without sophisticated proxy servers and DHCP servers. They do not support advanced routing features, however, and dynamic routing protocol support is limited to RIP. Although the performance of these devices is, predictably, rather minimal, these routers meet the needs in their intended environment admirably.

You can find more information on the 600 series at http://www.cisco.com/warp/public/cc/pd/rt/600rt/prodlit/index.shtml.

The 700 Series

The 700 series of routers are SOHO ISDN routers, and they are typically used to share an ISDN connection among multiple devices. The base model in this series, the 761, includes a single ISDN port and a single Ethernet port for connectivity. It includes a feature set similar to the 600 series for easy client configuration and sharing of the connection.

The higher-end models, the 775 and 776, also include a four-port Ethernet hub already integrated into the router, and two POTS ports for sharing the ISDN line among fax machines and POTS telephones. The only difference in the 775 and 776 models is that the 775 requires an S/T ISDN interface, whereas the 776 can use both S/T and U interfaces. Figure 12-1 provides an example of a 700 series router.

All models of the 700 series perform excellently for sharing a SOHO ISDN connection, as long as advanced routing features are not required. For more information on the 700 series, check out http://www.cisco.com/warp/public/cc/pd/rt/700/prodlit/index.shtml.

The 800 Series

The 800 series is an advanced line of SOHO routers for environments that require some advanced routing and firewall features, but that do not require the performance (or price) of the higher-end branch-router lines. The base models in this series (801–804)

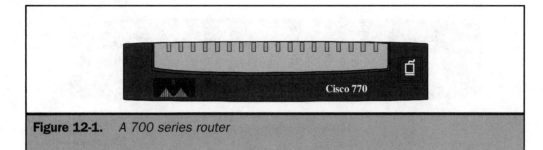

Figure 12-1.   *A 700 series router*

include one ISDN connection, and either one or four Ethernet ports. The 803 and 804 models also include two POTS ports for standard POTS device connectivity, such as the 775 and 776 routers.

The standard feature set on these models includes all of the features from the 600 and 700 series, plus support for access lists, queuing, and compression. Also, all routers in this series include advanced ISDN features such as Multilink Point-to-Point Protocol (MPPP or ML-PPP) and Bandwidth Allocation Control Protocol (BACP) used to bring up, bond, and release multiple ISDN B-channel connections dynamically based on demand. Optional feature sets for these routers can also add support for VPNs, firewall features, IPX routing support, and the EIGRP routing protocol.

The specialized models in the 800 series include the 805, 826, 827, and 827-4V routers. The 805 model is essentially the same as the 801 model, except the 805 includes a single synchronous serial port capable of up to 512 Kbps instead of an ISDN port. This allows the 805 to be used for a variety of lower-speed WAN connections.

The 826 model is an ISDN router very similar to the 802 model, but with the addition of features for slightly larger offices (such as small branches), including a faster processor and QoS support. The 827 model is nearly identical to the 826, except instead of ISDN, it includes an ADSL port. Finally, the 827-4V model includes all of the features of the standard 827, but adds four POTS telephone ports and corresponding PBX and voice features to allow Voice over IP (VoIP) for up to four simultaneous connections through a single DSL line. Figure 12-2 provides an example of an 800 series router.

Although the performance of the 800 series will not break any records (Cisco recommends the 800 series for environments with 20 users or less), it is adequate for its intended purpose, and it is slightly faster than the 600 and 700 series. All in all, this is an excellent router line for small environments that need a few of the more advanced features, but that don't yet need to step up to the branch routers (like the 1700 or 2600 series).

For more information on the 800 series of routers, check out http://www.cisco.com/warp/public/cc/pd/rt/800/prodlit/index.shtml.

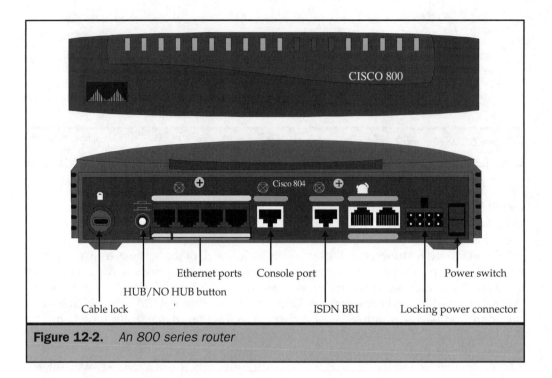

Figure 12-2.   *An 800 series router*

The 1000 Series

The 1000 series of routers are multiprotocol SOHO routers used to connect remote locations that may need AppleTalk or IPX support instead of (or in addition to) IP support. These routers are fairly simple; they include only basic routing features, similar to the 600 series. However, the embedded multiprotocol support allows these routers to be used in situations where the 600 series could not be used.

The 1000 series includes three models for different needs. The 1003 and 1004 models include an ISDN port (S/T on the 1003 and U on the 1004) and a single Ethernet port. The 1005 includes a single serial port that can be used either as a high-speed synchronous serial port (for WAN connections up to 2.048 Mbps), or as an asynchronous serial port (typically for dial-up applications). Because of its support for high-speed WAN connections, the 1005 model is also used in IP-only environments that require slightly more speed than the 800 series can provide, but it doesn't need most of the advanced features included in the 800 series. Figure 12-3 shows a 1000 series router.

All told, the 1000 series is an excellent choice for the SOHO user base that needs multi-protocol routing at an affordable price. You can find more information on the 1000 series at http://www.cisco.com/warp/public/cc/pd/rt/1000/prodlit/index.shtml.

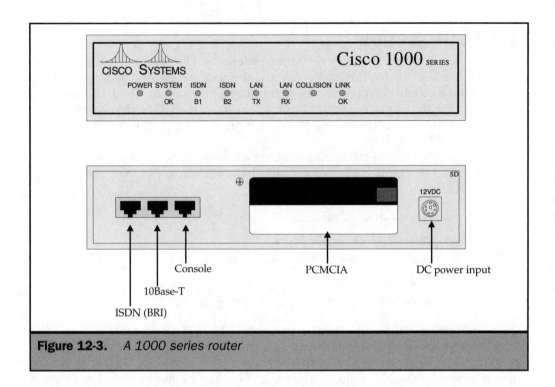

Figure 12-3. *A 1000 series router*

The 1400 Series

The 1400 series of routers are Cisco's high-end SOHO ADSL routers, designed for offices that require advanced features in an ADSL router, but that do not require the feature set, modularity, or cost associated with the 2600 series. The 1400 series supports the same feature sets as the 800 series, including access lists, firewall IOS capability, VPNs, and multiprotocol routing, but also includes additional QoS features such as Committed Access Rate (CAR), Weighted Fair Queuing (WFQ), and Resource Reservation Protocol (RSVP).

The 1400 series consists of two models: the 1401 and the 1417. The 1401 includes a single ATM 25 interface used to connect to an external ADSL modem, whereas the 1417 includes an integrated ADSL DMT Issue-2 modem for ADSL. Other than these differences, the models are identical.

In short, the 1400 series is a feature-rich router for SOHO ADSL networks that are a bit too advanced for the 600 series, but that don't yet require the power and modularity of the 2600 series. For more information on the 1400 series, please visit http://www.cisco.com/warp/public/cc/pd/rt/1400/prodlit/index.shtml.

The 1600 Series

The 1600 series of routers are Cisco's modular series of SOHO routers. They are designed for offices that require modularity and the ability to support a variety of interfaces, but they still don't require the performance provided by Cisco's branch router lines. The 1600 series supports all of the features of the 1400 series, but also includes support for RADIUS and enhanced multiprotocol support, including AppleTalk and SNA support.

As modular routers, the 1600 series has a variety of interface cards available. All models include a single interface card slot, which can contain any of the following interface cards:

■ A 1-port T1 with an integrated CSU/DSU

■ A 1-port serial interface (2.048 Mbps)

■ A 1-port 56 Kbps interface with an integrated CSU/DSU

■ A 1-port ISDN BRI interface card

■ A 1-port ISDN BRI interface card with an integrated NT-1

The 1600 series includes five models: the 1601, 1602, 1603, 1604, and 1605. The 1601 includes a single serial interface and dual Ethernet interfaces (RJ-45 and AUI). The 1602 includes a single 56-Kbps interface with an integrated CSU/DSU and dual Ethernet interfaces. The 1603 includes a single ISDN BRI S/T interface with dual Ethernet interfaces. The 1603 includes a single ISDN BRI S/T interface with dual Ethernet interfaces. The 1604 includes a single ISDN BRI U interface with an integrated NT-1, one S bus ISDN interface for ISDN phones, and dual Ethernet interfaces. The 1605 includes only three Ethernet interfaces (two RJ-45 and one AUI). Figure 12-4 shows a 1600 series router.

The number of models available, when combined with the wide variety of interface cards available, make the 1600 series a logical choice for SOHO environments that require diverse connectivity. For more information on the 1600 series, visit http://www.cisco.com/warp/public/cc/pd/rt/1600/prodlit/index.shtml.

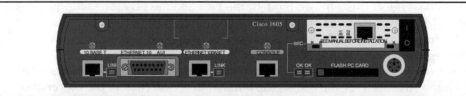

Figure 12-4.   *A 1600 series router*

The 1700 Series

The 1700 series is Cisco's entry-level branch office router line. The 1700 series is designed for branch offices that require comprehensive routing support, a variety of interface options, multiprotocol capability, voice integration, and VPN/firewall capability, but do not require the additional performance and features of the higher-end lines.

The 1700 series offers essentially the same feature sets available on the 1600 series, including DHCP support, NAT, QoS, access lists, VPN/firewall features, and multi-protocol (IP, IPX, AppleTalk, and SNA) routing capabilities. In addition, the 1700 series comes with a higher-performance processor than the 1600 series; includes greater support for routing protocols (including RIP, IGRP, EIGRP, OSPF, and BGP); and, in the 1750 models, includes support for voice features.

The 1700 series consists of four models: the 1720, 1750, 1750-2V, and 1750-4V. Because the 1700 series is modular, all of these models support the same interface cards. They also share these interface cards with the 1600, 2600, and 3600 series routers, which allows you greater investment protection if you need to upgrade in the future, as well as making it simple to keep shelf spares. The interface cards available for the 1700 series include the following:

- A 1-port serial interface (synchronous and asynchronous; 2.048 Mbps)
- A 2-port serial interface (synchronous and asynchronous; 2.048 Mbps)
- A 2-port low-speed (128 Kbps) serial interface (synchronous and asynchronous)
- A 1-port T1 interface with an integrated CSU/DSU
- A 1-port 56-Kbps interface with an integrated CSU/DSU
- A 1-port ISDN BRI interface card
- A 1-port ISDN BRI interface card with an integrated NT-1
- A 1-port ADSL interface
- A 1-port Ethernet interface (10 Mbps)

In addition, the 1750 routers support a two-port voice interface card.

The four models in the 1700 router line vary slightly in interface-card support and feature sets. The differences in these models are detailed here:

- **1720**   Includes a single 10/100 Ethernet interface, dual WIC slots, and a single VPN module expansion slot. Available IOS feature sets include support for DHCP, NAT/PAT, SNMP, access lists, advanced queuing, QoS, TACACS+, RADIUS, compression, VPN, IPSEC, firewall services, IPX, SNA, and AppleTalk.

- **1750**   Includes a single 10/100 Ethernet interface, three WIC slots, and a single VPN module expansion slot. Available IOS feature sets include support for all of the features of the 1720, as well as voice/fax integration features (with the addition of a voice-upgrade kit).

- **1750-2V** Includes a single 10/100 Ethernet interface, three WIC slots, and a single VPN module expansion slot. Available IOS feature sets include support for all of the features of the 1720, as well as voice/fax integration features, using a DSP that supports up to two voice/fax interface cards.

- **1750-4V** Includes a single 10/100 Ethernet interface, three WIC slots, and a single VPN module expansion slot. Available IOS feature sets include support for all of the features of the 1720 as well as voice/fax integration features, using a DSP that supports up to four voice/fax interface cards.

With the modularity available in the 1700 series, along with the capability for voice and data integration, the 1700 series is an obvious choice for small branch offices that require a versatile routing and voice solution. More information about the 1700 series can be found at http://www.cisco.com/warp/public/cc/pd/rt/1700/prodlit/index.shtml.

The 2500 Series

The 2500 series is an old standby in the Cisco router lineup. 2500 series routers are fixed-configuration routers ideally suited for midsize branch offices that require advanced routing and multiprotocol support in a simple, fixed router.

The 2500 series supports extensive feature-set options, including support for all of the features available in the 1700 series (except for voice capability). In addition to the vast routing support available in the 1700 series, the 2500 series offers support for additional network protocol suites, including DECNet, OSI, VINES, and XNS.

The 2500 series comes in a dizzying array of models to fit nearly any branch routing need. Figure 12-5 shows an example of a 2500 series router.

The following list briefly describes the available models in this series:

- **2501** Base Ethernet model in the line. Includes one AUI 10-Mbps Ethernet port and dual 2.048-Mbps serial ports (DB-60 connectors).

- **2502** Base Token Ring model in the line. Includes one DB-9 4/16-Mbps Token Ring port and dual 2.048-Mbps serial ports (DB-60 connectors).

- **2503** ISDN BRI/Ethernet/serial router. Includes one AUI 10-Mbps Ethernet port, one RJ-45 ISDN BRI port, and dual 2.048-Mbps serial ports (DB-60 connectors).

- **2504** ISDN BRI/Token Ring/serial router. Includes one DB-9 16/4-Mbps Token Ring port, one RJ-45 ISDN BRI port, and dual 2.048-Mbps serial ports (DB-60 connectors).

- **2505** Combination Ethernet hub/serial router. Includes eight RJ-45 10-Mbps Ethernet and dual 2.048-Mbps serial ports (DB-60 connectors).

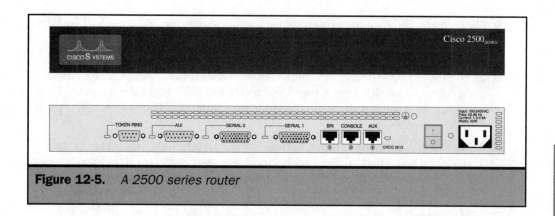

Figure 12-5.   *A 2500 series router*

- **2507**   Combination Ethernet hub/serial router. Includes 16 RJ-45 10-Mbps Ethernet and dual 2.048-Mbps serial ports (DB-60 connectors).

- **2509**   High-density serial/Ethernet router. Includes one AUI 10-Mbps Ethernet, eight low-speed asynchronous serial ports, and dual 2.048-Mbps serial ports (DB-60 connectors).

- **2510**   High-density serial/Token Ring router. Includes one DB-9 4/16-Mbps Token Ring port, eight low-speed asynchronous serial ports, and dual 2.048-Mbps serial ports (DB-60 connectors).

- **2511**   High-density serial/Ethernet router. Includes one AUI 10-Mbps Ethernet, 16 low-speed asynchronous serial ports, and dual 2.048-Mbps serial ports (DB-60 connectors).

- **2512**   High-density serial/Token Ring router. Includes one DB-9 4/16-Mbps Token Ring port, 16 low-speed asynchronous serial ports, and dual 2.048-Mbps serial ports (DB-60 connectors).

- **2513**   Dual LAN Token Ring/Ethernet/serial router. Includes one DB-9 4/16-Mbps Token Ring port, one AUI 10-Mbps Ethernet port, and dual 2.048-Mbps serial ports (DB-60 connectors).

- **2514**   Dual LAN Ethernet/serial router. Includes two AUI 10-Mbps Ethernet ports and dual 2.048-Mbps serial ports (DB-60 connectors).

- **2515**   Dual LAN Token Ring/serial router. Includes two DB-9 4/16-Mbps Token Ring ports and dual 2.048-Mbps serial ports (DB-60 connectors).

- **2520**   ISDN BRI/Ethernet/serial router. Includes one AUI 10-Mbps Ethernet port, one RJ-45 ISDN BRI port, dual low-speed asynchronous serial ports (DB-60 connectors), and dual 2.048-Mbps serial ports (DB-60 connectors).

- **2521**   ISDN BRI/Token Ring/serial router. Includes one DB-9 16/4-Mbps Token Ring port, one RJ-45 ISDN BRI port, dual low-speed asynchronous serial ports (DB-60 connectors), and dual 2.048-Mbps serial ports (DB-60 connectors).

- **2522**   ISDN BRI/Ethernet/high-density serial router. Includes one AUI 10-Mbps Ethernet port, one RJ-45 ISDN BRI port, eight low-speed asynchronous serial ports (DB-60 connectors), and dual 2.048-Mbps serial ports (DB-60 connectors).

- **2523**   ISDN BRI/Token Ring/high-density serial router. Includes one DB-9 16/4-Mbps Token Ring port, one RJ-45 ISDN BRI port, eight low-speed asynchronous serial ports (DB-60 connectors), and dual 2.048-Mbps serial ports (DB-60 connectors).

- **2524**   Modular Ethernet router. Includes one AUI 10-Mbps Ethernet port, dual 2.048-Mbps serial ports (DB-60 connectors), and two WIC card slots that can each contain a single-port T1 interface with integrated CSU/DSU, a single-port 56-Kbps interface with integrated CSU/DSU, or a single ISDN BRI interface.

- **2525**   Modular Token Ring router. Includes one DB-9 16/4-Mbps Token Ring port, dual 2.048-Mbps serial ports (DB-60 connectors), and two WIC card slots that can each contain a single-port T1 interface with integrated CSU/DSU, a single-port 56-Kbps interface with integrated CSU/DSU, or a single ISDN BRI interface.

With the wide variety of models available, it is easy to see why the 2500 series has enjoyed such a long and prosperous life in many branch offices. Nearly any scenario, from remote access server to Token Ring/Ethernet/WAN bridge, can be performed by this versatile line. Unfortunately for fans of the 2500 series, Cisco has begun the end-of-sale (EOS) for many of the routers in this line as of this writing. The 2500 series is being steadily replaced by the newer modular 2600 series; but for quite some time to come, the 2500 will remain active and productive in many branch office scenarios. For more information about the 2500 series, visit http://www.cisco.com/warp/public/cc/pd/ rt/2500/prodlit/index.shtml.

The 2600 Series

The 2600 series is Cisco's high-end line of branch office routers. They are ideally suited for branch offices that require advanced features and extensive modularity in a moderate-performing, reasonably priced router.

As for feature sets, the 2600 series supports all of the features of the 2500 series plus the addition of voice features similar to those found in the 1700 series. In addition, the 2600 series includes three different levels of processing power for better flexibility in demanding environments.

All of the 2600 series of routers include a single network module slot, dual WIC slots, and a single advanced integration module slot. Each network module slot can contain a variety of network modules. The following list includes available network modules shared with the 3600 series:

- A 1-port DS3 ATM network module
- A 1-port E3 ATM network module

- A 16-port asynchronous serial network module
- A 32-port asynchronous serial network module
- A 4-port synchronous/asynchronous serial network module
- An 8-port synchronous/asynchronous serial network module
- A 4-port ISDN BRI network module
- A 4-port ISDN BRI network module with integrated NT-1
- An 8-port ISDN BRI network module
- An 8-port ISDN BRI network module with integrated NT-1
- A 1-port ISDN PRI network module
- A 1-port ISDN PRI network module with integrated CSU/DSU
- A 2-port ISDN PRI network module
- A 2-port ISDN PRI network module with integrated CSU/DSU
- A 1-port ISDN PRI E1 network module
- A 2-port ISDN PRI E1 network module
- A 1-port Fast Ethernet network module
- A 4-port Fast Ethernet network module
- A 4-port T1 ATM network module
- A 4-port E1 ATM network module
- A 1-port ATM 25 network module
- An 8-port T1 ATM network module
- An 8-port E1 ATM network module
- An 8-port analog modem network module
- A 16-port analog modem network module
- A 1-port, 12-channel voice/fax T1 network module
- A 1-port 24-channel voice/fax T1 network module
- A 2-port 48-channel voice/fax T1 network module
- A 1-port 24-channel-enhanced voice/fax T1 network module
- A 1-port 15-channel voice/fax E1 network module
- A 1-port 30-channel voice/fax E1 network module
- A 2-port 60-channel voice/fax E1 network module
- A 1-port 30-channel–enhanced voice/fax E1 network module
- A 2-WIC slot network module

Note that with the dual-WIC slot network module, you can potentially have up to four WIC slots in a 2600 series router. As for the WIC slots themselves, they can contain any of the following WAN interface cards:

- A 2-port voice interface card
- A 2-port T1 multiflex voice/WAN interface card
- A 2-port E1 multiflex voice/WAN interface card
- A 2-port G.703 multiflex voice/WAN interface card
- A 1-port synchronous serial WIC
- A 2-port synchronous serial WIC
- A 2-port synchronous/asynchronous serial WIC
- A 1-port ADSL WIC
- A 1-port 56-Kbps WIC
- A 1-port 56-Kbps WIC with integrated CSU/DSU
- A 1-port ISDN WIC
- A 1-port ISDN WIC with integrated NT-1
- A 1-port T1 with integrated CSU/DSU

Finally, the Advanced Integration Module in the 2600 series can contain either an 8-Mbps hardware compression module or a DES/3DES hardware VPN encryption module.

To illustrate the various module slots, the rear view of a 2600 series router is depicted in Figure 12-6.

As for the models in the line, the 2600 series includes the 2610, 2611, 2612, 2613, 2620, 2621, 2650, and 2651. The 261*x* models include a 40-MHz processor for a maximum forwarding rate of around 15,000 PPS; the 262*x* models include a 50-MHz processor, for a forwarding rate of around 25,000 PPS; and the 265*x* models include an 80-MHz processor for a forwarding rate of around 37,000 PPS.

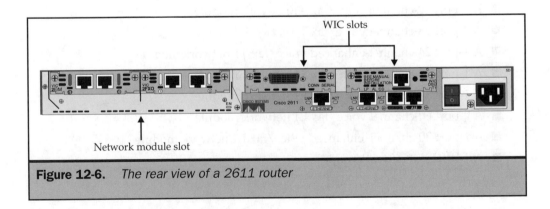

Figure 12-6. *The rear view of a 2611 router*

The 2610 model includes only a single RJ-45 10-Mbps Ethernet port; the 2611 model includes dual RJ-45 10-Mbps Ethernet ports; the 2612 model includes a single RJ-45 10-Mbps Ethernet port and one RJ-45 4/16-Mbps Token Ring port; the 2613 model includes a single RJ-45 4/16-Mbps Token Ring port; the 2620 model includes a single RJ-45 10/100-Mbps Ethernet port; the 2621 model includes dual RJ-45 10/100-Mbps Ethernet ports; the 2650 model includes a single RJ-45 10/100-Mbps Ethernet port; and the 2651 model includes dual RJ-45 10/100-Mbps Ethernet ports.

After considering the various performance, LAN, WAN, voice, and feature sets available for the 2600 series, it's easy to see how it is a perfect replacement for the 2500 series. It is also an excellent routing platform in nearly any branch office scenario. For more information about the 2600 series, visit http://www.cisco.com/warp/public/cc/pd/rt/2600/prodlit/index.shtml.

The 3600 Series

The 3600 series is Cisco's primary midsize corporate router line. These routers are suited to larger offices that require high-performance, feature-rich routing at a moderate price.

The feature sets available for the 3600 series are similar to those provided in the 2600 series, with the addition of AAA, making advanced authorization, authentication, and accounting services available.

The 3600 series includes three models: the 3620, 3640, and 3660. The 3620 includes two network module slots, supports 64MB of DRAM and 32MB of flash RAM, and forwards at around 50,000 PPS. The 3640 includes four network module slots, supports 128MB of DRAM and 32MB of flash RAM, and forwards at around 60,000 PPS. The 3640 includes six network module slots, supports 256MB of DRAM and 64MB of flash RAM, and forwards at around 70,000 PPS. The 3660 also includes two advanced integration module slots. All models in the 3600 series include PC Card slots for flash RAM PC Cards.

As for network modules, all of the 3600 series routers support the following (most of which are shared with the 2600 series):

- A 1-port OC-3 ATM network module
- A 1-port DS3 ATM network module
- A 1-port E3 ATM network module
- A 16-port asynchronous serial network module
- A 32-port asynchronous serial network module
- A 4-port synchronous/asynchronous serial network module
- An 8-port synchronous/asynchronous serial network module
- A 4-port ISDN BRI network module
- A 4-port ISDN BRI network module with integrated NT-1

- An 8-port ISDN BRI network module
- An 8-port ISDN BRI network module with integrated NT-1
- A 1-port ISDN PRI network module
- A 1-port ISDN PRI network module with integrated CSU/DSU
- A 2-port ISDN PRI network module
- A 2-port ISDN PRI network module with integrated CSU/DSU
- A 1-port ISDN PRI E1 network module
- A 2-port ISDN PRI E1 network module
- A 1-port Fast Ethernet, 1-port ISDN PRI E1 network module
- A 1-port Fast Ethernet, 2-port ISDN PRI E1 network module
- A 1-port Fast Ethernet, 1-port ISDN PRI network module
- A 1-port Fast Ethernet, 1-port ISDN PRI network module with integrated CSU/DSU
- A 1-port Fast Ethernet, 2-port ISDN PRI network module
- A 1-port Fast Ethernet, 2-port ISDN PRI network module with integrated CSU/DSU
- A 1-port Fast Ethernet network module
- A 4-port Fast Ethernet network module
- A 4-port T1 ATM network module
- A 4-port E1 ATM network module
- A 1-port ATM 25 network module
- An 8-port T1 ATM network module
- An 8-port E1 ATM network module
- An 8-port analog modem network module
- A 16-port analog modem network module
- A 6-port digital modem network module
- A 12-port digital modem network module
- An 18-port digital modem network module
- A 24-port digital modem network module
- A 30-port digital modem network module
- A 1-port 12-channel voice/fax T1 network module
- A 1-port 24-channel voice/fax T1 network module
- A 2-port 48-channel voice/fax T1 network module

- A 1-port 24-channel-enhanced voice/fax T1 network module
- A 1-port 15-channel voice/fax E1 network module
- A 1-port 30-channel voice/fax E1 network module
- A 2-port 60-channel voice/fax E1 network module
- A 1-port 30-channel–enhanced voice/fax E1 network module
- A 2-WIC slot network module
- A 1-port Ethernet, 2-WIC slot network module
- A 2-port Ethernet, 2-WIC slot network module
- A 1-port 10/100 Ethernet, 2-WIC slot network module
- A 2-port 10/100 Ethernet, 2-WIC slot network module
- A 1-port Ethernet, 1-Token Ring, 2-WIC slot network module
- A 1-port 10/100 Ethernet, 1-Token Ring, 2-WIC slot network module
- A 1-port HSSI network module
- A hardware compression network module
- A DES/3DES VPN hardware encryption module for 3620/3640
- A P2MP CPE wireless network module

As for interface cards, the 3600 series shares most of the 2600 series network modules, with the addition of a Direct Inward Dial (DID) voice interface card. The following list includes all of the WICs available for the 3600 series:

- A 2-port direct inward dial voice interface card
- A 2-port voice interface card
- A 2-port T1 multiflex voice/WAN interface card
- A 2-port E1 multiflex voice/WAN interface card
- A 2-port G.703 multiflex voice/WAN interface card
- A 1-port synchronous serial WIC
- A 2-port synchronous serial WIC
- A 2-port synchronous/asynchronous serial WIC
- A 1-port ADSL WIC
- A 1-port 56-Kbps WIC
- A 1-port 56-Kbps WIC with integrated CSU/DSU
- A 1-port ISDN WIC
- A 1-port ISDN WIC with integrated NT-1
- A 1-port T1 with integrated CSU/DSU

The advanced integration module slots in the 3660 can also contain an 8-Mbps hardware compression module, or a DES/3DES hardware VPN encryption module, or both.

After reviewing all of the features, performance, and versatility available in the 3600 series, it should be easy to see how it is uniquely positioned as a high-performance midsize routing platform, and that its unique features can benefit certain environments. For more information on the 3600 series, visit http://www.cisco.com/warp/public/cc/pd/rt/3600/prodlit/index.shtml.

The 7100 Series

The 7100 series is a niche line of routers. These routers are primarily designed to provide high-speed, hardware-assisted VPN services, in addition to functioning as a full-featured router. To this end, they are well suited to any environment able to take advantage of the massive number of VPN sessions that the device is able to support simultaneously.

The VPN features of the base model of 7100 router include hardware support for up to 2000 simultaneous Point-to-Point Tunneling Protocol (PPTP) sessions with an encryption/decryption rate allowing performance to reach up to 90 Mbps, allowing this device to support extensive quantities of standard Windows clients. With the installation of an optional integrated services module, the 7100 series can also support wire-speed Layer 2 Tunneling Protocol and IPSec sessions for up to 2000 clients at 90 Mbps, allowing it to support UNIX and Windows 2000 clients with a higher level of security.

The rest of the 7100 series' feature set is pretty similar to the 3600 series, including support for most routing protocols, diverse multiprotocol support, and advanced queuing and QoS features. The 7100 series does not, however, support voice traffic like the 3600 series does.

As for models, the 7100 series is divided into two main models, with several submodels of each model differentiated only by the WAN interface embedded in the particular model. The two primary models in the 7100 series are the 7120 and 7140. Both models come in a compact, two-slot chassis. All models come equipped with PC Card slots for flash RAM expansion.

The 7120 model is the base model of router in the series. The 7120 includes a MIPS 5000 processor capable of propelling the 7120 to a maximum forwarding rate of between 70,000 and 175,000 PPS (depending on the features enabled). The 7120 includes a single embedded 10/100 Fast Ethernet port, an embedded WAN port (a four-port serial interface, T1/E1, T3, E3, or OC-3 ATM interface), and a single LAN/WAN expansion slot for additional ports.

The 7140 model is the high-end model in the series. The 7140 includes a MIPS 7000 processor capable of propelling it to a maximum forwarding rate of between 150,000 and 300,000 PPS (again, depending on the features enabled). The 7140 includes dual embedded 10/100 Fast Ethernet ports, an embedded WAN port (which can include eight serial interfaces, eight T1/E1 interfaces, dual T3, dual E3, or dual OC-3 ATM interfaces), and a single LAN/WAN expansion slot for additional ports. In addition, the 7140 comes standard with redundant power supplies.

The LAN/WAN expansion slots in the 7100 series are not shared with any other model of router, but an extensive variety of expansion options are offered, including the following:

- A 4-port Ethernet
- An 8-port Ethernet
- A 5-port 10Base-FL Ethernet
- A 1-port Fast Ethernet
- A 1-port Gigabit Ethernet
- A 2-port Token Ring ISL (100Base-TX)
- A 2-port Token Ring ISL (100Base-FX)
- A 4-port 4/16 Mbps Token Ring
- A 4-port serial
- An 8-port serial
- A 1-port HSSI
- A 2-port HSSI
- A 1-port ATM DS3
- A 1-port ATM E3
- A 1-port ATM OC-3
- A 1-port T3 with integrated CSU/DSU
- A 2-port T3 with integrated CSU/DSU
- A 1-port E3 with integrated CSU/DSU
- A 2-port E3 with integrated CSU/DSU
- A 2-port T1 with integrated CSU/DSU
- A 4-port T1 with integrated CSU/DSU
- An 8-port T1 with integrated CSU/DSU
- A 2-port E1 with integrated CSU/DSU
- An 8-port E1 with integrated CSU/DSU
- A 4-port ISDN BRI (U interface)
- An 8-port ISDN BRI (S/T interface)

Looking at the options available, you can easily see that the 7100 series is a full-featured routing platform as well as a high-performance VPN device. This unique blend makes the 7100 series a fit for organizations that require both high-performance routing and high-performance VPN features in one unit, as well as for companies that need a dedicated VPN device. For more information about the 7100 series, visit http://www.cisco.com/warp/public/cc/pd/rt/7100/prodlit/index.shtml.

The 7200 Series

The 7200 series is designed to be a high-end enterprise router that also performs well in service-provider (i.e., ISP) networks. It includes a dizzying array of features, versatility, and a high-performance routing/switching engine to support a wide variety of functions.

The features available in the 7200 series include all of the features from both the 3600 and 7100 series (including extensive multiservice voice, video, and data support, as well as hardware VPN support), and add to this already impressive list by incorporating extensive support for WAN switching (including ATM, X.25, and Frame Relay switching), TDM support (to switch DS0 channels between physical DS1 and higher links), and CEF or PXF high-speed layer 3 switching.

The 7200 series comes in four distinct models: the 7204, 7204VXR, 7206, and 7206VXR. The primary difference in these models is in the number of slots in the chassis (the 7204 models include four slots, whereas the 7206 models include six slots), and the degree of multiservice support (only the VXR models include full multiservice support).

The 7200 series is a truly modular router line; the I/O controller, processing engine, interface cards, and chassis are all separate components, allowing you to build a custom 7200 series router to suit your needs.

The I/O controller in the 7200 series is responsible for directing and monitoring the flow of data across the midplane; it provides the console and auxiliary ports as well as the PC Card slots and, in some models, Fast or Gigabit Ethernet ports for uplinking to other switches and routers. The I/O controllers available for the 7200 series include a standard controller with no Ethernet ports, a controller with one Fast Ethernet port, a controller with dual 10/100 Fast Ethernet ports, and a controller with a single Gigabit Ethernet port and a single Ethernet port.

The processing engine on the 7200 series is responsible for the bulk of the switching and routing duties, as well as any optional functions such as encryption, compression, and QoS. The 7200 series supports three network processing engines (NPE): the NPE-225, NPE-300, and NPE-400. The primary difference in these engines is the performance offered and the degree of multiservice support.

The NPE-225 is the low end of the NPE family, and it is designed primarily for regional office or small service-provider duties. The NPE-225 supports a maximum forwarding rate of 225,000 PPS, and supports all 7200 chassis. It does not include multiservice support, but does include channel aggregation capabilities.

The NPE-300 is the midrange in the NPE family, and it is intended to be used in enterprise or midlevel service-provider networks. It supports forwarding rates up to

300,000 PPS, and enables the 1-Gbps bus in the VXR routers. Full multiservice support is included, and TDM capabilities are included in this model as well. The NPE-300 is available only in the VXR line of routers.

The NPE 400 is the high-end model of processing engine. It is designed primarily for service-provider duties, and it supports a maximum forwarding rate of 400,000 PPS. It supports all of the features of the NPE-300 and, like the NPE-300, is available for only the VXR routers.

The interface cards in the 7200 line are wide ranging, but are mostly for WAN support duties. Note that none of the interface card slots in the 7200 series are used for NPE or I/O controller duties, unlike the slots in some of the switch models. (See Figure 12-7 for clarification.) For this reason, all four or six slots (depending on the model) are available for interface cards.

The available interface cards for the 7200 series are the following:

- A 4-port Ethernet
- An 8-port Ethernet
- A 1-port 10/100 Fast Ethernet
- A 2-port 10/100 Fast Ethernet
- A 12-port Ethernet, 2-port Fast Ethernet Etherswitch
- A 4-port 4/16 Mbps Token Ring
- A 4-port serial
- A 4-port G.703 (E1) serial
- An 8-port serial
- A 1-port HSSI
- A 2-port HSSI T3 with integrated CSU/DSU
- A 2-port HSSI E3 with integrated CSU/DSU
- A 2-port T1
- A 4-port T1
- An 8-port T1

- A 2-port E1
- An 8-port E1
- A 1-port T3
- A 2-port T3
- A 1-port E3
- A 4-port ISDN BRI (U interface)
- An 8-port ISDN BRI (S/T interface)
- A 2-port T1/E1 voice
- A 2-port SONET OC-12
- A 1-port SONET OC-3
- An 8-port ATM T1
- An 8-port ATM E1
- A 1-port ATM DS3
- A 1-port ATM E3
- A 1-port ATM OC-3
- A 1-port ATM OC-12
- A 1-port mainframe channel port adapter

As you can see, with the variety of interfaces, I/O controllers, NPEs, and chassis options, the 7200 series serves well in a variety of high-performance roles. For more information on the 7200 series, check out this web site: http://www.cisco.com/warp/public/cc/pd/rt/7200/prodlit/index.shtml.

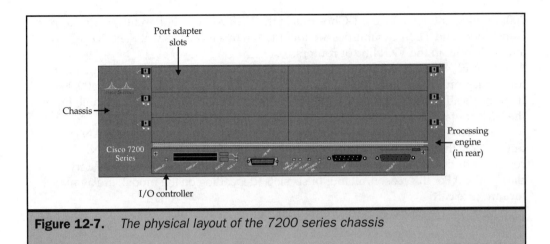

Figure 12-7. The physical layout of the 7200 series chassis

The 7500 Series

The Cisco 7500 series router line is designed for enterprise and service-provider networks that require extremely high-performance, modular, high-functionality routing solutions.

The 7500 series supports all of the features in the 7200 series, including the multiprotocol and multiservice support, and adds additional switching support. The 7500 series hardware architecture, however, allows for much higher levels of performance than what is available on the 7200 series. This allows the 7500 series to be used in situations in which the functionality of a 7200 would be adequate, but the speed is not (specifically, in very high-speed enterprise backbones or for midlevel ISP duties).

The 7500 series is made up of four major hardware components: the chassis, the Route Switch Processor (RSP), the Versatile Interface Processor (VIP), and the interface cards (called port adapters). Figure 12-8 depicts a 7500 series router.

The RSP is responsible for most of the functions in the 7500 series. These include running IOS, performing all routing updates, managing all routing/switching tables and caches, and handling all monitoring and management functions. Because of this, the routing features and performance in the 7500 series is almost directly related to the supported features and performance of the RSP.

Four models of RSP are currently available: the RSP1, RSP2, RSP4, and RSP8. The primary difference in these models is the amount of memory supported and the processor performance. Most RSPs support redundant RSPs in the same chassis. The most advanced RSP, as of this writing, is the RSP8. The RSP8 supports up to 256MB of RAM, and includes a 250-MHz RISC processor for high-performance environments.

The VIP, on the other hand, deals with most of the switching duties in the 7500 series. Each card slot in the 7500 series can contain a VIP, and each VIP can contain up to two port adapters (PAs).

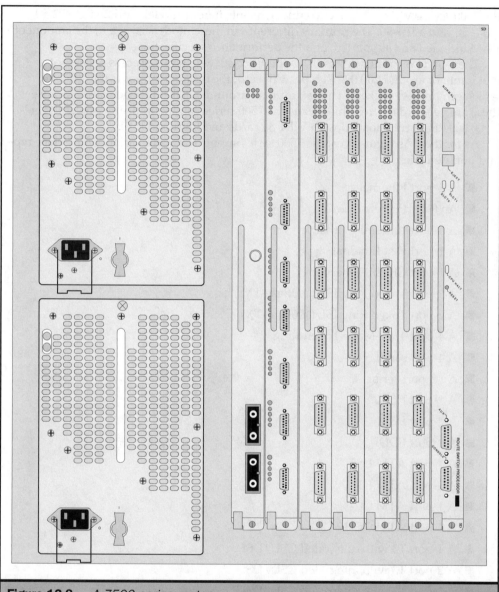

Figure 12-8.   *A 7500 series router*

The VIP includes its own processor and ASICs for switching functions and, therefore, reduces the load on the RSP by performing most functions through switching ASICs in the VIP card itself. For this reason, the VIP has a drastic impact on performance in the 7500 series. VIP cards are currently available in four models: VIP2-40, VIP2-50, VIP4-50, and VIP4-80. The primary difference in these models is, again, the amount of RAM included and supported, and the performance specifications.

The worst performer in this list is the VIP2-40, which can process packets at a maximum of around 200,000 PPS per VIP card; the best performer is the VIP2-80, which can process packets at a maximum of around 420,000 PPS per card. Because these are per-card figures, this means the 7500 series can theoretically process over 2,000,000 PPS if fully loaded with VIP4-80 cards and high-end PAs.

As for the PAs themselves, the 7500 series supports a wide variety of interface types and port densities:

- A 4-port Ethernet

- An 8-port Ethernet

- A 5-port 10Base-FL Ethernet

- A 1-port Fast Ethernet

- A 1-port Gigabit Ethernet

- A 2-port Token Ring ISL (100Base-TX)

- A 2-port Token Ring ISL (100Base-FX)

- A 4-port 4/16-Mbps Token Ring

- A 4-port serial

- An 8-port serial

- A 1-port HSSI

- A 2-port HSSI

- A 1-port FDDI

- A 1-port ATM DS3

- A 1-port ATM E3

- A 1-port ATM OC-3

- A 1-port ATM OC-12

- A 1-port T3 with integrated CSU/DSU

- A 2-port T3 with integrated CSU/DSU

- A 1-port E3 with integrated CSU/DSU

- A 2-port E3 with integrated CSU/DSU

- A 2-port T1 with integrated CSU/DSU

- A 4-port T1 with integrated CSU/DSU

- An 8-port T1 with integrated CSU/DSU

- A 2-port E1 with integrated CSU/DSU

- An 8-port E1 with integrated CSU/DSU

- A 4-port ISDN BRI (U interface)

- An 8-port ISDN BRI (S/T interface)

- A 1-port STM-1

- A hardware encryption service adapter

- A hardware compression service adapter

Now that we have examined all of the components in the 7500 series, let's examine the individual models in the line. The 7500 series consists of four models: the 7505, 7507, 7513, and 7576. The primary difference in all of these models (except the 7576) is the number of slots in the chassis. The 7505 has five total slots, four of which can be used for interface processors. (One slot is for the RSM.) The 7507 has seven total slots, five of which can be used for interface processors. (Two slots are for the RSM and redundant RSM.) The 7513 has 13 total slots, 11 of which can be used for interface processors. (Two slots are for the RSM and redundant RSM.)

The 7576 outwardly appears to be the same as the 7513, but the 7576 is actually designed to support two RSMs running concurrently, making it basically two routers in one chassis. For this, the backplane on the 7576 has been increased to double the capacity of the 7513 to 4 Gbps. The 7576 supports the same number of slots and usable slots as the 7513.

Overall, the 7500 series (also called the 7500 Advanced Router System or ARS) is an intelligent combination of router and high-end layer 3 switch that serves up high-performance, feature-rich IP services for enterprise and service-provider networks. For more information on the 7500 series, visit http://www.cisco.com/warp/public/cc/pd/rt/7500/prodlit/index.shtml.

The 10000 Series

The Cisco 10000 Edge Services Router (ESR) is an ultra-high-speed router designed primarily for service-provider networks that require high numbers of aggregate channel terminations. A fully equipped 10000 ESR can support thousands of channelized T1 speed connections (around 20,000 per chassis using channelized OC-12 cards). Even with all of these simultaneous connections, the 10000 series supports full IP services on every channel, allowing it to offer an unprecedented level of performance for ISP environments.

The 10000 series supports all of the features common in the 7500 series, except for multiservice support, but outperforms the 7500 series in all respects, due to the advanced PXF architecture of the 10000 series.

The 10000 series is composed of three primary components: the chassis, the Performance Routing Engine (PRE), and interface cards.

The chassis of the 10000 series is available in three versions: the 10005, 10008, and 10000. The 10005 comes with five total slots, two of which are dedicated to the primary and backup PREs, allowing three slots for interface cards. The 10008 comes with eight total slots, two of which are dedicated to the primary and backup PREs, allowing six slots for interface cards. Finally, the 10000 comes with ten total slots, two of which are dedicated to the primary and backup PREs, allowing eight slots for interface cards.

The PRE modules for the 10000 series control all of the routing and switching functions, as well as running the IOS on the 10000 series. Unlike the controlling processing engine in most other lines of Cisco routers, the PRE in the 10000 series does not use a single processor. Instead, it uses 16 individual custom processors to perform routing and switching functions at over 2,000,000 PPS. Also, all of the processors in the PRE have software-upgradeable microcode, meaning that enabling support for new features does not require purchasing a new PRE. Only one PRE module is available for the 10000 series.

The interface cards (called *line cards* in this series) in the 10000 series are not as widely varying as in some of the other lines of routers, reflecting this device's singular purpose. The available line cards for the 10000 series are the following:

- A 1-port Gigabit Ethernet
- A 6-port channelized T3
- A 4-port channelized OC-3
- A 1-port channelized OC-12

The 10000 series ESR is a high-performance, service-provider routing system rivaled only by the 12000 series in channel density or performance. For more information on the 10000 series, visit http://www.cisco.com/warp/public/cc/pd/rt/10000/prodlit/index.shtml.

The 12000 Series

The 12000 series Gigabit Switch Router (GSR) is the top of the Cisco routing line. These routers are designed for the highest network performance in the core of service-provider networks. Due to this, they support switching fabric speeds and forwarding rate unrivalled by any other Cisco router line (or switching line, for that matter), while at the same time supporting the vast array of features found in the 10000 series.

The components of the 12000 series follow the pattern set by most of the high-end Cisco routers. The chassis provides the foundation, backplane, and slot capacities; the Gigabit Route Processor (GRP) provides the routing logic; and, like the 7500 series, the individual interfaces provide the switching logic. Figure 12-9 provides an example of a 12000 series router.

The 12000 series is composed of six router models: the 12008, 12012, 12016, 12406, 12410, and 12416. The primary difference in these models is in the interface slots supported and the backplane speed.

The 12008 GSR includes eight slots, one of which is dedicated to the GRP. The other seven slots can be used solely for line cards, or six of them may be used for line cards and the seventh used for a redundant GRP. The 12012 follows the same pattern, with

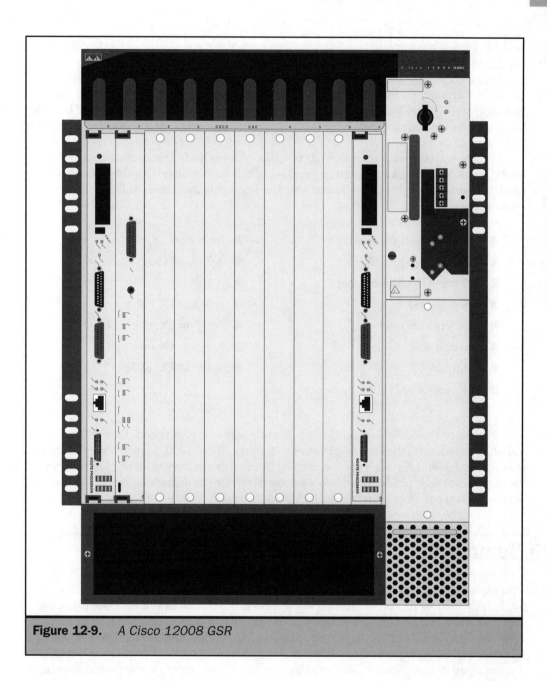

Figure 12-9.   *A Cisco 12008 GSR*

12 slots total, one of which is reserved for the GRP, and the other 11 can be used for line cards (or you can use ten for line cards and one for a redundant GRP). The 12016 has 16 slots, up to 15 of which may be used for line cards. All of the 120*xx* models

include a 2.5-Gbps per-card backplane. This backplane speed allows the fastest of the 120xx line (the 12016) to forward at a maximum of around 60,000,000 PPS.

The 124xx routers include a 10-Gbps per-card backplane. This allows the 12416 to forward at around 375,000,000 PPS, with the other 124xx models trailing a bit behind this figure. The 12406 includes 6 total slots (5 maximum for line cards), the 12410 includes 10 total slots (9 maximum for line cards), and the 12416 includes 16 total slots (15 maximum for line cards).

The line cards available for the 12000 series, although a bit more varied than the 10000 series, still reflect the singular purpose of this router line, with the slowest line card available supporting T3 connections. The line cards available for the 12000 series include the following:

- An 8-port Fast Ethernet
- A 1-port Gigabit Ethernet
- A 3-port Gigabit Ethernet
- A 6-port DS3
- A 12-port DS3
- A 6-port E3
- A 12-port E3
- A 4-port OC-3

- An 8-port OC-3
- A 16-port OC-3
- A 1-port OC-12
- A 4-port OC-12
- A 1-port OC-48c
- A 4-port OC-48c
- A 1-port OC-192c

Obviously, the 12000 series is built to handle some serious speed. Due to the capabilities, performance, and—of course—the cost of the 12000 series, it is relegated mostly to high-end service-provider duties; but in this environment, the 12000 series excels like no other product in Cisco's lineup. For more information on the 12000 series, visit http://www.cisco.com/warp/public/cc/pd/rt/12000/prodlit/index.shtml.

Summary

This chapter examined each major routing line in detail. This information will help you recognize the features and capabilities of the routers you will need to interact with daily in most routed environments. As always, you can find the most up-to-date information by using the following URL for Cisco's web site. However, the timesaving overview presented in this chapter should help you understand which line is most appropriate for your particular situation.

For more information on Cisco's routing line, check out http://www.cisco.com/warp/public/44/jump/routers.shtml.

The Complete Reference

Cisco

Chapter 13

Access Servers, Cache Engines, and Security Products

This chapter covers the basics of the rest of the major Cisco product lines. As with the previous chapters, the goal is to provide you with a basic understanding of the purposes and uses of these devices. Also, as with the previous chapters, you can find the most up-to-date information on Cisco's web site by using the URLs provided. Finally, realize that these are (for the most part) niche products, some of which you may never see in real life. However, they are still an important part of Cisco's product lineup and deserve some recognition for the part they can play in a modern network.

Access Servers

Access servers are basically self-contained remote access server products. They are designed to provide a very compact, low-cost-per-port alternative to PC-based remote access servers for environments (such as service providers and enterprise networks) that are required to support a lot of simultaneous remote access connections. Therefore, the terminology used and features supported are a bit different from other products you have seen thus far. To begin our discussion of these differences, let's start with hardware.

Access servers are a bit more like purpose-built routers than anything else. (In fact, the 2500 series access servers actually *are* purpose-built routers.) Their primary features are provided through software, making the processor and RAM supported of primary concern.

Supporting dial-up users and port densities is also important. Some access servers, such as the 2500 series, offer asynchronous serial ports. This approach means that you need to purchase external modems (nothing special, just standard external modems, but make sure they are not "win" modems) and plug them into the serial ports on the access server.

Other access servers, such as the AS5300, use integrated modem banks to support remote access users. These modems don't typically have physical ports, however. These servers use channelized trunk lines (like a T1) and split each channel (24 per T1) off inside the access server. Then they feed each channel to an internal modem for the demodulation portion of the communication, and then the modem forwards the raw data to the system processing unit.

As for the IOS, this varies slightly depending on the model of access server, although most use a standard-style IOS. The commands used for access server configuration may take some getting used to, and they are a bit harder to understand than the point-and-click interfaces that most PC-based remote access servers provide. However, again, once you master the commands, you can typically configure a Cisco access server in a fraction of the time it takes to configure a PC-based access server. Also, because they are purpose-built devices, the reliability and performance of Cisco access servers is typically much higher than PC-based solutions.

Let's take a look at the individual access server lines.

AS Series Universal Access Servers

The AS series universal access servers are a line of self-contained, compact, and feature-rich remote access servers that offer a large port density in a very small footprint for providers who need to maximize ports-per-rack. This makes it a very commonly used device in ISP environments. The AS series also includes universal access gateways, which, in addition to supporting large quantities of direct-dial modem ports, also support high-capacity voice T connections and features. For our purposes, we will concentrate solely on the access servers in this series.

The AS series includes two primary lines of access server: the AS5300 and the AS5800. The AS5300 is shown in Figure 13-1 and the AS5800 is shown in Figure 13-2.

The AS5300 is a compact, two-rack space unit including a reasonably fast 150-MHz RISC processor, 64MB of primary DRAM, 16MB of packet buffer DRAM, and 16MB of flash memory. The AS5300 has a rich feature set, including support for modem, ISDN, and VoIP (Voice over IP) connections and extensive remote access protocol support (SLIP, PPP, X-Remote, ATCP, and just about every other access protocol in existence).

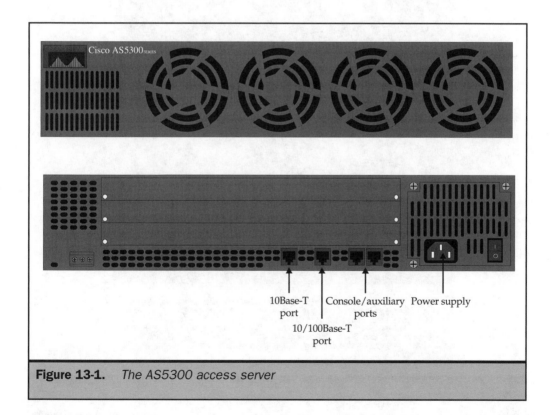

Figure 13-1.   *The AS5300 access server*

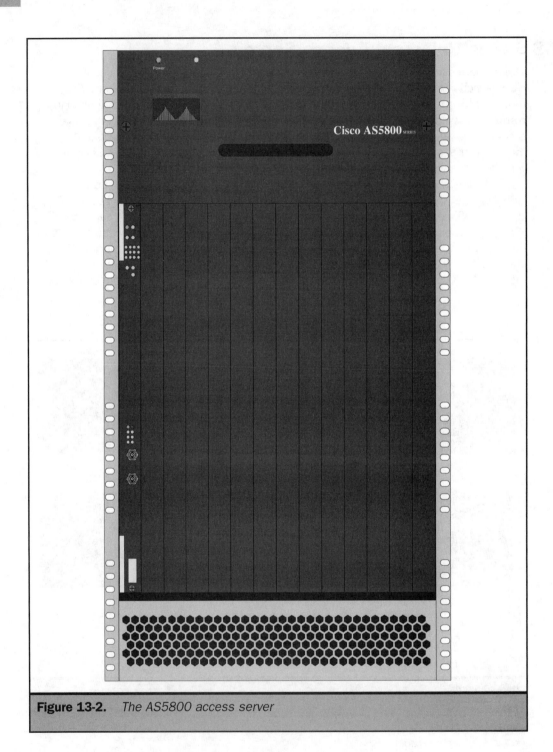

Figure 13-2. *The AS5800 access server*

As for port densities, the AS5300 contains three slots for interface cards. One slot is dedicated for WAN connections using either a four-port T1/E1 PRI card or an eight-port T1/E1 serial card. The other two ports are available for analog modem cards, and they can contain from 6 to 120 modems per card using high-density MICA carrier cards and Microcomm 6- or 12-port modem modules. This brings the total port density for analog modems on the AS5300 to 240 per two-rack unit chassis: impressive indeed. This allows a single AS5300 to support between 500 and 2400 users in most ISP point of presence (POP) environments, as individual users connect at different times throughout the day.

Performance for the AS5300 is around 100 PPS per port, which is adequate for most needs, and puts the total performance for the chassis at around 250,000 PPS. In addition, the reliability is around 500,000 hours mean time between failures (MTBF). A 500,000 MTBF figure gives the AS5300 about a 2-percent chance of failure per year, if run 24-7, 365 days a year.

All things considered, the AS5300 is a reliable, high-density access server solution that provides adequate performance in most remote access environments. You can find more information about the AS5300 at http://www.cisco.com/warp/public/cc/pd/as/as5300/prodlit/index.shtml.

The AS5800, on the other hand, is an access server *system*, and not a single unit. It consists of a dial shelf and a router shelf. The dial shelf includes 14 line card slots, each of which can contain 12-port T1/E1, single-port T3, or 144-port modem cards. In addition, one or two card slots are used for controller cards. This means the maximum density for modem connectivity on the AS5800 is 1440 ports. This should allow a large ISP to support up to 12,000 users per AS5800. In addition, all of the modem ports in the AS5800 are DSP controlled; they can be upgraded with a simple software procedure, making it unnecessary to replace modem cards to upgrade to new functionality.

The router shelf on the AS5800 contains a 7206VXR router, providing exceptional routing performance (around 300,000 PPS) and extensive WAN/LAN interface connectivity options.

As for features, the AS5800 supports all of the features of the AS5300, but it adds the additional functionality of the 7200 VXR series routers and the ability to switch and support pure voice calls. In addition, the AS5800 is designed for redundancy and provides for an MTBF of over 515,000 hours (1.7 percent chance of failure per year, or 99.999 percent uptime, also called "5nines" uptime).

These features and options make the AS5800 a true carrier-class solution, well suited to ISPs in need of higher port densities and enhanced feature sets. For more information on the AS5800, visit http://www.cisco.com/warp/public/cc/pd/as/as5800/prodlit/index.shtml.

AccessPath Access Servers

The AccessPath access server line is designed for extremely high-port-density environments, such as those found at midrange and top-tier ISPs. The AccessPath line consists of three primary models: the AccessPath-TS3, the AccessPath-UGS4, and the AccessPath-VS3. We will concentrate on the AccessPath-TS3 because it is the standard access server in the line. The AccessPath-UGS4 is a gateway. It provides voice, data, fax, and wireless services. It is a bit beyond the scope of our discussion. The AccessPath-VS3 is primarily for voice and fax communications, so it, too, is a bit out of our realm.

The AccessPath-TS3 is an ultra-high-density, ultra-high-performance, carrier-class modular access platform. Like the AS5800, it does not consist of a single unit. Rather, it is composed of four shelves. The system includes dial shelves (for modem ports), a system controller, a router shelf, and an optional switch interconnect.

The AccessPath-TS3 is not really a single unit but, rather, a combination of other stand-alone units with a special configuration allowing them all to be managed through one interface.

For the dial shelves, AS5300 access servers are used. For the system controller, a 3640 router is used. For the router shelf, a 7206 router is used; and for the optional switch interconnect, a 5002 switch is used. All of these are connected by special cabling and controlled as a single unit by special Cisco software called Cisco Access Manager (CAM). Figure 13-3 shows the AccessPath-TS3.

The compartmentalized design of the AccessPath system allows for a high degree of modularity and interface options but still allows the device to be managed as if it were a single integrated unit. For port densities, the AccessPath TS3 supports the following port configurations:

- Support for up to 168 T1/E1 trunk connections
- Support for over 4000 simultaneous POTS dial-up or ISDN BRI modem connections
- Support for standard 7200 series WAN data interfaces, including FDDI, ATM, HSSI, and serial connections (redundancy is supported).

These configurations allow a single AccessPath-TS3 to support up to 40,000 individual dial-up users in an ISP POP scenario. Of course, all of these ports don't come cheap, and the AccessPath-TS3 is *not* a compact unit. One entire rack is needed for a standard AccessPath-TS3 configuration. (Luckily, Cisco sells the full rack configuration already assembled, as well.)

As for reliability, the AccessPath-TS3 includes optional redundancy for nearly every component in the system. This allows the AccessPath-TS3 to achieve greater than 6nines, or 99.99998 percent uptime, which translates into an average of less than six seconds of downtime per year! Obviously, the AccessPath-TS3 is an extremely reliable system. Although expensive, it provides ISPs and other service providers with one of the most comprehensive, full-featured, reliable, high-performance, high-port-density remote access systems currently available. For more information on the AccessPath-TS3, visit http://www.cisco.com/warp/public/cc/pd/as/apsz/prodlit/index.shtml.

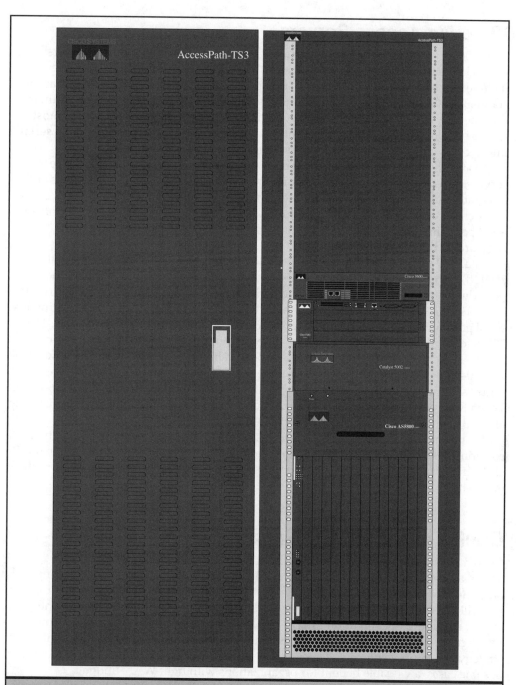

Figure 13-3.   *An AccessPath TS3 system*

2500 Series Access Servers

The 2500 series access server line is actually just purpose-built 2500 series routers with a large number of asynchronous serial interfaces. Therefore, these access servers include exactly the same performance, reliability, and feature-set specifications as the standard 2500 series routers. Unfortunately, they are also reaching the end of their lifetime and will most likely be replaced with modular 2600 series routers in most future installations. (A 2600 series router equipped with 8 or 16 asynchronous serial interfaces is equivalent to a 2500 series access server.)

Cisco calls the 2500 series access servers "general purpose, low-density" access servers, and that is a fairly accurate assessment. They include 8 or 16 serial ports and integrate routing capability, making them an ideal solution for very small POPs or low-density corporate remote access duties. With low per-port pricing and reasonably good performance, they are typically both cheaper and better performing than comparable PC-based solutions as well.

The 2500 series access server line is made up of five standard models: the 2509, 2509 RJ, 2511, 2511 RJ, and 2512. The 2509 model includes one AUI 10-Mbps Ethernet port, dual synchronous serial ports, and eight asynchronous serial ports using an octal cable. The 2509 RJ model includes one AUI 10-Mbps Ethernet port, a single synchronous serial port, and eight asynchronous serial ports with individual RJ-45 interfaces. The 2511 model includes one AUI 10-Mbps Ethernet port, dual synchronous serial ports, and 16 asynchronous serial ports using an octal cable. The 2511 RJ model includes one AUI 10-Mbps Ethernet port, a single synchronous serial port, and 16 asynchronous serial ports with individual RJ-45 interfaces. Finally, the 2512 model includes one 4/16-Mbps Token Ring port, dual synchronous serial ports, and 16 asynchronous serial ports using an octal cable. Figure 13-4 shows a 2509 model.

In summary, the Cisco 2500 line of access servers are excellent choices in situations requiring cost-effective, low-end remote access solutions. For more information on the 2500 series, check out http://www.cisco.com/warp/public/cc/pd/as/2500as/prodlit/index.shtml.

Cache and Content Engines

Cisco *cache* and *content engines* are designed to speed up access to remote content and reduce network traffic across WAN links. They do this by performing caching on content located across WAN segments. Much like a proxy server, whenever a particular resource is accessed extensively (such as a web page), the cache or content engine stores the document on internal disks. Then the next request for the same resource can be retrieved from the disk, rather than through the WAN link. This strategy can improve perceived speed and reduce WAN traffic by up to 60 percent in some cases, a valuable benefit for organizations with tight WAN budgets.

The cache or content engine uses a protocol called Web Cache Communication Protocol (WCCP) to allow for even-more-advanced functionality. With this protocol,

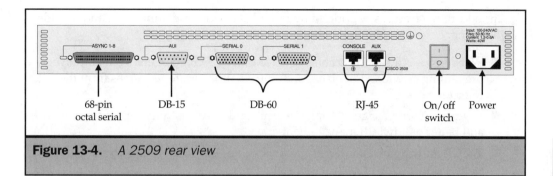

Figure 13-4. *A 2509 rear view*

a cache or content engine can be installed on the network and provide caching services with no change to the client configuration. It does this by communicating with Cisco routers that also support WCCP. When certain types of content are accessed, the router forwards the request to the cache or content engine. The cache or content engine then retrieves the document either from its hard disks or by requesting the document from the router, and then forwards the document to the client. If the cache/content engine needs to retrieve the information from the WAN, it may also cache the document to disk, allowing future accesses of the document to be retrieved directly from cache.

WCCP also allows cache/content engines to be *clustered,* or chained together as one logical entity, to improve both performance and fault tolerance. If one of the cache/content engines in the cluster fails, the others simply pick up the excess traffic and client requests can still be fulfilled. Also, if all of the members of the cluster fail, the router simply bypasses the cluster and retrieves the document directly. In short, WCCP allows the advanced functionality of caching to be completely transparent to end users.

Because cache and content engines perform largely the same function as file servers, the hardware resources most important to their function are similar to those of file servers. Physical DRAM and disk storage capacity and performance are of primary importance, with processor power running a close second. The performance of a cache or content engine is represented by maximum forwarding speed in megabits per second. Cisco has a cache or content engine solution providing optimal performance for nearly every environment, from SOHO branches to ISP backbones. With this, let's take a look at Cisco's individual product offerings.

Cache Engine 500 Series

The Cache Engine (CE) 500 series is Cisco's basic caching line. These devices support features for caching basic HTTP traffic, including the ability to bypass caching entirely for certain addresses and the ability to include the original header in the HTTP request (useful when retrieving dynamic content).

The additional features supported by the CE 500 series are numerous, and include the following:

- **Websense** Allows the cache engine to block certain content from employees, preventing download of objectionable or inappropriate resources from the Internet

- **Logging of web usage** Details Internet and WAN activity for future reference and better bandwidth management

- **WCCP multihomed router support** Enables multiple routers to share the same cache engine or cache engine cluster

- **Dynamic bypass** Allows clients to bypass the cache engine for sites that require authentication, preventing authentication errors

- **Overload bypass** Allows clients to dynamically bypass the cache engine when the engine is overloaded with traffic and response times are slow

- **Fault-tolerant disk support** Redundant drives are supported using mirroring (RAID 1)

- **Reverse proxy** Allows the cache engine to cache content located on internal web servers and provide that content to external clients, thereby reducing the load on internal web servers and hiding those servers from the outside world

- **Configurable data expiration** Allows you to fine-tune the *freshness* (how up-to-date the content is) of web data stored on the cache engine

The CE 500 series comes in four distinct models: the CE 505, CE 550, CE 570, and CE 590. All models contain essentially the same feature set. The primary difference in the models stems from their performance levels and hardware specifications. The CE 505 has a forwarding rate of about 1.5 Mbps, and it includes 128MB of SDRAM along with 9GB of hard disk space, which is upgradeable to 18GB. It is designed for small offices using T1 or E1 access links.

The CE 550 has a forwarding rate of about 11 Mbps, and it includes 256MB of SDRAM along with 18GB of hard disk space. It is designed for larger offices, with higher-speed (multiple T1 or fractional T3) links.

The CE 570 has a forwarding rate of about 22 Mbps, and it includes 512MB of SDRAM along with 36GB of hard disk space, which is upgradeable to 144GB using an external Cisco Storage Array. It is designed for regional offices and midlevel service providers with T3 or E3 links.

Finally, the CE 590 has a forwarding rate of about 45 Mbps. It includes 1GB of SDRAM along with 36GB of hard disk space, which is upgradeable to 252GB using an external Cisco Storage Array. It is designed for service provider and enterprise duty, with T3/E3 or higher-speed links.

All CE 500 models include dual auto-sensing 10/100 Fast Ethernet interfaces and take up one rack unit. They all also use SCSI disks. Figure 13-5 provides an example of a CE 500 series cache engine.

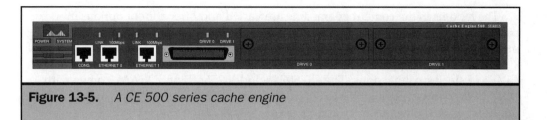

Figure 13-5.   *A CE 500 series cache engine*

As for reliability, with clustering, the CE 500 series can perform with 6nines or higher uptime ratings. With support for advanced features and a model to suit nearly any performance need, the CE 500 series of cache engines makes a major impact in any environment that requires extensive WAN usage. For more information on the CE 500 series of cache engines, visit http://www.cisco.com/warp/public/cc/pd/cxsr/500/prodlit/index.shtml.

Cisco Content Engines

The Cisco Content Engine line expands on the CE 500 line's functionality by offering caching for more service, higher-performance, and more content-control features. It is designed for organizations that need more than just HTTP caching services or higher speeds than those attainable with the CE 500 Cache Engine series.

All content engine models support the existing features provided by the cache engine line. They expand on these substantially, however, and offer support for the following additional features and protocols:

- **Content Delivery Network (CDN)**   Provides enhanced performance for streaming media, such as audio or video streaming, by caching this content and distributing it to clients as necessary. This allows for significantly reduced network traffic across WAN links and even across LAN backbones, as illustrated in Figure 13-6.

- **Wide protocol support**   Includes HTTP, FTP, and HTTPS.

- **WCCP flow protection**   Allows previously established sessions to continue regardless of changes (additions, removals, or failures) to the cache engines in the cluster.

- **Upstream proxy support**   Allows content engines to forward requests for needed information to upstream proxies or content engine clusters before retrieving the data directly.

- **Employee Internet Management (EIM)**   Allows employees access to objectionable content to be tracked and audited, as well as blocked.

The models in the content engine line are varied and offer a lot of options in performance and storage support to accommodate any sized environment. The models in the line are the CE 507, CE 507 AV, CE 560, CE 590, and CE 7320.

At the low end, the CE 507 model is available to support small branch offices with T1/E1 or slower connections. The CE 507 includes 256MB of SDRAM, comes with 18GB of disk space (expandable to 36GB), forwards at around 1.5 Mbps, and includes dual 10/100 Fast Ethernet interfaces. It uses one rack unit.

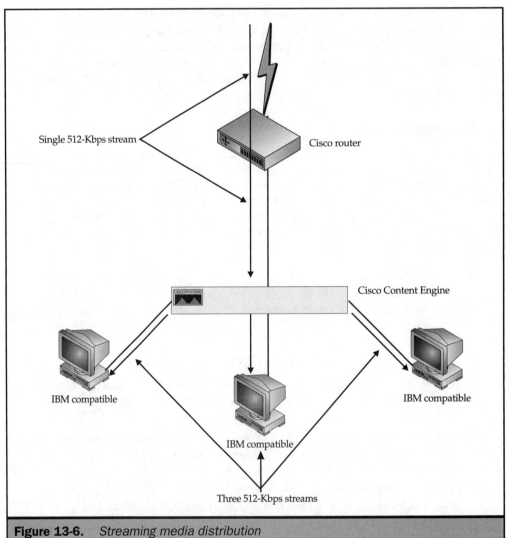

Single 512-Kbps stream

Cisco router

Cisco Content Engine

IBM compatible

IBM compatible

IBM compatible

Three 512-Kbps streams

Figure 13-6. *Streaming media distribution*

The CE 507 AV model is the same as the CE 507 model, except that it supports only 18GB of maximum storage capacity and includes a hardware MPEG decoder for video stream decoding.

In the midrange, the CE 560 model is designed for regional offices and comes standard with 512MB of SDRAM, 36GB of storage (expandable to 144GB through the addition of an external Cisco Storage Array), and dual 10/100 Fast Ethernet interfaces, and it forwards at around 20 Mbps. Like the CE 507, it also consumes one rack unit.

Toward the high end, the CE 590 model fills the need. It is designed for smaller service providers and enterprise networks, and it includes 1GB of SDRAM, 36GB of storage (expandable to 252GB through the addition of an external Cisco Storage Array), and dual 10/100 Fast Ethernet interfaces. It forwards at over 45 Mbps. Like the CE 507 and CE 560, it also consumes one rack unit.

Finally, for top-of-the-line caching needs, the CE 7320 steps in. The CE 7320 is designed purely for high-end service-provider duty, and it comes standard with 2GB of SDRAM, 180GB of storage (expandable to 396GB through the addition of an external Cisco Storage Array), dual Gigabit Ethernet interfaces, and quad 10/100 Fast Ethernet interfaces, and it forwards at over 155 Mbps! It requires seven rack units and some serious traffic!

All things considered, the Cisco content engine line provides demanding customers with many more options for reducing WAN bandwidth usage, and it is almost guaranteed to increase the efficiency of your network. For more information on the content engine line, visit http://www.cisco.com/warp/public/cc/pd/cxsr/ces/prodlit/.

Cisco Security Products

Cisco has a wide variety of security products available, ranging from software to monitor and control the entire network to hardware like PIX Firewall and IOS Firewall feature sets. This section details some of the more current Cisco security tools.

Cisco Secure Access Control Server

The Cisco Secure Access Control Server (ACS) system is a software application designed to centralize all authentication, authorization, and accounting (AAA) functions in your network. It allows administrators to administer *all* devices, including servers, routers, and switches, using one program and with one user account database. It does this by supporting nearly all popular user account database protocols, including LDAP (Windows 2000, Novell Netscape, and others), SAM support (Windows NT), TACACS+, RADIUS, and LEAP (Cisco Aeronet wireless devices).

Cisco Secure ACS is available for most popular operating systems, including Windows 2000, Windows NT, and Sun Solaris (UNIX).

With Cisco Secure ACS, you can control who can log on, what they can do, and what is audited during their network session, all from one interface. For more information on Cisco Secure ACS, visit http://www.cisco.com/warp/public/cc/pd/sqsw/sq/prodlit/index.shtml.

Cisco Secure Policy Manager

Cisco Secure Policy Manager (CSPM) is a software application designed to help you painlessly support and configure security settings on networks with large quantities of Cisco security products. Cisco Secure Policy Manager supports any device with the IOS Firewall or VPN-enabled feature sets, PIX Firewalls, and Cisco Secure Intrusion Detection System (CSID) sensors.

With Cisco Secure Policy Manager, an administrator can control VPN and other security settings (such as access list entries) from a centralized location using a GUI-based management console. These settings can then be applied to devices en masse with a few simple steps, and with very little knowledge of Cisco IOS. For more information about Cisco Secure Policy Manager, visit http://www.cisco.com/warp/public/cc/pd/sqsw/sqppmn/prodlit/index.shtml.

Cisco Secure Scanner

Cisco Secure Scanner is a network security–auditing tool designed to detect and verify network security holes before unauthorized intruders do. This software product scans nearly any TCP/IP-based device (including routers, firewalls, hosts, servers, and just about anything else with an IP address) both passively and actively to find vulnerabilities. It then verifies that the vulnerability exists and reports the results. It also includes a comprehensive security vulnerability database, detailing each security vulnerability along with tips on how to eliminate them, allowing even average network administrators to plug security holes without the assistance of a security expert.

Cisco Secure Scanner plays a large role in the Cisco security lineup; without testing, even the best-planned security policies may have vulnerabilities that are overlooked, at least until the network is breached. Cisco Secure Scanner has the ability to test and verify security problems in the following protocols:

- Domain Name Server (DNS) checks
- Finger service checks
- File Transfer Protocol (FTP) checks
- HTTP security checks
- NetBios system checks
- Network File System (NFS) checks
- Windows NT security checks
- POP server checks
- Rlogin Remote System checks

- Remote Procedure Call (RPC) checks
- Simple Mail Transport Protocol (SMTP) checks
- Simple Network Management Protocol (SNMP) checks
- Telnet service security checks
- Trivial File Transfer Protocol (TFTP) checks
- X Window System checks
- Server Message Block (SMB) checks
- Internet Mail Access Protocol (IMAP) checks
- Network News Transport Protocol (NNTP) checks
- Remote Shell (RSH) checks
- IDENT service checks
- RWHO service checks

Cisco Secure Scanner is available for both Windows NT and Solaris (both x86 and SPARC). For more information about Cisco Secure Scanner, visit http://www.cisco.com/warp/public/cc/pd/sqsw/nesn/prodlit/index.shtml.

Cisco IOS Firewall

The Cisco IOS Firewall feature set allows most standard Cisco routers, from the 800 series to the 7500 series, to leverage advanced security features in supporting networks, without or in addition to a dedicated hardware device (like a PIX Firewall).

The IOS Firewall feature set adds a number of new security tools to the Cisco routing line, including these:

- Context-Based Access Control (CBAC), a feature that enables per-application-based control of traffic. It does this by dynamically allowing sessions that were established from an internal address and denying sessions initiated from external hosts.

- Intrusion detection features to detect and report security violations and common attacks.

- TACACS+ and RADIUS dial-up authentication support.

- Denial-of-service (DOS) detection and prevention features to detect and eliminate packets involved in a DOS attack.

- Java applet blocking.

- VPNs, IPSec encryption, and QoS support.

- Real-time alerting features.

- Auditing capabilities to detail violations, including timestamps and address information.
- Basic and advanced access list features.
- Control over user access by IP address and interface.
- Time-based access lists.

These features turn an ordinary Cisco router into a formidable hardware firewall for networks that cannot afford a dedicated firewall device, or those that just need the additional security provided by this feature set. For more information, visit http://www.cisco.com/warp/public/cc/pd/iosw/ioft/iofwft/prodlit/index.shtml.

Cisco Secure Intrusion Detection System

The Cisco Secure Intrusion Detection System (IDS) is a hardware-based suite of devices designed to monitor the network for security breaches and eliminate the breach once it is detected. It does this by dynamically altering the access lists on affected Cisco devices once a security breach is detected, thereby shutting out the intruder but allowing all other operations to continue normally. This is almost like having your own security expert tracking packets and eliminating intruders, 24-7!

The Cisco Secure IDS can accurately detect and eliminate most major security breaches, including the following:

- Intrusion attempts, such as remote takeover (Backdoor Sub-seven, Back Orifice, and so on), failed login attempts, IP spoofing, and so on.
- DOS attacks, such as SYN floods and "ping of death" attacks.
- Discovery attempts, such as ping sweeps and port scans.
- Any attack using certain text strings in unencrypted traffic. For example, you could configure the Cisco Secure IDS to drop any e-mail containing the phrase "send this to all of your friends."

The Cisco Secure IDS is available in three models: the IDS 4210 for lower-end deployments, the IDS 4230 for high-end deployments, and the IDS module for Catalyst 6500 switches. The IDS 4210 and 4230 are PC-based systems, fully preconfigured and hardened to security vulnerabilities. The 4210 includes a 566-MHz Intel Celeron processor and 256MB of RAM, and it can support traffic flows of up to 45 Mbps (which should be plenty for nearly any Fast Ethernet network).

The 4230 includes dual 600-MHz Intel Pentium III processors and 512MB of RAM, and it can support traffic flows of up to 100 Mbps. The IDS network module includes a custom processor and 256MB of RAM, and can support traffic flows of up to 100 Mbps.

For most networks, one or more Cisco Secure IDS devices, combined with solid firewall products and intelligent security policies, should be more than adequate to help enforce security to a high degree. For more information on the Cisco Secure IDS, visit http://www.cisco.com/warp/public/cc/pd/sqsw/sqidsz/prodlit/index.shtml.

Cisco Secure PIX Firewall

The Cisco Secure PIX Firewall system is a high-end, dedicated hardware firewall device capable of protecting networks ranging in size from small offices to large enterprises.

At its heart, the Cisco PIX Firewall is simply a hardened PC with specialized firewall software. The wide availability of the components used in the PIX Firewall allow the device to be competitively priced, while the specialized software image installed on it allows it to be free of security vulnerabilities common to firewall solutions running on top of commercially available general-purpose operating systems.

As for features, the PIX Firewall includes a number of features designed to secure your network from most threats. The primary feature of the firewall is known as the Adaptive Security Algorithm (ASA), and it functions very similarly to the CBAC feature in the IOS Firewall, allowing connections established from the inside to function and denying attempts from the outside to intrude on a dynamic basis. The PIX also supports hot standby, allowing redundant PIX Firewalls to fail over to one another with no user disruption. Full NAT/PAT support is included, of course, as well as DOS suppression, and nearly every other feature in the IOS Firewall feature set is available.

The PIX Firewall is available in four models, all with differing performance and expansion capabilities. The 506 is the entry-level model in the line, followed by the 515, 525, and 535 models.

The 506 firewall is designed primarily for branch offices. It includes 32MB of DRAM, 8MB of flash RAM, and dual Ethernet (10 Mbps) NICs. It can forward at up to 8 Mbps for unencrypted traffic, and at up to 6 Mbps using DES (56-bit) or 3DES (168-bit) encryption.

The 515 model is designed for situations requiring significantly higher performance and interface expandability, like large regional offices. It includes either 32 or 64MB of DRAM, 16MB of flash RAM, dual 10/100 Mbps Fast Ethernet ports (expandable to a maximum of six), and a failover port for connection to a hot standby firewall. It can forward at up to 145 Mbps for unencrypted traffic, and at up to 10 Mbps using DES or 3DES encryption. Figure 13-7 shows an example of a PIX 515 model firewall.

The 525 model is designed for enterprise networks that require no-compromises high performance. It includes either 128 or 256MB of DRAM, 16MB of flash RAM, dual 10/100 Mbps Fast Ethernet ports (expandable to a maximum of eight), and a failover port. It can forward at up to 320 Mbps for unencrypted traffic, and at up to 70 Mbps using DES or 3DES encryption.

Finally, the 535 model is designed for the most intense security duties in service provider or enterprise networks. It includes either 512MB or 1GB of DRAM, 16MB of flash RAM, and a failover port. It has no Ethernet ports included, but single and quad

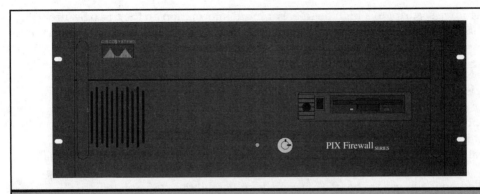

Figure 13-7.   *Front view of a PIX 515 firewall*

Fast Ethernet interfaces are available, as well as a single-port Gigabit Ethernet interface. With eight total expansion slots, the 535 can support up to eight total Gigabit or Fast Ethernet interfaces. As for performance, it can forward at up to 1.7 Gbps for unencrypted traffic, and at up to 95 Mbps using DES or 3DES encryption.

In the end, the PIX Firewall system is a platform supporting advanced network protection features required by many publicly accessible networks to prevent intrusion. For more information on the PIX Firewall, please visit http://www.cisco.com/warp/public/cc/pd/fw/sqfw500/prodlit/index.shtml.

Summary

This chapter examined the other lines of Cisco products, from access servers to security, and detailed the benefits of each product in Cisco's lineup. This is the final chapter detailing product lines. In the next chapter, you will begin to learn about the intricacies of the Cisco IOS.

Chapter 14

IOS Basics

T his chapter begins to build a foundation for understanding the standard Cisco IOS. You will take a look at the IOS modes of operation, syntax, storage, and organization to better understand how to navigate through the sometimes bewildering array of options within IOS. You can gain more out of this chapter by practicing these principles on a router or switch. If you don't have one, consider purchasing a functional used router or switch from an online auction. (An 800 series or 2500 series router and 1900 series switch are good choices.) Or you can purchase a quality simulator to help with deeper understanding of these concepts. (The best is probably direct from Cisco.) However, even if you do not use a router, switch, or simulator, you can still acquire a basic understanding of the concepts presented in this chapter simply by following the examples.

What Is the IOS, Anyway?

The IOS is the Internetworking Operating System, or the OS for your Cisco device. The IOS controls all of the functions of the device, from access lists to queuing, and provides the user interface (UI) for managing the device. Typically, the UI is a command line interface (CLI) in Cisco devices that resembles other popular CLIs such as DOS and C shell in UNIX. Like all CLIs, learning how to navigate and issue commands in the IOS is similar to learning a new language.

IOS Versions and Version Structure

The Cisco IOS has been through some major revisions since its initial release, and it seems like it is constantly being modified in some way. The most current version as of this writing is 12.2, and you are likely to see anywhere from version 11 to version 12 in most environments today.

Keep in mind that the IOS is typically model specific. Each IOS image also has a specific feature set. Typically, adding additional features to a router or switch simply means downloading a new IOS image (although, in some cases, additional hardware may be required).

These differences in IOS images are reflected in Cisco's IOS naming convention. The primary IOS version numbers can look fairly complex, but they aren't that difficult to understand once you know the method behind Cisco's madness. Note that while this method does not apply to all IOS releases, it does work for most of them. (A full study of IOS release trains is beyond the scope of this book.)

To start off, let's take a look at the cycle Cisco uses for IOS releases. Cisco releases the IOS in a Limited Deployment (LD) phase. This begins when the model is shipped to the first customer and continues up until around a year.

At this point, the product is considered to be in the General Deployment (GD) phase. This is where the product is considered to be undergoing final revisions to code for features and stability, and where it is continually refined into its final version. Early Deployment (ED) releases typically fall into this phase.

Next, the product enters the Mature Maintenance (MM) phase, where no new features are introduced, only bug fixes. The main release is built in this phase.

Finally, somewhere around two years after the initial release, the product begins the process of phasing out: first with end-of-sale (EOS); then with end-of-engineering (EOE), which means no more bug fixes; and, finally, with end-of-life (EOL), or death, for those who like it simple.

How Version Naming Works

Let's look at this process by taking version 12.0(3a)T as an example. First, the IOS is given a number for the primary version—in this case, *12.0*. Next, a maintenance revision is indicated in parenthesis *(3a)*. The *a* indicates that this revision has been rebuilt (usually due to a major bug). Finally, the ED field denotes the technology type for which the IOS is designed. In this case, it is *T*, which indicates that it is for *consolidated technology*, or multifunctional (typically modular) devices.

ED releases are not main IOS releases; they are used to add features and new functions to the IOS and are a bit buggy. However, in a lot of cases, you will end up using them anyway because they provide you with the advanced features not available with the main release.

Table 14-1 describes all of the standard ED fields.

Finally, Figure 14-1 illustrates this explanation visually.

However, even though you will generally be using the ED releases if you are on the bleeding edge, in some environments, you may be using what is called a *mature release*: a release without the ED, such as 12.1(3). This means that it is (usually) a more stable but perhaps less feature-rich version of the IOS than a newer ED version (like 12.1(3)T).

ED Code	Description
A	Access server
D	xDSL
E	Enterprise feature set
H	SDH/SONET
N	Multiservice (voice, video, and so on)
S	Service provider
T	Consolidated technology
W	ATM/WAN/L3 switching

Table 14-1. *ED Codes*

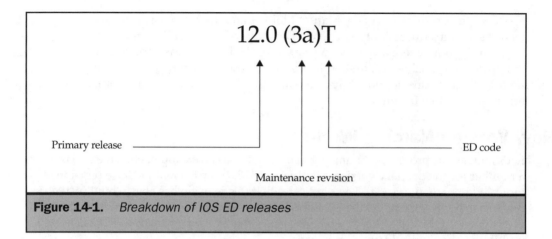

Figure 14-1.   *Breakdown of IOS ED releases*

As for file-naming conventions, Cisco takes the version of IOS and then makes it a whole lot more complex by adding all sorts of other specific information. (Hey, you're reading a Cisco book, what did you expect?)

To elucidate, the filename you would see on an IOS image might be something like c4500-js40_120-3t-mz. Pretty nasty, huh? Again, it really isn't all that bad, once you know the convention. Here's the breakdown:

- The *c4500* is for the platform, and, in this case, it tells you that this image is for a 4500 series router.

- The *js40* denotes the feature set (shown in Table 14-2), in this case, Enterprise Plus with 40-bit encryption.

- The *120-3t* is obviously the version (12.0(3)T, in this case).

- The *m* tells you that the software is to be executed from RAM.

- The *z* tells you that the file you have is zipped, or *compressed* (a decompressed file is denoted with a l) and must be decompressed into its executable .bin format before you can load it on your device.

Code	Feature Set
I	IP Only
IS	IP Plus
D	Desktop

Table 14-2.   *Feature Set Codes*

Code	Feature Set
DS	Desktop Plus
J	Enterprise
JS	Enterprise Plus
AJ	Enterprise/APPN
AJS	Enterprise/APPN Plus
P	Service Provider
G	ISDN
C	Communications Server
F	CFRAD
FIN	LANFRAD
B	AppleTalk
N	IP/IPX
R	IBM
F	Frame Relay
A	APPN (IBM)

Table 14-2.   *Feature Set Codes* (continued)

Figure 14-2 shows this entire filename layout.

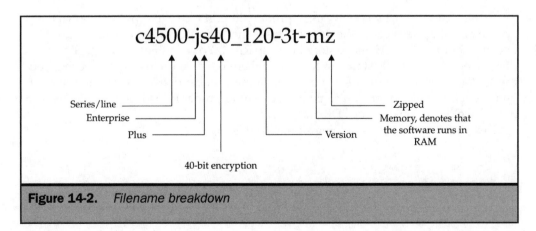

Figure 14-2.   *Filename breakdown*

Booting a Cisco Device

Booting a Cisco device is pretty easy. You simply plug in the power and hit the switch. However, a lot goes on behind the scenes that you should be aware of in case problems arise.

First, the device performs a power-on self test (POST), very similar to a PC. However, Cisco devices do not have beep codes; rather, they have flashing lights. While every line of device displays error codes a bit differently, they all have a key to the flash codes.

For instance, in a 1900 series switch, 16 different LEDs (one for each of the first 16 ports) are used to signify failures. If any component in the switch fails the POST tests, a port lights up amber on one or more of these LEDs. The port that turns amber indicates the component that failed. For example, if port 12 lights up as amber, it indicates that the forwarding engine has failed (a fatal error, meaning the switch cannot function until it is repaired). Therefore, you should know what your Cisco device normally displays when all is well. If it displays a different light code, you know something is most likely wrong.

Basic Cisco Device Configuration

Once the device has booted, you can begin configuring it. On a new device, configuration is usually performed by the console connection because the device does not yet have an IP address for Telnet configuration. Upon booting the device, however, a few details are worth paying attention to. These are illustrated and explained in the following example. (Note that all of the following displays are only for the standard IOS; most are from a 2600 series router.)

```
System Bootstrap, Version 11.3(2)XA4, RELEASE SOFTWARE (fc1)
Copyright (c) 1999 by cisco Systems, Inc.
TAC:Home:SW:IOS:Specials for info
C2600 platform with 32768 Kbytes of main memory
```

In this first section, two points are of primary importance. First, you are shown the system bootstrap version on the device. This is the "mini-IOS" that is actually used to boot the device to the IOS image stored in flash RAM. You can think of this as the bootup disk, except it is normally stored in ROM, and therefore difficult to change. Second, you can see the amount of system RAM installed. If this number does not match the amount you know is installed, then you could have bad or improper RAM installed. (Remember, most Cisco devices take special DRAM.)

Next, you should see something similar to this:

```
program load complete, entry point: 0x80008000, size: 0x403b9c
Self decompressing the image :
```

```
###################################################
#################################### [OK]
 Restricted Rights Legend
 Use, duplication, or disclosure by the Government is
subject to restrictions as set forth in subparagraph
(c) of the Commercial Computer Software - Restricted
Rights clause at FAR sec. 52.227-19 and subparagraph
(c) (1) (ii) of the Rights in Technical Data and Computer
Software clause at DFARS sec. 252.227-7013.
 cisco Systems, Inc.
 170 West Tasman Drive
 San Jose, California 95134-1706


Cisco Internetwork Operating System Software
IOS (tm) C2600 Software (C2600-I-M), Version 12.0(7)T,
RELEASE SOFTWARE (fc2)
Copyright (c) 1986-1999 by cisco Systems, Inc.
Compiled Tue 07-Dec-99 02:12 by phanguye
Image text-base: 0x80008088, data-base: 0x807AAF70
```

Again, two points are of primary importance. First, you see the *Self decompressing the image,* and then lots of pound (#) signs (some of which were removed in this example); and, finally, you hope, an OK. The router decompresses the image if the image is a self-decompressing compressed image, which is becoming common in newer routers. This functionality allows precious flash RAM space to be preserved by compressing
the image for storage, but then fully decompressing it for load into RAM at boot.

The second concern is near the end of the display. The *IOS c2600 software* prompt tells you which version of IOS you are loading from flash and booting the device with. In this case, the 2600 router is using version 12.0(7)T.

Finally, you will see more hardware-specific information on the next screen:

```
cisco 2611 (MPC860) processor (revision 0x203)
with 26624K/6144K bytes of memory
 .
Processor board ID JAD04360GH7 (4114038455)
M860 processor: part number 0, mask 49
Bridging software.
X.25 software, Version 3.0.0.
```

```
2 Ethernet/IEEE 802.3 interface(s)
1 Serial network interface(s)
32K bytes of non-volatile configuration memory.
8192K bytes of processor board System flash (Read/Write)

Press RETURN to get started!
```

First, you see the processor and a few more specifics about the RAM in your device. In the section that reads *26624K/6144K bytes of memory*, the first number is the amount of DRAM devoted to main memory (used for running the IOS, working space, and so on), and the second number is the amount of memory devoted to shared memory (used as buffers for all of your interfaces). These two numbers together should be equal to the total amount of DRAM in your system. Note that on some routers, the ratio of main to shared memory is a fixed distribution, while on others, it is changeable. Also note that in some of the high-end lines (like the 7500 series), each Versatile Interface Processor (VIP) may have its own packet or shared memory, separate from the main system memory.

Next, you see a bit about some of the software features installed and the interfaces installed. You simply need to make sure the interface types and counts displayed on this screen are in accordance with what you expect. After the interfaces is the amount of NVRAM and flash RAM installed. Finally, a prompt tells you to press RETURN to continue. Remember this, because when you first plug the console cable into an already booted device, you will not see anything (including this message) on the screen. You need to press ENTER to arrive at a prompt.

Note *If you miss the initial startup screen but you want to see the information presented there, the* show version *command displays the same information.*

On some Cisco devices (mostly routers), you will initially be greeted with a setup screen calling itself the *System Configuration Dialog*. This is like a text-only "wizard" that guides you through initial configuration by simply asking a series of questions. This process allows you to configure the device even if you have little or no experience with the IOS. The prompts you will see and explanations of what they mean are provided in the following discussion.

After pressing the ENTER key, you will usually be prompted with this:

```
- System Configuration Dialog -
At any point you may enter a question mark '?' for help.
Refer to the 'Getting Started' Guide for additional help.
Use ctrl-c to abort configuration dialog at any prompt.
Default settings are in square brackets '[]'.
Would you like to enter the initial configuration dialog? [yes]:
```

The *[yes]* means that the default answer to this question is yes, and if you simply press ENTER, that is what will be selected. If you select yes, you enter the "setup wizard" (a Microsoft term, but it fits) called *basic management setup*. If not, you will be sent directly to the CLI. Don't worry if you selected no and wanted to enter the wizard; you can simply reboot the device (as long as the device has a blank NVRAM, or startup, configuration), and you will get the same prompt. Alternatively, you can also enter enable mode (covered in "IOS Modes," later in this chapter) and type *setup* to return.

You will also notice that the question mark (?) key is used to get help. You will learn more about this concept later (in the "Command Help and Shortcuts" section), but remember it, because it will be your best friend (besides this book, of course). You can also press CTRL-C to completely exit back to the IOS at any time, should you decide you made a mistake. Your changes to the configuration do not take effect until the end of the process.

Assuming you chose yes, the next prompt you should see will be similar to this:

```
First, would you like to see the current interface summary? [yes]:
```

You are simply being asked if you would like to see the interfaces currently installed in the router and what the current settings on these interfaces are. Choosing yes will display something like this:

```
Any interface listed with OK? value "NO" does not
have a valid configuration.
Interface IP-Address OK? Method Status Protocol
TokenRing0 unassigned YES not set down down
Ethernet0 unassigned YES not set down down
Serial0 unassigned YES not set down down
Fddi0 unassigned YES not set down down
```

Notice that this information gives you the basics:

- What is the name of the interface?
- Has an IP address been assigned to it?
- Is it physically functional?
- How was it last configured?
- Is the interface active (in Ethernet terms, does it have a link)?
- Are any protocols active on the interface?

Most of the time, your only concern is that all of the interfaces are listed on this page that the router is supposed to have installed. Because you will typically use the setup tool only when the router is first installed, the IP address, method set, and so on, will all be blank. If the interface is not listed, the interface is either completely dead, or, possibly, a bug in the IOS is preventing detection of the interface (a known issue with some versions of IOS available for the 3660 router).

The next prompt you will be presented with should be as follows:

```
Configuring global parameters:
Enter host name [Router]:
```

Enter what you would like the DNS host name to be for the router. Note that the DNS host name will also become your prompt unless it is overridden with the *prompt* command (discussed further in Chapter 15). Note that entering a DNS host name here doesn't necessarily mean that the router host name can now be resolved on the PCs in your network (by typing *ping router,* for instance). For the name to be resolvable by other devices, you must enter an A (host) record for the name in your DNS server, or add the entry to the hosts file on applicable machines. The DNS host name simply tells the router what to call itself.

Next, you will see this message:

```
The enable secret is a password used to protect access to
privileged EXEC and configuration modes. This password, after
entered, becomes encrypted in the configuration.
Enter enable secret:
The enable password is used when you do not specify an
enable secret password, with some older software versions, and
some boot images.
Enter enable password:
```

The enable secret and enable password is almost like the administrator password for a device. By entering enable mode using the secret or password, you can modify the configuration and perform diagnostics that could potentially disrupt operations of the device. The difference between the enable secret and the enable password is that the enable secret is encrypted and not viewable in the configuration files, whereas the enable password is listed in plaintext. If both are set, then the secret is always used, unless you are using a really old version of the IOS that does not understand enable secrets.

When going through setup mode, the IOS will not allow you to set both of these passwords to the same name. It requires you to make them different. Enable password and enable secret are discussed in more detail in "IOS Modes," later in the chapter.

Next, you should see a prompt like this:

```
The virtual terminal password is used to protect
 access to the router over a network interface.
 Enter virtual terminal password:
```

This is your vty, or Telnet password. Whenever users attempt to access the router from Telnet, they will be required to enter this password, if it is set. This subject is discussed in more detail in Chapter 15. Next, you should see the following:

```
Configure SNMP Network Management? [yes]:
 Community string [public]:
Any interface listed with OK? value "NO" does not
have a valid configuration
Interface IP-Address OK? Method Status Protocol
BRI0 unassigned YES not set down down
Ethernet0 unassigned YES not set down down
Serial0 unassigned YES not set down down
Serial1 unassigned YES not set down down
Enter interface name used to connect to the
management network from the above interface summary: ethernet0
```

This output allows you to set up basic SNMP functionality on the router, including setting the community name and the interface from which you will manage the router. Next, you will see the following:

```
Configure IP? [yes]:
 Configure IGRP routing? [yes]:
 Your IGRP autonomous system number [1]: 1
```

This output allows you to enable IP on your interfaces. Note that the specific options you are given here may vary depending on the model of router and feature set of IOS. You may be asked to set up IPX, DECnet, AppleTalk, and other protocol suites if your router supports them. It may also ask to configure your routing protocols at this point (like in the previous example). (Chapters 23 through 27 discuss routing protocols in considerable detail.) Next, you will be presented with the following:

```
Configuring interface parameters: Ethernet0
Configure IP on this interface? [yes]:
 IP address for this interface: 192.168.1.1
 Subnet mask for this interface [255.255.255.0] :
 Class C network is 192.168.1.0, 24 subnet bits; mask is /24
```

This output allows you to perform basic IP configuration on all of your interfaces. Again, the specific options you are presented with may vary depending on the model of router and interfaces installed. Finally, it will build your configuration file, which will look something like this:

```
hostname router
enable secret 5 $1$HNfx$Nhj5AqtXt823hCEBf.JZt.
enable password test
line vty 0 4
password open
snmp-server community public
!
ip routing
!
interface BRI0
no ip address
!
interface Ethernet0
ip address 192.168.1.1 255.255.255.0
!
interface Serial0
no ip address
!
interface Serial1
no ip address
!
router igrp 1
network 192.168.1.0
!
end
```

This output simply details in IOS command format all of the changes that you requested. You are then prompted to save the configuration with the following:

```
[0] Go to the IOS command prompt without saving this config.
[1] Return back to the setup without saving this config.
[2] Save this configuration to nvram and exit.

Enter your selection [2]:
```

Choice 0 returns you to the IOS without making any changes. Choice 1 restarts setup without making changes (in case you made an error). Choice 2 saves the configuration and returns you to the IOS, which is typically what you want to do.

That's all there is to IOS setup mode. By answering a series of simple questions, you can set up the basic configuration of a router. The down side to setup mode is that you cannot configure anything complex or unique; and, once you learn the IOS, setup mode takes longer than just banging out the commands you need from the prompt. But for beginners, it is a very easy way to get the router up.

Working with the IOS

This section of the chapter looks at the basics of the IOS, including command structures, help, shortcuts, modes, and configuration files. These concepts are the fundamentals of learning to speak the IOS language. If you have Cisco products available, experimenting with the commands mentioned in this section will help you to understand the overall structure of the IOS. Also, if you are familiar with UNIX, the IOS in many ways resembles C shell, so you should feel right at home.

Command Structure

The IOS has a command structure that should be fairly simple to administrators familiar with common CLIs like DOS and UNIX. The first part of a command is the primary command, followed by one or more operators. Learning the command structure is simply a matter of learning how to speak the language of IOS. For instance, here's a common *show* command broken down into its component parts:

```
Router>Show ip interface Ethernet 0/0
```

This concept is fairly simple. The primary command, *show,* asks the router to display information for you. Unfortunately, the router cannot read your mind, so you will have to be a little more specific about the type of information you want to see. This is where the modifiers come in. The *ip* section informs the router that you want to see information specific to IP. The *interface* section informs the router that you want to see information specific to an interface. And the *Ethernet 0/0* section tells the router that the interface you are concerned about is the Ethernet interface in module slot 0 (the first slot), and that it is the first Ethernet port (port 0) on that module.

So, basically, what you said to the router was, "Show me everything you know about IP settings and statistics for the Ethernet interface located in module 0, port 0." Not too bad, huh? And the router rewards your good syntax by displaying exactly what you requested, like so:

```
Router>show ip interface Ethernet 0/0
Ethernet0/0 is up, line protocol is up
  Internet address is 10.1.1.12/8
```

```
Broadcast address is 255.255.255.255
Address determined by setup command
MTU is 1500 bytes
Helper address is not set
Directed broadcast forwarding is disabled
Outgoing access list is not set
Inbound access list is not set
Proxy ARP is enabled
Security level is default
Split horizon is enabled
ICMP redirects are always sent
ICMP unreachables are always sent
ICMP mask replies are never sent
IP fast switching is enabled
IP fast switching on the same interface is disabled
IP Flow switching is disabled
IP Fast switching turbo vector
IP multicast fast switching is enabled
IP multicast distributed fast switching is disabled
IP route-cache flags are Fast
Router Discovery is disabled
Proxy ARP is enabled
Security level is default
Split horizon is enabled
ICMP redirects are always sent
ICMP unreachables are always sent
ICMP mask replies are never sent
IP fast switching is enabled
IP fast switching on the same interface is disabled
IP Flow switching is disabled
IP Fast switching turbo vector
IP multicast fast switching is enabled
IP multicast distributed fast switching is disabled
IP route-cache flags are Fast
Router Discovery is disabled
IP output packet accounting is disabled
IP access violation accounting is disabled
TCP/IP header compression is disabled
RTP/IP header compression is disabled
Probe proxy name replies are disabled
Policy routing is disabled
Network address translation is disabled
```

CISCO TECHNOLOGY
OVERVIEW

```
WCCP Redirect outbound is disabled
WCCP Redirect exclude is disabled
BGP Policy Mapping is disabled
Router>
```

Unfortunately, the IOS is also like a three-year-old when it comes to comprehension. Tell it exactly what you want with the right syntax, and it will be more than willing to help (well, most of the time). Be vague, or speak to it with a Yoda-ism, and it will respond with a "huh?" or do something you *really* don't want it to do. For instance, the previous command *show ip interface Ethernet 0/0* worked like a charm, but being vague and using *show ip interface Ethernet* will get you a *% Incomplete command* response. On the other hand, giving it a Yoda-ism like *Ethernet show interface ip 0/0, yes* will get you this:

```
Router>Ethernet show interface ip 0/0, yes
         ^
% Invalid input detected at '^' marker.
```

This is the IOS equivalent of "What you talkin' about, fool?" So how do you know what the proper syntax is? Well, a number of tools can help you with that. First, IOS has online help, which is discussed in the next section, "Command Help and Shortcuts." Second, I have provided command references for the most common IOS commands (in Chapters 15, 16, and 17) as well as a complete index of the over 500 commands used in this book (in the Appendix) to help you with specific commands. And, finally, Cisco posts full command references for most models of routers and switches on their web site at www.cisco.com.

However, the best teacher is experience, or specifically, explosions when you mess something up. There's nothing like getting chewed out by the resident CCIE when you make a mistake to plant "Hey, that was probably a really bad idea" in your mind more firmly.

Anyway, if you don't have access to a router lab, as I mentioned at the beginning of this chapter, I strongly suggest that you purchase a few routers and switches, or, at the very least, a good simulator. The more practice you get with the IOS (or pseudo-IOS), the better you will understand the command syntax. And, unfortunately, getting your practice in on production routers is generally not a good idea (especially if it is preceded by "Hey ya'll, watch this!").

Command Help and Shortcuts

As mentioned in the preceding section, the IOS comes with online help to give you a hand when you are unsure about syntax. It also includes some useful shortcuts to make your life easier when issuing the same or similar commands many times.

As far as the help is concerned, the main help command you want to be aware of is the question mark. This command is extremely useful in a variety of situations. For instance, if you are at a router prompt and can't remember which command does what you are looking for, you can simply issue the question mark and get a full listing of all of the commands available from the mode you are in, as shown in this example:

```
Router>?
Exec commands:
 clear Reset functions
 disable Turn off privileged commands
 disconnect Disconnect an existing network connection
 enable Turn on privileged commands
 exit Exit from the EXEC
 lock Lock the terminal
 login Log in as a particular user
 logout Exit from the EXEC
 name-connection Name an existing network connection
 ping Send echo messages
 resume Resume an active network connection
 set Set system parameter (not config)
 show Show running system information
 systat Display information about terminal lines
 terminal Set terminal line parameters
 traceroute Trace route to destination
 where List active connections
 access-enable Create a temporary Access-List entry
 access-profile Apply user-profile to interface
 connect Open a terminal connection
 help Description of the interactive help system
 mrinfo Request neighbor and version information from a multicast
 router
 mstat Show statistics after multiple multicast traceroutes
 mtrace Trace reverse multicast path from destination to source
 pad Open a X.29 PAD connection
 ppp Start IETF Point-to-Point Protocol (PPP)
 rlogin Open an rlogin connection
 slip Start Serial-line IP (SLIP)
 telnet Open a telnet connection
 tunnel Open a tunnel connection
 udptn Open an udptn connection
 x28 Become an X.28 PAD
 x3 Set X.3 parameters on PAD
```

Now, if you happen to know that the command started with a letter, like *s*, you can perform a question-mark command just to retrieve all commands starting with *s*, like so:

```
Router>s?
*s=show set show slip systat
```

Say that now you know the command is *show*, but you can't remember the syntax for it. You can work through the syntax by issuing a question mark after each portion of the command and retrieving valid entries for the next portion of the command, like so:

```
Router>show ?
 backup Backup status
 c2600 Show c2600 information
 cca CCA information
 cdapi CDAPI information
 cef Cisco Express Forwarding
 class-map Show QoS Class Map
 clock Display the system clock
 compress Show compression statistics
 connection Show Connection
 dialer Dialer parameters and statistics
 exception exception informations
 flash: display information about flash: file system
 history Display the session command history
 hosts IP domain-name, lookup style, nameservers, and host table
 location Display the system location
 modemcap Show Modem Capabilities database
 policy-map Show QoS Policy Map
 ppp PPP parameters and statistics
 queue Show queue contents
 queuing Show queuing configuration
 radius Shows radius information
 rmon rmon statistics
 rtr Response Time Reporter (RTR)
 sessions Information about Telnet connections
 snmp snmp statistics
 tacacs Shows tacacs+ server statistics
 template Template information
 terminal Display terminal configuration parameters
 traffic-shape traffic rate shaping configuration
 users Display information about terminal lines
 version System hardware and software status

Router>show ip ?
 accounting The active IP accounting database
 aliases IP alias table
```

```
arp IP ARP table
as-path-access-list List AS path access lists
bgp BGP information
cache IP fast-switching route cache
cef Cisco Express Forwarding
community-list List community-list
dhcp Show items in the DHCP database
drp Director response protocol
dvmrp DVMRP information
egp EGP connections and statistics
eigrp IP-EIGRP show commands
extcommunity-list List extended community-list
flow NetFlow switching
igmp IGMP information
interface Interface status and configuration
ip IP information
irdp ICMP Router Discovery Protocol
local IP local options
masks Masks associated with a network
mcache IP multicast fast-switching cache
mrm IP Multicast Routing Monitor information
mroute IP multicast routing table
msdp Multicast Source Discovery Protocol (MSDP)
mtag IP Multicast Tagswitching TIB
nhrp NHRP information
ospf OSPF information
pim PIM information
policy Policy routing
prefix-list List IP prefix lists
protocols IP routing protocol process parameters and statistics
redirects IP redirects
rip IP RIP show commands
route IP routing table
rpf Display RPF information for multicast source
rsvp RSVP information
rtp RTP/UDP/IP header-compression statistics
sdr Session Directory (SDPv2) cache
sockets Open IP sockets
tcp TCP/IP header-compression statistics
traffic IP protocol statistics
vrf VPN Routing/Forwarding instance information
Router>show ip interface ?
```

```
Ethernet IEEE 802.3
Null Null interface
Serial Serial
brief Brief summary of IP status and configuration
| Output modifiers
<cr>
Router>show ip interface ethernet ?
<0-1> Ethernet interface number
Router>show ip interface ethernet 0?
/
Router>show ip interface ethernet 0/?
<0-1> Ethernet interface number
Router>show ip interface ethernet 0/0
Ethernet0/0 is administratively down, line protocol is down
Internet protocol processing disabled
```

By using the online help, you are able to retrieve the syntax for the router command, even though you may not have known it. The down side is that you have to sort through *tons* of commands and modifiers to find the one you want. But overall, this tool is pretty handy, and it is likely to be your best buddy for some time to come.

In addition to the question mark, you may have noticed the *help* command in the command listing. The *help* command in IOS is practically useless. Its only purpose is to help you learn how to use help. In other words, it tells you how to use the question-mark command, like so:

```
Router>help
Help may be requested at any point in a command by entering
a question mark '?'. If nothing matches, the help list will
be empty and you must backup until entering a '?' shows the
available options.
Two styles of help are provided:
1. Full help is available when you are ready to enter a
 command argument (e.g. 'show ?') and describes each possible
 argument.
2. Partial help is provided when an abbreviated argument is entered
 and you want to know what arguments match the input
 (e.g. 'show pr?'.)
```

This output is basically what we just went through. However, the IOS does provide some other useful shortcuts. The first, and probably most useful, is the history buffer. This shortcut is similar to DOSkey; it simply allows you to recall previous command entries by using the up and down arrow keys (or CTRL-P and CTRL-N, if you don't have the arrow keys). The up arrow (CTRL-P) recalls the previous command, and the down arrow (CTRL-N) recalls the next command in the list. By default, the IOS remembers ten commands, but you can modify this number by using the *terminal history size* command (explained further in Chapter 15), and set it as high as 256 (although this is definitely not recommended because you will spend so much time pressing up and down that you might as well type it over anyway).

Similarly, the IOS allows you to use the left arrow and right arrow keys (or CTRL-B and CTRL-F, respectively) to move forward or backward within a command without erasing characters. This trick is useful when you want to change something in the middle of a command without changing the end of the command.

In addition to the arrow keys, you can use the TAB key to complete a command once you have typed enough of the command for the IOS to recognize it as unique. For instance, you may type *sh* and press the TAB key, and IOS will turn it into *show* for you. This would be pretty useful, except the IOS uses the command as it is, as long as the command is unique. In other words, you don't ever really have to type the complete command; simply entering *sh* will suffice. So for the command *show ip Interface Ethernet 0/0*, all you really have to type is *sh ip int eth 0/0*. (Watching someone configure a router in the IOS can look like an interesting form of shorthand!)

On top of all of these shortcuts, a few others, while significantly less useful in day-to-day operations, still come in handy occasionally. The CTRL-A command moves you back to the first character in the command. The CTRL-E command moves you to the last character in the command. ESC-B moves you back one word in the command, and ESC-F moves you forward one word in the command. Finally, CTRL-R creates a new command prompt that is a duplicate of the last command prompt typed, including any commands typed. This command is mainly useful when a message from the router interrupts your typing and you can't remember where in the command statement you were.

One final note: you can negate any configuration command in the IOS by simply retyping the command with a *no* in front of it. For instance, if you use the command *ip address 192.168.1.1 255.255.255.0*, setting the IP address of an interface to 192.168.1.1 with a 255.255.255.0 mask, removing the IP address from the interface is as simple as typing *no ip address 192.168.1.1 255.255.255.0*.

IOS Modes

The IOS is built on modes of operation. The primary modes in IOS are based on privilege levels. The IOS includes 16 privilege levels, from 0 to 15; however, only 3 are used by default. The rest require special configuration to be used.

Privilege level 0 is rarely used. The only commands issuable from this level are *help, disable, enable, exit,* and *logout,* making it pretty much useless.

Privilege level 1 is the first mode you will be presented with upon logging into the router. This mode is called user privileged mode, sometimes called simply user mode. In this mode, you may not issue any commands that would cause the device grief, meaning that you cannot perform any sort of debug (advanced diagnostics, covered in later chapters), certain informational *(show)* commands are unavailable, and you may not reconfigure the device in any way. This mode is meant to be used simply when you need basic information about the device's operation. While in this mode, your prompt will appear like this:

```
Router>
```

The greater than (>) symbol is the clue as to the mode you are in. This symbol lets you know that you are in user mode.

The third mode of operation is system privileged mode, sometimes called simply privileged mode or enable mode. This mode is privilege level 15, and it allows you to access all commands usable by the device. You enter enable mode by typing *enable,* at which point, you are prompted for a password. This password will be either the enable secret or the enable password. If the enable secret is set and understood by the router, then it will be used. If not, the enable password will be used.

Tip   *You are generally better off using the enable secret instead of the enable password because most versions of the IOS currently in use understand the secret, and it is not stored in configuration files as plaintext.*

Once you successfully enter enable mode, your prompt will change to the pound (#) sign , as illustrated here:

```
Router>enable
Password:
Router#
```

At this point, you can issue any standard command, such as detailed *show* commands and *debug* commands. (*Debug commands* are special commands used to perform advanced diagnostics on the system.) You can also use a few configuration commands, such as *clear* (to clear many things, from the ARP cache to X.25 VCs), *clock* (to set the system time), *reload* (to reboot the device), *copy* (to copy configurations and IOS images), and *erase* (to delete images). However, most configuration tasks take place in a fourth mode: configure mode.

You reach configure mode from enable mode by typing the command *configure* and then specifying how you would like to configure the device. For instance, *configure terminal* means that you are going to configure the device by typing configuration commands in manually. This is the typical method of device configuration. Once you enter configure mode, the prompt changes again, as the following output illustrates:

```
Router#configure terminal
Enter configuration commands, one per line. End with CNTL/Z.
Router(config)#
```

This mode is also known as global config mode because all commands issued here affect the entire device. A separate mode, known as interface config (although it is not always for an interface), is used to set interface-specific parameters. To enter interface config mode, specify which interface you would like to configure by using the *interface* command, as follows:

```
Router(config)#interface Ethernet 0/0
Router(config-if)#
```

Note that interface config is not just for interfaces. It is also used for routing protocols, named access lists, and other commands. Some of these are shown in the following output:

```
Router(config)#ip access-list standard test
Router(config-std-nacl)#exit
Router(config)#router rip
Router(config-router)#exit
Router(config)#sub
Router(config)#subscriber-policy 1
Router(config-policy)#exit
Router(config)#
```

Notice that in all cases in which entering a command causes a mode change, the prompt changes to reflect the new mode. Also notice that to exit out of the interface config, you use the *exit* command. If you want to exit out of configure mode entirely (and back to enable mode), you can use CTRL-Z or the *end* command. These commands are good to keep up with because you cannot view your configurations from configure mode. You must return to enable mode. (No *show* commands exist in configure mode.)

Unlike most operating systems, where you can issue any command but access will be denied if the command is not available to you, in the IOS, commands and command lists are mode specific. That is, when in user mode, any unavailable commands will simply not appear in a command list, and entering these commands results in a "huh?" from the router. In fact, just remember that if you are sure of the command, but when you enter the command you get an error, chances are you are in the wrong mode.

IOS Configuration Files

The IOS uses IOS configuration files to determine interface, protocol, security, and other parameters. The IOS configuration files are similar to the DOS config.sys file. They are basic text files that simply issue commands and set environment variables in the device, telling it what to do and how to do it. Most Cisco devices follow a boot pattern similar to DOS.

First, a bootstrap program in ROM on the router begins the POST and basic startup processes. This is analogous to the BIOS functions on a PC. Next, the bootstrap program calls the IOS from its stored location (usually flash RAM), decompresses it if necessary, and loads it into RAM. (On some routers, the IOS is run directly from flash RAM, but this operation is not typical.) This process is similar to the basic boot process from a floppy disk or other storage media in DOS. The IOS loads the stored configuration file (called the startup-config) from NVRAM, which is then applied to the configuration running in RAM (called the running-config), and this configures the device according to your specifications.

If you make changes to the configuration while the device is running, most Cisco devices store this information in RAM, or the running-config (although there are exceptions; 1900 switches, for example, save to both the startup and running-config when changes are made). Therefore, unless you tell the device to save the configuration, when the router is reloaded, it loses these changes. This is in place mostly so that a bad mistake can be overcome by simply powering off the router.

Saving changes made in the router is a matter of simply copying the configuration from RAM (running-config) to NVRAM (startup-config). You can perform this task with a fairly simple command: *copy*. The *copy* command syntax is similar to that of DOS. The basic format is *copy (from) (to)*. So if you wanted to save your running-config to your startup-config, you would enter the command as follows:

```
Router#copy running-config startup-config
Destination filename [startup-config]?
Building configuration...
[OK]
Router#
```

Note that you can also copy these to an FTP or TFTP server. This is extremely useful because the files are very small (less than 32KB in most cases), so transfer time and storage space is not a major concern; therefore, you are provided with a backup of the device's configuration on the off chance that something is lost.

Even more, the configuration files are actually text files with nothing more complex than a list of commands, so you can go into the text file with a standard text editor and copy, paste, and modify to your heart's content. Then you can copy the file back into the router in much less time than it would take to type all of the commands in manually.

CISCO TECHNOLOGY OVERVIEW

The down side to this approach is that if you add a new command or extensively modify existing commands, you must know the IOS command structure and syntax very well indeed because Notepad won't tell you when you are wrong.

Cisco offers a free, generally useful (even if a bit buggy) Windows-based TFTP server on their web site, but a Cisco Connection Online (CCO) login is required to download it. If you are unable to acquire Cisco's TFTP server, a good Windows-based TFTP server is available for free at http://infodeli.3com.com/software/utilities_for_windows_32_bit.htm.

The syntax for copying to a TFTP server is simple. You issue the command *copy (source) TFTP*, where the source is either running-config or startup-config, and the IOS prompts you for all of the specifics, as the following example illustrates:

```
Router#copy running-config tftp
Address or name of remote host []? 192.168.1.1
Destination filename [router-confg]?
Write file router-confg on host 192.168.1.1? [confirm] y
#
Writing router-confg!!! [OK]
Router#
```

However, you need to be aware of a few issues with the *copy* command. First, anything you copy into NVRAM performs an overwrite of NVRAM. As a result, a copy to NVRAM is always a replace operation, regardless of the source. Copying up to TFTP or FTP is also an overwrite. Copies to RAM (running-config), however, are merge operations. This means that if the setting from the source is compatible with the destination, both are retained. If not, however, the source overwrites the destination.

In other words, if you are copying from NVRAM to RAM, and your IP address in NVRAM is set to 10.0.0.1, while your IP address in RAM is set to 192.168.1.1, then your IP address after the copy will be 10.0.0.1. However, if your IP address is *not* set (blank) in NVRAM, then after the copy, your IP address will still be 192.168.1.1. The bottom line? If you want to wipe the RAM clean and have exactly the same configuration in RAM that is present in NVRAM, the easiest way to do that is to reboot the router.

The Flash RAM File System

As mentioned previously in the "Booting a Cisco Device" section, the IOS image for your Cisco device is typically stored in flash RAM. Therefore, the Cisco IOS treats flash RAM more like a file system than RAM. Keep in mind that, just like a disk, flash RAM stores information in files. The IOS files, called *images,* take up a certain amount of space in flash RAM; and, just like a disk, you can store only so many megabytes of data on a flash RAM file system.

The amount of space an IOS image consumes depends mostly on the feature set and device model, but it is typically from 4MB to 16MB per image. Most routers and switches come standard with between 8MB and 32MB, however, and some support as much as 256MB of flash RAM. (Consult Chapters 9 through 12 for more details on memory support for each individual model.) If you want to upgrade to a high-level feature set, you might have to upgrade the flash RAM in the device as well.

Relative to configuration, the standard process of upgrading the IOS is simple and uses the *copy* command. To copy an image from a TFTP server to your router, for instance, you simply type the command *copy tftp flash* and follow the prompts. To copy the image from your router to the TFTP server, you perform the reverse of the previous command, typing *copy flash tftp*.

Note that if you are copying down a flash image and do not have enough free space in the flash file system, the IOS prompts you to overwrite the existing files in the flash. Once you choose yes at this prompt, you are committed: it deletes the image in flash *before* it pulls down the new image. If you happen to have a failure during the download at this point, the router reboots to a minimal version of the IOS stored in ROM known as ROM Monitor (Rommon).

Rommon is a "lite" version of IOS with a few simple commands designed to help you solve problems when major errors occur with the IOS image in flash. Rommon is automatically activated whenever the IOS image in flash is erased, but a router can be booted to it manually if needed. (See the next section for more details.)

A simpler solution to this problem on the higher-end routers is to use the embedded PC Card slots to store a copy of the IOS image you want to upgrade to, and then boot the router from this image. This scenario is covered in the section "Configuration Registers, Boot Options, and Password Recovery," later in the chapter.

Syslog Messages

Occasionally, while working at the console, you may be interrupted by a message from the IOS, informing you of some (usually trivial, but occasionally important) event. These messages are known as syslog messages, and the IOS uses them to inform you of events the IOS deems important. These messages generally appear in the following manner:

```
00:00:08: %LINK-3-UPDOWN: Interface Ethernet0/0,
changed state to up
00:00:08: %LINK-3-UPDOWN: Interface Ethernet0/1,
changed state to up
00:00:08: %LINK-3-UPDOWN: Interface Serial0/0, changed state to up
00:00:09: %SYS-5-CONFIG_I: Configured from memory by console
00:00:09: %LINEPROTO-5-UPDOWN: Line protocol on
Interface Ethernet0/0, changed state to up
00:00:09: %LINEPROTO-5-UPDOWN: Line protocol on
```

```
Interface Ethernet0/1, changed state to down
00:00:09: %LINEPROTO-5-UPDOWN: Line protocol on
Interface Serial0/0, changed state to up
00:00:09: %SYS-5-RESTART: System restarted --
00:00:11: %LINK-5-CHANGED: Interface Ethernet0/1,
changed state to administratively down
```

In this case, these messages informed you of links and line protocols on your links transitioning from up to down, and vice versa, as well as letting you know that the system had been restarted. Unfortunately, these events are typical when the router is rebooted, and they are more of a nuisance in this case than anything else. It's like your NT server popping up a message, interrupting your tasks every time a new event is written to the event log. The interruption gets tedious very quickly. Luckily, by default, these messages appear only on the console; even there, you can use some tools to make them a bit more bearable.

A global configuration command called *logging* is potentially the most useful tool at your disposal. The following output shows the operators available for this command:

```
Router(config)#logging ?
  Hostname or A.B.C.D IP address of the logging host
  buffered Set buffered logging parameters
  console Set console logging level
  facility Facility parameter for syslog messages
  history Configure syslog history table
  monitor Set terminal line (monitor) logging level
  on Enable logging to all supported destinations
  source-interface Specify interface for source address in logging
  transactions
  trap Set syslog server logging level
```

The *logging [hostname or IP address]* command allows you to save syslog messages to a syslog server for archival purposes. This does, of course, require that a syslog server application be running on the destination machine. Entering the command is fairly simple; but before you tell the system to save these messages to a syslog server, you need to make sure you enter the *no logging console* command, which halts all sending of syslog messages to the console port. This restriction is required because logging to both locations has been known to cause the router to hang.

The *logging buffered* command tells the IOS to copy all syslog messages to RAM on the router for future reference. This command is especially useful when logging into the router from a connection other than the console because, by default, syslog messages are not displayed on Telnet or aux connections. By using the *show logging* command, you can view these syslog messages at your convenience.

You can use the *logging console* command to set the level of logging to the console. Simply type *logging console* to enable logging to the console. However, the following optional statements can help you control how much information is sent to the console:

```
Router(config)#logging console ?
 <0-7> Logging severity level
 alerts Immediate action needed (severity=1)
 critical Critical conditions (severity=2)
 debugging Debugging messages (severity=7)
 emergencies System is unusable (severity=0)
 errors Error conditions (severity=3)
 guaranteed Guarantee console messages
 informational Informational messages (severity=6)
 notifications Normal but significant conditions (severity=5)
 warnings Warning conditions (severity=4)
```

Typically, you would want to know about a severity of 4 (warning) or higher, but the other messages are just annoying. To tell the router to display syslog messages with a severity of only 4 or higher, type *logging console 4* or *logging console warnings*. If you do this, however, remember to set the logging back to the default (severity 7, or debugging) before attempting to perform *debug* commands (advanced diagnostic commands, covered in detail in Chapter 15).

You use the *logging facility* command in conjunction with your syslog server configuration to specify how the message is processed. You need this command only when logging to a syslog server.

You can use the *logging history* command to both control the total number of syslog messages retained and to control what severity of messages are retained, independent from the severity levels enabled on the console connection. To set the number of messages retained, simply type *logging history size [size in messages]*. The size can be from 1 to 500. To control what severity of messages are retained, use the *logging history [severity]* command. The severity levels are identical to the severity levels used in the *logging console* command.

The *logging monitor* command sets up logging to terminal sessions other than the console session, such as Telnet and aux connections. In other respects, it is the same as the *logging console* command.

The *logging on* command simply enables logging to destinations other than the console. The *no logging on* command disables all logging (including syslog servers) except to the console. Typically, you will leave this setting on.

The *logging source-interface* command sets the source IP address of all syslog packets sent from the router. This is useful in high-security environments where syslog messages are not accepted by the server unless they originate from a specific IP.

The *logging trap* command sets the severity level for sending out SNMP traps to the configured management station. This allows you to specify that only certain high-importance messages are sent to your SNMP management station for analysis. The severity levels and syntax are the same as the *logging console* command.

Finally, the *logging synchronous* command can be particularly useful when you want to see all of the syslog messages, but you do not want them interrupting your command entry. This line configuration command (reached for the console port by first typing *line console 0* at the global configuration prompt) requires that you press the ENTER key before syslog messages are shown, meaning that your commands will not be interrupted when you receive a syslog message.

Configuration Registers, Boot Options, and Password Recovery

Sometimes a router has more than one boot image available, or someone has locked you out of a router and then left the company. In these cases, you need to understand how to use some of the tools the IOS provides and how to use Rommon to help solve these problems.

For your first problem, telling the router which image to boot from, the IOS can actually be quite helpful. Typically, you can use the *boot system* command to tell the router not only which IOS image to use, but also where to retrieve it from. For instance, the global config mode command *boot system flash [filename]* tells the router to boot by a certain name from an image in the flash file system. So *boot system flash c2600-i-mz.120-7.T* instructs the router to boot from a flash image named c2600-i-mz.120-7.T. If you are unsure of the name of the file, the *show flash* command gives you a list of all files stored in the flash file system.

Using the *boot system* command, you can also tell the router to boot from ROM (Rommon), or to boot from an image stored on a TFTP server. To get the router to boot to ROM, simply issue the *boot system rom* command. To get the router to boot from an image on a TFTP server, issue the *boot system tftp [filename] [ip address]* command. Fairly simple, right? Ah, but what if you attempted to download a new version of the IOS onto your router, and a junior network admin (now unemployed) happened to kick your cable out of the TFTP server in the middle of the transfer? Well, then it gets a bit more complicated.

The easiest method of solving this problem, assuming your router has a PC Card slot, and another (working) router of the same model also has a PC Card slot, is to load the file onto a PC Card and boot the router from that. Barring this option, the easiest way to perform this task is to boot into Rommon (which will happen automatically if a usable IOS cannot be located) and redownload the image. Rommon can be different depending on the model of router, however, so this process can get a little tricky. In most routers, this task can be done through a TFTP program within Rommon. On the 2600 series, for instance, the command syntax and requirements are as follows:

```
rommon 1 > tftpdnld ?

usage: tftpdnld [-r]
 Use this command for disaster recovery
only to recover an image via TFTP.
 Monitor variables are used to set up parameters for the transfer.
 (Syntax: "VARIABLE_NAME=value"
and use "set" to show current variables.)
 "ctrl-c" or "break" stops the transfer before flash erase begins.

 The following variables are REQUIRED to be set for tftpdnld:
 IP_ADDRESS: The IP address for this unit
 IP_SUBNET_MASK: The subnet mask for this unit
 DEFAULT_GATEWAY: The default gateway for this unit
 TFTP_SERVER: The IP address of the server to fetch from
 TFTP_FILE: The filename to fetch

 The following variables are OPTIONAL:
 TFTP_VERBOSE: Print setting.
0=quiet, 1=progress(default), 2=verbose
 TFTP_RETRY_COUNT: Retry count for ARP and TFTP (default=7)
 TFTP_TIMEOUT: Overall timeout of operation in seconds
(default=7200)
 TFTP_CHECKSUM: Perform checksum test on image,
0=no, 1=yes (default=1)

 Command line options:
 -r: do not write flash, load to DRAM only and launch image
rommon 2 >
```

Notice that the tftpdnld utility also allows you to simply launch the file directly from memory, rather than erasing the flash and launching it from there. This option can be useful in certain situations.

As far as the differences in Rommon are concerned, just remember that the online help is still available in Rommon; so if these commands don't work as expected, use help (and Cisco's web site) to determine the correct command and syntax.

In some routers, the tftpdnld option may not be available. The router may not have a *tftpdnld* command (or an equivalent). This requires that you use the *rommon xmodem* command to download the flash update over the console connection by using the following procedure:

```
rommon 1 > xmodem -y c2600-i-mz.120-7.T
Ready to receive file c2600-i-mz.120-7.T...
Download Complete!
```

The first line tells the router to download the file c2600-i-mz.120-7.T from the console port. The -y flag tells the router to use the *ymodem* protocol, which uses a 16-bit Cyclic Redundancy Check (CRC). (This is recommended.) Without the -y flag, the router simply transfers the file using *xmodem*, which uses only an 8-bit CRC.

Note *Be sure the transfer speed on your serial port is set to 9600. Some routers will exhibit errors if the speed is higher than this.*

When the console says "Ready to receive..." you need to begin the transfer using the *xmodem* protocol (if you didn't use the -y flag) or the *ymodem* protocol (if you did use the -y flag) in your terminal program. (HyperTerminal, which ships standard with Windows 9x, can use both protocols.) Once the download completes, you must reboot the router (either by power cycle or by using the *reset* command).

Finally, imagine that you have been locked out of a router and need to get in. You can force the router to ignore configuration settings in NVRAM (including passwords) by changing the configuration register.

The *configuration register* is a setting stored in NVRAM that lets the Cisco device know which settings to use upon boot, similar to the CMOS settings for the BIOS on a PC. Unfortunately, these settings are a bit more complicated than they are on a PC. The configuration register is always displayed as four hexadecimal digits, and the default setting is typically 0x2102. The *0x* part just lets you know it is in hex; the rest is the actual setting. The easiest way to understand the configuration register is to convert the hex value into binary and match them up with a key.

Hexadecimal is base 16, meaning that one of 16 values can be represented with a hexadecimal digit. The digits range from 0 through F, with 0 representing decimal 0, and F representing decimal 15. To convert hex into binary, you simply take each hex digit, convert it into its decimal equivalent, and then convert that into binary. Because each hex digit represents four binary digits (four binary digits are required to make a number between 0 and 15), you simply map it out with one hex digit equaling four binary digits.

To help with this, Table 14-3 shows conversions from hex to binary to decimal.

Hex	Decimal	Binary
0	0	0000
1	1	0001
2	2	0010
3	3	0011

Table 14-3. *Hex/Binary/Decimal Conversion Chart*

Hex	Decimal	Binary
4	4	0100
5	5	0101
6	6	0110
7	7	0111
8	8	1000
9	9	1001
A	10	1010
B	11	1011
C	12	1100
D	13	1101
E	14	1110
F	15	1111

Table 14-3. *Hex/Binary/Decimal Conversion Chart* (continued)

So, if you want to convert the hex number 0x1C01 into binary, first you split it up into four individual digits. Then you convert each individual digit into binary, and then you put the binary in the correct order again. This is shown in Figure 14-3.

Now, as for taking the hex code in the configuration register, converting it, and using a key, this should be fairly simple. You need to figure out what the binary is for the default settings. You go through the same process you just went through and end up with 0010000100000010, as shown in Figure 14-4.

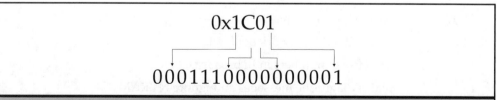

0x1C01

0001110000000001

Figure 14-3. *Hex-to-binary conversions*

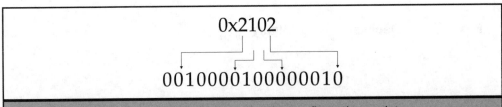

Figure 14-4. *Hex conversion into binary for the configuration register*

You then examine the key for the configuration register to see what each bit means. The key is shown in Table 14-4 and in Figure 14-5.

So, once you apply the key to the binary, you get the result shown in Figure 14-6. The default sets the router to boot to ROM if network boot fails (bit 13), disables the break function in bit 8, and sets the router to boot normally (bits 0–3).

Back to the original problem: how do you get the router to ignore the password? Well, the short answer is that you need to set bit six in your configuration register,

Bit	Meaning
0–3	If value = 0, boot to Rommon; if value = 1, boot to EPROM; if value 2–15, boot normally
4	Reserved
5	Reserved
6	Ignore NVRAM if set
7	OEM bit
8	Disable break function
9	Reserved
10	Node portion for IP broadcasts set to all zeros (instead of all ones)
11–12	Console port line speed
13	Boot to ROM if network boot fails
14	No network numbers in IP broadcasts
15	Enable diagnostic messages and ignore NVRAM

Table 14-4. *Configuration Register Key*

CISCO TECHNOLOGY
OVERVIEW

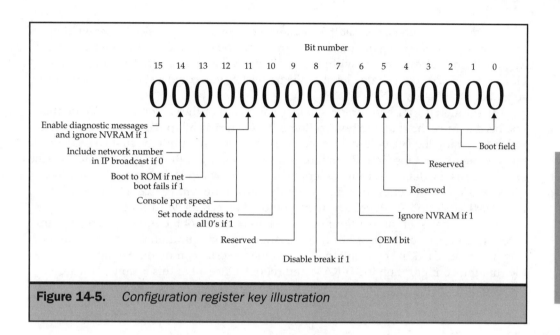

Figure 14-5. *Configuration register key illustration*

which causes the router to ignore NVRAM configuration. However, doing this is sometimes easier said than done, so here comes the long answer.

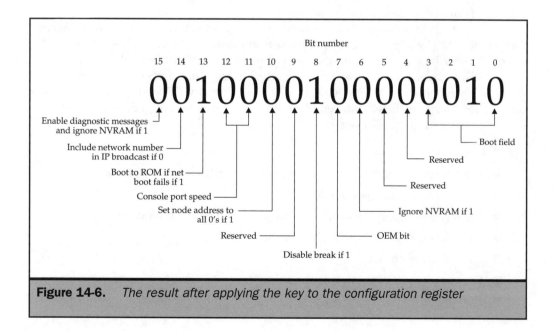

Figure 14-6. *The result after applying the key to the configuration register*

In some older routers, you set the configuration register by physically setting jumpers inside the router. However, on most newer routers, you can do this by using a command in global configuration mode within the IOS (the *configuration-register* command). However, if you could get into global config in the IOS, you could just reset the password, so that doesn't help much.

The solution is to use Rommon's *confreg* command or the *o/r* command. To do this, you need to force the router to boot to Rommon. On nearly all routers, you perform this task by pressing the BREAK key from a console connection within the first 60 seconds of booting the router. Once into Rommon, you set the configuration register to 0x2142 (the exact same as the default setting, except bit six is now set to 1). In the 1600, 2600, 3600, 4500, 7200, and 7500 series routers, you do this by typing *confreg 0x2142*. In the 2000, 2500, 3000, 4000, and 7000 series routers, you do it by typing *o/r 0x2142*.

Once you have set this configuration, you need to reboot the router. Then you can enter the router, reset the password, and change the configuration register back to its original value of 0x2102 by using the *configuration-register* command. Finally, you copy the running-configuration to NVRAM and reboot. Once all of this is done, you are finished and can log in to the router normally.

Cisco Discovery Protocol

Cisco Discovery Protocol (CDP) is designed to help you locate and identify which Cisco devices a given device is directly connected to. Its primary purpose is informative, allowing you to determine who your directly connected neighbors are, what they are capable of, and what their IP addresses are.

CDP operates by including information in layer 2 packets in a Sub-Network Access Protocol (SNAP) header. Using layer 2 packets to transport the CDP data makes CDP not only upper-layer-protocol independent (it works over IP, IPX, AppleTalk, and so on), but also allows it to run over any layer 2 technology that supports SNAP headers (Token Ring, Ethernet, ATM, Frame Relay, and most others). CDP sends periodic messages on all LAN interfaces (this is the default setting, but it can be changed, as you will see in a moment), listing the following:

- Host device's name
- The address of the interface on which the advertisement is being sent
- The platform (series) of the device
- The device's capabilities (bridge, router, switch, host, or repeater)
- The interface number on the host device
- The holddown timer
- The version of IOS the device is running
- The version of CDP being used (Version 2 is the most common on modern devices.)

By default, CDP sends these messages once every 60 seconds. A holddown timer is set on each device when it hears a CDP message. This timer detects when the neighboring device is down or otherwise unresponsive. When the holddown timer expires, the remote device is considered to be down and is removed from the list of CDP neighbors. The default holddown timer setting is 180 seconds, or three times the advertisement interval.

CDP is a Cisco proprietary protocol, so it is useful only for discovering other Cisco devices. It is also not propagated past the first device that receives it. In other words, in the example in Figure 14-7, Modena sends an advertisement to Maranello, letting Maranello become aware of Modena's capabilities. Maranello sends an advertisement to Dino, letting Dino know about Maranello's capabilities; however, Maranello will *not* propagate Modena's advertisement to Dino. And when Dino sends his advertisement to Maranello, Maranello will *not* propagate this information to Modena. So, in the end, only Maranello will be aware of both Dino and Modena.

You can view all of the information CDP provides by using a few simple *show* commands. The *show cdp* command gives you the CDP advertisement interval, the holddown timer value, the version of CDP being used, and whether CDP is enabled. (CDP is enabled on all LAN interfaces by default.) This is shown in the following output:

```
Router>sh cdp
Global CDP information:
 Sending CDP packets every 60 seconds
 Sending a holdtime value of 180 seconds
 Sending CDPv2 advertisements is enabled
Router>
```

The *show cdp interface* command gives you basic information about which interfaces are employing CDP, the encapsulation used on those interfaces, the advertisement interval, and the holddown timer settings. To see a single interface, you can simply

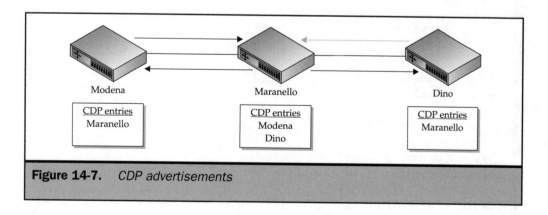

Figure 14-7. *CDP advertisements*

CISCO TECHNOLOGY OVERVIEW

append the interface name and number to the end of this command, for example, *show cdp interface Ethernet 0/0.* The following output illustrates:

```
Router>sh cdp interface
Ethernet0/0 is up, line protocol is up
 Encapsulation ARPA
 Sending CDP packets every 60 seconds
 Holdtime is 180 seconds
Serial0/0 is up, line protocol is up
 Encapsulation FRAME-RELAY
 Sending CDP packets every 60 seconds
 Holdtime is 180 seconds
Ethernet0/1 is administratively down, line protocol is down
 Encapsulation ARPA
 Sending CDP packets every 60 seconds
 Holdtime is 180 seconds
Router>sh cdp interface ethernet 0/0
Ethernet0/0 is up, line protocol is up
 Encapsulation ARPA
 Sending CDP packets every 60 seconds
 Holdtime is 180 seconds
Router>
```

To see information about your neighboring Cisco device, a few other commands come in handy. The *show cdp neighbor* command tells you basic information about each directly connected device. For more detail, you can simply append the detail to this command, making it *show cdp neighbor detail.* An example follows:

```
Router>show cdp neighbor
Capability Codes: R - Router, T - Trans Bridge,
B - Source Route Bridge
 S - Switch, H - Host, I - IGMP, r - Repeater

Device ID Local Intrfce Holdtme Capability Platform Port ID
SerialRouter1 Ser 0/0 140 R 3620 Ser 0/1
1900 Eth 0/0 134 T S 1900 2
Router>show cdp neighbor detail
-------------------------
Device ID: SerialRouter1
Entry address(es):
 IP address: 169.254.2.1
Platform: cisco 3620, Capabilities: Router
```

```
Interface: Serial0/0, Port ID (outgoing port): Serial0/1
Holdtime : 134 sec

Version :
Cisco Internetwork Operating System Software
IOS (tm) 3600 Software (C3620-D-M), Version 12.1(3a)T1,
RELEASE SOFTWARE (fc1)
Copyright (c) 1986-2000 by cisco Systems, Inc.
Compiled Fri 28-Jul-00 16:31 by ccai

advertisement version: 2

-------------------------
Device ID: 1900
Entry address(es):
 IP address: 10.1.1.1
Platform: cisco 1900, Capabilities: Trans-Bridge Switch
Interface: Ethernet0/0, Port ID (outgoing port): 2
Holdtime : 127 sec

Version :
V9.00

advertisement version: 2
Protocol Hello: OUI=0x00000C, Protocol ID=0x0112;
payload len=25, value=0000000
0FFFFFFFF0101050000000000000000003E305C600FF
VTP Management Domain: ''
Duplex: half

Router>
```

In addition, to see detailed information about just one neighboring device, you can use the *show cdp entry [name]* command. One small note: the name used in this command is case sensitive. If you use the wrong capitalization (in this case, if you use serialrouter1 instead of SerialRouter1), you will not get the expected result. This command is illustrated here:

```
Router>show cdp entry SerialRouter1
-------------------------
Device ID: SerialRouter1
```

```
Entry address(es):
 IP address: 169.254.2.1
Platform: cisco 3620, Capabilities: Router
Interface: Serial0/0, Port ID (outgoing port): Serial0/1
Holdtime : 159 sec

Version :
Cisco Internetwork Operating System Software
IOS (tm) 3600 Software (C3620-D-M), Version 12.1(3a)T1,
RELEASE SOFTWARE (fc1)
Copyright (c) 1986-2000 by cisco Systems, Inc.
Compiled Fri 28-Jul-00 16:31 by ccai

advertisement version: 2
Router>
```

As for CDP configuration, you need to do very little to enable CDP. As mentioned previously, CDP is enabled by default on all LAN interfaces. To enable CDP on WAN interfaces, simply use the interface configuration command *cdp enable*. To disable CDP on an interface, use the *no* version of the command *(no cdp enable)*. These versions are shown in the following example:

```
Router>show cdp neighbor
Capability Codes: R - Router, T - Trans Bridge,
B - Source Route Bridge
 S - Switch, H - Host, I - IGMP, r - Repeater

Device ID Local Intrfce Holdtme Capability Platform Port ID
1900 Eth 0/0 163 T S 1900 2
```

Here, only one device is advertising to you, 1900, using the Ethernet connection:

```
Router>show cdp interface
Ethernet0/0 is up, line protocol is up
 Encapsulation ARPA
 Sending CDP packets every 60 seconds
 Holdtime is 180 seconds
Ethernet0/1 is administratively down, line protocol is down
 Encapsulation ARPA
 Sending CDP packets every 60 seconds
 Holdtime is 180 seconds
```

After viewing the CDP interfaces, notice that only the LAN interfaces are advertising, which is the default configuration:

```
Router>enable
Password:
Router#config terminal
Enter configuration commands, one per line. End with CNTL/Z.
Router(config)#interface serial 0/0
Router(config-if)#cdp enable
Router(config-if)#^Z
Router#
1w1d: %SYS-5-CONFIG_I: Configured from console by console
```

Here, you enable CDP advertisements and enable listening for neighbors on your serial 0/0 interface.

```
Router#show cdp interface
Ethernet0/0 is up, line protocol is up
 Encapsulation ARPA
 Sending CDP packets every 60 seconds
 Holdtime is 180 seconds
Serial0/0 is up, line protocol is up
 Encapsulation FRAME-RELAY
 Sending CDP packets every 60 seconds
 Holdtime is 180 seconds
Ethernet0/1 is administratively down, line protocol is down
 Encapsulation ARPA
 Sending CDP packets every 60 seconds
 Holdtime is 180 seconds
Router#show cdp neighbor
Capability Codes: R - Router, T - Trans Bridge,
B - Source Route Bridge
 S - Switch, H - Host, I - IGMP, r - Repeater

Device ID Local Intrfce Holdtme Capability Platform Port ID
1900 Eth 0/0 176 T S 1900 2
```

Here, you check your CDP interface list to make sure the serial 0/0 interface is listed. Once you are satisfied, you attempt to see whether any new neighbors have been found

CISCO TECHNOLOGY OVERVIEW

and notice that none have. This is because you have just enabled CDP on the interface.
You need to wait up to 60 more seconds to discover a neighbor over serial 0/0:

```
Router#show cdp neighbor
Capability Codes: R - Router, T - Trans Bridge,
B - Source Route Bridge
 S - Switch, H - Host, I - IGMP, r - Repeater

Device ID Local Intrfce Holdtme Capability Platform Port ID
SerialRouter1 Ser 0/0 173 R 3620 Ser 0/1
1900 Eth 0/0 168 T S 1900 2
```

A few seconds later, SerialRouter1 shows up on your serial 0/0 interface.

```
Router#config terminal
Enter configuration commands, one per line. End with CNTL/Z.
Router(config)#interface serial 0/0
Router(config-if)#no cdp enable
Router(config-if)#^Z
1w1d: %SYS-5-CONFIG_I: Configured from console by console
Router#show cdp interface
Ethernet0/0 is up, line protocol is up
 Encapsulation ARPA
 Sending CDP packets every 60 seconds
 Holdtime is 180 seconds
Ethernet0/1 is administratively down, line protocol is down
 Encapsulation ARPA
 Sending CDP packets every 60 seconds
 Holdtime is 180 seconds
```

You then disable CDP and view your CDP interface list to make sure that serial 0/0
was removed. Once this is done, view the CDP neighbor list to see whether SerialRouter1
has been removed.

```
Router#show cdp neighbor
Capability Codes: R - Router, T - Trans Bridge,
B - Source Route Bridge
 S - Switch, H - Host, I - IGMP, r - Repeater

Device ID Local Intrfce Holdtme Capability Platform Port ID
SerialRouter1 Ser 0/0 133 R 3620 Ser 0/1
1900 Eth 0/0 129 T S 1900 2
```

Exactly as expected, the entry is still in the list because the holddown timer must expire before the entry is removed. You repeat the command 44 seconds later:

```
Router#show cdp neighbor
Capability Codes: R - Router, T - Trans Bridge,
B - Source Route Bridge
 S - Switch, H - Host, I - IGMP, r - Repeater

Device ID Local Intrfce Holdtme Capability Platform Port ID
SerialRouter1 Ser 0/0 89 R 3620 Ser 0/1
1900 Eth 0/0 145 T S 1900 2
```

Notice that the holddown timer entry for SerialRouter1 has been reduced by 44 seconds, while the holddown timer for 1900 has reset (due to your router receiving an update from 1900 35 seconds ago).

```
Router#show cdp neighbor
Capability Codes: R - Router, T - Trans Bridge,
B - Source Route Bridge
 S - Switch, H - Host, I - IGMP, r - Repeater

Device ID Local Intrfce Holdtme Capability Platform Port ID
SerialRouter1 Ser 0/0 15 R 3620 Ser 0/1
1900 Eth 0/0 131 T S 1900 2
```

SerialRouter1's holddown timer has nearly expired 74 seconds later:

```
Router#show cdp neighbor
Capability Codes: R - Router, T - Trans Bridge,
B - Source Route Bridge
 S - Switch, H - Host, I - IGMP, r - Repeater

Device ID Local Intrfce Holdtme Capability Platform Port ID
SerialRouter1 Ser 0/0 0 R 3620 Ser 0/1
1900 Eth 0/0 174 T S 1900 2
```

It does expire 15 seconds later. It will still be listed for a few more seconds until it can be removed.

```
Router#show cdp neighbor
Capability Codes: R - Router, T - Trans Bridge,
```

```
B - Source Route Bridge
 S - Switch, H - Host, I - IGMP, r - Repeater

Device ID Local Intrfce Holdtme Capability Platform Port ID
1900 Eth 0/0 165 T S 1900 2
Router#
```

And, finally, it is timed out of the table.

You can also enable or disable CDP globally on the router by using the *cdp run* and *no cdp run* global configuration commands, respectively. You can also globally set the holddown timer and update interval for the router with the *cdp holdtime [time in seconds]* and *cdp timer [time in seconds]* global configuration commands, although you should rarely, if ever, need to use them.

Note *In most environments, CDP is disabled entirely for security reasons.*

Summary

This chapter examined the basic operational principles behind the Cisco IOS and provided a few basic commands to help you with your tasks. Keep the tips and tricks presented in this chapter (especially the shortcuts!) in mind when performing any IOS configuration task. The next chapter delves more fully into the IOS by examining some of the most common IOS commands and their uses.

The
Complete
Reference

Chapter 15

Standard IOS
Commands: Part 1

This chapter discusses some of the most common generic commands in the standard IOS. By "generic," I mean we will cover only commands that are not covered in later chapters. For a full command reference, please examine Appendix A. The set-based IOS is also not covered in this chapter. (That is the focus of Chapter 17.) However, the standard IOS is applicable to most Cisco devices, many routers and switches included. Unfortunately, there can be some slight differences in these commands on certain platforms. The commands outlined in this chapter should be valid on most, if not all, Cisco devices.

This chapter is organized into two main parts: one for user mode and one for enable mode. In these sections, each command is examined individually in alphabetical order, with full examples when necessary. Global configuration mode and interface configuration mode are examined in the next chapter.

Common User Mode Commands

These are commands common to the most basic mode of device operation: user mode. Remember, you can tell which mode you are in based on the prompt. User mode prompts appear with the greater than symbol (>), like so: Router>. The commands discussed in this section are *connect, disconnect, enable, exit, name-connection, ping, resume, rlogin, show, telnet, terminal,* and *traceroute.*

The connect Command

This command is used to establish a terminal (Telnet) session to another device. In user mode, this command is identical to the *telnet* command, and, in truth, neither command is needed to establish a Telnet session. To establish a Telnet session, the easiest method is to simply type the name (if DNS or the local hosts table is configured) or IP address of the remote device at the prompt. The only case in which you can't use this method is when the name of the remote device is the same as an IOS command (which should rarely occur).

Syntax

Again, the easiest method to Telnet is to simply type the name or IP of the remote host from the prompt, like so:

```
ACCESSSERVER1>Router
Translating "Router"
Trying Router (10.1.1.1)... Open

Router>
```

However, if your remote host were named the same thing as an IOS command, you would need to type the command in the following manner: *connect [remote host]*, like so:

```
ACCESSSERVER1>Connect show
Translating "show"
Trying show (10.1.1.2)... Open

show>
```

Special Considerations

In some situations, you may wish to keep several Telnet sessions open at once on the same router. You can accomplish this by using a special command keystroke to exit the Telnet sessions. If you use the keystroke CTRL-SHIFT-6 and then type *x*, you will "suspend" your active Telnet session and go back to the host device. You may then Telnet into another device without losing your first connection. For instance, the following example uses a router called accessserver1 to connect to two other routers simultaneously:

```
ACCESSSERVER1>Router
Translating "Router"
Trying ROUTER (10.1.1.1)... Open

Router>
Ctrl-Shift-6    x
ACCESSSERVER1>OtherRouter
Translating "OtherRouter"
Trying OTHERROUTER (10.1.1.3)... Open

OTHERROUTER>
Ctrl-Shift-6    x
ACCESSSERVER1>
```

For more information about viewing, disconnecting, and resuming suspended Telnet sessions, see the following sections later in this chapter: "The show sessions Command," "The disconnect Command," and "The resume Command."

The disconnect Command

The *disconnect* command terminates a Telnet connection. Note that you must do this on suspended sessions; you cannot disconnect an active session. In other words, if you are Telneting into a router, typing *disconnect* will do you absolutely no good (although typing *exit* works like a charm). However, if you Telnet from that router into another router and then suspend that Telnet session (CTRL-SHIFT-6 x), you will be back at the first router. At this point, typing *disconnect* terminates your connection to the remote router.

Syntax

The base command *disconnect* disconnects the most recently used session. You may also type *disconnect [number of session]* to disconnect a specific session. To see the session numbers, you must use the *show sessions* command. See the following section for examples.

Usage Examples

In the following example, I will connect from a single router, AccessServer1, to three other routers, RouterA, RouterB, and RouterC. I will suspend each session with the three routers and then disconnect them one at a time.

```
ACCESSSERVER1>RouterA
Translating "RouterA"
Trying ROUTERA (10.1.1.1)... Open

ROUTERA>
ACCESSSERVER1>RouterB
Translating "RouterB"
Trying ROUTERB (10.1.1.2)... Open

ROUTERB>
ACCESSSERVER1>RouterC
Translating "RouterC"
Trying ROUTERC (10.1.1.3)... Open

ROUTERC>
ACCESSSERVER1>show sessions
Conn Host            Address          Byte   Idle Conn Name
   1 RouterA         10.1.1.1            0      0 RouterA
   2 RouterB         10.1.1.2            0      0 RouterB
*  3 RouterC         10.1.1.3            0      0 RouterC

ACCESSSERVER1>disconnect 1
Closing connection to RouterA [confirm]
ACCESSSERVER1>disconnect 3
Closing connection to RouterC [confirm]
ACCESSSERVER1>disconnect 2
Closing connection to RouterB [confirm]
```

The enable Command

This command simply takes you to privileged exec mode (enable mode) on the device.

Syntax

Simply type *enable* and supply the password for standard (level 15) enable mode access. Optionally, you can define the privilege level to log in to by using the *enable [0–15]* parameter; but unless other privilege levels besides the default 0, 1, and 15 have been defined, this will be useless.

The exit Command

This command logs you off the device, closing any Telnet sessions you have established and clearing your command history.

Syntax

Simply type *exit*.

The name-connection Command

This command allows you to assign a name to a suspended Telnet connection. This strategy makes it easier to remember which connection is which, but it is typically more trouble than it is worth.

Syntax

Type *name-connection*. The router will ask for the number of the connection you wish to name, and then it will ask what you want to name the connection. Once you have finished naming the connections, you can perform session control commands (such as *disconnect* and *resume*) on the session by name, as shown in the following example:

```
ACCESSSERVER1>name-connection
Connection number: 1
Enter logical name: RouterA
Connection 1 to RouterA will be named "ROUTERA" [confirm]
ACCESSSERVER1>disconnect routera
Closing connection to RouterA [confirm]
ACCESSSERVER1>
```

The ping Command

Possibly one of the most useful connectivity testing commands is the ubiquitous *ping* command. In the IOS, *ping* works like it does in nearly every other OS—by sending

Internet Control Messaging Protocol (ICMP) echo request packets and receiving echo replies. However, in the IOS, *ping* may take some getting used to because the codes are a bit unique. In user mode, it is what is known as a standard ping. With a standard ping, you can specify only the host to ping. The enable mode also includes an extended *ping* command, which can specify much more than simply an IP address. For now, we will concentrate on the user mode version of *ping*. (The "Common Enable Mode Commands" section discusses *extended ping*, later in this chapter.)

When you *ping* another device, you receive codes back for the results. Table 15-1 lists these codes.

Syntax

Type *ping [IP address/host name]*.

Usage Example

The following is an example of a *ping* that failed on its first attempt and then successfully received replies for the other four attempts.

```
RouterA>ping 10.1.1.1

Type escape sequence to abort.
Sending 5, 100-byte ICMP Echos to 10.1.1.1, timeout is 2 seconds:
.!!!!
Success rate is 80 percent (4/5), round-trip min/avg/max = 4/6/8 ms
RouterA>
```

Code	Meaning
!	Ping successful
.	No reply
U	Destination unreachable ("generic" unreachable)
N	Network unreachable
P	Port unreachable
Q	Source quench received
M	Packet undeliverable due to DF bit being set (could not fragment packet to send)
?	Unknown packet received in response

Table 15-1. *Ping Codes*

The resume Command

This command resumes a previously suspended Telnet session.

Syntax

Type *resume* and the name (if configured with the *name-connection* command) or number of the session, like so: *resume [name/number]*. You may also type *resume* to resume the last active connection.

The rlogin Command

This command allows you to connect to a remote UNIX system using the rlogin protocol. The *rlogin* command is similar to *telnet,* except that a username needs to be specified when initiating the connection.

Syntax

Use *rlogin* by typing *rlogin [IP address/host name]*. Optionally, you may also use the *–l username* or */user username* options to specify the username for rlogin. Both options perform the same function, but the *–l* option is the UNIX syntax, and the */user* option is the standard IOS syntax. IOS supports both. If no username is specified, rlogin defaults to using the local username. Sessions of rlogin can also be suspended, resumed, and disconnected in the same manner as Telnet sessions.

The show Command

The *show* command is potentially the most heavily used command in IOS. It is the primary viewing command, and nearly any part of an IOS device can be inspected by using it. Because *show* is so versatile and comprehensive, it is discussed in a slightly different format than the other commands. Also note that some of the *show* commands listed under this heading are usable only in privilege exec (enable) mode. They are listed here in the user mode section to reduce confusion. Those that are only privilege mode are identified as such in the command description.

Syntax

First, we will look at the standard syntax employed by all *show* commands, and then we will look a bit deeper into each individual *show* command. Standard *show* command syntax is *show [target] [modifiers]*. The target field specifies the individual *show* command, as in *show clock*. Modifiers (of which there may be several) can pinpoint the target more specifically in the case of a broad topic (such as *show ip route*), or specify the depth of information retrieved (such as *show ip interface brief*). This is the basic formula that all of the *show* commands tend to follow.

The show aliases Command   This command lists all aliases in the current mode. *Aliases* are pseudo-commands that invoke real commands with just one or two keys.

For instance, the default alias for *show* is *s*. The syntax for this command is *show aliases*. Optionally, you may append the mode to the end of the command to see the aliases for a mode other than the one you are currently in. For instance, *show aliases exec* shows you all of the aliases for privileged exec (enable) mode. An example of the *show aliases* command follows:

```
2600B#show aliases
Exec mode aliases:
  h               help
  lo              logout
  p               ping
  r               resume
  s               show
  u               undebug
  un              undebug
  w               where
```

The show arp Command   This command shows the Address Resolution Protocol (ARP) table for the entire device, regardless of the upper-layer protocol (as opposed to *show ip arp*, for instance, which shows only the ARP table for IP). The syntax for the command is simply *show arp*. An example of the *show arp* command follows:

```
2600B#show arp
Protocol  Address      Age (min)  Hardware Addr   Type  Interface
Internet  10.1.1.1          148   0003.e305.c600  ARPA  Ethernet0/0
Internet  10.1.1.200          -   0003.6b40.3700  ARPA  Ethernet0/0
```

The show async Command   This command shows information related to asynchronous serial connections. It has two modifiers: *bootp* and *status*. The *show async bootp* command shows all extended data (such as boot file information and time server addresses) sent in response to *bootp* requests. The *show async status* command shows statistics for asynchronous traffic, as well as all current asynchronous connections. Note that both of these commands are available only in privileged exec mode. An example of the *show async* command follows:

```
2600B#show async status
Async protocol statistics:
  Rcvd: 0 packets, 0 bytes
        0 format errors, 0 checksum errors, 0 overrun
  Sent: 0 packets, 0 bytes, 0 dropped
No lines currently configured for async protocol usage
```

The show cdp Command   As mentioned in the Chapter 14, the *show cdp* command is used to view all directly connected Cisco Discovery Protocol (CDP) neighbors, as well as to view basic CDP configuration information. The basic *show cdp* command shows your CDP configuration. Various modifiers can be used to show more specific information, as listed here:

- *show cdp entry [name]*   Shows you detailed information about all neighbors. It has the following optional modifiers:
  - *show cdp entry [name] protocol*   Shows you the layer 3 addressing of the entry.
  - *show cdp entry [name] version*   Shows you the remote device's IOS version.
- *show cdp interface [interface type and number]*   Shows you the CDP configuration for each interface.
- *show cdp neighbors*   Shows a summary of all of your neighbors. It has the following optional modifiers:
  - *show cdp neighbors [interface type]*   Allows you to see directly connected neighbors based on the connection type (Ethernet, serial, and so on).
  - *show cdp neighbors detail*   Shows you detailed information (similar to the *show cdp entry* command) on all neighbors.
- *show cdp traffic*   Shows you traffic statistics for CDP.

The show clock Command   This command, as you might have guessed, shows you the time and date set on the router. It also shows whether the time is believed to be correct or not, based on the device's time synchronization settings. The output of this command when the time is not synchronized appears as one of the following:

```
Router>show clock
Router>*11:13:33.110 EST Mon Jul 9 2001
```

In this case, the asterisk (*) means that the router assumes the time is not correct, usually because no time synchronization settings are in place.

```
Router>show clock
Router>.11:13:33.110 EST Mon Jul 9 2001
```

In this case, the period (.) means that the router assumes the time is correct; but it is not yet synchronized with its Network Time Protocol (NTP) source, so it could be wrong. If the router were synchronized, the display would appear like this:

```
Router>show clock
Router>11:13:33.110 EST Mon Jul 9 2001
```

CISCO TECHNOLOGY OVERVIEW

To display the NTP settings for your device, use the *show clock detail* command, as follows:

```
Router>show clock detail
Router>11:13:33.110 EST Mon Jul 9 2001
       Time source is NTP
```

The show debugging Command   *Debugging* is a form of advanced diagnostics with which you can view very detailed information about what occurs inside your Cisco device. (For more information on what can be monitored, please refer to "Common Enable Mode Commands," later in this chapter.) The *show debugging* command shows which items you happen to be debugging at any given time. This command is somewhat useful when you are troubleshooting a system that someone else has already been troubleshooting; you never know which debugs he or she may already have running. An example of this command is shown in this section: "The debug Command."

The show dhcp Command   This command is used mostly for serial or ISDN connections using Point-to-Point Protocol (PPP). It displays IP address lease information and known Dynamic Host Configuration Protocol (DHCP) servers. The *show dhcp lease* form of this command shows information on all DHCP leases on all interfaces. The *show dhcp server* command shows all known DHCP servers and statistics on leases given and released. An example of the *show dhcp* command follows:

```
2600B#show dhcp server
   DHCP server: ANY (255.255.255.255)
   Leases:    0
   Offers:    0      Requests: 0      Acks: 0      Naks: 0
   Declines:  0      Releases: 0      Bad:  0
```

The show diag Command   The primary use of the *show diag* command is to display diagnostic information about the modules installed in a router. Although a lot of the information displayed in the output is useful only to Cisco engineers, some of it can be extremely useful to network administrators (such as the detailed descriptions, including model numbers, of the modules and cards installed). The syntax for this command is *show diag [slot number]*. If you type *show diag*, it lists all information about all slots.

In addition, the higher-end routers (like the 12000GSR series) also include optional *detail* and *summary* modifiers that are appended to the end of the command. Note that this command can be used only in privileged mode. An example of the command for a 3620 router follows:

```
SerialRouter1#show diag 0
Slot 0:
```

```
NM-2FE2W Port adapter, 6 ports
Port adapter is analyzed
Port adapter insertion time unknown
EEPROM contents at hardware discovery:
Hardware Revision          : 1.0
Part Number                : 800-04797-01
Board Revision             : E0
Deviation Number           : 0-6158
Fab Version                : 04
PCB Serial Number          : JAB0422090L
RMA Test History           : 00
RMA Number                 : 0-0-0-0
RMA History                : 00
EEPROM format version 4
EEPROM contents (hex):
  >CONTENT SUPPRESSED DUE TO SIZE<


WIC Slot 0:
Serial 2T (12in1)
Hardware revision 1.0          Board revision D0
Serial number     22920279     Part number   800-03181-01
Test history      0x0          RMA number    00-00-00
Connector type    PCI
EEPROM format version 1
EEPROM contents (hex):
  >CONTENT SUPPRESSED DUE TO SIZE<


WIC Slot 1:
Serial 2T (12in1)
Hardware revision 1.0          Board revision D0
Serial number     22886858     Part number   800-03181-01
Test history      0x0          RMA number    00-00-00
Connector type    PCI
EEPROM format version 1
EEPROM contents (hex):
  >CONTENT SUPPRESSED DUE TO SIZE<
```

The show environment Command   Available in full form only on the higher-end routers (7000, 7200, 7500, and 12000 series), this is still a useful command. Limited support for this command is provided for some of the lower-end routers, beginning with the 3620. This command displays environmental statistics such as temperature, voltage, fan speed, power supply status, and so on. It updates once per minute and saves the latest statistics in NVRAM on the router in case a power failure or problem forces a shutdown. The syntax of this command is *show environment [optional modifiers]*. A basic *show environment* command gives you a simple summary of the system, as follows:

```
Router> show environment
Environmental Statistics
   Environmental status as of 13:17:39 UTC Thu Jun 6 1996
   Data is 7 second(s) old, refresh in 53 second(s)
    All Environmental Measurements are within specifications
```

Adding the *all* modifier gives you more detailed information about the entire system, as follows (from a 7000 series router):

```
BigOleRouter# show environment all
Environmental Statistics
   Environmental status as of 15:32:23 EST Mon Jul 9 2001
   Data is 42 second(s) old, refresh in 18 second(s)
   WARNING: Lower Power Supply is NON-OPERATIONAL
   Lower Power Supply:700W, OFF Upper Power Supply: 700W, ON
   Intermittent Powerfail(s): 4 Last on 15:11:54 EST Mon Jul 9 2001
   +12 volts measured at 12.21(V)
   +5 volts measured at 5.06(V)
   -12 volts measured at -12.20(V)
   +24 volts measured at 23.96(V)
   Airflow temperature measured at 36(C)
   Inlet temperature measured at 21(C)
```

You can also use additional modifiers to find out more specific information about a given subsystem: *fans, hardware, LEDs, power supply, temperatures,* and *voltages.* All of these modifiers are, of course, appended to the end of the command. Note also that this command must be run from privileged mode.

The show flash Command   This command can show detailed information about not only what is in the flash file system and the amount of free space left, but also how many flash SIMM/DIMM slots are in use. The most useful form of this command is the *show flash: all* command. This shows you all available flash information, like so (output from a 3620 router):

```
SerialRouter1>show flash: all
Partition   Size    Used     Free      Bank-Size  State       Copy Mode
   1        24576K  5808K    18767K    4096K      Read/Write  Direct
```

```
           +   4096K
           +   8192K
           +   8192K
System flash directory:
File  Length    Name/status
        addr      fcksum  ccksum
   1   5947868   c3620-d-mz.121-3a.T1.bin
        0x40       0x6802  0x6802
[5947932 bytes used, 19217892 available, 25165824 total]
24576K bytes of processor board System flash (Read/Write)
     Chip    Bank    Code    Size    Name
      1       1      01D5    1024KB   AMD    29F080
      2       1      01D5    1024KB   AMD    29F080
      3       1      01D5    1024KB   AMD    29F080
      4       1      01D5    1024KB   AMD    29F080
      1       2      01D5    1024KB   AMD    29F080
      2       2      01D5    1024KB   AMD    29F080
      3       2      01D5    1024KB   AMD    29F080
      4       2      01D5    1024KB   AMD    29F080
      1       3      89A0    2048KB   INTEL  28F016SA
      2       3      89A0    2048KB   INTEL  28F016SA
      3       3      89A0    2048KB   INTEL  28F016SA
      4       3      89A0    2048KB   INTEL  28F016SA
      1       4      89A0    2048KB   INTEL  28F016SA
      2       4      89A0    2048KB   INTEL  28F016SA
      3       4      89A0    2048KB   INTEL  28F016SA
      4       4      89A0    2048KB   INTEL  28F016SA
```

Other optional modifiers are *chips* (to see SIMM chip and bank information), *detailed*, *summary*, and *err* (to display rewrites or errors).

The show history Command   The *show history* command simply shows the commands issued since logging into the router, or up to the maximum configured history buffer size. The command has no modifiers. Note that the history is different for user/enable and configuration modes. An example of the *show history* command follows:

```
2600B#show history
     show aliases
     sh async status
     show async status
     show async bootp
```

```
sh dhcp
sh dhcp serv
show dhcp server
show dhcp lease
show arp
show history
```

The show hosts Command   This command simply shows the hosts table, along with the default domain and the lookup method (static or DNS) for the router. The syntax for the command is *show hosts [modifier]*, and the only modifier is an optional name (in case you want to look up an individual entry). An example of the *show hosts* command follows:

```
ACCESSSERVER1>show hosts
Default domain is not set
Name/address lookup uses static mappings

Host                    Flags          Age Type   Address(es)
2600C                   (perm, OK) 74   IP    10.1.1.1
2600P                   (perm, OK) **   IP    10.1.1.1
2600M                   (perm, OK) **   IP    10.1.1.1
2600A                   (perm, OK) 82   IP    10.1.1.1
2600G                   (perm, OK) **   IP    10.1.1.1
2600D                   (perm, OK) **   IP    10.1.1.1
2600B                   (perm, OK)  0   IP    10.1.1.1
2600N                   (perm, OK) **   IP    10.1.1.1
2600K                   (perm, OK) **   IP    10.1.1.1
2600F                   (perm, OK) **   IP    10.1.1.1
2600J                   (perm, OK) **   IP    10.1.1.1
2600O                   (perm, OK) **   IP    10.1.1.1
2600E                   (perm, OK) **   IP    10.1.1.1
2600L                   (perm, OK) **   IP    10.1.1.1
2600H                   (perm, OK) **   IP    10.1.1.1
2600I                   (perm, OK) **   IP    10.1.1.1
```

The show interfaces Command   This is one of the most commonly used *show* commands. Its job is to provide information on individual interfaces in the system that is protocol independent (layer 3 and above). The command is used primarily in the form *show interfaces*, which displays detailed information about all system interfaces, and in the form *show interface [interface type and number]*, which shows detailed information about a specific interface. Other options and modifiers for viewing queuing and bridging properties are available, but these are dependent on not only the type of interface, but also the type and model of the device. An example of the *show interface* command follows:

```
Router#show interface Ethernet 0/0
Ethernet0/0 is up, line protocol is up
  Hardware is AmdP2, address is 0003.6b40.3700 (bia 0003.6b40.3700)
  Internet address is 192.168.3.12/24
  MTU 1500 bytes, BW 10000 Kbit, DLY 1000 usec,
     reliability 255/255, txload 1/255, rxload 1/255
  Encapsulation ARPA, loopback not set
  Keepalive set (10 sec)
  ARP type: ARPA, ARP Timeout 04:00:00
  Last input 00:00:25, output 00:00:00, output hang never
  Last clearing of "show interface" counters never
  Queueing strategy: fifo
  Output queue 0/40, 0 drops; input queue 0/75, 0 drops
  5 minute input rate 0 bits/sec, 0 packets/sec
  5 minute output rate 0 bits/sec, 0 packets/sec
     2938 packets input, 399416 bytes, 0 no buffer
     Received 2938 broadcasts, 0 runts, 0 giants, 0 throttles
     0 input errors, 0 CRC, 0 frame, 0 overrun, 0 ignored
     0 input packets with dribble condition detected
     20583 packets output, 2010423 bytes, 0 underruns
     0 output errors, 0 collisions, 1 interface resets
     0 babbles, 0 late collision, 0 deferred
     0 lost carrier, 0 no carrier
     0 output buffer failures, 0 output buffers swapped out
Router#
```

The show ip Command   This command is more of a collection of subcommands than a command in and of itself. Because the *show ip* command consists of so many subcommands (over 50 on some routers), this section discusses only those that are commonly used and those not covered in later chapters.

- *show ip arp*   Similar to the *show arp* command, except this command shows only the ARP table for IP.

- *show ip interface*   Similar to the *show interface* command, except this command displays detailed IP-specific information about interfaces. The basic form of this command, *show ip interface*, shows details about all interfaces on the router. The brief form of the command, *show ip interface brief*, shows summarized information about all of the interfaces. Finally, the *show ip interface [interface name and number]* form of the command shows detailed information about a specific interface. All of these are shown in the following example (from a 2600 router):

```
2600A#show ip interface
Ethernet0/0 is up, line protocol is up
  Internet address is 10.1.1.100/24
```

```
          Broadcast address is 255.255.255.255
          Address determined by setup command
          MTU is 1500 bytes
          Helper address is not set
          Directed broadcast forwarding is disabled
          Outgoing access list is not set
          Inbound  access list is not set
          Proxy ARP is enabled
          Security level is default
          Split horizon is enabled
          ICMP redirects are always sent
          ICMP unreachables are always sent
          ICMP mask replies are never sent
          IP fast switching is enabled
          IP fast switching on the same interface is disabled
          IP Flow switching is disabled
          IP Fast switching turbo vector
          IP multicast fast switching is enabled
          IP multicast distributed fast switching is disabled
          IP route-cache flags are Fast
          Router Discovery is disabled
          IP output packet accounting is disabled
          IP access violation accounting is disabled
          TCP/IP header compression is disabled
          RTP/IP header compression is disabled
          Probe proxy name replies are disabled
          Policy routing is disabled
          Network address translation is disabled
          WCCP Redirect outbound is disabled
          WCCP Redirect exclude is disabled
          BGP Policy Mapping is disabled
        Serial0/0 is up, line protocol is up
          Internet address is 169.254.1.2/16
          Broadcast address is 255.255.255.255
          Address determined by non-volatile memory
          MTU is 1500 bytes
          Helper address is not set
          Directed broadcast forwarding is disabled
          Multicast reserved groups joined: 224.0.0.9
          >DETAIL REMOVED<
          BGP Policy Mapping is disabled
        Ethernet0/1 is administratively down, line protocol is down
```

```
   Internet address is 192.168.1.11/24
   Broadcast address is 255.255.255.255
   Address determined by non-volatile memory
   MTU is 1500 bytes
   >DETAIL REMOVED<
   BGP Policy Mapping is disabled
2600A#show ip interface brief
Interface      IP-Address    OK? Method Status    Protocol
Ethernet0/0    10.1.1.100    YES manual up           up


Serial0/0      169.254.1.2   YES NVRAM  up           up


Ethernet0/1    192.168.1.11  YES NVRAM  administratively down down

2600A#show ip interface Ethernet 0/0
Ethernet0/0 is up, line protocol is up
   Internet address is 10.1.1.100/24
   Broadcast address is 255.255.255.255
   >DETAIL REMOVED<
   BGP Policy Mapping is disabled
```

■ *show ip sockets*   Shows open sockets on your router. An example of its output
follows:

```
ACCESSSERVER1>show ip sockets
Proto    Remote    Port     Local    Port   In Out Stat TTY OutputIF
  17    --listen--         10.1.1.1     67    0   0  89   0
  17    --listen--         10.1.1.1  53114    0   0  89   0
ACCESSSERVER1>
```

■ *show ip traffic*   Shows detailed statistics about Internet Protocol (IP); Internet
Control Messaging Protocol (ICMP); and multicast traffic on the router, including
broadcast and error statistics. An example of this (from a 2600 series router) follows:

```
2600A#show ip traffic
IP statistics:
  Rcvd:  14549 total, 3252 local destination
         0 format errors, 0 checksum errors, 0 bad hop count
         0 unknown protocol, 2 not a gateway
         0 security failures, 0 bad options, 0 with options
  Opts:  0 end, 0 nop, 0 basic security, 0 loose source route
```

```
                     0 timestamp, 0 extended security, 0 record route
                     0 stream ID, 0 strict source route, 0 alert, 0 cipso
                     0 other
           Frags:    0 reassembled, 0 timeouts, 0 couldn't reassemble
                     0 fragmented, 0 couldn't fragment
           Bcast:    4 received, 3261 sent
           Mcast:    3239 received, 0 sent
           Sent:     3266 generated, 0 forwarded
           Drop:     5 encapsulation failed, 0 unresolved, 0 no adjacency
                     11295 no route, 0 unicast RPF, 0 forced drop

      ICMP statistics:
        Rcvd: 0 format errors, 0 checksum errors, 0 redirects, 0 unreachable
              0 echo, 9 echo reply, 0 mask requests, 0 mask replies, 0 quench
              0 parameter, 0 timestamp, 0 info request, 0 other
              0 irdp solicitations, 0 irdp advertisements
        Sent: 0 redirects, 0 unreachable, 10 echo, 0 echo reply
              0 mask requests, 0 mask replies, 0 quench, 0 timestamp
              0 info reply, 0 time exceeded, 0 parameter problem
              0 irdp solicitations, 14 irdp advertisements

      UDP statistics:
        Rcvd: 3244 total, 0 checksum errors, 3244 no port
        Sent: 3243 total, 0 forwarded broadcasts

      TCP statistics:
        Rcvd: 0 total, 0 checksum errors, 0 no port
        Sent: 0 total

      Probe statistics:
        Rcvd: 0 address requests, 0 address replies
              0 proxy name requests, 0 where-is requests, 0 other
        Sent: 0 address requests, 0 address replies (0 proxy)
              0 proxy name replies, 0 where-is replies

      EGP statistics:
        Rcvd: 0 total, 0 format errors, 0 checksum errors, 0 no listener
        Sent: 0 total

      IGRP statistics:
        Rcvd: 0 total, 0 checksum errors
        Sent: 0 total
```

```
OSPF statistics:
  Rcvd: 0 total, 0 checksum errors
        0 hello, 0 database desc, 0 link state req
        0 link state updates, 0 link state acks

  Sent: 0 total

IP-IGRP2 statistics:
  Rcvd: 0 total
  Sent: 0 total

PIMv2 statistics: Sent/Received
  Total: 0/0, 0 checksum errors, 0 format errors
  Registers: 0/0, Register Stops: 0/0,  Hellos: 0/0
  Join/Prunes: 0/0, Asserts: 0/0, grafts: 0/0
  Bootstraps: 0/0, Candidate_RP_Advertisements: 0/0

IGMP statistics: Sent/Received
  Total: 0/0, Format errors: 0/0, Checksum errors: 0/0
  Host Queries: 0/0, Host Reports: 0/0, Host Leaves: 0/0
  DVMRP: 0/0, PIM: 0/0

ARP statistics:
  Rcvd: 0 requests, 9 replies, 0 reverse, 0 other
  Sent: 7 requests, 3 replies (0 proxy), 0 reverse
```

The show ipx Command   Like the *show ip* command, *show ipx* is made up of several subcommands. Luckily, most of the IPX commands are very similar to the IP commands, making this fairly easy to understand. The *show ipx interfaces* command uses the same syntax and performs the same purpose for IPX as the *show ip interfaces* command for IP. By the same token, *show ipx traffic* is very similar to the *show ip traffic* command.

A few IPX commands, however, have no IP peers. One of these is the *show ipx servers* command. It shows the IPX servers that the router is aware of through SAP updates. Another is the *show ipx spx-spoof* command, which shows IPX keepalive spoofing information. An example of the *show ipx servers* command follows:

```
2948GL3>show ipx servers
Codes: S - Static, P - Periodic, E - EIGRP, N - NLSP,
H - Holddown, + = detail
U - Per-user static
2 Total IPX Servers
```

```
Table ordering is based on routing and server info

   Type Name                      Net      Address      Port    Route Hops Itf
 P   30C 105055DEB9DD0CA1NPID1       1.1050.55DE.b9DD:0CA1     1/00   1  BV1
 P   640 BIGNETWAREBOX           110E3721.0000.0000.0001:85E8   2/01   1  BV1
```

The show line Command   This command displays information about *lines,* a special type of interface in the Cisco IOS. The following interfaces on a router are typically considered lines (the IOS line name for the connection is provided as well): console port (console 0 or line 0), aux port (aux 0 or line 1, 17, or 65), asynchronous serial ports (lines 1–64), and Telnet ports (vty 0–181 or lines 18–199). This command shows you detailed information about the settings and configuration of these lines, as well as statistics on their use. The syntax for this command is *show line [modifier or line name and number].* The basic command, *show line,* shows you basic information on each of the lines. See the following example:

```
ACCESSSERVER1>show line
    Tty Typ    Tx/Rx      A Modem  Roty AccO AccI   Uses   Noise  Overruns Int
      0 CTY               -   -      -    -    -      3       1     0/0     -
 *    1 TTY    9600/9600  -   -      -    -    -     54    12664    0/0     -
      2 TTY    9600/9600  -   -      -    -    -     52      27     0/0     -
      3 TTY    9600/9600  -   -      -    -    -     40    50442    0/0     -
      4 TTY    9600/9600  -   -      -    -    -     31     1184    0/0     -
      5 TTY    9600/9600  -   -      -    -    -     25      14     0/0     -
      6 TTY    9600/9600  -   -      -    -    -     25      12     0/0     -
      7 TTY    9600/9600  -   -      -    -    -     13    51565    0/0     -
      8 TTY    9600/9600  -   -      -    -    -     13      10     0/0     -
      9 TTY    9600/9600  -   -      -    -    -     12      28     0/0     -
     10 TTY    9600/9600  -   -      -    -    -      9      44     0/0     -
     11 TTY    9600/9600  -   -      -    -    -     11     9886    0/0     -
     12 TTY    9600/9600  -   -      -    -    -      6       5     0/0     -
     13 TTY    9600/9600  -   -      -    -    -     12      64     0/0     -
     14 TTY    9600/9600  -   -      -    -    -     11       8     0/0     -
     15 TTY    9600/9600  -   -      -    -    -     10       7     0/0     -
     16 TTY    9600/9600  -   -      -    -    -     20       5     0/0     -
     17 AUX    9600/9600  -   -      -    -    -      0       0     0/0     -
 *   18 VTY               -   -      -    -    -    304       0     0/0     -
       >AND SO ON UNTIL<
    199 VTY               -   -      -    -    -      0       0     0/0     -

ACCESSSERVER1>
```

For basic usage information, the *show line summary* command provides you with usage information in a very concise report. An example of this follows:

```
ACCESSSERVER1>show line summary
        0: -u-- ---- ---- ---- -?u- ---- ---- ---- -???
       36: ???? ???? ???? ???? ???? ???? ???? ???? ????
       72: ???? ???? ???? ???? ???? ???? ???? ???? ????
      108: ???? ???? ???? ???? ???? ???? ???? ???? ????
      144: ???? ???? ???? ???? ???? ???? ???? ???? ????
      180: ???? ???? ???? ???? ????


   2 character mode users.         (U)
 168 lines never used              (?)
   2 total lines in use,    2 not authenticated (lowercase)
ACCESSSERVER1>
```

From this output, you can tell very quickly that line 1 and line 18 are currently in use by unauthenticated users, and all other lines are idle (and most of those have yet to be used). Finally, you can type *show line [line name and number]* to find out detailed information about a given line, like so:

```
ACCESSSERVER1>show line console 0
   Tty Typ   Tx/Rx    A Modem  Roty AccO AccI   Uses   Noise  Overruns Int
   0 CTY              -   -     -    -    -       3      1      0/0      -

Line 0, Location: "", Type: ""
Length: 24 lines, Width: 80 columns
Status: Ready
Capabilities: none
Modem state: Ready
Group codes:    0
Special Chars: Escape  Hold  Stop  Start  Disconnect  Activation
               ^^x     none  -     -      none
Timeouts:      Idle EXEC     Idle Session   Modem Answer  Session   Dispatch
               00:10:00        never                      none      not set
                             Idle Session Disconnect Warning
                               never
                             Login-sequence User Response
                             00:00:30
                             Autoselect Initial Wait
                               not set
```

```
Modem type is unknown.
Session limit is not set.
Time since activation: never
Editing is enabled.
History is enabled, history size is 10.
DNS resolution in show commands is enabled
Full user help is disabled
Allowed transports are lat pad v120 mop telnet rlogin nasi.
Preferred is lat.
No output characters are padded
No special data dispatching characters
ACCESSSERVER1>
```

The show logging Command   If you are logging system messages, this command shows you the messages currently in the log. It also shows the logging status (enabled or disabled) on all valid logging interfaces, as well as statistics on logging. An example of this command follows:

```
2600A>show logging
Syslog logging: enabled (0 messages dropped, 0 flushes, 0 overruns)
    Console logging: disabled
    Monitor logging: level debugging, 0 messages logged
    Buffer logging: level debugging, 9 messages logged
    Trap logging: level informational, 13 message lines logged

Log Buffer (4096 bytes):

00:00:08: %LINK-3-UPDOWN: Interface Ethernet0/0,
changed state to up
00:00:08: %LINK-3-UPDOWN: Interface Ethernet0/1,
changed state to up
00:00:08: %LINK-3-UPDOWN: Interface Serial0/0, changed state to up
00:00:09: %SYS-5-CONFIG_I: Configured from memory by console
00:00:09: %SYS-5-RESTART: System restarted --
Cisco Internetwork Operating System Software
IOS (tm) C2600 Software (C2600-I-M), Version 12.0(7)T,
RELEASE SOFTWARE (fc2)
Copyright (c) 1986-1999 by cisco Systems, Inc.
Compiled Tue 07-Dec-99 02:12 by phanguye
00:00:09: %LINEPROTO-5-UPDOWN: Line protocol on Interface Ethernet0/0,
 changed state to up
00:00:09: %LINEPROTO-5-UPDOWN: Line protocol on Interface Ethernet0/1,
 changed state to down
00:00:09: %LINEPROTO-5-UPDOWN: Line protocol on Interface Serial0/0,
changed state to up
```

```
00:00:11: %LINK-5-CHANGED: Interface Ethernet0/1,
changed state to administratively down
```

The *show logging history* command shows you statistics on logging and SNMP activity in general on the router as well. An example of this command follows:

```
2600A>show logging history
Syslog History Table:1 maximum table entries,
saving level warnings or higher
 6 messages ignored, 0 dropped, 0 recursion drops
 2 table entries flushed
SNMP notifications not enabled
    entry number 3 : LINK-3-UPDOWN
     Interface Serial0/0, changed state to up
     timestamp: 866
```

The show memory Command   The *show memory* command shows detailed information about both processor and I/O memory usage, including specific listings of which functions occupy the memory and the sections they occupy. This command is typically used more by Cisco engineers than network admins because understanding the output requires a programmer's understanding of the IOS. An example of the *show memory* command follows:

```
2600B#show memory summary
             Head    Total(b)     Used(b)     Free(b)   Lowest(b)  Largest(b)
Processor  80B58B54  15365292     2269272    13096020   12895720   12944036
      I/O  1A00000    6291456     1656280     4635176    4635176    4635132
Processor memory

Alloc PC        Size      Blocks      Bytes     What

0x800092E4       152          2        304      Init
0x80011C20       128         25       3200      RIF Cache
0x8001AF9C       416          1        416      IDB: Serial Info
>19 PAGES OF DETAIL REMOVED<
```

The show privilege Command   The *show privilege* command shows your current privilege level (0–15). The syntax is simply *show privilege,* with no modifiers. An example of the *show privilege* command follows:

```
2600B#show privilege
Current privilege level is 15
```

The show processes Command   This command is used more by Cisco engineers than network admins, but it is still useful to know in certain situations. The *show process* command shows memory use and CPU use for each individual function, or process, within the router. This allows you to pinpoint a performance problem by locating the process that is hogging CPU or RAM resources. The base command, *show processes*, shows you the average CPU utilization numbers for the entire router, as well as each running process and the resources they are consuming.

For determining CPU bottlenecks, the *show processes cpu* command shows the average and current processor usage of each individual process. To pinpoint memory bottlenecks, use *show processes memory*, although this command is typically less useful than the CPU variant. This command is available only in enable mode. An example of the *show processes* command follows:

```
2600B#sh processes cpu
CPU utilization for five seconds: 0%/0%; one minute: 0%; five minutes: 0%
 PID   Runtime(ms)   Invoked   uSecs    5Sec    1Min   5Min TTY Process
   1            24     87410        0   0.00%   0.00%  0.00%   0 Load Meter
   2          3789       766     4946   0.65%   0.35%  0.33%   0 Exec
   3        177735     44351     4007   0.00%   0.03%  0.00%   0 Check heaps
   4             0         1        0   0.00%   0.00%  0.00%   0 Chunk Manager
   5             4         1     4000   0.00%   0.00%  0.00%   0 Pool Manager
   6             0         2        0   0.00%   0.00%  0.00%   0 Timers
>DETAIL REMOVED<
```

The show protocols Command   The *show protocols* command tells you which layer 3 protocols are enabled and what the layer 3 addresses are for each interface. The syntax is *show protocols [optional interface name and number].* If you type *show protocols*, you will get the router's global parameters and each of the interface addresses. If you append the interface name and number, it will show you information concerning only that interface. An example of this command follows:

```
2600A#show protocols
Global values:
  Internet Protocol routing is enabled
Ethernet0/0 is up, line protocol is up
  Internet address is 10.1.1.100/24
Serial0/0 is up, line protocol is up
  Internet address is 169.254.1.2/16
Ethernet0/1 is administratively down, line protocol is down
  Internet address is 192.168.1.11/24
2600A#sh prot eth 0/0
```

```
Ethernet0/0 is up, line protocol is up
  Internet address is 10.1.1.100/24
```

The show running-config Command   Another extremely useful command, *show running-config*, shows the configuration in RAM (the configuration that the router is using at all times *after* boot). Remember, in IOS, any changes made to the configuration are placed in the running-config and must be saved to the startup-config to become permanent. This command shows the current configuration in running-config. The syntax for the command is *show running-config [optional interface name and number]*. Simply typing *show running-config* will list all configurations on the device, whereas typing *show running-config [interface name and number]* will show the configuration specific to an interface (which can be very useful on a high-density switch). This command is available only in enable mode. An example follows:

```
2600B#show running-config
Building configuration...

Current configuration:
!
! Last configuration change at 21:41:49 EST Sat Jul 14 2001
!
version 12.0
service timestamps debug datetime
service timestamps log uptime
no service password-encryption
no service compress-config
!
hostname 2600B
!
logging buffered 4096 debugging
logging console warnings
enable secret 5 $1$JPeK$XUmL9thAxfOu9Bay9r2Ef0
enable password ccna
!
!
!
!
!
clock timezone EST -4
ip subnet-zero
no ip domain-lookup
```

```
!
!
!
!
interface Ethernet0/0
 description Main Link
 ip address 10.1.1.200 255.255.255.0
 no ip directed-broadcast
!
interface Serial0/0
 ip address 169.254.2.2 255.255.255.0
 no ip directed-broadcast
 encapsulation frame-relay
 no ip mroute-cache
 no fair-queue
 frame-relay interface-dlci 200
 frame-relay lmi-type cisco
!
interface Ethernet0/1
 no ip address
 no ip directed-broadcast
 shutdown
!
ip classless
no ip http server
!
dialer-list 1 protocol ip permit
dialer-list 1 protocol ipx permit
!
line con 0
 exec-timeout 35791 0
 timeout login response 15
 password opensezme
 login
 transport input all
line aux 0
line vty 0 4
 password blank
 login
!
no scheduler allocate
end
```

The show sessions Command   As mentioned at the beginning of this chapter in the section "The disconnect Command," this command allows you to see the current Telnet connections you have established and subsequently suspended. The Telnet sessions established by other users are *not* shown. The syntax for this command is simply *show sessions.* An example of this command follows:

```
ACCESSSERVER1>show sessions
Conn Host             Address          Byte  Idle Conn Name
*  1 2600b            10.1.1.1            0     0 2600b
```

The show snmp Command   This command shows SNMP statistics, including errors and packets input and output. This command functions only if SNMP is enabled. The command's basic form is *show snmp.* The variants *show snmp pending* and *show snmp sessions* can also be used to show which communications are currently occurring between the router and SNMP management stations. The *pending* variant shows all pending (unanswered) requests from management stations to the router, and the *sessions* variant shows which communications between the router and the manager or managers are currently in use.

The show startup-config Command   This command simply shows you the startup-config (the config in NVRAM used at router boot) for the router. The syntax is *show startup-config,* with no modifiers. You must be in enable mode to issue this command. The output for this command is the same as the *show running-config* command, except it shows the startup-config.

The show tcp Command   This command shows detailed information on any open TCP connections to the router. (Note that this is *to* the router and not *through* the router.) Typically, you use this command to see the status and number of incoming Telnet sessions to the router. The basic form of this command is *show tcp,* which shows detailed TCP session information for all sessions on all lines. The *show tcp statistics* form of this command shows information on the number of sessions established since boot, as well as the number of packets and bytes transmitted and received. The *show tcp brief* form of this command simply lists the current sessions, along with the source and destination addresses and ports. Finally, the *show tcp [line name and number]* form of this command shows you details about TCP sessions on a given interface. An example of this command follows:

```
ACCESSSERVER1>show tcp brief
TCB        Local Address        Foreign Address        (state)
00584DB4   CCNA2600C.2002       CCNA2600C.24082        ESTAB
00584970   CCNA2600C.24082      CCNA2600C.2002         ESTAB
00585520   10.0.4.48.23         192.168.1.111.3358     ESTAB
```

The show tech-support Command This command attempts to put all information that may be needed by the Cisco Technical Assistance Center (TAC) in one listing. This command simply runs *show version, show running-config, show stacks, show interfaces, show controllers, show context, show diag, show cXXXX* (as in "show c2600"), *show process memory, show process cpu,* and *show buffers* all back to back. By default, *show tech-support* does not pause between each command, so make sure you have logging enabled for your terminal application because this information takes several screens to display.

Several modifiers to this command can be used to find tech support information on more specialized router features, but you shouldn't need to know most of these. (If Cisco TAC needs the information, they will tell you what to type.) One modifier, however, the *page* modifier, is useful to the network admin. This instructs the command to pause after each page of information, allowing you to view the information without needing to log it to a text file. To use this modifier, simply type *show tech-support page*.

The show terminal Command This command is the exact same as the *show line console 0* command mentioned earlier in the section "The show line Command." The syntax is simply *show terminal* with no modifiers.

The show version Command This command shows the version of IOS in use on the router, the version of IOS in ROM, the method of reload, the flash image used to boot, and the IOS capabilities, and it includes the uptime of the router at the time of execution. The syntax for the command is simply *show version.* An example of this command follows:

```
2600A#show version
Cisco Internetwork Operating System Software
IOS (tm) C2600 Software (C2600-I-M), Version 12.0(7)T,
RELEASE SOFTWARE (fc2)
Copyright (c) 1986-1999 by cisco Systems, Inc.
Compiled Tue 07-Dec-99 02:12 by phanguye
Image text-base: 0x80008088, data-base: 0x807AAF70

ROM: System Bootstrap, Version 11.3(2)XA4, RELEASE SOFTWARE (fc1)

2600A uptime is 12 hours, 32 minutes
System returned to ROM by reload
System image file is "flash:c2600-i-mz.120-7.T"

cisco 2611 (MPC860) processor (revision 0x203) with
26624K/6144K bytes of memory
.
Processor board ID JAD04360GH7 (4114038455)
```

```
M860 processor: part number 0, mask 49
Bridging software.
X.25 software, Version 3.0.0.
2 Ethernet/IEEE 802.3 interface(s)
1 Serial network interface(s)
32K bytes of non-volatile configuration memory.
8192K bytes of processor board System flash (Read/Write)

Configuration register is 0x2102
```

The telnet Command

The *telnet* command is used for the same purpose as the *connect* command discussed in the section "The connect Command" at the beginning of the chapter, but the *telnet* command also includes a wealth of options to allow you to customize your connection.

Syntax

The basic syntax of the command is *telnet [ip address/name]*, just like the *connect* command. The *telnet* command, however, has a number of optional modifiers that can be appended to the command to enable advanced features. First, *telnet* includes a */debug* option, which allows you to see exactly what is happening in the initial connection for troubleshooting purposes.

The second option is */encrypt kerberos*. If your router is equipped with the kerberized Telnet subsystem, this command allows you to connect to a remote host through an encrypted session.

Next, the */line* switch allows you to connect in line mode, which means that data is not sent to the remote host until you press the ENTER key. (This can save some bandwidth on slow connections.)

Next, the */source-interface* modifier allows you to specify the interface on your router to connect through.

Finally, you can use the port number as a modifier to specify a connection on a nonstandard port, or you can choose to connect to other ports besides Telnet by using the protocol as a modifier (for instance, *telnet 10.1.1.1 www* to connect using port 80). These optional protocols are listed in Table 15-2.

The terminal Command

The *terminal* command is an option-filled command that allows you to set most configuration requirements on your terminal session (Telnet, console, and so on). These include setting data bits, parity, speed, escape characters, stop bits, display width, and nearly anything else you might need.

Modifier	Description
<0-65535>	Port Number
bgp	Border Gateway Protocol (179)
chargen	Character generator (19)
cmd	Remote commands (rcmd, 514)
daytime	Daytime (13)
discard	Discard (9)
domain	Domain Name Service (53)
echo	Echo (7)
exec	Exec (rsh, 512)
finger	Finger (79)
ftp	File Transfer Protocol (21)
ftp-data	FTP data connections (used infrequently, 20)
gopher	Gopher (70)
hostname	NIC hostname server (101)
ident	Ident Protocol (113)
irc	Internet Relay Chat (194)
klogin	Kerberos login (543)
kshell	Kerberos shell (544)
login	Login (rlogin, 513)
lpd	Printer service (515)
nntp	Network News Transport Protocol (119)
pim-auto-rp	PIM Auto-RP (496)
pop2	Post Office Protocol v2 (109)
pop3	Post Office Protocol v3 (110)
smtp	Simple Mail Transport Protocol (25)
sunrpc	Sun Remote Procedure Call (111)

Table 15-2. *Telnet Port/Protocol Optional Modifiers*

Modifier	Description
syslog	Syslog (514)
tacacs	TAC Access Control System (49)
talk	Talk (517)
telnet	Telnet (23)
time	Time (37)
uucp	UNIX-to-UNIX Copy Program (540)
whois	Nicname (43)
www	World Wide Web (HTTP, 80)

Table 15-2.   *Telnet Port/Protocol Optional Modifiers* (continued)

Syntax

The basic syntax for the *terminal* command is *terminal [option] [suboptions or value]*. For instance, the command *terminal ip netmask-format bit-count* tells the terminal to display network masks in the /24 CIDR bit-count notation instead of the 255.255.255.0 decimal notation. Table 15-3 provides the full list of primary options available for the *terminal* command. Please note that changing some of these (such as *databits*) can cause the session to display weird characters if you do not also change the setting on your Telnet application.

Primary Options for the *terminal* Command	Description
autohangup	Automatically hangs up when last connection closes
data-character-bits	Sets the size of characters being handled
databits	Sets the number of data bits per character
default	Sets a command to its defaults
dispatch-character	Defines the dispatch character

Table 15-3.   *Primary* terminal *Command Options*

Primary Options for the *terminal* Command	Description
dispatch-timeout	Sets the dispatch timer
domain-lookup	Enables domain lookups in *show* commands
download	Puts line into download mode
editing	Enables command line editing
escape-character	Changes the current line's escape character
exec-character-bits	Sets the size of characters to the command *exec*
flowcontrol	Sets the flow control
full-help	Provides help to unprivileged users
help	Describes the interactive help system
history	Enables and controls the command history function
hold-character	Defines the hold character
international	Enables international 8-bit character support
ip	Provides IP options
length	Sets the number of lines on a screen
no	Negates a command or sets its defaults
notify	Informs users of output from concurrent sessions
padding	Sets the padding for a specified output character
parity	Sets the terminal parity
rxspeed	Sets the receive speed
special-character-bits	Sets the size of the escape (and other special) characters
speed	Sets the transmit and receive speeds
start-character	Defines the start character
stop-character	Defines the stop character
stopbits	Sets async line stop bits
telnet	Provides Telnet protocol–specific configuration

Table 15-3.   *Primary* terminal *Command Options* (continued)

Primary Options for the *terminal* Command	Description
terminal-type	Sets the terminal type
transport	Defines transport protocols for line
txspeed	Sets the transmit speeds
width	Sets width of the display terminal

Table 15-3.   *Primary* terminal *Command Options* (continued)

The traceroute Command

The final user mode command, *traceroute,* is also one of the most useful for connectivity troubleshooting. As discussed in Chapter 5, *traceroute* discovers the path you take from one point to another by sending packets with low Time to Live (TTL) settings and processing the ICMP time-exceeded messages that are sent in response. This allows you to map out the path a packet is taking, which is very useful when you have redundant paths because it helps pinpoint the problem.

Syntax

The syntax for a basic ip traceroute is *traceroute [ip address/name].* You can also perform a traceroute with other protocols (assuming your IOS feature set supports them) by typing *traceroute [protocol] [address].* An example of this command follows:

```
2948GL3>traceroute 216.240.148.171

Type escape sequence to abort.
Tracing the route to basic-argon.blanka.dreamhost.com
(216.240.148.171)

 1 172.16.1.14 4 msec 0 msec 4 msec
 2 172.32.253.1 0 msec 0 msec 4 msec
 3 ATM2-0.GW1.GSO1.ALTER.NET (157.130.59.197) 8 msec 12 msec 8 msec
 4 512.at-0-1-0.XR2.DCA8.ALTER.NET (152.63.36.130)
16 msec 16 msec 24 msec
 5 POS7-0.BR1.DCA8.ALTER.NET (146.188.162.213)
16 msec 16 msec 12 msec
 6 wdc-brdr-03.inet.qwest.net (205.171.4.69)
```

```
12 msec 12 msec 16 msec
 7 wdc-core-03.inet.qwest.net (205.171.24.69)
16 msec 12 msec 16 msec
 8 iah-core-01.inet.qwest.net (205.171.5.187)
52 msec 52 msec 48 msec
 9 lax-core-01.inet.qwest.net (205.171.5.164)
96 msec 92 msec 96 msec
10 lax-edge-08.inet.qwest.net (205.171.19.142)
92 msec 92 msec 92 msec
11 wvision-la-oc3-gsr1.webvision.com (63.149.192.126)
92 msec 92 msec 92 msec
12 * * *
13 * * *
14 basic-argon.blanka.dreamhost.com (216.240.148.171)
84 msec 88 msec 88 msec
```

Common Enable Mode Commands

These are commands that you can enter in enable or privileged exec mode of operation. Remember, you can tell this mode by the prompt. In enable mode, your prompt will end with a pound sign (#), like so: Router#. Note that we will cover only commands that are not discussed in later chapters; and unless the command has additional features, we will not discuss commands that can be entered from user mode. All of the commands from user mode are valid in enable mode as well. The following commands are discussed in this section: *clear, clock, configure, debug, delete, dir, disable, erase, lock, ping, reload, send, setup, squeeze, test, undelete, where,* and *write.*

The clear Command

This command is used to erase values, typically statistical values for many of the *show* commands. For instance, typing *clear cdp counters* erases the CDP statistics, as shown in the following example:

```
2600B#show cdp traffic
CDP counters :
        Total packets output: 5201, Input: 5195
        Hdr syntax: 0, Chksum error: 0, Encaps failed: 0
        No memory: 0, Invalid packet: 0, Fragmented: 0
        CDP version 1 advertisements output: 0, Input: 0
        CDP version 2 advertisements output: 5201, Input: 5195
2600B#clear cdp counters
2600B#show cdp traffic
```

```
CDP counters :
        Total packets output: 0, Input: 0
        Hdr syntax: 0, Chksum error: 0, Encaps failed: 0
        No memory: 0, Invalid packet: 0, Fragmented: 0
        CDP version 1 advertisements output: 0, Input: 0
        CDP version 2 advertisements output: 0, Input: 0
2600B#
```

Syntax

The basic syntax for the *clear* command is *clear [target] [modifiers]*. Because the *clear* command has so many possible targets, we will concern ourselves with only the most common targets. These are described in Table 15-4.

The clock Command

This command, as you might have guessed, sets the system clock, including the date. This often-overlooked feature allows the router to assign time and date stamps to

Common *clear* Command Targets	Description
access-list	Clears access list statistical information
arp-cache	Clears the entire ARP cache
bridge	Resets bridge forwarding cache
cdp	Resets cdp information
counters	Clears counters on one or all interfaces
frame-relay-inarp	Clears inverse ARP entries from the map table
host	Deletes host table entries
interface	Clears the hardware logic on an interface
ip	IP
line	Resets a terminal line
logging	Clears logging buffer
tcp	Clears a TCP connection or statistics

Table 15-4.   *Common* clear *Targets*

syslog messages and other messages, something that is extremely useful in some cases (like for organizations that require high security).

Syntax

The syntax for this command is very simple. Type *clock [time in the format of hh:mm:ss military time] [month] [day] [year in yyyy format]*. So, if you wanted to set the system clock for 3:00 P.M. on June 9, 2001, you would type *clock 15:00:00 june 9 2001*.

The configure Command

This command initiates global configuration mode. When this happens, your prompt will change from Router# to Router(config)#.

Syntax

The syntax for this command is *configure [method]*. The valid methods are the following:

- *terminal*   Allows you to type in the configuration commands (These commands are covered in Chapter 16.)
- *memory*   Executes the commands in the startup-config
- *network*   Executes commands from a configuration file on a TFTP server
- *overwrite-network*   Overwrites the startup-config from a configuration file on a TFTP sever (same as the *copy tftp startup-config* command)

With all of the methods except the overwrite-network *method, running-config is modified. To save the changes for the next reload, you must issue the* copy running-config startup-config *command.*

The debug Command

This is a major command—one that possibly has more options than any other in the IOS. The basic principle behind the *debug* command is to give additional information about router operation for troubleshooting purposes. To this end, a *debug* command exists for nearly every operation the device is likely to perform.

Syntax

The basic syntax for this command is *debug [target] [subtarget or modifiers]*. The *all* target instructs the device to enable all possible debugging. Other common valid targets are described in Table 15-5, with the most useful of these discussed next.

Of the targets in Table 15-5, one stands out as useful in most situations: the *ip* target. It can help you solve most of your IP connectivity problems. Because *ip* is such a broad

Target	Description
aaa	Authentication, authorization, and accounting (AAA)
access-expression	Boolean access expression
adjacency	Adjacency
all	Enables all debugging
arp	IP ARP and HP Probe transactions
async	Async interface information
callback	Callback activity
cdp	CDP information
chat	Chat scripts activity
compress	COMPRESS traffic
confmodem	Modem configuration database
conn	Connection Manager information
custom-queue	Custom output queuing
dhcp	DHCP client activity
dialer	Dial on Demand
domain	Domain Name System
dxi	atm-dxi information
eigrp	EIGRP Protocol information
entry	Incoming queue entries
ethernet-interface	Ethernet network interface events
fastethernet	Fast Ethernet interface information
frame-relay	Frame Relay
interface	interface
ip	IP information
lapb	LAPB protocol transactions

Table 15-5.   *Common Debug Targets*

CISCO TECHNOLOGY OVERVIEW

Target	Description
lex	LAN Extender protocol
list	Set interface and/or access list for the next *debug* command
modem	Modem control/process activation
ntp	NTP information
nvram	Debug NVRAM behavior
packet	Log unknown packets
ppp	PPP information
priority	Priority output queuing
radius	RADIUS protocol
serial	Serial interface information
snmp	SNMP information
spantree	Spanning Tree information
standby	Hot standby protocol
tacacs	TACACS authentication and authorization
tbridge	Transparent bridging
telnet	Incoming Telnet connections
tftp	TFTP debugging
token	Token Ring information
tunnel	Generic Tunnel Interface

Table 15-5.   *Common Debug Targets*

target, it has subtargets to allow you to further specify which information you are after. These subtargets are described in Table 15-6, with the most useful of these discussed next.

Of the subtargets listed in Table 15-6, one stands out as extremely useful: *debug ip packet*. As a general debugging tool, this one is hard to beat. It shows all IP packets traversing the router, giving you the source and destination information, the type of

Subtarget	Description
bgp	BGP information
cache	IP cache operations
cef	IP CEF operations
cgmp	CGMP (Cisco Gateway Multicast Protocol) activity
dhcp	Dynamic Host Configuration Protocol
drp	Director Response Protocol
dvmrp	DVMRP (Distance Vector Multicast Routing Protocol) activity
egp	EGP (Exterior Gateway Protocol) information
eigrp	IP-EIGRP information
error	IP error debugging
flow	IP flow switching operations
ftp	FTP dialogue
http	HTTP connections
icmp	ICMP transactions
igmp	IGMP protocol activity
igrp	IGRP information
interface	IP interface configuration changes
mbgp	MBGP information
mcache	IP multicast cache operations
mpacket	IP multicast packet debugging
mrm	IP Multicast Routing Monitor
mrouting	IP multicast routing table activity
msdp	Multicast Source Discovery Protocol (MSDP)
mtag	IP multicast tagswitching activity

Table 15-6. *IP Subtargets*

CISCO TECHNOLOGY OVERVIEW

Subtarget	Description
nat	NAT events
ospf	OSPF information
packet	General IP debugging and IPSO security transactions
peer	IP peer address activity
pim	PIM (Protocol Independent Multicast) Protocol activity
policy	Policy routing
rip	RIP protocol transactions
routing	Routing table events
rsvp	RSVP (Resource Reservation Protocol) activity
rtp	RTP (Real-time Transport Protocol) information
sd	Session Directory (SD)
security	IP security options
socket	Socket event
tcp	TCP information
tempacl	IP temporary ACL (Access Control List)
trigger-authentication	Trigger authentication
udp	UDP-based transactions
wccp	WCCP (Web Cache Communication Protocol) information

Table 15-6. *IP Subtargets* (continued)

packet, and any errors that the packet encountered. For even more information, use the *detail* modifier, as in *debug ip packet detail.* An example of the output from this command follows:

```
2600B#show logging
Syslog logging: enabled (0 messages dropped, 0 flushes, 0 overruns)
    Console logging: level warnings, 12 messages logged
```

```
      Monitor logging: level debugging, 0 messages logged
      Buffer logging: level debugging, 59 messages logged
      Trap logging: level informational, 35 message lines logged

Log Buffer (4096 bytes):


3d16h: %SYS-5-CONFIG_I: Configured from console by console
3d16h: IP: s=10.1.1.200 (local), d=10.1.1.1 (Ethernet0/0), len 100, sending
3d16h:     ICMP type=8, code=0
3d16h: IP: s=10.1.1.1 (Ethernet0/0), d=10.1.1.200 (Ethernet0/0), len 100,
rcvd 3
3d16h:     ICMP type=0, code=0
3d16h: IP: s=10.1.1.200 (local), d=10.1.1.1 (Ethernet0/0), len 100, sending
3d16h:     ICMP type=8, code=0
3d16h: IP: s=10.1.1.1 (Ethernet0/0), d=10.1.1.200 (Ethernet0/0), len 100,
rcvd 3
3d16h:     ICMP type=0, code=0
3d16h: IP: s=10.1.1.200 (local), d=10.1.1.1 (Ethernet0/0), len 100, sending
3d16h:     ICMP type=8, code=0
3d16h: IP: s=10.1.1.1 (Ethernet0/0), d=10.1.1.200 (Ethernet0/0), len 100,
rcvd 3
3d16h:     ICMP type=0, code=0
3d16h: IP: s=10.1.1.200 (local), d=10.1.1.1 (Ethernet0/0), len 100, sending
3d16h:     ICMP type=8, code=0
3d16h: IP: s=10.1.1.1 (Ethernet0/0), d=10.1.1.200 (Ethernet0/0), len 100,
rcvd 3
3d16h:     ICMP type=0, code=0
3d16h: IP: s=10.1.1.200 (local), d=10.1.1.1 (Ethernet0/0), len 100, sending
3d16h:     ICMP type=8, code=0
3d16h: IP: s=10.1.1.1 (Ethernet0/0), d=10.1.1.200 (Ethernet0/0), len 100,
rcvd 3
3d16h:     ICMP type=0, code=0
3d16h: IP: s=10.1.1.100 (Ethernet0/0), d=255.255.255.255, len 36, rcvd 0
3d16h:     ICMP type=9, code=0
3d16h: IP: s=10.1.1.200 (local), d=132.0.0.1, len 100, unroutable
3d16h:     ICMP type=8, code=0
3d16h: IP: s=10.1.1.200 (local), d=132.0.0.1, len 100, unroutable
3d16h:     ICMP type=8, code=0
3d16h: IP: s=10.1.1.200 (local), d=132.0.0.1, len 100, unroutable
3d16h:     ICMP type=8, code=0
3d16h: IP: s=10.1.1.200 (local), d=132.0.0.1, len 100, unroutable
3d16h:     ICMP type=8, code=0
3d16h: IP: s=10.1.1.200 (local), d=132.0.0.1, len 100, unroutable
3d16h:     ICMP type=8, code=0
```

The first point you should notice about this output is the timestamp to the far left of each entry. This tells you when this event occurred relative to router uptime. (In other words, all of these events occurred 3 days and 18 hours after the last reboot.) Next, you see IP *s:[number]* and *d:[number]*. These are the source and destination addresses of the packets in question. In the first part of the debug, you can see a ping being sent from the local router to host 10.1.1.1 out of the Ethernet 0/0 interface. Then you see the responses to these pings.

Toward the end of the debug, you can see the router attempting to ping 132.0.0.1 and deciding (before sending it out of the interface) that the packet is unroutable. If this was your problem, you should recognize immediately that the router needs a path to that location, a condition you will learn how to remedy in Chapter 22.

Special Considerations

The primary point to keep in mind about debugs is that they eat processor power. In fact, they can eat so much processor power that they can sever your session with the router. First, unless you are on a *very* lightly used router, do not use the *debug all* command. Second, never use debugs unless you need to troubleshoot a problem—and even then, try to use the most specific debug possible. Third, memorize the commands *undebug all* and *no debug all*. Type these commands in before you begin any debug.

This strategy accomplishes two goals: it keeps any previous debugs from using additional processor power, and it allows you to use the recall keystrokes to immediately pull the command back up and execute it. This approach can be extremely useful when debugging from a console connection because, by default, debugs pop up as syslog messages to the console, making it nearly impossible to type in new commands.

The delete Command

At first blush, the *delete* command might seem to be fairly simple. It deletes stuff, right? Well, not exactly. The *delete* command simply marks files as deleted; it doesn't actually get rid of them. This is because when a file is deleted, you are still allowed to go in and undelete it (just in case you made a really stupid mistake). To actually delete files, you have to use the *squeeze* or *erase* command. However, using the *delete* command is usually the first step in deleting a file.

Syntax

The syntax for this command is *delete [file system]: [filename]*. For instance, if you wanted to mark a file called "alternative" as deleted out of flash on a 2600 router, you would type *delete flash:alternative*. Note that this command marks files as deleted only from flash. You have to specify the file system because on some routers (like the 3600 series), there is a primary (internal) flash file system, as well as dual PC Card slots that can contain flash file systems. In these routers, *flash* refers to the internal flash memory, whereas *slot0* and *slot1* are used to refer to the PC Card flash memory.

Special Considerations

The *squeeze* command is used to actually delete a file from flash, whereas the *undelete* command is used to recover a file marked as deleted. (A file can be deleted and recovered up to 15 times.) However, on some routers (those with class C flash file systems, typically lower-end routers like the 2600 and 3600 series), the *squeeze* and *undelete* commands do not exist. This makes the *delete* command kind of useless on these routers because it doesn't really delete the files, and there is no way to recover them either. Basically, after deletion, they just sit there wasting space, unusable.

Frankly, I think Cisco should either add in the needed commands (which would require changing the file system type) or remove the *delete* command from the IOS on these models, but of course, no one asked me. (Don't worry, I'm not bitter!) Anyway, on these routers, I strongly suggest that you do not use the *delete* command because it does more harm than good. Instead, you will have to use the *erase* command, which deletes *all* files in the flash, and then copy the files you want to keep back in from a TFTP server.

The dir Command

This command is used to show the index numbers of files in the flash file system. The index numbers are used for the *undelete* command.

Syntax

The syntax for this command is *dir [file system]*. Valid file systems for this command are /all, flash, NVRAM, system, xmodem, and ymodem. An example of the output of this command follows:

```
2600B#dir /all
Directory of flash:/

  1   -rw-     4209848                <no date>  c2600-i-mz.120-7.T
  2   -rw-          814    Jul 09 2001 21:45:36  [startup]

8388608 bytes total (4177816 bytes free)
```

The disable Command

This command simply takes you back to user mode from enable mode. There are no modifiers. Just type *disable*.

The erase Command

This command is used on routers with class B and C flash file systems to recover space in the flash. It can also be used to erase the NVRAM and startup-config. Note that when the *erase* command is used, *all* files in the file system are erased. This makes it a bit impractical; unfortunately, on certain routers, you don't have a choice.

Syntax

The syntax for this command is *erase [file system]*. Valid file systems for this command are flash, slot0, slot1, NVRAM, and startup-config.

The lock Command

This command is used to lock a terminal session, similar to locking an NT or Windows 2000 workstation. Upon locking the session with the simple command *lock* (it has no modifiers), you are prompted for a password. Once you enter the password twice, the screen will display "locked." If anyone enters a keystroke, a password prompt comes up, and they must enter your password to arrive at a router prompt.

For this command to work, you must first enter the lockable *interface configuration mode command on the line you wish to lock. You will learn more about the* lockable *command in Chapter 16.*

The ping Command

In enable mode, you can perform what is known as an *extended* (or *privileged*) *ping*. In this mode of the *ping* command, you can set a number of options that are not available in the normal ping, including the datagram size, the DF bit setting, the ICMP option code, and many others. The syntax for an extended *ping* command is simply *ping* with no IP address. Then follow the prompts as they appear. The following example illustrates the capabilities of an extended ping:

```
2600B#ping
Protocol [ip]:
Target IP address: 10.1.1.1
Repeat count [5]: 1
Datagram size [100]:
Timeout in seconds [2]:
Extended commands [n]: y
Source address or interface:
Type of service [0]:
Set DF bit in IP header? [no]: y
Validate reply data? [no]:
Data pattern [0xABCD]:
Loose, Strict, Record, Timestamp, Verbose[none]: v
Loose, Strict, Record, Timestamp, Verbose[V]:
Sweep range of sizes [n]: y
```

```
Sweep min size [36]: 1400
Sweep max size [18024]: 1600
Sweep interval [1]: 64
Type escape sequence to abort.
Sending 4, [1400..1600]-byte ICMP Echos to 10.1.1.1,
timeout is 2 seconds:
Reply to request 0 (8 ms) (size 1400)
Reply to request 1 (12 ms) (size 1464)
Request 2 timed out (size 1528)
Request 3 timed out (size 1592)
Success rate is 50 percent (2/4),
round-trip min/avg/max = 8/10/12 ms
```

Notice that I told it to set the DF bit and to sweep a range of sizes, from 1400 bytes to 1600 bytes in 64-byte increments. Notice that the packets under 1500 bytes received a reply, but all of the ones over 1500 bytes failed. This is because the network connecting these devices is Ethernet, and the MTU for Ethernet is 1500 bytes. The device (a 1900 switch, in this case) at the other end simply threw the packets away without inspecting them due to the MTU infraction.

The reload Command

This command simply reboots the router (a warm boot, like restarting a PC). You can also specify why the router is reloaded or specify that it reload at a specific time.

Syntax

The *reload* command syntax is *reload [modifiers]*. The valid modifiers are *at* for setting a reload time and date, *in* for setting an interval of time before a reload, and an open field (called LINE in the IOS) allowing you to specify a reason for the reload. The first modifier, *at*, has the following syntax: *at [hh:mm month day reason]*. Therefore, if you wanted to reload at 12:45 A.M. on July 10 for testing purposes, the command (and the result) would be similar to the following example:

```
2600B#reload at 00:45 july 10 testing
Reload scheduled for 00:45:00 UTC Tue Jul 10 2001
(in 1 minute 53 seconds)
Reload reason: testing
Proceed with reload? [confirm]
2600B#
***
*** --- SHUTDOWN in 0:01:00 ---
```

```
***


***
*** --- SHUTDOWN NOW ---
***

2600B#
System Bootstrap, Version 11.3(2)XA4, RELEASE SOFTWARE (fc1)
Copyright (c) 1999 by cisco Systems, Inc.
TAC:Home:SW:IOS:Specials for info
C2600 platform with 32768 Kbytes of main memory

program load complete, entry point: 0x80008000, size: 0x403b9c
Self decompressing the image :
#################################################
>REMOVED DUE TO SIZE<
8192K bytes of processor board System flash (Read/Write)

Press RETURN to get started!
```

The second option, the *in* modifier, has this syntax: *in [hhh:mm reason]*. So, if you wanted the router to reload in two hours for no good reason, you would type *reload in 2:00 no good reason*.

The final option, the reason, can be used by itself or, as you've already seen, in conjunction with either of the other two commands. By itself, the syntax is *reload [reason]*. So you might type *reload I just felt like it*.

The send Command

This command allows you to send a message to one or all lines on a device. This allows you to inform others who may be connected of an impending reload (or better, mess with their heads a little).

Syntax

The syntax for this command is pretty simple. Just type *send [line name and number]* and follow the prompts. This is best shown in the following example:

```
2600B#send ?
  *        All tty lines
  <0-70>   Send a message to a specific line
```

```
   aux      Auxiliary line
   console  Primary terminal line
   tty      Terminal controller
   vty      Virtual terminal
2600B#send console 0
Enter message, end with CTRL/Z; abort with CTRL/C:
You have been tracked!
Cease and desist your activities!
Remain in your location, We will be with you momentarily!
^Z
Send message? [confirm]
2600B#


***
***
*** Message from tty0 to tty0:
***
You have been tracked!
Cease and desist your activities!
Remain in your location, We will be with you momentarily!
```

The setup Command

As mentioned in Chapter 14, this command takes you to the Setup Wizard. The syntax for this command is simply *setup*.

The squeeze Command

As you already know, this command removes files in the flash that have been deleted. Note that once this is done, all of the files marked for deletion are unrecoverable. It also removes any files with an error condition (such as those that failed to complete due to a download error). This command is valid only on routers with class A or B flash file systems.

Syntax

The syntax for this command is simply *squeeze [file system]*.

The test Command

This command causes the router to perform a series of tests on itself to assess operational status. However, these tests are relatively simple (the interface tests, for example, are just loopback tests), so even Cisco warns against using them as a final word on hardware status.

Syntax

The syntax for this command is *test [target]*. The valid targets vary a bit depending on the router line; but for most routers, they are *aim* (advanced interface module), *interfaces*, *memory* (for NVRAM), and *pas* (port adapters).

The undelete Command

On routers that have class A or B flash file systems, this command removes the *deleted* marker from files, allowing them to be used again. Of course, this strategy does not work if the file system has been squeezed.

Syntax

The syntax for this command is *undelete [index] [file system]*.

The where Command

This command is the older version of the *show sessions* command. The output is the same. The syntax for this command is simply *where* (no modifiers).

The write Command

This is an older command used to erase and display configuration files. There are four versions of this command: *write erase, write memory, write network*, and *write terminal*. The *write erase* command performs the same function as the *erase startup-config* command. The *write memory* command performs the same function as the *copy running-config startup-config* command. The *write network* command performs the same function as the *copy startup-config tftp* command. Finally, the *write term* command performs the same function as the *show startup-config* command. Aside from tests, you are unlikely to ever need to know these older commands.

Summary

This chapter examined the most common user mode and enable mode generic commands. This material, along with the next two chapters, provides the foundation for our exploration of specialized router and switch configurations in later chapters.

Chapter 16

Standard IOS Commands: Part 2

L ike the previous chapter, this chapter discusses some of the most common "generic" commands in the standard IOS. The focus here is on global- and interface-specific configuration mode commands. The commands covered are only those that are both commonly used and not covered later.

Like in the preceding chapter, this chapter is organized into two parts: one for global configuration mode and one for interface configuration mode. In these sections, each command is examined individually in alphabetical order, with full examples when necessary.

Common Global Configuration Mode Commands

These commands are common to the global configuration mode of device operation. In global configuration mode, the changes you make affect the entire device. Remember, you can tell which mode you are in based on the prompt. Global configuration mode prompts appear with the (config) at the end, like so: Router(config)#. The following commands are discussed in this section: *alias, arp, banner, boot, cdp, clock, config-register, default, enable, end, exit, hostname, interface, ip, line, logging, privilege, prompt,* and *service.*

The alias Command

The *alias* command sets aliases in the IOS. As mentioned in Chapter 15, aliases allow you to type one or two letters and complete a command. With the *alias* command, you can define what the aliases are for each mode.

> **Note**
>
> *Remember from Chapter 14 that you do not have to type the full IOS command in most cases, regardless of the aliases configured. You simply need to type enough of the command to allow the IOS to distinguish it from other similar commands. For instance,* show running-config *can be shortened to* sh ru, *and* shutdown *can be shortened to* shut.

Syntax

The syntax for the *alias* command is *alias [mode] [alias] [command]*. So, if you wanted to set a new alias for the *disable* command, and you wanted to make the alias key 1, you would type *alias exec 1 disable*. This example and its results are demonstrated in the following output:

```
2600B#show alias
Exec mode aliases:
 h help
 lo logout
 p ping
```

```
 r resume
 s show
 u undebug
 un undebug
 w where


2600B#configure terminal
Enter configuration commands, one per line. End with CNTL/Z.
2600B(config)#alias exec 1 disable
2600B(config)#^Z
2600B#show alias
Exec mode aliases:
 h help
 lo logout
 p ping
 r resume
 s show
 u undebug
 un undebug
 w where
 1 disable

2600B#1
2600B>
```

The arp Command

The *arp* command allows you to add static entries into the ARP table. ARP is
typically dynamic, so the *arp* command isn't normally needed; but you may need
to use it for a device that doesn't respond to ARP requests. (This is rare but possible
in high-security situations.) This command can also help to reduce broadcast traffic
in some cases (because the router doesn't need to broadcast to resolve the IP address
to a MAC address).

Syntax

The syntax for this command is *arp [IP address] [MAC address] [ARP type] [interface
name and number]*. The *ARP type* field is used to specify the type of ARP entry—ARPA
(normal for Ethernet), SAP, SMDS (for switched multimegabit data service), or
SNAP (FDDI and Token Ring). Also note that the MAC address is in the format of
0000.0000.0000 (in hex). So, if you wanted to add a static ARP entry for a server at
10.1.1.1 with a MAC address of 01-23-45-AB-CD-EF connected to your router's first
Ethernet interface, you would type *arp 10.1.1.1 0123.45AB.CDEF arpa Ethernet 0/0*.

The banner Command

The *banner* command displays a message to users or other administrators. The command has several variants, allowing you to set a banner for individual lines, for logins on any line, or for logins on a specific line. One of the primary uses of a banner is to provide a legal statement to would-be hackers, notifying them of possible criminal action from unauthorized use of equipment or systems.

Syntax

The syntax for this command is fairly simple. You type *banner [type]*, then a *delimiting character* (a character that you will not use in the message, like the pound (#) sign), and then press ENTER. The IOS prompts you to enter your banner, one line at a time, and use the delimiting character to signal the end of the message.

As for the valid banner types, the most useful is probably the message of the day (MOTD) banner. This banner displays any time anyone connects the router through a console or Telnet before the login message, making it extremely useful for conveying messages that apply to all users (such as security messages).

Other valid banner types include the following:

- **The exec banner**   Applies to all users except reverse Telnet users, and is displayed after the login

- **The incoming banner**   Applies to reverse Telnet, or users who are entering the router through an asynchronous serial connection, and is displayed after login

- **The login banner**   Displays before a username/password login appears, if you are using user-based authentication for the router

- **The prompt-timeout banner**   Allows you to change the message displayed when a user times out, for instance, when taking too long to type the enable password

- **The SLIP/PPP banner**   Displays for SLIP/PPP connections.

The following provides a basic example of banner use:

```
2600B(config)#banner motd #
Enter TEXT message. End with the character '#'.
Hello! Welcome to the ACME research corporation. If you are an
unauthorized user, please log off. Federal law prohibits
unauthorized use of this device. Thank you, and have a nice day.
#
2600B(config)#banner exec #
Enter TEXT message. End with the character '#'.
```

```
WARNING!!! You have now entered NORAD! We are tracking your
movements on Satellite, and you will be promptly terminated if you
persist!
#
2600B(config)#banner prompt-timeout #
Enter TEXT message. End with the character '#'.
Sorry dude, you like, took too long and stuff...
#
2600B(config)#^Z
2600B#exit

2600B con0 is now available

Press RETURN to get started.


Hello! Welcome to the ACME research corporation. If you are an
unauthorized user, please log off. Federal law prohibits
unauthorized use of this device. Thank you, and have a nice day.

User Access Verification

Password:


WARNING!!! You have now entered NORAD! We are tracking your
movements on Satellite, and you will be promptly terminated if you
persist!

2600B>enable
Password:
Sorry dude, you like, took too long and stuff...
```

The boot Command

The *boot* command sets the *bootup* method on most routers. You can use the *config-register* command for the same purpose, but, typically, the *boot* command is easier.

Syntax

Basic syntax of the *boot* command is *boot [target] [modifiers and values]*. There are six primary targets: *bootstrap, buffersize, config, host, network,* and *system.*

The bootstrap Target The *bootstrap* target sets the bootstrap image for initial boot. Normally, the bootstrap image is pulled from ROM; but in certain cases, it may be desirable to use an alternate image. To use the *bootstrap* target, type *boot bootstrap [flash | tftp] [filename]*.

The buffersize Target The *buffersize* target sets the size of the system buffer used to read the startup-config file. Normally, the system sets the buffer size equal to the size of NVRAM memory available on the router. Therefore, as long as you are having the router use the startup-config file from NVRAM, you shouldn't need to use this target. However, if you are loading a configuration file from a TFTP server at bootup (with the *boot host* or *boot network* commands), and the size of the configuration file is larger than the amount of NVRAM in your system (for instance, if the configuration file was 48KB and you had 32KB of NVRAM), you would need to set this target to an amount equal to or greater than the size of the configuration file. To use this target, type *boot buffersize [size in bytes]*.

The config Target The *config* target sets the location the router uses to load the startup-config file, either NVRAM or flash. This target is available only on routers with class A flash file systems (usually higher-end routers). To use this target, type *boot config [flash | NVRAM] [filename]*. The default is, of course, to boot from NVRAM.

The host Target The *host* target sets the router to load a startup-config file from a remote network host, using TFTP, RCP, or FTP. TFTP is the most commonly used host, and the syntax for setting this option is fairly simple: *boot host tftp: [//computer name or IP/directory/filename]*. Note that for this target to function, you must use the *service config* command (discussed later in the chapter in the section "The service Command") to enable this function.

Also note that the file specified with this target is considered to be the individual host configuration file. When using configurations in this manner, the device first loads the general network configuration file (specified with the *boot network* command); and then it applies the individual host configuration file, if present. Because in essence, it is copying these configurations to RAM, these are *merge operations,* meaning that if conflict exists between settings in the configuration files, the one that is applied *last* is the setting that is retained. In this manner, the settings configured with the *boot network* command are applied, containing general configuration information for the entire enterprise. Then anything specific to the device itself (using the *boot host* command) is applied, overwriting the more generic settings if necessary.

The network Target The *network* target, as described previously, also sets the router to use a startup-config file from a remote server. However, this target sets the generic startup-config file used by all routers. The syntax for this target is identical to the *boot host* command: *boot network [tftp | rcp | ftp]: [//computer name or IP/directory/filename]*.

The system Target   Finally, the *system* target tells the router which IOS image to use for boot. By default, the router typically uses the first available bootable image in flash. By using this target, you can specify that the router boot from a TFTP server instead. The syntax for this target is *boot system [flash | tftp | ftp | rcp] [filename (optional for flash)] [IP address (not used for flash)]*.

Note also that you can specify multiple boot system commands, allowing for fault tolerance. You do this by entering multiple boot system statements. The IOS processes these in order; so, if the first statement fails, it will try the second, and so on. For example, if you entered *boot system tftp bootfile.bin 10.1.1.1*, and then entered *boot system flash bootfile.bin* and—finally—*boot system flash*, the router would first attempt to retrieve the 2600.bin from your TFTP server. If that failed, it would try to retrieve 2600.bin from flash. If that also failed, it would retrieve the first executable file found in flash.

The cdp Command

The *cdp* command sets the basic operating configuration of Cisco Discovery Protocol (CDP). By using this command, you can set the interval between advertisements; the holdtime value; the version of CDP advertisements; and, of course, basic enabling or disabling of CDP.

In many environments, CDP is actually disabled. This cuts down on layer 2 traffic, as well as increasing security.

Syntax

The basic syntax of this command is *cdp [target] [value]*. The valid targets for this command are *advertise-v2, holdtime, timer,* and *run*.

The advertise-v2 Target   The *advertise-v2* target sets the version of CDP used for advertisements. By default, version 2 is used on newer Cisco devices. To force CDP to advertise using version 1, use the *no* version of this command.

The holdtime Target   The *holdtime* target sets the holddown timer that *other* devices should use for *this* device's CDP advertisements. This time is typically set to a multiple (by default, three) of the advertisement timer. The default setting is 180 seconds. The syntax for this target is *cdp holdtime [seconds]*. To return the *holdtime* to the default settings, use the *no* version of this target.

The timer Target   The *timer* target sets the interval between this device's CDP advertisements. The syntax for this target is *cdp timer [seconds]*. The default time is 60 seconds, and you can revert to the default by using the *no* version of the target.

The run Target   Finally, the *run* target just tells the device to enable CDP. (CDP is enabled by default.) To disable CDP, use the *no* version of this target.

The clock Command

The *clock* global configuration command (not to be confused with the *enable* command by the same name) configures basic clock functionality, namely, the time zone and whether to update for daylight saving time.

Syntax

The *clock* command follows the standard syntax *clock [target] [value]*. The two valid targets for this command are *timezone* and *summer-time*.

The timezone Target   The *timezone* target, predictably, sets the time zone for the router. The default is Coordinated Universal Time (UTC). The syntax for this command is *clock timezone [zone] [hour offset from UTC] [minute offset from UTC]*.

The summer-time Target   The *summer-time* target informs the device to update for daylight saving time. For the United States, the syntax for this command is simply *clock summer-time [zone] recurring*. For other countries, you need to know the specific dates or days when daylight saving time starts and ends. In these cases, to enable daylight saving to take effect on specific days (like the first day of August), you must type *clock summer-time [zone] recurring [week of month to begin] [day of month to begin] [month to begin] [hh:mm to begin] [week of month to end] [day of month to end] [month to end] [hh:mm to end] [time offset]*. To set daylight saving time to begin and end on a specific date every year, type *clock summer-time [zone] date [date to begin] [month to begin] [hh:mm to begin] [date to end] [month to end] [hh:mm to end] [time offset]*.

The config-register Command

As mentioned in Chapter 14, you use the *config-register* command to set the configuration register on a newer Cisco device.

Syntax

The syntax for this command is *config-register [hex value]*. Remember to include the *0x* in front of the hex value (as in 0x2102). For a recap of configuration register values, refer back to Chapter 14.

The default Command

This command simply sets configuration settings back to their default values. For many commands this is unnecessary because, in many cases, using the *no* version of a command sets it back to its default value.

Syntax

The basic syntax for this command is *default [target] [subtarget]*. For instance, if you wanted to set the CDP timer back to its default values, you could type *default cdp timer*; typing *no cdp timer* performs the same function, however. The valid targets for this command include nearly every command you can issue from global configuration mode.

The enable Command

Unlike the *enable* user mode command, you use this command to configure the enable password and secret.

Syntax

The syntax for this command is *enable [target] [modifiers or value]*. There are four valid targets in most versions of the IOS, but this discussion concentrates on only two of them. The two primary targets for this command are *password* and *secret*.

The password Target   The *password* target is typically used to specify an unencrypted password, but it can also be used to specify encrypted (secret) passwords as well. The basic syntax for this target (for setting the standard, unencrypted passwords) is *enable password [password]*. To set the encrypted password with this command, type *enable password 7 [password]*. The 7 informs the IOS that you will be setting the encrypted (or secret) password.

Note that you do not type in the password encrypted. (That would be pretty difficult!) You type the password in plaintext, and the IOS encrypts it before saving it to the configuration. Also note that just because the password is encrypted in the configuration file doesn't mean that the password is encrypted when sending it over an unencrypted connection (like a standard Telnet); it just encrypts the secret in the configuration files so that you can't just go in and view it outright.

The secret Target   The *secret* target works similarly except that, by default, this command sets the encrypted password. The basic syntax for this target is *enable secret [secret password]*. To set the unencrypted password using this command, type *enable secret 0 [password]*. An example of the use of these commands follows:

```
2600B(config)#enable password ccna
2600B(config)#enable secret ccie
2600B(config)#^Z
2600B#show running-config
Building configuration...

Current configuration:
!
version 12.0
!
hostname 2600B
```

```
!
logging buffered 4096 debugging
logging console warnings
enable secret 5 $1$JPeK$XUmL9thAxfOu9Bay9r2Ef0
enable password ccna
>OUTPUT TRUNCATED<
```

The end Command

This command takes you completely out of configuration mode and back to enable mode. It has no modifiers.

The exit Command

Use the *exit* command to drop back one level within configuration mode. In other words, if you are in interface configuration mode and you want to return to global configuration mode, using the *exit* command will take you there. (If you use the *end* command, it returns you directly to enable mode.) This process is illustrated in the following example:

```
2600B#configure terminal
Enter configuration commands, one per line. End with CNTL/Z.
2600B(config)#interface Ethernet 0/0
2600B(config-if)#exit
2600B(config)#interface Ethernet 0/0
2600B(config-if)#end
2600B#configure terminal
Enter configuration commands, one per line. End with CNTL/Z.
2600B(config)#interface ethernet 0/0
2600B(config-if)#^Z (Ctrl-Z)
2600B#
```

The *exit* command has no modifiers.

The hostname Command

The *hostname* command, quite predictably, sets the host name for the router. This command also has the effect of changing the system prompt to match the host name, but you can override this by using the *prompt* command. (It's a *really* bad idea to do so, however. If you apply a host name to the router, but the prompt states something different, you could be making incorrect assumptions about the name of the router. In general, just use the *hostname* command and ignore the *prompt* command.)

Syntax

The syntax for this command is *hostname [name]*.

Special Considerations

Note that this command does *not* automatically make the router resolvable by host name. To be able to resolve the router's host name to an IP address, you have to take the same steps as you would with any other host. (Add an A record to the DNS server or add the name/IP address mapping to the hosts file or hosts table.)

The interface Command

Use this command to enter interface configuration mode.

Syntax

The syntax for this command is *interface [interface name and number]*.

Special Considerations

The only tricky aspect of this command is the interface numbering scheme. In Chapters 9–13, you learned that the interface numbering differs depending on device type (router or switch) and individual product line. Here, these differences become important. To recap, routers typically begin interface numbering at zero, whereas switches generally begin at one. Also, modular routers and switches may number their interfaces in the *module/slot/port* or *module/port* scheme. Keep these points in mind when working on modular devices; and, if you are confused, refer back to the previous chapters for guidance.

The ip Command

The *ip* command is made up of many subcommands that perform a wide variety of roles, their only common bond being that they all perform IP-specific roles. Use the *ip* command to set IP addresses, default gateways, DHCP parameters, access lists, and more. This chapter covers only a small number of these commands. (Most of the rest are covered in later chapters, most notably, Chapters 18 and 22.)

Syntax

The basic syntax of this command is *ip [target] [subtarget, modifier, or value]*. This chapter is concerned with only five targets: *domain-list, domain-lookup, host, http,* and *name-server*.

The domain-list Target   You use the *domain-list* target to specify a list of domains to be appended to partial FQDN lookups. For instance, if you typed *ping server1*, the router would attempt to resolve the name by using its host table and configured DNS server, but it would need to "assume" a domain name somehow. The *domain-list* target tells the router which domains to assume that *server1* is a member of. If you tell it to try corp.com, then company.com, and then thisbiglongnameoverhere.com, it will first try to resolve server1.corp.com. If that fails, it will try server1.company.com, and so on. The syntax for this target is *ip domain-list [domain]*. You enter each domain you want to try in the order you want them to be used. A separate *ip domain-list* command is used to add each additional domain to the list.

The domain-lookup Target   The *domain-lookup* target tells the router to perform DNS resolution. The syntax for the command in this case is *ip domain-lookup*. This feature is typically more annoying than anything because every mistyped command you enter requires you to wait until the DNS resolution times out. This idiosyncrasy occurs because, if the router doesn't know you are typing a command, it assumes—at least in enable or user mode—that you are trying to connect to a host by that name.

   For example, if you type *contfgur*, the router responds with this: Translating "contfgur"...domain server (255.255.255.255). This forces you to wait what seems like an eternity (it's really only about a minute) before you can type the command again, correctly this time. You remove this "feature" by simply using the *no* version of this command or by entering a DNS server address with the *ip name-server* command.

The host Target   You use the *host* target to add entries into the device's static hosts table. This table is used before attempting to resolve a name using DNS (assuming DNS resolution is enabled). The syntax for this command is *ip host [hostname] [IP address]*. You enter each address-to-name mapping on a separate line, using a separate command for each.

The http Target   The *http* target sets the configuration for the internal HTTP server used to provide the optional web-based interface on most Cisco devices. This target has four subtargets:

- *access-class*   Controls access to the HTTP server using access lists. The syntax for this target and subtarget is *ip http access-class [access list number]*.

- *authentication*   Sets the method of authentication for access to the HTTP server. The authentication options are AAA, Enable, Local, and TACACS. Enable is the default (meaning you simply enter the enable secret or password with no username to gain access). The syntax for this target and subtarget is *ip http authentication [type]*.

- *port*   Sets the port used to connect to the HTTP server. The default is 80. If the server is enabled, it is recommended that you change this to a high-numbered port for security reasons. To connect to the server on an odd port number (like 14857), you must type the URL in your browser like so: *http://10.1.1.1:14857*. The syntax for this target and subtarget is *ip http port [port number]*.

■ *server* Enables the HTTP server. On most routers, the server is disabled by default. To enable the server, type *ip http server*. To disable the server, use the *no* form of this command.

The name-server Target Finally, the *name-server* target specifies a DNS server for the router to use. The syntax for this command is *ip name-server [IP address for preferred server] [IP address for secondary server] [IP address for tertiary server] [etc.]*. You can specify up to six DNS servers with this command.

The line Command

Use this command to enter interface configuration mode for a line. To recap, a *line* in Cisco is typically an asynchronous serial connection, including the console and aux ports. The prompt you will see when entering interface config mode for a line is Router(config-line)#.

Syntax

The syntax for this command is *line [target] [line number(s)]*. For example, to enter interface config mode for the console port, type *line console 0*. You could also just type the base line number, in this case, *line 0*. (Remember, you can view the line numbers by using the *show line* command.) You can also configure multiple lines at the same time by typing *line [type] [first line number] [last line number]*. For instance, if you wanted to configure your first five Telnet (VTY) lines at once, you would type *line vty 0 4*.

The logging Command

This command configures the logging parameters for your Cisco device. Chapter 14 thoroughly covered this command and its syntax, complete with examples.

The privilege Command

This command allows you to set the privilege level required to enter certain modes of router operation. This feature is useful when using username/password authentication methods because you can apply different levels of access to different users.

Syntax

The syntax for this command is *privilege [mode] [level]*. So, for example, if you wanted to change the privilege level required to enter interface config mode from 15 to 7, you would type *privilege interface 7*. When using this command, remember that, by default, user mode is level 1 and enable mode is level 15. For this command to function properly, user accounts must be set up and given different privilege levels.

The prompt Command

This command changes the system prompt. (Remember, so does the *hostname* command, but it also sets the router's host name.) Using this command instead of or combined with the *hostname* command is generally unwise because it tends to get a bit confusing.

Syntax

The syntax for this command is *prompt [text for prompt]*.

The service Command

The *service* command is another one of those large commands that encompasses several functions. With this command, you can compress the configuration files, enable time-stamping of debugs, encrypt all passwords in configuration files, and perform other assorted functions.

Syntax

The basic syntax for this command is *service [target] [modifier or value]*. This chapter is concerned with only a few specific targets: *compress-config, config, password-encryption, tcp-small-servers, timestamp,* and *udp-small-servers.* Each of these are addressed in the sections that follow.

The compress-config Target   This target allows you to compress the configuration files automatically when saving them. This feature is useful if you have a very large configuration file that will not fit in NVRAM. Once you enable this command, you need to perform a *copy running-config startup-config.* You will see a compression message stating what the original and final compressed size is. (It typically compresses the file to around 70 percent of its original size.) The IOS automatically decompresses the file upon boot. The syntax for this target is *service compress-config.* To remove the compression, use the *no* form of the command and resave the configuration.

The config Target   This target enables the device to use the *boot host* and *boot network* commands to load the config files from a network server. The syntax for this command is *service config.* This target is disabled by default, unless the NVRAM in the device is faulty.

The password-encryption Target   This target encrypts all passwords stored in config files on your router, including the standard enable password, as well as the console passwords, aux passwords, Telnet passwords, and all other passwords stored in the router's configuration files. Note that you must resave the config files after enabling this target to get the benefit. The syntax for this target is simply *service password-encryption.*

The tcp-small-servers Target   This target enables the router to respond to simple TCP services, such as ICMP echo requests (pings). This target is useful primarily because, if it is not enabled, pings to the router will not be answered. This feature can be positive or negative, depending on your needs; but typically, you want pings to your internal routers to be answered. The syntax for this target is *service tcp-small-servers.*

The timestamp Target   This target allows the router to timestamp (with hours; minutes; seconds; and, optionally, milliseconds) each debug or syslog message. The syntax for this target is a bit more complicated than most service targets. The format is *service timestamp [debug | log] [date | uptime] [options].* If you choose the *debug* option in the first section, the router timestamps debug messages but not syslog messages. If you choose the *log* option in the first section, the router timestamps syslog messages but not debug messages. If you want to timestamp both, simply enter the command twice (once for each type of message).

After that, you can choose to log in either uptime format or date format. If you choose the uptime format, the router timestamps the messages relative to the last reboot in HHHH:MM:SS format. If you choose the date format, the router timestamps the messages relative to its configured UTC time in the format MMM DD HH:MM:SS. If you choose the date format, you can also choose a few options: *msec, localtime,* and *show-timezone.* The *msec* option enables timestamping down to the millisecond. The *localtime* option formats the timestamp relative to the locally configured time zone. The *show-timezone* option displays the time zone used in the timestamp. The following example illustrates:

```
2600B#no debug all
All possible debugging has been turned off
2600B#debug ip packet
IP packet debugging is on
2600B#ping 10.1.1.1
 Type escape sequence to abort.
Sending 5, 100-byte ICMP Echos to 10.1.1.1, timeout is 2 seconds:
!!!!!
Success rate is 100 percent (5/5),
round-trip min/avg/max = 4/4/8 ms
2600B#show logging
Syslog logging: enabled (0 messages dropped, 0 flushes, 0 overruns)
 Console logging: level warnings, 3 messages logged
 Monitor logging: level debugging, 0 messages logged
 Buffer logging: level debugging, 153 messages logged
 Trap logging: level informational, 53 message lines logged
```

```
Log Buffer (4096 bytes):

%SYS-5-CONFIG_I: Configured from console by console
IP: s=10.1.1.200 (local), d=10.1.1.1 (Ethernet0/0), len 100,
sending
>Debug format before timestamp command<
>OUTPUT TRUNCATED<
2600B#configure terminal
Enter configuration commands, one per line. End with CNTL/Z.
2600B(config)#service timestamps debug uptime
2600B(config)#^Z
2600B#ping 10.1.1.1

Type escape sequence to abort.
Sending 5, 100-byte ICMP Echos to 10.1.1.1, timeout is 2 seconds:
!!!!!
Success rate is 100 percent (5/5),
round-trip min/avg/max = 4/7/8 ms
2600B#show logging
Syslog logging: enabled (0 messages dropped, 0 flushes, 0 overruns)
 Console logging: level warnings, 3 messages logged
 Monitor logging: level debugging, 0 messages logged
 Buffer logging: level debugging, 188 messages logged
 Trap logging: level informational, 54 message lines logged

Log Buffer (4096 bytes):

%SYS-5-CONFIG_I: Configured from console by console
IP: s=10.1.1.200 (local), d=10.1.1.1 (Ethernet0/0), len 100,
sending
>Debug format before timestamp command<
>OUTPUT TRUNCATED<
%SYS-5-CONFIG_I: Configured from console by console
2d22h: IP: s=169.254.2.1 (Serial0/0), d=224.0.0.9, len 232,
unroutable
>Debug format after timestamp uptime command<
>OUTPUT TRUNCATED<
2600B#configure terminal
Enter configuration commands, one per line. End with CNTL/Z.
2600B(config)#service timestamps debug datetime
2600B(config)#^Z
2600B#ping 10.1.1.1
```

```
Type escape sequence to abort.
Sending 5, 100-byte ICMP Echos to 10.1.1.1, timeout is 2 seconds:
!!!!!
Success rate is 100 percent (5/5),
round-trip min/avg/max = 4/5/8 ms
2600B#show logging
Syslog logging: enabled (0 messages dropped, 0 flushes, 0 overruns)
 Console logging: level warnings, 3 messages logged
 Monitor logging: level debugging, 0 messages logged
 Buffer logging: level debugging, 225 messages logged
 Trap logging: level informational, 55 message lines logged

Log Buffer (4096 bytes):

%SYS-5-CONFIG_I: Configured from console by console
IP: s=10.1.1.200 (local), d=10.1.1.1 (Ethernet0/0), len 100,
sending
>Debug format before timestamp command<
>OUTPUT TRUNCATED<
%SYS-5-CONFIG_I: Configured from console by console
2d22h: IP: s=169.254.2.1 (Serial0/0), d=224.0.0.9, len 232,
unroutable
>Debug format after timestamp uptime command<
>OUTPUT TRUNCATED<
%SYS-5-CONFIG_I: Configured from console by console
July 12 22:45:00: IP: s=10.1.1.200 (local), d=10.1.1.1 (Ethernet0/0),
len 100, sending
>Debug format after timestamp datetime command (UTC time,
time zone not shown)<
>OUTPUT TRUNCATED<
2600B#configure terminal
Enter configuration commands, one per line. End with CNTL/Z.
2600B(config)#service timestamps debug datetime msec localtime
show-timezone
2600B(config)#^Z
2600B#ping 10.1.1.1

Type escape sequence to abort.
Sending 5, 100-byte ICMP Echos to 10.1.1.1, timeout is 2 seconds:
!!!!!
Success rate is 100 percent (5/5),
round-trip min/avg/max = 4/6/8 ms
```

```
2600B#show logging
Syslog logging: enabled (0 messages dropped, 0 flushes, 0 overruns)
 Console logging: level warnings, 3 messages logged
 Monitor logging: level debugging, 0 messages logged
 Buffer logging: level debugging, 260 messages logged
 Trap logging: level informational, 56 message lines logged

Log Buffer (4096 bytes):
%SYS-5-CONFIG_I: Configured from console by console
IP: s=10.1.1.200 (local), d=10.1.1.1 (Ethernet0/0), len 100,
sending
>Debug format before timestamp command<
>OUTPUT TRUNCATED<
%SYS-5-CONFIG_I: Configured from console by console
2d22h: IP: s=169.254.2.1 (Serial0/0), d=224.0.0.9, len 232,
unroutable
>Debug format after timestamp uptime command<
>OUTPUT TRUNCATED<
%SYS-5-CONFIG_I: Configured from console by console
July 12 22:45:00: IP: s=10.1.1.200 (local), d=10.1.1.1 (Ethernet0/0),
len 100, sending
>Debug format after timestamp datetime command (UTC time,
time zone not shown)<
>OUTPUT TRUNCATED<
%SYS-5-CONFIG_I: Configured from console by console
July 12 18:45:15.434 EST: IP: s=10.1.1.200 (local),
d=10.1.1.1 (Ethernet0/0), len 100, sending
>Debug format after timestamp datetime msec localtime show-timezone
command (EST time, time zone shown)<
>OUTPUT TRUNCATED<
```

The udp-small-servers Target   Like the *tcp-small-servers* target, this target enables the router to respond to simple User Datagram Protocol (UDP) services, such as ICMP echo requests (pings). The syntax for this command is *service udp-small-servers*.

Common Interface Configuration Mode Commands

These commands are common to the interface configuration mode of device operation. In interface configuration mode, the changes you make affect only the selected interface or process. Again, you can tell which mode you are in based on the prompt. Interface configuration mode prompts appear with (config-if) at the end, like so: Router(config-if)#. Note that the commands available in interface configuration mode may differ somewhat depending on the device (router, switch, access server, and so on) and the specific type of interface (serial, Ethernet, and so on). These specifics are covered in later chapters (primarily, Chapters 18, 20, and 22). The commands discussed in this section apply to most standard IOS–based devices, and include the following: *cdp, description, full-duplex, half-duplex, ip, logging, loopback, mac-address, mtu,* and *shutdown.*

The cdp Command

This command simply enables or disables CDP on an interface. If CDP is enabled on an interface, advertisements are sent *and* listened for. If not, they are neither sent nor listened for.

Syntax

The syntax for this command is *cdp enable.* To disable CDP, use the *no* version of this command.

The description Command

This command sets a text description on an interface, which can be used to explain the settings on an interface, describe which network the interface links, or do anything else you find useful. This description shows up if a *show interface* is performed. An example of this follows:

```
2600B#show interface Ethernet 0/0
Ethernet0/0 is up, line protocol is up
  Hardware is AmdP2, address is 0003.6b40.3700 (bia 0003.6b40.3700)
  Description: Link to primary Ethernet network
  Internet address is 10.1.1.200/24
  MTU 1500 bytes, BW 10000 Kbit, DLY 1000 usec,
>OUTPUT TRUNCATED<
```

Syntax

The syntax for this command is *description [description text].*

The full-duplex and half-duplex Commands

As you might have guessed, these commands set the Ethernet interfaces on a router to either full- or half-duplex. Note that these commands will not function on all routers because some low-end routers are not capable of full-duplex operation.

Syntax

The syntax for these commands is *full-duplex* or *half-duplex*.

The ip Command

Like the *ip* global configuration commands, the *ip* interface configuration command encompasses a lot of different features, most of which are specific and discussed in later chapters (primarily Chapters 18 through 26).

Syntax

The basic syntax of this command is *ip [target] [subtarget, modifier, or value]*. This chapter covers only six targets: *address, mask-reply, mtu, redirects, unreachables*, and *verify*.

The address Target   The *address* target sets the IP address of the interface on a router or layer 3 switch. (Layer 2 switches do not have per-interface IP addresses; this is discussed further in Chapter 18.) The syntax for this target is *ip address [address] [mask] [secondary (optional)]*. The only tricky part about this command is the *secondary* option.

If *secondary* is appended to the end of this command, it makes the specified IP address a secondary address, meaning that two IP addresses are now bound to a single interface. Because this is for a router or layer 3 switch, both addresses are required to be on different subnets, meaning that the interface is now connected to two logical, but one physical, subnet. This feature is most useful when you have two addressing schemes running on the same physical link (which is rare), or in interfaces that support virtual circuits (VCs). (For VCs, secondary addresses are usable, but not recommended; subinterfaces, discussed in Chapter 22, are recommended in this case.)

The mask-reply Target   The *mask-reply* target enables ICMP mask replies on the selected interface (mentioned in Chapter 5). The syntax for this target is *ip mask-reply*.

The mtu Target   The *mtu* target sets the Maximum Transmission Unit (MTU) for the IP protocol (and *only* the IP protocol) on the interface. This target is fairly useless in most cases; typically, the MTU is defined by the layer 2 protocol and not the layer 3 protocol. The only point to keep in mind about this target is that if both the *mtu* command and the *ip mtu* command are set, the *ip mtu* command takes precedence. However, if the *mtu* command is set after the *ip mtu* command, the *ip mtu* command will be reset to the same value as the *mtu* command. The syntax for this command is *ip mtu [size in bytes]*.

The redirects and unreachables Targets   The *redirects* and *unreachables* targets enable ICMP redirect and unreachable messages on the selected interface (mentioned in Chapter 5). The syntax for these targets is *ip redirects* and *ip unreachables*.

The verify Target   Finally, the *verify* target enables per-packet verification of the source address listed in the packet. This is used to thwart some common denial of service (DOS) attacks (such as address spoofing). To enable this feature, type *ip verify unicast reverse-path.*

The logging Command

The *logging* interface configuration command (not to be confused with the *logging* global configuration command) allows you to enable or disable syslog messages for link state changes (when the link changes from up to down) on the specific interface.

Syntax

The syntax for this command to enable link state syslog messages is *logging event link-status.* To disable the messages, use the *no* form of the command. By default, link state syslog messages are enabled.

On serial interfaces, you may have two additional options related to VCs: *dlci-status-change* and *subif-status-change.* These options are used to send syslog messages whenever the state of a DLCI or subinterface changes.

The loopback Command

This command sets a loopback on the interface in question, causing every packet sent to be received back by the interface. The specifics of how this task is accomplished differ depending on the interface type, but the command is available for nearly all interfaces. The primary use of this command is to test basic interface functionality (in other words, to make sure the interface itself is functional at the physical layer). Note that on some interfaces (Ethernet, for example), enabling loopback makes the interface unusable for data traffic.

 Note   *Interface loopbacks are not software interfaces. This command physically places the port into a hardware loopback state.*

Syntax

The syntax for this command is *loopback.*

The mac-address Command

This command allows you to configure a custom MAC address on an interface. Typically, this feature is not required. It should be needed only if you have a MAC address conflict with another device (which is rare).

Syntax

The syntax for this command is *mac-address*.

The mtu command

This command sets the MTU on an interface. Typically, this feature is not required because the MTU values preset at the factory (shown in Table 16-1) are adequate in most cases. However, if you require a custom MTU setting, you can use this command to configure it. Note that unlike the *ip mtu* command, this command sets the MTU for all protocols on the interface.

Syntax

The syntax for this command is *mtu [size in bytes]*. The minimum value the IOS accepts is 64 bytes (ATM interfaces excluded), and the maximum size is 18,000 bytes.

The shutdown Command

The *shutdown* command takes an interface offline. This feature is useful when the interface is actually able to function, but, for some reason, you do not want the interface to transmit or receive data. Note that when an interface has been shut down, it displays "administratively down" in the show interfaces output, instead of "down," which displays when the interface is malfunctioning.

 By default, all router interfaces are in a shutdown state. To bring the interfaces online, issue the no *form of this command.*

Syntax

The syntax for this command is *shutdown*.

Interface Type	MTU
Ethernet	1500
Token Ring	4464
FDDI, ATM, and HSSI	4470
Serial	1500

Table 16-1. *The Default MTU Settings for Various Interface Types*

 **Summary**

This chapter examined the most common configuration mode commands for devices based on the standard IOS. Remember that although these commands apply to most devices, some exceptions do exist, depending on the specific model and revision of IOS. However, these commands should serve you well in most situations and will help tremendously as you work through the configuration examples in later chapters.

CISCO TECHNOLOGY
OVERVIEW

Chapter 17

Set-Based
IOS Commands

T his chapter again discusses some of the most common "generic" commands, but this time, in the set-based IOS, also known as *CatOS*. The set-based IOS is a bit different than the standard IOS. First, it is used only in high-end switches, so all of the commands are specific to switches. Second, switches (in most cases) tend to save configuration commands directly to nonvolatile RAM (NVRAM) (making all changes to the startup config). Finally, there are only two modes of operation (user and privileged), and there are only three primary commands (*set*, *clear*, and *show*) with a large number of subcommands.

For this reason, this chapter is organized alphabetically by command, although I have defined the valid modes for each command. Again, this chapter does not cover all commands, just those that are extremely useful and not explained in later chapters. Finally, if you have not had any experience with a CatOS-based switch, you may find it easier to begin with the *set* and *show* commands before reading the *clear* commands.

All configuration examples in this chapter were performed on a Catalyst 5500 switch. The commands may differ slightly on other switch lines (such as the 6500 series).

Common clear Commands

One of the three primary commands for a CatOS-based switch, the *clear* command is used to remove configuration settings. Note that in a CatOS-based switch, the *no* command is typically not used. The *clear* command is used instead to remove settings (most of the time; occasionally, *set* may also be used). Table 17-1 describes the commonly used *clear* command targets covered in this section.

Target	Description
alias	Clears aliases of commands
arp	Clears ARP table entries
.banner	Clears message of the day (MOTD) banner
boot	Clears booting environment variable
config	Clears configuration and resets system
counters	Clears MAC and port counters
ip	Clears IP settings; use the *clear ip help* command for more info

Table 17-1. *The clear Command Targets Discussed in This Chapter*

Target	Description
log	Clears log information
logging	Clears system logging information
timezone	Clears the time zone

Table 17-1.   *The* clear *Command Targets Discussed in This Chapter* (continued)

The basic syntax for all *clear* commands is *clear [target] [modifiers]*. The specific modifiers will be discussed in the sections on the individual *clear* command targets that follow.

The clear alias Command

This command removes command aliases. The syntax for the command is either *clear alias [alias name]* or *clear alias [all]*. The *clear alias [alias name]* version clears a single alias, whereas the *clear alias [all]* version clears all aliases. An example of this command follows:

```
Cat5K (enable) clear alias all
Command alias table cleared. (1)
```

The clear arp Command

The *clear arp* command clears the switch's ARP table. This command has several modifiers: *all, [ip address], dynamic, static,* and *permanent*. The *all* modifier clears all entries. The *[ip address]* modifier removes a specific ARP entry for a particular IP address. The *dynamic* modifier clears dynamically learned ARP entries. The *static* modifier clears all statically (manually) entered ARP entries that are not permanent. And, finally, the *permanent* modifier removes all manually entered ARP entries that are saved into NVRAM. The following example shows the use of some of these modifiers:

```
Cat5K (enable) clear arp 10.1.1.1
ARP entry deleted.
Cat5K (enable) clear arp dynamic
Dynamic ARP entries cleared. (1)
Cat5K (enable)
```

The clear banner Command

This command clears the message of the day (MOTD) banner on the switch. The syntax for this command is *clear banner motd*. An example follows:

```
Cat5K (enable) clear banner motd
MOTD banner cleared
Cat5K (enable)
```

The clear boot Command

The *clear boot* command clears the bootup configuration of the router (which is set with the *set boot* command, discussed in the section "The set boot Command," later in this chapter). This includes the ability to clear the specified boot image and NVRAM settings. The command comes in three different forms: *clear boot auto-config, clear boot system all*, and *clear boot system flash*.

The clear boot auto-config Command

The *clear boot auto-config* command clears the *config-file* variable in the router, removing any configuration file listings for the startup configuration of the switch. This causes the switch to use the default configuration file, if it exists, or default settings if it does not exist. The syntax for the command is *clear boot auto-config [module]*. The *module* modifier is used to specify the supervisor engine to clear (either the primary in slot 1 or the redundant in slot 2 for a Catalyst 5500).

The clear boot system all Command

The *clear boot system all* command clears the entire boot environment variable, including config file settings and CatOS image load settings. This causes the switch to revert to using the first available flash image and the default config file. If a boot image cannot be located, the switch boots to ROM Monitor (Rommon). The syntax for this command is *clear boot system all [module]*.

The clear boot system flash Command

The *clear boot system flash* command clears the flash boot image preference for the switch. This causes the switch to use the first available CatOS image at boot. If a boot image cannot be located, the switch boots to Rommon. The syntax for this command is *clear boot system flash [boot device (bootflash | slot0 | slot1)] [module]*.

The clear config Command

This command erases the configuration file for the switch. It can also be used to erase SNMP/RMON configuration. The syntax for this command is *clear config [module | rmon | snmp | all]*. The *snmp* and *rmon* modifiers clear the SNMP and RMON configurations, respectively, for the entire switch. The *module* modifier specifies clearing a specific module. The *all* modifier clears all switch configurations for all modules and slots, including management IP addresses and SNMP/RMON configurations. An example of this command follows:

```
Cat5K> (enable) clear config all

This command will clear all configuration in NVRAM.

This command will cause ifIndex to be reassigned on the next system startup.

Do you want to continue (y/n) [n]? y

Releasing IP address...Done

......

................

....................................................


System configuration cleared.

Cat5K> (enable)
```

The clear counters Command

This command clears the statistical MAC address and port counters (viewed with the *show mac* and *show port* commands). The syntax for this command is *clear counters*. An example of this command follows:

```
Cat5K (enable) show mac

Port      Rcv-Unicast          Rcv-Multicast        Rcv-Broadcast
--------  -------------------- -------------------- --------------------
 1/1                        0                    0                    0
 1/2                        0                    0                    0
 2/1                       81                   93                    0
 >DETAIL OMITTED<
Cat5K (enable) clear counters
This command will reset all MAC and port counters reported in CLI and SNMP.
Do you want to continue (y/n) [n]? y
MAC and Port counters cleared.
Cat5K (enable) show mac
```

```
Port      Rcv-Unicast          Rcv-Multicast         Rcv-Broadcast
--------  --------------------  --------------------  --------------------
 1/1                        0                     0                     0
 1/2                        0                     0                     0
 2/1                        3                     3                     0
>DETAIL OMITTED<
```

The clear ip Command

This command clears certain IP settings. The syntax is *clear ip [target] [modifiers]*. The valid targets for this command are *alias*, *dns*, *permit*, and *route*.

The alias Target

The *alias* target removes one or all IP aliases (configured with the *set ip alias* command). IP aliases are similar to host names, and they are used as a shortcut to typing a remote host's IP address. If both a DNS host name and alias are the same (in other words, if a host name is set to "gremlin" and an alias is also set to "gremlin"), the alias overrides the host name. The syntax for this target is *clear ip alias [alias name | all]*.

The dns Target

The *dns* target clears the default DNS domain and removes one or all of the DNS servers in the search order. To clear the default DNS domain, the syntax is *clear ip dns domain*. To clear a single server from the search order, type *clear ip dns server [server ip address]*; and to clear all DNS servers, type *clear ip dns server all*.

The permit Target

The *permit* target clears the IP permit list of one or all IP addresses. The IP permit list is used to allow access to the switch from one or more IP addresses using Telnet or SNMP. The syntax for this target is *clear ip permit [ip address | all] [subnet mask] [telnet | snmp | all]*.

The route Target

Finally, the *route* target removes one or all routing table entries in the switch. The syntax for this target is *clear ip route [destination | all] [gateway]*.

The clear log Command

This command clears the system error and command logs. The system error log (discussed further in the section "The show log Command," later in this chapter) logs errors that occur in the switch, whereas the command log stores command history. Clearing the error log allows you to more easily view newer errors because you do not need to sort through the old errors. Clearing the command log allows you to more easily locate commands with the shortcut keys.

The syntax for clearing the system error log is *clear log [module]*. If a module is not specified, all error messages for all modules are cleared.

The syntax for clearing the command log is *clear log command [module]*. Again, if a module is not specified, all command histories for all modules are cleared.

The clear logging Command

This command clears syslog messages, similar to the standard IOS command of the same name. The syntax for this command is *clear logging buffer* to clear the syslog messages stored on the switch. You can also use it to remove servers from the list of syslog servers by typing *clear logging server [server ip address]*.

The clear timezone Command

This command sets the time zone on the switch back to its default time zone (UTC). The syntax for this command is *clear timezone*.

The configure Command

This command downloads and executes commands from a configuration file on a TFTP or RCP server. This works similarly to the *configure network* command on a standard IOS-based device.

Syntax

The command has two variants: *configure [server name or ip address] [file] [optional RCP]* and *configure network*.

The *configure [server name or ip address]* variant configures the device with a single command. If this is typed with the optional *rcp* modifier, the device is configured using RCP. Otherwise, TFTP is used. An example of this variant follows:

```
Cat5K> (enable) configure 192.168.1.1 5500.cfg

Configure using 5500.cfg from 192.168.1.1 (y/n) [n]? y


/
Done.  Finished Network Download.  (302 bytes)
>> set ip alias CCNA1900A 192.168.1.11
IP alias added.
>> set prompt 5500>
>> set password
Enter old password: cisco
```

```
Enter new password: ccie

Retype new password: ccie

Password changed.
Cat5K> (enable)
```

The second variant of this command, *configure network,* prompts you for the IP address of the remote server and the filename of the configuration file. Other than this difference, it operates identically to the *configure [server name or ip address]* variant.

The copy Command

The *copy* command on a CatOS-based switch is used to copy configuration and flash images to and from the device.

Syntax

The basic syntax for this command is the same as for a standard IOS–based device: *copy [source] [destination].* However, on a CatOS-based switch, the source and destination targets can be slightly different. The valid targets on CatOS switches include *[file path], tftp, rcp, flash, config, cfg1,* and *cfg2.*

The [file path] Target

The *[file path]* target designates a specific file as the source or destination. The syntax of this target is *[module]/[device]:filename.* The device can be bootflash (internal flash memory), slot0, slot1, or tftp. So, for example, if you wanted to use a file called tempconfig.cfg on the slot0 flash file system on the first supervisor module, the path would be module1/ slot0:tempconfig.cfg.

The tftp and rcp Targets

The *tftp* and *rcp* targets specify a TFTP or RCP server. The syntax is *tftp/rcp:[file name].* You will be prompted for the server from which to copy the file, as shown in the following example:

```
Cat5K> (enable) copy config tftp:Cat5K.cfg

IP address or name of remote host []? 192.168.1.1
```

```
Upload configuration to imgFile:Cat5k.cfg (y/n) [n]? y

. . . . . . . . .
 .
 /
Configuration has been copied successfully. (1172 bytes).
Cat5K> (enable)
```

The flash Target

The *flash* target, predictably, copies the file to or from flash memory. The syntax is *flash*. The switch asks for the flash device to copy to or from, as shown in the following example:

```
Cat5K> (enable) copy config flash

Flash device [bootflash]? slot1:

Name of file to copy to [configFile]? Cat5K.cfg

Upload configuration to slot1:Cat5K.cfg
1247362 bytes available on device slot1, proceed (y/n) [n]? y

. . . . . . . . .
 .
 /
Configuration has been copied successfully. (1172 bytes).
Cat5K> (enable)
```

The config, cfg1, and cfg2 Targets

Finally, the *config*, *cfg1*, and *cfg2* targets designate a specific configuration file to copy to or from. The *config* target is the actual configuration file for the switch. The *cfg1* and *cfg2* targets are used on switches with Supervisor Engine IIG and IIIG, and they are secondary configuration files stored in flash memory.

The delete Command

The *delete* command completely deletes a secondary configuration file from the flash of a CatOS-based switch.

Syntax

The syntax for this command is *delete [module number]/[device]:[filename]*. An example of this command is shown here:

```
Cat5K> (enable) delete 1/slot0:5500.cfg

Cat5K> (enable)
```

The history Command

This command works similarly to the *show history* command on a standard IOS-based device. It shows the command history since boot (by default, 20 commands).

Syntax

The syntax for this command is *history*. The output is a bit different than an IOS-based switch, as shown here:

```
Cat5K (enable) history
      1 enable
      2 delete cfg1
      3 history
```

The ping Command

This command issues a standard ping (ICMP echoes) to a remote device, similar to the *ping* command in a standard IOS–based device.

Syntax

The syntax for this command is *ping [ip address]* for a standard ping, and *ping -s [ip address] [packet size] [packet count]* for an extended ping. See the example that follows:

```
Cat5K (enable) ping 10.1.1.1
10.1.1.1 is alive
Cat5K (enable) ping
Usage: ping <host>
       ping -s <host> [packet_size] [packet_count]
       (host is ipalias or IP address in dot notation: a.b.c.d)
       (packet_size range: 56 .. 1472)
Cat5K (enable) ping -s 10.1.1.1 64 5
```

```
PING 10.1.1.1: 64 data bytes
72 bytes from 10.1.1.1: icmp_seq=0. time=8 ms
72 bytes from 10.1.1.1: icmp_seq=1. time=8 ms
72 bytes from 10.1.1.1: icmp_seq=2. time=5 ms
72 bytes from 10.1.1.1: icmp_seq=3. time=8 ms
72 bytes from 10.1.1.1: icmp_seq=4. time=5 ms


----10.1.1.1 PING Statistics----
5 packets transmitted, 5 packets received, 0% packet loss
round-trip (ms)  min/avg/max = 5/6/8
Cat5K (enable)
```

The quit Command

The *quit* command is used to exit a session on the switch.

Syntax

The syntax for this command is *quit*.

The reset Command

This command forces a reload of the software on a module. By using the *reset* command, you can reset individual modules or the entire system, and even schedule resets for later.

Syntax

The basic syntax for this command is *reset [module number or system]*. So, to reset module 2, you would type *reset 2*. To reset the entire system, you would type *reset system*.

To reset the switch later, you can use the *reset [at or in] [hh:mm] [mm:dd] [reason]* command. If you use the *at* subtarget, this allows you to schedule the reset at a specific time and date. If you use the *in* subtarget, you can schedule the reset after a specific period of time has passed.

Finally, the *reset cancel* command cancels a scheduled reset.

The following examples show the use of the *reset* command:

```
Cat5K (enable) reset 4
Unsaved configuration on module 4 will be lost
Do you want to continue (y/n) [n]? y
2001 Jul 22 13:46:01 %SYS-5-MOD_RESET:Module 4 reset from Console//
Resetting module 4...
```

```
Cat5K (enable) 2001 Jul 22 13:47:26
%SYS-5-MOD_OK:Module 4 is online

Cat5K (enable) reset at 12:00 08/01 test
Reset schedule at Wed Aug 1 2001, 12:00:00.
Reset reason: test

Proceed with scheduled reset (y/n) [n]? y

Reset schedule for Wed Aug 1 2001, 12:00:00
(in 9 days 22 hours 11 minutes 26 seconds).
Reset reason: test
Cat5K (enable) reset cancel
Reset cancelled.
Cat5K (enable) 2001 Jul 22 13:48:58
%SYS-1-SYS_SCHEDRESETCANCEL:Scheduled reset cancelled by user

Cat5K (enable) reset in 12:00 test
Reset schedule in 12 hours 0 minute.
Reset reason: test

Proceed with scheduled reset (y/n) [n]? y

Reset schedule for Mon Jul 23 2001, 01:49:22
(in 0 day 12 hours 0 minute 0 second).
Reset reason: test
Cat5K (enable) reset cancel
Reset cancelled.
Cat5K (enable) 2001 Jul 22 13:49:28
%SYS-1-SYS_SCHEDRESETCANCEL:Scheduled reset cancelled by user
```

The session Command

The *session* command is used to enter a CLI session with intelligent modules, like an LS1010 ATM module or a 5500 RSFC. These modules have their own IOS and system images, and they are configured separately from the main switch.

Syntax

The syntax for this command is *session [module number]*. An example of the command follows:

```
Cat5K (enable) session 4
Trying ATM-4...
Connected to ATM-4.
Escape character is '^]'.

ATM>
```

Common set Commands

The *set* command, along with the *clear* and *show* commands, is one of the three primary commands in a CatOS switch. The *set* command, in general, is used to configure features in the switch. As such, it is made up of a lot of subcommands, many of which are covered here, but some of which are covered in later chapters. Table 17-2 describes the *set* command targets covered in this chapter.

Command Targets	Description
alias	Sets alias for command
arp	Sets ARP table entry
banner	Sets message of the day banner
boot	Sets booting environment variable
cdp	Sets cdp parameters
enablepass	Sets privilege mode password
interface	Sets network interface configuration
ip	Sets IP parameters
length	Sets number of lines on screen
logging	Sets system logging configuration information
logout	Sets number of minutes before automatic logout
module	Sets module configuration
password	Sets console password
port	Sets port features
prompt	Sets prompt
summertime	Sets summertime

Table 17-2. *The set Command Targets Discussed in This Chapter*

Command Targets	Description
system	Sets system information
time	Sets time
timezone	Sets timezone

Table 17-2. *The* set *Command Targets Discussed in This Chapter* (continued)

The basic syntax for all *set* commands is *set [target] [modifiers or value]*. The specific syntax for each target is discussed in the following sections.

The set alias Command

The *set alias* command sets a command alias in a CatOS switch. The syntax for this command is *set alias [alias] [command including parameters]*. For instance, if you wanted to set an alias for the *clear banner motd* command called *clban*, you would type *set alias clban clear banner motd*. This is shown in the following example:

```
Cat5K (enable) set alias clban clear banner motd
Command alias added.
Cat5K (enable) clban
MOTD banner cleared
Cat5K (enable)
```

The set arp Command

The *set arp* command adds entries into the ARP cache. Typically, this is not required because ARP dynamically resolves IP addresses to MAC addresses. However, it can be useful in situations in which ARP broadcast reduction is needed for commonly accessed hosts. In these cases, you can enter a static or permanent ARP entry to significantly reduce these broadcasts. The down side to static and permanent entries is that if the MAC address for a host changes (for example, if a network card in a server is replaced), the host is unreachable from the switch until the entry is either cleared or updated.

Along these lines, three types of ARP entries are defined:

- **Dynamic** The normal entry type. Ages out of the table after a short period of time and must be reacquired through the normal broadcast mechanism.
- **Static** Entered manually. Does not time out of the table, but is lost if the switch is reset.
- **Permanent** Entered manually. Does not time out and is retained in NVRAM.

The basic syntax for the *set arp* command is *set arp [type] [MAC address] [IP address]*. However, the *set arp* command can also be used to set the default ARP aging time. The syntax to set the aging time is *set arp agingtime [time in seconds]*. The default setting is 1200 seconds, and the range is from 0 seconds (aging disabled) to 1 million seconds.

The set banner Command

This command sets the MOTD banner. The syntax is *set banner motd [delimiting character]*, nearly identical to the *banner motd* command on a standard IOS-based device.

The set boot Command

This command configures the various boot options on a CatOS switch. The command consists of four subcommands: *auto-config, config-register, sync-now*, and *system*. Each of these is a bit different, so they are discussed individually in the following syntax sections.

Syntax for the set boot auto-config Subcommand

The *set boot auto-config* subcommand sets the startup configuration options on the switch. On a switch with a Supervisor Engine IIG or IIIG, this command has the following formats: *set boot auto-config [configuration file list]* and *set boot auto-config [recurring | nonrecurring]*.

The *set boot auto-config [configuration file list]* format specifies the configuration file used after a switch reset. By default, the current configuration (represented as simply *config*) is reused. The options on the IIG and IIG switches are either *cfg1* or *cfg2*, or both. The cfg1 and cfg2 files are configuration files stored in flash on the switch. (Use the *copy* command to create these files from a current configuration.)

These files are used to configure the switch only at boot or after using the *copy cfg1 config* or *copy cfg2 config* commands. By using the *set boot auto-config* command, you can force the switch to use one of these configuration files at reset. To specify that the switch uses both files after a reset, use the syntax *set boot auto-config [first choice];[second choice]*. For example, if you wanted the switch to use cfg2 if available, but use cfg1 if cfg2 is corrupted or has been deleted, you would type *set boot auto-config cfg2;cfg1*.

For a switch with a supervisor engine other than the IIG or IIIG, the syntax for this command is *set boot auto-config [device]:[filename]*. The device is bootflash, slot0, or slot1, and the filename is the name of the configuration file (for example, Cat5k.cfg). You can also choose to use multiple files by appending them to the command. So, if you wanted the switch to boot from a file called Cat5K.cfg in slot 0 if possible, but to boot from a backup configuration file called oldconfig.cfg in bootflash if the Cat5K.cfg file is unavailable, you would type *set boot auto-config slot0:Cat5K.cfg;bootflash:oldconfig.cfg*.

The *set boot auto-config [recurring | nonrecurring]* subcommand is used on switches with Supervisor Engines IIG or IIIG to determine whether the configuration register is cleared after each reset. If you choose *nonrecurring*, the configuration register is cleared, and the next reboot will be from the config (not from the cfg1 or cfg2 file you specified with the *set boot auto-config [configuration file]* command). This setting is the default. If you choose *recurring,* the configuration register will not be cleared, and your settings will remain.

Syntax for the set boot config-register Subcommand

The *set boot config-register* subcommand sets the configuration register from within the CatOS (rather than from Rommon). It has two primary syntax options: *set boot config-register [hex value]* and *set boot config-register [boot option] [value].*

The *set boot config-register [hex value]* syntax allows you to set the configuration register by simply typing in the hex value you wish to change it to. This is fast but rather user unfriendly. The default value for this field is *0x10F,* which sets the following parameters:

- Sets the *boot* method to *system* (allowing configuration commands to determine the boot image)

- Sets the console port to *9600 baud*

- Disables the *ignore-config* variable (allowing configuration in NVRAM to be used at reset; enabling this value clears the NVRAM at reload, giving the switch a blank configuration)

- Sets *auto-config* to *nonrecurring*

The *set boot config-register [option] [value]* syntax is typically easier to use, although somewhat more time-consuming. With this format, you simply specify the name of the option to change (*boot, baud, ignore-config,* or *auto-config*) and the value you want to change it to. For the *boot* option, the possible values are *rommon* (to cause the switch to boot directly into Rommon), *bootflash* (to cause the system to boot directly to the first image it can find), or *system* (to allow configuration files to determine the boot image). For the *baud* option, the possible values are *1200, 2400, 4800,* or *9600.* For the *ignore-config* option, the values are simply *enable* or *disable.* Finally, for the *auto-config* option, the values are simply *recurring* or *nonrecurring.*

Syntax for the set boot sync-now Subcommand

This subcommand initiates synchronization of the configuration files between the primary and standby supervisor engines in a CatOS switch. This allows you to keep the same configuration for both engines without manually copying the files. The syntax for this subcommand is simply *set boot sync-now.*

Syntax for the set boot system Subcommand

This command sets the image the switch boots from. The syntax for the command is *set boot system [device]:[filename] [prepend] [module number].* The device and filename follow the same conventions discussed in earlier commands (such as the *copy* command). The prepend flag is optional, and it tells the switch to place this entry at the top of the list of boot images. (Otherwise, boot images are processed in the order entered, and the first one found is used.) The module number states for which module you are specifying the boot image. This is also optional, and if left out, is assumed to be 1.

The set cdp Command

This command sets CDP preferences. Using this command, CDP can be enabled or disabled, the holdtime and interval can be set, and the version can be defined.

The basic syntax for this command is *set cdp [target] [modifier or value]*. The valid targets for this command are *enable/disable, holdtime, interval,* and *version*.

The *enable/disable* target enables or disables CDP either globally or for specific modules and ports. If you enter *set cdp enable* or *set cdp disable*, CDP is enabled or disabled globally.

The set enablepass Command

This command sets the enable password on CatOS switches. Note that in a CatOS switch, only one enable password exists, and it is automatically encrypted (like the enable secret on a standard IOS device).

The syntax for this command is *set enablepass*. The switch then prompts you for the old password and the new password. This is shown in the following example:

```
Cat5K (enable) set enablepass
Enter old password: cisco
Enter new password: ccie
Retype new password: ccie
Password changed.
Cat5K (enable)
```

Note that, by default, neither the enable password nor the console password are set.

The set interface Command

This command is used to configure the in-band (Ethernet-based or Telnet) and out-of-band (serial-based or console) interfaces on a CatOS switch. With this command, the interfaces can be enabled or disabled, the IP addresses set, and the virtual LAN (VLAN) number set for the in-band interface. (VLANs are discussed in Chapter 19.)

The basic syntax for this command is *set interface [sc0 | sl0] [options or value]*. For the sc0 (in-band) interface, the primary options are *[up | down], [ip address],* and *dhcp*. If you use the *set interface sc0 [up | down]* form, it enables or disables the interface. If you use *set interface sc0 [ip address] [subnet mask] [optional broadcast address]*, it sets the IP address and, optionally, the broadcast address on the interface. To also set the VLAN number, insert it into the syntax before the *[ip address]* section, like so: *set interface sc0 [vlan number] [ip address] [subnet mask] [optional broadcast address]*. The final form is *set interface sc0 dhcp [release | renew]*, which releases or renews a DHCP-configured IP address.

For the sl0 interface, the options are *[up | down]* and *[slip address]*. The *[up | down]* option enables or disables the interface, whereas the *[slip address]* option sets the IP address of both the console port and the directly connected host (which is not generally required or used). To use the *[slip address]* option, type *set interface sl0 [console ip address] [remote host ip address]*.

The set ip Command

This command sets several IP options on the switch, including IP aliases, DNS parameters, HTTP parameters, and ICMP parameters.

The *set ip* command's basic syntax is *set ip [target] [modifier or value]*. This command has five primary targets: *alias, dns, http, redirect,* and *unreachable*.

The alias Target

The *alias* target sets IP aliases (similar to host names) on the switch. Thus, you can use a simple name (like 2600A) to ping or Telnet into a device from the switch CLI. The syntax for this target is *set ip alias [alias] [ip address]*.

The dns Target

The *dns* target sets DNS preferences on the switch, specifically to disable or enable DNS, as well as to set the name servers and default domain. To globally enable or disable DNS, type *set ip dns [enable | disable]*. By default, DNS is disabled. To set the DNS servers for the switch, type *set ip dns server [ip address] [primary (optional)]*. If the optional *primary* modifier is used, the specified server is set as the primary server. Otherwise, servers are processed in the order entered. Finally, to set the default DNS domain name, type *set ip dns domain [domain name]*.

The http Target

The *http* target configures the internal HTTP management server, including globally enabling/disabling the server and setting the port for the switch to listen on. To enable or disable the server, type *set ip http server [enable | disable]*. The default setting is disabled. To set the port number, type *set ip http port [port number]*. The default port number is, of course, 80.

The redirect Target

The *redirect* target enables the issuance of ICMP redirect messages from the switch. By default, ICMP redirects are enabled. The syntax for this target is *set ip redirect [enable | disable]*.

The unreachable Target

Finally, the *unreachable* target globally enables or disables the sending of ICMP-unreachable messages. By default, this target is enabled, and ICMP unreachables are sent. The syntax for this command is *set ip unreachable [enable | disable]*.

The set length Command

The *set length* command sets the number of lines shown on the terminal screen before pausing with a more prompt. At the more prompt, output can be halted by pressing CTRL-C or Q, you can scroll down one line at a time by pressing ENTER, or you can scroll down an entire screen by pressing SPACEBAR.

To set the number of lines to any number between 1 and 512 lines per screen, type *set length [number of lines]*. To cause output to never pause, set the length to *0*. Finally, to set the length back to its default setting (24 lines per screen), type *set length default*.

The set logging Command

This command sets logging preferences for the switch, similar to the *logging* command on an IOS-based switch. The primary options for this command include buffering, console logging, syslog logging depth, server logging, and timestamping of syslog messages.

The basic syntax for this command is *set logging [target] [option or value]*. The primary targets for this command are *buffer, history, console, level, server,* and *timestamp.*

The buffer and history Targets

The *buffer* and *history* targets both perform the same task: they determine how many commands are retained in the syslog buffer. The default setting retains the maximum number of messages, 500. The syntax for these targets is *set logging buffer [number of messages]* and *set logging history [number of messages]*.

The console Target

The *console* target enables or disables logging to the console port. The syntax for this target is *set logging console [enable | disable]*.

The level Target

The *level* target sets the logging level for syslog messages. With this target, the logging level is individually definable for each facility (type of service) on the switch. Table 17-3 shows the default logging levels for various facilities.

Facility	Logging Level
System	5 (Notifications)
Dynamic Trunk Protocol	5 (Notifications)
Port Aggregation Protocol	5 (Notifications)

Table 17-3. *Default Logging Facilities and Levels*

Facility	Logging Level
Management	5 (Notifications)
Multilayer Switching	5 (Notifications)
All others	2 (Critical)

Table 17-3. *Default Logging Facilities and Levels* (continued)

The facilities available for the *level* target are shown in Table 17-4.

Facility CatOS Name	Facility Description
cdp	Cisco Discovery Protocol
cops	Common Open Policy Service
drip	Dual Ring Protocol
dtp	Dynamic Trunk Protocol
earl	Encoded Address Recognition Logic
fddi	Fiber Distributed Data Interface
filesys	File system
gvrp	GARP VLAN Registration Protocol
ip	Internet Protocol
kernel	Kernel
mcast	Multicast
mgmt	Management
mls	Multilayer switching
pagp	Port Aggregation Protocol
protfilt	Protocol filter
pruning	VTP pruning

Table 17-4. *Facility Types*

Facility CatOS Name	Facility Description
qos	Quality of Service
radius	Remote Access Dial-In User Service
security	Security
snmp	Simple Network Management Protocol
spantree	Spanning Tree Protocol
sys	System
tac	Terminal Access Controller
tcp	Transmission Control Protocol
telnet	Terminal Emulation Protocol
tftp	Trivial File Transfer Protocol
udld	User Datagram Protocol
vtp	Virtual Terminal Protocol

Table 17-4. *Facility Types* (continued)

Table 17-5 shows the logging levels of the *level* target.

Severity Level	Severity Type	Description
0	Emergencies	System unusable
1	Alerts	Immediate action required
2	Critical	Critical condition
3	Errors	Error conditions
4	Warnings	Warning conditions
5	Notifications	Normal bug significant condition
6	Informational	Informational messages
7	Debugging	Debugging messages

Table 17-5. *Syslog Severity Levels*

CISCO TECHNOLOGY
OVERVIEW

The syntax for the *level* target is *set logging level [facility] [level].* Type *set logging level default* to return the switch to its default values.

The server Target

The *server* target enables or disables sending syslog messages to a server, sets the syslog servers for the switch, sets the logging severity levels, and sets the format of the syslog messages. This target comes in four variations: *set logging server [enable | disable], set logging server [syslog server ip address], set logging server facility [facility type],* and *set logging server level [level to log].*

The *set logging server [enable | disable]* and *set logging server [syslog server ip address]* targets simply enable or disable logging to a syslog server globally, and they set the default syslog server.

The *set logging server facility [facility type]* variation sets the logging format to either syslog or local0 – local7. Your syslog server type determines the facility type you need to use.

Finally, the *set logging server level [level to log]* variation of this target sets the level of messages to send to the syslog server. Note that the base logging levels and IOS features (facilities) to log are still set globally for the entire switch by using the *set logging level* command. This means that the setting for logging to a syslog server must be equal to or greater than the severity level set globally for the switch. Otherwise, no syslog messages will be sent to the server.

The timestamp Target

The last target, *timestamp,* is used to enable or disable timestamping of syslog messages. The default setting for timestamping is *enabled.* The syntax for this target is *set logging timestamp [enable | disable].*

The set logout Command

This command configures how long a session can remain idle before it is forcibly terminated. This allows you to automatically close idle Telnet or console sessions, which improves security by making it more difficult for someone to access an unattended session. The syntax for this command is *set logout [time in minutes].* The valid values are from 1 to 10,000 minutes (which should be plenty!). The default value is 20 minutes. A setting of 0 disables auto logout entirely.

The set module Command

This command is used to enable, disable, or name a module in the switch. This command has two variations: *set module [enable | disable] [module number(s)]* and *set module name [module number] [name].* The *set module [enable | disable] [module number(s)]* form of the command enables or disables one or more modules. To specify more than one module, use commas or dashes to separate the module numbers. For instance, typing *set module disable 2,4,5* disables modules 2, 4, and 5. The *set module enable 2-4* command enables modules 2, 3, and 4. The default setting for all modules is *enabled*.

The *set module name [module number] [name]* form of the command sets a name for the module that is viewable in *show* command output. This is useful to quickly pinpoint various modules in the switch. An example of this command follows:

```
Cat5K (enable) set module name 1 SupervisorIIIG
Module name set.
Cat5K (enable) show module 1
Mod Slot Ports Module-Type              Model               Status
--- ---- ----- ------------------------ ------------------- ------
1   1    2     1000BaseX Supervisor IIIG WS-X5550            ok

Mod Module-Name        Serial-Num
--- ------------------ --------------------
1   SupervisorIIIG     12323434567

Mod MAC-Address(es)                            Hw    Fw         Sw
--- ------------------------------------------ ----- ---------- ------
1   00-03-e4-9f-d8-00 to 00-03-e4-9f-db-ff 1.2  5.1(1)     5.4(4)
Cat5K (enable)
```

The set password Command

This command sets the console password. By default, no password is configured. The syntax for this command is *set password [password]*. The password can be from 0 to 30 characters, is case sensitive, and can include spaces.

The set port Command

This command is used to set various port properties. The basic syntax is *set port [target] [subtarget, modifiers or value]*. Table 17-6 describes the primary targets for this command.

Target/Subcommand	Description
set port disable	Disables a port
set port enable	Enables a port
set port duplex	Sets port transmission type (full-/half- duplex)

Table 17-6.   set port *Subcommands*

Target/Subcommand	Description
set port flowcontrol	Sets port traffic flow control
set port name	Sets port name
set port negotiation	Sets port flow control negotiation
set port speed	Sets port transmission speed (4/10/16/100 Mbps)

Table 17-6.   set port *Subcommands* (continued)

The set port disable and set port enable Targets

The *set port disable* and *set port enable* targets simply enable or disable the specified port. The syntax for these commands is *set port [enable | disable] [module number]/[port number]*. So, to enable port 4 on module 2, you would type *set port enable 2/4*. You can also specify multiple ports on multiple modules by appending the comma or dash to the first port number and entering additional ports or modules. For example, *set port enable 2/1,3/1* enables both port 1 on module 2 and port 1 on module 3. Similarly, *set port enable 2/4-2/20* enables ports 4 through 20 on module 2. The default setting for all ports is *enabled*.

The set port duplex Target

The *set port duplex* target configures the duplex settings on a port. The syntax for this target is *set port duplex [module number]/[port number] [full | half]*. Multiple modules and ports can be configured at once by using the dash and comma to separate the values. For instance, *set port 3/1-3/24 full* sets ports 1 through 24 on module 3 to full-duplex. The default setting for Ethernet ports is half-duplex.

The set port flowcontrol Target

The *set port flowcontrol* target enables or disables flow control on a full-duplex Ethernet port. Flow control is negotiable and can be set for both the transmitting and receiving wires individually. The syntax for this target is *set port flowcontrol [module number]/[port number] [receive | send] [on | off | desirable]*. Again, multiple modules and ports can be configured at once by using the dash and comma to separate the values. The *[on | off | desirable]* modifier forces the port to use or not use flow control if set to *on* or *off*, and allows the port to negotiate to on or off depending on support from the remote device if set to *desirable*.

The set port name Target

The *set port name* target simply names the port. This target is sometimes useful to remind you of what the port is connected to. The syntax for this target is *set port name [module number]/[port number] [name].* If no name is entered, the previously configured name is cleared. By default, no ports are named.

The set port negotiation Target

The *set port negotiation* target allows you to enable or disable the Link Negotiation Protocol on Gigabit Ethernet interfaces. The Link Negotiation Protocol allows the port to automatically negotiate flow control and duplex settings on the port. Negotiation is enabled by default on Gigabit links. The syntax for this command is *set port negotiation [module number]/[port number] [enable | disable].*

The set port speed Target

Finally, the *set port speed* target sets the line speed on a given 10/100 Ethernet or 4/16 Token Ring port. The syntax for this target is *set port speed [module number]/[port number] [4 | 10 | 16 | 100 | auto].* The *auto* parameter sets the port to autonegotiate port speed. By default, all dual-speed ports are set to *auto.*

The set prompt Command

This command sets the prompt shown in the switch CLI. The syntax is *set prompt [prompt].* Like the standard IOS, if a system name is set using the *set system name* command, that name is used as the prompt. However, if a different prompt is configured using the *set prompt* command, the *set prompt* value is used.

The set summertime Command

The *set summertime* command informs the switch to update for daylight saving time. For the United States, the easiest way to make use of this command is to type *set summertime enable [zone].* (To disable, simply type *set summertime disable.*) For other countries, you need to know the specific dates or days when daylight saving time starts and ends. To enable daylight saving to take effect on specific days, like the first day of August, type *set summertime recurring [week of month to begin] [day of month to begin] [month to begin] [hh:mm to begin] [week of month to end] [day of month to end] [month to end] [hh:mm to end] [time offset].* To set daylight saving time to begin and end on a specific date every year, type *set summertime date [date to begin] [month to begin] [hh:mm to begin] [date to end] [month to end] [hh:mm to end] [time offset].*

The set system Command

This command sets various basic system parameters, including the baud rate for the console port, the contact person, the country code, the location, and the system name.

To set the console port baud rate, type *set system baud [speed].* The permitted values for the speed are *600, 1200, 2400, 4800, 9600, 19,200,* and *38,400.* The default value is *9600 baud.*

To set the system contact person, type *set system contact [contact name]*. This information is displayed in various *show* commands but is not useful otherwise. If you leave the contact name empty, the system contact is cleared.

To set the system country code, type *set system countrycode [two-digit ISO-3166 country code]*. For the United States, this code is US.

To enter text describing the physical location of the switch, use the *set system location [location]* command. This syntax is used in *show* commands to display the physical location of the switch. To clear this value, simply enter the command with a blank location field.

Finally, to set the system name, type *set system name [name]*. This syntax will also be used as the prompt unless the *set prompt* command is used.

The set time Command

The *set time* command sets the system time. The syntax for this command is *set time [day] [mm/dd/yy] [hh:mm:ss]*. The *[day]* section specifies the day of the week (Monday, Tuesday, Wednesday, and so on), and the *[hh:mm:ss]* section is the time in 24-hour time.

The set timezone Command

This command sets the time zone offsets from UTC. The syntax is *set timezone [name] [offset in hours] [offset in minutes]*.

Common show Commands

The *show* command serves exactly the same purpose on a CatOS-based switch as on a standard IOS–based device. The commands are a bit different, but they generally serve the same purpose. Rather than explaining each of these commands individually, for the commands that were covered in the previous chapter, I will simply describe the syntax and basic use of these commands in Table 17-7. The commands that are specific to the CatOS switches are described in more detail after the table.

Show Target	Description	Syntax (Optional Parameters)
alias	Shows aliases for commands	*show alias [name]*
arp	Shows ARP table	*show arp [ip address] [hostname]*
boot	Shows booting environment variables	*show boot [module number]*

Table 17-7.   *Common* show *Commands*

Show Target	Description	Syntax (Optional Parameters)
cdp	Shows Cisco Discovery Protocol information	*show cdp [neighbors | port]*
config	Shows system configuration	*show config [all | system | module]*
file	Displays contents of file	*show file [device]:[filename]*
flash	Shows system flash information	*show flash [flash device | devices | all | chips | filesys]*
interface	Shows sc0 and sl0 interface information	*show interface*
ip	Shows IP information	*show ip [alias | dns | http]*
log	Shows log information	*show log [module number]*
logging	Shows system logging information	*show logging [buffer]*
mac	Shows MAC information	*show mac [module number]*
module	Shows module info	*show module [module number]*
netstat	Shows network statistics	*show netstat [interface | icmp | ip | stats | tcp | udp]*
port	Shows port information	*show port [numerous optional parameters, described later]*
proc	Shows CPU and processes	*show proc [cpu | mem]*
reset	Shows schedule reset information	*show reset*
summertime	Shows state of summertime information	*show summertime*
system	Shows system information (mainly environmental information, similar to the show environment command)	*show system*
tech-support	Shows system information for tech support	*show tech-support [config | memory | module | port]*

Table 17-7. *Common* show *Commands* (continued)

Show Target	Description	Syntax (Optional Parameters)
test	Shows results of diagnostic tests	*show test [diaglevel	packetbuffer	module number]*
time	Shows time of day	*show time*
timezone	Shows the current time zone and offset	*show timezone*
traffic	Shows traffic information	*show traffic*
users	Shows active admin sessions	*show users*
version	Shows version information	*show version [module number]*

Table 17-7. *Common* show *Commands* (continued)

The show boot Command

This command is used to show boot environment variables, as the following example illustrates (unlike the others, this one is taken from a Catalyst 6500):

```
Cat6500> show boot
BOOT variable = bootflash:cat6000-sup.5-5-4.bin,1;
CONFIG_FILE variable = slot0:switch.cfg

Configuration register is 0x102
ignore-config: disabled
auto-config: non-recurring, overwrite, sync disabled
console baud: 9600
boot: image specified by the boot system commands

Cat6500>
```

The show config Command

This command shows the current switch configuration. Remember, in most switches, the *running config* and *startup config* commands are the same. The standard version of this command simply shows the nondefault configuration. If the *show config all* version is used, the entire switch configuration is shown. If the *show config system* version of this command is used, system-specific information is shown (like the configuration register value). Finally, the *show config [module number]* version shows the configuration for a specific module. See the following examples of these commands:

```
Cat5K (enable) show config
This command shows non-default configurations only.
Use 'show config all' to show both default and
non-default configurations.
......
...............

...............

..

begin
!
# ***** NON-DEFAULT CONFIGURATION *****
!
!
#time: Wed Jul 25 2001, 20:14:18
!
#version 5.4(4)
!
set enablepass $2$0o8Z$5LMNlAart.RBqpzBGlzXm0
set prompt Cat5K
set logout 0
!
#errordetection
set errordetection inband enable
set errordetection memory enable
!
#system
set system name Cat5500
!
#frame distribution method
set port channel all distribution mac both
!
#ip
set interface sc0 1 10.1.1.254/255.255.255.0 10.1.1.255
```

```
!
#command alias
set alias clban clear banner motd
!
#set boot command
set boot auto-config recurring
clear boot auto-config
!
#mls
set mls nde disable
!
#port channel
set port channel 3/1-4 22
set port channel 3/5-8 23
set port channel 3/9-12 24
set port channel 3/13-16 25
set port channel 3/17-20 26
set port channel 3/21-24 27
!
# default port status is enable
!
!
#module 1 : 2-port 1000BaseX Supervisor IIIG
set module name 1 SupervisorIIIG
!
#module 2 : 24-port 10/100BaseTX Ethernet
!
#module 3 : 24-port 10/100BaseTX Ethernet
!
#module 4 : 2-port MM OC-3 Dual-Phy ATM
!
#module 5 empty
!
#module 6 empty
!
#module 7 empty
!
#module 8 empty
!
#module 9 empty
!
#module 10 empty
```

```
!
#module 11 empty
!
#module 12 empty
!
#module 13 empty
!
#module 15 empty
!
#module 16 empty
end
Cat5K (enable) show config all
......
..............
..............
...............
.............
..

begin
!
# ***** ALL (DEFAULT and NON-DEFAULT) CONFIGURATION *****
!
!
#time: Wed Jul 25 2001, 20:14:29
!
#version 5.4(4)
!
set password $2$FMFQ$HfZR5DUszVHIRhrz4h6V70
set enablepass $2$0o8Z$5LMNlAart.RBqpzBGlzXm0
set prompt Cat5K
set length 24 default
set logout 0
set banner motd ^C^C
!
#test
set test diaglevel minimal
set test packetbuffer sun 03:30
set test packetbuffer enable
!
#errordetection
set errordetection inband enable
```

```
set errordetection memory enable
!
#system
set system baud 9600
set system modem disable
set system name Cat5500
set system location
set system contact
set system countrycode
set traffic monitor 100
!
#frame distribution method
>DETAIL OMITTED<
end
Cat5K (enable) show config system
This command shows non-default configurations only.
Use 'show config system all' to show both default and
non-default configurations
.
......
..

begin
!
# ***** NON-DEFAULT CONFIGURATION *****
!
!
#time: Wed Jul 25 2001, 20:15:15
!
#version 5.4(4)
!
set enablepass $2$0o8Z$5LMNlAart.RBqpzBGlzXm0
set prompt Cat5K
set logout 0
!
#errordetection
set errordetection inband enable
set errordetection memory enable
!
#system
set system name Cat5500
!
```

```
#frame distribution method
set port channel all distribution mac both
!
#ip
set interface sc0 1 10.1.1.254/255.255.255.0 10.1.1.255

!
#command alias
set alias clban clear banner motd
!
#set boot command
set boot auto-config recurring
clear boot auto-config
!
#mls
set mls nde disable
!
#port channel
set port channel 3/1-4 22
set port channel 3/5-8 23
set port channel 3/9-12 24
set port channel 3/13-16 25
set port channel 3/17-20 26
set port channel 3/21-24 27
end
Cat5K (enable) show config 2
This command shows non-default configurations only.
Use 'show config <mod> all' to show both default and
non-default configurations.

begin
!
# ***** NON-DEFAULT CONFIGURATION *****
!
!
#time: Wed Jul 25 2001, 20:15:31
!
# default port status is enable
!
!
#module 2 : 24-port 10/100BaseTX Ethernet
end
```

```
Cat5K (enable) show config 4
This command shows non-default configurations only.
Use 'show config <mod> all' to show both default and
non-default configurations.

begin
!
# ***** NON-DEFAULT CONFIGURATION *****
!
!
#time: Wed Jul 25 2001, 20:15:37
!
# default port status is enable
!
!
#module 4 : 2-port MM OC-3 Dual-Phy ATM
end
```

The show file Command

This command shows information on files contained in the flash file system. This command has two variations: one for the Catalyst 5500 switches with the Supervisor Engine IIG or IIIG, and one for all other switches. On the 5500 with the G supervisor engines, the command syntax is *show file [cfg1 | cfg2]*:

```
Cat5K (enable) show file cfg1

begin
!
# ***** ALL (DEFAULT and NON-DEFAULT) CONFIGURATION *****
!
!
#time: Sun Jul 22 2001, 17:25:10
!
#version 5.4(4)
!
set password $2$FMFQ$HfZR5DUszVHIRhrz4h6V70
set enablepass $2$vX0Q$gSbNcUYHgN24gMyrW5U7E0
set prompt Cat5500
>DETAIL OMITTED<
```

On all other switches, the command syntax is *show file [device]:[filename]*. This is shown in the following example:

```
Cat5K2> (enable) show file slot0:cfgfile


begin
#version 5.3(0.11)BOU-Eng
!
set password $1$FMFQ$HfZR5DUszVHIRhrz4h6V70
set enablepass $1$FMFQ$HfZR5DUszVHIRhrz4h6V70
set prompt Cat5K2>
set length 24 default
...
Cat5K2> (enable)
```

The show log Command

This command shows the error log for one or all modules. The error log also includes useful nonerror-related information, including the reset count and history, as well as a history of NVRAM configuration changes. The output from this command is shown in the following example:

```
Cat5K (enable) show log

Network Management Processor (ACTIVE NMP) Log:
  Reset count:   18
  Re-boot History:    Jul 23 2001 16:29:40 3, Jul 23 2001 16:19:46 3
                      Jul 22 2001 17:55:22 3, Jul 22 2001 17:52:03 3
                      Jul 22 2001 17:48:53 3, Jul 22 2001 17:45:02 3
                      Jul 22 2001 17:39:18 3, Jul 22 2001 17:33:59 3
                      Jul 22 2001 17:29:26 3, Jul 22 2001 17:25:49 3
  Bootrom Checksum Failures:      0    UART Failures:                0
  Flash Checksum Failures:        0    Flash Program Failures:       0
  Power Supply 1 Failures:        0    Power Supply 2 Failures:      0
  Swapped to CLKA:                0    Swapped to CLKB:              0
  Swapped to Processor 1:         0    Swapped to Processor 2:       0
  DRAM Failures:                  0

  Exceptions:                     0

  Last software reset by user: 7/23/2001,16:29:32

Heap Memory Log:
```

```
Corrupted Block = none

NVRAM log:

01. 7/22/2001,17:19:18: initBootNvram:Auto config detected
02. 7/22/2001,17:21:23: StartupConfig:Auto config started
03. 7/22/2001,17:26:13: initBootNvram:Auto config detected
04. 7/22/2001,17:28:18: StartupConfig:Auto config started
05. 7/22/2001,17:34:24: initBootNvram:Auto config detected
06. 7/22/2001,17:36:28: StartupConfig:Auto config started
07. 7/22/2001,17:45:27: initBootNvram:Auto config detected
08. 7/22/2001,17:47:31: StartupConfig:Auto config started
09. 7/22/2001,17:52:28: initBootNvram:Auto config detected
10. 7/22/2001,17:54:32: StartupConfig:Auto config started
11. 7/23/2001,16:20:11: initBootNvram:Auto config detected
12. 7/23/2001,16:22:15: StartupConfig:Auto config started

Module 2 Log:
  Reset Count:    18
  Reset History: Mon Jul 23 2001, 16:31:17
                 Mon Jul 23 2001, 16:21:23
                 Sun Jul 22 2001, 17:56:59
                 Sun Jul 22 2001, 17:53:40


Module 3 Log:
  Reset Count:    18
  Reset History: Mon Jul 23 2001, 16:31:21
                 Mon Jul 23 2001, 16:21:27
                 Sun Jul 22 2001, 17:57:03
                 Sun Jul 22 2001, 17:53:44


Module 4 Log:
  Reset Count:    20
  Reset History: Mon Jul 23 2001, 16:32:09
                 Mon Jul 23 2001, 16:22:14
                 Sun Jul 22 2001, 17:57:50
                 Sun Jul 22 2001, 17:54:32
```

The show mac Command

This command shows MAC port counters on the switch, which list the received layer 2 packets and sort them by type (unicast, multicast, or broadcast). The output of this command is shown here:

```
Cat5K (enable) show mac

Port      Rcv-Unicast           Rcv-Multicast          Rcv-Broadcast
--------  --------------------  --------------------   ----------------
 1/1                        0                     0                   0
 1/2                        0                     0                   0
 2/1                    94516                 96698                 357
 2/2                        0                     0                   0
 2/3                        0                     0                   0
 2/4                        0                     0                   0
>DETAIL OMITTED<
```

The show module Command

This command shows basic information on the modules installed in the switch,
including the MAC addresses, serial numbers, and names. The output of this
command is shown here:

```
Cat5K (enable) show module
Mod Slot Ports Module-Type               Model               Status
--- ---- ----- ------------------------- ------------------- ------
1   1    2     1000BaseX Supervisor IIIG WS-X5550            ok
2   2    24    10/100BaseTX Ethernet     WS-X5224            ok
3   3    24    10/100BaseTX Ethernet     WS-X5225R           ok
4   4    2     MM OC-3 Dual-Phy ATM      WS-X5167            ok


Mod Module-Name         Serial-Num
--- ------------------- --------------------
1   SupervisorIIIG      30802010212
2                       55577000223
3                       35500470207
4                       00213402185


Mod MAC-Address(es)                           Hw     Fw         Sw
--- ----------------------------------------- ------ ---------- ------
1   00-03-e4-9f-d8-00 to 00-03-e4-9f-db-ff 1.2    5.1(1)     5.4(4)
2   00-01-c9-ac-cd-d0 to 00-01-c9-ac-cd-e7 1.5    3.1(1)     5.4(4)
3   00-01-63-56-12-d8 to 00-01-63-56-12-ef 3.3    4.3(1)     5.4(4)
4   00-10-7b-21-18-40 to 00-10-7b-21-18-47 3.0    5.1(1)     12.0(4a)W5(10),
Cat5K (enable)
```

The show netstat Command

This command shows a wide variety of complied network statistics, including ICMP, IP, TCP, and management statistics. The output of several versions of this command is shown in the following example:

```
Cat5K (enable) show netstat ?
  icmp                        Show ICMP statistics
  interface                   Show interface statistics
  ip                          Show IP statistics
  routes                      Show IP routing table
  stats                       Show TCP, UDP, IP, ICMP statistics
  tcp                         Show TCP statistics
  udp                         Show UDP statistics
  <cr>
Cat5K (enable) show netstat
Active Internet connections (including servers)
Proto Recv-Q Send-Q  Local Address           Foreign Address          (state)
tcp        0      0  *.7161                  *.*                      LISTEN
tcp        0      0  *.23                    *.*                      LISTEN
udp        0      0  *.*                     *.*
udp        0      0  *.161                   *.*
Cat5K (enable) show netstat icmp
icmp:
        Redirect enabled
        0 calls to icmp_error
        0 errors not generated 'cuz old message was icmp
        0 messages with bad code fields
        0 messages < minimum length
        0 bad checksums
        0 messages with bad length
        Input histogram:
                #9: 359
        0 message responses generated
Cat5K (enable) show netstat ip
ip:
        3541 total packets received
        0 bad header checksums
        0 with size smaller than minimum
        0 with data size < data length
        0 with header length < data size
        0 with data length < header length
        0 fragments received
        0 fragments dropped (dup or out of space)
        0 fragments dropped after timeout
        0 packets forwarded
        0 packets not forwardable
```

```
        0 redirects sent
Cat5K (enable) show netstat tcp
tcp:
        30 packets sent
                2 data packets (100 bytes)
                0 data packets (0 bytes) retransmitted
                2 ack-only packets (1 delayed)
                0 URG only packets
                0 window probe packets
                26 window update packets
                0 control packets
        3183 packets received
                3 acks (for 101 bytes)
                3145 duplicate acks
                0 acks for unsent data
                34 packets (46468 bytes) received in-sequence
                0 completely duplicate packets (0 bytes)
                0 packets with some dup. data (0 bytes duped)
                0 out-of-order packets (0 bytes)
                0 packets (0 bytes) of data after window
                0 window probes
                0 window update packets
                0 packets received after close
                0 discarded for bad checksums
                0 discarded for bad header offset fields
                0 discarded because packet too short
        0 connection requests
        1 connection accept
        1 connection established (including accepts)
        0 connections closed (including 0 drops)
        0 embryonic connections dropped
        3 segments updated rtt (of 3 attempts)
        0 retransmit timeouts
                0 connections dropped by rexmit timeout
        0 persist timeouts
        3145 keepalive timeouts
                3145 keepalive probes sent
                0 connections dropped by keepalive
Cat5K (enable) show netstat udp
udp:
        0 incomplete headers
        0 bad data length fields
        0 bad checksums
        0 socket overflows
        0 no such ports
Cat5K (enable) show netstat interface
Interface          InPackets InErrors OutPackets OutErrors
```

```
sl0                         0        0           0           0
sc0                    100957        0      733484           0
Interface Rcv-Octet              Xmit-Octet
--------- --------------------   --------------------
sc0        7961870               63835448
sl0              0               0
Interface Rcv-Unicast            Xmit-Unicast
--------- --------------------   --------------------
sc0         100539               733416
sl0              0               0
```

The show port Command

This command is extremely versatile and is used to show nearly anything you might want to know about a port. Because the useful targets for this command vary based on the actual type of port, and most of the options are self-explanatory, rather than attempting to explain every possible option, I instead provide some detailed examples of its use with explanations following each example.

```
Cat5K (enable) show port ?
  broadcast                    Show port broadcast information
  cdp                          Show port CDP information
  capabilities                 Show port capabilities
  channel                      Show port channel information
  counters                     Show port counters
  fddi                         Show port FDDI information
  flowcontrol                  Show port traffic flowcontrol
  filter                       Show Token Ring port filtering information
  ifindex                      Show port IfIndex information
  mac                          Show port MAC counters
  negotiation                  Show port flowcontrol negotiation
  protocol                     Show port protocol membership
  qos                          Show port QoS information
  security                     Show port security information
  spantree                     Show port spantree information
  status                       Show port status
  trap                         Show port trap information
  trunk                        Show port trunk information
  <mod>                        Module number
  <mod/port>                   Module number and Port number(s)
  <cr>
```

This example simply displays the command help, which includes a short description of what each target does. With a large command like *show port,* online help can be your best friend.

```
Cat5K (enable) show port
Port  Name              Status      Vlan        Level  Duplex Speed Type
----- ----------------- ----------- ----------- ------ ------ ----- ------------
 1/1                    notconnect 1            normal  full  1000 No GBIC
 1/2                    notconnect 1            normal  full  1000 No GBIC
 2/1                    connected  1            normal a-half a-10 10/100BaseTX
 2/2                    connected  1            normal a-half a-10 10/100BaseTX
 2/3                    notconnect 1            normal  auto  auto 10/100BaseTX
  >DETAIL OMITTED<
 2/24                   notconnect 1            normal  auto  auto 10/100BaseTX
 3/1                    notconnect 1            normal  auto  auto 10/100BaseTX
  >DETAIL OMITTED<
 3/24                   notconnect 1            normal  auto  auto 10/100BaseTX
 4/1                    notconnect trunk        normal  full   155 OC3 MMF ATM
 4/2                    notconnect trunk        normal  full   155 OC3 MMF ATM

Port  Security Violation Shutdown-Time Age-Time Max-Addr Trap     IfIndex
----- -------- --------- ------------- -------- -------- -------- -------
 2/1  disabled shutdown              0        0        1 disabled      28
  >DETAIL OMITTED<
 3/24 disabled shutdown              0        0        1 disabled      75

Port  Num-Addr Secure-Src-Addr   Age-Left Last-Src-Addr  Shutdown/Time-Left
----- -------- ----------------- -------- -------------- ------------------
 2/1         0                 -        -              -        -          -
  >DETAIL OMITTED<
 3/24        0                 -        -              -        -          -

Port  Trap     IfIndex
----- -------- -------
 1/1  disabled 3
 1/2  disabled 4

Port  ifIndex
----- -------
 4/1  78
 4/2  79

Port     Broadcast-Limit Broadcast-Drop
-------- --------------- --------------
 1/1                   -              0
  >DETAIL OMITTED<
 3/24                  -              0
```

Port	Send FlowControl		Receive FlowControl		RxPause	TxPause	Unsupported
	admin	oper	admin	oper			opcodes
1/1	desired	off	off	off	0	0	0
1/2	desired	off	off	off	0	0	0
3/1	off	off	on	on	0	0	0
>DETAIL OMITTED<							
3/24	off	off	on	on	0	0	0

Port	Align-Err	FCS-Err	Xmit-Err	Rcv-Err	UnderSize
1/1	0	0	0	0	0
1/2	0	0	0	0	0
>DETAIL OMITTED<					
3/24	0	0	0	0	0

Port	Single-Col	Multi-Coll	Late-Coll	Excess-Col	Carri-Sen	Runts	Giants
1/1	0	0	0	0	0	0	0
>DETAIL OMITTED<							
3/24	0	0	0	0	0	0	0

```
Last-Time-Cleared
--------------------------
Mon Jul 23 2001, 16:30:14

Cat5K (enable) show port cdp
CDP                : enabled
Message Interval   : 60
Hold Time          : 180
Version            : V2
```

Port	CDP Status
1/1	enabled
1/2	enabled
>DETAIL OMITTED<	
4/1-2	enabled

This example uses a basic *show port* command, which prompts a 30-page output, most of which I have omitted due to space constraints. This includes all of the *show port* targets for all interfaces. To see the output for a single port or module, append the module and port number to the end of the command.

```
Cat5K (enable) show port capabilities
Model                   WS-X5550
Port                    1/1
Type                    No GBIC
Speed                   1000
Duplex                  full
Trunk encap type        802.1Q,ISL
Trunk mode              on,off,desirable,auto,nonegotiate
Channel                 no
Broadcast suppression   percentage(0-100)
Flow control
receive-(off,on,desired),send-(off,on,desired)
Security                no
Membership              static
Fast start              yes
QOS scheduling          rx-(none),tx-(none)
CoS rewrite             no
ToS rewrite             no
Rewrite                 no
UDLD                    yes
SPAN                    source,destination

----------------------------------------------------------------

Model                   WS-X5550
Port                    1/2
Type                    No GBIC
Speed                   1000
Duplex                  full
Trunk encap type        802.1Q,ISL
Trunk mode              on,off,desirable,auto,nonegotiate
Channel                 no
Broadcast suppression   percentage(0-100)
Flow control
receive-(off,on,desired),send-(off,on,desired)
Security                no
Membership              static
Fast start              yes
QOS scheduling          rx-(none),tx-(none)
CoS rewrite             no
ToS rewrite             no
Rewrite                 no
UDLD                    yes
SPAN                    source,destination
```

```
-------------------------------------------------------------
Model                    WS-X5224
Port                     2/1
Type                     10/100BaseTX
Speed                    auto,10,100
Duplex                   half,full
Trunk encap type         no
Trunk mode               off
Channel                  no
Broadcast suppression    pps(0-150000),percentage(0-100)
Flow control             no
Security                 yes
Membership               static,dynamic
Fast start               yes
QOS scheduling           rx-(none),tx-(none)
CoS rewrite              no
ToS rewrite              no
Rewrite                  no
UDLD                     yes
SPAN                     source,destination


-------------------------------------------------------------
Model                    WS-X5224
Port                     2/2
 >DETAIL OMITTED<
```

This example uses a *show port capabilities* command. This shows the port specifics, including speeds, duplex, type, QoS settings, encapsulations, and other details. Again, the base form of the command shows all of this information for every port. To see the output for a single port or module, append the module and port number to the end of the command.

```
Cat5K (enable) show port counters 4/1

Last-Time-Cleared
-------------------------
Mon Jul 23 2001, 16:30:14
Cat5K (enable) show port 2/1
Port  Name             Status     Vlan       Level  Duplex Speed Type
----- ---------------- ---------- ---------- ------ ------ ----- ------------
 2/1                   connected  1          normal a-half a-10  10/100BaseTX
```

```
Port  Security Violation Shutdown-Time Age-Time Max-Addr Trap     IfIndex
----- -------- --------- ------------- -------- -------- -------- -------
2/1   disabled shutdown              0        0        1 disabled      28

Port  Num-Addr Secure-Src-Addr   Age-Left Last-Src-Addr  Shutdown/Time-Left
----- -------- ---------------   -------- -------------  ------------------
2/1          0               -          -             -          -        -

Port     Broadcast-Limit Broadcast-Drop
-------- --------------- --------------
2/1                    -              0

Port  Align-Err FCS-Err   Xmit-Err   Rcv-Err    UnderSize
----- --------- --------- ---------- ---------- ---------
2/1           0         0          0          0         0

Port  Single-Col Multi-Coll Late-Coll  Excess-Col Carri-Sen Runts  Giants
----- ---------- ---------- ---------- ---------- --------- ------ ---------
2/1            0          0          0          0         0      0         0

Last-Time-Cleared
------------------------
Mon Jul 23 2001, 16:30:14
```

The final example shows the counters for a specific port, in this case, an ATM OC-3 port on module 4 and a 10/100 Ethernet port on module 2. This shows statistical data for these ports (or simply the last clear time for the ATM module), including error statistics and security violations.

The show test Command

This command shows detailed information on the tests performed by the switch and the results of those tests. Remember, because the CatOS switches are typically higher-end switches, redundancy and reliability are primary concerns. The *show test* command lets you view all of the detailed self-tests that the switch performs, as shown in the following examples:

```
Cat5K (enable) show test

Diagnostic mode: minimal    (mode at next reset: minimal)

Environmental Status (. = Pass, F = Fail, U = Unknown, N = Not Present)
  PS (3.3V):   N   PS (12V): .   PS (24V):   .   PS1: .     PS2: .
  PS1 Fan:      .   PS2 Fan : .   Clock(A/B): A   Chassis-Ser-EEPROM: .
  Temperature: .   Fan:        .
```

```
Module 1 : 2-port 1000BaseX Supervisor IIIG
Network Management Processor (NMP) Status: (. = Pass, F = Fail, U = Unknown)
  ROM:   .   Flash-EEPROM: .   Ser-EEPROM: .   NVRAM: .   MCP Comm: .

  EARL III Status :
        DisableIndexLearnTest:          .
        DontLearnTest:                  .
        DisableNewLearnTest:            .
        ConditionalLearnTest:           .
        MonitorColorFloodTest:          .
        EarlTrapTest:                   .
        StaticMacAndTypeTest:           .
        BadDvlanTest:                   .
        BadBpduTest:                    .
        IndexMatchTest:                 .
        ProtocolTypeTest:               .
        IgmpTest:                       .
        SourceMissTest:                 .
        SourceModifiedTest:             .
        ArpaToArpaShortcutTest:         .
        ArpaToSnapShortcutTest:         .
        SnapToArpaShortcutTest:         .
        SnapToSnapShortcutTest:         .
        SoftwareShortcutTest:           .
        MulticastExpansionTest:         .
        DontShortcutTest:               .
        ShortcutTableFullTest:          .
        ArpaToArpaShortcutTest(IPX): .
        ArpaToSnapShortcutTest(IPX): .
        ArpaToSapShortcutTest(IPX):  .
        ArpaToRawShortcutTest(IPX):  .
        SnapToArpaShortcutTest(IPX): .
        SnapToSnapShortcutTest(IPX): .
        SnapToSapShortcutTest(IPX):  .
        SnapToRawShortcutTest(IPX):  .
        SapToArpaShortcutTest(IPX):  .
        SapToSnapShortcutTest(IPX):  .
        SapToSapShortcutTest(IPX):   .
        SapToRawShortcutTest(IPX):   .
        RawToArpaShortcutTest(IPX):  .
        RawToSnapShortcutTest(IPX):  .
        RawToSapShortcutTest(IPX):   .
        RawToRawShortcutTest(IPX):   .
        SoftwareShortcutTest(IPX):   .
        DontShortcutTest(IPX):       .
        ShortcutTableFullTest(IPX):  .
```

```
Line Card Diag Status for Module 1  (. = Pass, F = Fail, N = N/A)
 CPU          : .    Sprom    : .    Bootcsum : .    Archsum  : N
 RAM          : .    LTL      : .    CBL      : .    DPRAM    : .    ARBITER:N
 Phoenix      : .    Pkt Bufs : .    Repeater : N    FLASH    : N
 Phoenix      : . TrafficMeter: . UplinkSprom : . PhoenixSprom: .

 PHOENIX Status :
  Ports T1  R1  T2  R2
 --------------------
          .   .   .   .

 GIGMAC Status :
  Ports 1  2
 -----------
          .  .

 PHOENIX Port Status :
  Ports 9    17   18    19    20    21    22
      INBAND A->B B->A B->C C->B A->C C->A
 ------------------------------------------
          .    .    .    .     .     .     .

 Packet Buffer Status :
  Ports 1  2
 -----------
          .  .

 PHOENIX Packet Buffer Status :
  Ports INBAND A<->B B<->C A<->C
 -------------------------------
          .     .      .     .

 Loopback Status [Reported by Module 1] :
  Ports  1  2  9
 --------------
           .  .  .


Cat5K (enable) show test diaglevel
Diagnostic mode at last bootup : minimal
Diagnostic mode at next reset  : minimal

Cat5K (enable) show test packetbuffer status
Last packet buffer test details
Test Type           : never run
Test Started        : 00:00:00 Month 00 0000
Test Finished       : 00:00:00 Month 00 0000
```

The show traffic Command

This command shows the traffic statistics for the switching bus or busses in a CatOS switch. This command can be very useful for identifying a backplane oversubscription. The output is simple and easy to read, as shown in the following example (from a Catalyst 5500, with three switching backplanes):

```
Cat5K (enable) show traffic
Threshold: 100%
Switching-Bus Traffic Peak Peak-Time
------------- ------- ---- ------------------------
A               0%      0% Mon Jul 23 2001, 16:30:14
B               0%      0% Mon Jul 23 2001, 16:30:14
C               0%      0% Mon Jul 23 2001, 16:30:14
```

Summary

This chapter examined the most common generic commands for devices based on the CatOS, or set-based, IOS. Although most of your focus will be on standard IOS devices, the commands presented in this chapter will help ground you in the command use for the higher-end CatOS-based switches, allowing you to work with nearly any modern Cisco device.

The Complete Reference

Cisco

Part III

Cisco LAN Switching

In this part of the book, we discuss all aspects of Cisco LAN switching, from basic switch configuration to MLS. Chapter 18 covers basic configuration tasks for both the standard and set-based IOS. Chapter 19 examines standard layer 2 switching, including STP and VLAN configuration. In Chapter 20, we explore layer 3 switching using MLS; and, finally, in Chapter 21, we delve into layer 4 switching using SLB.

Chapter 18

Basic Switch Configuration

his chapter covers some of the basic configuration tasks common to nearly every switch. Through examples and explanations, we will take a blank new standard IOS switch and a blank new CatOS-based switch (a Catalyst 5500), and configure everything required for basic connectivity and management. The standard IOS switches used are a Catalyst 1924 and Catalyst 2948G Layer 3. Even though, technically, the 1924 runs a custom OS, it is very similar to standard IOS.

Configuring the General System Information

One of the first tasks involved with configuring a switch is setting basic information, such as the switch's name, management IP address, and the time. (The administrative contact and location variables come in handy in a large environment, but you seldom need to set them in smaller environments.) We begin this section by configuring a 1900 switch and then move into configuring the 5500 switch.

Standard IOS Configuration

The standard IOS devices are the main focus of this book, so we will begin our configuration with a standard IOS switch. Immediately after booting the switch, you will be greeted with one of two prompts: either a setup mode prompt or a menu prompt. The following example shows a setup mode prompt from a Catalyst 2948 switch:

```
        --- System Configuration Dialog ---

Continue with configuration dialog? [yes/no]:
```

An example of the menu-based prompt from a 1924 switch is as follows:

```
Catalyst 1900 Management Console
Copyright (c) Cisco Systems, Inc.  1993-1999
All rights reserved.
Enterprise Edition Software
Ethernet Address:        00-03-E3-05-C6-00

PCA Number:              73-3121-04
PCA Serial Number:       FCG1726A875
Model Number:            WS-C1924-EN
System Serial Number:    FCD1263B6DS
Power Supply S/N:        PHI273845HM
PCB Serial Number:       FAB0945C243,73-3121-04
-------------------------------------------------
```

```
1 user(s) now active on Management Console.

        User Interface Menu

    [M] Menus
    [K] Command Line
    [I] IP Configuration
    [P] Console Password

Enter Selection:
```

The [K] *option is available only on the Enterprise Edition 1900 series switching software. If you have the Standard Edition, you must use the menu system.*

Either way, you want to get to the command prompt. On the standard IOS switches, enter *no* to the setup prompt; and on the 1900 series, issue *[K]*. A standard user mode prompt will appear, similar to the one shown here:

```
        User Interface Menu

    [M] Menus
    [K] Command Line
    [I] IP Configuration
    [P] Console Password

Enter Selection:   K


    CLI session with the switch is open.
    To end the CLI session, enter [Exit].

 >
```

First, you need to get into privileged mode, and then you need to get into global configuration mode because all of the system configuration commands are available in global config mode. The password on a new switch is not set, so all you need to do is issue the *enable* command and the *configure terminal* command, as shown here:

```
>enab
#conf t
```

```
Enter configuration commands, one per line.  End with CNTL/Z
(config)#
```

Note   *I used the abbreviations* enab *and* conf t *for the* enable *command and the* configure terminal *command, respectively. This is to get you used to the idea of using command abbreviations because that is what you will most likely see in real life.*

Once in global config mode, set the switch name and IP address by using the *hostname* and *ip address* commands, as shown here:

```
(config)#hostname 1900A
1900A(config)#ip address 10.1.10.1 255.255.255.0
1900A(config)#
```

Note   *In most cases, the switch will not tell you when you get a command correct—only when you get it wrong. This scheme can take some adaptation if you are more familiar with the DOS CLI. Use* show *commands to verify your entries.*

Next, you may want to go back and verify your settings. Return to enable mode by using CTRL-Z or typing *EXIT*, and run the appropriate *show* commands (in this case, *show ip* and *show running-config*), as displayed in the following example:

```
1900A(config)# ^Z
1900A#show ip
IP Address: 10.1.10.1
Subnet Mask: 255.255.255.0
Default Gateway: 0.0.0.0
Management VLAN:  1
Domain name:
Name server 1: 0.0.0.0
Name server 2: 0.0.0.0
HTTP server : Enabled
HTTP port :  80
RIP : Enabled
1900A#show running-config
Building configuration...
Current configuration:
!
!
!
!
```

```
!
!
!
hostname "1900A"
!
!
!
!
ip address 10.1.10.1 255.255.255.0
!
!
!
!
!
interface Ethernet 0/1

!

--More--
```

If you made a mistake, you can simply reissue the command (in the case of the *hostname* command), or you can use the *no* version of the command and retype the appropriate command (in the case of the *ip address* command). The following example illustrates:

```
1900A(config)#no ip address
1900A(config)#ip address 10.1.1.1 255.255.255.0
1900A(config)#hostname cat1924A
cat1924A(config)#
```

Note

On the 1900 series and some other switches, using the no ip address *command is not required to set a new system IP address. However, this habit is good to get into because on a router, it* is *required.*

Now you might want to set the time on the switch: you can do so manually or by using the Network Time Protocol (NTP). NTP allows you to synchronize all of your switches and routers to a central time server, making sure that all devices issue logging messages and other timestamped information by using the same "clock." NTP is typically a much better approach to time synchronization, and, luckily, it isn't very difficult to configure. Unfortunately, lower-end switches, such as the 1900 series, may not support NTP. In fact, the 1900 series does not support setting the time and date at all! It timestamps messages with the elapsed time since the last reboot.

On higher-end switches, you generally want to enable NTP and configure the switch to synchronize its clock with a centralized NTP server. The server can be a stand-alone, dedicated device (like the DATUM Tymserve), a PC running NTP server software, or even another router or switch. Typically, the most accurate results are from dedicated devices; but in a pinch, you can have all of your Cisco devices sync to a central switch or router. For a standard IOS device, all you need to enable synchronization to a central time server is the *ntp server [ip address of NTP server]* command.

Although it isn't suggested, you can set up a switch as an NTP server by using the *ntp master [stratum]* command. The *stratum* is a number from 1 to 15 that, in essence, rates the "trustworthiness" of the server. The lower the stratum number, the more trustworthy the server. (Although this explanation is a bit of an oversimplification, it works well in most scenarios.) Therefore, you need to be very careful when assigning a stratum of 1 to any device, especially a router or switch, because they could potentially override a more accurate time server.

To configure a switch's time and date manually, use the *clock set* command, as described in Chapter 15. The following example shows both NTP configuration and manual configuration:

```
2948G#show clock
*18:37:04.104 UTC Mon Jul 30 2001
2948G#clock set 9:00:00 Jan 1 2001
2948G#show clock
09:00:04.123 UTC Mon Jan 1 2001
2948G#configure terminal
Enter configuration commands, one per line.  End with CNTL/Z.
2948G(config)#ntp server 192.168.1.2
2948G(config)#^Z
2948G#show clock
.09:01:30.651 UTC Mon Jan 1 2001
2948G#sh clock
18:37:04.104 EST Mon Jul 30 2001
```

Note *You may have noticed that the switch took some time to synchronize with the NTP server. In some cases, it may take up to 30 minutes to achieve synchronization.*

As discussed at the beginning of the chapter, setting the location and administrative contact are useful in a large organization that has many of the same switches, but it is generally unnecessary in smaller organizations. To set the administrative contact, use the *snmp-server contact* command, as shown in this example:

```
cat1924A(config)#snmp-server contact John Doe
```

To set the location for the switch, use the *snmp-server location* command, as shown next:

```
cat1924A(config)#snmp-server location Corp HQ Closet 5
```

When finished, the *show snmp contact* and *show snmp location* commands should display your settings, like so:

```
cat1924A#show snmp contact
John Doe
cat1924A#show snmp location
Corp HQ Closet 5
cat1924A#
```

CatOS Configuration

To configure the general system information on a CatOS-based switch, the commands are a bit different. First, when you boot the switch, you will be greeted with a simple prompt, like this:

```
Console>
```

You need to type *enable* to set switch parameters. Because no default enable password is set, the switch simply enters into enable mode with no further typing required, as shown here:

```
Console enable
Console (enable)
```

To set the name on a CatOS switch, use the *set system name* command. To set the IP address for the in-band (Ethernet) management interface, use the *set interface sc0* command. Both of these are shown next:

```
5500A (enable) set system name Cat5K
System name set.
Cat5K (enable) set interface sc0 10.1.1.254 255.255.255.0
Interface sc0 IP address and netmask set.
Cat5K (enable) show interface
sl0: flags=51<UP,POINTOPOINT,RUNNING>
        slip 0.0.0.0 dest 0.0.0.0
sc0: flags=63<UP,BROADCAST,RUNNING>
vlan 1 inet 10.1.1.254 netmask 255.255.255.0 broadcast 10.1.1.255
Cat5K (enable)
```

CISCO LAN SWITCHING

You can set the date and time on a CatOS switch either manually or by using NTP, like with the standard IOS switches. To use NTP, issue the *set ntp server* command. To set the time manually, use the *set time* command. Both of these commands are shown in the following example:

```
Cat5K (enable) set ntp server 10.1.1.1
NTP server 10.1.1.1 added
Cat5K (enable) show ntp

Current time: Tue Jul 31 2001, 02:42:25
Timezone: '', offset from UTC is 0 hours
Summertime: '', disabled
Last NTP update:
Broadcast client mode: disabled
Broadcast delay: 3000 microseconds
Client mode: disabled
Authentication: disabled

NTP-Server                                    Server Key
--------------------------------------    ----------
10.1.1.1                                         -

Key Number    Mode
----------    ---------
Cat5K (enable) set time wednesday 8/1/2001 14:44:00
Wed Aug 1 2001, 14:44:00
Cat5K (enable) show time
Wed Aug 1 2001, 14:44:06
Cat5K (enable)
```

Finally, to set the contact and location information for the switch, use the *set system contact* command and the *set system location* command. You can then view this information with the *show system* command, like so:

```
Cat5K (enable) set system contact John Doe
System contact set.
Cat5K (enable) set system location Corp HQ Closet 5
System location set.
Cat5K (enable) show system
PS1-Status PS2-Status
---------- ----------
ok         ok
```

```
Fan-Status Temp-Alarm Sys-Status Uptime d,h:m:s Logout
---------- ---------- ---------- -------------- ---------
ok         off        ok           8,19:42:30   none

PS1-Type      PS2-Type
------------ ------------
WS-C5508      WS-C5508

Modem   Baud  Traffic Peak Peak-Time
------- ----- ------- ---- ------------------------
disable 9600  0%       0% Mon Jul 23 2001, 16:30:14

System Name            System Location          System Contact
---------------------- ------------------------ --------------
Cat5K                  Corp HQ Closet 5         John Doe
Cat5K (enable)
```

Configuring Logging and Passwords

Logging can be a very important tool when troubleshooting problems on any switch; however, the nearly endless barrage of syslog messages to the console and the default setting of unbuffered logging can lead to some headaches. Therefore, setting the logging parameters of your switch is the next priority. In addition, setting the enable, console, and telnet passwords is also a high priority because you never know who will want to hack your switches.

Standard IOS Configuration

Configuring the logging parameters on the standard IOS is a fairly simple process. First, you generally want to make sure syslog messages are buffered, allowing you to view them from a remote (i.e., nonconsole) connection. The two methods of buffering are to a syslog server and to the local device. To buffer to a syslog server, you must first install and configure a syslog server application on a network host. Then you can enter the *logging <IP address>* command to enable logging to the server. This is shown in the following example:

```
2948GL3(config)#logging 10.1.1.1
2948GL3(config)#^Z
2948GL3#show logging
Syslog logging: enabled (0 messages dropped, 0 flushes, 0 overruns)
```

```
     Console logging: disabled
     Monitor logging: level debugging, 0 messages logged
     Buffer logging: level debugging, 11 messages logged
     Trap logging: level informational, 15 message lines logged
          Logging to 10.1.1.1, 2 message lines logged
 Log Buffer (4096 bytes):

>DETAIL REMOVED<
2948GL3#
```

To enable buffered logging to the switch itself, use the *logging buffered* command. Optionally, you can specify the level of messages to log, but I typically leave it at the default. An example of enabling buffered logging is shown here:

```
2948GL3(config)#logging buffered
2948GL3(config)#^Z
2948GL3#show logging
Syslog logging: enabled (0 messages dropped, 0 flushes, 0 overruns)
     Console logging: disabled
     Monitor logging: level debugging, 0 messages logged
     Buffer logging: level debugging, 1 messages logged
     Trap logging: level informational, 16 message lines logged
          Logging to 10.1.1.1, 3 message lines logged
Log Buffer (4096 bytes):

01:43:50: %SYS-5-CONFIG_I: Configured from console by console
```

Next, you want to eliminate syslog messages from interrupting your work when you are using the console connection. The easiest way to do this is to simply disable console logging by using the *no logging console* command. However, if you want to see the really important messages but eliminate the "informative" (read: annoying) messages, you can also use the *logging console [level]* command. I normally disable console logging entirely; but if I do leave it enabled, I typically set it at the warning level (severity level 4). An example of both methods is shown next:

```
2948GL3(config)#logging console warning
2948GL3(config)#^Z
2948GL3#show logging
Syslog logging: enabled (0 messages dropped, 0 flushes, 0 overruns)
     Console logging: level warnings, 0 messages logged
```

```
     Monitor logging: level debugging, 0 messages logged
     Buffer logging: level debugging, 2 messages logged
     Trap logging: level informational, 17 message lines logged
          Logging to 10.1.1.1, 4 message lines logged


2948GL3#conf t
Enter configuration commands, one per line.  End with CNTL/Z.
2948GL3(config)#no logging console
2948GL3(config)#^Z
2948GL3#show logging
Syslog logging: enabled (0 messages dropped, 0 flushes, 0 overruns)
     Console logging: disabled
     Monitor logging: level debugging, 0 messages logged
     Buffer logging: level debugging, 3 messages logged
     Trap logging: level informational, 18 message lines logged
          Logging to 10.1.1.1, 5 message lines logged


2948GL3#
```

Finally, you need to set your passwords for switch management, starting with the most important: the enable secret. Although enable mode technically has two passwords that you can set, the password (unencrypted) and the secret (encrypted), I usually set only the secret because the enable password is used only if the secret does not exist or if you are using a *very* old version of IOS.

If you do choose to set the enable password, you should *always* set it to be different than the secret, because the enable password can be viewed as plaintext in archived configuration files on your network servers. Besides, if the unencrypted password is the same as your encrypted secret, what's the point of encrypting the secret? This is a moot point in most versions of IOS anyway, because the IOS normally won't allow the password and secret to be the same.

To set the enable password, use the *enable password* command. To set the secret, use the *enable secret* command. These commands are shown in the next example:

```
2948GL3(config)#enable password whatchamacallit
2948GL3(config)#enable secret whosiewhatsit
2948GL3(config)#^Z
2948GL3#show running-config
Building configuration...

Current configuration:
```

```
!
version 12.0
service timestamps debug uptime
service timestamps log uptime
no service password-encryption
!
hostname 2948GL3
!
logging buffered 4096 debugging
no logging console
enable secret 5 $1$2X9L$3pFJSSCmDEK/W.IYUKHTE1
enable password whatchamacallit
!
>DETAIL OMITTED<
```

As for the Telnet, console, and aux connections, you want to set passwords on these to provide as much security as possible. These passwords are not encrypted, however, so do not reuse the *enable secret* command for these passwords.

To set the passwords for these ports, you need to be in line configuration mode, a special form of interface config mode. This mode allows you to change the parameters specific to these ports. To enter line configuration mode, type *line [con | aux | tty | vty] [number]*. To enter line config mode for the console port, type *line console 0*. The aux port is *line aux 0*. For TTY (asynchronous serial lines) and VTY (Telnet) ports, you can select multiple ports to configure at once by typing *line [vty | tty] [first port] [last port]*. For instance, if you wanted to select Telnet ports 0 through 4 (the first five ports), you would type *line vty 0 4*.

In line config mode, two commands are important: *password* and *login*. You use the *password* command to set the password. You use the *login* command to force a login upon connect. (Otherwise, the password will be set, but there will not be a prompt to log in—making the password useless.) The following example shows how to use these commands:

```
2948GL3(config)#line vty 0 4
2948GL3(config-line)#password tricksare4kids
2948GL3(config-line)# no login
2948GL3(config-line)#^Z
2948GL3#
>ATTEMPT TO CONNECT FROM A CONNECTED 5500<
Cat5K (enable) telnet 10.1.1.100
```

```
Trying 10.1.1.100...
Connected to 10.1.1.100.
Escape character is '^]'.

2948GL3>exit
>NOTICE THE PASSWORD WAS NOT REQUIRED<

2948GL3(config)#line vty 0 4
2948GL3(config-line)#password tricksare4kids
2948GL3(config-line)# login
2948GL3(config-line)#^Z
2948GL3#

>ATTEMPT TO CONNECT FROM A CONNECTED 5500<
Cat5K (enable) telnet 10.1.1.100
Trying 10.1.1.100...
Connected to 10.1.1.100.
Escape character is '^]'.


User Access Verification

Password:
2948GL3>
>NOTICE THAT, THIS TIME, THE PASSWORD WAS REQUIRED<
```

| Note | *Although you can set different passwords for each line, it is not recommended. This recommendation is especially true of Telnet (VTY) lines because you never know (and there is no way to specify) which line you will get when you Telnet in.* |

CatOS Configuration

Except for completely different commands, CatOS-based switch logging configuration is fairly similar to standard IOS logging configuration. Again, you want to set logging buffering and eliminate console syslog messages. To buffer your syslog messages to the local switch, you don't need to do anything. By default, up to 500 syslog messages are automatically buffered. To enable logging to a syslog server, use the *set logging server enable* and *set logging server [IP address]* commands. The *set logging server enable* command enables logging to syslog servers, whereas the *set logging server [IP address]* command tells the switch which server or servers to log messages to. The following example shows these commands:

```
Cat5K (enable) set logging server enable
System logging messages will be sent
to the configured syslog servers.
Cat5K (enable) set logging server 10.1.1.1
10.1.1.1 added to system logging server table.
Cat5K (enable) show logging

Logging buffer size:            500
       timestamp option:        enabled
Logging history size:           1
Logging console:                enabled
Logging server:                 enabled
{10.1.1.1}
>DETAIL REMOVED<
```

To disable logging to the console, you use the *set logging console disable* command.

Luckily, setting passwords for a CatOS switch is a bit simpler than for a standard IOS-based switch. CatOS switches have only one enable password (and it *is* encrypted) and only one user mode password. To set the enable password, use the *set enablepass* command; and to set the user mode password, use the *set password* command. The next example shows these commands:

```
Cat5K (enable) set password
Enter old password:nunyabiznes
Enter new password:yawontguessthis1
Retype new password: yawontguessthis1
Password changed.
Cat5K (enable) set enablepass
Enter old password:blank
Enter new password:password
Retype new password:password
Password changed.
Cat5K (enable)
```

Configuring Boot Parameters

With boot parameters, you need to be concerned about only two items: the boot image and the boot configuration. The next two sections detail the configuration commands for both of these boot options.

Standard IOS Configuration

By default, most standard IOS switches boot from the first available image in flash and use the configuration stored in NVRAM. In most situations, this boot order is exactly what is needed, and no further configuration is required. However, in some high-availability environments, you may want to specify an additional IOS image or configuration file to provide for redundancy in case of a flash RAM failure or configuration corruption.

For the boot image, you configure the boot file by using the *boot system* command. By using multiple *boot system* commands, you can specify a list of boot images. If one or more of the images is corrupt, the switch uses the next image listed. For instance, if you enter *boot system tftp cat2948g.bin 10.1.1.1*, then *boot system tftp cat2948g.bin*, then *boot system flash cat2948g.bin*, and—finally—*boot system flash*, the switch first attempts to boot by using the image cat2948g.bin from the TFTP server 10.1.1.1.

If this image or server is not available, the switch next attempts to retrieve the same boot image from any available TFTP server. (If a server IP address is not specified, it broadcasts for one.) If this also fails, the switch attempts to boot from cat2948g.bin in flash RAM. If that file does not exist or is corrupt, it finally boots from the first bootable image found in flash. This boot configuration produces a very high level of redundancy.

For the bootup configuration, you use the *service config* command, along with the *boot network* and *boot host* commands. The *service config* command tells the switch to look for a network configuration file. Without this command, the switch boots by using the config in NVRAM, ignoring the *boot host* and *boot network* commands. Once you enter the *service config* command, use the *boot network* command to specify a configuration file with settings that apply to all switches, and use the *boot host* command to specify a configuration file specific to an individual switch. If the network configuration files cannot be found, the switch boots by using the configuration in NVRAM. See the following example:

```
CAT2948GL3(config)#service config
CAT2948GL3(config)#boot network allhosts.cfg 10.1.1.200
CAT2948GL3(config)#boot host 2948gL3.cfg 10.1.1.200
CAT2948GL3(config)#
>AFTER NEXT RELOAD<
```

```
%Error opening tftp://10.1.1.200/allhosts.cfg (Timed out)
%Error opening tftp://10.1.1.200/2948gL3.cfg (Timed out)
Press RETURN to get started!
CAT2948GL3>
```

CatOS Configuration

In many ways, configuring boot parameters on a CatOS-based switch is easier than it is on a standard IOS-based switch. To configure your boot image, you need to issue the *set boot system flash* command. Entering multiple *set boot system flash* commands provides a list of image files that the switch tries in the order you enter them. For increased redundancy, use images in both the system flash and PC Card flash. For example, if you wanted the switch to boot from a file called cat5500.bin, and you wanted it to look first in the system flash and then in PC Card slot 0, you would issue the following commands:

```
Cat5K> (enable) set boot system flash bootflash:cat5500.bin
BOOT variable = bootflash:cat5500.bin;
Cat5K> (enable) set boot system flash slot0:cat5500.bin
BOOT variable = bootflash:cat5500.bin;slot0:cat5500.bin;
```

To set the configuration file, issue the *set boot auto-config* command. If you issue multiple *set boot auto-config* commands, you can configure the switch to use the configuration files in the order entered. For instance, if you enter *set boot auto-config slot0:Cat5K.cfg* and *set boot auto-config slot1:5500.cfg*, the switch first looks in the slot0 file system for Cat5K.cfg and then in slot1 for 5500.cfg.

Saving the Configuration

With switch configuration files, you need to be concerned about two tasks: saving the files to NVRAM and saving the files to a server for storage.

In most switches, IOS or CatOS, the running configuration is automatically saved to NVRAM. However, in some switches, such as 2948G Layer 3, saving the running config to the startup config is required. To do this, issue the *copy running-config startup-config* command (or simply *cop ru sta*).

You should save the file to a TFTP server on all switches as soon as possible. This strategy allows you to recover from configuration mistakes fairly quickly and easily. On a standard IOS-based switch, the command for this is usually *copy startup-config tftp*. (The 1900 series is a bit different and uses the *copy nvram tftp://server/file* command.) These two commands are illustrated next.

```
>COPY FROM A 2948G SWITCH<
2948GL3>enable
Password:
Password:
2948GL3#copy startup-config tftp
Address or name of remote host []? 192.168.1.1
Destination filename [startup-config]?
!!!!!!!!!!!!!!!!!!!!!!!!!!!!!!!!!!!!!!!!!!!!!!!!!!!!!!!!!!!!!!!!!!!!!!
Upload to server done
Copy took 00:00:23 [hh:mm:ss]
>COPY FROM A 1900 SWITCH<
cat1924A#copy nvram tftp://192.168.1.1/config.cfg
!!!!!!!!!!!!
Upload successful
Copy took 00:00:03 [hh:mm:ss]
```

On a CatOS-based switch, the command is *copy module#/device:filename tftp:filename*, as discussed in Chapter 17. This following example illustrates:

```
Console> (enable) copy config tftp:lab2.cfg all

IP address or name of remote host [172.20.22.7]? y

Upload configuration to imgFile:lab2.cfg (y/n) [n]? y


.........
  .
  /
Configuration has been copied successfully. (10299 bytes).
Console> (enable)
Cat5K2> (enable) copy 1/slot0:cat5k.cfg tftp:cat5k.cfg

IP address or name of remote host [192.168.1.1]? y

Upload configuration to imgFile:cat5k.cfg (y/n) [n]? y


.........
  .
  /
Configuration has been copied successfully. (7574 bytes).
Cat5K2> (enable)
```

Summary

This chapter covered basic initial switch configuration for both standard IOS switches and CatOS switches. Think of this as the introduction to the next three chapters, which examine more advanced configurations for layer 2, layer 3, and layer 4 switches.

The Complete Reference

Cisco

Chapter 19

Layer 2 Switching

his chapter covers layer 2 switching technologies that are common to all Cisco switching devices. It also discusses common technologies related to nearly all layer 2 switching devices from any vendor, including Spanning Tree Protocol (STP) and virtual LANs (VLANs). You'll also learn about Cisco-specific enhancements to layer 2 switching. Finally, you will explore configuration and troubleshooting using both standard IOS and CatOS commands on Catalyst 1900, 2948G, and 5500 switches.

Spanning Tree Protocol (STP)

Spanning Tree Protocol (STP) is used to provide for loopfree redundant switching topologies in switched networks. You need STP because of a simple problem inherent in bridging and repeating. As you may recall from Chapter 2, a *repeater* (hub) simply repeats all signals received out all ports. This situation leads to a problem with building fault-tolerant repeated networks, known as *looping*.

Consider the example network topology in Figure 19-1. In this example, four hubs are configured in a full mesh. If Hub 1 receives a packet from PC 1, it will repeat it out all ports.

As a result, Hubs 2–4 will receive the same packet and—that's right—repeat the packet out all ports. Therefore, nine copies of the exact same packet now exist (which, incidentally, had already reached the designated host, Server 1) traversing the network. This situation is shown in Figure 19-2.

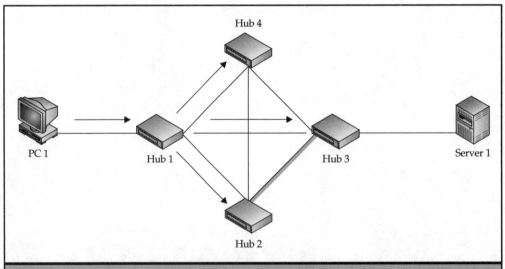

Figure 19-1. *Initial repetition of a frame from PC 1 destined for Server 1*

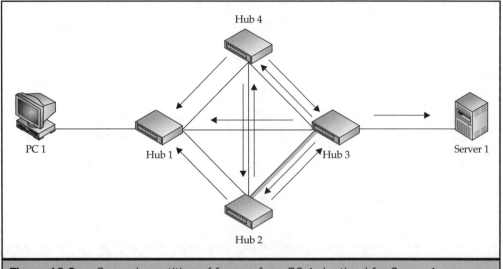

Figure 19-2.   *Second repetition of frames from PC 1 destined for Server 1*

This process will go on forever, exponentially increasing with each repetition. In this situation, assuming a 64-byte frame size and *no* other traffic on the network (both best-case scenarios and very unlikely), the replicated packets would take only about one minute to consume *all* of the network bandwidth. This situation is obviously undesirable, and it is the primary reason that you can't build a fault-tolerant Ethernet using simple hubs.

Layer 2 switches don't have this specific problem (remembering that, as long as they have an entry for the destination in the MAC table, they forward frames only out of the specific port to which the destination host is connected). However, switches do suffer from a similar problem with broadcasts.

Recall from Chapter 2 that if an Ethernet switch receives a broadcast (and in some cases, a multicast), it is *required* to forward the broadcast out all ports (except the source port, of course), because a broadcast is destined for all ports. This problem doesn't exist in a nonredundant (physical star) topology, as shown in Figure 19-3. In the physical star topology, there is no danger of a loop because the switch has no redundant paths in which to forward a frame.

In a mesh topology, however, the switch repeats the frame to all other redundantly connected switches, causing the frame to be repeated indefinitely (or until all bandwidth is consumed). This dilemma is the same looping problem in a redundant (full or hybrid physical mesh) topology, like the one shown in Figure 19-4.

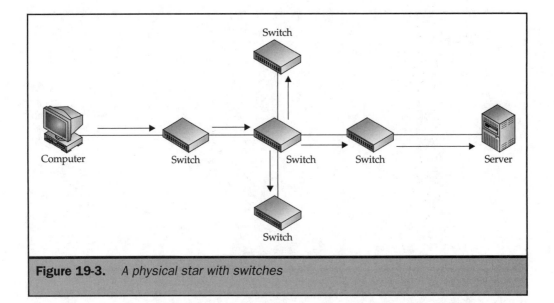

Figure 19-3.   *A physical star with switches*

Enter STP. It automatically determines where redundant links are present and blocks data traffic on those ports until the primary link fails (in which case, it automatically opens the redundant ports). Figure 19-5 shows the mesh topology using STP.

Now that you know why STP is used, let's concentrate on how it works.

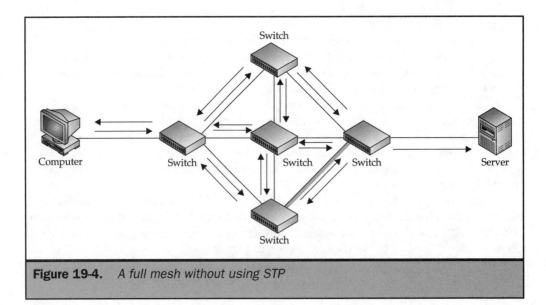

Figure 19-4.   *A full mesh without using STP*

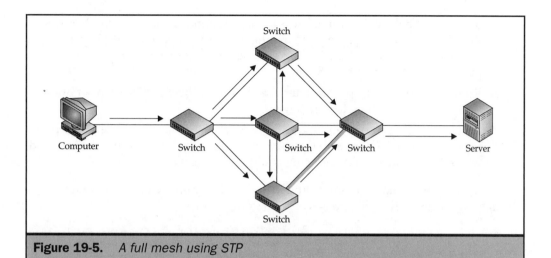

Figure 19-5.   *A full mesh using STP*

How STP Works

STP works by determining what the "best" path is through the network and allowing data traffic to be sent only along that path. This functionality is fairly similar to the main functionality of dynamic routing protocols, but, luckily, STP is simpler than most routing protocols.

Understanding basic STP functionality requires an understanding of some new terms and rules. Let's begin with the basic terms:

- **Port states**   STP defines several port states: blocking, listening, learning, forwarding, and disabled. For right now, you need to be concerned with only three: blocking (which disables data traffic on the port but still allows STP messages to be transmitted and received), forwarding (which allows all traffic), and disabled (which allows no traffic). The disabled state might be used when a link is not present on the port or the port has been administratively disabled.

- **Root bridge (RB)**   The root bridge is the center of the STP topology. The basic idea is that a path through the root bridge is the best path to use to get from one side of the network to the other. All ports on the root bridge are always in a forwarding state.

- **Designated bridge (DB)**   The designated bridge is the one with the shortest path (lowest cost) to a given network segment. This bridge is also sometimes called a *parent switch*.

- **Bridge Protocol Data Unit (BPDU) or Configuration Bridge Protocol Data Unit (CBPDU)**   STP sends these frames to communicate with other bridges in the STP topology. Their purpose is to carry information about the STP topology to all other bridges that are participating in the same topology. BPDUs are sent periodically by all switches in the STP topology, but they are not propagated past the first switch to hear them.

- **Root port (RP)**   This port is determined to be the best path to the root bridge. This port is always in a forwarding state.

- **Port cost**   The cost of a port indicates how much STP prefers a particular path out a given port. STP uses this cost to determine which port is the root port and which bridge on a given segment should be the designated bridge.

STP uses the Spanning Tree Algorithm (STA) to determine what the best path throughout the network is. STA relies on a few simple rules:

- The root bridge places all ports in a forwarding state (except those that are disabled administratively or not connected).

- The bridge with the lowest cost to a given segment is considered the designated bridge for that segment. The designated bridge for a given segment places the port connected to that segment in a forwarding state.

- All bridges have a single port that is considered to have the lowest cost (shortest path) to the root bridge. This port is considered to be the root port and is placed in a forwarding state.

- All other ports are placed in a blocking state.

Now that you understand the rules and terms, let's look at how STP determines the optimal topology and maintains that topology. Figure 19-6 shows a sample network topology.

First, a root bridge must be selected. This task is performed through a process known as an *election*. When the switches are powered up, they have no knowledge of the other switches or topology of the network. Therefore, the switch doesn't know which one is the best switch for the root bridge role. Therefore, it simply assumes that it is the best switch for the root bridge role, and it begins advertising itself (using BPDUs) as the root bridge out all nondisabled ports.

The BPDUs contain only a few pieces of information:

- **The root bridge's bridge ID**   The bridge ID is usually the MAC address for the bridge.

- **The root bridge's root priority**   This priority can be set by the administrator to assure that one bridge is likely to become the root. This 16-bit field is set to 32,768 by default, or 0x8000 (hex).

- **The bridge's bridge ID**   This ID identifies the bridge that is sending this BPDU, usually the MAC address of the sending bridge.

■ **The cost to the root bridge from the advertising bridge**   This cost is known as the *path cost*. It tells the receiving bridge how "far" from the root bridge the advertising bridge is.

■ **The sending bridge's port ID and port cost**   This information tells the receiving bridge which port on the sending bridge this advertisement was sent from, and what the cost is on that port. By default, the switch usually sets this number to a value equaling 1000 divided by the speed of the port in megabits per second. For example, the default cost for most 10Base-T interfaces is 100.

■ **Various timers**   These timers (max age, hello, and forward delay) determine how quickly the STP topology responds to link failures. They are discussed individually later in this section of the chapter.

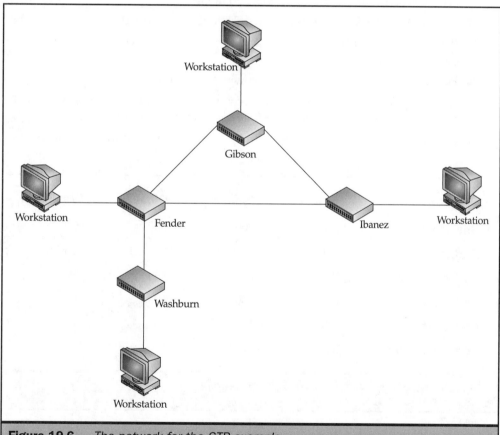

Figure 19-6.   *The network for the STP example*

Using the example shown in Figure 19-6, let's see how this election process occurs. Table 19-1 shows the STP configuration.

Figure 19-7 displays this information graphically. Note that, in this example, the MAC addresses and priorities are simplified, and we will power up all the switches at the same time.

Because each bridge initially thinks that it is the best candidate for the root bridge role (let's begin with Fender, in this case), it sends out an advertisement containing the following information:

- **Root bridge ID**   Because Fender thinks it should be the root bridge, it sends its MAC address, simplified in this example to 001.

- **The root bridge's bridge priority**   Because Fender thinks it should be the root bridge, it sends its priority, 12 (or 0xC in hex).

- **Bridge ID**   This will be Fender's MAC address, simplified to 001.

- **Path cost**   Because Fender thinks it's the root, this cost is 0. (Lower costs are better.)

- **Port ID**   This ID is set to the port on which the information was sent.

- **Port cost**   This cost is set to the cost of whichever port the advertisement was sent on—in this case, 100 for ports 2, 3, and 4, and 10 for port 1.

- **Timers**   This is the root bridge's (in this case, Fender's) configured timers.

	Fender	**Ibanez**	**Gibson**	**Washburn**
Bridge ID	001	900	F00	050
Bridge priority	12 (0xC)	5 (0x5)	1 (0x1)	15 (0xF)
Port cost, Port 1	10	10	100	100
Port cost, Port 2	100	100	10	100
Port cost, Port 3	100	100	10	N/A
Port cost, Port 4	100	N/A	N/A	N/A

Table 19-1.   *Bridge Configuration for the Example*

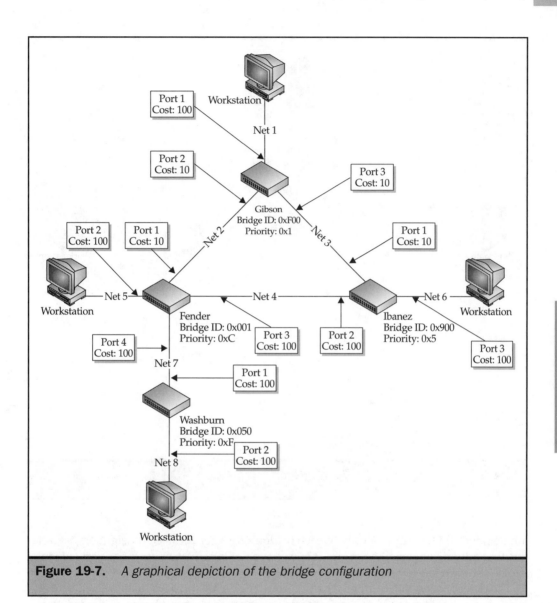

Figure 19-7. *A graphical depiction of the bridge configuration*

Figure 19-8 shows the initial advertisement.

The other bridges (Washburn, Ibanez, and Gibson) also think they are the root bridge, so their initial BPDUs contain information proclaiming that they are the root.

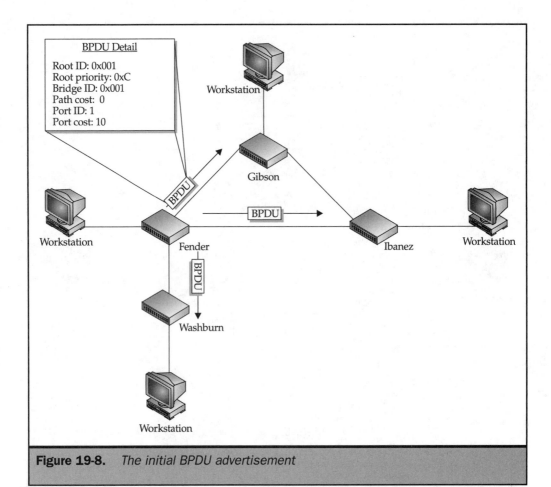

BPDU Detail

Root ID: 0x001
Root priority: 0xC
Bridge ID: 0x001
Path cost: 0
Port ID: 1
Port cost: 10

Figure 19-8. *The initial BPDU advertisement*

Remember, BPDUs are sent only to each bridge's directly connected neighbors; so, after the first advertisement, Gibson and Ibanez have not heard Washburn's advertisement, and Washburn has not heard Ibanez or Gibson's advertisements.

To determine which bridge is the best root bridge, the bridges take the root bridge priority and root bridge MAC address from each advertisement heard and compare these to all other advertisements, including their own. The priority and MAC address are combined, as shown in Figure 19-9, and the lowest total wins.

At Fender (we use Fender in this case because it is the only bridge that has heard all advertisements), after combining each advertised MAC address and priority for this example, we get the results shown in Table 19-2.

Obviously, the lowest total priority is Gibson, at 7936 (or 0x1F00). Consequently, as far as Fender and Ibanez are concerned, Gibson has won the election and is the root

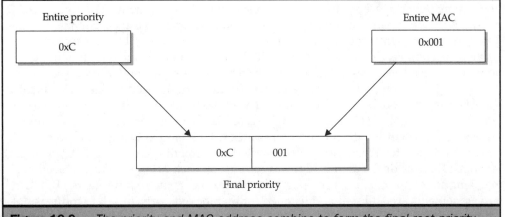

Figure 19-9. *The priority and MAC address combine to form the final root priority.*

bridge. However, Washburn has not yet heard an advertisement from Gibson; all it has heard is an advertisement from Fender. As far as Washburn is concerned, the root bridge should be Fender. Luckily, with the election process, once Fender determines that Gibson is a better candidate, it begins to advertise Gibson as the root bridge in all future BPDUs. So, for right now, Washburn thinks Fender is the root, but at the next advertisement, it knows that Gibson is the best candidate and updates its STP information appropriately.

After a root bridge is elected, all switches begin the process of determining which port should be the designated bridge for each segment and which port should be the root port. The designated bridge for a segment is the bridge with the best cost to that segment. Costs are cumulative, so all costs for all bridges along a given path are added together. In this case, the designated bridge for segments 1, 2, and 3 is Gibson, because Gibson obviously has the best cost for these segments. For segments 5 and 7, Fender is

Advertised Root	Fender	Ibanez	Gibson	Washburn
Advertised root total priority (priority plus MAC) in hex	0xC001	0x5900	0x1F00	0xF050
Advertised root total priority (priority plus MAC) in decimal	49,153	22,784	7936	61,520

Table 19-2. *Root Bridge Advertisements Heard at Fender*

the designated bridge, because it has the best cost to these segments. For segment 6, Ibanez is the designated bridge, and for segment 8, Washburn is the designated bridge. Only segment 4 is unaccounted for. A problem occurs, however, because the cost for both Ibanez and Fender to segment 4 is 110. In the case of a tie, the switch with the lowest bridge ID becomes the designated bridge. In this case, Fender has the lowest bridge ID (0×001 versus 0×900), so Fender becomes the designated bridge for segment 4.

Choosing the root port is a bit simpler. The root port on the switch is simply the port with the lowest total cost to the root. In this case, the root port for Fender is 1 (connected to segment 2, cost of 10), the root port for Ibanez is 1 (connected to segment 3, cost of 10), and the root port for Washburn is 1 (connected to segment 7, cost of 110). All ports that are not either the designated bridge or root port on each bridge are then placed in a blocking state. (In this case, only port 2 on Ibanez meets the requirements.) BPDU advertisements are still sent and listened for on blocked ports, but data traffic is not allowed. Because data traffic is now blocked on port 2 at Ibanez, loops cannot occur. Figure 19-10 shows the final STP topology for this example.

Once the topology is configured, switches continue to issue BPDUs at scheduled intervals. These scheduled BPDU messages allow the switches to react to topology changes as they occur. This functionality is controlled by timers on the switches. These timers are as follows:

- **Hello**   This timer determines how often the bridge will issue a BPDU. Two seconds is the default.

- **Max age**   This timer determines the longest a bridge will wait for a hello message from a neighbor before determining that an error has occurred. Twenty seconds is the default.

- **Forward delay**   This timer determines how long the bridge will wait to transition from a listening state to a learning state. The listening and learning states are short-term states designed to add an additional delay in the transition from blocking to forwarding. This delay gives the switch a bit more time to make sure that it's not going to cause a bridging loop before entering forwarding mode. While in the listening state, the switch listens to BPDUs but will not transfer any data or add MAC addresses to its table. In the learning state, the switch learns MAC addresses but still will not transmit data. The default forward delay is 15 seconds.

If an error occurs, the bridge will know of the problem and respond because of these timers. If the bridge is directly connected to the link, it typically will not need to wait for the max age timer to expire. It will have firsthand knowledge of the link failure. One way or another, when an error occurs, STP on the sensing bridge will send BPDUs along the root path to inform the root bridge of the topology change. Once the root bridge is aware of the change, it sends BPDUs to all switches requesting a recalculation of the spanning tree (the spanning tree is the topology "map" that STP builds by listening to BPDUs).

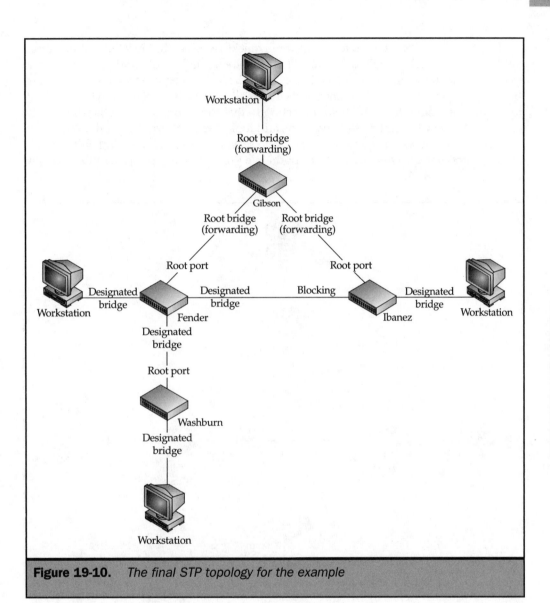

Figure 19-10.   *The final STP topology for the example*

The down side to Spanning Tree Protocol is that it takes time to calculate the spanning tree and transition port states. This delay can cause serious problems for some protocols, such as BOOTP, that may need an active link immediately upon system boot. To help alleviate these problems, Cisco includes some manually configurable enhancements to spanning tree: Backbonefast, Uplinkfast, and Portfast.

Backbonefast allows the switch to converge quickly when an indirect link failure occurs on the primary backbone link. An *indirect link failure* is a failure of another switch's link to the root, as shown in Figure 19-11. In this case, the switch begins receiving higher-cost BPDUs on its root port than on other ports. Normally, the switch would still have to wait for its max age timer to expire before reacting to this condition; but with Backbonefast enabled, the switch immediately responds and looks for an alternative path to the root.

Uplinkfast allows an access-layer switch to react and resolve direct link failures almost immediately. The switch places ports in a group that are receiving BPDUs from

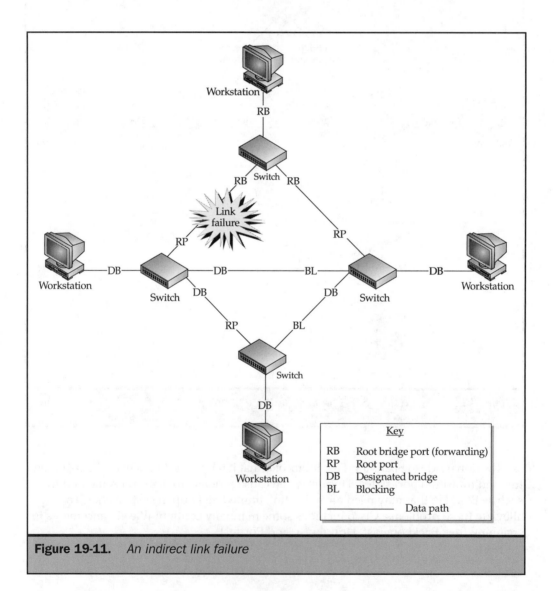

Figure 19-11. *An indirect link failure*

upstream switches. If the primary path to the root experiences a direct link failure, the switch responds by setting one of the other ports in the group to forwarding mode almost immediately (less than five seconds), without transitioning through listening and learning states. This functionality is extremely useful for access switches because it allows them to very quickly reconfigure their spanning tree when the primary link fails; however, enabling Uplinkfast automatically adds 3000 to the cost for all ports and sets the priority of the bridge to 49,152. These changes occur because enabling Uplinkfast on distribution-layer or core switches can cause looping and lost data. By increasing the cost and priority of the switch, the switch is unlikely to be used as a backbone switch for STP.

Portfast is used on access switches to allow a switch port to enter forwarding mode immediately, rather than transitioning through states normally. Portfast is useful in workstations in which the protocols require connectivity almost immediately upon bootup. For example, if you are using BOOTP for a diskless workstation, it may broadcast within the first ten seconds after powering the machine on. This quick broadcast is problematic with STP because the default forward delay prevents connectivity. However, when using Portfast, the switch port immediately begins forwarding frames, and BOOTP functions properly.

 Be sure to enable Portfast only on links directly connected to network hosts, never on links connected to other switches. Enabling Portfast on switch links can cause loops to occur because the switch does not use the forward delay to detect possible loops.

Basic STP Configuration

STP configuration for a given switch varies depending on the environment and switch type (access, distribution, or core) in question. You first need to determine whether STP is even necessary. In most small environments, redundant paths between switches may not even exist. In these cases, bandwidth can be spared and performance slightly improved by disabling STP entirely.

In a larger or more redundant environment, you will need to leave STP enabled and ask yourself a few questions concerning the network layout to determine the proper base STP configuration, such as the following:

- Are any non-Cisco switches present?
- What is the network diameter (number of switches through which any frame must pass)?
- Which switch would make the best STP root?
- Should Backbonefast be enabled?
- Which switches should have Uplinkfast enabled?
- Which access links should have Portfast enabled?

First, let's look at how to disable STP entirely. In a smaller environment, disabling STP can provide a small performance improvement, but you need to make absolutely sure that no loops are present in the topology. (The topology must be a true physical star.) To disable spanning tree on a standard IOS-based switch (2948G and 3500 series), issue the *no spanning-tree [vlan list]* command. On the 1900 series, issue the *no spantree [vlan list]* command; and on a CatOS-based switch, issue the *set spantree disable all* command. Spanning tree should be enabled by default, but if you need to re-enable it, simply issue the opposite command (*spanning-tree [vlan list]*, *spantree [vlan list]*, or *set spantree enable all*).

> *On some versions of the CatOS supervisor engines, a VLAN list is supported after the* set spantree enable *or* set spantree disable *command. The examples used in this chapter are from a Catalyst 5500 with the Supervisor IIIG engine, which supports only enabling or disabling of spanning tree for* all *VLANs (rather than for specific VLANs).*

The following output shows examples of these commands for all the switches mentioned:

```
>1900 Switch Configuration<

1924(config)# no spantree vlan 1 2 3 4
1924(config)# spantree vlan 1 2 3 4


>3500 Switch Configuration<

3508(config)# no spanning-tree vlan 200
3508(config)# spanning-tree vlan 200


>5500 Switch Configuration<

Cat5K (enable) set spantree disable all
Spantree disabled.
Cat5K (enable) set spantree enable all
Spantree enabled.
```

After you choose to enable spanning tree, you must determine the proper STP configuration. First, determine whether any non-Cisco switches are present in the environment. This information lets you know which Cisco proprietary enhancements you can use, such as Fast Etherchannel and Per VLAN Spanning Tree (PVST.) Per VLAN Spanning Tree is discussed in the section "VLAN STP Considerations," later in this chapter.

Next, determine the switch diameter. By setting the switch diameter, you cause STP to recalculate timer values to help ensure optimal performance. This technique is the primary, and typically the best, way of setting the timer values. Manual modification of the timers is not suggested, because improper settings can reduce performance or cause looping.

To calculate the diameter, examine the number of switches between any two hosts in the network. The default STP diameter is seven, which is also the maximum setting.

In most networks, this number can be reduced to four or less to improve performance. (Remember, the Ethernet specification allows only four repeaters, including switches, between any two hosts.)

To configure the STP diameter on a standard IOS switch, use the *spanning-tree [vlan list] root primary diameter [value]* command. Note that you can configure the diameter only on the STP root, because it is the switch that controls the spanning tree timers. The *root primary* parameter in this command sets the switch's priority to 8192, almost ensuring that it will become the STP root. On a CatOS-based switch, use the command *set spantree root [vlan list] dia [diameter value]*.

As part of this process, you will by necessity be forced to choose the STP root. The STP root bridge should generally be a fast switch near or at the center of your network topology. Remember, the majority of your network traffic will likely flow through the root bridge, so you do not want the root bridge to be an underpowered or poorly positioned switch.

Figure 19-12 shows a good example of a root bridge selection and the resulting traffic paths. You can choose the root bridge by either using the commands just outlined

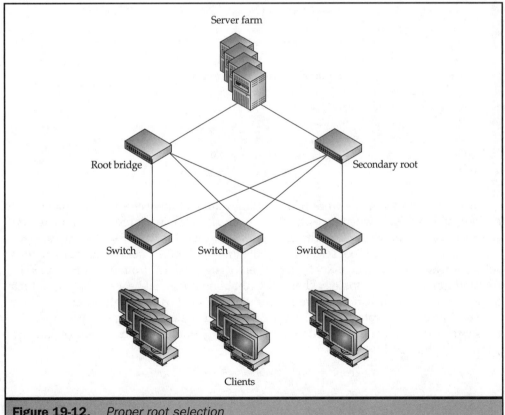

Figure 19-12.   *Proper root selection*

(*spanning-tree [vlan list] root* or *set spantree root*), or by setting the priority manually by using the *spanning-tree priority [value]* or *set spantree priority [value]* commands.

You should also choose a secondary root bridge in case the primary root bridge fails. You can use the *spanning-tree [vlan list] root secondary* or *set spantree root secondary* command to make secondary root bridge selection happen automatically (setting the priority to 16,384). You can, of course, manually set the priority as well by using the *spanning-tree priority [value]* or *set spantree priority [value]* commands and entering a value that is less than the default (32,768) but higher than the primary root's priority.

An example of setting a switch to be the primary root, along with setting the priority manually, is provided for both a standard IOS and CatOS-based switch:

```
>3500 Switch Configuration<
3508(config)# spanning-tree vlan 1 root primary diameter 7
3508(config)# spanning-tree vlan 1 priority 100
>5500 Switch Configuration<
Cat5K (enable) set spantree root 1 dia 7
VLAN 1 bridge priority set to 8192.
VLAN 1 bridge max aging time set to 20.
VLAN 1 bridge hello time set to 2.
VLAN 1 bridge forward delay set to 15.
Switch is now the root switch for active VLAN 1.
Cat5K (enable) set spantree root secondary 1 dia 7
VLAN 1 bridge priority set to 16384.
VLAN 1 bridge max aging time set to 20.
VLAN 1 bridge hello time set to 2.
VLAN 1 bridge forward delay set to 15.
Cat5K (enable) set spantree priority 100
Spantree 1 bridge priority set to 100.
```

In addition, in a large network with multiple administrative entities, you may wish to enable a special Cisco switching feature, *root guard.* Because in a large environment, you can never be sure of what other administrators will configure, it is best in some cases to guard against unwanted root selection when another administrator configures a better priority on an inferior switch. Root guard keeps a rogue switch from becoming the STP root and forcing an STP topology change by disabling ports that hear of a better root than the current root. For instance, Figure 19-13 shows a proper STP configuration.

Then, an administrator adds a new switch and sets the spanning tree priority for the switch to 0. Under normal situations, this setting would force the topology change shown in Figure 19-14.

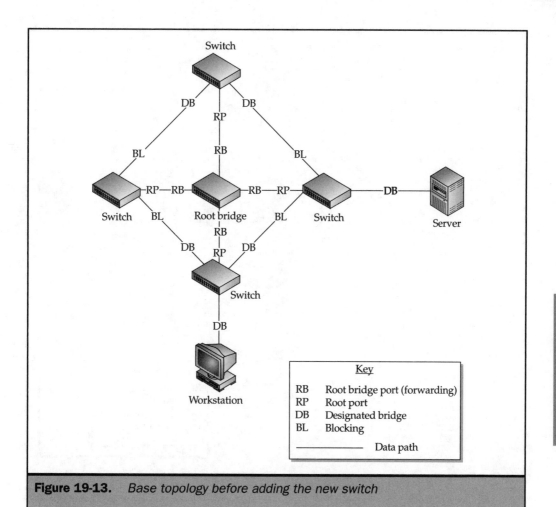

Figure 19-13.   *Base topology before adding the new switch*

However, if you had enabled root guard on the ports the administrator was connecting to, as well as all ports along the new switch's path back to the correct root, root guard would have blocked these ports, disabling data traffic to the new switch until such time as the administrator set the priority back to something reasonable. Figure 19-15 shows this situation.

To configure root guard on a standard IOS switch, use the *spanning-tree guard root* command in interface config mode for the specific interface that is required. On a

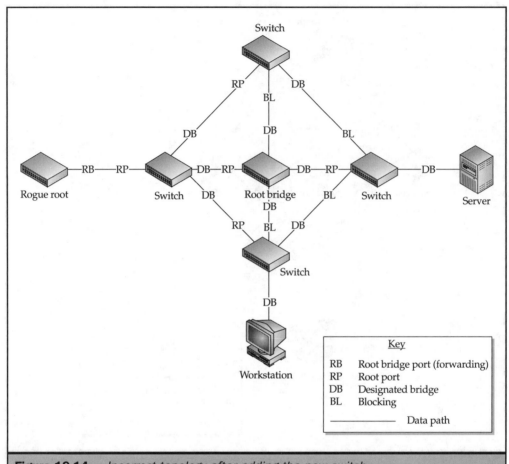

Figure 19-14. *Incorrect topology after adding the new switch*

CatOS-based switch, use the *set spantree guard root [module/port]* command. These
commands are shown in the following output:

```
>3500 Switch Configuration<
3508(config)#interface GigabitEthernet 0/8
3508(config-if)#spanning-tree rootguard
>5500 Switch Configuration<
Cat5K> (enable) set spantree guard root 3/12
Rootguard on port 3/12 is enabled.
Warning!! Enabling rootguard may result in a topology change.
```

```
Cat5K> (enable) set spantree guard none 3/12
Rootguard on port 3/12 is disabled.
```

Note *Be very careful about where you enable root guard. If you enable it on all ports or improper ports, you might lose data and have inconsistent STP configurations.*

Choosing which switches to enable Backbonefast on is a bit easier. Typically, all backbone ports on all switches can benefit from Backbonefast. To enable Backbonefast on a standard IOS-based switch, use the *spanning-tree backbonefast* command. On a CatOS-based switch, use the *set spantree backbonefast* command. The only caveat in

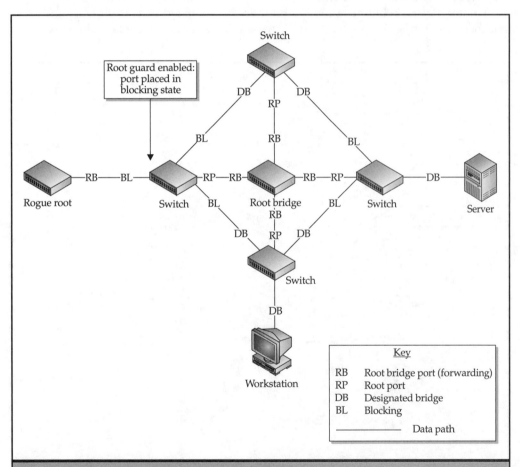

Figure 19-15. *Where root guard needs to be enabled to protect the topology*

using Backbonefast is that for it to function, Backbonefast must be enabled on *all* switches in the topology.

For Uplinkfast, the decision is nearly as simple as Backbonefast. Uplinkfast should typically be enabled only on access-layer switches, as shown in Figure 19-16. Enabling Uplinkfast on core or distribution switches can cause the STP topology to become inconsistent and hamper performance. However, on access-layer switches, Uplinkfast can reduce the time to reconfigure the STP topology after a direct link failure from 30 seconds to less than 5 seconds, which can help eliminate buffer overflows and lost data due to STP delays after link failures.

To enable Uplinkfast on a standard IOS switch, simply enter the command *spanning-tree uplinkfast* in global config mode. For CatOS-based switches, use the *set spantree uplinkfast enable* command.

Finally, Portfast configuration is a bit more complicated than Uplinkfast and Backbonefast configuration. First, unlike the other two enhancements, Portfast is

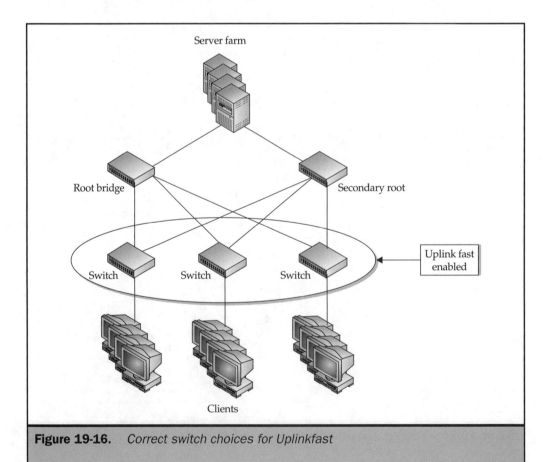

Figure 19-16. *Correct switch choices for Uplinkfast*

port specific, requiring that you manually configure it on a per-port basis. Also, using Portfast on links to other switches can cause serious STP problems, so you must be careful when deciding on the ports to enable for Portfast. However, when used correctly, Portfast can transition a port into the forwarding state almost immediately upon connection. This quick response allows hosts that require immediate network access upon boot to receive that access with little or no drama.

Basic Portfast selection and configuration is fairly simple. First, determine which links are access links for network hosts. For instance, in the example in Figure 19-17, Ports 1 and 3 on Corvette and Ports 4 and 5 on Viper should have Portfast enabled. Once the proper ports have been selected, use the *spanning-tree portfast* interface config command (on standard IOS switches) or the *set spantree portfast [module/port] enable* command (on CatOS switches).

Next, you may want to enable an additional Cisco feature for Portfast, called BPDU guard. *BPDU guard* disables any Portfast-configured port if a BPDU is received on that port. It performs this action because, due to the removal of the waiting period for a port to transition to a forwarding state, loops can be caused if a Portfast port is connected to another switch. If a port is connected to another switch, spanning tree BPDUs should begin entering the port. BPDU guard listens for these BPDUs, and if it detects one, it will immediately disable the port to prevent a loop. To enable BPDU guard on a standard IOS switch, use the *spanning-tree portfast bpdu-guard* global config command. On a CatOS switch, use the *set spantree portfast bpdu-guard* command.

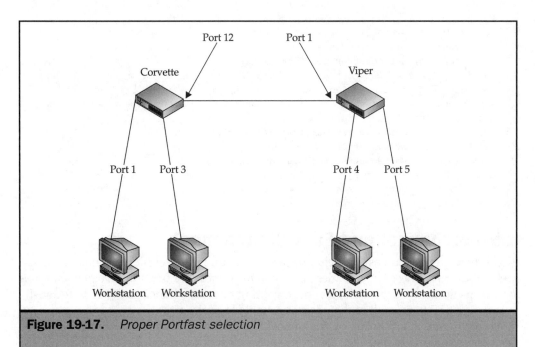

Figure 19-17.   *Proper Portfast selection*

Basic Backbonefast, Uplinkfast, and Portfast configuration commands for both Catalyst 5500 and 3500 switches are shown in the following output:

```
>3500 Switch Configuration<
3508(config)# spanning-tree backbonefast
3508(config)# spanning-tree uplinkfast
3508(config)#interface GigabitEthernet 0/8
3508(config-if)# spanning-tree portfast
>5500 Switch Configuration<
Cat5K (enable) set spantree uplinkfast enable
VLANs 1-1005 bridge priority set to 49152.
The port cost and portvlancost of all ports set to above 3000.
Station update rate set to 15 packets/100ms.
uplinkfast all-protocols field set to off.
uplinkfast enabled for bridge.
Cat5K (enable) set spantree uplinkfast disable
uplinkfast disabled for bridge.
Use clear spantree uplinkfast to return stp parameters to default.
Cat5K (enable) clear spantree uplinkfast
This command will cause all portcosts, portvlancosts, and the
bridge priority on all vlans to be set to default.
Do you want to continue (y/n) [n]? y
VLANs 1-1005 bridge priority set to 32768.
The port cost of all bridge ports set to default value.
The portvlancost of all bridge ports set to default value.
uplinkfast all-protocols field set to off.
uplinkfast disabled for bridge.
Cat5K (enable) set spantree portfast 3/1 enable
Warning: Spantree port fast start should only be enabled on ports connected
to a single host. Connecting hubs, concentrators, switches, bridges, etc. to
a fast start port can cause temporary spanning tree loops. Use with caution.
Spantree port 3/1 fast start enabled.
Cat5K (enable) set spantree portfast 3/1 disable
Spantree port 3/1 fast start disabled.
Cat5K (enable) set spantree backbonefast enable
Backbonefast enabled for all VLANs.
Cat5K (enable) set spantree backbonefast disable
Backbonefast disabled for all VLANs.
```

Fast Etherchannel and Fast Gigachannel

Fast Etherchannel and Fast Gigachannel are two extremely useful Cisco enhancements to standard STP. These technologies allow you to "bundle" or bond up to eight 100-Mbps or 1-Gbps ports into a larger logical port, gaining increased bandwidth (up to 8 Gbps!) by causing STP to treat all of the ports as a single logical port.

To configure a port group on a standard IOS–based switch, use the *port group [number]* command in interface config mode for each of the ports. On a CatOS-based switch, use the *set port channel [module/port list] mode on* command. By setting a number of ports to the same group, you configure one larger logical port.

Troubleshooting STP

STP problems are common, but with proper configuration, most can be avoided. STP problems typically fall into one of the following categories:

- Bridging loops
- Inefficient data transfer
- Slow convergence after a link failure
- BOOTP/DHCP problems

Although all of these problems can be significant, bridging loops are perhaps the most disastrous. Because this problem is exactly the one STP is designed to solve, encountering a bridging loop should be a rare occurrence. Nevertheless, it will still happen occasionally, usually due to an improper configuration. To solve this problem, you must first determine the cause. Common causes of bridging loops are the following:

- *Insertion of a bridge that does not support STP or has STP disabled.* In this case, simply enable STP or remove the redundant switch links from the offending switch.

- *Improper Portfast configuration.* If Portfast is enabled on an uplink, the switch can potentially begin forwarding on the uplink port when it should be blocking. To resolve this problem, disable Portfast on the link.

- *Improper timer configuration.* If the timers are configured for improper values (either manually or by setting the diameter incorrectly), the switch may not have enough time to determine that a port should be placed in a blocking state. To resolve this problem, avoid setting timers manually, and use the correct diameter setting.

Inefficient data transfer can be the result of a poorly optimized STP topology, which leads to improper root selection and poor data paths through the switch fabric. For example, in Figure 19-18, a root was chosen at the network edge. This situation leads to a very inefficient tree and places an access-layer switch at the root of the topology, causing poor performance.

To resolve this problem, make sure the proper priorities are set on your STP root bridge and secondary root.

Slow layer 2 network convergence times are typically the result of improperly configured timers—in this case, timers that are set to be too long. By default, the diameter for a Cisco switch is set to seven. In networks smaller than this, the diameter should be adjusted to accurately reflect the topology.

CISCO LAN SWITCHING

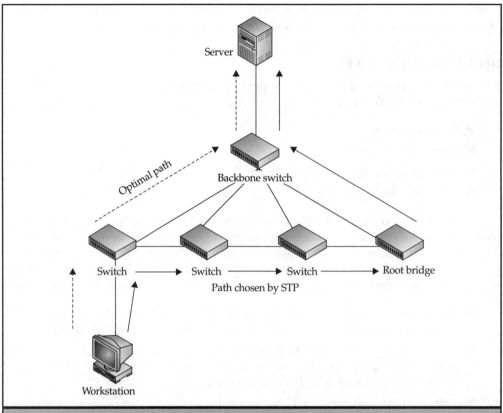

Figure 19-18.   *Poor root selection leading to inefficient data paths through the fabric*

BOOTP and DHCP problems can be an issue if the default switch configuration is used in access-layer switches with clients that require immediate network access (such as diskless workstations and some imaging software). To eliminate this problem, enable Portfast on the necessary ports on the access-layer switches, or disable STP entirely if no redundant links exist from the switch into the rest of the switch fabric.

Finally, several *show* and *debug* commands are useful in troubleshooting your STP problems. Table 19-3 describes the standard IOS commands.

Table 19-4 describes the CatOS *show* commands.

Command	Description
show spanning-tree	Shows detailed information on all interfaces and VLANs
show spanning-tree interface [number]	Shows detailed information on a specific interface
show spanning-tree vlan [number]	Shows detailed information on a specific VLAN
debug spantree events	Debugs topology changes
debug spantree tree	Debugs BPDUs

Table 19-3.   *Useful STP-Related Standard IOS* show *and* debug *Commands*

Command	Description
show spantree	Shows summary information
show spantree backbonefast	Shows spanning tree Backbonefast
show spantree blockedports	Shows ports that are blocked
show spantree portstate	Shows spanning tree state of a Token Ring port
show spantree portvlancost	Shows spanning tree port VLAN cost
show spantree statistics	Shows spanning tree statistic information
show spantree summary	Shows spanning tree summary
show spantree uplinkfast	Shows spanning tree Uplinkfast
show spantree [mod/port]	Shows information on a specific port or ports
show spantree [vlan]	Shows information on a specific VLAN number

Table 19-4.   *Useful STP-Related CatOS* show *Commands*

Virtual LANs (VLANs)

Virtual LANs are a major benefit to layer 2 switching. By using VLANs, you can split
a single switch into multiple broadcast domains. Note that a layer 2 switch still can't
route the packets, but using VLANs does allow the separation to be done with much
less hardware than is normally required. For instance, imagine that you have an
environment with two departments and a server, all of which must be separated into
different broadcast domains. Without using VLANs, you would need three switches
and a router, as shown in Figure 19-19.

With VLANs, however, you can split these into separate broadcast domains on a
single switch, allowing you to use a single switch and a single router. See Figure 19-20.

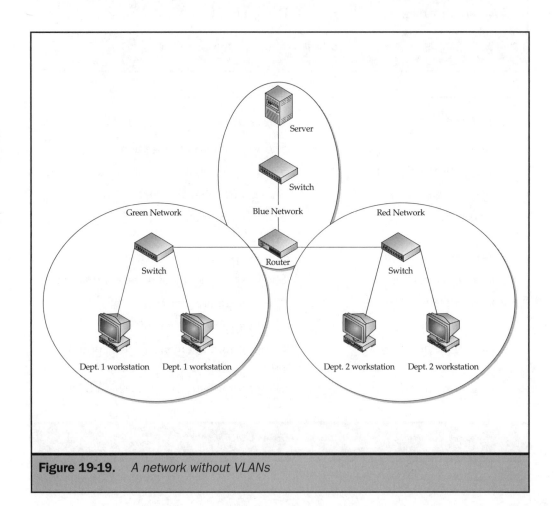

Figure 19-19.   *A network without VLANs*

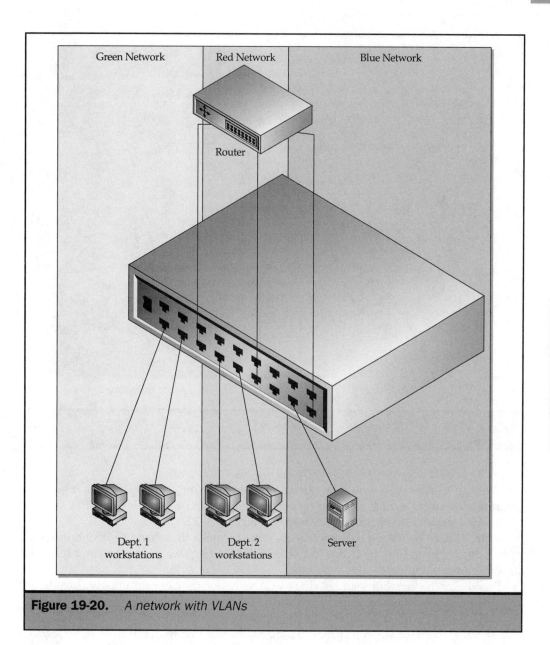

Figure 19-20. *A network with VLANs*

VLANs also allow you to complete user migrations in record time. For example, Figure 19-21 shows a group of users (Green Network) that needs to move to another building.

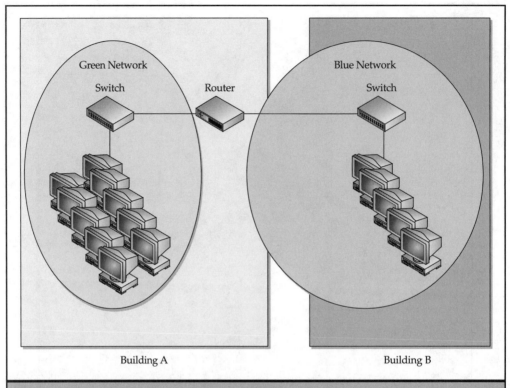

Figure 19-21.   *The base network for a migration example*

Without VLANs, you would be forced to make some major equipment changes, migrating the users' switch and adding a new router to the destination building (which currently does not have a router), as shown in Figure 19-22.

With VLANs, as shown in Figure 19-23, you can make this change with a minimum of fuss by simply adding the Green Network to the existing switch, saving you a bundle in both money and effort.

Because of the versatility and savings VLANs allow, they are commonly considered to be one of the most useful improvements made to layer 2 switching.

The basic principle of VLANs is fairly simple. You take individual ports on a switch and separate them into distinct broadcast domains. Just like with multiple switches, to get from one broadcast domain to another requires the use of a router. However, once you dig below the surface, a number of more advanced issues crop up—including VLAN membership (color) determination, VLAN tagging, use of VLAN Trunking Protocol (VTP), and STP considerations. But first, let's see how to define a VLAN.

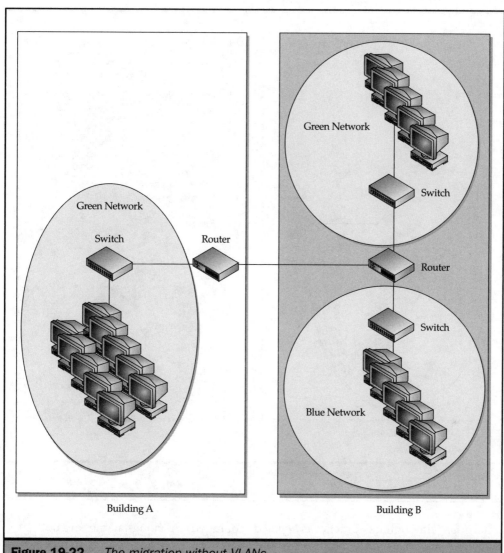

Figure 19-22.   *The migration without VLANs*

Defining a VLAN

On the most basic level, defining a VLAN simply consists of assigning the VLAN a
name and number. If trunking (covered in the "VLAN Tagging" section, later in the
chapter) is not used, this name and number can be different on each switch in the

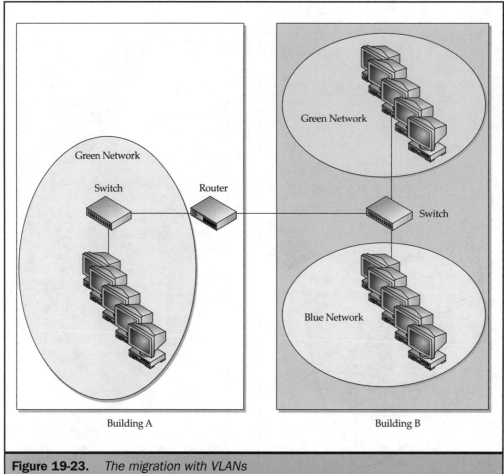

Figure 19-23. *The migration with VLANs*

enterprise (although this practice is *highly* discouraged). Although names are not actually required, typically, you will want to name the VLAN something that makes it easy for you to remember why the VLAN exists, like "executive" for the executive users VLAN.

Naming helps when you need to reconfigure the VLANs. (Otherwise, in a large environment, you will need charts or other documentation to help you remember what the VLAN was used for in the first place.) Several default VLANs may exist in a Cisco switch, depending on the media types the switch supports. In all cases, VLAN 1 is the default Ethernet VLAN, and, by default, all Ethernet ports on the switch are members of this VLAN.

VLAN Membership

VLAN membership (commonly called *color*) defines which VLAN a given device is a member of. VLAN membership can be defined either statically or dynamically.

Statically defined VLANs are the most common. When VLAN membership is statically defined, a device's VLAN membership is based on the switch port to which it is connected. This setup is the simplest way of configuring membership to a VLAN.

Dynamically defined VLAN membership is based on device MAC addresses. This setup a bit more complicated to configure, but it allows you to move hosts at will and have their VLAN membership "follow them around," which can be very useful for companies that relocate employees often or environments with lots of traveling laptop users.

VLAN Tagging

VLAN tagging is a process whereby a frame's VLAN membership is identified so that other switches and routers can tell from which VLAN the frame originated. As a result, you can send frames from many VLANs over the same physical port, a process known as *trunking*. This extremely useful enhancement allows you to conserve ports on switches and routers by making a single port a member of multiple VLANs. For instance, in Figure 19-24, three VLANs are configured on two separate switches and then connected to a router for routing between the VLANs, all performed without frame tagging. This setup requires a total of 12 ports (three per switch for interconnection between switches, plus three on one switch for the router connection and three ports on the router). This configuration is a bit wasteful and expensive.

In contrast, Figure 19-25 shows the same VLAN configuration and functionality, but using only four ports (one per switch for switch interconnection, plus one for the router connection and a single router port).

A couple of issues exist with trunking, however. First, over Ethernet, you can perform trunking only with Fast Ethernet or Gigabit links. Second, VLAN tagging requires modification of the Ethernet frame, meaning that all switches participating in VLAN tagging (at least with Cisco's Inter Switch Link (ISL) protocol) must support VLAN tagging. Finally, you can perform VLAN tagging in only one of two incompatible manners, using either the ISL (proprietary) or IEEE 802.1q (multivendor) protocols. Switches must, of course, be able to support the specific protocol used as well.

If you use 802.1q as your frame-tagging method, you allow for multivendor interoperability, as well as retain the standard Ethernet frame format, allowing switches that do not understand or support 802.1q to continue to forward the frame without registering an error. With 802.1q, the standard Ethernet frame is simply modified to carry the VLAN tagging information. This tagging methodology is useful if you need to extend a segment using a cheap (or old) switch. 802.1q also supports other vendor's switches, making it useful in non-Cisco shops as well.

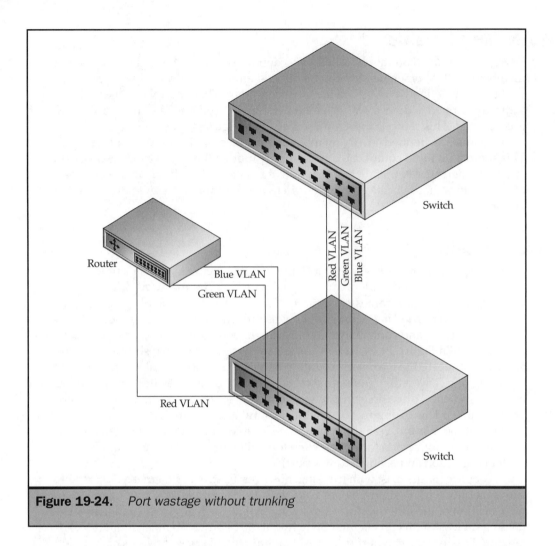

Figure 19-24. *Port wastage without trunking*

To top off all of this good news, 802.1q also adds only four bytes to the original frame, making it extremely efficient as well. The only real down side to using the 802.1q tagging method is that it is not supported in older Cisco switches; and even if it is supported, the specific switch software revision determines whether autonegotiation of trunk ports is supported. (Autonegotiation using Dynamic Trunking Protocol (DTP) is discussed in the "VLAN Configuration" section, later in the chapter.)

Note *As of this writing, most Cisco switch products support both ISL and 802.1q, and they also support DTP for trunk negotiation. For the Catalyst supervisor engine in the higher-end CatOS switches, 802.1q support was provided in release 4.1, and DTP support was provided in release 4.2.*

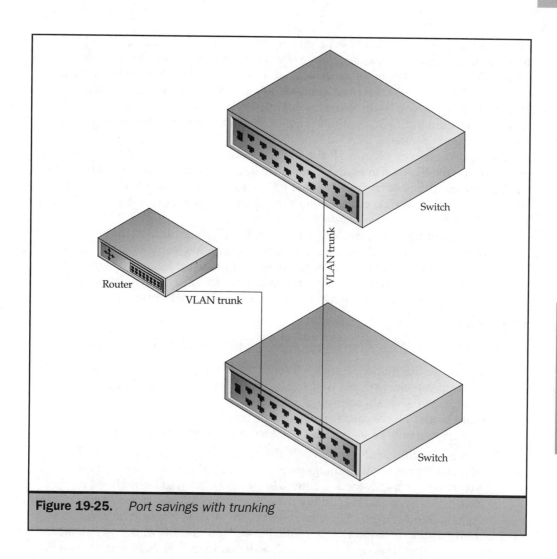

Figure 19-25. *Port savings with trunking*

For ISL, the news isn't as cheery. First, ISL will not work with any other vendor's products, because ISL re-encapsulates the base Ethernet frame. This limitation turns it into a frame that other switches may not understand. Second, ISL adds a whopping 30 bytes to each frame it encapsulates, making it rather inefficient. The only real good side to ISL is that it is supported in most past and current Cisco switching and routing software revisions, and it allows you to use Per VLAN Spanning Tree (PVST).

VLAN Trunking Protocol (VTP)

VTP is a useful addition to VLAN management that allows you to automatically assign VLAN definitions to multiple switches in a network. To truly appreciate the usefulness of this addition, put yourself in the shoes of a network engineer for a large, diverse network. Imagine you have 500 switches and over 100 VLANs defined. For VLANs to work over trunked links as advertised, the VLAN number needs to be identical on all participating switches in the enterprise, and you must pay attention to how you will know that "this VLAN is for executives, and this VLAN is for peons," and so on. With just numbers, you begin to see the complexity involved in configuring VLANs in such a large environment. Imagine if a user port were placed in the wrong VLAN because someone confused VLAN 151 with VLAN 115!

To help solve this problem, VTP automatically pushes out your VLAN definitions for you, allowing the VLAN names and numbers to be set on a single switch and propagated throughout the entire enterprise. Note that VTP does *not* propagate VLAN memberships to all switches (that would be disastrous in most environments), just the definition (name, number, and other basic information).

To accomplish this feat, first, VTP advertises VLAN configuration information out all trunk ports when enabled. Thus, neighboring switches learn about the VLANs present in the network and their configuration. Then these switches propagate the VLAN information to other switches, and so on. In addition, this process has a few more facets: VTP modes, VTP domains, VTP versions, and VTP pruning.

VTP Modes

VTP can run on a switch in one of three modes: client, server, and transparent:

- **Client**   In this mode, the switch listens for and propagates VTP advertisements for its management domain (discussed further in the next section). It makes changes to its VLAN configuration based on these advertisements. While in client mode, VLAN changes may not be made directly to the switch. VLAN changes can only be made using VTP while the switch is in client mode.

- **Server**   In this mode, the switch also listens for and propagates VTP advertisements for its management domain, but it also *generates* new advertisements. This mode allows VLAN information to be modified on the switch directly, including additions and deletions of VLANs from the management domain. Modifying a management domain's VTP configuration causes the configuration revision number (like the version number of the VTP database) to be updated. This update causes all switches in the management domain to update their VTP configuration with the new information. Typically, only one or two VTP servers should exist in each management domain, and configuration rights to these switches should be guarded carefully; otherwise, mistakes may be propagated across the entire management domain. (And if the configuration revision number is high enough, these mistakes can be costly and time-consuming to repair. This situation, and solutions to it, are covered in the "Troubleshooting LANs" section.)

- **Transparent**   In transparent mode, VTP information is forwarded, but the VLAN configurations contained within these advertisements are ignored. VLAN configuration can be directly modified on a switch in transparent mode, but these configuration modifications apply only to the local switch.

VTP Management Domains

VTP configuration is controlled through VTP management domains. A *management domain* is a specific name that all switches that will participate in VTP must be a member of. In other words, if a switch is not a member of a VTP domain, it will not get the VLAN information sent to that domain. Switches can be a member of only one VTP domain. In addition, for a VTP domain to operate properly, a couple of conditions must be met.

First, the VTP domain must be *contiguous*: for VTP information to flow between switches, the switches must have a continuous VTP domain without any breaks. For instance, in Figure 19-26, VTP advertisements will never reach Diablo, because it has no link to the other switches in its domain.

However, if you configure a link from Countach to Diablo, as shown in Figure 19-27, you will have VTP connectivity.

Second, VTP information is propagated only across trunk links. Therefore, if the trunking is not configured correctly or the trunked link fails, VTP information ceases to flow.

If these conditions are not met, VTP will not function properly. Therefore, it is typically easiest to configure all switches as part of the same VTP management domain.

VTP Versions

VTP has two versions that you can use: version 1 and version 2. These versions are, of course, completely incompatible with one another, and the world will come to a dramatic end if you configure both in the same network. (OK, the world will be fine, but be careful nonetheless.) VTP version 2 simply includes some additional features

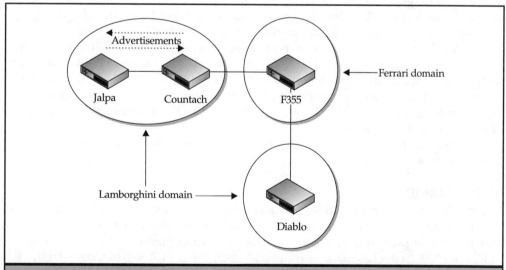

Figure 19-26.   *VTP information will not be propagated in the Lamborghini domain because it is not contiguous.*

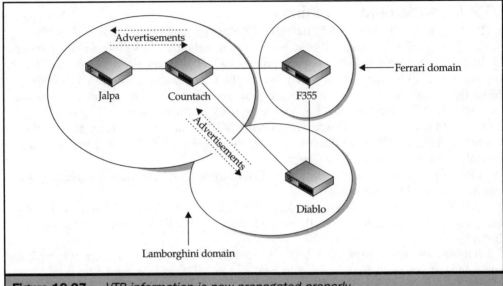

Figure 19-27. *VTP information is now propagated properly.*

not present in version 1, only two of which will be of any real importance in most networks: Token Ring VLAN support and multidomain transparent mode support.

Token Ring VLAN support allows a switch to use VTP to propagate VLAN information for your Token Ring environment. This functionality is of minimal use in most modern networks. Multidomain transparent mode VTP support, however, is useful in any environment that uses multiple VTP domains.

In a VTP version 1 switch, transparent mode switches propagate VTP information only if the management domain of the VTP advertisement matches their own. In other words, if your transparent mode switch is a member of the Corp VTP domain, and it receives an advertisement for the Org domain, it will not pass this advertisement on to any other switches. With VTP version 2, however, all VTP advertisements, regardless of the domain, are propagated.

Typically, it is easier to set all switches to use version 2, as long as version 2 is supported on all switches.

VTP Pruning

Finally, VTP includes an exceptional enhancement known as *VTP pruning*, which allows VTP to automatically determine which VLANs have members on a given switch, and removes (or prunes) unnecessary broadcast traffic from those switches. For instance, in Figure 19-28, Tama has members only on VLAN 1 and VLAN 4, but, by default, it receives all of the broadcast traffic for VLAN 2 and 3 as well. If VTP pruning were

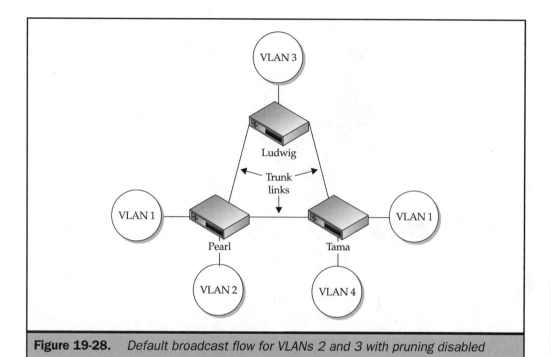

Figure 19-28.   *Default broadcast flow for VLANs 2 and 3 with pruning disabled*

enabled, broadcast traffic for VLANs 2 and 3 would be *pruned,* or never sent down the trunk connection to Tama, because it doesn't have any members on those VLANs.

This enhancement can obviously lead to significant broadcast traffic reduction in networks with concentrated user populations, like the network shown in Figure 19-29, because the broadcast traffic can be better contained.

VLAN STP Considerations

Another issue with VLANs is how they make use of STP. Like any switched network, as long as redundant links are present, STP needs to be enabled. The problem with VLANs is that STP actually has multiple broadcast domains to contend with. Three methods exist for STP functionality on VLANs: Common Spanning Tree (CST), Per VLAN Spanning Tree (PVST), and Per VLAN Spanning Tree Plus (PVST+).

CST is used with the IEEE's 802.1q protocol, and it simply sets up a single spanning tree for the entire switching fabric, regardless of the number of VLANs supported. A single STP topology is chosen for the entire physical network, and all VLANs are forced to use this topology. The up side is that this method takes minimal switch resources to compute the STP topology. Only one STP root exists, and STP recomputations should be minimal. The down side is that all VLANs must use the same (typically huge)

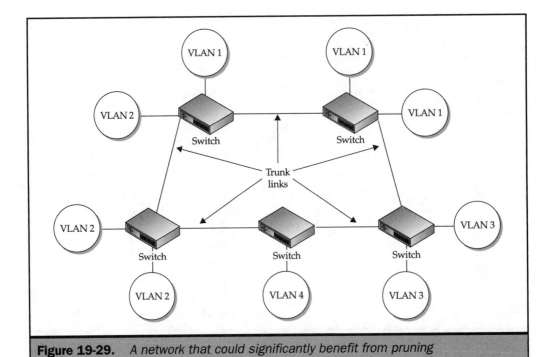

Figure 19-29. *A network that could significantly benefit from pruning*

topology, leading to the possibility that less than optimal paths will be chosen, as shown in Figure 19-30. In a large environment, the topology may also take a long time to converge.

PVST is used with Cisco's ISL protocol, and it uses a separate STP topology for each VLAN. This method has the advantage of optimal path selection (as shown in Figure 19-31) and minimal convergence time. The down side is that multiple STP roots must be used, multiple STP topologies must be calculated, and BPDUs must be sent for each topology, all of which lead to increased switch resource and bandwidth consumption.

PVST+ is an enhancement to PVST that allows interoperability between a CST network and a PVST network. Basically, PVST+ maps PVST spanning trees to the CST spanning tree, sort of like a "trunking gateway." This method can lead to some weird STP topologies, and it is generally much easier and more efficient to stick with PVST if you use ISL or CST if you use 802.1q.

VLAN Configuration

The first step in configuring VLANs is to define the logical topology you require. Configuring your VLANs is a fairly simple process if you have a good design to work

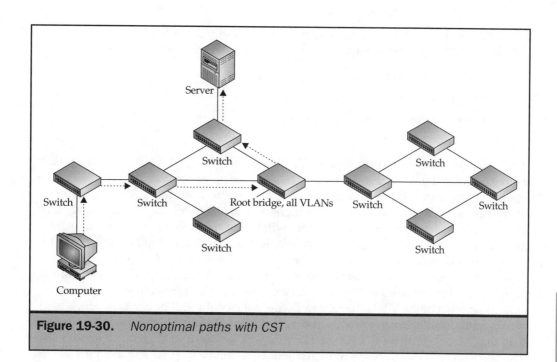

Figure 19-30.   *Nonoptimal paths with CST*

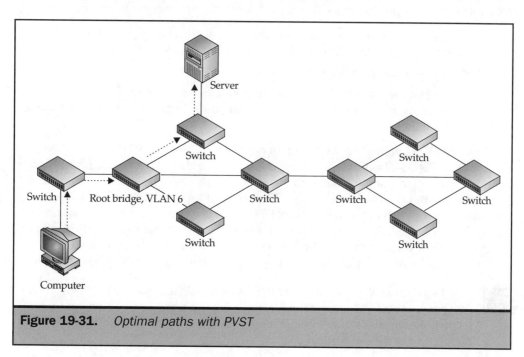

Figure 19-31.   *Optimal paths with PVST*

from. Although network design is beyond the scope of this book, once you have decided on a design, implementation consists of following four basic steps:

1. Enable and configure VTP.
2. Define the VLANs.
3. Assign VLAN membership.
4. Configure trunking.

You should enable and configure VTP before you define VLANs, because VLANs cannot be defined unless the switch is in VTP server or transparent mode. When installing other VTP-capable switches into the network, be extremely careful. By default, switches are typically configured for VTP server mode with a blank default domain name. If no domain name is assigned to a switch, the switch automatically uses the first domain name it hears in VTP advertisements. (From then on, however, it retains this name.)

This functionality can lead to an issue where you install an additional switch into a network that just happens to ship from the factory with a very high configuration revision number. Remember that VTP uses this number to figure out what the most current revision of the VTP database is. Because the configuration for this switch is blank and it defaults to server mode, it may begin advertising a higher revision number than the one currently used, overwriting your VTP configuration on all switches in the network with its "newer," blank configuration. To eliminate this problem, before installing a new switch into a production VTP-enabled network, perform the following steps:

1. Clear the switch configuration using the command *clear config all* (a CatOS switch), *erase config* (a standard IOS switch), or *delete nvram* (a 1900/2820 series switch).
2. Power cycle the switch.
3. Verify that the switch's VTP configuration revision number is zero by using the *show vtp domain* or *show vtp status* commands. Repeat steps if the number is higher than zero.

To enable VTP, define the VTP domain name and enable the VTP mode on the switch. On a CatOS-based switch, set the domain using the *set vtp domain [domain name]* command; and on a standard IOS switch, use the *vtp domain [domain name]* command.

At this point, you should also set a VTP password for the domain. If you set a password, for a switch to participate in VTP for that domain, it must have the correct password configured. Setting the domain password ensures that rogue switches do not enter the topology and overwrite the VTP configuration. To set the password, use the *set vtp password [password]* (CatOS) or *vtp password [password]* (standard IOS) command.

Note *To set VTP parameters on a standard IOS-based switch, you must be in VLAN configuration mode. To enter this mode, type* vlan database *from a privileged mode prompt. Also note that to apply any changes made in this mode, you must exit from the mode using the* exit *command.*

The following output shows these commands:

```
>3500 Configuration<
3508# vlan database
3508(vlan)# vtp domain Thisone password secret
Domain name set to Thisone .
Setting device VLAN database password to secret.
3508(vlan)# vtp server
Setting device to VTP SERVER mode.
3508(vlan)#exit
APPLY completed.
Exiting....


>5500 Configuration<
Cat5K (enable) set vtp domain Mydomain mode server passwd knowme
pruning enable v2 enable
Generating MD5 secret for the password ....
This command will enable the pruning function in the entire
management domain.
All devices in the management domain should be pruning-capable
before enabling.
Do you want to continue (y/n) [n]? y
This command will enable the version 2 function in the entire
management domain.
All devices in the management domain should be version2-capable
before enabling.
Do you want to continue (y/n) [n]? y
VTP domain Mydomain modified
Cat5K (enable)
```

If you decide not to use VTP, you can skip the domain name and password configuration, but you still need to set the switch's VTP mode to transparent to ensure that you can create VLANs as necessary without causing any undue configuration overwrites on other switches. Setting the mode on all switches to transparent effectively disables VTP.

To set the VTP mode, use the *set vtp mode [client | server | transparent]* (CatOS) or *vtp [client | server | transparent]* commands.

Once the mode is set, you will typically want to set the VTP to version 2 if all of your switches support version 2. To do this, use the command *set vtp v2 enable* (CatOS) or *vtp v2-mode* (standard IOS).

If you set v.2 on any switch in the domain, you must make sure to set it on all switches in the domain to ensure proper VTP functionality.

Finally, you will typically want to enable VTP pruning if you enable VTP. There are really no drawbacks to using pruning, so unless all of your switches are set to transparent mode, you should probably enable pruning. To enable pruning on the entire management domain, from the VTP server switch, issue the command *set vtp pruneeligible [vlan list]* (CatOS) or *vtp pruning* (standard IOS). You normally shouldn't need to do this because, by default, switches are typically set to prune VLANs 2-1000. To check which VLANs are eligible for pruning, use the *show vtp domain* or *show vtp status* commands.

Once you have configured VTP, you need to define your VLANs. To perform this task, set the VLAN names and numbers using the *set vlan [number] name [name]* (CatOS) or *vlan [number] name [name]* (standard IOS) command. If you are using VTP, remember to perform this task on the VTP server.

Once you have defined your VLANs, you will then need to assign members to the VLANs. To assign static VLAN members (based on port number), use the *set vlan [vlan number] [mod/port list]* (CatOS) or *switchport access [vlan number]* (standard IOS) command.

Both of these commands for 5500 and 3500 switches are shown in the following output:

```
>3500 Configuration<
3508(vlan)#vlan 200 name Classroom
VLAN 200 added:
 Name: Classroom
3508(vlan)#exit
APPLY completed.
Exiting....
3508#configure terminal
Enter configuration commands, one per line. End with CNTL/Z.
3508(config)#interface GigabitEthernet 0/2
3508(config-if)#switchport access vlan 200

>5500 Configuration<
Cat5K (enable) set vlan 200 name Classroom
Vlan 200 configuration successful
Cat5K (enable) set vlan 200 3/1-10
VLAN 200 modified.
VLAN 1 modified.
VLAN Mod/Ports
---- ----------------------
200 3/1-10
Cat5K (enable)
```

To assign VLAN membership dynamically (based on MAC address), you must use a VLAN Membership Policy Server (VMPS). Unfortunately, VMPS configuration could be a chapter in and of itself and is beyond the scope of this book.

> **Note**   *For more information on VMPS configuration, visit the following web addresses:*
> *http://www.cisco.com/univercd/cc/td/doc/product/lan/c2900xl/29_35xu/scg/kivlan.htm*
> *http://www.cisco.com/univercd/cc/td/doc/product/lan/c2900xl/29_35xp/eescg/mascvmps*
> *.htm*

Finally, you must configure trunking for the switch. Before getting into the specific command required for this, however, you need to learn about *trunk negotiation*. Trunk negotiation using Dynamic ISL (DISL) or Dynamic Trunking Protocol (DTP) allows two ports to negotiate whether to become trunking ports. This negotiation is used to allow you to configure only one side of the trunk. The other side can autoconfigure itself to the required parameters. Personally, I prefer to manually configure my trunk ports, because autonegotiation typically doesn't save enough time to be worthwhile (and has been known to cause major headaches).

Autoconfiguration works by using negotiation frames between two ports. If the settings on the ports are compatible, the ports will enter a trunking state. The ports can be set to one of the following five states:

- **On**   This state cuts trunking on. In this state, negotiation does not occur, but DTP frames are still issued (meaning that if both sides do not support DTP, errors may occur).
- **Off**   This state cuts trunking off.
- **Auto**   This it the default setting. The port will negotiate trunking status and encapsulation (ISL or 802.1q). The port will become a trunk if the other side is set to either on or desirable.
- **Desirable**   With this state, the port will become a trunk if the other side is set to on, auto, or desirable.
- **Nonegotiate**   With this state, the port will be a trunk, and it will not issue DTP frames to negotiate any parameters. This state is useful when switches do not support the same Dynamic Trunking Protocol.

Typically, it is easiest to just set both sides to either on or off. However, if you set one side to either on or desirable, the other side will automatically start trunking, because the default setting for all trunk-capable (100 Mbps or faster Ethernet) ports is auto.

Note   *As of this writing, 2900 and 3500 series switches (i.e., the standard IOS switches) do not support autonegotiation. For these switches to connect to a 1900 series or 5500 series switch, you will need to set the trunking mode on the 1900 or 5500 to nonegotiate.*

If you choose to use autoconfiguration, set the trunk port on one side to desirable, nonegotiate, or on. To do this for a 1900 series switch, use the command *trunk [desirable | on | auto]* from interface config mode on either of the Fast Ethernet interfaces. On a 5500 series switch, use the *set trunk [module/port] [desirable | on | auto]* command. The following output shows these commands:

```
>1900 Configuration<
1900F(config-if)#trunk on

>5500 Configuration<
Cat5K (enable) set trunk 3/1 desirable isl
Port(s) 3/1 trunk mode set to desirable.
Port(s) 3/1 trunk type set to isl.
Cat5K (enable) clear trunk 3/1
Port(s) 3/1 trunk mode set to auto.
Port(s) 3/1 trunk type set to negotiate.
```

To eliminate negotiation (and the possibility of autonegotiation failure), use the *trunk nonegotiate* command on a 1900 switch, or the *set trunk [module/port] nonegotiate* command on a 5500 switch. The 2900 and 3500 series will never negotiate, so the command to enable trunking on these switches is *switchport mode trunk.*

To set the VLANs transported across your trunk links, use the command *trunk-vlan [vlan list]* (1900 series), *set trunk [module/port] [vlan list]* (CatOS), or *switchport allowed vlan [add | remove] [vlan list]* (2900 or 3500 series).

Once these parameters are configured correctly on both sides, your switches should trunk correctly. Use the *show trunk* command to verify trunking functionality, as shown here:

```
Cat5K (enable) show trunk detail
* - indicates vtp domain mismatch
    Port        Mode Encapsulation      Status Native vlan
-------- ----------- -------------- ----------- ----------
    4/1-2        on          lane    trunking            1
```

```
Port                                                  Vlans allowed on trunk
-------- ---------------------------------------------------------------
4/1-2                                                            1-1005

Port                    Vlans allowed and active in management domain
-------- ---------------------------------------------------------------
4/1-2

Port            Vlans in spanning tree forwarding state and not pruned
-------- ---------------------------------------------------------------
4/1-2
```

Note *For trunked connections to routers, additional configuration on the routers is necessary. Chapter 22 covers router trunk configuration.*

Troubleshooting VLANs

Most VLAN problems are actually routing problems, because communication between VLANs is not possible without routing. Chapter 22 covers these routing issues in more detail. Other than routing problems, the major cause of VLAN woes is usually a configuration error. When troubleshooting VLAN configuration problems, look for these common issues:

- DTP/DISL autoconfiguration mismatches
- VLAN numbering mismatches over trunk links
- Trunking encapsulation problems (i.e., ISL on one side and 802.1q on the other)
- STP loops
- VLAN membership misconfigurations
- VTP domain names or passwords being typed incorrectly
- VTP domains not being contiguous
- VTP versions not matching

Several *show* and *debug* commands can help you determine the root cause of your VLAN issues. These commands are described in Table 19-5 for standard IOS switches. Table 19-6 describes the *show* and *debug* commands for CatOS switches.

CISCO LAN SWITCHING

Command	Description
show vlan	Shows all VLANs
show vlan brief	Shows a summary of all VLANs
show vlan id [vlan number]	Shows details on a specific VLAN
show vlan name [vlan name]	Shows details on a specific VLAN
show vtp status	Shows information about your VTP domain, such as the name and revision number
show vtp counters	Shows statistical information on VTP
show interface type [number]	Shows basic interface information
show interface type [number] switchport	Shows interface VLAN membership and trunking
*show [changes	current	proposed]*	From VLAN config mode. Shows database changes, and current config and proposed changes to database
debug sw-vlan vtp	Debugs VTP packets

Table 19-5. *Useful VLAN-Related Standard IOS* show *and* debug *Commands*

Command	Description
show vlan	Shows a summary of all VLANs
show vlan trunk	Shows which VLANs are being trunked
show vlan [vlan]	Shows details on a specific VLAN
show vtp domain	Shows information about your VTP domain, such as the name and revision number
show vtp statistics	Shows statistical information on VTP
show port trunk	Shows which ports are trunking
show port [module/port]	Shows port information (including trunking status)

Table 19-6. *Useful VLAN-Related CatOS* show *and* debug *Commands*

Summary

This chapter covered layer 2 switching technologies from both a theoretical and a practical perspective. You learned how switches use STP and VLANs to eliminate bridging loops and ease administration. The next chapter examines layer 3 switching and how it is used to combine the speed of switching with the power of routing.

Chapter 20

Layer 3 Switching

This chapter covers layer 3 switching functionality and details how it is used to provide the best in high-performance LAN routing. It begins with a brief look at what makes a layer 3 switch tick and then takes a look at specific layer 3 switching devices. Finally, the chapter covers different layer 3 switching methods and route processors, and ends with configuration guidelines and tips.

Note *Routing is covered more specifically in Part IV of this book, so this chapter does not cover routing protocols, static configuration, or other topics that, although they do apply to layer 3 switching, are covered in later chapters. Instead, this chapter covers topics specific only to layer 3 switching and leaves the explanations and configuration commands for routing to later chapters.*

Layer 3 Switching Overview

Layer 3 switching is essentially very fast routing. As mentioned in Chapter 9, layer 3 switches are essentially routers that perform their calculations in application-specific integrated circuits (ASICs) rather than in software. As a result, a bit of flexibility is removed from the device (upgrading routing processes is a matter of switching out hardware, not just software), but the speed of these devices is increased to wire speed.

The real difference between a router and a layer 3 switch in *most* environments (ISPs and some enterprise networks excluded) lies in the device's role in the network. Layer 3 switches are typically used in the LAN or MAN to provide high-speed data transfer for local clients, whereas routers are used to provide lower-speed WAN access. See Figure 20-1.

Because routers have more flexibility in both their feature sets and their interfaces, using software-based processing works out pretty well. Also, most WAN connections are run to branch offices and tend to be slower than 100 Mbps, so the performance reduction inherent in a router is not such a major issue.

However, layer 3 switches are still very useful and powerful devices that can increase network performance by segregating your environment into broadcast domains, just like routers. In addition, they can significantly ease configuration headaches (and sometimes even the expense) when using VLANs by allowing you to use the switches to route between VLANs.

From this standpoint, a layer 3 switch uses VLANs just like a layer 2 switch: you assign VLAN memberships to split ports into logical networks. The layer 3 switch, however, can route between these networks on its own and (usually) at wire speed, meaning that you do not have to configure the "router on a stick" or worry about trunking between the switch and the router (reducing port requirements in the process).

However, cramming all of these features into a single device does lead to some interesting configuration requirements. From the configuration end, two primary types of layer 3 switches exist: those with the route processor configuration integrated into

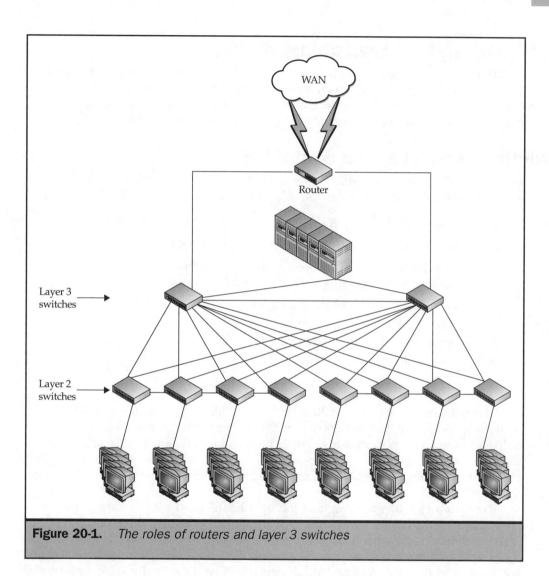

Figure 20-1.   *The roles of routers and layer 3 switches*

the switching command set (like the 2948G L3 and 6000/6500 series), and those for which a separate IOS and command set are used (like the 5000/5500 series).

In addition, layer 3 switches can be classified by the layer 3 switching method they use. This chapter concentrates on Cisco's most common layer 3 switching method, known as *multilayer switching (MLS)*. This switching technology is used in nearly every layer 3 switch Cisco makes, including the 2948G L3, 4908G L3, the 5000/5500 series, and the 6000/6500 series, and it's the method you are most likely to use in real life.

How Layer 3 Switching Works

In this section of the chapter, we will discuss how layer 3 switching works. First, we will cover the difference between layer 3 switching and routing. Then, we will cover MLS's basic data-forwarding paradigm. Finally, we will take a look at the interactions between switching devices and routing devices in a multilayer-switching environment.

Routing Versus Layer 3 Switching

First, you need to know that MLS makes a distinction between routing and layer 3 switching. Basically, when a router routes a packet, it performs the following actions:

- Receives a packet destined for its MAC address, but a remote IP address
- Looks in its routing table to determine where to send the packet
- Modifies the source MAC address in the packet, removing the sender's MAC address and inserting the router's MAC address
- Modifies the destination MAC address in the packet, removing its MAC address and inserting the next hop MAC address
- Decrements the IP TTL by at least 1
- Recomputes the frame and IP checksums
- Forwards the packet out the appropriate interface

A layer 3 switch does things a little differently. First, layer 3 switching is based on the understanding that packets do not live in a vacuum. In other words, you are unlikely to send a single packet to a remote host, and that remote host is unlikely to send only one packet back. In most cases, you will both transmit and receive several packets to and from a remote host. Routers (in the classical definition) do not follow this principle.

Every packet a router receives goes through the full process of looking in the table, changing the headers, and being forwarded. This process is time-consuming. A layer 3 switch, on the other hand, believes in "route once, switch many." In other words, if you send 4,000 packets from your PC to a specific remote host (say, www.cisco.com), you have to do all of that routing work for only the first packet in the communication. The rest just use the information gained from routing the first packet. Thus, the entire communication (except for the first packet) is *much* faster than if you routed each individual packet.

Note   *The classical definition of routing is known as* process switching. *By default, most Cisco routers are configured to process switch packets. However, some models are also able to use other switching mechanisms (such as fast switching and Cisco Express Forwarding) that more closely resemble the layer 3 switching "route once, switch many" rule. For more information on switching mechanisms in routers, please visit http://www.cisco.com/univercd/cc/td/doc/product/software/ios121/121cgcr/switch_c/xcprt1/xcdovips.htm.*

MLS Data Forwarding

In Cisco-ese, the communication stream is known as a *flow*. A flow can be described in three ways:

- A communication to a given destination (known as destination IP)

- A communication to a given destination from a specific host (known as source destination IP)

- A communication to a given destination from a specific host using specific layer 4 protocols and port numbers (known as IP flow)

Note   *Flows are unidirectional. That is, if PC1 sends a packet to PC2, the packet stream from PC1 to PC2 is a single flow. If PC2 responds back to PC1, that is classified as a separate flow.*

The type of flow used by the switch (known as the MLS-SE, or *multilayer switching-switching engine*) is known as the *flow mask*. In most cases, the switch uses the destination IP mask. However, if you are using access lists (discussed later in this section, as well as in Chapter 27), or if you need more specific accounting or logging data (in other words, you want to know who is eating up all of your bandwidth and which application they are using), you may choose to use one of the other masks.

The flow mask is primarily used to determine *which* information about flows is entered into the switching cache (known as the *MLS cache*). For instance, in Figure 20-2, PC1 (192.168.1.1) is communicating with Server1 (10.1.1.1) through a layer 3 switch. For the first packet, the layer 3 switch notices that the destination MAC is for the router, so it looks in its MLS cache for the destination IP address.

Because an MLS entry was not found for the destination IP address, it forwards the packet to the router (known as the MLS-RP, or *multilayer switching-route processor*) for processing (shown in Figure 20-3).

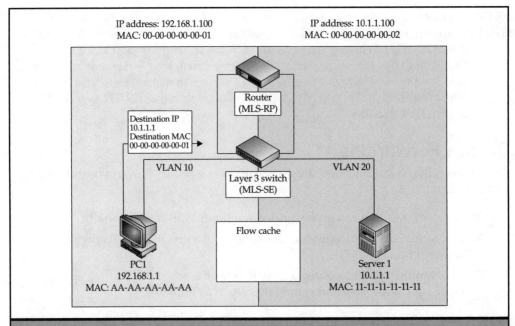

Figure 20-2. The MLS-SE cannot find the destination in the MLS cache.

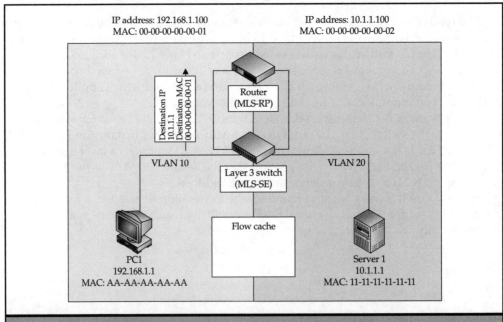

Figure 20-3. The MLS-SE forwards the packet to the MLS-RP.

The router routes the packet and forwards it back to the MLS-SE (switch), which adds the destination to the MLS cache (shown in Figure 20-4). Because this example is using only the destination IP flow mask, the only information added is the destination IP, destination MAC address, and destination port.

When PC1 sends the next packet, the MLS-SE looks in its cache, finds the entry for the destination, and rewrites the packet with the new information, without all of the additional overhead associated with routing the packet. See Figure 20-5.

Consequently, when PC 2 sends a packet to Server1, the communication will match the flow already entered in the table; and it will be switched, not routed, to the destination (shown in Figure 20-6).

So, the basic process for layer 3 switching is as follows:

■ The switch examines the destination MAC address. If it is for any of the switch's (MLS-SE) configured routers (MLS-RP), layer 3 processing begins. If not, the packet is layer 2 switched.

■ If the packet is to be layer 3 processed, the switch looks for an entry in the MLS cache for the destination IP address.

■ If an MLS entry is found, the packet is rewritten using the information in the entry, the TTL is decremented, and the checksums are recomputed.

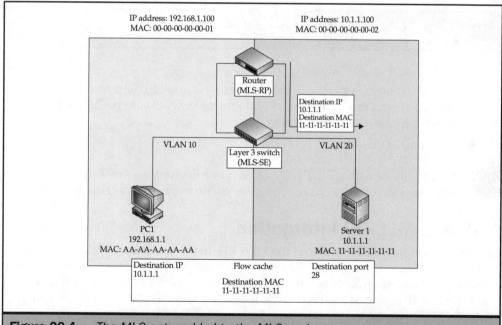

Figure 20-4.   *The MLS entry added to the MLS cache*

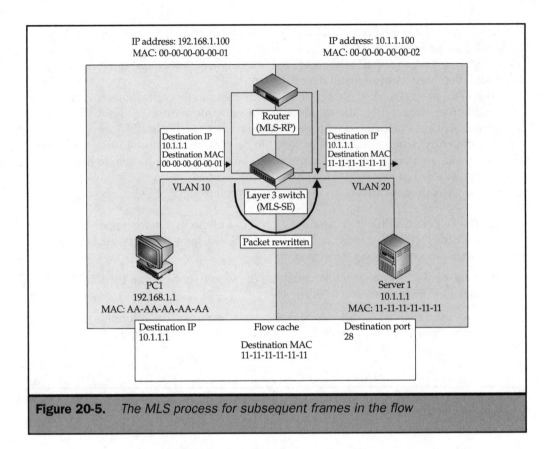

Figure 20-5. *The MLS process for subsequent frames in the flow*

- If an MLS entry is not found, the frame is forwarded to the appropriate MLS-RP; and it routes the packet and returns the packet to the MLS-SE, which writes a new cache entry.

- The frame is forwarded.

 In MLS, only the destination IP is checked for a match, regardless of the flow mask used. This point is important because some Cisco documentation about this subject can be misleading.

MLS-SE and MLS-RP Interaction

So how do the MLS-SEs know about the MLS-RPs in the first place? Well, it turns out that you must manually add MLS-RPs once to a given MLS-SE. However, all MLS-RPs will also send advertisements out at regular intervals (every 15 seconds, by default) using the CGMP multicast address. These messages, known as Multilayer Switching Protocol (MLSP) messages, contain information about the MLS-RPs' known routes, MAC address information, and access lists.

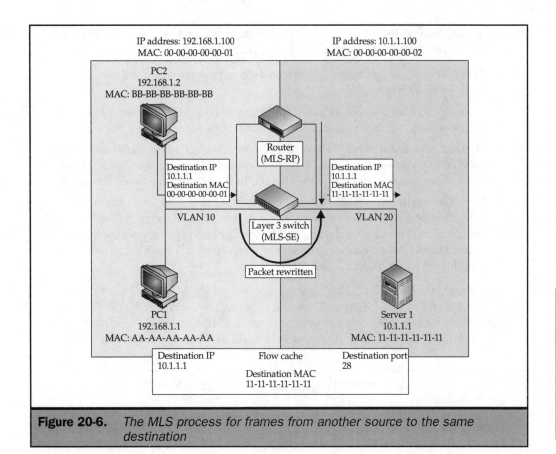

IP address: 192.168.1.100
MAC: 00-00-00-00-00-01

IP address: 10.1.1.100
MAC: 00-00-00-00-00-02

PC2
192.168.1.2
MAC: BB-BB-BB-BB-BB-BB

Router
(MLS-RP)

Destination IP
10.1.1.1
Destination MAC
00-00-00-00-00-01

Destination IP
10.1.1.1
Destination MAC
11-11-11-11-11-11

VLAN 10

VLAN 20

Layer 3 switch
(MLS-SE)

Packet rewritten

PC1
192.168.1.1
MAC: AA-AA-AA-AA-AA

Server 1
10.1.1.1
MAC: 11-11-11-11-11-11

Destination IP
10.1.1.1

Flow cache

Destination MAC
11-11-11-11-11-11

Destination port
28

Figure 20-6.   *The MLS process for frames from another source to the same
destination*

This information allows the MLS-SEs to automatically learn all of the important
details they need (like MAC addresses and flow masks) from the MLS-RPs. The
MLS-SEs in the network listen to this address; but all non-MLS switches simply forward
the information, so it doesn't take up processor time on switches that do not need the
information. MLSP is also the enabler for MLS, which means that for MLS to work
properly, MLSP must be enabled on the MLS-RPs.

The catch is that all of the MLS-SEs and all of the MLS-RPs that are going to know
about each other must be in the same VTP domain.

> **Note**   *You can sort of get around the VTP domain issue by not making your switches members
> of any VTP domain. If the VTP domain field on your MLS-SEs is blank, you do not have
> to configure a VTP domain on the MLS-RPs. (Their VTP domain will be blank as well.)*

Also, access lists need special care in MLS. First, a brief explanation of access lists is
in order. (For a more thorough discussion, please refer to Chapter 27.) Access lists (or

ACLs, for *access control lists*) are used to allow or deny access based on certain criteria. The following list describes the four types of ACLs:

- **Standard**   This type filters (allows or denies) packets based only on source address or addresses.

- **Extended**   This type filters based on protocol, source and destination IP address(es), port numbers, and other parameters.

- **Dynamic (lock and key)**   This type allows you to provide access lists that are specific to individual users, which requires that the user log in to the router or firewall before access is allowed. This method of access is a bit cumbersome.

- **Reflexive**   This type filters based on information similar to an extended access list, but reflexive access lists allow or deny connections based on session information. This allows the firewall to configure temporary access list entries automatically—for instance, to allow any return traffic from a session established by an internal client.

The real issue with ACLs is that they are applied on the MLS-RP; so if a packet is MLS-switched by the MLS-SE, it will never see the ACL—and it may be improperly allowed. This problem is best shown with an example. In Figure 20-7, the MLS-RP has an outbound ACL set to allow FTP access to host 10.1.1.1 but deny all other traffic.

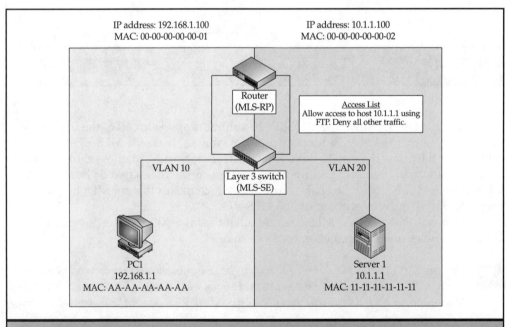

Figure 20-7.   *The network configuration for the MLS ACL example*

Host 192.168.1.1 sends an FTP packet to 10.1.1.1. The MLS-SE forwards this packet to the MLS-RP because it has no entry in its MLS cache for the destination. The MLS-RP checks the ACL and allows the packet, forwarding it back to the MLS-SE. The MLS-SE adds the entry to its MLS cache and forwards the packet. Now, if host 192.168.1.1 sends a Telnet packet to 10.1.1.1, it will be allowed because the MLS-SE examines the MLS cache for the destination IP (and *only* the destination IP, regardless of the flow mask used), and it will match. Therefore, the packet will never be forwarded to the MLS-RP, which means the ACL will not be enforced.

You can use flow masks to help reduce this problem (although not the way you would think). Whenever a standard ACL is configured on an MLS-RP, it sets its flow mask to source destination IP. Whenever an extended ACL is set on the MLS-RP, it changes its mask to IP flow. The MLS-SEs in the environment can use only one flow mask, and they always use the most specific flow mask configured on any of its configured MLS-RPs, so the MLS-SE will change its flow mask to the most specific one encountered.

> **Note**   *Some switches (the 6000 series included) do not support external MLS-RPs, and therefore do not use MLSP. For this reason, an ACL change on the MLS-RP in these switches will not change the flow mask automatically.*

If the flow mask is changed, the change causes all entries in the MLS cache to be removed. This functionality causes all packets to be sent to the MLS-RP for processing. The basic idea is that if the communication is denied, the packet will never be forwarded back to the MLS-SE; it will simply be dropped. Therefore, the MLS-SE should not add the denied entry to its MLS cache, and the communication will be (correctly) denied. This process breaks down in the previous example. Because the MLS-SE does *not* use the flow mask to help it match flows (it always matches based *only* on the destination IP), if a client is allowed access using one protocol but denied access for other protocols, the issue may arise where the client is incorrectly allowed access to the remote host.

> **Note**   *The 6000 series with the Policy Feature Card (PFC) does not suffer from the same problem with incorrectly allowing access.*

A few additional restrictions apply when using ACLs with MLS:

- Only standard and extended ACLs may be used. (Dynamic and reflexive ACLs are not supported.)
- ACLs may be set only on outbound interfaces on the MLS-RP. Inbound access lists are not supported.

> **Note**   *You can use an inbound access list on the 6000 series by using the command* mls rp ip input-acl *in IOS 12.0(7)XE or later.*

A few other restrictions apply on MLS, such as hardware and software requirements. From the switching side, the requirements are as follows:

- A Catalyst 5000/5500 series switch with
    - Supervisor Engine 4.1(1) or later
    - Supervisor Engine II G or III G, or Supervisor Engine III or III F with a NetFlow Feature Card (NFFC) or NFFC II
- A Catalyst 6000/6500 series switch with a Multilayer Switch Feature Card (MSFC)

From the routing side, the requirements are the following:

- RSM, RSFC, or an external Cisco 7500, 7200, 4700, 4500, or 3600 series router
- Cisco IOS release 12.0(3c)W5(8a) or later on the Route Switch Feature Card (RSFC)
- Cisco IOS release 12.0(2) or later on Cisco 3600 series routers
- Cisco IOS release 11.3(2)WA4(4) or later on the Route Switch Module (RSM), or Cisco 7500, 7200, 4700, and 4500 series routers

In addition, you cannot use MLS if the following items are configured on the MLS-RP:

- IP Security (IPSec encryption commands and options) disables MLS on the given router interface.
- Any compression applied to an interface disables MLS for that interface.
- NAT enabled on an interface disables MLS for that interface.
- QoS features on an interface may disable MLS, specifically, Committed Access Rate (CAR).

Layer 3 Switching Configuration

Because the switches with the integrated routing and switching command sets and engines (like the 2948G L3 or 4908G L3) are typically the easiest to configure, let's concentrate on them first.

These switches perform the routing functions internally, so the first step in configuring these devices is to decide which routing/switching mode to use: Integrated Routing and Bridging (IRB) or Concurrent Routing and Bridging (CRB). Typically, you will want to use IRB, but let me explain the differences between the two so you understand why.

IRB Versus CRB

With CRB, you can both route and bridge a given protocol on a given switch, but no communication can occur between routed clients and bridged clients (at least, not without adding another router). For example, with CRB, you can set up the configuration shown in Figure 20-8.

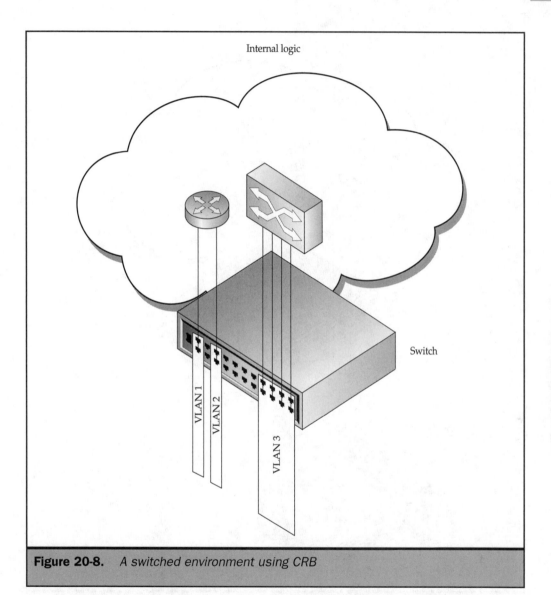

Figure 20-8. *A switched environment using CRB*

In this scenario, VLAN 1 can communicate with VLAN 2, but no connectivity exists between VLAN 1 and VLAN 3, or VLAN 2 and VLAN 3. To achieve connectivity between these networks, a separate router (and some additional routes) needs to be added, as shown in Figure 20-9.

With IRB, on the other hand, you could route between the sets of bridged interfaces without any additional hardware. This task is accomplished through specialized

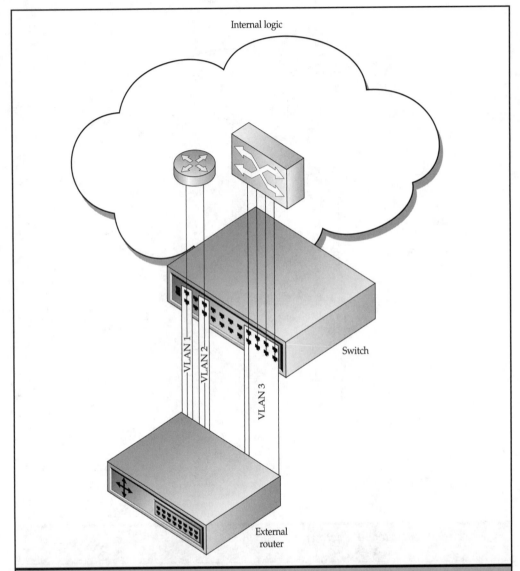

Figure 20-9. *The changes required to enable routing between VLAN 3 and the other VLANs using CRB*

virtual interfaces known as *bridge virtual interfaces (BVIs)*. The BVIs are like special internal ports that connect to all other BVI ports, as shown in Figure 20-10.

In Figure 20-10, each BVI is connected to the other BVIs. Each BVI has its own IP address, which is what the clients would use as their default gateway. When a client

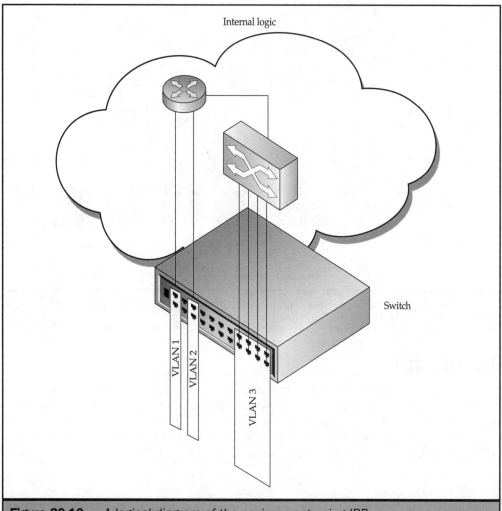

Figure 20-10.   *A logical diagram of the environment using IRB*

wants to communicate with another VLAN, the client simply forwards the packet to the BVI IP address for its subnet, and the switch routes between the BVIs. This functionality makes it extremely easy for the network admin to configure multiple VLANs and have the switch route between them. All you have to do is configure the ports for the appropriate BVI and enable routing between the BVIs.

For the full effect, let's take a look at the configuration commands for CRB versus IRB. The next two sections will use the environment shown in Figure 20-11 as an example to demonstrate the differences between CRB and IRB.

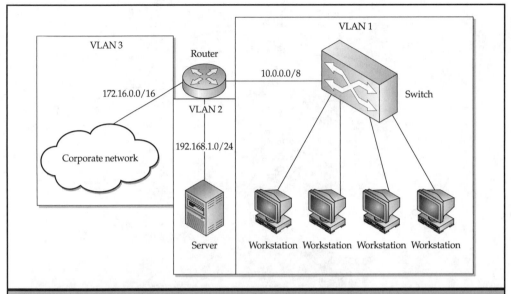

Figure 20-11. *The proposed environment, independent of the bridging modes*

Configuring CRB

To configure the example environment shown in Figure 20-11 with CRB, you must first add a few additional devices, as shown in Figure 20-12: a hub (or simple switch) and a router.

Once you perform this task, you enter the following commands on the switch (the commands are explained in the text after the example):

```
Switch# config t
Switch(config)# bridge crb
Switch(config)# interface e 1
Switch(config-if)# ip address 172.16.1.1 255.255.0.0
Switch(config-if)# bridge-group 3
Switch(config-if)# interface e 3
Switch(config-if)# ip address 192.168.1.1 255.255.255.0
Switch(config-if)# bridge-group 2
Switch(config-if)# interface e 6
Switch(config-if)# bridge-group 1
Switch(config-if)# interface e 7
Switch(config-if)# bridge-group 1
Switch(config-if)# interface e 8
Switch(config-if)# bridge-group 1
```

```
Switch(config-if)# interface e 9
Switch(config-if)# bridge-group 1
Switch(config-if)# interface e 10
Switch(config-if)# bridge-group 1
Switch(config-if)# ^Z
Switch(config)# bridge 1 protocol ieee
Switch(config)# bridge 2 protocol ieee
Switch(config)# bridge 2 route ip
Switch(config)# bridge 3 protocol ieee
Switch(config)# bridge 3 route ip
```

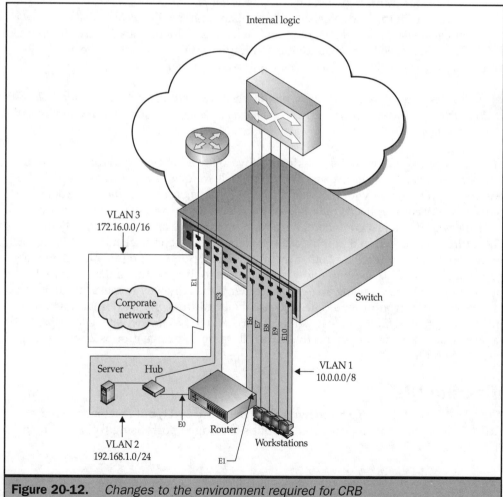

Figure 20-12. *Changes to the environment required for CRB*

The first bolded command, *bridge crb,* enables CRB on the switch. The second bolded command, *ip address,* gives the e1 interface a valid IP address for its subnet (required because you will be routing over this interface). You then assign the interface to a bridge group (in this case, 3). The bridge group number you choose doesn't really matter, as long as you remember that the bridge group is the logical bridging or routing interface.

The same is done for the e3 interface for the same reasons. You give it a different bridge group number to denote in the switch that they are members of separate logical networks. By putting the interfaces in separate bridge groups, you are telling the switch not to bridge traffic between the interfaces.

You then switch to the e6 interface and configure e6–e10 to be members of bridge group 1. You do not give these interfaces IP addresses, because they will not (directly) participate in routing. You set them all to the same bridge group to ensure that the switch will bridge traffic between this set of interfaces.

Finally, you tell the bridge groups which Spanning Tree Protocol to use with the *bridge [group number] protocol ieee* command (which configures spanning tree to use the IEEE 802.1D STP specification), and you configure routing between bridge groups 2 and 3 with the *bridge [group number] route [protocol]* commands.

> **Note**   *For these scenarios to work, a few routing commands must be entered on both the layer 3 switch and the external router. These commands, along with static routing, are covered in Chapter 22.*

At this point, you might be thinking, "Why do I need the external router? Couldn't I just issue the *bridge 1 route ip* command and enable routing using bridge group 1?" The answer is no, at least not with CRB. For the *bridge [group number] route [protocol]* command to work, an IP address needs to be set for the clients to use as the default gateway.

However, if you just go in and give a random interface an IP address, only the client directly attached to that interface would be able to send packets, and that client would not be able to reach any other members of its bridge group (only hosts on other subnets) because CRB does not allow bridging *and* routing on the same interface. In other words, with CRB, using an external router is the only way to meet the goal. The clients on bridge group 1 would need to use the IP address of the external router's e1 interface as their default gateway. All traffic for bridge group 1 will have to be bridged (at least from the switch's perspective), and all traffic on bridge groups 2 and 3 will have to be routed.

Configuring IRB

Now, let's look at the scenario shown earlier in Figure 20-11, but this time we'll use IRB. With IRB, you can logically route groups of bridged ports using BVIs, as shown in Figure 20-13.

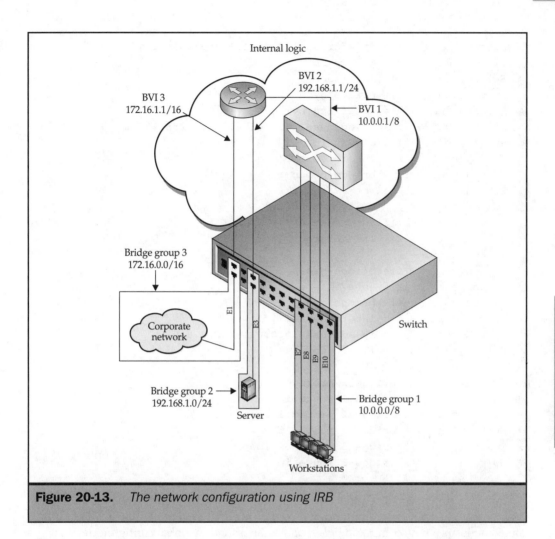

Figure 20-13.   *The network configuration using IRB*

Wow, that's a much simpler physical configuration, right? Luckily, the software configuration is actually easier as well—because with IRB, you do not have an external router to worry about. For this example, the configuration is as follows:

```
Switch# config t
Switch(config)# bridge irb
Switch(config)# mls rp ip
Switch(config)# interface e 1
Switch(config-if)# bridge-group 3
Switch(config-if)# interface bvi 3
```

```
Switch(config-if)# ip address 172.16.1.1 255.255.0.0
Switch(config-if)# mls rp ip
Switch(config-if)# interface e 3
Switch(config-if)# bridge-group 2
Switch(config-if)# interface bvi 2
Switch(config-if)# ip address 192.168.1.1 255.255.255.0
Switch(config-if)# mls rp ip
Switch(config-if)# interface e 6
Switch(config-if)# bridge-group 1
Switch(config-if)# interface e 7
Switch(config-if)# bridge-group 1
Switch(config-if)# interface e 8
Switch(config-if)# bridge-group 1
Switch(config-if)# interface e 9
Switch(config-if)# bridge-group 1
Switch(config-if)# interface e 10
Switch(config-if)# bridge-group 1
Switch(config-if)# interface bvi 1
Switch(config-if)# ip address 10.0.0.1 255.0.0.0
Switch(config-if)# mls rp ip
Switch(config-if)# ^Z
Switch(config)# bridge 1 protocol ieee
Switch(config)# bridge 1 route ip
Switch(config)# bridge 2 protocol ieee
Switch(config)# bridge 2 route ip
Switch(config)# bridge 3 protocol ieee
Switch(config)# bridge 3 route ip
```

In this example, you just add a few commands to the CRB config. The first difference is obviously the *bridge irb* command, which tells the switch to use IRB instead of CRB. Then you issue a new command, *mls rp ip*. When issued in global config mode, this command globally enables multilayer switching on the switch. (You still need to enable it on each BVI, however.)

As you will see later in this section, the mls rp ip *command also enables MLSP on a router. Therefore, it must be configured both globally and for each interface that will be participating in MLS on all MLS-RPs in the network.*

Next, you add interface e1 to bridge group 3, just like before; but instead of giving the interface an IP address, you configure the BVI for the bridge group to which it belongs with the IP address. To perform this task, you issue the command *interface bvi [bridge group number]*. Then you will go into interface config mode for the BVI for the given bridge group (in this case, 3). There, you issue the IP address command, just like normal. However, you also add an additional command, *mls rp ip*, to allow this BVI to be used as a routing interface. You duplicate these steps for BVIs 2 and 1, and then you issue the *bridge [group number] route ip* command to enable routing on all bridge groups. Now you have your functional, fully routed network all contained in one very powerful layer 3 switch. Note that from a client standpoint, the IP address configured for the BVI for a given bridge group is used as the default gateway.

In addition to being less of a headache for the administrator, another advantage of using IRB is flexibility. In this scenario, you have four ports unused on the switch. With IRB, you can use these ports as necessary to expand your bridge groups. For example, if you wanted to add a new server to bridge group 2, as shown in Figure 20-14, you

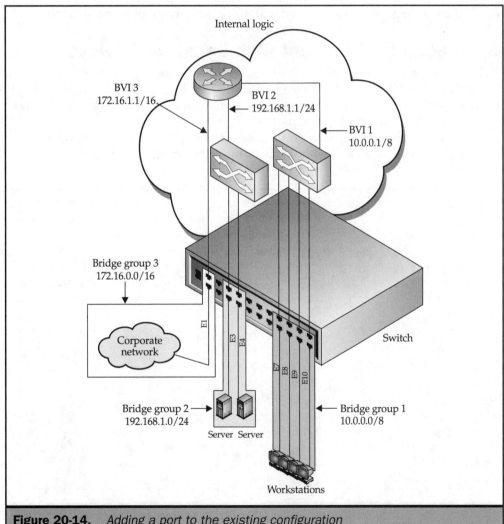

Figure 20-14.   *Adding a port to the existing configuration*

would simply add an available port (such as e4) to the bridge group, and you would be done.

Adding the additional port to the bridge group is as simple as issuing the following commands:

```
Switch(config)# interface e 4
Switch(config-if)# bridge-group 2
```

Truthfully, once you get used to the idea of using layer 3 switching (especially with IRB in a consolidated device), you will have difficulty imagining how to configure or handle the wiring nightmares involved with using separate routers and switches.

Configuring an Environment with Separate Devices

Unfortunately, even some MLS-capable switches require using separate devices for the routing and switching duties (like the 5000/5500 series). Configuration for these devices is a bit more complicated and time-consuming than what you have seen thus far, but it involves the same basic steps.

To see the configuration process for these switches, let's take another example network (shown in Figure 20-15) and walk through the configuration.

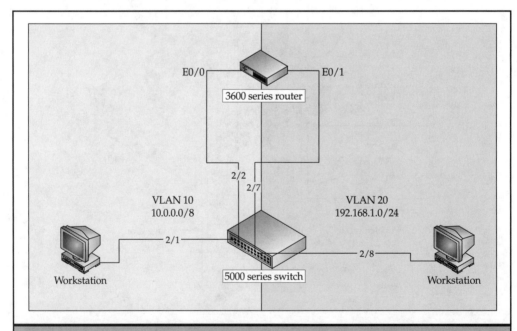

Figure 20-15. *A simple configuration for basic layer 3 switching*

First, you need to configure the router. Issue the following commands (again, explained in the text that follows the configuration):

```
3600(config)# mls rp ip
3600(config)# interface fa 0/0
3600(config-if)# ip address 10.0.0.1 255.0.0.0
3600(config-if)# mls rp vtp-domain small
3600(config-if)# mls rp ip
3600(config-if)# mls rp management-interface
3600(config-if)# mls rp vlan-id 10
3600(config-if)# interface fa 0/1
3600(config-if)# ip address 192.168.1.1 255.255.255.0
3600(config-if)# mls rp vtp-domain small
3600(config-if)# mls rp ip
3600(config-if)# mls rp vlan-id 20
```

Note *All interfaces on a router are administratively disabled (shutdown) by default. While not shown in this example, you may need to enable these interfaces by using the* no shutdown *interface config mode command to actually use the interface. For more information on the* shutdown *command, refer to Chapter 16.*

As for this configuration, the first command, *mls rp ip*, simply globally enables MLS and MLSP on the router. After that, you switch to the fa 0/0 interface (the first Fast Ethernet port) and give it an IP address. Then you use the *mls rp vtp-domain [domain name]* command to set the VTP domain for this interface. (Remember, the VTP domain must be the same for both the MLS-RP and MLS-SE.) You then issue the *mls rp ip* command to enable MLS on this specific interface. Then you issue the *mls rp management-interface* command to set this interface as the management interface.

The management interface is the one that MLSP messages will be sent out. You will need to configure at least one of these; otherwise, MLSP messages will not be sent, and MLS will not function. You can set more than one interface as a management interface, but doing so increases both processor overhead on the router and network overhead, because hello messages will be advertised out all configured management interfaces. Finally, you issue the *mls rp vlan-id [vlan number]* command. This command tells the MLS-RP which VLAN this interface is associated with on the MLS-SE.

Note *Make sure you enter the* mls rp vtp-domain *command before entering the* mls rp ip *interface config command, the* mls rp vlan-id *command, or the* mls rp management-interface *command. Failure to do so will result in the interface in question being placed in the "null" domain (meaning MLS will not function unless the switch has no VTP domain configured). To remove the interface from the "null" domain, you must perform the* no *versions of all of these commands and then retype them all.*

Once you finish configuring fa 0/0, you switch to fa 0/1 and perform the same steps, except for the *mls rp management-interface* command. That's it; now you have configured your MLS-RP.

Now for your switch (the MLS-SE). For a 5000 series switch, you perform the following commands:

```
Cat5K> (enable) set mls enable
Multilayer switching is enabled
Cat5K> (enable) set VTP domain small mode server
VTP domain small modified
Cat5K> (enable) set vlan 10 2/1-4

VLAN 20 modified.
VLAN 1 modified.
VLAN  Mod/Ports
---- ----------------------
10    2/1-4
Cat5K> (enable) set vlan 20 2/5-8

VLAN 20 modified.
VLAN 1 modified.
VLAN  Mod/Ports
---- ----------------------
10    2/5-8
Cat5K> (enable) set mls include ip 10.0.0.1
IP Multilayer switching is enabled for router 10.0.0.1
```

This configuration is fairly straightforward. First, you enable MLS globally on the switch with the *set mls enable* command. If your switch supports MLS, this setting should be the default; but I have included the command just in case. You then issue the *set vtp* command (discussed in the preceding chapter) to set the vtp domain and mode. You then add interfaces 2/1–2/4 to VLAN 10 and interfaces 2/5–2/8 to VLAN 20, following the criteria in the example. Finally, you use the *set mls include [protocol] [address of MLS-RP]* command to add the 3600 as a MLS-RP for the switch. This completes the configuration. Again, not too difficult, once you understand the process, but it does get a little trickier when you begin trunking VLANs.

Configuring an Environment with a Trunked Connection

The next task is to perform this configuration using only a single trunked connection to the router, as shown in Figure 20-16.

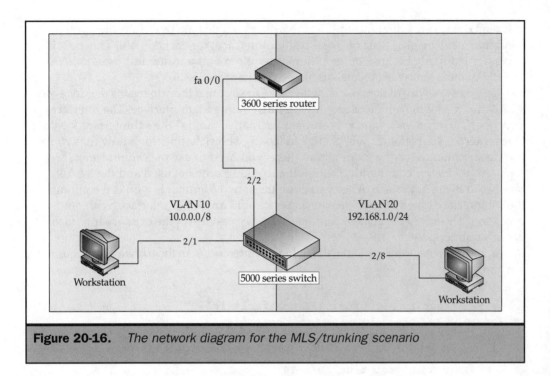

Figure 20-16.   *The network diagram for the MLS/trunking scenario*

In this case, you have to do a little more work to get MLS up and running. You again start by configuring the router, as shown in the following output:

```
3600(config)# mls rp ip
3600(config-if)# interface fa 0/0
3600(config-if)# mls rp vtp-domain small
3600(config-if)# interface fa 0/0.10
3600(config-subif)# encapsulation isl 10
3600(config-subif)# ip address 10.0.0.1 255.0.0.0
3600(config-subif)# mls rp ip
3600(config-subif)# mls rp management-interface
3600(config-subif)# interface fa 0/0.20
3600(config-subif)# encapsulation isl 20
3600(config-subif)# ip address 192.168.1.1 255.255.255.0
3600(config-subif)# mls rp ip
```

The primary difference between this example and the previous one is in the use of subinterfaces and the *encapsulation* command. In the preceding chapter, you saw how to configure VLAN trunking on switches; but on a router, the configuration process is a bit different. The function is the same as on a switch; the trunking process just lets you send data destined for multiple VLANs over a single link. But in a router, you must configure a *subinterface* (sort of like a pseudointerface) for each VLAN. This task is necessary primarily because of the different requirements a router imposes on VLANs (mostly due to routing protocols, discussed in Chapters 23–26).

Regardless of why routers use subinterfaces, the fact that they do means that for every VLAN that you will trunk to a router, you must define a subinterface. The interface *[type] [port] . [subinterface number]* command actually "builds" the subinterface for you. Once entered, your prompt will change to (config-subif), letting you know that you are now in subinterface config mode. From there, you need to use the *encapsulation [isl | dot1q] [vlan number]* command to assign the trunking protocol used and the VLAN number to the subinterface. After you issue these two commands, you can configure the subinterface exactly the same way as you would any other interface, with one exception. The *mls rp vtp-domain* command can be set only on the primary interface, not on subinterfaces.

As for the switch configuration, the only real difference is in the trunking configuration, as shown here:

```
Cat5K> (enable) set trunk 2/2 desirable isl
Port(s)  2/2 trunk mode set to desirable.
Port(s)  2/2 trunk type set to isl.

Cat5K> (enable) set trunk 2/2 10
Adding vlans 10 to allowed list.
Port(s) 2/2 allowed vlans modified to 1,10.

Cat5K> (enable) set trunk 2/2 20
Adding vlans 20 to allowed list.
Port(s) 2/2 allowed vlans modified to 1,10,20.
Cat5K> (enable) set mls enable
Multilayer switching is enabled
Cat5K> (enable) set VTP domain small mode server
VTP domain small modified
Cat5K> (enable) set vlan 10 2/1,3-4

VLAN 20 modified.
VLAN 1 modified.
VLAN  Mod/Ports
----  ----------------------
10    2/1,3-4
Cat5K> (enable) set vlan 20 2/5-8

VLAN 20 modified.
VLAN 1 modified.
```

```
VLAN  Mod/Ports
----  ----------------------
10    2/5-8
Cat5K> (enable) set mls include ip 10.0.0.1
IP Multilayer switching is enabled for router 10.0.0.1
```

Again, the only real difference here is in the configuration of the trunk link. You use the *set trunk* command (described in the preceding chapter) to enable trunking using ISL, set the mode to desirable, and allow trunking for VLANs 10 and 20 over link 2/2. The setup really isn't that difficult once you understand the process.

Finally, you configure the same basic environment using an RSM in a 5500 series switch. The RSM is just a "router on a card" that you insert into the 5500 to enable integrated routing in the 5500. Notice that I did not say layer 3 switching, but *routing*. The 5500 (with the correct supervisor engine or NFFC) can already perform layer 3 switching. Adding the RSM just puts the router inside the box, and it allows you to configure the entire environment from one interface. That being said, you will notice that you actually have to "connect" to the RSM using the *session* command (discussed in Chapter 17), and then use standard IOS commands to configure the RSM. It's really like having two separate devices without all of the wires. Figure 20-17 shows the layout for this configuration.

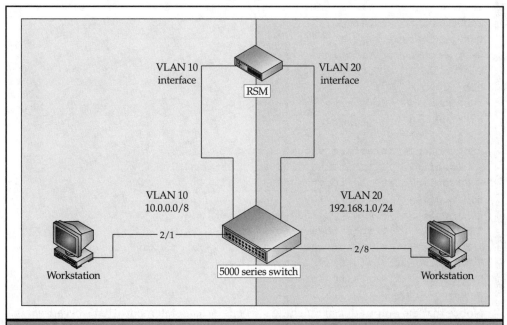

Figure 20-17. *A logical network diagram when using the RSM in a 5500*

The configuration commands follow:

```
Cat5K> (enable) set mls enable
Multilayer switching is enabled
Cat5K> (enable) set VTP domain small mode server
VTP domain small modified
Cat5K> (enable) set vlan 10 2/1-4

VLAN 20 modified.
VLAN 1 modified.
VLAN  Mod/Ports
----  ----------------------
10    2/1-4
Cat5K> (enable) set vlan 20 2/5-8

VLAN 20 modified.
VLAN 1 modified.
VLAN  Mod/Ports
----  ----------------------
10    2/5-8
Cat5K> (enable) session 3
Trying Router-3...
Connected to Router-3.
Escape character is '^]'.
User Access Verification
Password:
RSM> enable
Password:
RSM# config t
RSM(config)# mls rp ip
RSM(config)# interface vlan10
RSM(config-if)# ip address 10.0.0.1 255.0.0.0
RSM(config-if)# mls rp vtp-domain small
RSM(config-if)# mls rp ip
RSM(config-if)# mls rp management-interface
RSM(config-if)# interface vlan20
RSM(config-if)# ip address 192.168.1.1 255.255.255.0
RSM(config-if)# mls rp vtp-domain small
RSM(config-if)# mls rp ip
```

Notice that the output doesn't include a *set mls include* statement. That's because it is supplied automatically for an RSM installed in a switch. The rest of the steps are similar to the original configuration example using an external router (from the "Configuring an Environment with Separate Devices" section earlier in the chapter), except for the use of the *session* command to connect to the RSM, and then you use the *interface vlan [vlan number]* command to configure the individual virtual connections from the RSM to each VLAN on the switch. Otherwise, the commands are practically identical.

At this point, you've learned about layer 3 switching configuration for both the standard IOS–based and CatOS-based switches. Although you shouldn't have too many problems with layer 3 switching if you follow the guidelines in this section, you still need to know what to do when something goes wrong. Let's delve into how to optimize and troubleshoot layer 3 switching.

Layer 3 Switching Troubleshooting and Optimization

The first step in solving your layer 3 problems is to make sure you've done the simple stuff. From that angle, Cisco offers a list of issues that they suggest you check before you do anything else:

- Make sure your equipment meets the minimum requirements for MLS (discussed in the section "Layer 3 Switching Configuration," earlier in this chapter).

- Find out whether the topology supports MLS. Specifically, are the hosts in two separate logical networks, and does the same MLS-SE see the packet before and after it passes through the MLS-RP?

- Make sure the *mls rp ip* statement is entered on the MLS-RP both globally and on each interface that needs to support MLS.

- Make sure you've informed the MLS-SE of the MLS-RP using the *set mls include* command. Note that this requirement is necessary only if the MLS-RP is external to the switch.

- Make sure VTP is enabled, and that both the MLS-SE and MLS-RP are in the same VTP domain.

- Make sure the flow masks on the MLS-SE and MLS-RP match. Typically, this should not be an issue because, unless you manually configure the masks on the MLS-SE, the most specific flow mask heard will be used automatically. Also, if access lists are being used, make sure that the communication is allowed in the access list.

- Make sure no options (like NAT, encryption, or compression) that disable MLS are enabled on any of the interfaces used for MLS on the MLS-RP.

- Make sure that you do not have duplicate MAC addresses on the network, and that no bridging loops exist in the network. If you have either of these problems, you may continually get "MLS Too Many Moves" errors.

After you've ensured that these problems are not the cause of your errors, you must use *show* commands to determine what *is* the cause. On a standard IOS switch, the *show* and *debug* commands listed in Table 20-1 apply to MLS.

Command	Description
show mls flowmask	Shows the current flow mask used by the switch
show mls ip [target] [subtarget]	Shows various IP MLS information (explained in the text following the table)
show mls rp [target] [subtarget]	Shows various MLS information (available only on switches with integrated routing, like the 2948 and 4908; explained in the text following the table)
debug mls rp [target]	Enables MLS debugging (explained in the text following the table)

Table 20-1.   *Useful MLS* show *and* debug *Commands for the Standard IOS*

The first command in Table 20-1 is fairly straightforward. The *show mls flowmask* command simply shows the flow mask currently in use on the MLS-SE, like so:

```
Cat6K# show mls flowmask
current ip flowmask for unicast:  destination only
current ipx flowmask for unicast: destination only
```

This command is fairly useful if you think you may have a flow mask mismatch on a 6000/6500 series switch.

The next useful *show* command in the previous table is *show mls ip*. This command is the primary source for detailed MLS information, and it has lots of options. Table 20-2 describes the possible targets and subtargets for this command.

The output from the *show mls ip any* command is shown next. (Note that, due to the width of the output from this command, the lines wrap.)

```
Cat6K# show mls ip any
DstIP           SrcIP           DstVlan-DstMAC          Pkts         Bytes
---------------------------------------------------------------------------
SrcDstPorts SrcDstEncap Age    LastSeen
------------------------------------------
172.16.1.1      0.0.0.0          10   : 0000.1111.2222   5            397
2 /2 ,2 /8  ARPA,ARPA   42    11:23:00
  Number of Entries Found = 1
Cat6K#
```

Target	Subtarget	Description
any	N/A	Displays all MLS information
destination	[host name or IP address]	Displays MLS entries for a specific destination address
detail	N/A	Displays detailed information for complete flow masks
flow	[tcp or udp]	Displays flow information for the specific layer 4 protocol
interface	[type and number]	Displays MLS information specific to the interface
macd	[MAC address]	Displays MLS information related to a specific MAC destination address
macs	[MAC address]	Displays MLS information related to a specific MAC source address
slot	[number]	Displays information about MLS on a specific slot
source	[host name or IP address]	Displays MLS entries for a specific source address
count	N/A	Displays the total number of MLS entries

Table 20-2.   *The* show mls ip *Command Targets and Subtargets*

The *show mls rp* command is similar to the *show mls ip* command, except for the 2948G L3 and 4908G L3 switch lines. It is also used on MLS-RPs (external routers and RSMs) to display information on MLS. Table 20-3 describes the targets and subtargets for this command.

The following is an example of the *show mls rp* command output (from a 2948G L3):

```
2948GL3#show mls rp ?
  interface    fcp interface status and configuration
  ip           ip
  ipx          ipx
  vtp-domain   vtp-domain
  <cr>
2948GL3#show mls rp
ip multilayer switching is globally enabled
```

```
ipx multilayer switching is globally enabled
ipx mls inbound acl override is globally disabled
mls id is 0001.0002.0003
mls ip address 192.168.1.1
mls ip flow mask is destination
mls ipx flow mask is destination
number of domains configured for mls 1

vlan domain name: lab
   current ip flow mask: destination
   ip current/next global purge: true/false
   ip current/next purge count: 0/0
   current ipx flow mask: destination
   ipx current/next global purge: true/false
   ipx current/next purge count: 0/0
   current sequence number: 1396533568
   current/maximum retry count: 0/10
   current domain state: no-change
   domain uptime: 8w3d
   keepalive timer expires in 10 seconds
   retry timer not running
   change timer not running
   fcp subblock count = 14
    0 management interface(s) currently defined:
    9 mac-vlan(s) configured for multi-layer switching
    9 mac-vlan(s) enabled for ip multi-layer switching:
      mac 0001.0002.0003
        vlan id(s)
        1    5    14    15
      mac 0001.0002.0004
        vlan id(s)
        1    11    12    13    15
    8 mac-vlan(s) enabled for ipx multi-layer switching:
      mac 0001.0002.0004
        vlan id(s)
        1    14    15
      mac 0001.0002.0005
        vlan id(s)
        1    11    12    13    15
   router currently aware of following 0 switch(es):
     no switch id's currently exists in domain
 2948GL3#
```

Target	Subtarget	Description
interface	*[type and number]*	Displays MLS information specific to the interface
ip	N/A	Displays MLS information specific to IP
ipx	N/A	Displays MLS information specific to IPX
vtp-domain	*[domain name]*	Displays MLS information specific to the VTP domain

Table 20-3. *The* show mls rp *Command Targets and Subtargets*

Finally, the *debug mls rp* command is useful for seeing the guts of MLS operation; but like with any *debug* command, you should enable it only when absolutely necessary for troubleshooting—and even then, use the most specific target possible. The valid targets for this command are *all* (writes all possible MLS information to the system log), *error* (writes error events to the system log), *ipx* (writes only IPX MLS information to the system log), *locator* (writes information about which switches are switching MLPS packets), and *packet* (writes statistical packet counts to the log).

For CatOS-based switches, the commands are fairly similar—just a little more inclusive. Table 20-4 describes useful MLS-related CatOS *show* commands.

Command	Description
show mls ip	Displays information about MLS for IP
show mls debug	Displays MLS debugging information (highly technical)
show mls entry	Displays specific MLS switching table entries (described in more detail in the text that follows)
show mls include [ip | ipx]	Displays information on configured MLS-RPs
show mls rp [ip | ipx] [address of MLS-RP]	Displays information about a specific MLS-RP
show mls statistics	Displays MLS statistical information (described in more detail in the text that follows)

Table 20-4. *Useful MLS-Related CatOS* show *Commands*

Most of these commands are fairly straightforward. The *show mls ip* command shows most of the basic MLS info, including basic statistics, as you can see in the following command output:

```
Cat5K> show mls ip
Total packets switched = 1012
Total Active MLS entries = 15
IP Multilayer switching enabled
IP Multilayer switching aging time = 256 seconds
IP Multilayer switching fast aging time = 0 seconds, packet threshold =0
IP Current flow mask is Destination flow
Configured flow mask is Destination flow
Active IP MLS entries = 0
Netflow Data Export disabled
IP MLS-RP IP     MLS-RP ID     XTAG MLS-RP MAC-Vlans
---------------- ------------  ---- -------------------------------
172.16.1.1       000000000001 10    00-00-00-00-00-02  2-5
Cat5K>
```

The second command, *show MLS entry*, is highly configurable and allows you to view specific information about a given MLS entry or set of entries. Table 20-5 describes the most common targets and subtargets for this command.

Target	Subtarget	Description
ip	*destination [ip address or subnet]*	Displays MLS information about specific IP destinations
	source [ip address or subnet]	Displays MLS information about specific IP sources
ipx	*destination [ipx address or network]*	Displays MLS information about specific IPX destinations
	source [ipx address or network]	Displays MLS information about specific IPX sources
rp	*[IP address of MLS-RP]*	Displays MLS information associated with a given MLS-RP

Table 20-5. *Common* show mls entry *Command Targets and Subtargets*

An example of the *show mls entry* output is shown here:

```
Cat5K> show mls entry ip
                Last Used        Last    Used
Destination IP  Source IP        Prot DstPrt SrcPrt Destination Mac  Vlan Port
--------------  ---------------  ---- ------ ------ ----------------
IP MLS-RP 172.16.1.1:
10.1.1.1        172.16.1.2       TCP   80    41602  00-00-aa-bb-cc-dd  5  2/5
```

The next command, *show mls include,* simply shows all of the configured MLS-RPs for this switch. An example of its output follows:

```
Cat5K> show mls include ip
Included IP MLS-RP
----------------------------------------
172.16.1.1
```

If you want more detailed information about a particular MLS-RP, you can use the *show mls rp* command, as shown here:

```
Cat5K> show mls rp ip 172.16.1.1

MLS-RP IP        MLS-RP ID       XTAG  MLS-RP MAC-        Vlans
---------------------------------------------------------------
172.16.1.1       0010298a0c09    2     00-00-00-00-00-01  1,5
```

Finally, if you want to view detailed statistical information about MLS (such as which protocol is using up the most bandwidth), you can use the *show mls statistics* command. This command is also very configurable, including several targets and subtargets, the most useful of which are detailed in Table 20-6.

Note *The port and protocol statistics will be unavailable unless you are using the IP flow (full flow) mask.*

This command is especially useful for evaluating and tuning your MLS environment's performance. The following output shows many of the permutations of this command:

```
Cat5K>(enable) show mls statistics protocol
Protocol  TotalFlows  TotalPackets   Total Bytes
-------   ----------  --------------  ------------
Telnet    5           97              6221
WWW       74          479             93423
```

```
SMTP      35          961              139733
DNS       85          912              141036
Total     199         2449             380413
Cat5K> show mls statistics rp
Total packets switched = 2449
Active IP MLS entries = 0
Total packets exported= 0
                              Total switched
MLS-RP IP       MLS-RP ID     packets    bytes
--------------- ------------- ---------- ------------
172.16.1.1    000000000001 2449     380413
Cat5K> show mls statistics entry destination 10.0.0.1
               Last Used        Last    Used
Destination IP Source IP        Prot DstPrt SrcPrt Stat-Pkts  Stat-Bytes
--------------- --------------- ---- ------ ------ ---------- ---------------
10.0.0.1       172.16.1.2      6    46451 80      15           12543
```

Now that you've seen some of the *show* and *debug* commands related to MLS, if you are still having problems, you should be better equipped to locate and resolve those problems. Just remember to keep an eye out for those eight common MLS issues listed

Target	Subtarget	Description
protocol	N/A	Sorts the MLS statistics based on protocol used
ip	*rp [IP address of MLS-RP]*	Displays IP MLS statistics associated with a given MLS-RP
ipx	*rp [IP address of MLS-RP]*	Displays IPX MLS statistics associated with a given MLS-RP
entry [ip | ipx]	*destination [address or network]*	Displays MLS statistics about specific destination addresses
	source [address or network]	Displays MLS statistics about specific source addresses
	flow [protocol | source port | destination port]	Displays MLS statistics about a protocol for a given entry or entries

Table 20-6. *Common* show mls statistics *Targets and Subtargets*

at the beginning of this section. If you are having problems with MLS, chances are, one of those issues is to blame.

> **Tip** *Sometimes "MLS errors" are actually routing errors. If you have verified that the MLS environment is correctly configured, examine your routing configuration for errors.*

If you determine that you do have an MLS misconfiguration, you must reconfigure the device. To do this, you may need to use one of the commands listed in Table 20-7 to remove the current MLS configuration.

Once you have MLS up and running, you need to think about how to tune MLS to achieve optimal performance. Follow a couple of basic tips to ensure the highest performance MLS environment possible.

First, attempt to reduce your flow masks, if possible. The more a switch has to look at a packet, the higher the latency. To ensure that the least specific possible flow mask is being used, first, avoid manually setting the flow mask on your switches. The flow mask is to be used primarily to enforce access lists; and if access lists are enabled on your routers, they will inform all of the switches and change the flow mask for you. Configuring the flow mask on your switches manually can cause problems and reduce performance. Second, use access lists on your MLS-RPs only when absolutely necessary. Again, configuring access lists on your MLS-RPs will increase the specificity of the flow mask used in the domain, which will reduce performance.

> **Note** *The Catalyst 6000 and 6500 series switches have hardware support for layer 4 switching, and therefore do not suffer from increased flow mask granularity nearly as much as the 5000/5500 series does.*

Command	Description	IOS Type
set mls disable	Disables MLS globally	CatOS
clear mls entry	Removes one or more entries from the flow cache	CatOS
clear mls include	Removes one or more MLS-RP entries	CatOS
no mls rp ip	Disables MLS globally or on specific router interfaces	Standard IOS
no mls rp management-interface	Removes the interface from the list of management interfaces	Standard IOS

Table 20-7. *Commands to Remove MLS Configurations*

Next, try to keep to a minimum the number of flows that the switches need to keep up with. The more flows in the flow cache, the longer it takes to match packets to flows, and the higher the latency. While the flow cache can support up to 128,000 entries on most platforms, Cisco recommends that you allow it to reach a maximum of only 32,000 entries. This restriction is because, on most platforms, at 32,000 entries, the layer 3 switching process slows down to about the same speed as standard routing at 32,000 entries. If layer 3 switching isn't any faster than routing, then there isn't much point in using it!

To control the size of the flow cache, you can use one of two methods. The first method is to set the global time period that a flow is retained, called the *aging time*, to a lower value. The default value is 256 seconds. (The minimum value permitted is 8 seconds, and the maximum value permitted is 2032 seconds.) Therefore, as long as a single packet is sent within 256 seconds of the last packet, the flow will remain in the cache. The performance of your routers and switches, as well as the amount and type of traffic prevalent on your network, will determine whether this time is optimal. You can increase or reduce the aging time by using the *set mls agingtime [value in seconds]* command on a CatOS-based switch, or the *mls aging normal [time in seconds]* command on a standard IOS–based switch. Note that MLS will accept only a multiple of 8 seconds for the aging time, so if you entered 180 seconds, the switch would automatically adjust this time 184 seconds. Typically, the default aging time is fine, but your mileage may vary.

The second method of controlling the cache size (and, typically, the better method) is to enable and configure *fast aging*, which removes flow entries faster than normal that are briefly used from the cache. For instance, when using a User Datagram Protocol (UDP) application, typically, the flow is very short lived. (In some cases, only one packet may be transmitted per flow.) In a situation in which the flow will be used only once and then the application establishes a new flow, there is no benefit to caching that flow. However, by default, all flows are cached, so these flows will take up space in the cache. By enabling fast aging, you define a threshold and a fast aging time limit. If a number of packets equal to or greater than the threshold are not sent within the time limit specified, the entry is aged out of the cache early. This allows the switch to identify short-lived flows and remove them at a faster rate to make room for valid flows.

To configure fast aging, use the command *set mls agingtime fast [time in seconds] [threshold]* on a CatOS switch, or the *mls aging fast threshold [number of packets] time [time in seconds]* command on a standard IOS–based switch. Cisco recommends that if you enable fast aging, you begin with the maximum time value of 128 and a low threshold. This aging configuration allows you to see what, if any, impact fast aging has on your cache size without inadvertently overloading your routers.

If you need to view the configured aging settings on your switch, use the *show mls* command on a CatOS-based switch or the *show mls aging* command on a standard IOS–based switch.

Finally, try to keep topology changes to a minimum. When a MLS-RP is removed from the network, all of the entries for that device must be removed, leading to a temporary performance hit that can take you by surprise if you aren't prepared. When using MLS, make sure you adequately prepare for planned topology changes, and attempt to remove MLS-RPs from the network only during periods of relative inactivity.

Summary

This chapter covered layer 3 switching with MLS on both CatOS and standard IOS–based switches. It also examined the router configuration commands required to enable MLS. At this point, you should be able to recognize the performance benefits of layer 3 switching, and you should be able to configure and troubleshoot this useful technology in most environments.

The Complete Reference

Cisco

Chapter 21

Layer 4 Switching

This chapter covers layer 4 switching from both a theoretical and practical standpoint. Note that because upper-layer switching techniques are still relatively new, various vendors define layer 4 switching a bit differently. Obviously, because this book is about all things Cisco, I will be concentrating on Cisco's definition of layer 4 switching functionality; but I will touch on some of the paradigms supported by other vendors to some degree as well because Cisco may include these functions in the future.

Layer 4 Switching Overview

Layer 4 switching is an interesting concept. The basic idea is that different applications have different needs; so by identifying a given application in a network communication, access can be optimized based on the application, as well as the network and MAC addresses. Because most network protocol suites (TCP/IP and IPX/SPX included) identify the application with layer 4 headers to some degree, layer 4 information can be used to "switch" traffic flows based on the application. (Well, that's the idea, anyway; unfortunately, with some of today's complex applications, this functionality doesn't work as advertised.)

Using sockets (in TCP/IP, the combination of the IP address, port number, and transport protocol), a layer 4 switch attempts to identify each individual flow and modify the path for that flow. Now, I will be the first to admit that this definition is a little bit ambiguous. (What exactly does "modify the path for that flow" mean, for instance?) But it is purposely ambiguous, because each vendor has a different idea of how to improve performance using layer 4 information. Therefore, I will simply detail three of the more common methods of switching based on layer 4 information: server load balancing (SLB), multilayer switching (MLS), and congestion management.

SLB

Server load balancing (SLB) is one of the techniques this chapter concentrates on the most. Supported by a few Cisco products (most notably the 4840G SLB switch and Cisco LocalDirector products), this method takes the layer 4 information, identifies flows, and balances those flows across a cluster of servers (sometimes called a *server farm*). This strategy allows traffic to high-usage applications like web servers to be spread across several physical servers intelligently. The benefits primarily are increased redundancy and improved performance for the client.

MLS

Multilayer switching (MLS) is discussed in Chapter 20 as the technology of choice for layer 3 switching in Cisco products; but in the 6000/6500 series, with the inclusion of the Policy Feature Card (PFC), MLS can also be used to forward traffic selectively based on layer 4 information. This task is performed based on the flow mask; like with the PFC, the flows can be matched based on layer 4 information as well as layer 3 information. (Remember that without the PFC, you could specify the IP flow mask, which includes layer 4 information, but the switch searches only for a destination IP address match.) The primary benefit of this additional MLS functionality is that it allows access lists to be more correctly applied to the traffic stream.

 The PFC in a 6000 series switch also allows for other sophisticated QoS functions, including SLB. For more information on the PFC, please visit http://www.cisco.com/warp/public/cc/pd/si/casi/ca6000/prodlit/6kpfc_ds.htm.

Congestion Management

A fancy phrase for queuing, *congestion management* is a broad term that encompasses several different approaches to prioritizing traffic based on layer 3 and 4 information. This method is the other form of layer 4 switching covered in detail in this chapter. The basic idea behind congestion management is that different types of traffic are assigned to different prioritized queues inside the router or switch. Thus, the higher-priority traffic takes precedence and is forwarded first when the link is oversubscribed. Several vendors, including Cisco, support this particular method of switching.

How Layer 4 Switching Works: SLB

SLB works by creating virtual server farms and displaying them to the outside world as a single IP address. SLB then handles which real server is contacted regarding the client's request using its load-balancing algorithm. Thus, SLB controls the loads on the real servers dynamically based on traffic loads. Unlike other methods of load balancing (such as round-robin DNS), SLB enables dynamic response to varying server loads.

SLB works based on a few fundamental rules:

- Virtual servers are defined on the SLB device. These virtual servers are configured with the IP addresses that remote clients use to connect to the resource.

- The IP address for the virtual server is translated into an IP address of the real server deemed best able to respond to the request by the SLB device.

- The real server address is chosen from a pool of real server addresses defined for the specific application that the client is requesting. In this manner, multiple servers are displayed as a single server to the client.

- Because the pools of real servers are definable by application, a single virtual server IP address can be used to redirect traffic to several pools of real servers, with different real servers serving different requests.

- SLB is transparent to the client and *can* be transparent to the internal server pools.

- SLB can also increase security for the server pools by enforcing protocol restrictions, as well as performing more complex packet analysis (like SYN flood protection).

- SLB can operate as a NAT server for the clients, the servers, both, or neither.

To examine how SLB works in more detail, let's take the example network shown in Figure 21-1, and go through the process as a remote client requests a web page. Note that in this example, NAT is enabled on the 4840 for the servers.

First, the client (Client1) attempts to connect to the target web site, www.corp.com, and the DNS server resolves this name to the IP address of the virtual server configured on the SLB device (69.4.1.1). The SLB device (4840G) then looks at its listing of real servers for this application (Port 80, HTTP) and forwards the packet to the server it believes is the least busy—in this case, Server1 (10.1.1.1)—by modifying the destination IP to 10.1.1.1. (You will learn more about how the least busy server is determined later

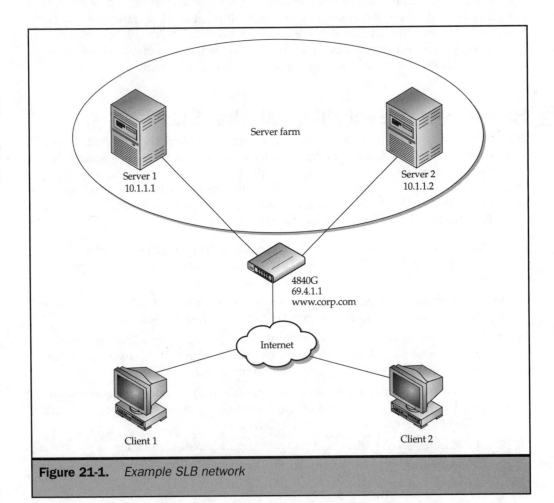

Figure 21-1. *Example SLB network*

in this section.) Server1 responds, and the 4840 forwards the packet back to Client1, this time modifying the source IP from 10.1.1.1 to 69.4.1.1 so that Client1 can't tell which server is actually responding.

While this process is occurring, let's assume that Client2 is also attempting to reach www.corp.com. Client2 visits the exact same IP address as Client1 (69.4.1.1), but this time, the 4840 decides that Server2 is best able to handle the request. The 4840 modifies the destination IP to 10.1.1.2 and forwards the packet. Server2 responds, and the 4840 again modifies the IP address in the packet to 10.1.1.1 to keep the client unaware of this process.

Now that you have seen basic SLB functioning, you're ready to learn about some of the more advanced functionality.

First, let's examine the load-balancing mechanisms available with SLB. It offers two different load-balancing mechanisms to suit your particular needs: weighted round-robin and weighted least connections. Weighted round-robin is the simplest.

With a *nonweighted round-robin* algorithm, the pool of servers is simply accessed in a circular fashion, with no particular priority given to one server over another. Therefore, if you had three servers, it sends the first request to Server1, the second to Server2, and the third to Server3, and then restarts with Server1. With *weighted round-robin*, however, you could apply a weight of four to Server1, five to Server2, and one to Server3. As a result, the first four connections go to Server1, the next five go to Server2, and the last connection goes to Server3.

This strategy is a very simple method of weighing access to your servers, however, and its primary downfall is the same as any other simple round-robin technique: if a server fails, that server has to be accessed *x* number of times before the algorithm switches the connection to a new server. A second problem with this technique is that time and session length are not taken into account. Sure, maybe Server1 should be able to respond to four times as many connections as Server3, but what if they all exist at the same time? Or what if they are all activated during a short time (like within milliseconds of each other)? Weighted round-robin cannot take these factors into account, and, therefore, slow response times can still be a problem in certain situations. It is simple to design and implement, however, and works well for applications with fairly static data access and connection establishment requirements.

Weighted least connections, on the other hand, is a more complicated, and consequently more robust, algorithm. Weighted least connections examines current connections established to all of the servers in the pool and assigns new connections based on the current traffic load of the server, as well as the server's configured weight. Although this process sounds complicated, the math behind the algorithm is really fairly simple; just add up the total of all of the weights, and each server gets its corresponding percentage of the total. So, in the previous example, the weight of the servers was four for Server1, five for Server2, and one for Server3. Four plus five plus one equals ten, so Server1 (weight of four) gets 40 percent of the traffic, Server2 (weight of five) gets 50 percent of the traffic, and Server3 (weight of one) gets 10 percent of the traffic.

CISCO LAN SWITCHING

If you started with these three servers and no connections were established, the first connection to come in goes to the server that was farthest away from its configured total—in this case, Server2 (50 percent away from the configured traffic total). The second connection then goes to Server1 because it is 40 percent away from its configured total, Server2 is at its configured total (two connections divided by one connection to Server2 equals 50 percent of all connections allocated to Server2), and Server3 is only 10 percent away from its configured total. The third connection then goes to Server2 because Server1 is only 6.66 percent away from its configured total. (Three connections divided by one connection to Server1 equals 33.33 percent; 40 percent minus 33 percent equals 6.66 percent.) Server3 is still only 10 percent away from its configured total, but Server2 is now 16.66 percent away from its configured total. (Three connections divided by one connection to Server2 equals 33.33 percent; 50 percent minus 33 percent equals 16.66 percent.)

On the next connection, Server1 receives the connection because it is now 15 percent away from its configured total. (Four connections divided by one connection to Server1 equals 25 percent; 40 percent minus 25 percent equals 15 percent.) Server2 is at its configured total (four connections divided by two connections to Server2 equals 50 percent; 50 percent minus 50 percent equals 0 percent), and Server3 is still only 10 percent away from its configured total.

Somewhere around the sixth connection, Server3 finally gets a connection, but you get the idea. This algorithm deals very well with the issue of having several connections open at the same time, but it is still susceptible to the problem in which several connections are established in a short time. It is also still susceptible to the problem in which a server dies, and packets are still forwarded to that server instead of other servers in the pool. To deal with these problems, a few additional functions need to be employed.

To eliminate the issues caused by large quantities of connections being sent to a server in a short time, two features of SLB are used: slow start and maximum connections.

The *slow start feature* limits the number of connections that can be sent to a server over a short time. This restriction helps reduce errors when adding a new server to the pool because, otherwise, it will be sent a large number of connections in a short time.

The *maximum connections feature* puts limits on the maximum number of connections to a single server at any given time. This restriction keeps a server from being overloaded in situations in which the hardware is reaching its capacity.

To help reduce errors when a server fails, an additional feature, automatic failure detection, is employed. When configured, automatic failure detection keeps track of responses to client requests from each server. If a server fails to respond to a client request, automatic failure detection increments a failure counter. Once the failure counter reaches a certain configured threshold, SLB removes the server from the pool, assuring that no new connections are sent to that server. A retry timer is then set, and after that timer expires, SLB sends the next available connection to the server. If the connection succeeds, the server is again added to the pool; but if the connection fails, the retry timer is reset.

This functionality is called *auto unfail*, and it is extremely handy when a server simply times out on one or more requests (but isn't actually down). With auto unfail, the server is removed from the pool only after failing a specified number of requests, and it is automatically added back in when it begins responding normally.

Another feature that helps eliminate problems when a server is just overloaded or slow to respond is *TCP session reassignment.* Remember from Chapter 5 that when a client wants to connect to a server using TCP, it sends a packet with the SYN bit set. This signals the request to establish a session. The SLB device tracks SYN bits sent to each server; and if a server does not respond after a configurable number of connection establishment attempts (marked by multiple SYN packets being sent from the same client), SLB automatically sends the next SYN request to a different server.

To allow for servers that have very specialized "tuning" for a specific application, SLB supports the use of "per application" server farms known as *port-bound servers.* This feature allows you to configure different server pools to support different protocols while using the same external IP address. For instance, Figure 21-2 shows one external

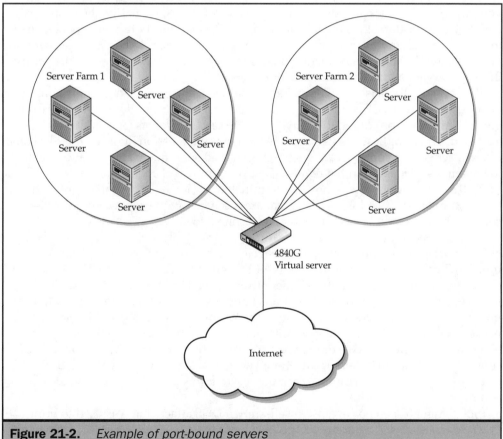

Figure 21-2. *Example of port-bound servers*

IP address that all remote clients use to connect to all offered network services. However, client requests destined for port 80 or 443 (HTTP and HTTPS) go to Server Farm 1, while client requests for port 20 and 21 (FTP) go to Server Farm 2.

For even more advanced performance tuning and monitoring, SLB includes a feature known as *Dynamic Feedback Protocol (DFP)*. DFP monitors servers using either one or more *agents* (software designed to track performance statistics on devices), HTTP probes, or both, and alerts SLB to any performance problems (including, with DFP, general server slowdowns due to memory leaks, high RAM use, hardware issues, and so on). The DFP functionality allows SLB to dynamically adjust the weights of the servers in a pool to deal with unexpected performance issues, including either under- or overuse of a given server. *HTTP probes*, on the other hand, allow for a very simple mechanism to verify that HTTP servers and firewalls are up and responding to requests. With an HTTP probe, a request is sent to the configured web servers at specified intervals. If a configurable positive status code is returned ($4xx$ is the default), the server is assumed to be functioning.

SLB, although solving many of the problems associated with consolidating high volumes of traffic, does introduce its own new set of issues. Primarily, the problems with SLB are attributable to the fact that most TCP/IP applications were not designed to enable several servers to handle a given stream of requests. As a result, a server may not understand a subsequent client request that relies on a previous request because a different server handled the initial client request.

To deal with these issues, SLB makes use of two features: "sticky" connections and delayed removal of TCP connection contexts. *Sticky connections* allows a request from a given client to an application to always go to the same server, as long as a timer is not exceeded. Thus, applications that must rely on previous session information (like some types of active web content) can function because all subsequent client requests will go to the same server as the initial request. Delayed removal of connection contexts allows the SLB switch to delay the removal of terminated sessions from its tables until a delay timer has expired. Therefore, packets that are received out of order can be processed correctly by the server before the session is terminated.

SLB also includes some anti-DOS features, most notably, the ability to perform NAT for both clients and servers, and *Synguard* (SYN flood protection). The NAT functionality of SLB is similar to other NAT servers, and it conforms to RFC 1631. SLB can perform translation for both the client (translating public client addresses into internal, private IPs) and the servers (translating the server's public or private address into the NAT server's IP address). When enabled for the client side, NAT ensures that packets back to the client are always automatically sent back through the SLB device. (Otherwise, you must make sure the servers are configured to point back to the NAT server, typically as their default gateway.) When enabled for the server side, SLB NAT can ensure that the clients cannot reach the server farms directly, increasing the overall security of the server farms.

SLB Synguard defends against a well-known DOS attack that was used to bring down several web sites (most notably, CNN.com) in early 2000. A *SYN flood* is when a device sends a packet with the SYN bit set and with a fake return address. The end host attempts to keep track of the partially open connection, which consumes resources. If enough SYN packets of this type are sent in a short time, the device may run out of resources (typically RAM) and crash. Synguard in SLB attempts to eliminate this problem by rejecting any SYN packets over a configurable threshold (measured in unanswered SYN requests) within a configurable time (measured in milliseconds).

> **Note**   *Synguard can protect devices located only* behind *it in the network. Routers in front of it need to be protected using access lists or firewall devices.*

Finally, what if you are configuring one box to keep your servers clustered so that you have redundancy, but the SLB switch fails? Cisco has a solution for this problem as well. You can achieve SLB redundancy through two methods: Hot Standby Router Protocol (HSRP) and SLB stateful backup.

The basic idea of *HSRP* is that it clusters your routers (or, in this case, your SLB switches—"clustering your clusterers," you could say) by assigning the group of devices a virtual IP and MAC address, similar to SLB. The major differences between SLB and HSRP lie in the implementation and logic. (HSRP is implemented on the devices in the "farm" instead of on other devices, and HSRP does not offer any type of load balancing—only redundancy.)

> **Note**   *Due to page constraints, HSRP is not covered in detail in this book. However, for basic HSRP information, you can visit the following address: http://www .cisco.com/ warp/ public/cc/so/cuso/epso/entdes/hsrp_wp.htm.*

SLB *stateful backup* works in conjunction with HSRP to make sure the backup switches in the HSRP cluster have the most recent SLB information from the primary switch. This functionality allows for minimal data loss if the primary device fails and HSRP has to failover to the backup switch.

All things considered, SLB is a robust method of controlling access to your servers and providing for complex load balancing in high-traffic, high-availability environments.

How Layer 4 Switching Works: Congestion Management

As you already learned, congestion management is just complex queuing based on layer 2, layer 3, or layer 4 information. Although queuing policies are not technically considered layer 4 switching (they are actually QoS specifications), they fit best in the discussion of this subject because they can use layer 4 information to make switching or routing decisions.

Queuing is simply a useful way of prioritizing and dedicating link bandwidth to a specific user or application. In other words, queuing will not perform load balancing, or any other actions that will improve anything other than *perceived* link performance. Note that I said *perceived* link performance. Queuing, in and of itself, does not actually increase available bandwidth; it simply assigns it differently. Because of these limitations, let's begin by examining situations in which an advanced queuing method should and should not be used:

- There must be congestion for queuing to do any good. If you are not experiencing congestion at any time, queuing will not help.

- There must not be *constant* congestion. If the link is constantly congested, queuing will most likely not help, and may actually worsen the situation. In these cases, your only options are to remove services that are using the bandwidth or increase the bandwidth.

- If you have applications with time constraints that are consistently timing out, queuing can help by allocating a higher priority to those types of traffic, but this comes at the expense of other traffic.

- Queuing can ensure that no one user or application monopolizes the bandwidth by dedicating a specific amount of bandwidth to each user and application.

- Queuing is typically useful only for WAN links. For LAN links, the performance reduction in the router or switch typically offsets the usefulness of queuing.

- Queuing is most useful for low-speed (E1 or lower) WAN links. Again, this is because as packets per second (PPS) rates increase, the resources required for queuing also increase, reducing its effectiveness.

- To accurately determine the need and usefulness of queuing, you must have detailed network use information. At minimum, you should know your bandwidth use during peak periods, the bandwidth breakdown by application, and which packets are being dropped due to queue overflows and lack of bandwidth.

Now, let's take a look at the various common queuing methods and their uses.

First in, first out (FIFO) queuing is the most commonly used and simplest of all queuing methods. Basically, with FIFO queuing, packets are forwarded in the order in which they are received, with no regard of which packets are more important than others. This method is the default queuing method for links at E1 speed or higher on nearly all Cisco routers and switches.

Weighted fair queuing (WFQ) is the default queuing method on serial links running at E1 speeds or less. WFQ dynamically assigns traffic to higher- or lower-priority queues in an attempt to equalize bandwidth use between applications and users. To better understand this, let's take an example of FIFO queuing with three traffic streams: an FTP data transfer, a Telnet session from Client A, and a Telnet session from Client B, as shown in Figure 21-3.

Due to the constraints of printed media, the figures in this section are presented in black and white. For the original color images, visit my web site at http://www.alfageek.com/mybooks.htm.

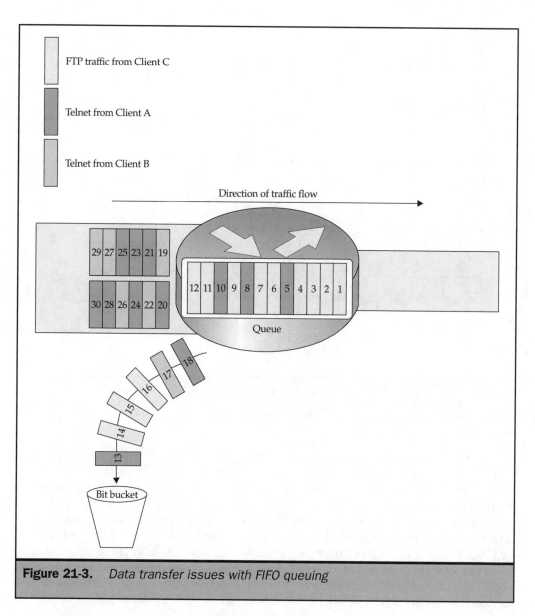

Figure 21-3. *Data transfer issues with FIFO queuing*

In this example, with FIFO, the traffic is sent in the order it arrives, and the Telnet transfer from Client B is dropped because the FTP transfer fills the queue to capacity before the Telnet packet from Client B arrives at the router. However, if you were using WFQ, the Telnet transfers would be assigned a higher priority than the FTP transfer, regardless of the order in which they were received; and if packets needed to be dropped, they would be dropped from the FTP flow. This concept is best explained by walking through Figures 21-4 through 21-8.

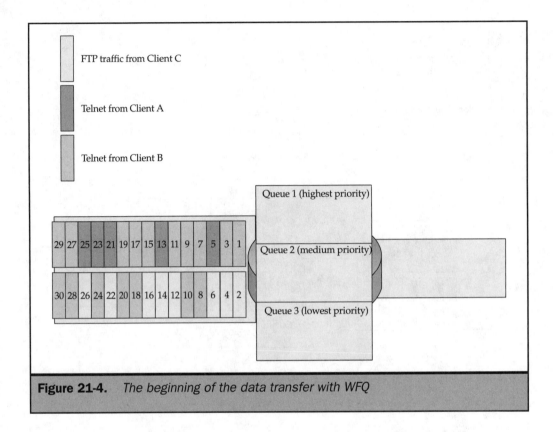

FTP traffic from Client C

Telnet from Client A

Telnet from Client B

Queue 1 (highest priority)

29 27 25 23 21 19 17 15 13 11 9 7 5 3 1

Queue 2 (medium priority)

30 28 26 24 22 20 18 16 14 12 10 8 6 4 2

Queue 3 (lowest priority)

Figure 21-4.   *The beginning of the data transfer with WFQ*

Figure 21-4 shows the beginning of the data transfer. Notice that the diagram actually includes three separate queues. With WFQ, each flow is assigned to a dynamically created queue. Each queue then has a set number of packets that it will accept. When the queue for a given flow fills up, packets for that flow *may* be dropped. For flows that are classified as "high bandwidth," like FTP, the additional packets will be dropped once the queue fills. Low-bandwidth applications (like Telnet), however, are allowed by FWQ to exceed their limit to a degree; even though you may have the same number of packets in an FTP queue as you do in a Telnet queue, because Telnet is a low-bandwidth application, the average packet size for Telnet is much smaller than FTP. To bias WFQ toward low-bandwidth applications, they are occasionally allowed to exceed their queue limits.

There is a configurable limit to the number of dynamic queues the system will create. Each queue also has a configurable number of packets that it will accept. Both of these topics are discussed in greater detail in the section "Layer 4 Switching Configuration: Congestion Management," later in the chapter.

In this example, assume that the incoming speed is twice the outgoing speed. For each successive figure, ten packets will enter the router, but only five will be allowed to

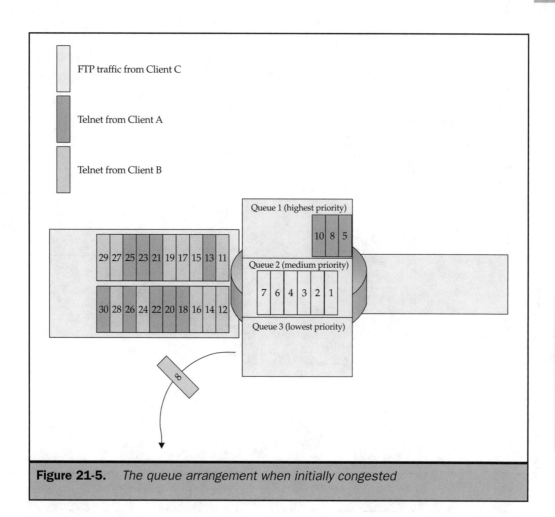

Figure 21-5.   *The queue arrangement when initially congested*

exit. In Figure 21-5, the ten packets have entered and (most) have been queued. Notice that the Telnet traffic was inserted into the first queue (high priority).

When emptying the queues, WFQ will go round-robin from the first to the last queue, so whatever is in the top queue is transmitted first. Telnet traffic is considered high priority (because it is low bandwidth), so it is placed above the FTP traffic in the top queue to give it higher transmission priority. Notice that packet 9 for the FTP session was discarded, even though an entire queue is open (Queue 3). This occurs because the queue assigned to this FTP flow (Queue 2) is full. WFQ does not assign the packet to a new queue because, if it did, the FTP session from this user would be using an unfair amount of the queue space.

In Figure 21-6, five packets have been transmitted: 5, 1, 8, 2, and 10. Notice that these packets were transmitted round-robin from the two queues that were filled

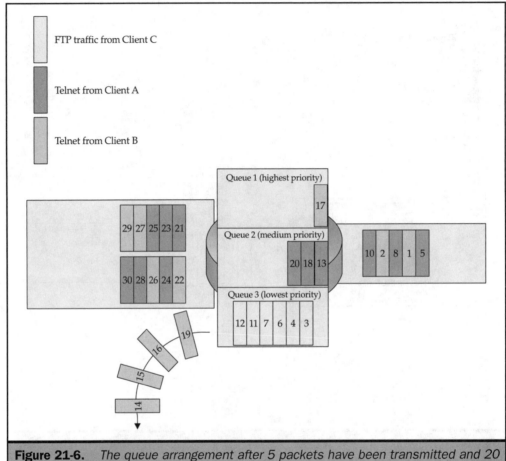

Figure 21-6. *The queue arrangement after 5 packets have been transmitted and 20 total packets have been received*

(Queues 1 and 2) in Figure 21-5. In other words, packet 5 from Queue 1 was transmitted first, then packet 1 from Queue 2, then packet 8 from Queue 1, and so on.

Again, this round-robin queuing strategy is used to ensure fair transmission for all flows. In this example, the incoming traffic moves at twice the speed as the outgoing traffic (and the packets are all equally sized); therefore, ten packets (packets 11 through 20) have attempted to enter the router. Notice that packet 17 (the first packet in Client B's Telnet session) was inserted into the top queue; and packets 20, 18, and 13 from Client A's Telnet session were moved down into Queue 2. This was done because, at this time, client B's Telnet session is using less bandwidth than Client A's Telnet session, making Client B's Telnet session higher priority for WFQ. WFQ prioritizes flows based on inverse bandwidth use; in other words, the lower the bandwidth use

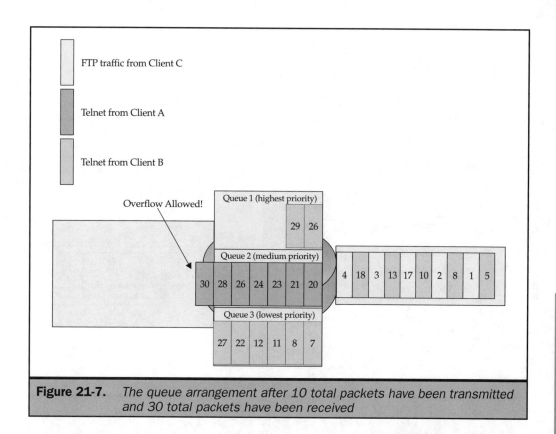

Figure 21-7. *The queue arrangement after 10 total packets have been transmitted and 30 total packets have been received*

of a flow, the higher the priority WFQ assigns the flow. Finally, the FTP traffic was dropped into the last queue and quickly filled the queue to capacity. Because packets 11 and 12 took all available space in the FTP session's queue, packets 14, 15, 16, and 19 for the FTP transfer were dropped.

In Figure 21-7, five more packets have been transmitted and the final ten packets have been received. Notice that the Telnet session for Client A managed to exhaust its queue in this transfer. However, unlike the FTP traffic, the packets were not dropped because WFQ allows low-bandwidth traffic like Telnet to go over the allocated queue space occasionally. As a result, no packets from Client A's Telnet session are lost.

In the final figure for this example, Figure 21-8, five more packets have been transmitted. Notice that this was again done in a round-robin fashion—Queue 1 was emptied, allowing WFQ to adjust the queues, once again giving the Telnet session from Client A top priority. Also, notice that the discarded packets from the FTP session for Client C have now been retransmitted. In reality, they wouldn't come in one lump chunk as illustrated here, but the TCP retransmit timer on Client C would eventually time out and retransmit the dropped packets, allowing Client C's FTP session to complete successfully, with the user (or the administrator) none the wiser. And this

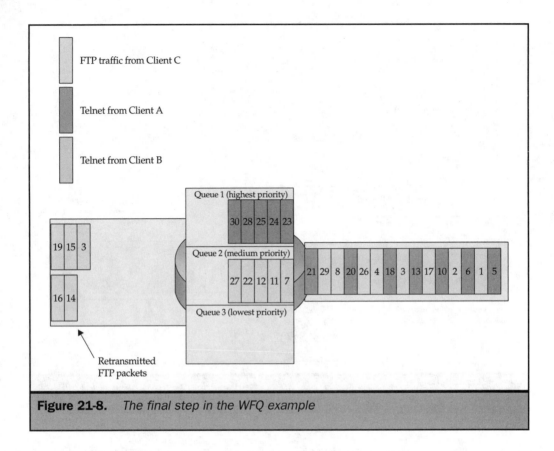

Figure 21-8. *The final step in the WFQ example*

time, because the queue will be empty by the time the packets are received, the packets should be successfully transmitted.

Note *I chose to explain the actual operation of WFQ, rather than simplifying it to the point of technical inaccuracy. Unfortunately, the discussion is a bit tricky to follow, but you will benefit in the long run.*

Now you can see the primary goal of WFQ: to provide for a balanced data transfer strategy while allowing no single application or user to monopolize the connection. Luckily, basic WFQ is all dynamic and automatic; so, in truth, no configuration (or understanding) is necessary to implement WFQ. However, understanding WFQ's operation will help you better determine when it is the best queuing strategy for your organization.

To this end, you should also understand that WFQ automatically assigns packets to separate queues based on IP precedence. IP precedence has eight possible values (0 through 7), or precedence levels, that can be used to classify traffic. The higher the

IP precedence is, the more important the packet is. Because the IP precedence field is just a value in the IP header, it is included in every IP packet. Normally, this value is set to 0, which is classified as "routine" traffic. Using Policy Based Routing (PBR) or Committed Access Rate (CAR), you can set the IP precedence bit on packets as they enter your network (based on details like source IP address, protocol used, and port numbers), and then let WFQ use these bits to partition network bandwidth as necessary.

WFQ schedules higher-precedence packets in higher-priority queues, ensuring that the higher-priority packets are given preferred use of the link. WFQ also assigns an amount of bandwidth proportionate to the precedence level of the packet flow, ensuring that higher-precedence packets get more bandwidth. It does all of this, however, without letting even high-priority traffic overuse the line by assigning queue and bandwidth limits to all queues. Thus, no traffic flow gets sent at the total expense of other traffic flows. With WFQ, *all* traffic gets *some* use of the line, but higher-precedence packets get a larger than normal share of the bandwidth.

If you are looking for a little more control over WFQ's decision process, Cisco does allow you to specify additional classes of service using class-based weighted fair queuing (CBWFQ). Using CBWFQ, you can specify specific properties of a packet (like specific protocols, or specific source or destination IP addresses) that cause it to be inserted into a special class queue. For each class of traffic you define (you can define up to 64 custom classes), a special queue is assigned. For each queue, you define a policy (define the properties of the queue) by setting the following parameters:

- **Packet limit** The packet limit for the queue controls how many packets can be in the queue at once. When this number is exceeded, CBWFQ begins dropping packets that match the class.

- **Bandwidth** This parameter determines how much bandwidth the class is guaranteed during times of congestion.

Once your classes and policies are configured, you apply them to one or more interfaces. If a packet matches a given class, the properties for that class are then applied. If a packet matches no classes, it goes into a default queue and is processed using normal WFQ parameters after all other queues have been processed.

You'll learn more about this topic in the "CBWFQ Configuration" section, later in this chapter. For now, you just need to know when WFQ and CBWFQ can help. Use WFQ in the following situations:

- When you need access to be equalized among all users and applications, with no one user or application monopolizing the available bandwidth

- When you need a fully automated, dynamic queuing strategy that requires minimal administrative effort and dynamically reconfigures itself to suit transient needs

- When you do not need to configure more than six classes of prioritized traffic

> **Note**   *Even though IP precedence specifies eight precedence levels, the two highest levels (levels 6 and 7) are reserved for network functions (like routing updates) and cannot be used to classify user traffic.*

- When you do not need to ensure that one or more types of traffic is always sent, even at the complete expense of others

CBWFQ is appropriate in the following situations:

- When you need access to be equalized among all users and applications, with no one user or application monopolizing the available bandwidth
- When you need to classify packets based on specialized packet properties, and IP precedence is not used to classify traffic
- When you need to specify more than six individual queues
- When the administrative overhead involved in configuring and maintaining CBWFQ is not prohibitive
- When you do not need to ensure that one or more types of traffic is always sent, even at the complete expense of others

Fortunately, in the discussion of queuing, WFQ and CBWFQ are probably the two most detailed and complex of the queuing methods. However, they are also the most forgiving and dynamic of the queuing methods, and typically the most useful.

Custom queuing (CQ), on the other hand, is fairly simple. It classifies traffic similarly to CBWFQ, placing the traffic into one of 17 queues. However, CQ statically defines the bandwidth and queuing logic, making it able to more specifically define bandwidth use, but also reducing its flexibility to transient traffic needs. To illustrate this concept, let's look at how CQ works.

First, you must assign traffic to queues. You perform this task similarly to how you do it when using CBWFQ—by using packet details such as IP addresses (source and destination), protocol types, port numbers, and so on. In this classification process, you choose which queue to place each type of traffic in, using queues 1 through 16. (Queue 0 is reserved by the system for signaling packets, keepalives, and other system necessities.) For each queue, you specify a byte count. These byte counts specify the amount of traffic that will be sent from each queue before moving on to the next queue.

CQ goes in order, from Queue 0 to Queue 16, sending all packets in each queue until it reaches the byte count threshold for the queue, or when all packets in the queue have been delivered. By specifying which types of traffic go into which queues, and the byte count limits on each queue, you can specify the priority of traffic (for low-latency applications) and the total amount of bandwidth each traffic type will use. Finally, you specify the packet limit for each queue. When a queue reaches its configured packet limit, it discards any further packets attempting to enter that particular queue. Configuring the packet limit allows you to ensure that queues do not stay perpetually full.

The real trick to CQ is setting the byte limits. By using the byte limits properly, you can specify the specific amount of bandwidth allocated to a particular class of traffic. For instance, if your total outgoing bandwidth is 768 Kbps, that is the figure by which you divide your bandwidth. You then figure out how many traffic classes you need and what the bandwidth requirements are for each class of traffic. If class 1 traffic requires 25 percent of the bandwidth, class 2 requires 10 percent, and all other traffic should get the remaining bandwidth, you have a basic calculation. In this case, 192 Kbps should go to class 1, 77 Kbps should go to class 2, and 499 Kbps to all other traffic.

However, configuring CQ is not as simple as setting a 192,000-byte limit on Queue 1, a 77,000-byte limit on Queue 2, and a 499,000-byte limit on Queue 3. If you did that, the latency would be a bit high because each queue is processed in order, and CQ will not switch from one queue to another until either all bytes up to the byte limit have been transferred or the queue is empty. For instance, if Queue 2 and Queue 3 were full, Queue 1 would have to wait at least 750 milliseconds (0.75 seconds) to be serviced—an eternity in computer terms. To put that time in perspective, right now, over a modem connection, a ping to yahoo.com takes me approximately 150 ms. That means, through 12 routers, it took less than 75 ms for my packet to reach the server and less than 75 ms for it to return. Obviously, a 750-ms wait (just on the sending side; once you count the return packet, it becomes a 1.5-second wait) is unacceptable.

> **Tip** *The following example contains my own calculations for figuring out how to set byte limits in CQ. Although I think this way works better for understanding* what *you are doing, Cisco actually has a quicker method to perform the same task, which is detailed at the following web address: http://www.cisco.com/ univercd/cc/td/doc/ product/software/ ios121/121cgcr/qos_c/qcprt2/qcdconmg.htm#xtocid840033.*

A better way to set byte limits is to chop the total bandwidth into smaller time slices and use those to determine the byte limits. For instance, if you chopped the bandwidth up into 1/40th of a second, Queue 1 would get a byte limit of 4800 (384,000 divided by 40 equals 4800), Queue 2 would get 1925 (77,000 divided by 40), and Queue 3 would get 12,475 (499,000 divided by 40). This arrangement would give Queue 1 a wait time of about 19 ms if both Queues 2 and 3 were full.

This scenario is much better, but in reality, it should probably be even lower. The problem with setting the queue significantly lower than this is that, even though you set the limit in bytes, the router can send only full packets. So, if your packet sizes were always 1500 bytes (the maximum size for Ethernet), when Queue 2 is processed, it will actually send 3000 bytes instead of 1925. Applying the same logic to Queue 1 and Queue 3, you end up with 6000 bytes for Queue 1 and 13,500 bytes for Queue 3, giving you a bandwidth split of 26.6 percent for Queue 1, 13.3 percent for Queue 2, and 60 percent for Queue 3, which isn't exactly desirable. So, to correctly set the byte limits, you need to take a look at what the average packet size is for each queue.

Say, for instance, that the application used for Queue 1 has an average packet size of 1000 bytes, Queue 2 has an average packet size of 1200 bytes, and Queue 3 has an average packet size of 800 bytes. Well, to figure the correct ratio out, you need to figure out what the ratio of packets is for each queue. In other words, how many packets do you want sent out of Queue 1, as compared to packets from Queue 2 and Queue 3? Well, based on the bandwidth breakdown, you want 192,000 bytes allocated for Queue 1, which is 192 packets. You want 77,000 bytes from Queue 2, which is about 64 packets (64.1, rounded down so that the result is even), and 499,000 bytes from Queue 3, which is about 624 packets (623.7, rounded up so that the result is even).

Tip *For this calculation to work, the packet numbers must be even.*

Now you need to figure out what the byte size is if you divide the second interval into the smallest possible pieces, and the easiest strategy is to find the highest common denominator (HCD) for all of the packet results. (Break out the calculator!) In this case, the HCD for these numbers (192, 64, and 624) is 16, which means that you will be splitting each second into 16 pieces. You then take the HCD and divide the number of packets by it, which gives you 12 for Queue 1 (192 divided by 16), 4 for Queue 2 (64 divided by 16), and 39 for Queue 3 (624 divided by 16). So, for every 12 packets sent from Queue 1, 4 will be sent from Queue 2 and 39 will be sent from Queue 3. You then multiply the number of packets for each queue by the average size of packets for that queue. This calculation results in a 12,000-byte limit for Queue 1, a 4800-byte limit for Queue 2, and a 31,200-byte limit for Queue 3. It also keeps your ratios very close to the original bandwidth requirement because Queue 1 is assigned 192 Kbps, Queue 2 is assigned 76.8 Kbps, and Queue 3 is assigned 499.2 Kbps.

This strategy works out pretty well, except each second is divided into only 16 parts, which means your wait time for Queue 1 could be as high as 40 ms. If you could afford to move the bandwidth requirements a little, you could reduce Queue 3's needs to 608 packets per second (486.4 Kbps), and you could then change the ratios to 6/2/19 (Queue 1/Queue 2/Queue 3) and split each second into 32 parts. This strategy would give you a total of only 755.2 Kbps used, however, so the router would actually cycle through each queue slightly more than 32 times per second.

Tip *Luckily, if you hate number games, you don't have to worry about packet size nearly as much with IOS 12.1 or later. In IOS 12.1, the CQ algorithm was changed slightly, and it puts a queue in a deficit if it has to be sent over its byte limit in a given cycle. In other words, if CQ has to send 3000 bytes instead of 2000 for a given queue due to packet sizing, it puts that queue in a 1000-byte deficit for the next cycle, and it attempts to transfer only 1000 bytes instead of 2000.*

CQ is a good choice for congestion management if you need to assign a very specific amount of bandwidth to a given application but do not want that application to impact the performance of other applications (and vice versa). The down side of CQ is that it requires a lot of work to implement properly, and it is not dynamic. Also, CQ does not classify traffic within a queue (user 1 FTP verses user 2 FTP), so one user of a protocol could potentially monopolize the bandwidth for all users of that protocol.

Finally, priority queuing (PQ) simply classifies traffic as one of four types and assigns the traffic to queues based on the traffic's classification. The four traffic classifications that PQ uses are high, medium, normal, and low. You choose how traffic is classified similarly to the process used for CBWFQ or CQ, by choosing one or more packet characteristics that cause the packet to be entered into a specific queue.

PQ's queuing mechanism is designed to ensure that high-priority traffic is *always* sent ahead of low-priority traffic. The logic is fairly simple. When able to send a packet, every queue is examined in order, starting with the high-priority queue and ending with the low-priority queue. The first packet PQ finds gets sent. Consequently, if a packet is in the high queue, it is sent immediately, regardless of the other queues. If another packet arrives in the high-priority queue while this packet is being sent, it is sent next. As a result, low-priority (and sometimes even normal-priority) traffic may *never* be sent. For this reason, you must be very careful when you classify traffic using PQ.

As for its advantages, PQ's only real advantage is also its biggest disadvantage. You can use PQ to make sure a certain type of traffic is *always* sent, even at the expense of *all* other traffic. Unfortunately, this method is a very extreme way of configuring queuing—and if you use it, you must use it with great care.

> **Note** *You can combine PQ with CAR or traffic shaping to extend its usefulness by having the transmission rate for high-priority traffic reduced before it reaches the PQ device. This strategy can help ensure that all traffic types receive some service, but, typically, I find it easier to simply use another queuing method.*

Layer 4 Switching Configuration: SLB

This section covers how to get basic SLB running on an SLB-capable switch. Note that a 4840 switch is used in all of these configuration examples. If you are performing SLB configuration on another supported platform, the commands may differ slightly.

Basic SLB Configuration

To configure SLB, you'll walk through the commands used in an example configuration. Figure 21-9 shows the network diagram for the example.

> **Note** *In the following configuration steps, it is assumed that basic layer 3 switch configuration (including IP address assignment) has been done, so commands for those functions are not shown. For more information on basic switch configuration, refer to Chapters 18, 19, and 20.*

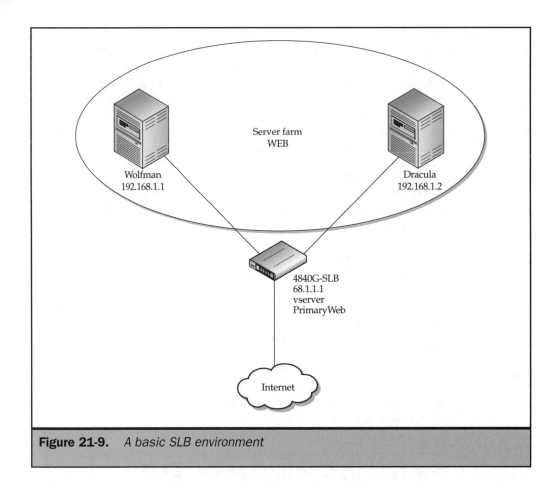

Figure 21-9.   *A basic SLB environment*

The first step in configuring SLB is to perform initial configuration of the server farm and assign a virtual IP address to the farm. During this process, you also enable NAT for both the clients and the servers to enable the basic configuration in the figure. The configuration is shown in the following output (with detailed explanations provided in the text that follows):

```
4840G-SLB# config t
Enter configuration commands, one per line.  End with CNTL/Z.
4840G-SLB(config)# ip slb natpool websurfers 10.1.1.1 10.1.1.254
netmask 255.255.255.0
4840G-SLB(config)# ip slb serverfarm WEB
4840G-SLB(config-slb-sfarm)# nat client websurfers
4840G-SLB(config-slb-sfarm)# nat server
4840G-SLB(config-slb-sfarm)# real 192.168.1.1
```

```
4840G-SLB(config-slb-real)# inservice
4840G-SLB(config-slb-real)# exit
4840G-SLB(config-slb-sfarm)# real 192.168.1.2
4840G-SLB(config-slb-real)# inservice
4840G-SLB(config-slb-real)# end
4840G-SLB# config t
Enter configuration commands, one per line.  End with CNTL/Z.
4840G-SLB(config)# ip slb vserver PrimaryWeb
4840G-SLB(config-slb-vserver)# virtual 68.1.1.1 tcp www https
4840G-SLB(config-slb-vserver)# serverfarm WEB
4840G-SLB(config-slb-vserver)# inservice
```

The first SLB command in this example, *ip slb natpool websurfers 10.1.1.1 10.1.1.254 netmask 255.255.255.0*, defines the NAT pool for your client connections to the servers. The syntax for this command is *ip slb natpool [pool name] [first IP address in pool] [last IP address in pool] netmask [subnet mask for pool]*. So, this example translates client IP addresses into an address in the range of 10.1.1.1 through 10.1.1.254, with a 24-bit mask.

The next command you issue is *ip slb serverfarm WEB*. This command initializes and names your server farm, and it puts you into server farm configuration mode. The syntax for this command is *ip slb serverfarm [farm name]*.

You then issue the commands *nat client websurfers* and *nat server*. The *nat client* command enables NAT for your client connections to this pool of servers using the NAT pool *websurfers* defined earlier. The *nat server* command enables translation of your server's private IP addresses into public IP addresses when they are forwarded back to the clients.

Next, you issue *real 192.168.1.1.* This command configures your first web server, Wolfman, to be a member of this server farm, and enters real server config mode. Once in real server config mode, you issue the *inservice* command, which enables Wolfman, allowing it to participate in SLB for this server farm.

Tip   *You must enable your real and virtual servers using the* inservice *command, or they will not be used. You must issue this command any time you modify the configuration of the servers to re-enable them.*

You then perform the same configuration steps for Dracula, entering it into your WEB server farm and enabling it with the *inservice* command.

Finally, you go back to global config mode and configure your virtual server. The first command you issue is *ip slb vserver PrimaryWeb*. This command creates and names your virtual server, PrimaryWeb, and inserts you into virtual server configuration mode.

Once in virtual server config mode, you issue the *virtual 68.1.1.1 tcp www https* command. This command sets your virtual server's IP address and configures it to load balance HTTP and HTTPS connections. This command actually has a few additional options that deserve discussion. First, the syntax for this command is *virtual [virtual server IP address] [tcp | udp] [protocols to load balance] service (optional) [service name (optional)].* The *virtual server IP address* section is the address your clients use to connect to the virtual server. The *tcp | udp* section tells the SLB switch which protocol this server farm should support. The *protocols to load balance* section can be port numbers (separated by spaces) or names of the protocols (dns, ftp, https, nntp, pop2, pop3, smtp, telnet, or www).

The final *service* option allows you to specify that a specific port be bound to another for protocols that use multiple ports. This ensures that all ports from the same client using this protocol are sent to the same SLB server for processing. If you specify the *service* option, you must specify a protocol to which it applies. Currently, only FTP is supported as a protocol for service coupling. (It uses ports 20 and 21.)

After specifying the IP address and protocols that this virtual server will support, you set the server farm that this virtual server will load balance for (WEB) with the *serverfarm WEB* command. You then enable the virtual server with the *inservice* command. That's it. Now you have a very basic but functional SLB environment.

Advanced SLB Configuration

Most likely, however, you will want to configure some of the optional SLB features to protect and improve the performance of your servers. First, you set weights for your servers so that SLB can accurately evaluate their performance potential, and change the load-balancing mechanism to weighted least connections to ensure better performance under various situations. You also want to limit the maximum connections to each server, and enable TCP session reassignment, failure detection, and auto unfail to improve performance. To increase security, you enable Synguard; and, finally, because you are taking online orders with these servers and using dynamic content, you enable sticky connections. The configuration commands for these tasks are shown in the following output (with explanations following the example):

```
4840G-SLB(config)# ip slb serverfarm WEB
4840G-SLB(config-slb-sfarm)# predictor leastconns
4840G-SLB(config-slb-sfarm)# real 192.168.1.1
4840G-SLB(config-slb-real)# weight 32
4840G-SLB(config-slb-real)# maxconns 200
4840G-SLB(config-slb-real)# reassign 2
4840G-SLB(config-slb-real)# faildetect numconns 10 numclients 5
4840G-SLB(config-slb-real)# retry 90
4840G-SLB(config-slb-real)# inservice
4840G-SLB(config-slb-real)# exit
4840G-SLB(config-slb-sfarm)# real 192.168.1.2
4840G-SLB(config-slb-real)# weight 64
4840G-SLB(config-slb-real)# maxconns 400
4840G-SLB(config-slb-real)# reassign 4
4840G-SLB(config-slb-real)# faildetect numconns 20 numclients 10
```

```
4840G-SLB(config-slb-real)# retry 45
4840G-SLB(config-slb-real)# inservice
4840G-SLB(config-slb-real)# end
4840G-SLB# config t
Enter configuration commands, one per line.  End with CNTL/Z.
4840G-SLB(config)# ip slb vserver PrimaryWeb
4840G-SLB(config-slb-vserver)# synguard 100 200
4840G-SLB(config-slb-vserver)# sticky 90
4840G-SLB(config-slb-vserver)# inservice
```

First, you enter server farm config mode. Then you issue the *predictor leastconns* command. This command configures the predictor (SLB algorithm) to use weighted least connections. If, in the future, you wanted to set the switch back to weighted round-robin, you would issue either the *no predictor* command or the *predictor roundrobin* command. (They both set the predictor algorithm back to the default round-robin setting.)

Next, you enter real server config mode for Wolfman and set the weight to 32, using the *weight 32* command. This weight is four times the default value of 8, so any new servers entering the environment that do not have the weight configured are unlikely to take a large portion of the load. Valid ranges for the *weight* command are from 1 to 255.

Tip   *Avoid using weight values that are lower than the default of 8. Otherwise, new servers for which weight has not been configured might take an inordinate amount of the load.*

Next, you issue the *maxconns 200* command, which sets the maximum number of concurrent connections to this server to be 200. The ranges for this value are from 1 to 4 billion. (Now *that's* a big server!) The default setting is the maximum of 4 billion.

The *reassign 2* command tells the switch to assign the connection to a new server if two consecutive SYN packets from the same host go unanswered. The default value for this setting is 3, and the range is from 1 to 4.

After the *reassign* command, you issue the *faildetect numconns 10 numclients 5* command. The *numconns 10* setting tells the SLB switch that if ten consecutive sessions to this server fail, to consider the server down and remove it from the server pool. The *numclients 5* setting tells SLB that if five consecutive sessions from the same client fail, to also consider the server down and remove it from the server pool. The *numclients* setting is optional. The default value for the *numconns* setting is 8, and the range is from 1 to 255. The *numclients* setting defaults to the same value as the *numconns* command if the *numconns* setting is 8 or less, and to 8 if the *numconns* setting is over 8. The range for the *numclients* setting is from 1 to 8.

The last step for Wolfman is to configure the auto unfail feature. You perform this task using the *retry [value in seconds]* command. In this case, you set the retry interval to 90 seconds. So, if the server is "failed" by the *faildetect* command, auto unfail sends a

CISCO LAN SWITCHING

packet to Wolfman every 90 seconds, hoping for a response so Wolfman can be added back to the pool. The default setting for this command is 60 seconds, and the valid range is from 1 to 3600.

Next, you configure Dracula using the same commands, but setting different values, because Dracula is more powerful than Wolfman. (No, really! Have you ever seen his teeth?)

Finally, you configure Synguard using the *synguard 100 200* command, and you configure sticky connections using the *sticky 90* command.

The *synguard* command syntax is *synguard [number of invalid SYN packets] [interval in ms]*. In this case, you have configured Synguard to drop new SYN packets if 100 SYN packets are outstanding in a 200-ms time period. The default value for this command is 0 for the SYN threshold (cutting Synguard off) and 100 ms for the interval. Valid values for the SYN threshold are from 0 to 4 billion, and 50 ms to 5000 ms for the interval.

The *sticky 90* command enables sticky connections, and it configures the virtual server to send any new session from a previously connected client to the same server it had last, as long as the client's new session is attempted within 90 seconds of the termination of the previous session. The default setting for the *sticky* command is 0 (off), and the valid range is from 0 to 65,535 seconds (about 18 hours).

For the final configuration example, let's configure the virtual server to restrict access to the real servers to allow access only from your public address space (68.1.1.0/24). To perform this task, you issue the *client [ip address] [mask]* command, as follows:

```
4840G-SLB(config)# ip slb vserver PrimaryWeb
4840G-SLB(config-slb-vserver)# client 68.1.1.0 255.255.255.0
```

This configuration allows only clients with IP addresses from 68.1.1.1 through 68.1.1.254 access to the server pool. The default setting for this command is *client 0.0.0.0 0.0.0.0*, which allows all IP addresses access.

 You should not *use the* client *command as a substitute for a firewall. Think of the* client *command as a way to provide extra security, in addition to standard firewall protection.*

That's it. SLB configuration for most simple environments is actually incredibly easy. However, in more complex environments with lots of port-bound servers configured for multiple virtual servers, and SLB redundancy configured on two or more SLB switches, configuration can get a bit more complicated. If you need to configure any of these optional features, please visit the following addresses for complete configuration steps:

- **Catalyst 4840G SLB Configuration Guide** http://www.cisco.com/univercd/cc/td/doc/product/l3sw/4840g/ios_12/120_10/soft_fet/4840slb.htm
- **Catalyst 4840G SLB Redundancy Configuration Guide** http://www.cisco.com/univercd/cc/td/doc/product/l3sw/4840g/ios_12/120_10/soft_fet/4840red.htm

■ **Catalyst 4840G SLB Command Reference (IOS 12.0)** http://
www.cisco.com/univercd/cc/td/doc/product/l3sw/4840g/ios_12/
120_10/soft_fet/4840cmds.htm

Troubleshooting SLB

Troubleshooting SLB usually consists of merely checking your configuration for errors.
Like with layer 3 switching, you should verify a few points before making any drastic
modifications:

■ If NAT is *not* configured for your real servers (using the *nat server* command),
you need to ensure that the real servers use the same IP address (as a secondary
or loopback IP address) as the SLB virtual server. Without NAT, the SLB switch
sends packets to the server pool using the MAC address of the chosen server
but the IP address that is assigned to the virtual server. This mode of SLB
operation (the non-NAT mode) is known as *dispatched mode*.

■ If NAT is *not* configured for your real servers (if you are using dispatched mode),
make sure that no routers exist between the SLB switch and the server pool.

■ If NAT *is* configured for the real servers (called *directed mode*), and routers are
between the SLB switch and the server pool, ensure that the routers are able to
route packets from the SLB switch to the server pool and back.

■ If NAT *is* configured for the clients, and routers are between the SLB switch and
the server pool, ensure that the routers are able to route packets that use the
client address pool from the SLB switch to the server pool and back.

■ Verify that your server IP configuration is valid.

■ Verify that the SLB switch IP configuration is valid.

■ If you are not using NAT, ensure that the server pool is routing client packets
through the SLB switch, and not through other routers in the network. If the
servers route the packets through another router or layer 3 switch, the SLB
switch may incorrectly believe that connections to the server pool are failing.

■ Ensure that optional configuration parameters like Synguard, automatic failure
detection, and session reassignment are not set to extreme values. To eliminate
these as possibilities, use the *no* form of the commands to revert the switch to
the default settings.

■ Ensure that the clients that are attempting to connect are not prevented from
connecting with the *client* command.

■ Ensure that you enter the *inservice* command after *every* configuration change to
either the virtual or real servers.

To help determine the sources of your problems, Table 21-1 describes some useful
SLB-related *show* commands.

Command	Description
show ip slb conns [vserver (virtual server name)] [client (ip-address)] [detail]	Displays connections handled by SLB (either all connections or only those handled by a specific virtual server or from a specific client)
show ip slb reals [vserver (virtual server name)] [detail]	Displays configured real servers, including connection statistics
show ip slb serverfarms [name (serverfarm name)] [detail]	Displays information about server farms
show ip slb stats	Displays SLB statistics.
show ip slb sticky [client (ip-address)]	Displays current sticky connections
show ip slb vservers [name (virtual server name)] [detail]	Displays configured virtual servers, including connection statistics

Table 21-1. *Useful SLB* show *Commands*

Layer 4 Switching Configuration: Congestion Management

Congestion management configuration is a bit trickier than SLB configuration. Like you learned earlier in the "How Layer 4 Switching Works: Congestion Management" section of the chapter, congestion management encompasses several queuing strategies, each with their own way of performing tasks. This section examines configuration tasks for WFQ, CBWFQ, CQ, and PQ. Again, assume that basic configuration has already been performed.

WFQ Configuration

Luckily, WFQ is mostly automatic, so you need to do very little to enable and configure this queuing strategy. In fact, if you have a router interface running at 2.048 Mbps or less, WFQ is already enabled by default.

WFQ is set per interface or ATM PVC. WFQ is not supported on interfaces with the following encapsulations: Link Access Procedure Balanced (LAPB), X.25, and Synchronous Data Link Control (SDLC).

Configuring WFQ actually consists of entering a single command—*fair-queue [congestive discard threshold] [dynamic queues] [reservable queues]*—in interface config mode for the interface for which you wish to configure WFQ. The only mandatory part of this command is *fair-queue*. All other sections are optional.

The *congestive discard threshold* section is the number of messages each queue will hold. After reaching this number of packets, the queue is full and new packets are dropped. The range for the *congestive discard threshold* option is from 16 to 4096 in powers of 2, and the default setting is 64.

The *dynamic queues* setting controls how many queues are allowed in WFQ. By default, this figure is calculated based on the bandwidth of the interface. For links at 64 Kbps or lower, WFQ uses 16 dynamic queues by default. From 64 Kbps to 128 Kbps, 32 queues are used. From 128 Kbps to 256 Kbps, 64 queues are used. From 256 Kbps to 512 Kbps, 128 queues are used; and for any link over 512 Kbps, 256 queues are used by default. The valid values for this setting are 16, 32, 64, 128, 256, 512, 1024, 2048, and 4096.

Note
ATM PVCs have different WFQ settings for each PVC. The default settings for these queues are a bit different than for other interfaces.

Finally, the *reserved queues* option sets how many queues are reserved for RSVP packets. The default setting is 0, and the valid range is from 0 to 1000.

Tip
Typically, the default settings for congestive discard and queues are fine. In most cases, enabling extreme values for either of these settings does not help significantly, unless your link is either extremely slow or extremely fast.

As an example of WFQ configuration, the following command sets WFQ to set a congestive discard of 128 on all queues and allow up to 256 dynamic queues:

```
3600 (config)# interface serial 0/0
3600 (config-if)# fair-queue 128 256
```

CBWFQ Configuration

Class-based WFQ is a bit more complicated to configure than standard WFQ. Configuring CBWFQ basically consists of three steps:

1. Define and configure your class maps.

2. Configure your class policy and policy maps.

3. Apply the policy map to an interface.

The first step in this process is to configure your class maps. *Class maps* are the criteria that determine which packets will be assigned to a given class. By building class maps, you are essentially defining the classes you will use to classify your traffic.

To build class maps, you use the *class-map [name of class]* command from global config mode. This command enters you into class map config mode, as shown here:

```
3600 (config)# class-map testclass
3600 (config-cmap)#
```

You then assign types of packets to this class using the *match* command, which can be used to match packets based on protocol used, input interface (the interface on which the packet arrives), IP precedence, and access lists (discussed in Chapter 27). You choose which criteria to base the match on by using different targets in the *match* command. Table 21-2 describes the common targets for the *match* command, with further explanations following the table.

Note *You can match a class map using only one* match *command. For instance, if you enter* match access-group 1 *and then enter* match protocol ip, *the only command that will be used is the last one typed, in this case,* match protocol ip. *Therefore, most* match *commands are configured to use access lists, because each access list can be configured to match multiple items. For more information on access lists, please refer to Chapter 27.*

The *match access-group* command can match packets based on specific (already configured) named or numbered access lists. To use this command, from class map config mode, type the command followed by the appropriate named or numbered access list. (For named access lists, you must specify the *name* subtarget.) For instance, the following configuration matches packets based on access list 50:

```
3600 (config)# class-map testclass
3600 (config-cmap)# match access-group 50
```

And the following command matches packets based on named access list "executives":

```
3600 (config-cmap)# match access-group name executives
```

Command	Description
match access-group [access list number | name (access list name)]	Matches packets based on a named or numbered access list
match input-interface [interface name and number]	Matches packets based on the interface on which the packet was received
match ip-precedence [up to four precedence values, separated by spaces]	Matches packets based on the IP precedence of the packet
match protocol [protocol name]	Matches packets based on the protocol used

Table 21-2. *Common* match *Command Targets*

The *match input-interface* command matches packets based on the interface on which they arrive. The syntax for this command is very simple: you specify the name and number of the interface after entering *match input-interface*. For instance, the following example matches all packets entering the first Fast Ethernet port:

```
3600 (config-cmap)# match input-interface FastEthernet 0/0
```

The *match ip precedence* command matches packets based on their IP precedence bits. This is useful if other routers along the path have set these bits to signify important traffic. The syntax for this command is *match ip-precedence*, with a listing of up to four precedence values after the command. For instance, the following command matches packets with an IP precedence setting of 2 through 5:

```
3600 (config-cmap)# match ip-precedence 2 3 4 5
```

Finally, the *match protocol* command matches packets based on a specified layer 2 or layer 3 protocol. The common protocols that IOS recognizes include ARP, CDP, IP, and IPX. For example, the following command matches all ARP packets:

```
3600 (config-cmap)# match protocol arp
```

Tip   *Unless your router supports Network-Based Application Recognition (NBAR), the* match protocol *command cannot be used to match upper-layer traffic, which severely limits its usefulness. If your router does not support NBAR, you may still match upper-layer traffic using access lists. Unfortunately, NBAR is beyond the scope of this book. For more information on NBAR, please refer to http://www.cisco.com/warp/ public/cc/so/neso/ienesv/cxne/nbar_ov.htm.*

After building your class maps, the next step is to build and configure policy maps. *Policy maps* are used to set the queuing requirements for one or more classes. Think of a policy map as the container for your classes. By configuring the policy map with bandwidth requirements and packet limits for each class map, you set the queuing parameters for your classes. The first command you use to configure a policy map is the *policy-map [policy name]* global config command, which creates the specified policy map and switches you to policy map config mode, like so:

```
3600 (config)# policy-map slowwanlink
3600 (config-pmap)#
```

Tip *I typically name my policy maps to conform with the name of the interface on which I plan to enable the policy map. Because bandwidth allocations are set with the* policy map *command, this naming convention helps keep me from setting the wrong policy on the wrong interface (and using the wrong amount of bandwidth).*

Once in policy map config mode, you need to assign classes to the policy and set the queuing parameters for each class. To assign a class to a policy, use the *class [class name]* command. This command also switches you to policy map-class config mode, like so:

```
3600 (config-pmap)# class testclass
3600 (config-pmap-c)#
```

Once in policy map-class config mode, you may configure the bandwidth for the class by using the *bandwidth [bandwidth in Kbps]* command, like so:

```
3600 (config-pmap-c)# bandwidth 512
```

In this case, the bandwidth for the class *testclass* has been set to 512 Kbps.

Note *The total bandwidth configured for all classes in a policy map should not exceed 75 percent of the total bandwidth for the link. If you configure more than this amount, the IOS will not let you apply the policy to the interface. You can change the 75 percent limit using the* max-reserved-bandwidth *command, but I don't recommend it because it will reduce the amount of bandwidth available for core functions like signaling and routing updates.*

Tip *You do not have to use all of the link's available bandwidth. For any bandwidth not specifically assigned to a class, CBWFQ distributes the bandwidth to each class weighted by that class's bandwidth setting. In other words, if on a 1.544-Mbps link, you assign 512 Kbps to class 1 and 256 Kbps to class 2, out of the remaining 384 Kbps (remember, IOS uses 25 percent, or 384 Kbps in this case, for system functions), 288 Kbps is assigned to class 1 and 96 Kbps to class 2.*

After configuring the bandwidth for the class, you should apply a queue limit to the class. The queue limit specifies the maximum number of packets that may be in the queue at one time. To set the queue limit, use the *queue-limit [number of packets]* command, like so:

```
3600 (config-pmap-c)# queue-limit 32
```

The default queue limit for most platforms is the maximum of 64.

Once you configure all of your defined classes within the policy map, you will need to configure the default queue. The default queue (commonly called the *class-default* class) is the queue that is used when packets match none of the configured queues. The default queue operates in the exact same manner as basic WFQ: it creates a separate queue for each flow and uses the WFQ algorithm to distribute bandwidth to each queue, as discussed earlier in the "WFQ Configuration" section. For this reason, configuration of the default class is a bit different than configuring a standard class. First, from policy map config mode, you enter the command *class class-default*, like so:

```
3600 (config-pmap)# class class-default
3600 (config-pmap-c)#
```

You then enter the *fair-queue [number of queues]* command to set the maximum number of dynamic queues assigned to the default queue. After this, you can enter the *queue-limit [number of packets]* command to set the maximum size of the default queue. (The default of 64 is typically sufficient, however.) The following example sets the number of dynamic queues to 16 and the queue limit to 32:

```
3600 (config-pmap)# class class-default
3600 (config-pmap-c)# fair-queue 16
3600 (config-pmap-c)# queue-limit 32
```

The last step in configuring CBWFQ is to apply the policy to an interface, which enables CBWFQ on that interface. You perform this task with the *service-policy output [name of policy]* interface config mode command, like so:

```
3600 (config)# interface serial 0/0
3600 (config-if)# service-policy output slowwanlink
```

That's all there is to configuring basic CBWFQ. To bring it all together, let's take a look at an example network in Figure 21-10 and walk through the configuration. In this example, we want to set all packets from the executive subnet (interface FastEthernet 0/0) to use 25 percent of the available WAN bandwidth, and all packets from the server subnet (interface Fast Ethernet 0/1) to use 40 percent of the available WAN bandwidth, and leave the other 10 percent unallocated.

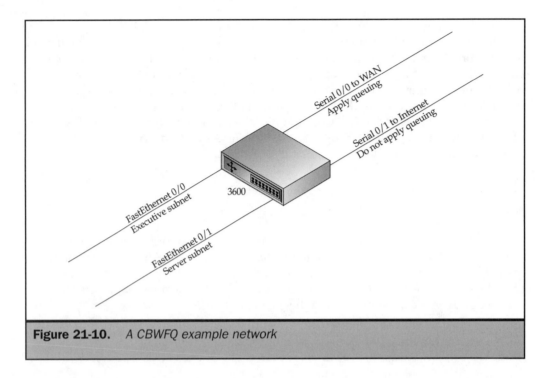

Figure 21-10. *A CBWFQ example network*

First, create your classes and set the matching parameters, shown in the following output:

```
3600 (config)# class-map executive
3600 (config-cmap)# match input-interface fastethernet 0/0
3600 (config-cmap)# class-map server
3600 (config-cmap)# match input-interface fastethernet 0/1
```

Next, create the policy map, add the classes to it, configure the queue properties of the classes, and configure the default class:

```
3600 (config)# policy-map WAN
3600 (config-pmap)# class executive
3600 (config-pmap-c)# bandwidth 384
3600 (config-pmap-c)# queue-limit 40
3600 (config-pmap)# class server
3600 (config-pmap-c)# bandwidth 768
3600 (config-pmap-c)# queue-limit 60
3600 (config-pmap)# class class-default
3600 (config-pmap-c)# fair-queue 16
```

Finally, assign the policy to the serial 0/0 interface, like so:

```
3600 (config)# interface serial 0/0
3600 (config-if)# service-policy output WAN
```

Note   *Keep in mind that queuing statements are* interface specific. *They do* not *apply to interfaces on which they are not configured.*

CQ Configuration

CQ configuration is, believe it or not, a little less complicated than CBWFQ configuration. Basic CQ configuration consists of three steps (which happen to be almost exactly the reverse of CBWFQ):

1. Create the queue list and assign it to interfaces.
2. Configure the queuing properties of each queue in the list.
3. Configure matching for each queue.

The first step is to create the queue list itself. The *queue list* is the "container" that holds a given queue configuration. You specify queuing parameters for each of the 16 queues in your queue list, and apply all of them to an interface by applying the queue list to the interface. You may have up to 16 queue lists active on the router at any time, and each queue list can have up to 16 user-defined queues (plus one system queue). Only one queue list can be active on each interface at any time.

To create the queue list, use the *custom-queue-list [number of queue list]* interface config mode command, like so:

```
3600 (config)# interface serial 0/0
3600 (config-if)# custom-queue-list 1
```

Next, configure the queuing properties (packet limit and byte count) for each of the queues to which you wish to assign special traffic. You perform this task with the *queue-list [queue list number] queue [queue number]* global config mode command. To configure the packet limit, use the syntax *queue-list [queue list number] queue [queue number] limit [maximum packets]*, like so:

```
3600 (config)# queue-list 1 queue 1 limit 40
```

This configures the first queue in Queue List 1 for a limit of 40 packets. The default setting for the packet limit is 20. The valid range is from 0 to 32,767.

Setting a queue limit of 0 allows for an unlimited queue size.

To configure the byte count, use the *queue-list [queue list number] queue [queue number] byte-count [number of bytes]* command, like so:

```
3600 (config)# queue-list 1 queue 1 byte-count 1000
```

This configures the first queue in list 1 for a byte count of 1000 bytes per cycle. The default byte count for a given queue is 1500 bytes.

Finally, you specify which properties a packet must have to be matched for each of your queues, and set a "default" rule for all packets that do not match any of the configured properties.

To configure matching, you use the *queue-list [list number] interface [interface name and number] [queue number]* or *queue-list [list number] protocol [protocol name] [queue number] queue-keyword [modifier] [value]* command. These two commands allow you to match packets using almost the same criteria that are used with CBWFQ. The difference is in how they are configured.

In CBWFQ, you enter a special configuration mode to configure matching for each class, and you may have only one match statement per class. In CQ, however, you configure matching directly from global config mode, and you are allowed to enter as many statements as you like. CQ processes statements in the order in which they are entered, so the first command that matches a packet determines the queue in which the packet is placed. Therefore, be careful when entering *queue-list* statements.

A good way to ensure that your queue list matches packets correctly is to always begin configuring your matching statements with the most specific match, and work your way down to the least specific match. This strategy works well for most situations.

The *queue-list [list number] interface [interface name and number] [queue number]* command classifies packets based on the input interface, just like the *match input-interface* command in CBWFQ. For instance, to match packets entering the first Ethernet interface and classify them for Queue 2 in list 1, you would issue this command:

```
3600 (config)# queue-list 1 interface ethernet 0/0 2
```

Try to keep the number of queue-list *matching statements to a minimum. Every time a packet needs to pass through CQ, every command in the queue list must be parsed from top to bottom. The more commands you enter, the longer this process will take, and the higher your router's processor use and latency will be.*

The protocol variant is a bit more complicated. First, if you just want to match the base protocol (IP and IPX are the only common protocols supported), you leave off the *queue-keyword [modifier] [value]* section of the command. For instance, if you just want to match all IP packets and assign them to Queue 2 in list 1, you enter the following:

```
3600 (config)# queue-list 1 protocol ip 2
```

However, this queue list really isn't all that useful. To get the full power out of this command, you add the *queue-keyword [modifier] [value]* section on the end of the command. The common modifiers you can use for this command are *lt, gt, list, tcp,* and *udp.* The *lt* and *gt* modifiers stand for *less than* and *greater than,* and they are used to match packets with byte sizes less than or greater than the value entered. For instance, the following statement matches packets of less than 500 bytes and enters them into Queue 2 of list 1:

```
3600 (config)# queue-list 1 protocol ip 2 queue-keyword lt 500
```

The list modifier specifies an access list to use for matching. The following statement attempts to match IP packets to access list 150. If they match the access list, they will be inserted into Queue 2 of list 1:

```
3600 (config)# queue-list 1 protocol ip 2 queue-keyword list 150
```

The *tcp* and *udp* modifiers are used to match packets that enter on a specific port. For instance, the following command enters all HTTP packets (TCP port 80) into Queue 2 of list 1:

```
3600 (config)# queue-list 1 protocol ip 2 queue-keyword tcp 80
```

Note *Unfortunately, you cannot specify multiple port numbers or a range of port numbers using the* tcp *or* udp *modifiers. If you want to match a range of addresses, you can either enter multiple* queue-list *commands (only useful for small ranges) or use an access list.*

Finally, your last command should define the default queue, telling CQ where to place a packet that meets no other criteria. You accomplish this goal using the *queue-list [list number] default-queue [queue number]* command. For instance, the following command sets Queue 6 to be the default queue for list 1. (The default setting is that Queue 1 is the default queue for each list.)

```
3600 (config)# queue-list 1 default-queue 6
```

To bring this all together, let's walk through a sample configuration by using the network diagram in Figure 21-11.

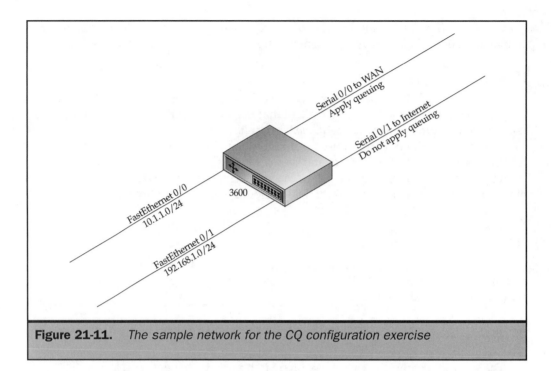

Figure 21-11.   *The sample network for the CQ configuration exercise*

In this example, the requirements are that any packet with a byte count of less than 200 bytes should be placed in Queue 1, any packet that matches access list 101 (which matches all TCP traffic from network 10.1.1.0) should be placed in Queue 2, the default queue should be Queue 3, and any packet using TCP port 20 or 21 (FTP) should be placed in Queue 4. The limits and byte counts are as follows:

■ **Queue 1**   Limit: 80; byte count: 400
■ **Queue 2**   Limit: 40; byte count: 4000
■ **Queue 3**   Limit: 30; byte count: 3000
■ **Queue 4**   Limit: 20; byte count: 1500

First, you configure your serial interface with the queue list:

```
3600 (config)# interface serial 0/0
3600 (config-if)# custom-queue-list 1
```

Next, you build your four queues and configure them:

```
3600 (config)# queue-list 1 queue 1 limit 80
```

```
3600 (config)# queue-list 1 queue 1 byte-count 400
3600 (config)# queue-list 1 queue 2 limit 40
3600 (config)# queue-list 1 queue 2 byte-count 4000
3600 (config)# queue-list 1 queue 3 limit 30
3600 (config)# queue-list 1 queue 3 byte-count 3000
3600 (config)# queue-list 1 queue 4 limit 20
3600 (config)# queue-list 1 queue 4 byte-count 1500
```

Finally, you set up your *queue-list* matching statements. Pay careful attention to the order, and notice how it flows from most to least specific. Specifically, notice that the FTP statement is first, making sure that FTP traffic is always entered into Queue 4. Then notice the byte count statement for any traffic that is under 200 bytes. Finally, you enter the *access list* matching statement and the *default-queue* statement to finalize your requirements:

```
3600 (config)# queue-list 1 protocol ip 4 queue-keyword tcp 20
3600 (config)# queue-list 1 protocol ip 4 queue-keyword tcp 21
3600 (config)# queue-list 1 protocol ip 1 queue-keyword lt 200
3600 (config)# queue-list 1 protocol ip 2 queue-keyword list 101
3600 (config)# queue-list 1 default-queue 3
```

Be aware that this statement has no effect on the serial 0/1 interface on the 3600, regardless of where the packet originated from or the port used. For this statement to apply to the serial 0/1 interface, you would need to add a priority group statement to that interface.

PQ Configuration

Configuring PQ is like simplified CQ configuration. If you remember from previously in the "How Layer 4 Switching Works: Congestion Management" section of this chapter, PQ simply uses four queues (high, medium, normal, and low), and it goes through them in order every time a packet is to be sent. Packets in higher-priority queues are *always* sent before packets in lower-priority interfaces, so the real trick to configuring PQ is to be *very* careful about which packets you classify above normal priority. Beyond that, configuring PQ consists of two steps:

1. Configure the matching statements and the queue properties.

2. Assign PQ to one or more interfaces.

Like CQ, PQ defines up to 16 queue lists within the IOS, each with all four queues, but PQ calls these *priority lists.* Not surprisingly, the command you use for most of your configuration is *priority-list*. The *priority-list* command has four variants: *priority-list interface, priority-list protocol, priority-list queue-limit,* and *priority-list default*. Because you need to set up your matching commands first, let's examine the *priority-list interface* and *priority-list protocol* commands.

The *priority-list interface* command classifies packets based on the input interface. Its syntax is *priority-list [list number] interface [interface name and number] [high | medium | normal | low]*. So if you want all packets entering your first Ethernet interface to be inserted into the high-priority queue for list 1, you enter this command:

```
3600 (config)# priority-list 1 interface Ethernet 0/0 high
```

The *priority-list protocol* command classifies packets based on the protocol used and optional modifiers. Its syntax is *priority-list [list number] protocol [protocol name] [high | medium | normal | low] queue-keyword [modifier] [value]*. The *queue-keyword* section of this command functions exactly like the *queue-keyword* section of the *queue-list* command for CQ (meaning that it is not required, and the modifiers and values operate in exactly the same manner). So, if you want all IP packets with a packet size larger than 1000 bytes to be entered into the low-priority queue for list 1, you enter the following command:

```
3600 (config)# priority-list 1 protocol ip low queue-keyword gt 1000
```

Note *Just like CQ, the matching statements for PQ are parsed in the order in which they are entered. Therefore, make sure the most specific statements are first, and try to keep the total number of statements as low as possible.*

The *priority-list queue-limit* variant sets the size in packets of each queue for PQ. The syntax for this command is *priority-list [list number] queue-limit [high limit] [medium limit] [normal limit] [low limit]*. So, if you want to set the packet limits for the high queue to 30, the medium queue to 40, the normal queue to 50, and the low queue to 60 for list 1, you enter the following command:

```
3600 (config)# priority-list 1 queue-limit 30 40 50 60
```

The default values for the queues are the following: high is 20, medium is 40, normal is 60, and low is 80.

The *priority-list default* command sets the default classification for packets that match none of the other *priority-list* statements. The syntax for the command is *priority-list [list number] default [high | medium | normal | low]*. The following command sets the default classification to low for list 1:

```
3600 (config)# priority-list 1 default low
```

After configuring all of your matching statements, setting queue limits, and configuring the default queue, the only task left is to assign PQ to an interface. You do this using the *priority-group [group number]* interface config mode command, like so:

```
3600 (config)# interface serial 0/0
3600 (config-if)# priority-group 1
```

Again, to put this into perspective, let's examine a sample network (shown in Figure 21-12) and configure PQ. In this case, the criteria stipulates that all packets exiting serial 0/0 that entered from the executive subnet (Fast Ethernet 0/0) should be given medium priority, all packets exiting serial 0/0 for a custom application running on TCP port 6000 should be given high priority, all packets exiting serial 0/0 that

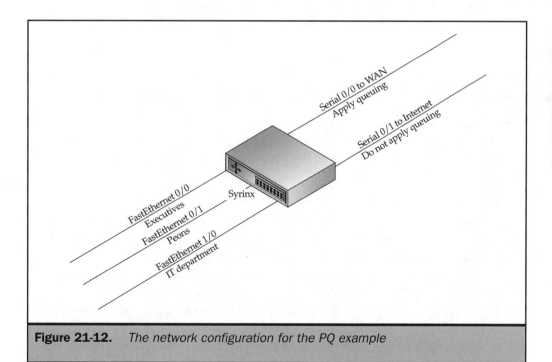

Figure 21-12. *The network configuration for the PQ example*

entered from the peons' subnet (Fast Ethernet 0/1) should be given low priority, and all other traffic should be at normal priority.

```
Syrinx (config)# priority-list 1 protocol ip high queue-keyword tcp 6000
Syrinx (config)# priority-list 1 interface fastethernet 0/0 medium
Syrinx (config)# priority-list 1 interface fastethernet 0/1 low
Syrinx (config)# priority-list 1 interface default normal
Syrinx (config)# interface serial 0/0
Syrinx (config-if)# priority-group 1
```

Once again, note the order of the matching statements. You set the matching statement for the high-priority traffic first so that even the peons get to use the custom application; then the two interface statements; and, finally, the default statement. That's all there is to configuring PQ.

Note *Just like in the CQ example, be aware that this statement has no effect on the serial 0/1 interface on Syrinx, regardless of where the packet originated from or which port is used. For PQ to apply to the serial 0/1 interface, you must add a priority group statement to that interface.*

Verifying and Troubleshooting Congestion Management

In most cases, problems with congestion management are simply due to improper configurations. However, you should use the following guidelines for what to look for if you are having problems with congestion management:

- Make sure the device performing the queuing is not underpowered.
- Ensure that your link is not constantly oversubscribed. Queuing cannot help with this situation (unless you wish to eliminate some of your traffic entirely).
- Ensure that queuing is configured on the correct interface.
- If you are performing matching with access lists, ensure that the access lists are correctly configured. (Access lists are discussed in Chapter 27.)
- For problems with basic WFQ, ensure that WFQ's congestive discard threshold and dynamic queue parameters are not set to extreme values.
- For problems with CBWFQ, ensure that the combined total of all of your bandwidth statements for the policy are configured to use no more than 75 percent of the total bandwidth for the interface to which you have the policy applied.
- For CQ and PQ, ensure that you have the correct default queue set.
- For CQ and PQ, ensure that your matching statements are entered in the proper order.

Finally, to help you identify your problem areas, as well as verify the queuing configuration, Table 21-3 provides a list of useful *show* commands related to queuing.

Command	Optional Modifiers	Description
show policy-map	None	Shows configurations for all CBWFQ policy maps, including all configured classes.
show policy-map [policy name]	*class [class name]*	Shows configurations for a single CBWFQ policy map, including all configured classes. Can also show information on a single class with the optional *class* keyword.
show policy-map interface [interface name and number]	*vc [vci/vpi] dlci [number]*	Shows configuration for a single CBWFQ interface, including statistics. Can also show information on a single ATM or Frame Relay VC with the optional *vc* or *dlci* keywords.
show queue [interface name and number]	*vc [vci/vpi]*	Shows queuing information for a single interface, including statistics. Can also show information on a single ATM VC with the optional *vc* keyword.
show queuing	*[custom	fair	priority]*	Shows either all queuing statistics, or just queuing statistics for a particular queuing strategy for all interfaces.

Table 21-3.   *Useful Congestion Management* show *Commands*

Command	Optional Modifiers	Description
show queuing interface [interface name and number]	*vc [vci/vpi]*	Shows queuing statistics for a single interface. Can also show information on a single ATM VC with the optional *vc* keyword.

Table 21-3. *Useful Congestion Management* show *Commands* (continued)

Summary

This chapter covered layer 4 switching with SLB and congestion management strategies. It also covered configuration and troubleshooting at a basic level for both of these technologies. You should now be able to configure basic SLB (even though it is fairly rare at this point), as well as congestion management using WFQ, CBWFQ, CQ, or PQ. Although understanding or configuring SLB may not be required in your present environment, in a few years, SLB will most likely be a commonly used technology—and it never hurts to be prepared. As for congestion management, in most environments, a good working knowledge of queuing methods is invaluable in reducing latency for bandwidth-intensive or near–real-time applications, and your knowledge of these queuing methods will most likely serve you well for some time to come.

The Complete Reference

Part IV

Cisco Routing

In this part, we cover the most common internal routing configurations in use today. We begin with Chapter 22, where we discuss the basics of routing and the benefits of static routing. Chapter 23 discusses possibly the most common internal distance-vector routing protocol, RIP. Chapter 24 discusses IGRP, a Cisco proprietary routing protocol that overcomes some of the shortcomings of RIP while retaining simplicity.

In Chapter 25, we discuss IGRP's more scalable big brother protocol, EIGRP. In Chapter 26, we discuss the alternative to EIGRP for highly scalable internal routing requirements—OSPF. Finally, in Chapter 27, we discuss the role of access lists in modern routed environments.

Chapter 22

Understanding Routing

This chapter examines routing as a whole and takes a look at what makes both dynamic and static routing tick from a theoretical perspective. You will learn about configurations and situations in which static routing is appropriate. The chapter then switches gears a bit and covers the basics of configuring a router, regardless of the routing paradigm used. Finally, you will learn about the intricacies of configuring static routing and look at troubleshooting static routing in common environments.

How Routing Works

Chapter 6 covered the basic functionality of routing, but I will go through a quick refresher just in case I managed to put you to sleep before you got to that part. (I know, I know, you are protesting right now that no one could *possibly* sleep through my book—but I like to remain humble.)

Anyhow, basic routing works like this. The clients perform the ANDing process with the source IP address (hopefully, their IP address) and the destination IP address. If the result is the same, the client knows that it should be able to find the destination device's MAC address using an Address Resolution Protocol (ARP) broadcast. If the result is different, the client knows that the destination is remote, and it needs to use a router to reach the destination client. Assuming that a router is required, the client examines the IP address entered for the default gateway in its configuration. It then determines the MAC address for the default gateway using an ARP broadcast. Once it has the router's MAC address, it builds the packet using the final destination device's IP address, but using the router's MAC address. The router examines the packet; notices that its MAC address has been used, meaning that it needs to process the packet; and examines the destination IP address in the packet.

The router then looks through a table that lists all remote networks it is currently aware of, called a *routing table,* and attempts to find a path to the destination network in this table. If a path to the remote network *is* found, the router enters the MAC address of the next-hop device (either the next router along the path or the remote host itself) in the packet, and forwards the packet. If *no* path to the remote network is found (even a poor one, like a default route), the router returns an Internet Control Messaging Protocol (ICMP) unreachable message back to the source, informing the source, "Hey pal, you can't get there from here."

That's it. Not all that complex, huh? So what's all this hubbub about routing being this big monstrously complex topic? Well, unfortunately, the devil is in the details. To begin the discussion of this topic, let's take a look at the most basic form of routing: static routing.

How Static Routing Works

Static routing is based on the idea that, if there are any networks that you want the router to be aware of, you will manually enter those routes yourself, thank you very much. Static routing is typically pretty easy to understand and configure, at least on a small network, and is definitely the least complex method of routing. To examine static routing, let's take the basic network shown in Figure 22-1 and examine the routing process from the time the routers are first booted.

First, looking at the routing tables for each router, the first column listed is Destination. The only destination networks currently listed are those to which the routers are directly connected, the default setting when using static routing. If you want other networks to be listed in this table, you have to enter them manually. After the destination network is the Next Hop column, which tells the routers what the next hop router is for this destination network. This column is currently blank because no one has manually entered any routes; the only routes the routers know about are the ones to which they are directly connected.

Next is the Exit I/F column, which tells the router which interface to forward the packet out from. (This information isn't usually needed, but I have included it for illustrative purposes.) Finally, the Metric column is used when multiple routes are in the table for the same destination. The router uses the route with the lowest cost to any given

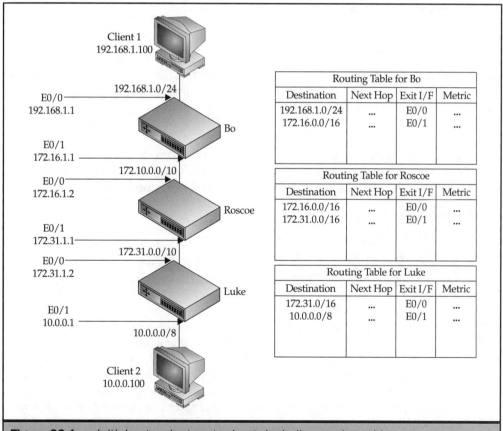

Figure 22-1.   *Initial network at router boot, including routing tables*

destination. In this example, the metric used is *hop count* (how many routers are between this router and the destination network); and because all routers are directly connected to the only networks they are currently aware of, the Metric columns are all blank.

Next, notice that each router interface is assigned an IP address. These addresses will be discussed more in a moment; but for right now, the IP address for Client 1's (192.168.1.100) default gateway is Bo's E0/0 interface (192.168.1.1), and Client 2 (10.0.0.100) has its default gateway set to 10.0.0.1 (Luke's E0/1 interface).

What happens when Client 1 sends a packet to Client 2? Client 1 performs the ANDing process and realizes that Client 2 is on a different network. Client 1 looks in its IP configuration and finds the default gateway entry for Bo (192.168.1.1). It then sends the packet out with the following parameters:

- **Destination MAC** 11-11-11-11-11-11 (Bo's E0/0 interface MAC address)
- **Source MAC** 00-00-00-00-00-01 (Client 1's MAC address)
- **Destination IP** 10.0.0.100 (Client 2)
- **Source IP** 192.168.1.100 (Client 1)
- **TTL** 128

Bo receives the packet on the E0/0 interface and examines the destination MAC address field. Noting that the destination MAC is his MAC address, he begins processing the packet by looking in his routing table for the destination network (10.0.0.0). After looking in the routing table, he realizes that he has no path to the specified destination network, so he forwards an ICMP packet back to Client 1 informing Client 1 that the destination is unreachable.

To resolve this communication's failure, you need to tell Bo how to reach 10.0.0.0. Well, Bo needs to send all packets for the 10.0.0.0 network to Roscoe, because Roscoe is the next router in the path. So, you add a static route to Bo with the following parameters:

- **Destination Network** 10.0.0.0/8
- **Next Hop** 172.16.1.2 (Roscoe's E0/0 interface)
- **Exit Interface** E0/1
- **Metric (hop count)** 2 (The packet must pass through two routers to reach the destination.)

Note *Because no alternative paths exist to the 10.0.0.0 network, setting the metric isn't really necessary. However, it is a good idea to go ahead and insert metrics to reduce administrative effort in case you change the network layout later. Also note, from a configuration perspective, static routes don't really have a metric. When you set what the IOS calls a metric on a static route, you are actually setting the administrative distance of the static route (discussed in the "Understanding Administrative Distance" section, later in the chapter). As long as you are using only static routes, you can treat the administrative distance the same as a metric, but you need to consider the differences when adding dynamic routing protocols to the network.*

After adding this information, your network configuration would resemble Figure 22-2. Now when Client 1 sends a packet to Client 2, it should get there, right? Let's walk through the process and see. Client 1 goes through the same process as in the previous example, and sends the packet to Bo using Bo's MAC address and Client 2's IP address. Bo examines the packet, finds the routing entry in the table, and forwards the packet to the next-hop router (Roscoe) with the following parameters:

- **Destination MAC** 22-22-22-22-22-22 (Roscoe's E0/0 interface MAC address)
- **Source MAC** 11-11-11-11-11-12 (Bo's E0/1 interface MAC address)
- **Destination IP** 10.0.0.100 (Client 2)
- **Source IP** 192.168.1.100 (Client 1)
- **TTL** 127 (Remember, each router is *required* to decrement the Time to Live by at least 1.)

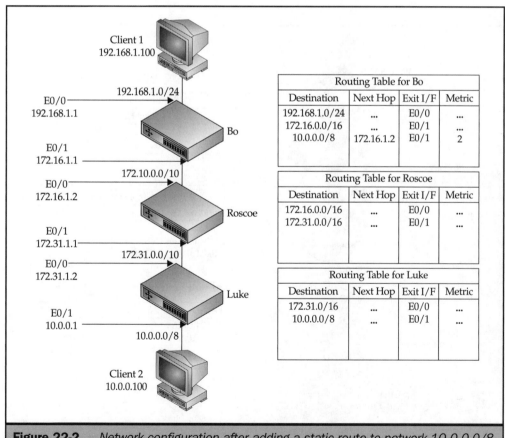

Figure 22-2. *Network configuration after adding a static route to network 10.0.0.0/8 for Bo*

So far, so good. Now Roscoe examines the packet details, looks for an entry for the destination network in his routing table, and finds—nothing. So Roscoe attempts to send an ICMP packet informing Client 1 that the destination is unreachable, but, looking in his routing table, he realizes that he doesn't even know how to get the packet back! So he simply drops the original packet, and Client 1 just has to wait for the transmission to time out (instead of almost immediately getting a Destination Unreachable message).

To solve this problem, you need to add a static route to Roscoe for network 10.0.0.0 as well, using the following parameters:

- **Destination Network**   10.0.0.0/8
- **Next Hop**   172.31.1.2 (Luke's E0/0 interface)
- **Exit Interface**   E0/1
- **Metric (hop count)**   1

These settings give you the configuration shown in Figure 22-3.

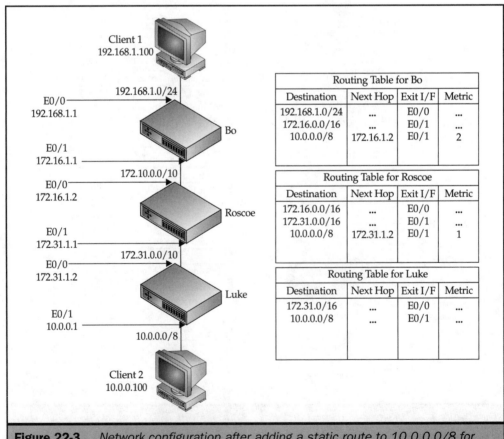

Figure 22-3.   *Network configuration after adding a static route to 10.0.0.0/8 for Roscoe*

Now, Client 1 sends the packet, and Bo looks in his routing table and forwards the packet to Roscoe. Roscoe looks in his routing table and forwards the packet to Luke. Luke looks in his routing table, realizes he is directly connected to 10.0.0.0/8, and forwards the packet directly to Client 2—*success!*

Now Client 2 attempts to respond, sending the packet to Luke. Luke looks at his routing table for a path to the 192.168.1.0/24 network and (here we go again) comes up empty. So, what do you have to do to actually get this thing to work? You must add a path to networks 10.0.0.0/8 and 192.168.1.0/24 for every router along the path, as shown in Figure 22-4.

Great, now Client 1 can reach Client 2. However, this layout has a few issues. For instance, even though Client 1 can send packets all the way to Client 2, Client 1 will get a Network Unreachable message when trying to ping 172.31.1.3 (Luke's E0/0 interface). Why? Because there is no route to the 172.31.0.0 network on Bo. Although this isn't an issue right now, if you were to add a server to the 172.31.0.0 network, Client 1 would not be able to reach it. (A similar issue exists with Client 2 and network 172.16.0.0.) So,

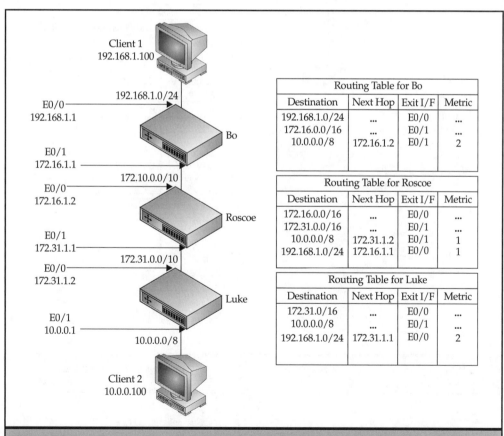

Figure 22-4. *The functional configuration, with paths to both 192.168.1.0/24 and 10.0.0.0/8 enabled on all routers*

CISCO ROUTING

to allow full connectivity to every network, you need to make sure every router has a route to every network, as shown in Figure 22-5.

At this point, you're probably thinking that there must be an easier way of doing this. There is: you can configure default routes on Bo and Luke. *Default routes* are routes that the router uses if it has no route entry for the destination network. You can configure a default route in a couple ways; but right now, we will use the 0.0.0.0 network. If you add a route to the router for 0.0.0.0 with a 0.0.0.0 (0-bit) mask, the router basically matches that address to everything. So, if it can't find a more specific route in the routing table, it will send the packet using the default route.

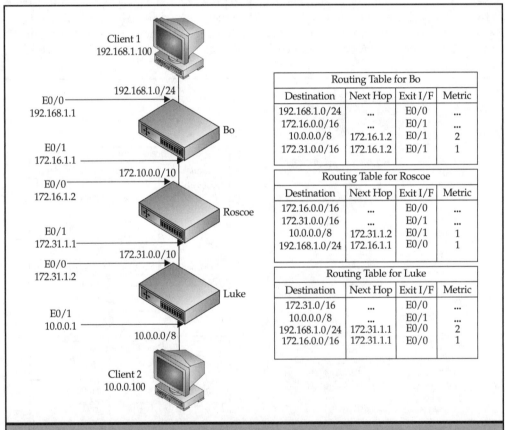

Figure 22-5. *A fully functional network configuration, with a route to every network on every router*

Note *Keep in mind that a router will use the most specific route it can find in the routing table. For instance, if you were trying to get a packet to 10.1.1.1 and the router had a route to 10.0.0.0/8, a route to 10.1.1.0/24, and a route to 0.0.0.0/0, it would use the 10.1.1.0/24 route. This rule is known as the longest match rule.*

To configure default routes for this network, you would add routes to the all 0's network for Bo and Luke, as shown in Figure 22-6.

Okay, now that you've seen how basic static routing works, let's take a look at a more complicated static routing implementation. Figure 22-7 adds a new router to the network, along with two additional links to provide some redundancy. The routing table for Figure 22-7 follows.

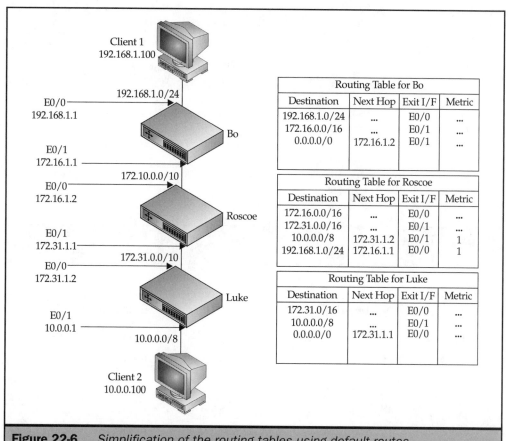

Figure 22-6. *Simplification of the routing tables using default routes*

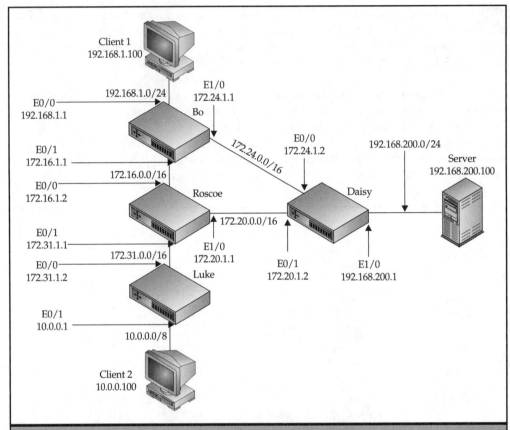

Figure 22-7. *A more complicated example including redundancy*

	Destination	Next Hop	Exit I/F	Metric
Daisy	192.168.200.0/24	—	E1/0	—
	172.24.0.0/16	—	E0/0	—
	172.20.0.0/16	—	E0/1	—
	192.168.1.0/24	172.24.1.1	E0/0	1
	192.168.1.0/24	172.20.1.2	E0/1	2
	172.16.0.0/16	172.20.1.1	E0/1	1
	172.16.0.0/16	172.24.1.1	E0/0	1

	Destination	Next Hop	Exit I/F	Metric
Daisy (continued)	172.31.0.0/16	172.20.1.1	E0/1	1
	172.31.0.0/16	172.24.1.1	E0/0	2
	10.0.0.0/8	172.20.1.1	E0/1	2
	10.0.0.0/8	172.24.1.1	E0/0	3
Bo	192.168.1.0/24	—	E0/1	—
	172.16.0.0/16	—	E0/1	—
	172.24.0.0/16	—	E0/1	—
	0.0.0.0/0	172.24.1.2	E1/0	—
	0.0.0.0/0	172.16.1.2	E0/1	—
Roscoe	172.16.0.0/16	—	E0/0	—
	172.31.0.0/16	—	E0/1	—
	172.20.0.0/16	—	E1/0	—
	10.0.0.0/8	172.31.1.2	E0/1	1
	192.168.1.0/24	172.16.1.1	E0/0	1
	192.168.1.0/24	172.20.1.2	E1/0	2
	192.168.200.0/24	172.20.1.2	E1/0	1
	192.168.200.0/24	172.16.1.1	E0/0	2
	172.24.0.0/16	172.16.1.1	E0/0	1
	172.24.0.0/16	172.20.1.2	E1/0	1
Luke	172.31.0.0/16	—	E0/0	—
	10.0.0.0/8	—	E0/1	—
	0.0.0.0/0	172.31.1.1	E0/0	—

This time, I already went in and added all the necessary routes. Notice the major changes to the routing table on Roscoe, as well as the large number of routes on the new router, Daisy. Also notice the addition of another default route on Bo. Let's go through these routers one at a time, starting with Bo.

Tip *The Cisco IOS normally load balances across up to six static routes.*

At Bo, the only real changes were the addition of the directly connected 172.24.0.0 network (Bo's E1/0 interface), and the addition of a second default route. With static routing, if more than one static route to the same network with the same metric is entered, the router load balances across all routes. For example, if you were to ping Client 2 from Bo, the first packet Bo would send would be through Roscoe, the second packet would be through Daisy, the third packet would be through Roscoe, and so on.

Note   *Load balancing across redundant static routes is known as* equal cost load balancing—*the links and performance of both routes should be roughly equal. If they are not, then one link will be overused while the other is underused. This problem is known as* pinhole congestion, *and it is discussed further in Chapter 23.*

By load balancing the static routes, you also get the advantage of redundancy. In Cisco routers, if the next hop address for any route (including default routes) becomes unreachable due to a local link failure (the router cannot sense a link on the exit interface), that route will be removed from the routing table until the link is back up. So, if Bo's 0/1 interface were to fail, Bo would remove the default route to Roscoe (172.16.1.2) and forward all packets to Daisy.

Note   *Redundancy with static routing applies only to* direct *link failures. A failure on a remote device cannot be detected with static routing, and will not cause the route to be removed from the table. Therefore, packets will intermittently fail to be delivered. (With two links, 50 percent would fail; with three links, 33 percent would fail—and so on.)*

At Roscoe, I have added routes to the new networks, as well as a redundant route to the 192.168.1.0 network. Note that on the redundant routes to each network, the metric is higher than the primary route. Because we are using hop count as the metric, this concept is fairly easy to understand. Roscoe uses the route with the lowest metric, and if that route fails, falls back and uses the higher metric route. For instance, to deliver packets to the 192.168.200.0 network, Roscoe uses the route through Daisy (172.20.1.2) because it has the lowest metric (it is the most direct route). However, if Roscoe's E1/0 interface fails, he removes the route and uses the route through Bo instead. Keep in mind that with static routing, load balancing across unequal cost paths does *not* occur. The router simply uses the path with the lowest metric.

Note   *In the Cisco IOS, the higher metric routes would not show in the routing table. However, if a direct link failure occurred, they would then be inserted into the routing table and would be used normally.*

Looking at Daisy, you can see that she has two paths to nearly everywhere. Again, notice how the metrics are used to determine which path she chooses as the best path. For instance, to deliver a packet to the 10.0.0.0 network, her preferred path is through

Roscoe (metric of 2). However, if her E0/1 interface were dead, she would send the packet to Bo instead. Unfortunately, that is where the load-balancing feature of static routing begins to cause problems with the redundancy feature. Because Bo simply has two equal-cost static routes, 50 percent of the time, he sends the packet right back to Daisy. Daisy then sends the packet back to Bo who—again, 50 percent of the time—sends it back to Daisy (as if they were playing "pass the packet").

Although most of the time the packet would *eventually* be delivered, occasionally the TTL on the packet would expire and it would be dropped. In addition, all of the packet retransmissions would use valuable router and link resources. A better routing-table design for Bo is shown in the next table (with the topology repeated in Figure 22-8 for ease of reference).

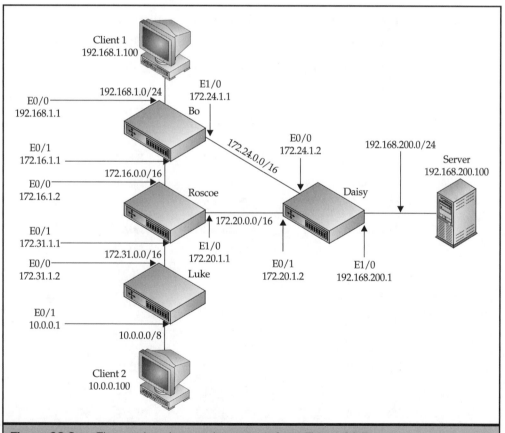

Figure 22-8. *The reprinted example network from Figure 22-7*

	Destination	Next Hop	Exit I/F	Metric
Daisy	192.168.200.0/24	—	E1/0	—
	172.24.0.0/16	—	E0/0	—
	172.20.0.0/16	—	E0/1	—
	192.168.1.0/24	172.24.1.1	E0/0	1
	192.168.1.0/24	172.20.1.2	E0/1	2
	172.16.0.0/16	172.20.1.1	E0/1	1
	172.16.0.0/16	172.24.1.1	E0/0	1
	172.31.0.0/16	172.20.1.1	E0/1	1
	172.31.0.0/16	172.24.1.1	E0/0	2
	10.0.0.0/8	172.20.1.1	E0/1	2
	10.0.0.0/8	172.24.1.1	E0/0	3
Bo	192.168.1.0/24	—	E0/0	—
	172.16.0.0/16	—	E0/1	—
	172.24.0.0/16	—	E1/0	—
	192.168.200.0/24	172.24.1.2	E1/0	1
	192.168.200.0/24	172.16.1.2	E0/1	2
	172.20.0.0/16	172.24.1.2	E1/0	1
	172.20.0.0/16	172.16.1.2	E0/1	1
	172.31.0.0/16	172.16.1.2	E0/1	1
	10.0.0.0/8	172.16.1.2	E0/1	1
Roscoe	172.16.0.0/16	—	E0/0	—
	172.31.0.0/16	—	E0/1	—
	172.20.0.0/16	—	E1/0	—
	10.0.0.0/8	172.31.1.2	E0/1	1
	192.168.1.0/24	172.16.1.1	E0/0	1
	192.168.1.0/24	172.20.1.2	E1/0	2
	192.168.200.0/24	172.20.1.2	E1/0	1
	192.168.200.0/24	172.16.1.1	E0/0	2
	172.24.0.0/16	172.16.1.1	E0/0	1
	172.24.0.0/16	172.20.1.2	E1/0	1

	Destination	Next Hop	Exit I/F	Metric
Luke	172.31.0.0/16	—	E0/0	—
	10.0.0.0/8	—	E0/1	—
	0.0.0.0/0	172.31.1.1	E0/0	—

With this configuration, if Daisy forwarded a packet to Bo for the 10.0.0.0 network, Bo would forward the packet to Roscoe 100 percent of the time.

Now you should understand what static routing is capable of. The bottom line is, if you have a relatively small and simple network, static routing is probably your best choice. If you have redundant links, you can still use static routing; but, in most cases, you would be better served by a dynamic routing protocol.

How Dynamic Routing Works

Dynamic routing, on the other hand, doesn't suffer from many of static routing's limitations. The basic idea of dynamic routing is that a special protocol, called a *routing protocol,* communicates routes between the routers in the topology. For instance, take a look at the network in Figure 22-9.

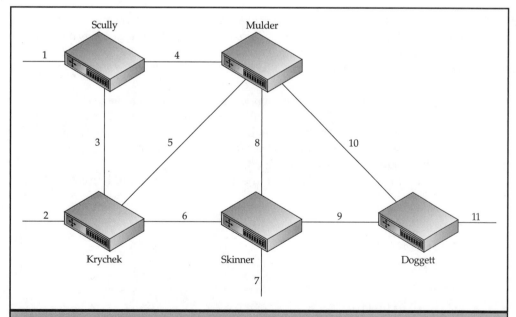

Figure 22-9. *The initial network for the dynamic routing example*

The known networks for Figure 22-9 are as follows:

- **Scully** 1, 4, 3
- **Mulder** 4, 5, 8, 10
- **Krychek** 2, 3, 5, 6
- **Skinner** 6, 7, 8, 9
- **Doggett** 9, 10, 11

This network has five routers in a partial (or hybrid) mesh. For the sake of brevity, I have listed only their known networks (rather than the entire routing table) and have simplified the networks down to a single number. At boot, each router would know about only the networks to which they are directly connected. With static routing, for each router to "learn" the routes to all networks in this topology, you would need to manually enter all primary and backup routes manually. This task would be a bit difficult, as well as be prone to the same packet-passing issues you saw earlier.

A dynamic routing protocol, on the other hand, automatically informs the other routers of all routes it knows about at regular intervals. These "information" packets are called *advertisements* or *routing updates,* depending on the specific protocol in question. These updates allow all routers to automatically learn about all routes (primary and backup).

Note *The following example has been greatly simplified to be non-protocol specific and to illustrate the basics of routing updates. The purpose is not to show you how dynamic routing protocols work in real life (that is covered in Chapters 23–26), but to give you a basic understanding of the routing update process.*

To help you see how dynamic routing works, let's walk through the update process with the example network. In Figure 22-10, all routers are sending out their initial advertisement to all other neighboring routers. So Scully, for example, sends an advertisement to Mulder and Krychek containing information about her known networks (at this point, only networks 1, 3, and 4).

At the time of the initial advertisement, the known networks for Figure 22-10 are as follows:

- **Scully** 1, 4, 3
- **Mulder** 4, 5, 8, 10
- **Krychek** 2, 3, 5, 6
- **Skinner** 6, 7, 8, 9
- **Doggett** 9, 10, 11

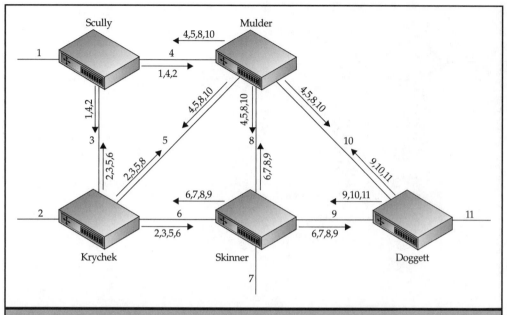

Figure 22-10. *The initial advertisement in the example network*

After receiving the updates, the routers update their respective tables and, at the next update interval, send out a new update containing the updated list of networks, as shown in Figure 22-11. The dynamically learned entries in the routing table are marked in italic type, and the routers whose networks were learned from are listed in parentheses.

At the time this second update is sent, the known networks for Figure 22-11 are as follows (with the router the network was learned from in parentheses):

■ **Scully**   1, 2 *(Krychek)*, 3, 4, 5 *(Mulder, Krychek)*, 6 *(Krychek)*, 8 *(Mulder)*, and 10 *(Mulder)*

■ **Mulder**   1 *(Scully)*, 2 *(Krychek)*, 3 *(Scully, Krychek)*, 4, 5, 6 *(Krychek, Skinner)*, 7 *(Skinner)*, 8, 9 *(Skinner, Doggett)*, 10, and 11 *(Doggett)*

■ **Krychek**   1 *(Scully)*, 2, 3, 4 *(Scully, Mulder)*, 5, 6, 7 *(Skinner)*, 8 *(Mulder, Skinner)*, 9 *(Skinner)*, and 10 *(Mulder)*

■ **Skinner**   2 *(Krychek)*, 3 *(Krychek)*, 4 *(Mulder)*, 5 *(Krychek, Mulder)*, 6, 7, 8, 9, 10 *(Mulder, Doggett)*, and 11 *(Doggett)*

■ **Doggett**   4 *(Mulder)*, 5 *(Mulder)*, 6 *(Skinner)*, 7 *(Skinner)*, 8 *(Mulder, Skinner)*, 9, 10, 11

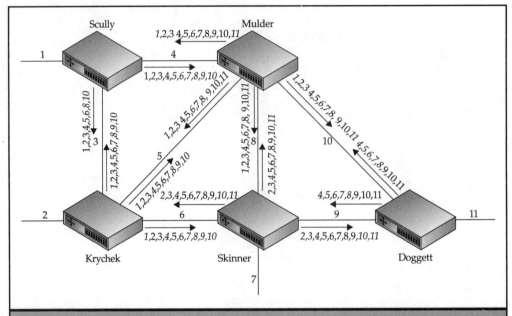

Figure 22-11. *The second update in the dynamic routing example.*

After looking at Figure 22-11, you are probably thinking that this thing just got very complicated really quick! Unfortunately, that complexity is one of the biggest drawbacks of dynamic routing (and one of the reasons I believe that the more you understand about networking, the more surprised you will be that it actually works *at all*!). Eventually, however, the network will settle down and every router will know of every path to every network. This state is known as *convergence*, and it is shown in Figure 22-12.

At full convergence, the known networks for Figure 22-12 are as follows (with the router the network was learned from in parentheses):

- **Scully** 1, 2 *(Krychek, Mulder)*, 3, 4, 5 *(Krychek, Mulder)*, 6 *(Krychek, Mulder)*, 8 *(Krychek, Mulder)*, 9 *(Krychek, Mulder)*, 10 *(Krychek, Mulder)*, and 11 *(Krychek, Mulder)*

- **Mulder** 1 *(Scully, Krychek, Skinner, Doggett)*, 2 *(Scully, Krychek, Skinner, Doggett)*, 3 *(Scully, Krychek, Skinner, Doggett)*, 4, 5, 6 *(Scully, Krychek, Skinner, Doggett)*, 7 *(Scully, Krychek, Skinner, Doggett)*, 8, 9 *(Scully, Krychek, Skinner, Doggett)*, 10, and 11 *(Scully, Krychek, Skinner, Doggett)*

- **Krychek** 1 *(Scully, Mulder, Skinner)*, 2, 3, 4 *(Scully, Mulder, Skinner)*, 5, 6, 7 *(Scully, Mulder, Skinner)*, 8 *(Scully, Mulder, Skinner)*, 9 *(Scully, Mulder, Skinner)*, 10 *(Scully, Mulder, Skinner)*, and 11 *(Scully, Mulder, Skinner)*

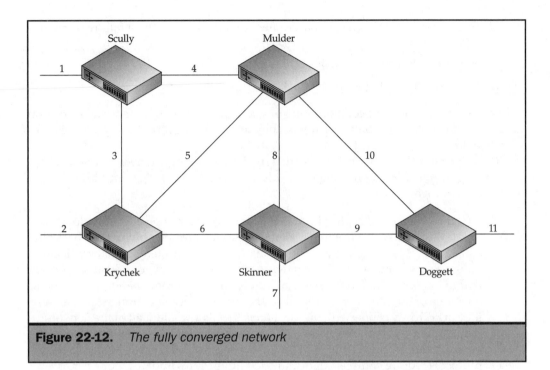

Figure 22-12.   *The fully converged network*

- **Skinner**   *1 (Krychek, Mulder, Doggett), 2 (Krychek, Mulder, Doggett), 3 (Krychek, Mulder, Doggett), 4 (Krychek, Mulder, Doggett), 5 (Krychek, Mulder, Doggett), 6, 7, 8, 9, 10 (Krychek, Mulder, Doggett), and 11 (Krychek, Mulder, Doggett)*

- **Doggett**   *1 (Mulder, Skinner), 2 (Mulder, Skinner), 3 (Mulder, Skinner), 4 (Mulder, Skinner), 5 (Mulder, Skinner), 6 (Mulder, Skinner), 7 (Mulder, Skinner), 8 (Mulder, Skinner),* 9, 10, *and* 11

Note   *If you have a sharp eye, you may also have noticed one of the other major drawbacks of dynamic routing: routing loops. Preventing routing loops is a major chore for most dynamic routing protocols, and all of them handle it a bit differently. You will learn about the various solutions to routing loop problems in later chapters.*

Now that you've seen the basics of what a routing protocol does, let's take a look at the goals of all routing protocols:

- To reduce administrative effort by dynamically filling the routing tables with routes to all networks

- When more than one route to a given network is available, either
  - To place the best route in the table, or
  - To place multiple routes in the table and load balance across the routes

- To automatically remove invalid routes from the table when a failure (direct or indirect) occurs

- If a better route is heard, to add that route to the table

- To eliminate routing loops as quickly as possible

These goals are the same for all routing protocols, regardless of the routing algorithm used. As you will see in upcoming chapters, sometimes meeting all of these goals is easier said than done.

Routing protocols are typically classified by the type of logic they use. The three primary algorithms used by routing protocols are listed here, along with a brief explanation of each:

- **Distance vector** This algorithm is the most common type of routing logic used today. Distance vector routing is sometimes called "routing by rumor." Basically, a distance vector protocol yells out to all directly connected neighbors, "Hey guys, I know about all of these networks!" The neighbors go, "Great, let me add that to my table. And, by the way, I know about all of these networks!"—and so on. None of the routers actually *know* about these networks (unless, of course, they are directly connected), they just *think* they know about all of the networks. If one router has bad information, it passes this invalid information on to all other routers without realizing that it is wrong. For this reason, distance vector protocols require complex algorithms to "doubt" any updates they hear to prevent large-scale routing loops. Distance vector protocols covered in this book include Routing Information Protocol (RIP) and Interior Gateway Routing Protocol (IGRP). I also consider Enhanced IGRP (EIGRP) to be a distance vector protocol (albeit, one on steroids), but Cisco classifies it as a "balanced hybrid" (probably because distance vector has such a bad rap).

- **Link state** Link state routing logic operates on the Dykstra Shortest Path First (SPF) algorithm, and it operates a bit differently than distance vector. Basically, link state protocols build a "map" of the network, so they inherently have a better idea of where everything is located than distance vector protocols do. This advantage makes link state protocols much more graceful and sophisticated, but it also makes them harder to understand and implement properly. Very few routing protocols use the link state logic, and the only one this book covers is Open Shortest Path First (OSPF).

- **Balanced hybrid** This algorithm is Cisco's term for EIGRP's routing logic. Personally, I think Cisco just made this designation up for marketing purposes; but because they did classify it this way, I will cover it in this manner. The idea behind the balanced hybrid designation is that it includes features of both distance vector and link state algorithms. For instance, the core logic behind

EIGRP is basically distance vector, but EIGRP transmits additional topology information in order to "map" the network, like link state. The only balanced hybrid protocol discussed in this book is EIGRP.

Here's a quick analogy to help describe the difference in these routing logic types. Pretend your routing protocol is a traveler lost in the desert, wandering in circles looking for water, and it comes across a fork in the path. If it were a distance vector protocol, it would read the sign that says "water—over here" and follow that path, not realizing that some kids were having fun and switched the sign. If it were link state, it would take the time to look at its meticulously drawn map, and (eventually) figure out the correct path. If it were balanced hybrid, it would ask for directions.

In addition to being classified by logic type, routing protocols are also classified by their intended use: either as an interior gateway protocol (IGP) or an exterior gateway protocol (EGP). IGPs are typically limited in the size of networks they can adequately support (although, with a little tweaking, some of them can support rather large networks); and they are, therefore, usually a bit simpler than EGPs. IGPs are designed to route within autonomous systems (ASs). An AS is a fancy way of saying a network controlled by a single administrative entity. LANs are a single AS, and so are most corporate WANs. The Internet, however, is a collection of several ASs. The AS concept is examined in more detail in later chapters. EGPs are built to support huge networks. An EGP's primary purpose is to route between ASs. Due to the complexities involved with EGPs, they will not be covered in this book.

Now let's take a look at a technique to deal with the common problem of having a router make routing decisions based on information from multiple dynamic routing protocols. The technique used by Cisco to combat this issue is called administrative distance, and it is described in the next section.

Understanding Administrative Distance

Administrative distance is an additional metric placed on each route to describe the "trustworthiness" of that route. Because of the differences in logic and metrics among the routing protocols, if you have more than one routing protocol in use, how is a router to know which one is giving it the most correct information? Cisco deals with this issue by assigning an administrative distance to each routing protocol and using that as the primary decision maker, with metric taken into account only after administrative distance has been examined. To see the usefulness of this feature, let's take a look at an example both with and without the use of administrative distance. Take a look at Figure 22-13 and walk through the route determination process without the use of administrative distance.

Without administrative distance, all Popeye has to go on is the metrics of each of the routes. Well, as you can see, that leads to a few problems, because Bluto (running

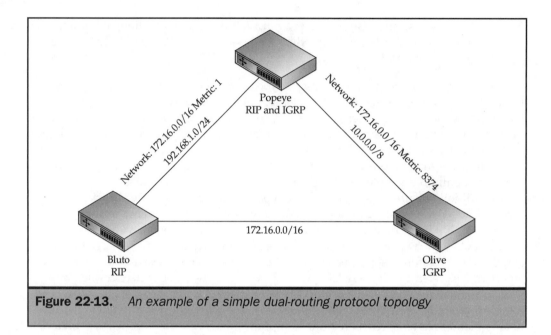

Figure 22-13.   *An example of a simple dual-routing protocol topology*

RIP) is advertising a metric of 1, whereas Olive (running IGRP) is advertising a metric of 8374. The metrics are so different because of the differences in how the two protocols calculate the metric. RIP uses a very simple metric: hop count. Because of the simplicity of RIP's metric, the valid range for a RIP metric is from 1 to 15. So, of course, Bluto advertises a metric of 1 for the 172.160.0.0 network.

IGRP, however, takes bandwidth, delay, load, and reliability into account. Thus, IGRP has to have a more granular metric structure, and an IGRP metric can range between 1 and 16.7 million. So, even though it looks large, a metric of 8000 and some change is actually a very low IGRP metric. Of course, without administrative distance (or some other function to validate the different metrics in use), Popeye has no way of knowing that the IGRP metric is better than the RIP metric, so he simply adds the best metric to his table. Popeye looks at the metrics similar to a price. As far as he is concerned, he can pay either $1 to get to network 172.16.0.0, or he can pay $8374. Which one do you think he will choose? Without administrative distance, Popeye would choose RIP over IGRP every time, even though RIP tends to give considerably less reliable information in most cases.

With administrative distance, both routes would be given a metric in the format of (Administrative Distance)/(Metric). Before looking at the metric for each of the routes, Popeye would look at the administrative distance. The route with the best administrative distance would be put in the table, and the other route would be effectively ignored. If both routes had the same administrative distance, the route with the best metric would be used. If the routes have the same administrative distance and the same metric, the router will either use the first route received or load balance across the routes (depending on how you have the router configured). In this case, the administrative distance for RIP is 120 (by default), and the administrative distance of IGRP is 100. Because IGRP has the lowest administrative distance, the IGRP route will be used and the RIP route ignored.

Table 22-1 displays the default administrative distances for Cisco routers.

Tip   *The default administrative distances can be changed in the IOS, although this practice is not recommended.*

Route Source	Administrative Distance
Directly connected	0
Static route	1
EIGRP summary route	5
External BGP	20
Internal EIGRP	90
IGRP	100
OSPF	110
Intermediate System-to-Intermediate System (IS-IS)	115
RIP	120
Exterior Gateway Protocol (EGP)	140
External EIGRP	170
Internal BGP	200
Unknown	255

Table 22-1. *Default Administrative Distances*

CISCO ROUTING

Basic Router Configuration: Interface Configuration

This section discusses basic router configuration tasks that you will use regardless of the router model. Because almost all routers use standard IOS, I will not cover configuration of parameters like name, location, administrative contact, boot options, and so on, because you already learned how to configure those items in Chapter 18 when you learned about standard IOS switch configuration. Instead, I will focus on configuring the primary difference in router and switch configuration: interface configuration. Because the technologies focused on in this book are Ethernet, Frame Relay, and ATM, I will discuss configuration steps for each of these interface types one at a time, beginning with the simplest: Ethernet.

 This book covers only synchronous serial interface configuration. For asynchronous serial configuration steps, including analog modem and ISDN configuration, please refer to http://www.cisco.com/univercd/cc/td/doc/product/software/ios122/122cgcr/fdial_c/index.htm.

Ethernet Configuration

Ethernet configuration is relatively easy. Configuring your Ethernet, Fast Ethernet, and Gigabit Ethernet interfaces consists of three to five basic steps (depending on whether you are configuring trunking and Fast Etherchannel). These steps are listed next:

1. Set the encapsulation.

2. Configure the speed and duplex.

3. Configure the IP address.

4. Configure trunking.

5. Configure Etherchannel.

Note *Remember, interfaces are named differently on different router lines. For the specific naming conventions used on specific router lines, please refer to the router classification chart in Chapter 11.*

Setting the Encapsulation

Setting the encapsulation is typically not necessary if you are using standard (DIX) Ethernet. The default encapsulation for (IP) Ethernet interfaces on Cisco routers is DIX encapsulation (which Cisco calls ARPA). For IPX interfaces, the default encapsulation is 802.3 (which Cisco calls Novell-Ether). Other encapsulations are typically used only in IPX. If you need to set the encapsulation, you will need to use one of the following interface configuration mode commands (depending on your specific model of router): *encapsulation [type]* or *ipx network [network number] encapsulation [type]*.

Remember, all interfaces on a router are in the shutdown state *by default. You must issue the* no shutdown *command from interface configuration mode for each interface to enable the interfaces.*

Table 22-2 details the valid encapsulation types.

Configuring the Speed and Duplex

Configuring the duplex on an Ethernet interface is pretty straightforward as well. You use the interface config mode command *duplex [half | full | auto].*

Not all interfaces are capable of full-duplex operation, but most will let you use the command even if they are not capable of the desired duplex mode. In most cases, the interface will continue to function normally (at half-duplex, of course), but you will get repeated Duplex Mismatch syslog messages.

To configure the speed on a 10/100 Ethernet interface, simply use the command *speed [10 | 100 | auto].*

Remember, it is recommended that you manually set the speed and duplex on your interfaces to avoid potential autonegotiation failures. Most 10/100 Ethernet interfaces default to auto for both speed and duplex.

Configuring the IP Address

Once again, configuring the IP address on an Ethernet interface is fairly basic. From interface config mode, you use the *ip address [address] [mask].* You can also set a secondary IP address using the *ip address [address] [mask] secondary* command. Secondary IP addresses can be useful if, for some reason, you have two layer 3 addressing mechanisms running over the same layer 2 network (i.e., VLANs are not used to separate the physical networks). Secondary addressing is not a recommended practice because it can lead to some rather strange problems, but it is possible on most Cisco routers.

CISCO ROUTING

Common Encapsulation Name	Cisco IOS Name
Ethernet II (also known as DIX Ethernet)	ARPA
Ethernet 802.3	Novell-Ether
Ethernet 802.2	SAP
Ethernet SNAP	SNAP

Table 22-2.   *Ethernet Encapsulation Types*

Configuring Trunking

If you need to configure ISL or 802.1q trunking (for instance, to allow a layer 2 switch to use a single router interface for inter-VLAN route processing), you can configure trunking on your router's Fast or Gigabit Ethernet interfaces by creating a subinterface for each VLAN to be supported.

Note *You can perform trunking only over Fast or Gigabit Ethernet interfaces.*

To create a subinterface, from global or interface config mode, type the *interface [number].[subinterface number]* command, as shown here:

```
3600 (config)# interface fastethernet 0/0.10
3600 (config-subif)#
```

Once you have created your subinterface, use the *encapsulation [isl | dot1q] [vlan]* command to apply the appropriate VLAN to the subinterface, like so:

```
3600 (config-subif)# encapsulation isl 10
```

Finally, apply an IP address to the subinterface (using the standard *ip address* command), and you are done.

Tip *At this point, you may want to apply a description of the purpose of each subinterface using the* description [descriptive text] *command, especially if you are configuring several subinterfaces. It's also a good idea to create the subinterface number based on the VLAN you plan to apply to it (for instance, Fa0/0.62 for VLAN 62), if possible.*

Configuring Etherchannel

To configure Fast Etherchannel or Fast Gigachannel on a router, you need to perform two steps: create the channel group and apply interfaces to the channel group.

Note *You can configure Fast Etherchannel or Fast Gigachannel only on high-end routers, like the 7200 series.*

To create the channel group, from global config mode, issue the *interface port-channel [number]* command. This will create the port channel and place you in interface config mode for the port channel. From there, you can apply the IP address, description, and so on, like you would with any other interface.

Note *If you want to perform trunking over the port channel, you need to create port channel subinterfaces and configure the subinterfaces just like you would do with any other interface. Also, if you are using trunking and subinterfaces, do* not *apply an IP address to the base interface.*

Once you have created the port channel, you need to apply actual interfaces to the channel group. You perform this task with the *channel-group [number]* interface config mode command.

Note *Do not configure IP addresses on the individual interfaces in a channel group, only on the primary port channel interface or its subinterfaces.*

To show you how ports are bound to a channel, the following output configures two Fast Ethernet ports on a 7500 series router with Fast Etherchannel, and then sets the port channel to trunk VLANs 2 and 3:

```
7500 (config)# interface port-channel 1
7500 (config-if)# interface fastethernet 1/0/0
7500 (config-if)# channel-group 1
7500 (config-if)# interface fastethernet 1/1/0
7500 (config-if)# channel-group 1
7500 (config-if)# interface port-channel 1.2
7500 (config-subif)# encapsulation isl 2
7500 (config-subif)# ip address 10.0.0.1 255.0.0.0
7500 (config-subif)# interface port-channel 1.3
7500 (config-subif)# encapsulation isl 3
7500 (config-subif)# ip address 192.168.1.1 255.255.255.0
```

Frame Relay Configuration

Frame Relay configuration is, of course, slightly more complicated than Ethernet configuration. For starters, there are two primary ways of physically connecting to a Frame Relay link, and each is configured a bit differently. In most environments, the most common method of connecting to a Frame Relay link is to use an external channel services unit/data services unit (CSU/DSU) connected to a synchronous serial port on the router. The other method of connecting involves using an internal CSU/DSU on a modular router. The Frame Relay configuration is the same for both methods; but if you use an internal CSU/DSU, you will need to configure it as well. For now, I'll concentrate on the basic steps to configure Frame Relay across a synchronous serial port.

There are three basic steps to configuring Frame Relay links:

1. Configure Frame Relay encapsulation.
2. Configure address mapping.
3. Configure Local Management Interface (LMI).

Configuring Frame Relay Encapsulation

Configuring Frame Relay encapsulation is typically pretty straightforward. First, you must know which type of encapsulation your provider is using—Cisco or IETF. (It will almost always be IETF.) Once you have this information, you simply enter interface config mode for your serial interface and issue the *encapsulation frame-relay [ietf]* command. If you are using Cisco encapsulation, you simply enter *encapsulation frame-relay*; if you are using the IETF frame type, use the *encapsulation frame-relay ietf* form of the command. That's it: encapsulation is set.

Frame Relay encapsulation types can also be set on a per-VC basis, but, personally, I have never seen the need to do so. If you set the encapsulation for the serial port itself, it automatically applies to all VCs for that port.

Configuring Frame Relay Address Mapping, Single PVC

With a single PVC, in most cases, you have to do very little to configure Frame Relay. With LMI autodetect and Inverse ARP, all you need to do is issue the *ip address* command on the serial interface. The router automatically determines the DLCI number, maps the IP address to it, and configures LMI. However, if your provider doesn't support LMI autodetect and Inverse ARP, you have a bit more work to do.

If you are using IOS 11.2 or later, LMI type autodetect and Inverse ARP (InARP) are enabled by default, so as long as you have only one PVC, you can basically just set the encapsulation, apply the correct IP address to the serial interface, and you're done.

First, you need to know the DLCI for your PVC. Once you have that information, go into interface config mode for your serial interface. From there, issue the *frame-relay interface-dlci [dlci number]* command. After you have configured the DLCI, you need to configure the ip address for the interface. Do this with the standard *ip address* interface config mode command. Finally, you need to map the DLCI to the IP address. To perform this mapping, use the *frame-relay map [layer 3 protocol] [layer 3 protocol address] [dlci] [broadcast] [ietf | cisco]* command.

The *layer 3 protocol* section of this command specifies the layer 3 protocol (typically, IP or IPX). The *layer 3 protocol address* section specifies the IP (or IPX) address you wish to map to this VC. The *dlci* section specifies the DLCI number to map the protocol to. The *IETF | Cisco* section specifies the encapsulation type to use for this VC (typically not needed; by default, the frame type specified with the *encapsulation* command will be used). Finally, the *broadcast* section specifies whether to "pretend" broadcast over this DLCI. Remember, Frame Relay is an NBMA network specification; so if you are using a routing protocol that broadcasts rather than multicasts routing updates (like RIP v.1), you will need to enable broadcast emulation to allow routing updates across the link.

To show you what is required to enable a single Frame Relay PVC over a serial interface, the following example manually configures the serial 0/0 interface on a 3600 series router. I will assign 10.1.1.1/8 as the IP address, and map DLCI 200 to this IP address. I will also enable broadcasts over the DLCI.

```
3620 (config)# interface serial 0/0
3620 (config-if)# encapsulation frame-relay ietf
3620 (config-if)# frame-relay interface-dlci 200
3620 (config-if)# ip address 10.1.1.1 255.0.0.0
3620 (config-if)# frame-relay map ip 10.1.1.1 200 broadcast
```

Note   *When configuring multiple PVCs, you can configure them all on the same serial interface, without the use of subinterfaces. However, this strategy causes major split horizon problems (discussed further in Chapter 23), so it is not recommended.*

Configuring Frame Relay Address Mapping, Multiple Point-to-Point PVCs

If you have multiple point-to-point PVC's, the router cannot autoconfigure nearly as much for you. With multiple PVCs over the same link, you must create serial subinterfaces, one per PVC, and then apply the VCs to the subinterfaces as desired.

Again, you need to know the DLCs for your PVCs; and for every DLCI, you need to create a subinterface. To create a point-to-point Frame Relay serial subinterface, use the *interface serial [number].[subinterface number] point-to-point* command. The additional *point-to-point* modifier on the end tells the router that a point-to-point (single DLCI) subinterface will be created. As usual, this command enters you into subinterface config mode, where you need to configure your individual PVC. From here, you once again use the *ip address* and *frame-relay interface-dlci* commands to assign an IP address and DLCI to the PVC.

If you are able to use Inverse ARP, this will be all you need to do. InARP performs the DLCI mapping for you. However, if you are not able to use InARP, you will also need to issue the *frame-relay map* command for each subinterface. To better illustrate this process, I will configure a 3600 series router to match the configuration of HQRouter in Figure 22-14. For this example, I will once again statically map the DLCIs.

Note   *When looking at these figures, remember that DLCI-swapping occurs during the cloud, so, normally, you will not have the same DLCI on one side of the connection as you do on the other.*

```
HQRouter (config)# interface serial 0/0
HQRouter (config-if)# encapsulation frame-relay ietf
HQRouter (config-if)# interface serial 0/0.1 point-to-point
HQRouter (config-subif)# frame-relay interface-dlci 100
HQRouter (config-subif)# ip address 192.168.1.1 255.255.255.0
HQRouter (config-subif)# frame-relay map ip 192.168.1.1 100 broadcast
HQRouter (config-subif)# interface serial 0/0.2 point-to-point
HQRouter (config-subif)# frame-relay interface-dlci 200
HQRouter (config-subif)# ip address 192.168.2.1 255.255.255.0
HQRouter (config-subif)# frame-relay map ip 192.168.2.1 200 broadcast
```

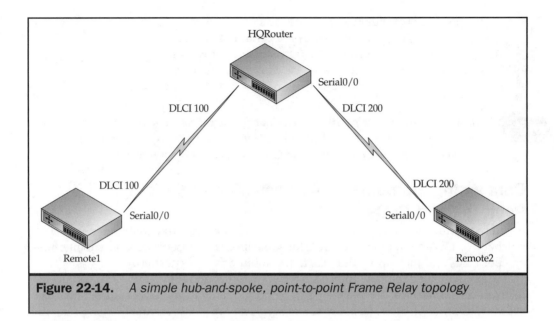

Figure 22-14. *A simple hub-and-spoke, point-to-point Frame Relay topology*

Configuring Frame Relay Address Mapping, Multipoint

Multipoint topologies are fairly rare, with the possible exception of OSPF networks (for reasons you will learn about in Chapter 26), but they still deserve some coverage. Remember from Chapter 3, a multipoint connection is one that operates like a layer 2 bus, where packets from one network address are sent down multiple VCs. Multipoint connections allow a router to send a packet to one logical network, and the packet gets split off to multiple routers on the same logical network. Multipoint configuration doesn't actually use any new commands; it just uses the ones you have already seen a little differently. To explain multipoint configuration, the following output takes the previous example network (shown in Figure 22-14) and configures it with multipoint Frame Relay at the HQRouter:

```
HQRouter (config)# interface serial 0/0
HQRouter (config-if)# encapsulation frame-relay ietf
HQRouter (config-if)# interface serial 0/0.1 multipoint
HQRouter (config-subif)# frame-relay interface-dlci 100
HQRouter (config-subif)# frame-relay interface-dlci 200
HQRouter (config-subif)# ip address 192.168.1.1 255.255.255.0
HQRouter (config-subif)# frame-relay map ip 192.168.1.1 100 broadcast
HQRouter (config-subif)# frame-relay map ip 192.168.1.1 200 broadcast
```

Notice that two DLCIs were mapped to the same logical network. Thus, HQRouter sends packets to both Remote1 and Remote2 any time it sends packets to the 192.168.1.0 network. The only real issue with this layout is the remote site configuration. Right now, with this partially meshed topology, Remote1 and Remote2 are simply configured with point-to-point interfaces back to HQRouter. For example, Remote1's configuration is as follows:

```
Remote1 (config)# interface serial 0/0
Remote1 (config-if)# encapsulation frame-relay ietf
Remote1 (config-if)# interface serial 0/0.1 point-to-point
Remote1 (config-subif)# frame-relay interface-dlci 100
Remote1 (config-subif)# ip address 192.168.1.2 255.255.255.0
Remote1 (config-subif)# frame-relay map ip 192.168.1.2 100 broadcast
```

The problem with this configuration is that, if Remote1 attempts to ping Remote2; the ping will fail. The failure is because Remote1 is not directly connected to Remote2; but as far as HQRouter is concerned, they are all on the same network. In other words, to ping 192.168.1.3, Remote1 sends a packet to Frame Relay DLCI 100, which comes in on DLCI 100 at HQRouter. Well, HQRouter isn't going to route the packet because, as far as it's concerned, a packet that came from network 192.168.1.0 and is destined for 192.168.1.0 shouldn't need routing. To enable Remote1 to ping Remote2, you need to either revert back to point-to-point or configure a new PVC between Remote1 and Remote2, as shown in Figure 22-15, and configure multipoint connections on Remote1 and Remote2.

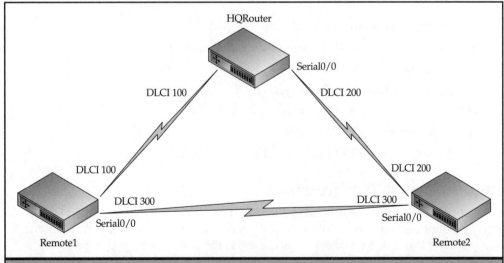

Figure 22-15. *The required full mesh topology to allow Remote1 and Remote 2 to communicate with each other*

Sometimes multipoint topologies, in which two remote sites cannot communicate with each other, only with the HQ, are desirable. Some companies may have explicit requirements for departmental autonomy and security between divisions or branches.

Configuring an Integrated CSU/DSU Module

Typically, configuring an internal CSU/DSU isn't all that complicated, unless, of course, you don't have the required information from your telco. (Unfortunately, you can't just make that stuff up.) The basic information you need to configure your internal CSU/DSU is listed in the following:

- The facility data link (FDL) mode (either att, ansi, or unused)
- The clock source (either internal or line)
- The data coding setting (inverted or normal)
- The framing type (either the extended super frame (ESF) or super frame (SF) type)
- The line build out (LBO) (can be none, –7.5db, or –15db)
- The linecode (AMI or B8ZS)
- The number of channels and speed (64 Kbps or 56 Kbps) of each channel if using a fractional link.

The commands required to set each of these parameters are shown in the following list, with the typical settings (for North America) bolded. All of these commands are entered from interface config mode on the serial interface that corresponds to the slot the CSU/DSU is installed in. (In other words, if the CSU/DSU was in slot 1 of module 1, you would enter *interface serial 1/1*.)

- *[no] service-module t1 fdl [att | ansi]*
- *[no] service-module t1 clock-source [internal | **line**]*
- *[no] service-module t1 data-coding [inverted | **normal**]*
- *[no] service-module t1 framing [esf | sf]*
- *[no] service-module t1 lbo [**none** | –7.5db | –15db]*
- *[no] service-module t1 linecode [ami | **b8zs**]*
- *[no] service-module t1 timeslots [**all** | range] speed [56 | **64**]*

Configuring an ATM Interface

Configuring an ATM PVC interface is, believe it or not, usually simpler than Frame Relay configuration. First, ATM can be configured using either a native ATM interface in a router or through a serial connection to an ATM DSU (ADSU). Using a serial connection, ATM configuration is fairly similar to Frame Relay configuration, and consists of the following steps:

1. Configure the serial interface.
2. Set the encapsulation.
3. Define the PVCs and perform address mapping.

ATM switched virtual circuits require very specialized knowledge of ATM infrastructures and ATM routing, and entire books have been devoted to the intricacies of their operation. The discussion of ATM SVCs is beyond the scope of this book.

Configuring an ATM PVC Using a Serial Connection

After cabling your ADSU to the serial port, you need to perform some basic serial configuration tasks. First, switch to interface config mode for the designated serial interface and apply the IP address for the ATM PVC to that interface using the standard *ip address* command. Next, set the encapsulation of the serial interface to ATM DXI by using the *encapsulation atm-dxi* command. After configuring the line encapsulation, you need to build the PVC and—optionally—configure the PVC's encapsulation. You perform this task with the *dxi pvc [vpi] [vci] [snap | nlpid | mux]* command. The *vpi* and *vci* fields in this command specify the VPI/VCI values, similar to the DLCI for Frame Relay. (Refer to Chapter 4 for more information.)

Note *This chapter covers only the user side of ATM UNI configuration (the user side being your router and the network side being the provider's or your backbone ATM switch). ATM switching is, unfortunately, beyond the scope of this book.*

The *snap | nlpid | mux* section of this command specifies the encapsulation that should be used for this PVC. Your provider should tell you which encapsulation is appropriate, but SNAP encapsulation is the default on most Cisco routers. The *mux* encapsulation should be used only if you wish to map a single protocol to the VC. The other encapsulation options allow you to map multiple protocols to the same VC.

Finally, you need to configure protocol mappings to the PVC by using the *dxi map [protocol] [address] [vpi] [vci] [broadcast]* command, similar to the *frame-relay map* command. All things considered, this entire process is nearly identical to the Frame Relay serial configuration. An example of serial configuration for an ATM PVC using VPI 10 VCI 11 and SNAP encapsulation is shown in the following output:

```
3600 (config)# interface serial 0/0
3600 (config-if)# ip address 10.0.0.1 255.0.0.0
3600 (config-if)# encapsulation atm-dxi
3600 (config-if)# dxi pvc 10 11
3600 (config-if)# dxi map ip 10.0.0.1 10 11 broadcast
```

CISCO ROUTING

Configuring an ATM PVC Using an Integrated ATM Interface

On some routers, you may be able to use an integrated ATM interface instead of the serial for your ATM connections. The integrated interfaces typically provide a big advantage in speed over serial interfaces. To configure an ATM PVC with an integrated ATM interface, you must perform the following steps:

1. Configure the PVC.
2. Map the PVC to a protocol address.

For the first step, you need to enter ATM interface configuration mode by typing *interface ATM [number]*. Once in ATM interface config mode, you issue the *ip address* command to configure an IP address, the same as with any other interface. You then issue the *pvc [name] [vpi/vci]* command to create the PVC and enter VC configuration mode. The *name* section of this command is optional, and is used only for informational purposes (similar to the *description* command). Finally, you configure protocol to PVC mapping using the *protocol [protocol] [address] [broadcast]* command. So, to configure an internal ATM interface with the same settings as the previous example, you would do the following:

```
3600 (config)# interface ATM 1/0
3600 (config-if)# ip address 10.0.0.1 255.0.0.0
3600 (config-if)# pvc test 10/11
3600 (config-if-atm-vc)# protocol ip 10.0.0.1 broadcast
```

Static Routing Configuration, Verification, and Troubleshooting

Configuring static routing is a very simple process involving only one major command. However, because in order for static routing to work, you must have everything else properly configured, this section discusses some troubleshooting tips for basic connectivity as well.

First, let's detail the major static routing command, *ip route*. You use this command to create standard static routes, including default routes. The syntax is *ip route [destination network] [subnet mask] [next hop address | interface] [administrative distance] [permanent]*. The destination network and subnet mask are pretty straightforward. The *next hop address | interface* section allows you to specify the next hop router or (optionally) the interface to use. In nearly all cases, you should use the next hop router IP address instead of specifying the interface.

The administrative distance, in addition to being useful in complex static routing environments with redundant paths, is also used to flag the static route for removal from the table if a dynamic route is heard. For instance, if you had a static route to the

10.0.0.0/8 network that you wanted removed if the router hears a RIP advertisement for the 10.0.0.0/8 network, you would set the administrative distance of the static route to 121 or higher.

Finally, you can flag the route as permanent, which means the route stays in the table even if the interface that is used for forwarding the route's packets goes down. This strategy is typically not recommended, because it can cause problems if you have static routes set up for redundant paths.

Note *Specifying the interface is not typically recommended. By specifying the interface, you are telling the router that it should use the route as if it were directly connected. In other words, rather than sending the packet to a remote router for forwarding, the router ARPs for the destination IP on the selected interface. If the router is not directly connected (if it were, there would be no need to use this command in the first place), the remote router needs to have proxy ARP enabled for this command to work.*

To create default routes using the *ip route* command, simply enter the command with the address and mask for the destination network set to 0.0.0.0, as in the following example:

```
2600A (config)# ip route 0.0.0.0 0.0.0.0 10.1.1.1
```

If you enter multiple default routes, the one with the lowest administrative distance is used. If they all have the same administrative distance, the last one entered is used.

Tip *All default routes entered are retained by the router. If you enter several, the last one entered will be used. If the interface that is used for that route fails, the previously entered default route will be used. This can be helpful if you want to configure default route redundancy. Note that the router does not load balance across default routes.*

One interesting point to keep in mind is that, for a default route to actually be used, the router must be operating in "classless" mode. Otherwise, the router will attempt to match a destination IP address only with its class-based network entry in the routing table, regardless of the subnet mask configured for the destination network in the routing table. Obviously, this functionality can also cause all kinds of problems with discontiguous subnets and complex masking. In most cases, this classfull behavior isn't an issue because, by default, Cisco routers come with *ip classless* enabled. However, if you have problems getting a default route to work, make sure that the *no ip classless* command is not entered. If it is, just go into global config mode and enter *ip classless*.

Note *You can also use the* ip default-network *command to configure a default route; but, typically, you use this command only with a routing protocol that does not support the all 0's route (mainly IGRP). The* ip default-network *command is discussed in more detail in Chapter 24.*

CISCO ROUTING

You also need to be aware of one key *show* command when using static routing: *show ip route.* This command displays your routing table, so you can see your changes after entering a static route. The syntax for this command is *show ip route [address] [mask] [longer-prefixes] [protocol].* The only required part of the command is *show ip route.* A simple *show ip route* command shows all networks in the routing table from all protocols, like so:

```
3620#show ip route
Codes: C - connected, S - static, I - IGRP, R - RIP, M - mobile, B - BGP
       D - EIGRP, EX - EIGRP external, O - OSPF, IA - OSPF inter area
       N1 - OSPF NSSA external type 1, N2 - OSPF NSSA external type 2
       E1 - OSPF external type 1, E2 - OSPF external type 2, E - EGP
       i - IS-IS, L1 - IS-IS level-1, L2 - IS-IS level-2,
       ia - IS-IS inter area
       * - candidate default, U - per-user static route, o - ODR
       P - periodic downloaded static route

Gateway of last resort is not set

     169.254.0.0/24 is subnetted, 8 subnets
C       169.254.16.0 is directly connected, Serial1/3
C       169.254.15.0 is directly connected, Serial1/2
C       169.254.14.0 is directly connected, Serial1/1
C       169.254.13.0 is directly connected, Serial1/0
C       169.254.12.0 is directly connected, Serial0/3
C       169.254.11.0 is directly connected, Serial0/2
C       169.254.10.0 is directly connected, Serial0/1
C       169.254.9.0 is directly connected, Serial0/0
S    10.0.0.0/8 [2/0] via 192.168.3.1
C    192.168.3.0/24 is directly connected, FastEthernet0/0
3620#
```

There are a couple of points to note about this output. First, the capability codes section at the top details which protocols are responsible for which routes. Right now, you are concerned only with *C* (directly connected) and *S* (static), but you will learn about the others in later chapters. Next, notice the section that says *gateway of last resort is not set.* Examining this line is one way to find out what your default route is. The default route will also have an asterisk (*) beside it. Next, notice that the output also shows you both the class-based network and all of the subnets for that network. Also, the capability code listed on the far left tells you how the router knows about this particular route.

Finally, notice that the static route shows a [2/0] before telling you what the next hop router is with *via www.xxx.yyy.zzz.* The [2/0] is the administrative distance/metric combination. In this case, the route to the 10.0.0.0 network has been entered with a nondefault administrative distance of 2; and the metric is, of course, 0. (Remember, you cannot set metrics on static routes.)

Now, let's say you want to see information only on the 169.254.0.0 network. For this, you would change the command to *show ip route 169.254.0.0,* like so:

```
3620>show ip route 169.254.0.0
Routing entry for 169.254.0.0/24, 8 known subnets
  Attached (8 connections)
  Redistributing via rip

C       169.254.16.0 is directly connected, Serial1/3
C       169.254.15.0 is directly connected, Serial1/2
C       169.254.14.0 is directly connected, Serial1/1
C       169.254.13.0 is directly connected, Serial1/0
C       169.254.12.0 is directly connected, Serial0/3
C       169.254.11.0 is directly connected, Serial0/2
C       169.254.10.0 is directly connected, Serial0/1
C       169.254.9.0 is directly connected, Serial0/0
3620>
```

If you wanted to see the entry for only a specific subnet of the 169.254.0.0 network, like 169.254.12.0, you would issue the command *show ip route 169.254.12.0,* like so:

```
3620>show ip route 169.254.12.0
Routing entry for 169.254.12.0/24
Known via "connected", distance 0, metric 0 (connected, via interface)
  Redistributing via rip
  Advertised by rip
  Routing Descriptor Blocks:
  * directly connected, via Serial0/3
      Route metric is 0, traffic share count is 1
3620>
```

Finally, if you wanted to see only a specific subset of the networks, like from 169.254.12.0 through 169.254.15.0, you would issue the *show ip route 169.254.12.0 255.255.252.0 longer-prefixes* command.

```
3620>show ip route 169.254.12.0 255.255.252.0 longer-prefixes
Codes: C - connected, S - static, I - IGRP, R - RIP, M - mobile, B - BGP
       D - EIGRP, EX - EIGRP external, O - OSPF, IA - OSPF inter area
```

```
            N1 - OSPF NSSA external type 1, N2 - OSPF NSSA external type 2
            E1 - OSPF external type 1, E2 - OSPF external type 2, E - EGP
            i - IS-IS, L1 - IS-IS level-1, L2 - IS-IS level-2,
            ia - IS-IS inter area
            * - candidate default, U - per-user static route, o - ODR
            P - periodic downloaded static route

Gateway of last resort is not set

        169.254.0.0/24 is subnetted, 8 subnets
C           169.254.15.0 is directly connected, Serial1/2
C           169.254.14.0 is directly connected, Serial1/1
C           169.254.13.0 is directly connected, Serial1/0
C           169.254.12.0 is directly connected, Serial0/3
3620>
```

The *longer-prefixes* option in this command tells it to match every route that the subnet mask you enter matches. For instance, 10.128.0.0 255.192.0.0 will match networks 10.128.0.0 through 10.191.0.0. If this range of addresses confuses you, you may need to perform the binary ANDing, as if you were subnetting, to see why the 10.128.0.0 with the 255.192.0.0 subnet matches that range of networks. (Refer to Chapter 6 for a detailed discussion of subnetting.)

Finally, the following troubleshooting tips might come in handy. Unless specifically stated, most of these tips are just general routing troubleshooting tips, and they are not necessarily specific to static routing.

■ First, ensure that the interface(s) you are using are up and running. You can perform this task with the *show interface* command (or just look for the link light on Ethernet).

Note   *There is an exception to this rule. If ISL trunking is configured on one side of a single physical connection (one individual cable) but not on the other, a link may be present, but you will get no connectivity. Ensure that there is no trunking misconfiguration by using the* show running-config *command on both sides of the connection before proceeding.*

■ Ensure that the interface is not disabled with the *shutdown* command. If the interface is shut down, use the *no shutdown* command to enable it.

■ After you have verified physical and datalink-layer functionality, verify that IP is properly bound to the interface. For an Ethernet interface, the easiest way to perform this task is to ping the interface IP address. A serial interface, however, will not respond to a ping from the local router (unless you install a loopback adapter). For serial interfaces, use the *show ip interface* command.

- Ensure that your IP configuration is correct. Also, if possible, try to determine whether the problem is with a specific host or all hosts. For instance, if you cannot ping a *specific* host, then that host is generally the source of the problem. However, if you cannot ping *any* hosts, then your configuration is most likely to blame.

- If you are having problems connecting to a destination that is several hops away, try performing a trace route (using the *trace* command) to determine where the failure is occurring. If the *trace* command returns any hops along the path, then look for where it fails because that will be the router you need to troubleshoot. If the *trace* command shows no hops along the path, then the problem is on your router.

- Use the *show ip route* command to ensure that you have a route to the remote network.

- If you are using a default route or static routing to reach the remote network, ensure that the designated router can reach the remote network.

- If you are using complex static routing with redundant paths, ensure that you do not have a routing loop. The easiest way to perform this task is generally to remove the redundant routes and attempt to connect.

Summary

This chapter examined the basic process of routing and looked at how static routing can be used to implement simple networks. It also discussed the basic principles and operation of dynamic routing, and examined what administrative distance is and when it is needed. The next chapter begins a detailed examination of routing protocols with one of the most widely supported protocols—RIP.

The Complete Reference

Cisco

Chapter 23

RIP Versions 1 and 2

This chapter examines routing with Routing Information Protocol (RIP) versions 1 and 2, arguably the most common dynamic interior IP routing protocol. You will learn the advantages and disadvantages of using RIP in most networks and take a look at how to configure, troubleshoot, and optimize RIP. The beginning of the chapter concentrates on RIP version 1, and later in the chapter, you'll learn about the improvements in version 2.

How RIP Works

RIP is a simple distance vector routing protocol intended for small- to medium-sized redundant networks. It is vendor independent, and version 1 is defined primarily in RFC 1058. This vendor independence gives RIP the advantage of having multivendor interoperability. (Even Microsoft supports RIP, assuming you need to use a Windows 2000 box as a router.) Also, because RIP is such a simple protocol, it is very easy to configure. Unfortunately, its simplicity also means it is prone to problems.

Basic RIP 1 Operation

The basic operation of RIP is very simple and follows some fairly basic rules:

- When a router boots, the only routes it is initially aware of are its directly connected networks.

- In RIP version 1, the router broadcasts information about all known networks to all directly connected networks. These broadcasts are known as *updates* or *advertisements*.

- RIP routers listen for RIP broadcasts. By listening to RIP broadcasts, RIP routers can become aware of networks they may not have direct knowledge of.

- RIP's metric is *hop count* (loosely defined as the number of routers in the path), and is advertised in RIP broadcasts for each network. RIP's maximum hop count is 15. A metric of 16 is considered infinite.

- Any route learned from a RIP router is assumed to be through that router. In other words, if Router A sends an update to Router B, Router B assumes that the next hop for the networks included in the update is Router A.

- Updates are sent at regular intervals.

To solidify these basic rules, let's walk through an example. Figure 23-1 has four routers and eight networks. I have also included the initial routing tables for all routers, but note that some information (such as the next hop address) has been left out to simplify the example. The beginning metrics in this case are all zero because RIP uses hop count as a metric, and, currently, all known networks are directly connected.

When Richard boots, he sends a broadcast about all known networks to Stan and Fran, advertising a metric of 1 for these networks. RIP does not "locate" or determine

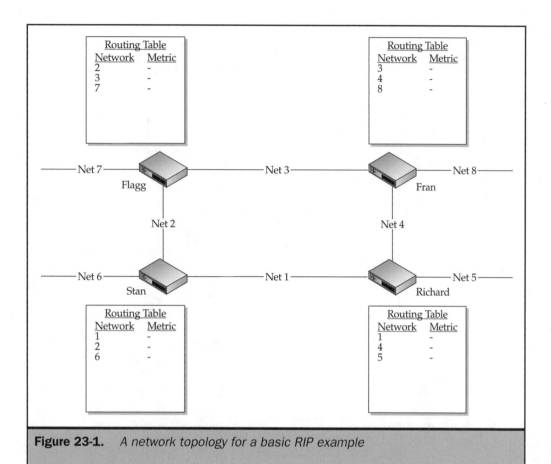

Routing Table

Network	Metric
2	-
3	-
7	-

Routing Table

Network	Metric
3	-
4	-
8	-

——Net 7—— Flagg ——Net 3—— Fran ——Net 8——

Net 2   Net 4

——Net 6—— Stan ——Net 1—— Richard ——Net 5——

Routing Table

Network	Metric
1	-
2	-
6	-

Routing Table

Network	Metric
1	-
4	-
5	-

Figure 23-1. *A network topology for a basic RIP example*

CISCO ROUTING

which RIP routers exist in the network, so Richard has no idea that a router does not exist on Net 5. Therefore, Richard also sends a broadcast to Net 5. Note that the lack of knowledge of other routers also means that Richard has no idea how many routers are connected to Nets 1 and 4, or if any are connected at all. He just broadcasts out and assumes someone is listening.

Stan and Fran add the new networks to their routing tables with a metric of 1 (using the advertised metric in the update). This example allows only one route to each destination network, so Stan and Fran will add to their routing tables only the best (lowest metric) route for each destination. Therefore, Stan will not change his routing table to include Net 1 as advertised by Richard, because he already has a better route to that destination in his table. Similarly, Fran will not add the new route for Net 4. Figure 23-2 shows this initial broadcast and the routing table additions it causes.

Fran and Stan then broadcast information about all of their known networks, including the new networks just learned from Richard, to all directly connected subnets. From this

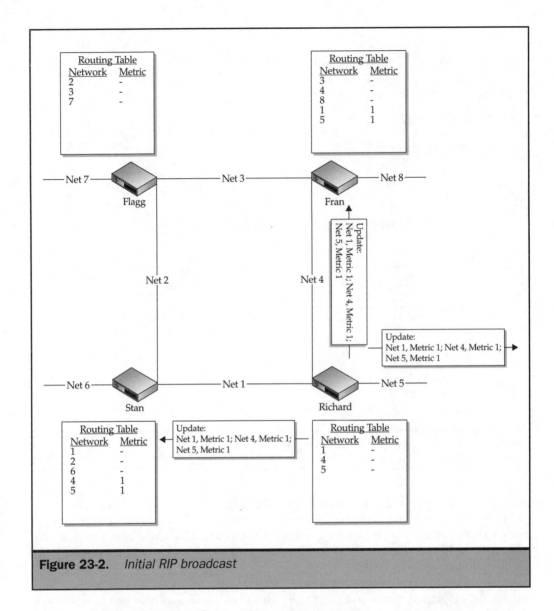

Figure 23-2.   *Initial RIP broadcast*

broadcast, Richard learns of Nets 3 and 8 (from Fran), and Nets 2 and 6 (from Stan). Flagg learns of all networks in the topology during this update process and adds them to his table. Note that because this example allows only one route for each network in the table, when Flagg hears of two equal-cost paths to the same network (for instance, the update about Net 5 with a metric of 2 from both Stan and Fran), he includes into his routing table only the first advertised route heard. Figure 23-3 shows the second broadcast and the routing table updates it caused.

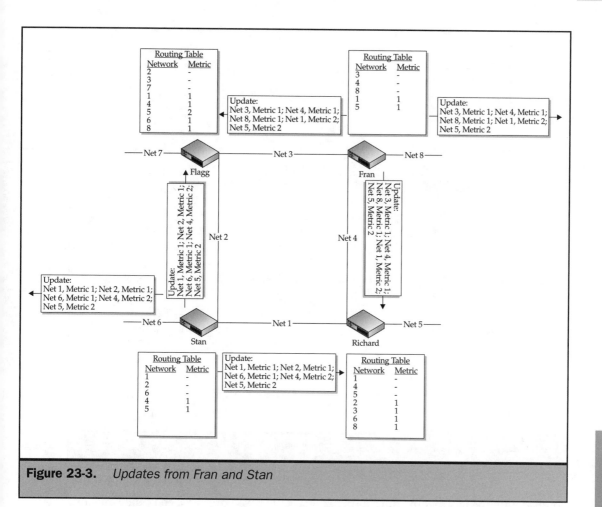

Figure 23-3. *Updates from Fran and Stan*

Finally, Flagg broadcasts information about his networks to Stan and Fran. At the next update interval, Stan and Fran broadcast these new routes to Richard, and convergence is achieved. This setup is shown in Figure 23-4.

This entire example, although a bit oversimplified, displays the basic operation of RIP. For a more complete understanding of RIP, you need to learn about a few other features: update timers, split horizon (with and without poison reverse), holddown timers, route poisoning, and triggered updates.

Update timers are used so that the router knows how long to wait before sending periodic updates. In RIP version 1, each update includes all routes (except those removed by split horizon), regardless of whether any changes have occurred since the last update. The periodic update process ensures that routers can determine whether other routers are down (almost like a pulse). But the short time period that RIP waits between updates,

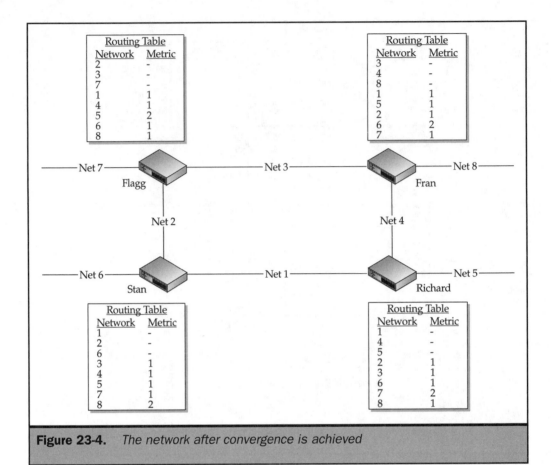

Figure 23-4. *The network after convergence is achieved*

coupled with the fact that the full routing table is advertised during each update, means that RIP updates can use large amounts of bandwidth in complex networks. The update timer is configurable on Cisco routers, but you must ensure that all routers have the same timer settings.

Split horizon is one of several features implemented in distance vector routing to reduce the possibility of routing loops. Basic split horizon ensures that routes learned through a given interface are never broadcast back out the same interface. In addition, the network associated with an interface is never advertised on that same interface. Split horizon with *poison reverse* reduces the convergence time associated with split horizon in the case of a routing loop by advertising an infinite metric on a given interface for routes learned through that interface. The examples following this brief explanation include both split horizon and split horizon with poison reverse, so don't get concerned if this discussion doesn't make sense yet.

Holddown timers are used to prevent loops in a complex topology by requiring that the RIP router wait a specified period of time (180 seconds, or 6 times the update interval, by default) before believing any new information about an updated route. Holddown timers are sort of like the reality check of distance vector routing. They are discussed further in the coming example as well.

Route poisoning is also used to reduce the possibility of routing loops. Without route poisoning, bad routing entries are removed from the table after the route timeout and route-flush timers expire. The *route timeout* timer (called the *invalid* timer in IOS) is used to determine when a route has failed. If an update has not heard about a given route before this timer expires, the route is considered invalid and enters holddown. The route, however, is still used (but no longer advertised) until the route-flush timer expires. Once the route-flush timer expires, the route is entirely removed from the routing table.

The default setting for the route timeout is 180 seconds, and the default setting for the route-flush timer is 240 seconds. Without route poisoning, a router simply stops advertising bad routes after the route timeout timer expires. The next router in line then needs to wait for its route timeout to expire, and so on. This means that in a large RIP network (10+ hops), it could take over 30 minutes for all routers to be notified of a bad route (which can eliminate the usefulness of holddown timers). Route poisoning works with triggered updates to reduce this time by advertising a route with an infinite metric (16 in RIP) once that route's timeout timer expires. Route poisoning, when combined with triggered updates, could reduce the convergence time for a 10+ hop network to less than 30 seconds.

Triggered updates are used to reduce both the possibility of routing loops and to reduce the network convergence time. If a link to a directly connected network fails, rather than waiting on an update timer to expire, RIP immediately advertises the failure (as an infinite distance, as per route poisoning). Also, once a route has been updated, RIP immediately advertises the updated route, rather than waiting for the update timer to expire. Finally, if an indirect failure occurs (i.e., the router doesn't hear about an advertised route and the timeout timer expires), triggered updates are used in conjunction with route poisoning to propagate the route failure quickly.

At this point, you are probably thinking that RIP is not so simple after all. In truth, it is really a very simple protocol. Being simple has its benefits, but the major drawback is that routing loops are very easy to come by. Because of routing loops, all of these additional features are required, which introduce the complexity into the equation.

To help you understand why all of these features are required, let's take a simple mesh network (shown in Figure 23-5) and examine the RIP update process without the benefit of any of the loop-prevention mechanisms.

Initially, the only routes that any routers are aware of are their directly connected networks. This means that Jack, Danny, and Hallorann would have routing tables similar to those shown in Table 23-1.

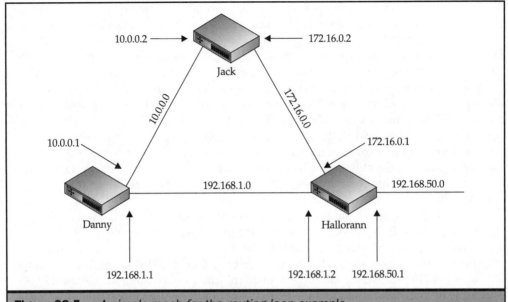

Figure 23-5.   *A simple mesh for the routing loop example*

The routers would begin sending updates as per the normal RIP update interval, and about 120 seconds later, would have converged to the routing tables shown in Table 23-2.

	Network	Next Hop	Metric
Jack	10.0.0.0	—	—
	172.16.0.0	—	—
Danny	10.0.0.0	—	—
	192.168.1.0	—	—
Hallorann	192.168.1.0	—	—
	172.16.0.0	—	—
	192.168.50.0	—	—

Table 23-1.   *Initial Routing Tables for Figure 23-5*

	Network	Next Hop	Metric
Jack	10.0.0.0	—	—
	172.16.0.0	—	—
	192.168.1.0	10.0.0.1	1
	192.168.50.0	172.16.0.1	1
Danny	10.0.0.0	—	—
	172.16.0.0	10.0.0.2	1
	192.168.1.0	—	—
	192.168.50.0	192.168.1.2	1
Hallorann	10.0.0.0	172.16.0.2	1
	172.16.0.0	—	—
	192.168.1.0	—	—
	192.168.50.0	—	—

Table 23-2.   *Routing Tables for Figure 23-5 After Convergence*

Now, imagine that Hallorann's link to the 192.168.50.0 network fails. Hallorann immediately removes the entry from his table (because he is no longer directly connected), but he does not inform Jack or Danny of the removal. (Remember, this example is not using any loop-prevention features.) Because the network failure was not advertised to Danny or Jack, they will have to wait until the timeout for the 192.168.50.0 network expires, which is 180 seconds, before they stop advertising the network. Because they haven't yet noticed that the 192.168.50.0 route is bad, they will send out an advertisement to all networks (remember, this example is not using split horizon), including their links to Hallorann, that have a route for the 192.168.50.0 network. This update causes Hallorann to enter a bad route to the 192.168.50.0 network using one of the other routers as his next hop (let's say Danny, in this case). The routing table for Hallorann now resembles the following table.

Network	Next Hop	Metric
10.0.0.0	172.16.0.2	1
172.16.0.0	—	—
192.168.1.0	—	—
192.168.50.0	192.168.1.1	2

At this point, there is a loop. Hallorann forwards packets destined for the 192.168.50.0 network to Danny, who forwards them to Hallorann, who forwards them to Danny, and so on, until the TTL expires. About three minutes later, however, Danny would recognize that the route through Hallorann is no longer valid (because he is no longer hearing single-hop metric advertisements from Hallorann), and he would remove it from his table. Jack would notice the same problem, and he would remove the initial route to 192.168.50.0 from his table as well.

However, the problem persists because Hallorann is still advertising the 192.168.50.0 network to both Jack and Danny with a metric of 3 (based on the bogus route through Danny he now has in his table). Then Jack and Danny add a metric 3 route to their tables that is directed to Hallorann, who is just going to send the packet back to Danny, which starts the problem all over again. Eventually, Hallorann's two-hop metric route to 192.168.50.0 will expire, and he will hear of a four-hop metric from Danny or Jack. But then Danny and Jack's three-hop metric route will expire, and they will hear of a five-hop metric route from Hallorann, and so on, until the metric for the route reaches 16 (about an hour later). This problem is aptly called *counting to infinity*, and it is probably the primary problem with distance vector routing.

To solve this problem (and the routing loops that caused the problem in the first place), let's add RIP's loop-prevention mechanisms into the puzzle, beginning with split horizon. If only split horizon had been enabled, when Hallorann's connection to the 192.168.50.0 network failed, he would not have heard of a bogus route to the 192.168.50.0 network from either Danny or Jack. Because Danny heard of the 192.168.50.0 network from the link to the 192.168.1.0 network, split horizon would prevent him from sending an update about the 192.168.50.0 network to the 192.168.1.0 network. Similarly, Jack would be prevented from sending an update down the 172.16.0.0 link because he initially heard the route on that link.

However, the problem still persists because for the next 180 seconds, Danny and Jack will continue advertising the 192.168.50.0 network *to each other* across the 10.0.0.0 link with a metric of 2. Suppose that Danny times out his route to the 192.168.50.0 network, but before Jack can time the network out, he manages to send an update to Danny about this network with the two-hop metric. Jack then times the route out of his table normally, but the damage has already been done. Danny now thinks he has a valid two-hop route to 192.168.50.0 through Jack. He then advertises this route with a three-hop metric to Hallorann, who adds it to his table and advertises the route with a four-hop metric to Jack, and so on. A routing loop still exists in all its glory. However, if you add holddown timers to the equation, you can finally solve the problem.

Note   *Split horizon with poison reverse improves on split horizon by ensuring that loops caused by split horizon failure are removed from the routing table quickly. If a router has a bogus route from a neighboring route, poison reverse ensures that the bogus route is removed by updating it to an infinite metric route. In most cases, you can consider poison reverse to be a more failsafe way of implementing split horizon.*

With holddown timers, when Danny's timeout for the 192.168.50.0 network initially expires, he places the route in holddown. For the next 180 seconds, any updates about the 192.168.50.0 network are ignored. By the time the route is removed from holddown,

Jack is aware of the failure of the route, and the loop is avoided. The problem with these loop-prevention mechanisms is time. In this case, it would take 180 seconds for Jack and Danny to recognize that the route to the 192.168.50.0 network is invalid. Even worse, if Hallorann's connection happened to come back up, it would take another 180 seconds before Jack or Danny would believe Hallorann's updates. Enter triggered updates and route poisoning.

With triggered updates and route poisoning, as soon as Hallorann's link fails, he immediately advertises (triggered update) an infinite metric (route poisoning) for the 192.168.50.0 network to both Danny and Jack. They immediately expire the route and enter holddown, which eliminates the initial route timeout wait of 180 seconds that normally occurs. They also immediately advertise the route to 192.168.50.0 with an infinite metric. Although propagating the infinite metric doesn't help much in this case, if there were several other routers between Jack and Danny, convergence time would reduce tremendously (with the default timers, around 180 seconds per additional hop).

Finally, if Hallorann's link to the 192.168.50.0 network manages to become functional again, Hallorann immediately advertises this information to Jack and Danny. If Jack and Danny's holddown timers had expired, they would then immediately advertise the new route to any other routers, eliminating the need to wait 30 seconds until the next update interval.

RIP 1 Advanced Topics

Now that you've seen the basic operation of RIP, let's examine some more advanced topics. First, let's examine RIP's ability to add more than one route to the routing table and perform load balancing.

By default, Cisco routers use up to four equal-cost routes and load balance between them. With RIP, if a router hears about a route to a network from two or more routers, and all are using the same metric, it performs equal-cost load balancing between all of the routes. For instance, in Figure 23-6, Roland hears of two routes to the 192.168.200.0 network, both with a metric of 1, from Flagg and Eddie. In a Cisco router, Roland adds both to the routing table and forwards packets in a round-robin fashion between both routes (sending 50 percent to Flagg and 50 percent to Eddie). This is great if the links are the same speed, but if not, it causes a problem known as pinhole congestion.

Pinhole congestion occurs when a router is not aware of a speed difference between two routes. This problem is most common in RIP, because RIP has no way of specifying link speed. (Its metric is solely concerned with hop count.) However, this problem can crop up in other protocols as well (usually due to poor configuration). If, in the previous example, the link speed for Roland's link to Flagg were 64 Kbps, and the link speed from Roland to Eddie were 1.544 Mbps (T1), Roland would use only 64 Kbps of the 1.544-Mbps link. This is pinhole congestion. The T1 link wouldn't actually be congested, but Roland wouldn't have any way of knowing that. Roland would simply send one packet using the 64-Kbps link, then another using the T1, then the next using the 64-Kbps link, and so on. Once the 64-Kbps link becomes saturated, no additional packets would be sent across the T1 either. With RIP, the only way to ensure that pinhole congestion does not occur is to ensure that all equal metric redundant links have the same bandwidth capacity.

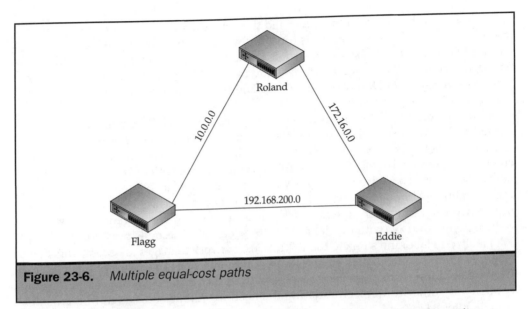

Figure 23-6.   *Multiple equal-cost paths*

Another concern with RIP version 1 is its inability to send masks with routing updates. As a result, you cannot use RIP version 1 in networks that rely on complex VLSM or CIDR subnetting. Because RIP version 1 does not send masks with its updates, it has to assume that a certain mask is used when it receives an update. The basic rules that RIP applies to updates received are as follows:

- If the update entering an interface is the same class of network as any of the router's interface IP addresses, RIP applies the mask used on the interface to the update. If, after applying the interface's mask to the update, bits in the host portion of the update are set, RIP inserts the route into the table with a 32-bit (255.255.255.255) mask, which means the route is for an individual host.

- If the update entering an interface is not the same class of network as any of the router's interface IP addresses, RIP applies the default class A, B, or C mask to the update.

The rules RIP applies to sent updates are as follows:

- When sending an update out an interface, if the class A, B, or C network address of the IP address applied to the interface that is sending the update is the same as the class A, B, or C network address of the route entries in the update, and the mask applied to the sending interface is the same as the mask used for the subnets in the update, RIP sends the subnet numbers in the update but does not send the masks.

- When sending an update out an interface, if the class A, B, or C network address of the IP address applied to the interface that is sending the update is the same as the class A, B, or C network address of the route entries in the update, and the mask applied to the sending interface is different from the mask used for the subnets in the update, RIP does not send the update.

- When sending an update out an interface, if the class A, B, or C network address of the IP address applied to the interface that is sending the update is different from the class A, B, or C network address of the route entries in the update, RIP summarizes the route as a route to the entire class A, B, or C network.

To show you how these rules work, let's take a look at the network shown in Figure 23-7.

In this example, Pennywise sends three routing entries to Silver: 10.0.0.0, 172.16.64.0, and 172.16.80.0. The update regarding the 10.14.0.0 network is summarized into a route about the entire 10.0.0.0 network, based on the sending rules for RIP. Silver applies the /20 mask to the other two routes for the 172.16.0.0 network in this case, because they match the mask used on the link the updates were received on.

However, if a /19 mask were being used on the link between Silver and Pennywise, Silver would apply the /19 mask to the 172.16.64.0 network and a /32 mask to the 172.16.80.0 entry. Like with a 19-bit mask, the entry has bits turned on in the host portion of the address. (If you are confused, drop back to binary and divide the address into network, host, and subnet sections using a 19-bit mask.) Note that this occurrence is very unlikely because, if Pennywise were using a /19 mask for the 172.16.192.0 link, he would have never sent the 172.16.64.0 or 172.16.80.0 entry in his update, as per the second RIP sending rule. Therefore, with a /20 mask on the 172.16.192.0 network, Silver will never send an update about the 172.16.128.0 network to Pennywise.

To keep life simple, the easiest solution to the no-subnet-masks-in-the-update issue is either to use the same subnet mask for all subnets of a given class A, B, or C network, or to use a routing protocol (like RIP version 2 or EIGRP) that supports VLSM.

Table 23-3 summarizes the advantages and disadvantages of RIP version 1.

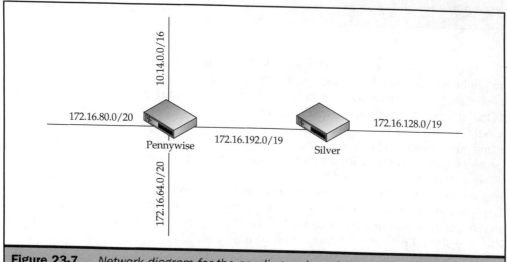

Figure 23-7.   *Network diagram for the sending and receiving rules example*

Advantage	Disadvantage
Very easy to understand and configure.	Inefficient (bandwidth intensive).
Almost guaranteed to be supported by all routers.	Slow convergence in larger networks.
Supports load balancing.	Supports only equal-cost load balancing Pinhole congestion can be a problem.
Generally loop free.	*Generally* loop free.
	Limited scalability (15-hop maximum).
	Does not take bandwidth, delay, or reliability into account for metrics.
	Does not support VLSM.
	If a network is large, complex, and prone to change, routers may never fully converge.
	Broadcasted updates can cause widespread waste of CPU cycles on hosts.
	Does not support authenticated updates, meaning a rogue router could potentially disrupt routes.

Table 23-3.   *Advantages and Disadvantages of RIP Version 1*

As you can see, the disadvantages of RIP far outweigh the advantages. Now, you might be wondering why anyone would *ever* use RIP if it's this problematic. I've actually been asking myself that question for some time now, and the best answer I can give you is that it's easy to configure, and it's always supported. In almost all cases, however, I would recommend EIGRP over RIP (if you are using Cisco routers) or OSPF (if you are using other routers). That is, of course, assuming that you actually need a dynamic routing protocol. (See Chapter 22 for a list of reasons to use static routing.)

In a lot of cases, RIP is used simply because the administrators involved in the network implementation don't understand routing well enough to add the correct static routes. It is easier for them to just let RIP figure everything out. If you are caught in this situation, I advise you to reexamine the need for RIP. In most cases, if you are small enough to use RIP, you should probably be using static routing instead.

Improvements with RIP 2

RIP version 2 (defined primarily in RFC 2453) improves on some of version 1's shortcomings without drastically changing the protocol. The following list details the most useful improvements to version 2:

- **VLSM support**   Subnet masks are transmitted with RIP 2 updates.
- **Multicasted updates**   Updates are multicasted rather than broadcasted, reducing CPU cycle wastage for non-RIP hosts.
- **Authentication support**   Cleartext authentication is supported for RFC-compliant routers. (MD5-encrypted authentication is also supported on Cisco routers.)

Version 2 still retains many of the problems of version 1, but it also retains the advantages of simplicity and nearly unanimous support. If you must use RIP, version 2 is typically the better choice.

Basic RIP Configuration

Luckily, RIP configuration is very simple. To configure RIP using the default settings requires only two commands: *router rip* and *network [network address]*.

The *router rip* command, when entered from global configuration mode, enables RIP globally and inserts you into router configuration mode, as shown here:

```
3600A(config)#router rip
3600A(config-router)#
```

Once in router config mode, you can enter your network commands to enable RIP routing for each class-based network individually. Note that the *network* command accomplishes three functions:

1. Routes belonging to the specific class-based network are advertised.
2. Updates are listened for on all interfaces belonging to the class-based network in question.
3. Updates are sent on all interfaces belonging to the class-based network in question.

The moment you enter the network command for a given network, all three of these functions are enabled. This is an all or nothing affair that you cannot selectively control. Also note that the command enables all three functions for an entire class-based network. In other words, if you have a router with 172.16.64.0/18 on E0 and 172.16.128.0/18 on S0, the command *network 172.16.64.0* automatically enables RIP on both E0 and S0. As far as the router is concerned, the network command you entered enabled RIP for the entire class B network and all subnets and interfaces on that network. Make sure that you remember this little detail when you are configuring RIP.

CISCO ROUTING

Finally, if you want to change the version of RIP being used, simply use either the *version [1 | 2]* or *ip rip [send version | receive version] [1 | 2]* command. You enter the *version* command from router config mode and set the version of RIP used on the router globally. If you want to change the version of RIP used for sending updates or receiving updates on a particular interface (for instance, when the interface is connected to a router that supports only RIP 1, but the rest of the environment supports RIP 2), use the *ip rip* command. The following example shows how to set the version to 1 for all updates sent out of the first Fast Ethernet interface:

```
3600A(config)#interface fastEthernet 0/0
3600A(config-if)#ip rip send version 1
```

Tip   *If you want to enable sending or receiving of both versions of RIP on a given interface, put both 1 and 2 in your* ip rip *command, like so:* ip rip send version 1 2. *(This is the default setting.)*

As you can see, getting RIP up and running using the defaults is not rocket science. However, if you want to optimize RIP's operation, you will be required to perform a few more steps, which are detailed in the next section.

Advanced RIP Configuration and Optimization

In a Cisco router, you can use a number of optional commands to gain greater control over RIP. Although none of these commands are required, they can be particularly helpful in certain situations. A list of optional configuration tasks is provided here:

- Configuring passive interfaces (disabling broadcasted updates)
- Configuring unicast updates
- Adding metric offsets to routes
- Adjusting RIP timers
- Disabling split horizon
- Setting the maximum number of paths
- Configuring authentication (RIP 2)
- Disabling autosummarization (RIP 2)

The following sections discuss each of these tasks one at a time, beginning with passive interfaces.

Configuring Passive Interfaces

A *passive interface* is an interface that does not broadcast routing updates, but is still advertised in routing updates and still listens for routing updates. If you remember the three tasks the *networks* command performs (advertises all networks in the class-based network, listens for updates on all interfaces in the class-based network, and broadcasts updates on all interfaces in the class-based network), the *passive interface* command shuts off the third task for a particular interface. This functionality is particularly useful in two common situations:

■ The network attached to the RIP interface includes only hosts, but it needs to be advertised to other routers.

■ For security or performance reasons, you wish to disable broadcasted routing updates and would rather selectively choose which routers receive unicast updates. (This topic is explained further in the next section, "Configuring Unicast Updates.")

The first situation is actually fairly common, so let's take a look at an example of a scenario that could use passive interfaces (shown in Figure 23-8) and walk through the RIP configuration of a router.

In this example, the 172.16.64.0/18 network (connected to Alan through the E0 interface) contains only hosts. RIP broadcasts on this network would be wasteful; but if you neglect to enter a network command for the 172.16.0.0 network, not only will the 172.16.64.0/18 network not be advertised to Polly or Leland, but the 172.16.128.0/18 network also won't be advertised. In addition, *no* networks would be advertised from Alan to Polly (the third rule of the *network* command), meaning that Polly would never learn of the 192.168.1.0/24 network, and neither the 172.16.64.0/18 network nor the 172.16.128.0/18 network would be advertised to Leland.

To solve this little dilemma, you need to use the *passive interface* command for the E0 interface (attached to the 172.16.64.0/18 network) on Alan. The *passive interface*

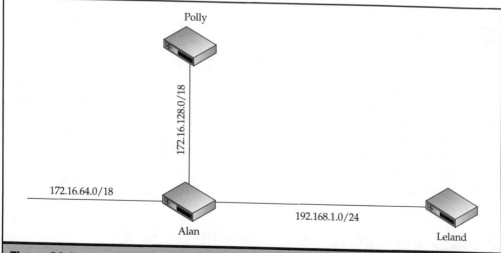

Figure 23-8.   *A network that needs a passive interface defined*

command's syntax is as follows: *passive-interface [interface type and number].* You enter the command from router config mode, as shown in the following example:

```
Alan (config)# router rip
Alan (config-router)# network 172.16.0.0
Alan (config-router)# network 192.168.1.0
Alan (config-router)# passive-interface Ethernet 0/0
```

After performing these configuration steps, Alan's RIP functionality will be as follows:

- All networks will be advertised to Polly and Leland.
- Routing updates will be listened for on all interfaces.
- Routing updates will not be advertised on Alan's E0/0 interface.

Configuring Unicast Updates

Although RIP 1 typically uses broadcasts for routing updates and RIP 2 normally uses multicasts, in some situations, you may need to enable unicast updates. One such situation is when you are sending updates over a link that does not support broadcasts (NBMA networks), such as Frame Relay. Another case is when you do not want broadcasts to waste CPU resources on clients attached to the same network as a router that needs to receive broadcasts. Finally, unicast updates are useful in situations where you require security between routers for routing updates. Because the updates are unicast to each router, in a switched network, normal hosts would not be able to use a sniffer to read the details of each RIP update. To see how unicast routing updates can be helpful, take a look at the simple internal network shown in Figure 23-9.

The broadcast keyword can be used on Frame Relay PVCs to ensure that broadcasts are sent to the PVC endpoint. If the broadcast keyword is used, unicast updates are not needed. If not, unicast updates should be configured on all routers participating in the VC.

In this example, if you were broadcasting updates, all of the hosts on the 192.168.1.0 network would receive these updates. Instead, you could just enter the IP address of Flagg on Peter, and vice versa, and then specify that the E0/0 interface on both routers is a passive interface. This strategy would ensure that all updates between Peter and Flagg are unicast, with no RIP broadcast traffic crossing the network.

To enable unicast updates, you need to use the *neighbor [ip address]* router configuration mode command. For the previous example, the configuration of Flagg is as follows:

```
Flagg (config)# router rip
Flagg (config-router)# network 192.168.1.0
Flagg (config-router)# neighbor 192.168.1.1
Flagg (config-router)# passive-interface Ethernet 0/0
```

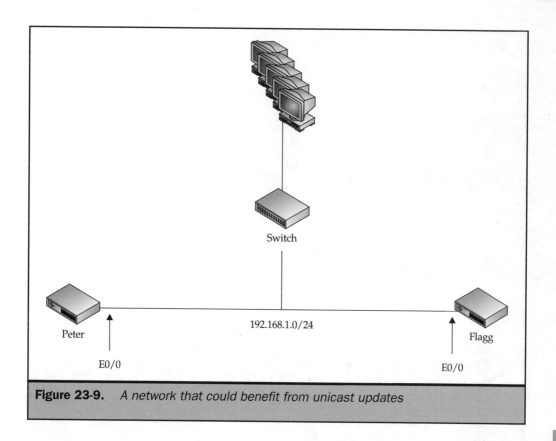

Figure 23-9. *A network that could benefit from unicast updates*

Adding a Metric Offset

Adding a metric offset to a route allows you to specify that the metric for routes coming from a certain router or network be increased by a specific amount. This functionality allows you to crudely specify that routes from one or more routers are less "preferred" than other routes. For instance, Figure 23-10 has two routes to the 10.0.0.0 network from Roland. With the route through Peter, you are connected to Alan through a T1, but with the route through Flagg, you are connected with a 128-Kbps link.

The problem with this situation is that RIP receives the same metric (two hops) for both routes, and therefore performs equal-cost load balancing across these two routes, leading to a buffer overflow on Flagg and causing indirect pinhole congestion. To remedy this problem, you could specify that all routes entering Roland's E0/0 interface are given a +1 metric offset. Thus, the route from Flagg would appear to cost more than the route from Peter, and Roland would use the route through Peter instead of performing load balancing.

To add a metric offset, you use the *offset-list [(optional) access list] [in | out] [offset] [(optional) interface type and number]* command. This command is a bit complicated, so I'll break each component down individually.

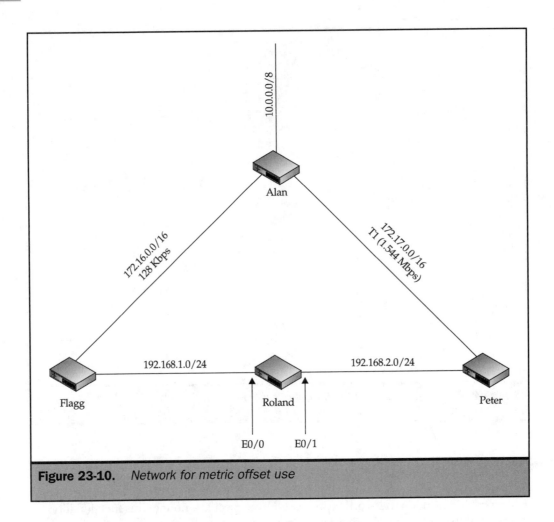

Figure 23-10. Network for metric offset use

The *[(optional) access list]* section is an optional component that you can use to add an offset to update entries that match the access list. For instance, if your access list matches all networks beginning with 192, then updates for the 192.168.1.0 network will have the offset added to their metrics.

Note *Chapter 27 discusses access lists in greater detail.*

The *[in | out]* section is used to define whether the offset applies to incoming updates or outgoing updates. The *[offset]* section defines the metric value that will be added to the metric on each route matching the *offset-list* statement. In RIP, this value should be between 1 and 14. Finally, the *[(optional) interface type and number]* section specifies that this offset list should be applied to route updates entering or leaving a specific interface.

To see how this command works, let's take the example shown in Figure 23-10 and apply an offset of 5 to all routes entering Roland's E0/0 interface. The required configuration is shown in the following output:

```
Roland (config)# router rip
Roland (config-router)# network 192.168.1.0
Roland (config-router)# network 192.168.2.0
Roland (config-router)# offset-list in 5 Ethernet 0/0
```

Adjusting RIP's Timers

RIP timer adjustment is useful if you want to optimize the convergence of your network. For instance, on a high-speed, high-bandwidth internal network (like a Fast Ethernet LAN), you may want to reduce the timers, allowing RIP to converge faster at the expense of bandwidth. On a WAN, however, you may want to increase the timers to reduce bandwidth use, at the expense of convergence time. One way or another, however, when you modify the timers, ensure that you set all routers to use the same timer values, and be careful to remember the relationships the timers share with each other. To simplify this process, the following table lists recommended timer multiples.

Timer	Multiple	Default (IP RIP)
Update	Base timer	30 seconds
Invalid	3x update	180 seconds
Holddown	3x update	180 seconds
Flush	Greater than invalid	240 seconds

Note *Always ensure that the timers are the same on all routers and that the timers are relative to each other. If you don't, routes may be lost, and routing loops may occur.*

If you have a lot of routes, setting the timers too low can result in high processor use on your routers. The default of 30 seconds for the update timer is fine for most WAN links; and because convergence is generally not an issue in LANs (because they should have relatively static configurations and should rarely fail), keeping the timers at the default is actually appropriate in many situations. However, if your network changes often (usually due to link failures), reducing the timers results in faster convergence.

To set RIP's timers, use the *timers basic [update in seconds] [invalid in seconds] [holddown in seconds] [flush in seconds]* router config mode command. For instance, to set the update timer to 15 seconds, the invalid timer to 45 seconds, the holddown timer to 55 seconds, and the flush timer to 90 seconds, you issue the following command:

```
Router (config-router)# timers basic 15 45 55 100
```

This configuration sets the router to send and expect to receive updates every 15 seconds; to declare a route bad after 45 seconds without an update and enter holddown; to remain in holddown for an additional 55 seconds; and then, at 100 seconds, to remove the route from the table.

Disabling Split Horizon

Before getting into how to disable split horizon, you should know that it is most definitely *not* recommended in most situations. In most cases, split horizon saves you a few headaches by preventing routing loops; however, if you have WAN links with multiple VCs on a single physical interface, split horizon can actually backfire on you. For instance, imagine that you have a single serial interface that has four Frame Relay VCs and four networks applied to it. With split horizon enabled, no updates received from *any* of the VCs on the serial link would be forwarded to any of the other VCs, as per split horizon's rules. Luckily, an easy solution exists to this problem. First, however, let's examine the IOS's default settings for split horizon over Frame Relay. (Split horizon is enabled by default for all LAN interfaces.) The default split horizon settings for Frame Relay are the following:

- If no subinterfaces are defined, split horizon is disabled.
- For point-to-point subinterfaces, split horizon is enabled.
- For multipoint subinterfaces, split horizon is disabled.

Note   *Split horizon is enabled by default for HDLC and PPP interfaces.*

Notice that split horizon is enabled on point-to-point subinterfaces. In this case, RIP treats each subinterface as a separate interface for split horizon. In other words, if a route is heard on S0/0.1, it will not be rebroadcast to S0/0.1, but it *will* be broadcast to S0/0.2.

For interfaces without subinterfaces, if only one VC is assigned to the interface, it is best to enable split horizon on the interface. If multiple VCs are assigned to the interface, you have three choices:

- Enable split horizon and statically configure any routes that are not propagated.
- Leave split horizon disabled, and filter out routes that could potentially cause loops using access lists (discussed in Chapter 27).
- Reconfigure the router to use subinterfaces.

The first two choices are a pain and are not really recommended. (Don't worry, though, that doesn't mean I won't tell you how to do it.) The third choice is preferable in almost all situations.

For the first choice, you need to determine which routes are not being propagated from one VC to another and statically add those routes to your routers. For instance, in the network shown in Figure 23-11, you would need to statically add a route to the 192.168.1.0 and 172.16.0.0 networks to Al's routing table, and a route to the 192.168.2.0

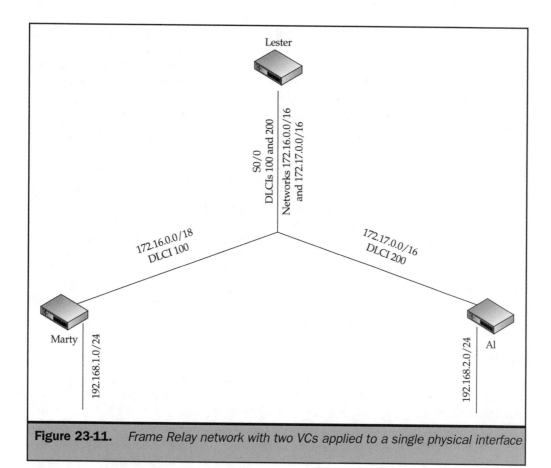

Figure 23-11. *Frame Relay network with two VCs applied to a single physical interface*

and 172.17.0.0 networks to Marty's table. You would add these routes because both of these VCs enter on Lester's S0/0 interface, so Lester will not propagate the routes back out the S0/0 interface (at least, not with a valid metric). This scenario is obviously useful only in small networks. (And if you are adding a lot of static routes, what exactly is RIP doing for you anyway?)

The second scenario involves keeping split horizon enabled and adding access list statements that filter bad routes out of updates entering each remote router. At each remote router, you would add a filter for each network to which the router is directly connected. For instance, in the previous example, you would need to filter out incoming updates for the 192.168.1.0 and 172.16.0.0 networks on Marty, to keep any bogus routes for these networks from entering Marty's routing table.

Note *If you are using multipoint subinterfaces, route filtering is almost a necessity to avoid routing loops.*

The final solution is to just reconfigure the router to use subinterfaces for each VC. This solution is typically the easiest, and is covered in detail in Chapter 22.

One way or another, enabling and disabling split horizon is as easy as entering the *ip split-horizon* (to enable split horizon) or *no ip split-horizon* command in interface config mode for each particular interface.

To perform filtering on an interface, you must perform two basic tasks:

1. Configure one or more access lists to match the update entries (networks) that you wish to remove from updates.

2. Configure an incoming or outgoing distribute list to use the access list.

Again, access lists are covered in Chapter 27, so I won't discuss the first step here. Once the access list is configured, use the *distribute-list [access list] [in | out] [(optional) interface type and number]* router configuration command to apply the filter either globally (by not specifying an interface) or to a specific interface. The *[in | out]* parameter is used to specify whether the filter should be applied to incoming updates or outgoing updates. When applying the list to an outgoing update, you may also specify that the update apply to a specific routing protocol and autonomous system (AS) by appending the protocol and AS number to the end of the statement (discussed further in the next chapter).

Setting the Maximum Number of Paths

From the previous discussions in this chapter, you should be aware of RIP's ability to perform load balancing across equal-cost paths and some of the problems this feature can cause. In the IOS, the default setting for all routing protocols *except* BGP is to load balance across up to four equal-cost paths. (Load balancing is disabled by default for BGP.) In some cases, this setting can lead to pinhole congestion. To adjust this figure to allow for more or less paths, you use the *maximum-paths [number of paths]* router configuration command for each routing protocol in use on the router. You may include up to six paths for IP, and the minimum is, of course, one.

Configuring Authentication

For RIP 2, you can ensure that to process an update from a neighboring router, the router must be authenticated. This authentication helps (somewhat) to ensure that routing updates are processed only if they come from "trusted" routers. For authentication to work, all routers must be using the same password. Enabling authentication in RIP 2 involves two steps:

1. Configure a key chain.

2. Enable plaintext or MD5 authentication.

The first step in this process is to configure a *key chain*: a list of authentication "keys," or passwords, that can be used to authenticate the routing protocol. To configure a key chain, you must first enter the command *key chain [name of chain]* from global config mode, which will put you into key-chain configuration mode for this specific key chain, like so:

```
3620B(config)#key chain test
3620B(config-keychai)#
```

You must then use the *key [key number]* command to begin configuring a specific key. The key number can be any number between 1 and 4 billion. The *key* command enters you into a special mode to configure this individual key, but your prompt will *not* change.

Next, use the *key-string [password]* command to set the password. By default, this password is used both to authenticate this router with other routers, as well as to authenticate other routers attempting to send an update to this router. You can modify this behavior (for instance, use one password to authenticate to other routers and use a different password for other routers to authenticate to your router with), but it is typically easier to use the same password for both authentication methods.

To set the times in which the router will accept the key, use the *accept-lifetime [start time] [end time]* command. The *[start time]* parameter is entered in hh:mm:ss month date year format and specifies the earliest time that this password will be accepted. The default is the beginning of time. (Well, as close as Cisco gets to it anyway; it's actually January 1, 1993.) The *[end time]* parameter can be specified in one of three ways: by using the *infinite* keyword, which means forever; by using the *duration* keyword followed by a number of seconds, which specifies how many seconds after the start time the password is accepted; or by using the same hh:mm:ss month date year format as the start time. The default for the end time is infinite.

To set the times in which the router will use the key string to authenticate with other routers, use the *send-lifetime [start time] [end time]* command. The start time and end time parameters for *send-lifetime* are the same as for *accept-lifetime*. The defaults are also the same (from the beginning of time through the end of time).

Because all of these commands are a bit complex, let's walk through an example in which you want to set the password used in the key chain to be *secret*, set the key to be used for both authentication to this router and from this router, and set the times allowed to be from now until the end of time:

```
3620B(config)#key chain test
3620B(config-keychai)# key 1
3620B(config-keychai)# key-string secret
```

That task was actually pretty easy because the defaults are that the string will be used for authentication from and to the router and the lifetimes are infinite. Let's look at a more complex example. In this example, let's set the key used to authenticate to this router to be *secret* and the key used to authenticate to other routers to be *terces*. Although

CISCO ROUTING

there isn't a specific setting to enable this function, clever use of the *accept-lifetime* and *send-lifetime* parameters can do the job. Look at the following example configuration:

```
3620B(config)#key chain test
3620B(config-keychai)# key 1
3620B(config-keychai)# key-string secret
3620B(config-keychai)#send-lifetime 00:00:00 jan 1 2035 duration 1
3620B(config-keychai)# key 2
3620B(config-keychai)# key-string terces
3620B(config-keychai)#accept-lifetime 00:00:00 jan 1 2035 duration 1
```

In this configuration, any routing protocol using this key chain will use the *terces* key (key 2) to authenticate with other routers at all times except for one second at midnight on January 1, 2035. Similarly, the router will accept only the *secret* key (key 1) at all times except for one second at midnight on January 1, 2035. Just remember that if you configure a router in this manner, all other routers using RIP to advertise routes to this router must have the reverse of this key chain configured. (Yes, this is a nightmare, which is why complicated key chains are rarely done.)

Once you have configured these parameters, your key is ready to be applied to RIP 2 authentication. Luckily, this process is fairly simple. First, you must apply the key chain to a specific interface by using the *ip rip authentication key-chain [name of key chain]* interface configuration mode command. Then, you set the authentication mode for the interface using the *ip rip authentication mode [md5 | text]* interface config mode command. Then authentication is configured.

Disabling Autosummarization

With RIP 2, routes are automatically summarized into the entire classful network address if it meets autosummarization requirements. The autosummarization requirements are basically as follows: if a route is advertised on an interface that has a different class-based network address than the route to be advertised, all subnets of the advertised network will be advertised as one entry for the entire class-based network.

In other words, if you are sending an update containing networks 172.16.64.0/18 and 172.16.128.0/18 on an interface with the IP address of 172.31.1.1/20, the advertisement will be for 172.16.0.0/16, *not* for each individual subnet. This feature can cause problems if you have a complex VLSM-based topology; but if not, it considerably reduces the number of routes that are required to be advertised and kept in the routing table.

However, if you have a complex VLSM topology, you can still have your cake and eat it, too. With IOS 12.1 and later, RIP on Cisco routers now supports a limited form of route aggregation, so you can manually summarize networks. For instance, examine the network shown in Figure 23-12.

In the example in Figure 23-12, the default settings of RIP summarize the routes from Gage to Lewis as 172.16.0.0/16. The problem is that Gage doesn't contain all of the subnets in the 172.16.0.0/16 network. In this case, the defaults would actually work

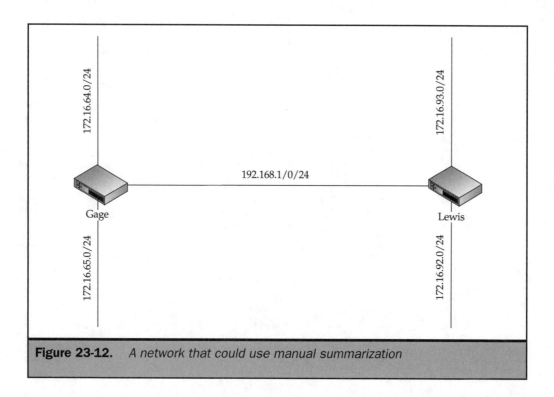

Figure 23-12. *A network that could use manual summarization*

because the router would send everything destined for the 172.16.0.0/16 network to the other router (except for packets destined for Gage's directly connected subnets, of course, because the most specific route to a destination wins). But if you added an additional router, as shown in Figure 23-13, now you have a problem.

In this scenario, using the defaults, Gage, Jud, and Lewis all advertise an autosummary route to the 172.16.0.0/16 network with a one-hop metric. With the default of four equal-cost paths to each network, on Gage, there would be two entries for the 172.16.0.0/16 network: one through Jud and one through Lewis. Gage would load balance between these two routes, meaning that if you sent something to 172.16.128.1, at first blush, it looks like it would have a 50-percent chance of succeeding. Actually, it would probably get there; it would just take a while and use a bunch of bandwidth.

Let's walk through this example so you see what I mean. First, let's assume that a user on 172.16.64.0/24 needs to get to 172.16.128.1. She sends her packet to Gage, who looks in his routing table, flips a coin, and sends the packet to Lewis. Lewis receives the packet and looks in his routing table, and the only route that matches is the 172.16.0.0/16 entry—which he has two of (one from Gage and one from Jud). Therefore, he flips a coin and forwards the packet back to Gage. Now, assuming that no one else has sent any packets through Gage to the 172.16.0.0 network while Gage and Lewis have been busy playing "pass the packet," Gage actually (*finally*) sends the packet to Jud, who gets it to its final destination.

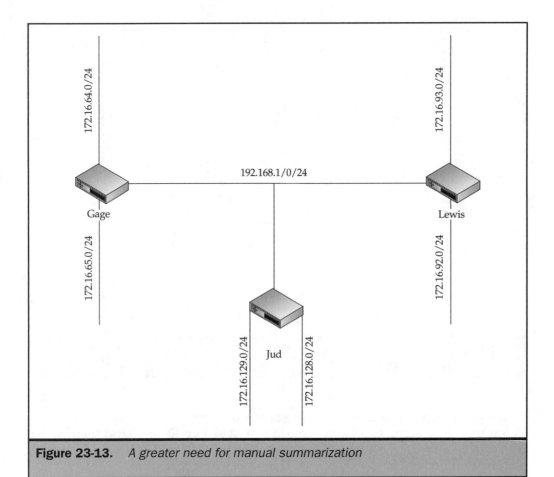

Figure 23-13. *A greater need for manual summarization*

Although this whole process *does* actually work (depending on how you define *work*), it is not exactly what you would call graceful. Also, the bigger the environment gets, and the more subnets of the 172.16.0.0 network you have, the worse this little problem becomes (sometimes to the point of expiring the TTL on the packet). So this configuration is obviously not recommended.

To solve this problem, you have two choices. The first is to disable autosummarization entirely and let the routers advertise the subnets individually. Although this choice is good in this case (because it means adding only four new routes to each router's table), in a larger environment, it could cause needless traffic and CPU overhead (due to the sheer size of the routing tables). In a larger environment, the second option—manual summarization—is a far better choice.

Although this book isn't meant to be test preparation, just in case you are looking to take the CCNA or CCNP tests, you will need to know that only EIGRP and OSPF can perform route aggregation. This case is not one in which the test and real life don't match, just a case of a new feature being added to the IOS that hasn't quite made it into the test yet.

Manual summarization, more commonly called *route aggregation,* allows you to specify how the routes are summarized to neighbors in advertisements. With manual summarization, you could advertise 172.16.64.0/23 from Gage, 172.16.92.0/23 from Lewis, and 172.16.128.0/23 from Jud, using only two entries in each router's table to reach all networks.

> **Tip**   *Route aggregation requires that you fully understand the binary math behind subnetting. If this topic is giving you trouble, refer to Chapter 6 and examine the "IP Addressing Basics" section in greater detail.*

To perform route aggregation, you first need to disable autosummarization using the *no auto-summary* router config mode command. Then you need to use the *ip summary-address rip [ip address] [mask]* interface config mode command on every interface where you want to summarize the routes. For instance, assuming Gage's E0/0 interface is the interface attached to the 192.168.1.0 network, you configure Gage for route aggregation as follows:

```
Gage (config-router)# no auto-summary
Gage (config-router)# exit
Gage (config)# interface Ethernet 0/0
Gage (config-if)# ip summary-address 172.16.64.0 255.255.254.0
```

That's it. Route aggregation really isn't hard, as long as you understand the binary behind it. Now that you've learned about nearly every common configuration option that RIP has to offer, let's take a look at some common troubleshooting tips.

Troubleshooting RIP

Like with all other networking issues, the first step in solving any problem is to determine the root cause. For this task, your *show* and *debug* commands should come in particularly handy. For *show* commands, the hands down most useful commands at your disposal are *show ip route* and *show running-config. Show ip route* shows you what is in your routing table (as described in Chapter 22), whereas *show running-config* verifies the RIP configuration. You can also use the *show ip protocols* command to show details about the running routing protocols on the router, including timers and summarization details, as shown in the following example:

```
Gage# show ip protocols

Routing Protocol is "rip"
  Sending updates every 30 seconds, next due in 25 seconds
Invalid after 180 seconds, hold down 180, flushed after 240
    Outgoing update filter list for all interfaces is
    Incoming update filter list for all interfaces is
```

CISCO ROUTING

```
Redistributing: rip
Default version control: send version 2, receive version 2
  Interface        Send  Recv  Triggered RIP  Key-chain
  Ethernet0/0       2     2
  Ethernet0/1       2     2
  Ethernet1/0       2     2
  Automatic network summarization is not in effect
Address Summarization:
  172.16.64.0/23 for Ethernet0/0
```

Show ip rip database is also useful for showing which networks are currently being advertised by RIP, as you can see in the following output:

```
CCNA2600C#sh ip rip database
10.0.0.0/8     auto-summary
10.1.1.0/24     directly connected, Ethernet0/0
169.254.0.0/16     auto-summary
169.254.0.0/16
    [1] via 10.1.1.14, 00:00:19, Ethernet0/0
172.31.0.0/16     auto-summary
172.31.0.0/16
    [2] via 10.1.1.14, 00:00:19, Ethernet0/0
```

The most useful debugging command is probably *debug ip rip*. Once enabled, RIP debugging shows you all RIP packets sent from or received by the router, as well as the associated error messages and networks sent in the update.

If you run into problems with RIP, be on the watch for a few common issues:

- VLSM is not supported in RIP 1. Remember, depending on the configuration, RIP 1 either summarizes VLSM subnets into one class-based network address or neglects to advertise the route entirely.

- Autosummary is enabled by default in RIP 2. If you are using RIP 2, ensure that your network topology can be autosummarized, or disable autosummarization.

- If authentication is enabled, all routers that will participate in the RIP network must use the same password(s).

- Non-Cisco routers may not support MD5 authentication.

- Split horizon can cause problems on Frame Relay PVCs if multiple PVCs are mapped to the same interface. Use subinterfaces with one PVC per subinterface to eliminate this problem.

- Split horizon must be disabled for multipoint PVCs. Make sure to include route filtering to prevent routing loops in multipoint topologies.

■ Ensure that timers on all routers in the topology match. If they do not, routing updates can fail and routing loops can form.

■ Ensure that the timer values are relative to each other.

■ Remember, RIP is prone to pinhole congestion when load balancing.

■ If updates are not being sent out an interface, ensure that the correct network statements are entered and that the interface is not configured as a passive interface.

■ If you are using unicast updates, ensure that the IP addresses of the remote routers are entered correctly.

■ Ensure that the problem is due to an issue involving RIP and not a physical or basic configuration problem (i.e., wrong IP address applied to an interface).

Again, remember that most errors are simply misconfigurations. Although these tips do not cover every problem you are likely to see with RIP (a list of bugs in different IOS versions *alone* would take a full chapter), they are the most common causes of RIP problems.

Summary

This chapter discussed RIP 1 and 2, and examined how to configure, optimize, and troubleshoot RIP in most environments. Although an old protocol, RIP is still very widely used, so this information will likely come in handy in a variety of situations. In addition, many of the conventions discussed in this chapter (like split horizon and timers) apply to nearly all routing protocols, and will be an important building block as you move into more complex routing protocols, beginning with IGRP in Chapter 24.

CISCO ROUTING

The Complete Reference

Cisco

Chapter 24

Interior Gateway Routing Protocol

his chapter examines routing with Interior Gateway Routing Protocol (IGRP). IGRP is a Cisco proprietary routing protocol that was designed to solve some of the scalability problems with RIP in larger, more complex networks. As such, it includes improvements on most of RIP's shortcomings, including timers and metrics; but, otherwise, it is basically "RIP on steroids." IGRP has been almost entirely replaced by its big-brother protocol Enhanced IGRP (EIGRP) in most modern networks, but you may still occasionally run across it. Luckily, it's so similar to RIP that very little of IGRP's operation requires additional explanation.

How IGRP Works

As previously mentioned, IGRP operation is very similar to RIP in many respects. IGRP is a distance vector routing protocol; and as such, it follows most of RIP 1's operational procedures, including the following:

- IGRP broadcasts routing tables to neighboring routers at predefined intervals.
- IGRP uses update, invalid, holddown, and flush timers.
- IGRP doesn't support VLSM.
- IGRP uses split horizon, triggered updates, and route poisoning.

The primary differences between IGRP and RIP include the following:

- IGRP uses autonomous systems (AS).
- IGRP supports a much more complex and flexible metric.
- IGRP can span networks of up to 255 hops.
- IGRP's load-balancing mechanism is different.
- IGRP uses longer timers.

Each of these topics are discussed in the following sections.

 Although IGRP can span up to 255 hops, the default (and recommended) limit is 100 hops.

Autonomous Systems

An IGRP autonomous system (AS) is what is known as a process domain. A *process domain* is an area in which a specific routing protocol instance is processed on all routers. Because IGRP accepts or sends updates to routers only within its own AS, you could have two (or more) separate process domains within a single routing domain. A *routing domain* is an area in which routing is internally controlled and monitored. Basically, your routing domain is your internal network. The connection to your ISP is the ISP's routing domain. Once a packet leaves your routing domain, it is your ISP's responsibility (as least until it leaves your ISP's routing domain). To examine the distinction between routing domains and process domains more closely, take a look at the network in Figure 24-1.

In this example, routing updates from AS 1 will not enter AS 2, and vice versa. Thus, scalability increases because you can have up to 255 hops in each AS. The border router, in the middle, is a member of both ASs. Therefore, it knows all routes to all destinations. In this case, you could just insert a default route into both AS 1 and AS 2 that points to the border router, and the border router would route packets both between process domains and between routing domains (to the Internet).

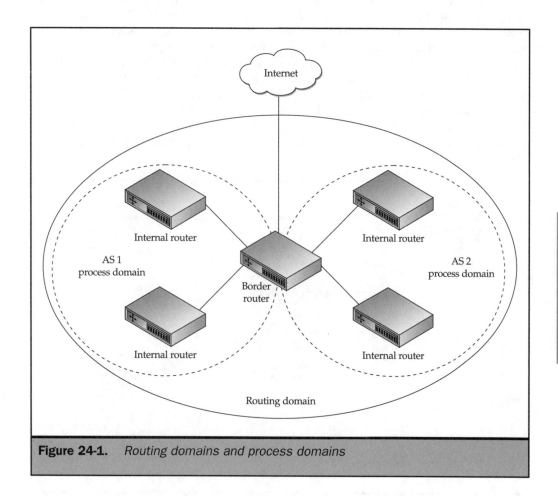

Figure 24-1.   *Routing domains and process domains*

In addition to distinguishing between ASs, IGRP also distinguishes routes depending on the network topology. IGRP defines three types of routes: internal, system, and external.

- An *internal* route is a route to a router that is a subnet of the connection between the sending router and the receiving router.

- A *system* route is a summary route to a major network.

- An *external* route is a default network. Like the default route, the default network is where the router sends packets when it cannot find a better route.

To solidify these concepts, take a look at another example network, shown in Figure 24-2.

In this example, if Corvette advertises 172.16.32.0/19 to Camaro, it will advertise the network as an *internal route* using the address 172.16.32.0 because the 172.16.32.0/19 network is a subnet of the 172.16.0.0 network—the same class-based network that is used in the link between Corvette and Camaro. When Corvette advertises 172.16.32.0/19 to Firebird, it will advertise a summary route (a *system* route in IGRP) for the entire 172.16.0.0 network because the link to Firebird is on the 192.168.1.0 network. When it advertises the 192.168.1.128/26 network to Firebird, however, it will be advertised as an internal route to 192.168.1.128. Finally, if Corvette were to forward the route to the 10.16.0.0 network to either Camaro or Firebird, it would advertise it as an external route. As a result, the default network on Camaro and Firebird would be

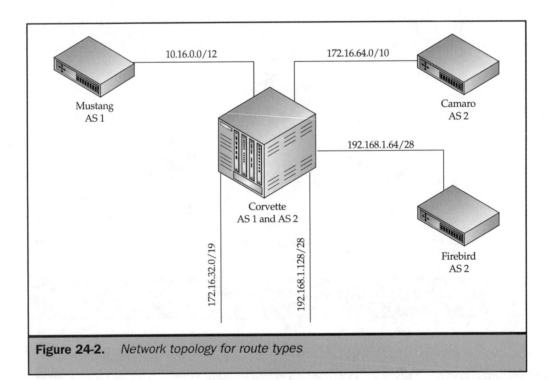

Figure 24-2. *Network topology for route types*

set to the 10.0.0.0 network, and any traffic that didn't match any other routes in their table would be forwarded there.

The basic IGRP rules for the three types of routes advertised are as follows:

- If a route is advertised to a subnet of the same major network as the route itself, the route is sent as an internal route and includes subnet information.

- If a route is advertised to a subnet of a different major network as the route itself, the route is sent as a system route and does not include subnet information.

- If the advertised route has been marked as a default route, the route is advertised as an external route.

Metric and Maximum Hops

IGRP uses a complex calculation to produce a single "composite" metric. The calculation IGRP uses includes bandwidth, delay, reliability, and load, but only bandwidth and delay are used by default. IGRP also tracks hop count and Maximum Transmission Unit (MTU), although neither are used in the metric calculation. *Bandwidth* in IGRP specifies the lowest-rated bandwidth used in the entire path, and it is manually defined. *Delay* is the amount of time it takes a single packet to reach the destination, assuming an uncongested network. Delay is also manually defined. *Reliability* is defined by the number of packets that arrive on the link undamaged. This calculation is measured (although it is not used, by default). Finally, *load* specifies the amount of traffic that crosses the link relative to the bandwidth, on average. In other words, load describes how under- or oversubscribed the line is. Load is measured, although, again, it is not used to calculate the metric by default.

> **Note** *MTU, while tracked, serves no real purpose in IGRP. Hops, on the other hand, although they do not influence metric calculation, are tracked to limit the size of the IGRP network. By default, a single IGRP AS can span up to 100 hops, but this number can be modified up to a maximum of 255 hops.*

IGRP still transmits metrics in routing updates, like RIP does, but these metrics are not used to determine the metric for the receiving router. Rather, IGRP also transmits the metric components (bandwidth, delay, load, and reliability) with each update. Consequently, each router can apply different weights to each metric component to arrive at an appropriate metric for that router. This functionality can be very useful in certain situations (one of which is explored later in this section).

So how do all of these pieces fit together to create a composite metric? Here's the formula:

$$\text{metric} = K1 \times Be + (K2 \times Be)/(256 - \text{load}) + K3 \times Dc) \times (K5/(\text{reliability} + K4).$$

It really isn't that complex once you understand the individual components. First, forget the formula for a minute and concentrate on the components.

The Bandwidth Component

Bandwidth is calculated as the lowest-rated bandwidth along the entire path, with a bit of a twist. First, the bandwidth rating used is the bandwidth of the slowest link along the entire route. In other words, because the value used is the lowest bandwidth of any link in the path, if you had a path consisting of a 1.544-Mbps link, a 1-Gbps link, and a 56-Kbps link, the bandwidth value used for metric calculation would be 56 Kbps.

Second, to make higher bandwidth "better" (lower) in the metric, the bandwidth rating is 10,000,000 divided by the actual bandwidth in Kbps of the slowest link. For instance, with the slowest link in the path being 56 Kbps, the final bandwidth would be 178,571. In contrast, a 1-Gbps link would have a bandwidth of 10. The process of dividing 10,000,000 by the bandwidth makes faster links "better" in terms of metric. This final bandwidth number is represented as Be in the formula.

The Delay Component

Delay is the *sum* of all of the delays along the path, expressed as units of ten microseconds. You could calculate the delay for each individual link and set them that way (pretty tedious). However, normally, you just set a specific value depending on the type of link. Setting the delay is covered in the "IGRP Configuration" section, later in the chapter; but be aware that although the delay is set on a link in microseconds, it is calculated for the metric in units of ten microseconds. In other words, if you set the delay of a link to be 10,000 microseconds, the calculated delay will be 1,000. This calculated delay figure is represented as Dc in the formula. Table 24-1 shows the recommended delay values for different link types.

Load and Reliability

Load is not normally used, but when it is, it is based on the exponentially weighted average of how saturated the link is during a given sample period. The sample period is normally five minutes, and the load calculation is updated every five seconds (in other words, every five seconds, the load calculation will change to reflect the load on the link in the previous five minutes). Load is measured from 1 to 255, with 1 being minimal use and 255 being fully saturated (100 percent used).

Link Type	Delay (in Microseconds)
100+ Mbps	100
Token Ring	630
10-Mbps Ethernet	1,000
Serial	20,000

Table 24-1.   *Recommended Delay Values*

Reliability is similar to load in that it is calculated using an exponentially weighted average that is taken from activity occurring in the previous five minutes and is updated every five seconds. It is also not normally used. Reliability, however, uses a reverse scale from load (I am convinced this was done solely to cause confusion), with 255 being completely error-free and 1 being unusable.

Calculating the Formula

Now that you have a better idea of how each component is calculated, take a look at that formula again:

metric = K1 × Be + (K2 × Be)/(256 – load) + K3 × Dc) × (K5/(reliability + K4).

Well, you now know what the Be and Dc are, but what about all those Ks? *K* is a constant, referred to as a *weight*. Weights allow you to assign different priorities to different metric components, which is useful when classifying traffic and routing packets based on the classification. For instance, for voice traffic, latency (delay) is most important, with bandwidth being secondary. For file transfers, however, delay is pretty unimportant, while bandwidth reigns supreme. The original idea was to use IGRP as a routing mechanism that could use packet classifications to choose the best route for specific types of traffic, but this concept was never fully implemented, so choosing weights ends up coming down to choosing which metric component you feel is most important.

Simple Metric Calculations   Usually, it is recommended that you simply use the defaults, which are K1 (bandwidth weight) = 1, K3 (delay weight) = 1, K2 (load weight) = 0, K4 (secondary reliability weight) = 0, and K5 (primary reliability weight) = 0. With the default weights, only K1 and K3 are used. The rest of the weights are set to zero, which, instead of being inserted as zero into the equation (and giving a resulting metric of zero for every route), means that part of the equation is removed. After removing the unused sections of the equation, the formula simplifies to this: metric = K1 × Be + K3 × Dc, or, even more simply (because the default weights for K1 and K3 are 1), metric = Be + Dc. With the default weights, metric determination is fairly simple. For instance, examine the network shown in Figure 24-3.

In this example, calculating the metric for the route from Miura to Net 4 is fairly simple. First, you find the lowest bandwidth in the path, which is the serial link between Urraco and Jalpa. You then divide 10,000,000 by the bandwidth in bps of that link to create the final bandwidth figure, 19,531. You then add up all of the delays on all of the *outgoing interfaces* to create a delay score of 40,630 (20,000 + 20,000 + 630). You then divide this delay score by 10 (because delay is calculated for the metric in units of ten microseconds) for a total of 4,063. Finally, you add the bandwidth to the delay to achieve a total metric of 23,594.

You need to be aware of a couple of points here. First, notice how the outgoing interface is used to achieve the delay score. This point is important, because it means that the path from Net 1 to Net 4 will have a metric of 23,594, but the return path has

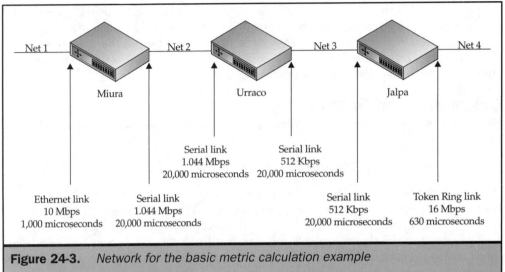

Figure 24-3. *Network for the basic metric calculation example*

a metric of 23,631 (19,531 (bandwidth) + 2,000 + 2,000 + 100 (delays)). Keep in mind that in a large, complex network, a packet might take a different path to return than it took to arrive. Second, notice how important bandwidth is in this example. Delay was responsible for 4,063 points (about 20 percent) on the metric, while bandwidth accounted for a whopping 19,531 (75 percent). The moral of the story is that, with the default weights, delay along the path typically has minimal impact compared to bandwidth.

If you wanted to reduce bandwidth's role in metric determination, you would either increase the delay weight (K3) or increase the delay settings for each link. Because you are already using the recommended delay figures, let's increase K3 and examine the results.

With this example, if you set K3 to equal 2, then your simplified formula would change to metric = Be + (2 × Dc). In this case, the bandwidth computation is 19,531, while the delay is 8126. That's not quite good enough. If you increase the weight to 4, however, you have a much better match. In this network, this would give you a metric of 35,783, with approximately a 50/50-percent split between bandwidth (19,531) and delay (16,252). This split is pretty good if you want delay to make a large impact on route selection. Now, in this network, if you added a slightly higher speed (like 1.544 Mbps) but a high-latency link (for instance, a VPN tunnel through another network) between Urraco and Jalpa, it would be not preferred over the standard serial link.

Okay, now that you have learned how to handle the weights and values for bandwidth and delay, what about load and reliability? Well, first, load is not typically used (and highly discouraged) because load is not controllable, and a transient high load measurement can cause a route to fail. However, because load can only double

Tips for Using Weights

Cisco recommends that you leave the weights at their default settings, mostly because if you don't understand the consequences involved with changing them, you might make poor routing choices. I generally break with Cisco a bit on this subject, however, and recommend that you understand not only the math behind metric calculation, but also the needs (latency vs. throughput) of your network applications. If you feel you have a firm grasp of these concepts, I suggest computing an optimal weighting scheme and testing the scheme in a lab environment to see whether you can obtain a performance improvement. If you cannot set up a lab (due to costs or other constraints), then choose a time of little or no activity and test then. One way or another, be very careful about changing weights in a production network until you are *sure* the results will be favorable.

Be aware that if you change the delay (K3) weight, that delay is the *sum* of all delays between the source and the destination. In a large network, you may actually have to increase the bandwidth (K1) weight instead of the delay (K3) weight, because the cumulative delays may be very large indeed. For instance, if you had 22 serial hops and a 512-Kbps slowest link, with the default K values, delay would be responsible for 44,000 points of the metric (2,000 × 22), while bandwidth is responsible for only 19,531 points.

the total metric (with a K2 value of 1, anyway), I don't typically consider this to be the main problem.

The primary problem with either load or reliability is that they tend to cause significant traffic through *flash,* or triggered, updates. These flash updates cause the remote routers to place the route in holddown and, in some cases, remove the route from the table entirely. (The best route can become the worst route very quickly, and with little or no warning.) To illustrate, change the K2 weight to 1 for the previous example, and examine the resulting formula and metric calculation. The formula that would result (with the default weights for K1 and K3) would be metric = bandwidth + (bandwidth/(256 − load) + delay. If the load calculation were the lowest possible (remember, load is not set, it is measured, so you have no control over this), the metric would be 19,531 + (19,531/255) + 4,063 = 23,671. This figure is almost the same as before, right? Yes, but what if the load rises to its maximum (255)? Then you have 19,531 + (19,531/1) + 4,063 = 43,125. This figure is a big difference.

A second side effect of this metric change is that it will cause a triggered update. And because load is updated every five seconds, if you had a really "bursty" line, load has the ability to cause triggered updates every five seconds, which is bad news. Furthermore, if you give load a high weight (by setting the K2 weight), like 1000, serious problems may result. With this kind of weight, the formula changes to this: bandwidth + (bandwidth × 1,000/(256 − load) + delay. When you plug in the numbers, it looks like this (with a load of 255): 19,531 + (19,531,250/1) + 4,063 = 19,554,844.

CISCO ROUTING

Pretty significant jump, huh? What's worse is that with a load of 1, it looks like this: 19,531 + (19,531,250/255) + 4,063 = 100,187. This leads to a metric that has a pretty wide range, all controlled by a factor that you can't manually modify. The end result is that problems tend to occur at the worst possible times.

Reliability can cause the same constant triggered updates problem as load, although it normally doesn't. Because most lines today are relatively error-free, you will likely see little or no difference in the metric due to reliability. However, because reliability multiplies the rest of the calculated metric, on an error-prone link, reliability has the ability to single-handedly cause a route to be removed because it can cause a metric to be higher than infinite (4,294,967,295).

Reliability also uses two weights, K4 and K5, to attempt to "scale" the calculation to be more or less powerful. For instance, in the previous example, with the bandwidth (K1) weight set to 1, the load (K2) weight set to 1, and the delay (K3) weight set to 1, if you set the reliability weights (K4 and K5) to 1, the resulting formula will be (bandwidth + (bandwidth/(256 – load) + delay) × (1/(reliability + 1)). With a maximum reliability score (255) and a minimal load (1), plugging in the numbers ends up looking like this: (19,531 + (19,531/255) + 4,063) × (1/(255 + 1)) = 92. Whoa, big jump, huh? In this case, maximum reliability cuts the metric down to practically nothing. However, with a minimum reliability score (1) and a maximal load (255), plugging in the numbers ends up looking like this: (19,531 + (19,531/1) + 4,063) × (1/(1+1)) = 21,563. As you can see, reliability can greatly impact your metric calculations. Therefore, I suggest that you do not use reliability in your metrics.

Advanced Metric Calculation So what do you do if you actually need to use all of the metric components to calculate your metrics? First, realize that *all* of the metrics are very rarely needed. But sometimes, you might need to use all of the metric variables in the metric calculation. Take the remote office network shown in Figure 24-4, for example.

Note *With the possible exception of the CCIE lab exam, you will most likely never need to understand IGRP or EIGRP metric calculation to the degree covered here, because almost no one goes to all of this trouble in real life. I am including the subject just so you have a better understanding of how the variables in a metric work together to create the total metric.*

In this case, a router at a distant location needs to have full redundancy in its connection to the network at HQ. In this location, let's say a large number of huge file transfers need to take place. In this scenario, you want the router to use the highest speed link possible for the route to HQ, but you want load on the link to be taken into account. In addition, because some of the links are not very reliable (the satellite link, for instance), you want reliability to be accounted for. Finally, delay, although important, is not nearly as important as unused bandwidth for this scenario.

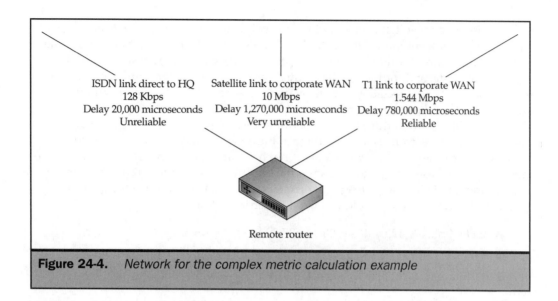

ISDN link direct to HQ
128 Kbps
Delay 20,000 microseconds
Unreliable

Satellite link to corporate WAN
10 Mbps
Delay 1,270,000 microseconds
Very unreliable

T1 link to corporate WAN
1.544 Mbps
Delay 780,000 microseconds
Reliable

Remote router

Figure 24-4.   *Network for the complex metric calculation example*

Tip   *If you want to perform these calculations, the best strategy is to build a spreadsheet that allows you to punch the numbers into the various variables until you achieve the desired results. Attempting to perform this math using pencil, paper, and a standard calculator is a frustrating and time-consuming process. For a free Excel spreadsheet that can be used to perform these calculations, visit http://www.alfageek.com/mybooks.htm.*

You want to make sure that the satellite link is used as the primary link if it has ample bandwidth and is reliable, because it is the fastest link. If there is a reliability issue or most of the bandwidth is being used on the satellite link, you want the T1 to be used. If the T1 is having reliability difficulties or is heavily overused, you want the ISDN link to be used. The problem is that, with the default weights, you would end up using the ISDN link much more than the other two links, which is *not* the idea.

Let's plug the math into the equation. (Note that, in this example, delay shown on the links in question is the *cumulative* delay for the route to HQ, *not* the individual delay settings for each link.) With default settings, the K1 and K3 values are 1, and K2, K4, and K5 are all 0, making the final equation simply bandwidth + delay = metric. In this scenario, the ISDN link would have a metric of 80,125 (78,125 + 2000), the satellite link would have a metric of 121,000 (1,000 + 120,000), and the T1 would have a metric of 84,477 (6,476.7 + 78,000). Therefore, the ISDN line would be preferred. Several solutions exist to this problem. (The Cisco suggested solution is to adjust the delays; but in this case, we will assume the delays are accurate.) However, the only solution that accomplishes all of the goals is to modify the weights.

To eliminate the initial problem of the ISDN line being preferred over the faster T1 and sat links, you could modify the delays on the sat and T1 links (again, this is the Cisco-recommended solution). But your problem is that, for the sat link, the delay figure is accurate; and for the T1, the delay figure is caused by the route having to be sent through 30+ routers, all with a delay of 20,000—leading to a massive cumulative delay. To reduce the delay figure on the T1, you would either need to purchase a PVC directly to the HQ (to reduce the number of hops in the path) or change the delay figures on *all* of the routers in the path (a solution with unintended consequences, because other routers in the WAN that do not have this specific problem also use these delay figures). However, if you change the bandwidth (K1) weight on the remote router, it will not adversely affect other routers and it will accomplish the goals. By changing the K1 to 8, you would change the metrics as follows:

- ISDN link: $(78{,}125 \times 8) + 2{,}000 = 627{,}000$
- Sat link: $(1{,}000 \times 8) + 120{,}000 = 128{,}000$
- T1 link: $(6476.7 \times 8) + 78{,}000 = 129{,}813$

Bingo. Initial problem solved. The sat link is now preferred, with the T1 second, and the ISDN is a long last. However, now you need to take load into account. If you change the load (K2) weight to a nonzero integer, you end up with the following formula: $(K1 \times bandwidth) + (K2 \times bandwidth)/(256 - load) + delay$. With a K2 of 1, your metrics would look like this under a minimal load:

- ISDN link: $(78{,}125 \times 8) + (78{,}125/(256 - 1)) + 2{,}000 = 627{,}306$
- Sat link: $(1{,}000 \times 8) + (1{,}000/(256 - 1)) + 120{,}000 = 128{,}004$
- T1 link: $(6{,}476.7 \times 8) + (6{,}476.7/(256 - 1)) + 78{,}000 = 129{,}839$

So, under minimal load, nothing really changes. Well, let's see what happens if you put the sat link under full load. Remember, the goal was to have the T1 be used if the sat link was saturated. Under full load for only the sat link, the metrics would be as follows:

- ISDN link: $(78{,}125 \times 8) + (70{,}000/(256 - 1)) + 2{,}000 = 627{,}306$
- Sat link: $(1{,}000 \times 8) + (1{,}000/(256 - 255)) + 120{,}000 = 129{,}000$
- T1 link: $(6{,}476.7 \times 8) + (6{,}476.7/(256 - 1)) + 78{,}000 = 129{,}839$

Close, but no cigar. Even under full load, the route through the sat link still has a lower metric than the route through the T1; therefore, the sat link will still be the primary link, even under times of heavy congestion. To solve this problem, you need to make the load a more important factor in the metric calculation by increasing the K2. To simply ensure that the route through the T1 will be the primary route if the sat link is fully saturated, you could just increase the K2 to 2. However, then the sat link will have to reach full saturation before the route through the T1 becomes the primary route.

Actually, if the sat link is more than 85-percent saturated (because the sat link is down to less than 1.5 Mbps at 85-percent saturation), and the T1 is unsaturated, the preferred primary route should be through the T1. To enable this setup, you need to increase the K2 to a level where the sat link at a load score of 217 (85 percent) becomes the secondary route. Unfortunately, because load is on a logarithmic scale (sort of like half of a bell curve), load doesn't really start significantly influencing the metric until it reaches 252 or higher.

To allow the route through the T1 to become the primary route once the sat link reaches 85-percent saturation, you have to increase the K2 to around 10,000. As a result, load become a *major* influence in route selection and, normally, would cause problems with excessive triggered updates. However, in this situation, the remote router is an edge device that doesn't re-advertise the routes to HQ back to any other routers (due to split horizon), so increasing the K2 won't cause any update or holddown problems. Once you increase the K2 to 10,000, the metrics with the sat link at a 217 load and the other links at 1 become the following:

- ISDN link: $(78,125 \times 8) + ((10,000 \times 70,000)/(256 - 1)) + 2,000 = 3,690,725$
- Sat link: $(1,000 \times 8) + ((10,000 \times 1,000)/(256 - 217)) + 120,000 = 384,410$
- T1 link: $(6,476.7 \times 8) + ((10,000 \times 6,476.7)/(256 - 1)) + 78,000 = 383,801$

Now let's see what happens if the T1 is saturated as well. (Remember, the goal was to have the ISDN link become the primary route if the T1 and the sat link were saturated.) Again, you are looking for the route through the ISDN line to become the primary if the sat link is over 99-percent saturated (253 load) *and* the T1 is over 92-percent saturated (235 load). The numbers, if both the T1 and sat link were saturated, are as follows:

- ISDN link: $(78,125 \times 8) + ((10,000 \times 70,000)/(256 - 1)) + 2,000 = 3,690,725$
- Sat link: $(1,000 \times 8) + ((10,000 \times 1,000)/(256 - 253)) + 120,000 = 3,461,333$
- T1 link: $(6,476.7 \times 8) + ((10,000 \times 6,476.7)/(256 - 235)) + 78,000 = 3,213,949$

You are very close, but still no cigar. At this level of saturation, the route through the T1 would still be the primary, the route through the sat link would be the secondary, and the ISDN would be a close third. So what do you need to do? That's right, increase the K2 yet again. Like before, due to the logarithmic nature of the load calculation, you will have to increase the K2 by a very large number to achieve your goals. In this case, you will have to increase K2 to around 300,000. Once you increase the K2 to 300,000, the metrics under the previous load values are as follows:

- ISDN link: $(78,125 \times 8) + ((300,000 \times 70,000)/(256 - 1)) + 2,000 = 92,538,765$
- Sat link: $(1,000 \times 8) + ((300,000 \times 1,000)/(256 - 253)) + 120,000 = 100,128,000$
- T1 link: $(6,476.7 \times 8) + ((300,000 \times 6,476.7)/(256 - 235)) + 78,000 = 92,653,870$

CISCO ROUTING

Whew! It's a good thing the infinite metric in IGRP is 4 billion! Now, let's add the final two weights (K4 and K5) and put reliability into the equation. In this case, you would like to use the sat link unless it is unreliable enough that the T1 will have a higher throughput (due to retransmissions required on the sat link). Again, with just reliability taken into account, the sat link would have to reach an 85-percent error rate for the route through the T1 to be faster. An 85-percent error rate on the sat link ends up as a 38 reliability rating. (Remember, reliability is backward, with 255 being the best and 1 being the worst.) Because the T1 is a reliable link, you don't need to be too concerned about reliability issues with it. Therefore, this equation will be a bit simpler than the load equation because you do not need to worry about failover to the ISDN due to T1 reliability.

> **Note** *In actuality, IGRP metrics are 24-bit, making the infinite metric 16.7 million, not 4 billion (which is the infinite value for EIGRP). However, this is a miniscule point in most environments, where the metrics typically never reach either maximum value. I chose to use 4 billion as the maximum value in this chapter in order to illustrate the use of extremely large metrics. Just keep in mind that metrics larger than 16 million are not actually valid in IGRP.*

Before modifying the reliability weights for the example, let's look at the details of the reliability calculation. When modifying the K4 and K5 weights, you need to realize the purpose of each and the mathematical relationship each has with the rest of the metric. The composite metric calculated up to this point is modified in its entirety by the reliability portion of the equation. Basically, the relationship boils down to this: (all of the other factors) × (reliability factors) = metric. So, reliability will modify everything else. As far as the reliability portion of the equation is concerned, K4 and K5 influence each other.

Just for review, the formula in the reliability portion of the equation is K5/(reliability + K4). You can use K4 to decrease the metric range based on reliability (in other words, K4 sets how important reliability is), and K5 to increase or decrease the final metric. For instance, if your base metric (before setting K4 and K5) is 500,000, you set K4 and K5 to be 1, and then you compute the metric at maximum reliability (255), your final formula will be $(500,000) \times (1/(255 + 1))$, for a final metric of 1,953. If you set K5 to be 2, however, your formula becomes $(500,000) \times (2/(255 + 1))$, for a final metric of 3,906. To understand K5, think of it as a metric *scalar*, meaning that it can scale the final metric up or down to match it to your environment. K4, on the other hand, tells you how important reliability is in the final metric calculation.

For instance, in the previous example, with a base metric of 500,000 and K5 and K4 values of 1, the metric with a reliability rating of 255 was 1,953. However, with a reliability rating of 1, the metric increases to a whopping 250,000! If you set K4 to be 1,000, however, the metric with a 255 reliability is 398, while the metric with a 1 reliability is 500. In this manner, K4 determines how much reliability matters in the final metric. You must be very careful when matching K4 and K5 in your metric calculations, because if you set K5 too low and K4 too high, you will end up with extremely low metrics. For instance, with a base metric of 500,000, if you set K4 = 1,000,000 and K5 = 1, your metric will always be 1, regardless of the reliability of the link.

Now that you have seen how K4 and K5 modify the metric, let's configure the reliability settings for our "Complex Metric Calculation" example. So far, in my example, I have decided on the following weights: K1 = 8, K2 = 300,000, K3 = 1. These weights provide the following metrics during periods of low congestion (load of 1):

- ISDN link: 92,538,765
- Sat link: 1,304,471
- T1 link: 7,749,442

These weights also provide the following metrics in times of total congestion (load of 255):

- ISDN link: 23,438,127,000 (unreachable, infinite distance)
- Sat link: 300,128,000
- T1 link: 1,943,134,995

Again, the main goal is to allow the route through the T1 to become the primary route if the sat link experiences error rates of greater than 85 percent (reliability rating of 38 or lower). Secondarily, you also want to reduce the metrics to numbers that are within the realm of mere mortals. Finally, you want to ensure that the best route will be chosen based on the combination of available bandwidth and reliability. For this exercise, I chose a K4 value of 5 and a K1 value of 1. When you plug those figures into the equation, with no congestion, if the sat link has a reliability rating of 38, the formulas work out as follows:

- ISDN link: $92,538,765 \times (1/255 + 5) = 355,918$
- Sat link: $1,304,471 \times (1/38 + 5) = 30,337$
- T1 link: $7,749,442 \times (1/255 + 5) = 29,806$

Notice that the route through the sat link becomes a higher-cost route than the T1 link if the reliability of the sat link drops to 38 or lower. If the sat link also had congestion, the metric differences would be even more pronounced. For instance, if the sat link were experiencing 50-percent congestion as well (load of 128), the metric for the sat link would be 57,483.

The final formula is $(8 \times \text{bandwidth} + (300,000 \times \text{bandwidth})/(256 - \text{load}) + \text{delay}) \times (1/\text{reliability} + 5)$. The final weights are K1 = 8, K2 = 300,000, K3 = 1, K4 = 5, and K5 = 1.

Load Balancing

IGRP load balances across equal-cost links, similar to RIP. However, due to IGRP's advanced metric, you are unlikely to ever get two completely equal metrics for the same destination (unless, of course, your weights are set incorrectly). Therefore, IGRP includes a function called *variance* to define how different two metrics can be and still be considered "equal cost." Variance defines how far off a redundant path to a destination can be and still be considered for load balancing. Variance is listed as a multiple of the lowest-cost path to the destination. For instance, if your router receives two routes to the 10.0.0.0 network and applies two different metrics to those routes, normally, only the lowest-cost route would be used. However, with variance, if the metric for the lowest-cost route were 8,000, and your variance were 2, any route to the 10.0.0.0 network that had a metric of 15,999 or lower would be accepted into the table, and IGRP would load balance between the two paths.

However, because the metrics are different, IGRP would not perform equal-cost load balancing, like RIP does. Rather, IGRP would send a percentage of traffic down each link, relative to the metrics on each route. For instance, if you had two paths to the same destination, one with a metric of 5,000, and the other with a metric of 15,000 (assuming a variance greater than 3), 75 percent of the traffic would be sent using the 5,000 metric route, and 25 percent would be sent using the 15,000 metric route. Variance allows IGRP to perform unequal-cost load balancing. However, the default setting for variance is 1, meaning that unequal-cost load balancing is disabled.

A few rules are associated with IGRP unequal-cost load balancing:

■ The maximum paths (as in the *maximum-paths* command discussed in Chapter 23) defines how many paths IGRP will load balance over. The default is 4.

■ All routes to a given destination must be less than the lowest-cost route multiplied by the variance. For example, if the lowest-cost route is 25,000 and the variance is 4, all routes to the same destination with a metric of 99,999 or less will be considered acceptable for load balancing.

■ The next hop router listed in the route entry must have a lower metric to the destination network than the lowest metric route to the destination network on the local router.

These rules are fairly simple except the third rule. To help explain this rule, take a look at the example network, shown in Figure 24-5.

In this example, assume you have set the variance to 4, and you are using the default weights (K1 and K3 = 1, all others = 0). Testarossa will hear of two routes to the 10.0.0.0/8 network—one from Mondial and one from GTO. The route from GTO will have the lowest metric, 8,577. The route from Mondial will have a metric of 15,131, which is higher than the metric from GTO but within the variance limit (34,307). However, Testarossa will not add the route from Mondial to the routing table and perform load balancing because of the third rule. Testarossa's lowest metric to the 10.0.0.0/8 network is 8,577, but Mondial's metric for the route it advertised to Testarossa for the 10.0.0.0/8 network is 15,121.

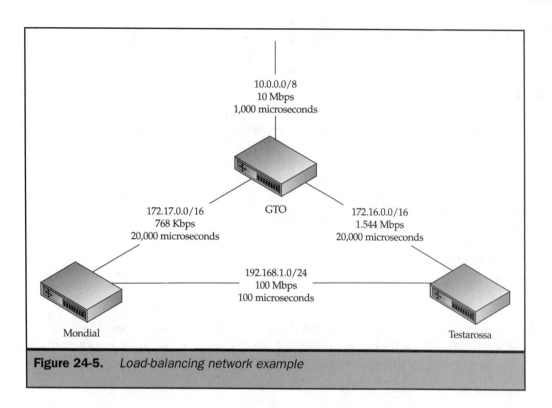

Figure 24-5.   *Load-balancing network example*

The third rule states that the route will not be used if the metric on the upstream router is not lower than the *best metric* on the downstream router. Consequently, the route through Mondial will not be used. The idea behind this rule is to keep routers from forwarding packets to other routers that are actually farther away from the destination, but this rule can cause problems in certain situations. In this case, Cisco recommends that you change either the bandwidth or delay settings on the problem router. However, if you change the bandwidth on Mondial to 1.5441 Mbps (the minimum required to bypass the rule), although Testarossa will begin to properly load balance across the link, it will do so at the wrong percentage. (It will send around 50 percent of all packets through Mondial.)

Even worse, Mondial will cease load balancing to the 10.0.0.0/8 network through Testarossa (because Mondial's best metric will be 8,576 and Testarossa's best metric will be 8,577). In this case, the bandwidth disparity is too wide to reduce the delay and achieve the desired result; but even if you could, you would end up with the same problem. The moral of the story is that IGRP load balances across unequal-cost routes— but it does so for the router with the worst metrics, not the routers with the best metrics. For instance, with the original settings, while Testarossa would not load balance, Mondial would. In fact, Mondial's lowest metric, in this case, was through Testarossa (with a metric of 8,587). The direct connection to GTO was a secondary route (with a metric of 15,121), and Mondial would send approximately 70 percent of the traffic through Testarossa and 30 percent of the traffic across the direct link.

Timers

IGRP uses fundamentally the same timers as RIP, but the timers are set to much longer intervals. Longer intervals have both strengths and weaknesses. As mentioned in the discussion of RIP, longer intervals mean that less overhead is required for routing updates, but they also mean that convergence takes longer. IGRP combats the convergence problems by using triggered updates, however, so convergence time is typically not a major issue (unless a router completely goes down). The default timers in IGRP are as follows:

- Update: 90 seconds
- Invalid: 270 seconds
- Holddown: 280 seconds
- Flush: 630 seconds

In addition, IGRP varies the update timer by up to 20 percent to prevent all of the routers from flooding the network with updates at exactly the same time. As a result, the actual update time (with the default timers) is between 72 and 90 seconds. Like RIP, timers in IGRP are set on each individual router, and you should take care to ensure that they are all the same.

IGRP Configuration

Now that you have seen how IGRP works, you're ready to learn about its configuration options. Like RIP, the minimal requirement to get IGRP running is the use of two statements: *router igrp* and *network*. However, because IGRP's metric takes a few other parameters into account, you will usually want to set some advanced parameters. Let's begin with the basics, however.

Basic Configuration

To configure IGRP, you first need to enable it and enter your network statements. To enable IGRP, simply type *router igrp [as number]* from global config mode, like so:

```
3660A (config)# router igrp 1
3660A (config-router)#
```

The AS number can be any number from 1 to 65,535, but you must be sure to use the same AS number for all routers in a given AS. You can also enable multiple IGRP processes on a given router by entering multiple *router igrp* statements and assigning different networks to different AS numbers. This topic is discussed further in the next section.

To tell IGRP which networks to broadcast updates on, listen for updates on, and advertise, you must use the *network [network number]* router configuration command, just like in RIP, like so:

```
3660A (config-router)# network 172.16.0.0
3660A (config-router)#
```

Just like RIP, this command accepts only the class-based network. If you enter a subnet of a class-based network, IGRP simply uses the entire class-based network.

Note   *Like RIP, you can also use the* passive-interface *command to control routing updates exiting the interface.*

Once you are finished with your network statements, you should make sure the bandwidth and delay settings on your interfaces are correctly set. You can see the bandwidth and delay settings with the *show interface* command. (Refer to Chapter 15 for a full description of the *show interface* command.)

```
3660A # show interface fastethernet 0/0
FastEthernet0/0 is up, line protocol is up
  Hardware is AmdFE, address is 0001.968e.fce1 (bia 0001.968e.fce1)
  Internet address is 172.16.0.200/16
  MTU 1500 bytes, BW 100000 Kbit, DLY 100 usec,
     reliability 255/255, txload 1/255, rxload 1/255
>DETAIL REMOVED<
```

The default settings for bandwidth are 10,000 Kbps for Ethernet, 100,000 Kbps for Fast Ethernet, 1,000,000 Kbps for Gigabit Ethernet, and 1,544 Kbps for serial links. The default delay settings are 1000 microseconds for Ethernet, 100 microseconds for Fast Ethernet, 10 microseconds for Gigabit Ethernet, and 20,000 microseconds for serial interfaces. To change the bandwidth on an interface, use the *bandwidth [speed in Kbps]* interface config mode command. To change the delay on an interface, use the *delay [tens of microseconds]* interface config mode command. Both of these commands are shown in the following output. (Note that the bandwidth is changed to 768 Kbps and the delay is changed to 10,000 microseconds.)

```
3660A (config)# interface serial 0/0
3660A (config-if)# bandwidth 768
3660A (config-if)# delay 1000
```

CISCO ROUTING

The bandwidth *and* delay *commands do not change the properties of the interface itself. In other words, setting a bandwidth statement for 5 Kbps on a 10-Mbps Ethernet interface does* not *change the speed of the interface. It simply tells routing protocols (IGRP, in this case) how the bandwidth parameter of the metric should be calculated.*

That's it. Once you have performed these steps, you have successfully set up basic IGRP. Now let's move on to some more advanced configuration topics.

Advanced Configuration

Advanced IGRP configuration covers a lot of the same ground as advanced RIP configuration, with a few new items of interest. First, you'll learn about the settings that are the same, and then you'll move on to IGRP-specific configurations. This section covers the following topics:

- Configuration tasks common to RIP and IGRP:
  - Configuring passive interfaces (disabling broadcasted updates)
  - Configuring unicast updates
  - Adding metric offsets to routes
  - Adjusting IGRP timers
  - Disabling split horizon
  - Setting the maximum number of paths
- IGRP-specific configuration tasks:
  - Setting variance
  - Controlling load balancing
  - Adjusting metric weights
  - Setting the network diameter
  - Setting a default network
- Route redistribution:
  - Redistributing IGRP into other IGRP ASs
  - Redistributing IGRP into RIP
  - Redistributing RIP into IGRP

Configuration Tasks Common to RIP and IGRP

The tasks covered in this section are common to both RIP and IGRP, and the commands used in IGRP do not differ significantly from those used in RIP. Therefore, this section simply provides a brief overview of each topic and the commands used. For full coverage of these commands, please refer to Chapter 23.

Configuring Passive Interfaces   Passive interfaces in IGRP are exactly the same as passive interfaces in RIP. Broadcasted updates are not sent out of a passive interface, but updates are listened for and the network(s) associated with the passive interface are advertised to other routers. You configure passive interfaces with the *passive-interface* router config mode command.

Configuring Unicast Updates   Unicast updates in IGRP, like RIP, are used to configure the IGRP router to advertise routes to specific IGRP routers using a unicast. This feature is usually combined with the passive interface feature to reduce network broadcasts. To configure unicast updates, use the *neighbor* router config mode command.

Adding Metric Offsets to Routes   Just like RIP, you can add a certain amount to all routes that match the criteria set forth in the offset-list. Note that, unlike RIP, IGRP on each router calculates its unique metric. The *offset-list* router configuration mode command simply adds a value to the final computed metric to increase the metric.

Adjusting IGRP Timers   IGRP timers are the same as RIP timers (except, of course, they are longer by default), except for one timer: sleeptime. IGRP allows an additional, optional sleeptime timer to be set for flash updates. The sleeptime timer postpones scheduled updates in the event that a flash update is sent. The sleeptime timer is normally disabled, but it can be set (in milliseconds) with the same command used to set all of the other IGRP timers: *timers basic*.

Disabling Split Horizon   Just like RIP, split horizon is enabled for most interfaces by default in IGRP. To disable split horizon for a specific interface, use the *no ip split-horizon* interface config mode command.

Setting the Maximum Number of Paths   Just like RIP, IGRP will, by default, perform load balancing over up to four equal-cost paths. You can set IGRP to load balance over a maximum of six paths using the *maximum-paths* router config mode command.

IGRP-Specific Configuration Tasks

IGRP-specific configuration tasks are generally fairly straightforward. They are tasks that cannot be set in RIP and are unique to IGRP and EIGRP. Luckily, all of these features have been discussed previously in this chapter, so now you just need to learn how to configure them.

Setting Variance   Variance, as mentioned previously, defines which redundant routes IGRP will consider for load-balancing purposes. Variance is set as a multiple of the lowest-cost route to a given network. By default, variance is set to 1, which effectively disables unequal-cost load balancing. (IGRP still load balances across up to four paths with *exactly* the same metric, however.) You set variance with the *variance [multiple]* router config mode command. The multiplier can be any value between

1 and 128, but you should be very careful about setting the multiplier too high. An example of setting the variance to be eight times the lowest metric is shown in the following output:

```
3660A (config-router)# variance 8
```

Tip   *If you are using load and/or reliability in your metric calculation, make sure to take the full range of possible metrics into account before setting the variance. Failure to do so can lead to routes being spontaneously dropped and/or added to the table based on transient metric changes.*

Controlling Load Balancing   In addition to setting the variance, you can also disable the load-balancing function of IGRP while allowing IGRP to enter multiple routes to a given destination in the routing table. This functionality is useful in situations when you want IGRP to use *only* the best path to a given destination, yet recover quickly in case that route fails. Normally, if a route fails and it is the only route in the table, the route would be placed in holddown, and you would need to wait for the holddown timer to elapse before the backup path could be learned of and used.

By using the *traffic-share* command, you can specify that IGRP enter multiple paths (that meet the load-balancing rules mentioned previously in the "Load Balancing" section of the chapter) into the routing table, but only use the best route for forwarding packets. If that route fails, IGRP immediately begins using the next best route. The syntax for the *traffic-share* router config mode command is *traffic-share [balanced | min]*. To allow load balancing, use the *traffic-share balanced* version of this command (the default setting); and to disable load balancing (but still enter alternate routes into the table), use the *traffic-share min* version of this command.

Adjusting Metric Weights   In addition to setting global load-balancing parameters, you may need to adjust the weights applied to the components used to define IGRP metrics. Any changes to the weights in IGRP apply only to the IGRP routing process for the specific AS it is changed for on that specific router. In other words, changes made to the weights on Router 1 for AS 5 do not apply to Router 2 and do not apply to AS 2 running on Router 1. The router config mode command that you use to set the IGRP weights is *metric weights [tos weight] [K1 weight] [K2 weight] [K3 weight] [K4 weight] [K5 weight]*.

All of the weight parameters can be set to any value between 0 and 4 billion except for TOS, which is always zero. (TOS was meant to be used to allow IGRP to do type of service routing, but was never fully implemented.) The default weights are K1 = 1, K2 = 0, K3 = 1, K4 = 0, and K5 = 0. The following example shows setting the weights for AS 2 to be K1 = 1, K2 = 2, K3 = 3, K4 = 4, and K5 = 5:

```
3660A (config)# router igrp 2
3660A (config-router)# metric weights 0 1 2 3 4 5
```

At the risk of sounding like a broken record, be very careful when changing the weights used for IGRP to nondefault settings. Modify the weights only when you have considered the impact of the change on metric calculations as well as load balancing on your routers, and have fully tested your theories to quantify the benefits.

Setting the Network Diameter   You may also want to change the *network diameter*, or the maximum number of routers between any two destinations. By default, the maximum network diameter in IGRP is 100, although this number can be increased up to 255 using the *metric maximum-hops [number of hops]* router config mode command. Realize, however, that IGRP performance suffers in extremely large networks. Rather than increasing the diameter, the general suggestion is to use multiple IGRP ASs and redistribute the routes (or change to EIGRP or OSPF). An example of setting the maximum network diameter to 200 hops is shown here:

```
3660A (config-router)# metric maximum-hops 200
```

Setting a Default Network   Finally, you will usually want to set a default network for IGRP. IGRP, unlike RIP, does not advertise the 0.0.0.0 network as the default route. IGRP instead uses a default network. If packets do not match any other route in the table, they are sent to the configured default network. In this way, the default network in IGRP and the default route in RIP are essentially the same. You set the default network for IGRP with the *ip default-network [network number]* global config mode command. You typically set the default network on only the border router for your IGRP network. IGRP then propagates this route to all other routers in the AS as an "external route," and the other routers will begin to use the route as their default network. An example of setting the default network to 172.16.0.0 is shown next:

```
3660A (config)# ip default-network 172.16.0.0
```

 Remember that the IP classless *command is required in order to use a default network or default route.*

Route Redistribution

Route redistribution is a commonly configured feature in complex, multiprotocol networks, and—in some cases—even single-protocol networks. Router redistribution allows you to advertise routes from one routing protocol or AS into another routing protocol or AS. The redistribution features in Cisco routers allow you to achieve full connectivity in environments with mixed routing protocols, and they give you an incredible amount of control over which routes are redistributed and how metrics are applied to those routes.

Regardless of whether you are redistributing between different routing protocols or between different ASs of the same protocol, route redistribution can be done in one of four basic ways:

■ One-way redistribution with a single border router

■ One-way redistribution with multiple border routers

■ Two-way redistribution with a single border router

■ Two-way redistribution with multiple border routers

For the first two scenarios, you will typically have what is known as a "stub" AS that must redistribute routes to the network core to achieve connectivity with other stub ASs. A *stub* AS (which is an adaptation of the OSPF term *stub area*, covered in Chapter 26) is an AS that does not have direct knowledge of any other ASs besides the backbone or *core* AS. All inter-AS routing for a stub AS must be transmitted through the core AS. To help explain this whole concept a bit better, let's examine the example network shown in Figures 24-6 and 24-7, and walk through configuring multi-AS IGRP redistribution with this example.

IGRP into IGRP: One-Way Redistribution

Figure 24-6 shows a network diagram from a high-level view. In this figure, the network is in a hub-and-spoke WAN topology, with each remote site having its own AS, and each AS is connected to the core AS (AS 1) only by redundant WAN links.

This topology is very common in hierarchical network designs, because it reduces router and network overhead over WAN links by splitting off each physical location into its own AS. (Remember, routing updates from one AS are not automatically propagated to other ASs.)

Figure 24-7 shows a closeup view of AS 10. AS 10 has two internal routers routing for the subnets of the contiguous 172.16.0.0 network. The routers are connected to each other with a direct link, as well as connected to each other using the border router.

Because all routers are directly connected using a subnet of the 172.16.0.0 network, they will advertise full subnet information to each other (an internal route in IGRP). IGRP doesn't advertise the masks, of course, which is why the subnets of this network are all using the /20 mask. Thus, every router in this LAN has full knowledge of every route in the LAN. However, they would have no knowledge of any of the other LANs, and except for the border router, they would have no knowledge of the core network, because they are in different ASs.

This arrangement is actually what you want. For the internal routers to have knowledge of the core and/or the other LANs, they would have to be receiving routing updates about these networks across the WAN link. You would also have to configure your timers to allow for convergence delays across the WAN. By splitting each office into its own AS, you can reduce routing updates across the WAN links, while at the same time providing for much lower timers on the LANs, allowing for faster convergence.

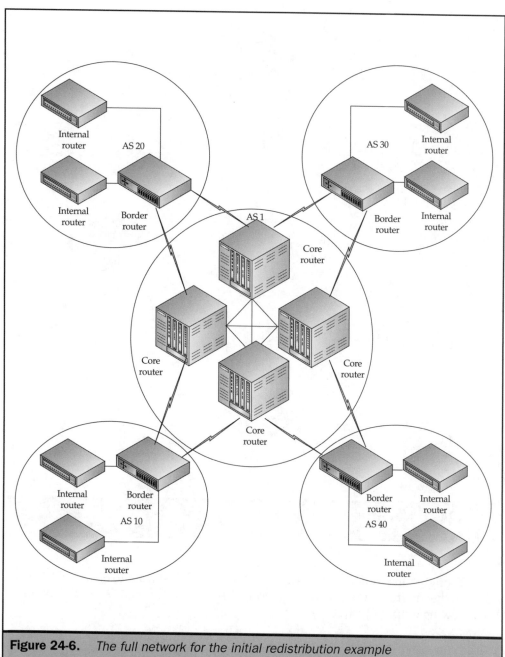

Figure 24-6. *The full network for the initial redistribution example*

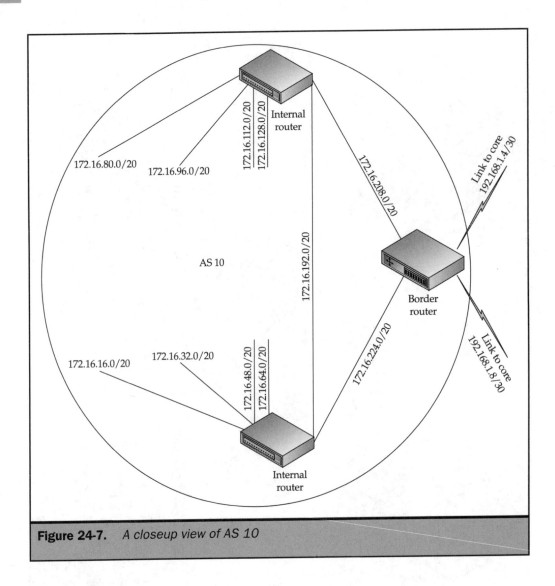

Figure 24-7. *A closeup view of AS 10*

To configure the routers in this scenario, you would first need to enable IGRP for AS 10 on the internal routers, and add a *network* statement for the 172.16.0.0 network. The following output illustrates:

```
AS10Internal(config)# router igrp 10
AS10Internal(config-router)# network 172.16.0.0
```

The border router, however, needs to be a member of both AS 10 and AS 1. For AS 10 on the border router, you would have a *network* statement for the 172.16.0.0 network *and* the 192.168.1.0 network. You add the 192.168.1.0 *network* statement because you need the internal routers to have a route to the 192.168.1.0 network, to use the 192.168.1.0 network as their default network. You could achieve the same results by redistributing AS 1 into AS 10, but in this situation, it is just easier to add the network to AS10.

For AS 1 on the border router, you would have a *network* statement only for the 192.168.1.0 network. You would also want to configure the links to the 192.168.1.0 subnets (serial 0/0 and 0/1) as passive interfaces for AS 10, because no members of AS 10 will be listening on the 192.168.1.0 network. Therefore, updates for AS 10 on the 192.168.1.0 network serve no purpose. This configuration is shown in the following output:

```
AS10Border(config)# router igrp 10
AS10Border(config-router)# network 172.16.0.0
AS10Border(config-router)# network 192.168.1.0
AS10Border(config-router)# passive-interface serial 0/0
AS10Border(config-router)# passive-interface serial 0/1
AS10Border(config-router)# router igrp 1
AS10Border(config-router)# network 192.168.1.0
```

You would then configure the 192.168.1.0 network as the default network using the standard *ip default-network* statement on the border router, as shown here:

```
AS10Border(config)# ip default-network 192.168.1.0
```

The fact that you set the *ip default-network* statement on the border router and not the internal routers is an important distinction, because the border router doesn't really *have* a 192.168.1.0 network. Instead, it has the 192.168.1.4 and 192.168.1.8 subnets of the 192.168.1.0 network. As a result, the border router can't use its own *default-network* statement.

However, due to IGRP's update rules, the 192.168.1.4 and 192.168.1.8 networks will not actually be advertised to the internal routers. IGRP instead sends a summary route (called a *system route* in IGRP) for the entire 192.168.1.0 network to the internal routers. (Refer back to the "Autonomous Systems" section if you are confused.) So, by setting the default network to be 192.168.1.0, you are actually changing the system route to the 192.168.1.0 network into an external route, and therefore setting the default network on all of the internal routers.

This strategy sounds a bit strange, I know—but believe me, it works (and it's much easier than typing the *ip default-network* statement on each individual internal router). For the final border router configuration task, you tell the border router to redistribute AS 10 to AS 1. To perform this task, you use a new command, the *redistribute* config mode command. Because this command is fairly complex, I will detail each piece of the command individually as it applies to IGRP, using the IOS online help as a guide.

```
AS10Border(config)# router igrp 1
AS10Border(config-router)#redistribute ?
  bgp         Border Gateway Protocol (BGP)
  connected   Connected
  egp         Exterior Gateway Protocol (EGP)
  eigrp       Enhanced Interior Gateway Routing Protocol (EIGRP)
  igrp        Interior Gateway Routing Protocol (IGRP)
  isis        ISO IS-IS
  iso-igrp    IGRP for OSI networks
  metric      Metric for redistributed routes
  mobile      Mobile routes
  odr         On Demand stub Routes
  ospf        Open Shortest Path First (OSPF)
  rip         Routing Information Protocol (RIP)
  route-map   Route map reference
  static      Static routes
  <cr>
```

The first point you should notice about this command is that I entered the command, in this case, for AS 1, not AS 10. You enter the *redistribute* command in router config mode for the AS you wish to redistribute the routes to (the destination AS), which, in this case, is AS 1. Second, notice that the primary target for this command is the routing protocol in question. Based on your choice of routing protocol, other options will begin to appear. For now, you are concerned with only the IGRP target; but notice that some of the other options allow you to choose not only other routing protocols, but static routes and connected networks as well.

Once you enter the *igrp* target, you need to enter the AS that you would like to redistribute (the source AS), as shown in the following output:

```
AS10Border(config-router)#redistribute igrp ?
  <1-65535>  Autonomous system number

AS10Border(config-router)#redistribute igrp 10 ?
  metric      Metric for redistributed routes
  route-map   Route map reference
  <cr>
```

Once you have chosen an AS to redistribute, you may simply press ENTER to continue, or you can choose to set the metric for all routes redistributed using this statement. This step is important, because this method is the only way IGRP has of assigning a metric to the routes you will redistribute. Typically, if you want to redistribute multiple ASs into this AS, all with the same base metric, it is easier to use the *default-metric* router config mode command, rather than including the metrics with each individual *redistribute* statement. The *default-metric* command instead specifies a base metric for all routes redistributed into this AS, regardless of source AS or routing protocol.

The syntax for the *default-metric* command is *default-metric [bandwidth in Kbps] [delay in tens of microseconds] [reliability] [load] [mtu in bytes]*. For instance, to set *default-metric* to equal a bandwidth of 512 Kbps, a delay of 20,000 microseconds, a reliability of 255, a load of 1, and an MTU of 1,500 bytes, you would enter the following command:

```
AS10Border(config-router)# default-metric 512 2000 255 1 1500
```

However, in this case (because you are redistributing only a single route), it is just as easy to enter the metrics from the *redistribute* statement. The syntax for the *redistribute* statement, in this case, is *redistribute igrp [as number] metric [bandwidth in Kbps] [delay in tens of microseconds] [reliability] [load] [mtu in bytes]*. An example of the completed *redistribute* statement for this network, using the same metric values as in the previous *default-metric* command, follows:

```
AS10Border(config)# router igrp 1
AS10Border(config-router)#redistribute igrp 10 metric 512 2000 255 1 1500
```

Note   *You must use either the* default-metric *command or add the metric using the* redistribute *command. If you do not, the routing protocol (with the sole exception of OSPF) will use an infinite metric (16 in RIP, –1 in IGRP) to advertise the route.*

You might have also noticed an option to use a route map statement in your *redistribute* statement. Route maps are normally used for policy-based routing; but they can also be used to selectively choose which routes are sent or accepted during redistribution, as well as to set metrics to different values based on properties (like destination network) of the update. Route map use in redistribution is beyond the scope of this book.

Tip   *If you are redistributing multiple ASs, you can also use the* default-metric *command to set the common metric; and for any AS that needs a better or worse metric applied, use the* redistribute *command to set the metric for that specific AS.*

After all of these steps, the final IGRP configuration for the border router appears as follows:

```
AS10Border(config)# router igrp 10
AS10Border(config-router)# network 172.16.0.0
```

```
AS10Border(config-router)# network 192.168.1.0
AS10Border(config-router)# passive-interface serial 0/0
AS10Border(config-router)# passive-interface serial 0/1
AS10Border(config)# ip default-network 192.168.1.0
AS10Border(config)# router igrp 1
AS10Border(config-router)# network 192.168.1.0
AS10Border(config-router)#redistribute igrp 10 metric 100000 100 255 1 1500
```

Notice that the default bandwidth metric component was set to 100 Mbps and the default delay metric component was set to 1,000 microseconds. This is because, at the point that the routes are redistributed into AS 1, the slowest link in the path is a Fast Ethernet connection.

On the core routers, the configuration takes a bit of a twist. First, you must look at how the core is set up. If the entire core is based on the subnetted 192.168.1.0 network, then the core configuration is fairly simple, but you will lose some of the benefits of using multiple ASs. For instance, imagine that the core is configured as shown in Figure 24-8.

For this configuration, you would simply have AS 1 configured and a network statement for the 192.168.1.0 network, like so:

```
Core(config)# router igrp 1
Core(config-router)# network 192.168.1.0
```

At this point, route redistribution would be fully functional (assuming, of course, that you performed all of these steps on all remote ASs). However, you would still be using an excessive amount of bandwidth on your WAN links. The reason for the excessive bandwidth use is that, because all of the routers involved in AS 1 (core and otherwise) are using subnets of the 192.168.1.0 network to communicate, all of the 192.168.1.0 subnets would be advertised down to your border routers.

To get around this problem, you could add two static routes to the 0.0.0.0 network on the border routers, one for each core connection, and then use the *passive-interface* command on the core routers for the interfaces connected to the border routers. This will keep the routing updates from the core from propagating down to the border routers, while still allowing the border routers to forward all traffic to the core. However, Cisco routers do not load balance over multiple static default routes; so if you add these static routes, you will lose your ability to load balance over the WAN links. The better solution is to change the IP structure in the core to something similar to what is shown in Figure 24-9.

Although changing the IP address scheme in and of itself would cut down slightly on the number of routes that are propagated down to the border routers, since the routes to other class-based networks will be sent as system (summary) routes to the

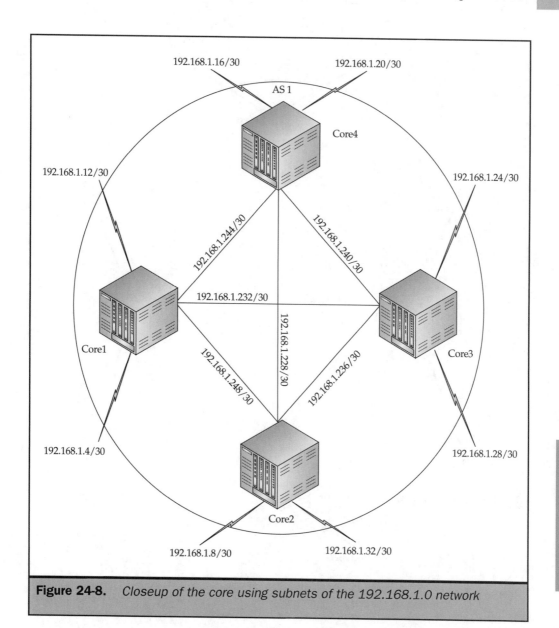

192.168.1.16/30   192.168.1.20/30

AS 1

Core4

192.168.1.12/30   192.168.1.24/30

192.168.1.244/30   192.168.1.240/30

192.168.1.232/30

192.168.1.228/30

Core1

192.168.1.248/30   192.168.1.236/30   Core3

192.168.1.4/30   192.168.1.28/30

Core2

192.168.1.8/30   192.168.1.32/30

Figure 24-8. *Closeup of the core using subnets of the 192.168.1.0 network*

border routers, it does not completely solve the problem. The ideal situation is to have each core router advertise only a single default (external) route to the border routers. To accomplish this task, you need to "filter" the routing updates sent from the core to the border routers by using the *distribute-list* command.

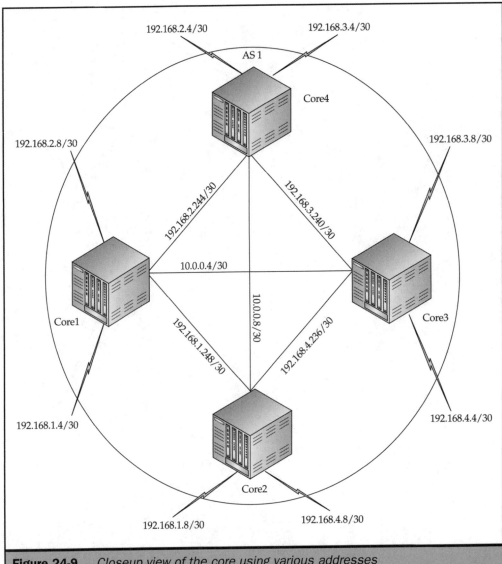

Figure 24-9. *Closeup view of the core using various addresses*

The *distribute-list* command uses standard access lists to deny or permit specific entries in the routing update. Because access lists aren't covered until Chapter 27, you have not learned about them in detail; but I will briefly explain access lists in the context of distribute lists so that you get the general idea.

Standard IP access lists (the kind used in this discussion) have basically four components: an access list number, a *permit* or *deny* statement, an IP address (or portion of an IP address), and a reverse mask. The number tells which access list the statement belongs to (you may have multiple statements per access list), the *permit* or *deny* statement tells whether the access list will permit or deny the routing entries matched by the access list statement, the IP address specifies the base IP address, and the reverse mask specifies the range of IP addresses the list will match.

The basic syntax of a standard IP *access-list* statement is *access-list [number of access list (1–99)] [permit | deny] [ip address] [reverse mask]*. The reverse mask portion is the complex part, but the basic rule is that any bits in the reverse mask that are binary 0 must *exactly* match the IP address entered in the *access-list* statement. For instance, the following access list, in the context of a distribute list, allows an update for the 192.168.1.0 network:

```
Core1(config)# access-list 1 permit 192.168.1.0 0.0.0.255
```

The reverse mask portion of this statement (the *0.0.0.255* section) tells the router that the first three octets of the IP address in the update must exactly match the first three octets of the IP address in the *access-list* statement. The permit allows the update. This statement by itself, however, would also deny all updates that do not match the statement. Access lists have an implicit *deny* statement at the end. Therefore, any update that did not match any of your configured *access-list* statements would be automatically denied.

Access-list statements sharing the same access-list number are also processed in the order entered, and only the first matching statement is used. This functionality can cause problems for the unwary. For example, if you wanted to allow all updates from the 192.168.1.0 network, deny all updates from any other 192.168 network, and then allow all other updates, there are three methods that you could use to formulate an access list to meet this criteria, as shown next:

Method 1:

```
Core1(config)# access-list 1 deny 192.168.0.0 0.0.255.255
Core1(config)# access-list 1 permit 192.168.1.0 0.0.0.255
Core1(config)# access-list 1 permit 0.0.0.0 255.255.255.255
```

Method 2:

```
Core1(config)# access-list 1 permit 0.0.0.0 255.255.255.255
Core1(config)# access-list 1 permit 192.168.1.0 0.0.0.255
Core1(config)# access-list 1 deny 192.168.0.0 0.0.255.255
```

CISCO ROUTING

Method 3:

```
Core1(config)# access-list 1 permit 192.168.1.0 0.0.0.255
Core1(config)# access-list 1 deny 192.168.0.0 0.0.255.255
Core1(config)# access-list 1 permit 0.0.0.0 255.255.255.255
```

Of these methods, only method 3 will work.

In method 1, the *deny* statement for the 192.168.0.0 network block is listed first. Therefore, if a route for the 192.168.1.0 network is checked against this list, it will match the first statement and be immediately denied. (Remember, access-list processing stops once a match, even a poor one, is found.)

In method 2, the first statement, *permit 0.0.0.0 255.255.255.255*, is basically the same as saying "permit everything." Because it is first, *all* routes will match this statement and be immediately allowed.

However, method 3 has the correct processing order. In this instance, all routes to the 192.168.1.0 network will be immediately allowed. If the route is not for the 192.168.1.0 network, but is a route for another network beginning with 192.168, the route will be dropped. Finally, all other routes are allowed.

Getting back to our problem, in this case, you would want to build an *access-list* statement that allows routes for the 10.0.0.0 network to be transmitted only to your border routers. I chose the 10.0.0.0 network because it is the only network that all core routers have in common. If you enter a *default-network 10.0.0.0* statement on your core routers, they will not use the statement (because they have no route for the 10.0.0.0 network, only routes for the 10.0.0.4/30 or 10.0.0.8/30 subnets of the 10.0.0.0 network). However, they *will* advertise the 10.0.0.0 network to the border routers as an external route, which will cause the border routers to use the 10.0.0.0 route as their default route.

To build your access list to deny all routing updates except updates advertising the 10.0.0.0 network, you would use the following *access-list* statement:

```
Core1(config)# access-list 1 permit 10.0.0.0 0.255.255.255
```

Then, you would need to configure your *distribute-list* statement. Distribute lists are router config mode commands that have the following basic syntax: *distribute-list [access-list number] [in | out] [(optional) interface] [(optional) routing protocol and AS number]*. To break this syntax down is fairly simple; each section of the command is described in the following list:

- *access-list number* is the access list that applies to this distribute list. Only one access list can be applied to a given distribute list.

- *in | out* specifies whether the distribute list should apply to incoming updates (updates from other routers) or outgoing updates (updates from this router).

- *(optional) interface,* if used, specifies that the distribute list applies only to a specific interface. Only one distribute list for incoming updates and one distribute list for outgoing updates can be applied to an interface. In other words, you may have a maximum of two distribute lists per interface, one for incoming updates and one for outgoing updates.

- *(optional) routing protocol and AS number,* if used, specifies that the distribute list applies only to a specific routing protocol and (optionally) a specific AS for that protocol. Only one distribute list for incoming updates and one distribute list for outgoing updates may be applied to any routing protocol or AS.

- If neither an interface nor a routing protocol/AS are chosen, the distribute list is applied to *all* interfaces and routing processes. This is called a global distribute list.

Because it is possible for different distribute lists to be applied to the routing protocol, to the interface, and globally, distribute lists are processed in the following order and based on the following rules:

1. Distribute lists applied to the interface that is either sending or receiving the update are checked. If no distribute list is applied to the interface, continue to step 2. If a distribute list *is* applied to the interface, and the routing update is allowed by the distribute list, it may continue to step 2. Otherwise, the route is put in the bit bucket.

2. Distribute lists applied to the specific routing protocol or AS that is either sending or receiving the update are checked. If no distribute list is applied to the routing protocol or AS, continue to step 3. If a distribute list *is* applied to the routing protocol or AS, and the routing update is allowed by the distribute list, it may continue to step 3. Otherwise, the route is put in the bit bucket.

3. Global distribute lists are checked. If there are no global distribute lists, the route can be processed by the routing protocol on the router. If a global distribute list *is* applied, and the routing update is allowed by the distribute list, the route can be processed by the routing protocol on the router. Otherwise, the route is put in the bit bucket.

Back to the example network. In this case, you do not want to apply the distribute list either to the routing protocol or globally, because you need updates for other networks besides the 10.0.0.0 network to be sent between the core routers. In this case, all you really want to do is apply the distribute list to the links to the border routers (on Core1, serial 0/0, and 0/1). To achieve this functionality, you build and apply the following access list and distribute list on Core1:

```
Core1(config)# access-list 1 permit 10.0.0.0 0.255.255.255
Core1(config)# router igrp 1
Core1(config-router)# distribute-list 1 out serial 0/0
Core1(config-router)# distribute-list 1 out serial 0/1
```

You then perform these same steps on all core routers. Once you have finished, the core routers will not forward any updates about networks other than the 10.0.0.0 networks to the border routers, but they will still listen for all routing updates on all interfaces and will still send updates about all known networks to each other.

Because this entire explanation has been a bit lengthy, let's take a look at all of the pertinent configurations for the routers in this example. If the reasons for adding any of these commands are not clear, refer back to the text in the example before moving on. The configurations for each type of router are shown next:

Internal:

```
AS10Internal(config)# router igrp 10
AS10Internal(config-router)# network 172.16.0.0
```

Border:

```
AS10Border(config)# router igrp 10
AS10Border(config-router)# network 172.16.0.0
AS10Border(config-router)# network 192.168.1.0
AS10Border(config-router)# passive-interface serial 0/0
AS10Border(config-router)# passive-interface serial 0/1
AS10Border(config)# ip default-network 192.168.1.0
AS10Border(config)# router igrp 1
AS10Border(config-router)# network 192.168.1.0
AS10Border(config-router)#redistribute igrp 10 metric 100000 100 255 1 1500
```

Core:

```
Core1(config)# router igrp 1
Core1(config-router)# network 192.168.1.0
Core1(config-router)# network 10.0.0.0
Core1(config)# access-list 1 permit 10.0.0.0 0.255.255.255
Core1(config)# router igrp 1
Core1(config-router)# distribute-list 1 out serial 0/0
Core1(config-router)# distribute-list 1 out serial 0/1
```

Whew, that was a pretty lengthy description for just redistribution from IGRP into IGRP using a single border router and one-way redistribution. Luckily, most of the information you will need for any other type of redistribution was included in that example. In fact, performing one-way redistribution of IGRP into IGRP with multiple routers simply involves performing the same steps as you performed in the previous example. For instance, to configure the example AS (AS 10) to use two border routers, you would just add another router and links, as shown in Figure 24-10, and mirror the configuration of the first border router (with changes to network numbers, of course) to the new router.

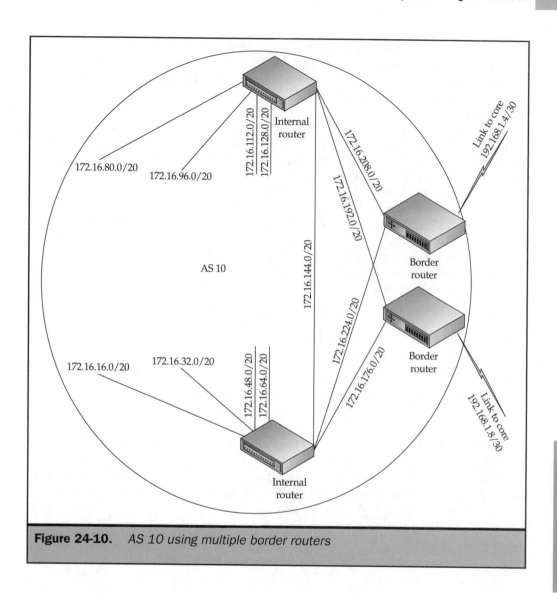

Figure 24-10. *AS 10 using multiple border routers*

With a multiborder router scenario like the one in this example, it would take a lot to kill connectivity! The disadvantage to all of this redundancy is cost, of course; but if you need high uptime, this strategy is the way to go.

IGRP into IGRP: Two-Way Redistribution   When redistributing different ASs of the same routing protocol, about the only reason you would want to perform two-way redistribution is when you want to allow the network to grow larger than the 255-hop limit—and even then, two-way redistribution just doesn't make much sense. If you need to perform two-way redistribution between multiple ASs of the same routing protocol (IGRP or otherwise), my advice is either to figure out a way to change all of the routers to use the same AS (because with two-way redistribution, you are not really gaining any bandwidth conservation benefits anyway), and use default routes and distribute lists to limit updates; or, alternatively, figure out a way to convert into a one-way redistribution scenario. For example, take a look at the network shown in Figure 24-11.

In this example, you might think that the only solution is to perform two-way redistribution because only one AS in the entire network could be classified as a stub AS. In addition, there is no definable "core" network that all traffic must go through, because this is not a hub-and-spoke topology. (Although this topology is actually a really strange mesh, the most technical definition for it is "an absolute mess.")

In the example in Figure 24-11, the easiest solution is to use one AS and not redistribute at all. If the network diameter (number of hops) or some other concern prevents you from using a single AS, you *could* still use one-way redistribution and default routes, but you are going to run into problems. For instance, take a look at the reconfigured network in Figure 24-12.

In Figure 24-12, redistribution will affect the ASs as follows:

- **AS 10** Includes routes from AS 10 and AS 20

- **AS 30** Includes routes from AS 20, AS 30, AS 70 (from AS 80), and AS 80

- **AS 40** Includes routes from AS 10, AS 20 (from AS 10 and AS 30), AS 30, AS 40, AS 50, AS 60 (from AS 50), AS 70 (from AS 50), and AS 80 (from AS 30)

- **AS 50** Includes routes from AS 50, AS 60, and AS 70

- **AS 80** Includes routes from AS 70 and AS 80

This configuration *will* work (at least until a link goes down); but you will lose some redundancy and load balancing, which is the whole reason we have this screwed up mesh topology instead of a hub-and-spoke in the first place. To understand this concept more fully, imagine that the link from AS 10 to AS 40 fails. If that one link were the

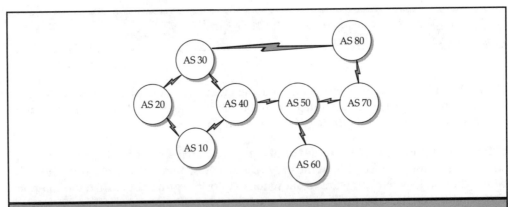

Figure 24-11. *A network that needs two-way redistribution*

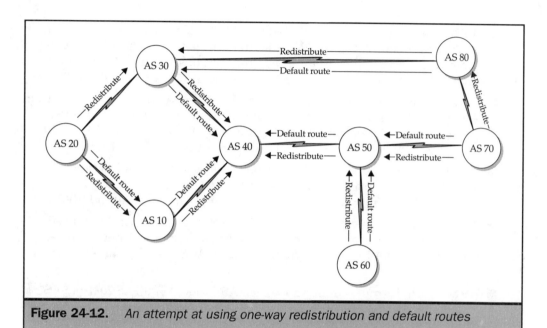

Figure 24-12. *An attempt at using one-way redistribution and default routes*

only link that failed, AS 10 *should* still be able to reach every network (including AS 40); but with this distribution scenario, the only networks AS 10 could reach are the ones in its AS and AS 20.

You could add an additional default route to AS 10 to point to a network in AS 20, and then AS 10 would send packets to AS 20 when the link to AS 40 failed, but AS 20 has no better idea of how to get to other ASs than AS 10 does. In this case, even adding another default route to AS 20 won't help because the router uses only one default route at a time, and its current default route (using the link to AS 10) is up. The only real solution is to perform two-way redistribution, as shown in Figure 24-13. Once you begin using two-way redistribution, your bandwidth usage for routing updates is going to skyrocket.

With two-way redistribution, the ASs will include the following routes:

- **AS 10**   Includes routes from AS 10, AS 20 (from AS 20 and AS 40), AS 30 (from AS 20 and AS 40), AS 40 (from AS 20 and AS 40), AS 50 (from AS 20 and AS 40), AS 60 (from AS 20 and AS 40), AS 70 (from AS 20 and AS 40), and AS 80 (from AS 20 and AS 40)

- **AS 20**   Includes routes from AS 10 (from AS 10 and AS 30), AS 20, AS 30 (from AS 10 and AS 30), AS 40 (from AS 10 and AS 30), AS 50 (from AS 10 and AS 30), AS 60 (from AS 10 and AS 30), AS 70 (from AS 10 and AS 30), and AS 80 (from AS 10 and AS 30)

- **AS 30** Includes routes from AS 10 (from AS 20, AS 80, and AS 40), AS 20 (from AS 20, AS 80, and AS 40), AS 30, AS 40 (from AS 20, AS 80, and AS 40), AS 50 (from AS 20, AS 80, and AS 40), AS 60 (from AS 20, AS 80, and AS 40), AS 70 (from AS 20, AS 80, and AS 40), and AS 80 (from AS 20, AS 80, and AS 40)

- **AS 40** Includes routes from AS 10 (from AS 10, AS 30, and AS 50), AS 20 (from AS 10, AS 30, and AS 50), AS 30 (from AS 10, AS 30, and AS 50), AS 40, AS 50 (from AS 10, AS 30, and AS 50), AS 60 (from AS 10, AS 30, and AS 50), AS 70 (from AS 10, AS 30, and AS 50), and AS 80 (from AS 10, AS 30, and AS 50)

- **AS 50** Includes routes from AS 10 (from AS 40 and AS 70), AS 20 (from AS 40 and AS 70), AS 30 (from AS 40 and AS 70), AS 40 (from AS 40 and AS 70), AS 50, AS 60, AS 70 (from AS 40 and AS 70), and AS 80 (from AS 40 and AS 70)

- **AS 70** Includes routes from AS 10 (from AS 50 and AS 80), AS 20 (from AS 50 and AS 80), AS 30 (from AS 50 and AS 80), AS 40 (from AS 50 and AS 80), AS 50 (from AS 50 and AS 80), AS 60 (from AS 50 and AS 80), AS 70, and AS 80 (from AS 50 and AS 80)

- **AS 80** Includes routes from AS 10 (from AS 30 and AS 70), AS 20 (from AS 30 and AS 70), AS 30 (from AS 30 and AS 70), AS 40 (from AS 30 and AS 70), AS 50 (from AS 30 and AS 70), AS 60 (from AS 30 and AS 70), AS 70 (from AS 30 and AS 70), and AS 80

The only real trick to two-way redistribution using a single routing protocol is metric settings. When using the *redistribute* command, you cannot specify which routes

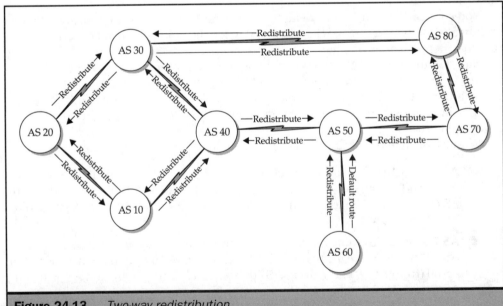

Figure 24-13. *Two-way redistribution*

get any particular metric (all redistributed routes are given the same metric), so issues can occur when re-redistributing routes. For instance, AS 40 will hear of routes within AS 20 from AS 10, AS 30, and AS 50. If all of the ASs are using the same metric for route redistribution, AS 40 will not know which path is the best path for packets to networks within AS 20. Based on IGRP's load-balancing mechanisms, the routers in AS 40 will attempt to load balance all packets to routes in AS 20 through the WAN links to AS 10, AS 30, *and* AS 50, not realizing that the path through AS 50 is a *much* longer path.

To solve this problem, you could increase metrics for redistributed routes from AS 50, but then AS 40 would think that the best path to routes within AS 60 is through AS 10 or AS 30 (because the *redistribute* command alone would assign the same metric to *all* redistributed routes). The only real solution to this dilemma is to use *route-map* statements to selectively set metrics on redistributed routes and give each route re-redistributed from each particular AS an appropriate metric. This solution is *extremely* time-consuming (especially if each AS has a lot of routes); but if you want to do two-way redistribution and choose the best path for packets (that is still a router's job, right?), it is your only choice.

Now that you have seen what a pain two-way redistribution is in a single routing protocol environment, you should have a better appreciation for single AS and hierarchical network designs. If you must keep a complex mesh-style topology, as shown in these examples, and you have too many hops to include all of the network routes in a single AS, I suggest changing the AS structure a little to allow for one-way redistribution and proper load balancing, as shown in the reconfigured network in Figure 24-14.

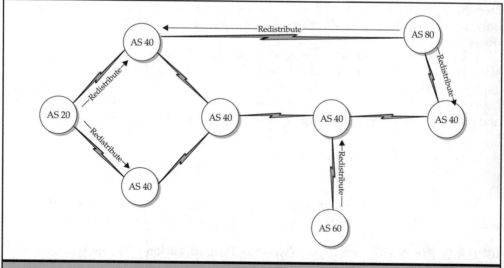

Figure 24-14.   *Changes to the AS structure to allow for one-way redistribution*

IGRP into RIP or RIP into IGRP: One-Way Redistribution Redistributing a routing protocol into another protocol is fairly common. There are several key reasons for redistribution among multiple protocols; but in this case (IGRP into RIP), the primary motivation is because of a merger or routing protocol migration. For example, let's say your company has been using another vendor's routers for years and using RIP as the primary routing protocol. However, you have acquired a smaller company that is using Cisco routers and IGRP for internal routing. You want to provide this new entity with access to your internal resources (and vice versa), as well as Internet access through your connection, without having to significantly reconfigure your internal network or the newly acquired network. Assume that you configure the network topology as shown in Figure 24-15.

In this network, the IGRP routers don't need to know about all of the RIP-based routes because you can simply set their default network to be the 192.168.200.0 network. However, the RIP routers will need to know how to get to all of the IGRP networks because you already have a 0.0.0.0 route in RIP so that all RIP routers can reach the Internet. So, in this example, you will need to set up redistribution and default routes as shown in Figure 24-16.

On the RIP side, you just add a default route (0.0.0.0) to RIP to point to the router that is connected to the Internet. On the IGRP side, you add the 192.168.200.0 network to the IGRP AS, and set that route as the default network. For the border router, you simply add both RIP and IGRP as routing protocols, include a *network* statement for RIP for only the 192.168.200.0 network, and include *network* statements for IGRP for the 10.0.0.0 and 192.168.200.0 networks. You then set the interface connected to the 192.168.200.0 network in IGRP as a passive interface. If the border router for RIP supported it (remember, it is a non-Cisco router), to reduce the number of updates crossing the WAN link, you set a distribute list on the link to the IGRP domain to allow only the default route to be advertised. Finally, you enable redistribution of the IGRP domain into RIP.

The only new configuration required in this scenario is the addition of a *redistribute* statement for redistribution into RIP. For RIP, the *redistribute* statement changes slightly to reflect RIP's metric, as follows: *redistribute [protocol and AS] [(optional) metric] [(optional) 1–15]*. For instance, to redistribute the IGRP routes into RIP for this example, and set the metric on all redistributed routes to two hops, you type the following:

```
IGRPborder(config)# router rip
IGRPborder(config-router)# redistribute igrp 1 metric 2
```

That's it. Because RIP is such a simple protocol, one-way redistribution with RIP is a fairly simple process. Two-way redistribution, however, is a bit more complex.

IGRP into RIP or RIP into IGRP: Two-Way Redistribution The most common reasons for performing two-way redistribution with multiple protocols are the same as one-way redistribution, just in more complex environments. For instance, take the network shown in Figure 24-17.

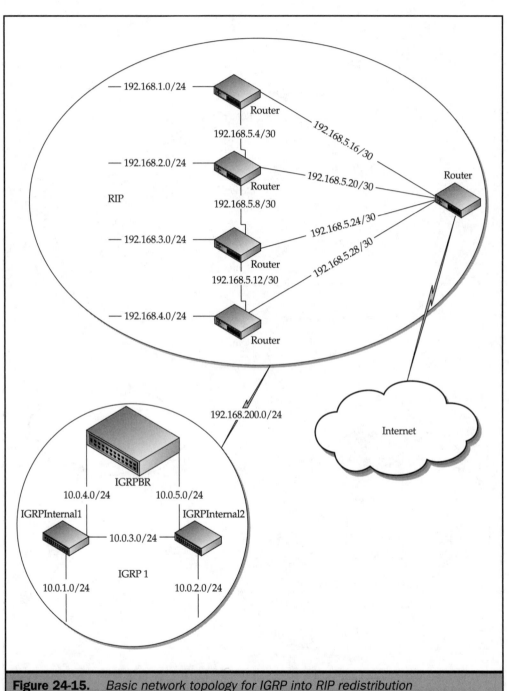

Figure 24-15. *Basic network topology for IGRP into RIP redistribution*

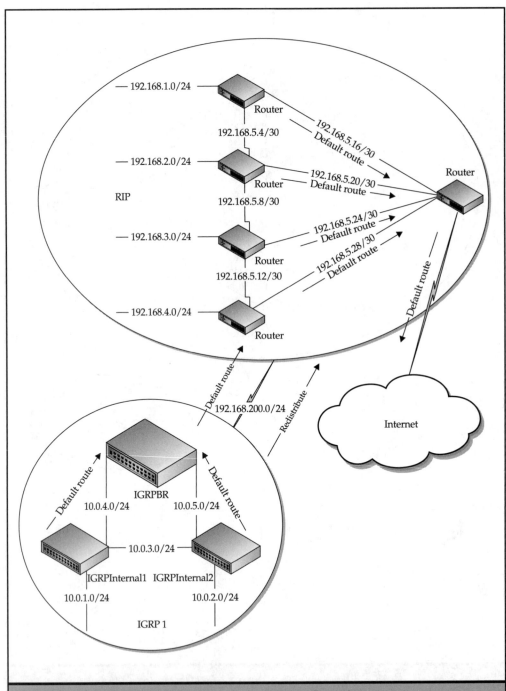

Figure 24-16. *Redistribution and default route configuration*

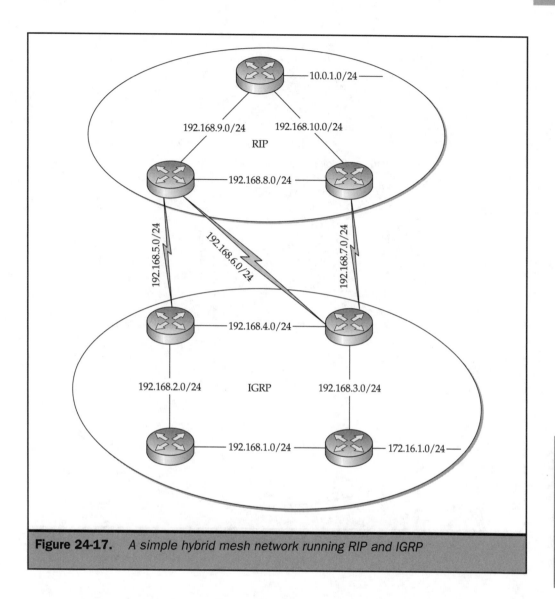

Figure 24-17.   *A simple hybrid mesh network running RIP and IGRP*

In this network, one-way redistribution with default routes used in the other routing protocol results in a loss of load balancing and redundancy capabilities. Two-way redistribution, however, allows both routing protocols to learn about all possible routes to each destination. However, two-way redistribution leads to a potential problem with routing loops and suboptimal paths. For instance, assume you are performing two-way redistribution, as shown in Figure 24-18.

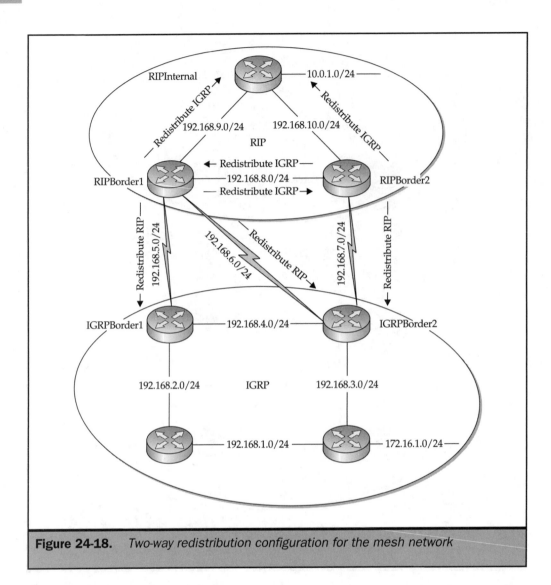

Figure 24-18. *Two-way redistribution configuration for the mesh network*

In this case, let's follow the route for network 10.0.1.0 as it is redistributed. The network will be advertised to RIP's border routers using RIP's metric (1 hop) and administrative distance (120), which applies an IGRP metric (let's say 4000) and administrative distance (100) and redistributes the route. Let's assume that, in this case, RIPBorder1 redistributes the route before RIPBorder2 does, as shown in Figure 24-19.

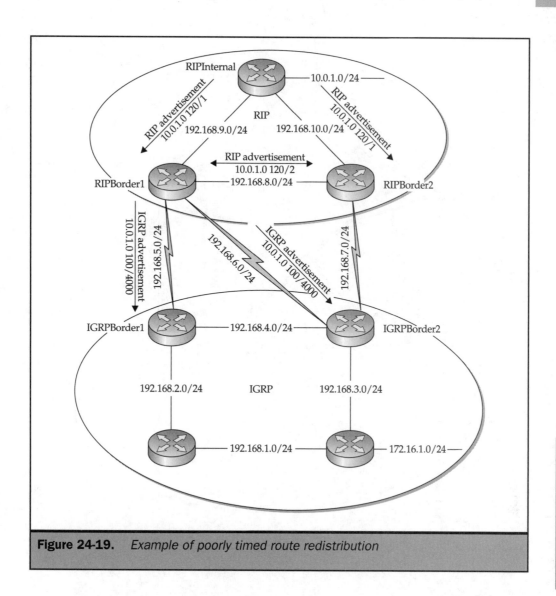

Figure 24-19.   *Example of poorly timed route redistribution*

Then IGRP begins forwarding updates to all IGRP routers in the AS, adding to the metric as the updates are sent. Because RIPBorder2 has still not sent the redistributed route to IGRPBorder2, IGRPBorder2 is not prevented by split horizon from advertising the route back to RIPBorder2, so it will send an update to RIPBorder2, as shown in Figure 24-20. RIPBorder2 sees the "new" IGRP route from IGRPBorder1 and uses it

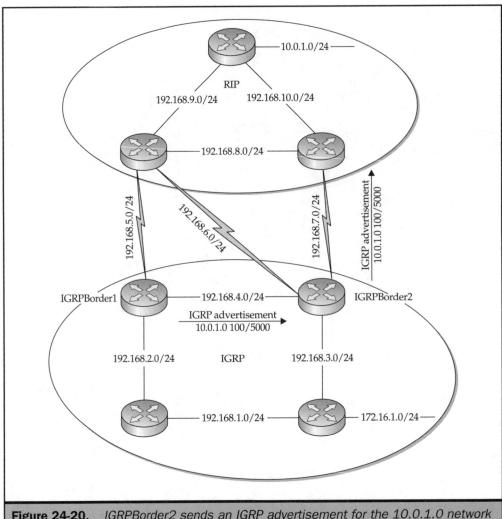

Figure 24-20. *IGRPBorder2 sends an IGRP advertisement for the 10.0.1.0 network to RIPBorder2.*

instead of the correct RIP route because the IGRP route has a lower administrative distance than the RIP route. Eventually, split horizon and holddown timers sort this mess out; but until they do (around ten minutes with the default timers), RIPBorder1 sends all packets destined for the 10.0.1.0 network to IGRPBorder2, which sends them to RIPBorder1, which (finally) sends them to RIPInternal. Even worse, in more complex environments, a routing loop may form between the two protocols, causing even more headache.

Several possible solutions exist to this problem (including changing the default administrative distances and reducing the timers), but the least troublesome solution is to use distribute lists to limit the updates that can be sent back across the WAN links from IGRPBorder1 and IGRPBorder2. For instance, the following distribute list, when applied to the WAN interfaces on IGRPBorder1 and IGRPBorder2, eliminates this problem entirely:

```
IGRPBorder2(config)# access-list 1 deny 10.0.0.0 0.255.255.255
IGRPBorder2(config)# access-list 1 deny 192.168.8.0 0.0.3.255
IGRPBorder2(config)# access-list 1 permit 0.0.0.0 255.255.255.255
IGRPBorder2(config)# router igrp 1
IGRPBorder2(config-router)# distribute-list 1 out serial 0/0
IGRPBorder2(config-router)# distribute-list 1 out serial 0/1
```

The *access-list* statements in this configuration are fairly simple, except for the second statement. The first statement denies routing updates that match the 10.0.0.0 – 10.255.255.255 IP address range from exiting the WAN links (serial 0/0 and 0/1) on IGRPBorder2. (They allow incoming packets, however, because of the *out* statement in the distribute list.) The second statement denies routing updates that match the 192.168.8.0–192.168.11.255 address range from exiting the WAN interfaces because, in the reverse mask, all bits must match except the last three bits in the third octet and the entire fourth octet. You need the second statement to prevent the same poor route selection problem from happening with the 192.168.8.0, 192.168.9.0, and 192.168.10.0 networks. Finally, the last statement allows all other updates.

The moral of the story is that if you are going to use two-way redistribution between routing protocols, you should ensure (using route maps or distribute lists) that routing loops and poor route selection cannot be caused by one routing protocol reredistributing a route into another routing protocol.

Troubleshooting IGRP

Most of the problems you will encounter with basic IGRP configurations are the same as the problems encountered with RIP. In addition, the commands you use to detect most IGRP problems are the same (or very similar) to those used for RIP. Rather than simply rehashing all of those issues, I will instead list the most common network problems; explain the causes; and, where needed, walk you through solving them. Some of the most common problems related to IGRP are as follows:

1. Updates for a subnet of a major network are not received.

2. Routes time out and are relearned at "random" intervals.

CISCO ROUTING

3. Load balancing does not work as expected.

4. Network convergence is slow.

5. Routing loops occur.

6. Certain routers never receive updates.

7. Default routes don't work as expected.

Problem 1: Discontiguous Networks and VLSM

If updates to a router for subnets of a major network are never received, chances are, you have what is known as a *discontiguous network*: a network that is subnetted and the subnets are separated by a completely different major network. For instance, look at the example network in Figure 24-21.

In this example, subnets of the 172.16.0.0 network are separated by the 10.0.0.0 network. If you remember from the "Autonomous Systems" section of the chapter, IGRP sends system (summary) routes about a network if it is sending a route to another router on a different major network than the route it is advertising. In other

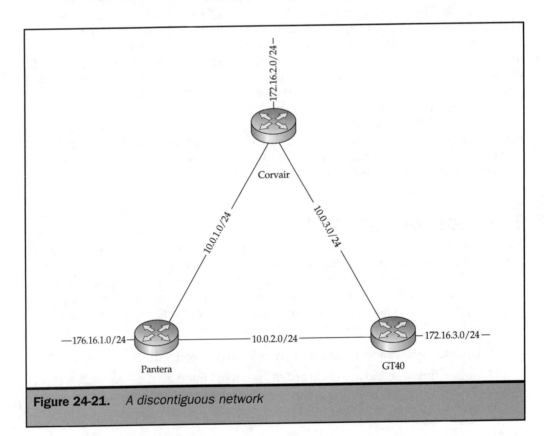

Figure 24-21.   *A discontiguous network*

words, because Corvair is connected to Pantera by the 10.0.0.0 network, it will advertise a summary route for the entire 172.16.0.0 network to Pantera, not the individual 172.16.2.0 subnet of the 172.16.0.0 network. Because of this problem, when you perform a *show ip route* command on Pantera, you will get something similar to the following:

```
3620#show ip route
Codes: C - connected, S - static, I - IGRP, R - RIP, M - mobile, B - BGP
       D - EIGRP, EX - EIGRP external, O - OSPF, IA - OSPF inter area
       N1 - OSPF NSSA external type 1, N2 - OSPF NSSA external type 2
       E1 - OSPF external type 1, E2 - OSPF external type 2, E - EGP
       i - IS-IS, L1 - IS-IS level-1, L2 - IS-IS level-2, ia - IS-IS
inter area
       * - candidate default, U - per-user static route, o - ODR
       P - periodic downloaded static route

Gateway of last resort is not set

     172.16.0.0/24 is subnetted, 1 subnets
C       172.16.1.0 is directly connected, Ethernet0/0

     10.0.0.0/24 is subnetted, 3 subnets
C       10.0.1.0 is directly connected, FastEthernet0/0
C       10.0.2.0 is directly connected, FastEthernet0/1
I       10.0.1.0 [100/4375] via 10.0.1.1
```

It looks as if you are not getting routes at all for the 172.16.0.0 subnets. Yet, if you look at the configuration for Pantera and Corvair, you would see this:

```
!
router igrp 1
 network 10.0.0.0
 network 172.16.0.0
>DETAIL REMOVED<
```

Perhaps you should check to see whether you are receiving updates; that seems like the most logical cause of the problem. To perform this task, you can use the *debug ip igrp events* and *debug ip igrp transactions* commands. The *debug ip igrp events* command shows sort of a quick and dirty summary of which IGRP updates are being sent and received, like so:

```
1w3d: IGRP: sending update to 255.255.255.255 via FastEthernet0/0 (10.0.1.2)
1w3d: IGRP: Update contains 2 interior, 1 system, and 0 exterior routes.
1w3d: IGRP: Total routes in update: 3
```

The *debug ip igrp transactions* command, however, shows very specific information about which network addresses are included in each update, like so:

```
1w3d:          network 172.16.0.0, metric=8476
1w3d:          subnet 10.0.2.0, metric=1501
1w3d:          subnet 10.0.3.0, metric=2601
```

When you combine the two commands, you get a very powerful tool for diagnosing routing update problems (or the lack thereof). In this case, after entering both of these commands, you would see the following:

```
1w3d: IGRP: sending update to 255.255.255.255 via FastEthernet0/0 (10.0.1.2)
1w3d:          network 172.16.0.0, metric=8476
1w3d:          subnet 10.0.2.0, metric=1501
1w3d:          subnet 10.0.3.0, metric=2601
1w3d: IGRP: Update contains 2 interior, 1 system, and 0 exterior routes.
1w3d: IGRP: Total routes in update: 3
1w3d: IGRP: received update from 10.0.1.1 on FastEthernet0/0
1w3d:          network 172.16.0.0, metric 8576 (neighbor 8476)
1w3d:          subnet 10.0.3.0, metric 2601 (neighbor 1501)
1w3d: IGRP: Update contains 1 interior, 1 system, and 0 exterior routes.
1w3d: IGRP: Total routes in update: 2
```

Well, Pantera is receiving *and* sending the updates, just like it's supposed to. So why isn't the output showing the routes to the subnets of the 172.16.0.0 network? The router is trying to protect you from problems. When it sees the summarized route for a network it *knows* is subnetted (because it *has* one of the subnets), it realizes that the other router must be incorrect. If the remote router had the *entire* 172.16.0.0 network (which is what the system route is stating, basically), you would have an address conflict because your router knows it has at least one subnet of that network. So, your router just ignores the advertisement entirely, which means you will never get the route to those subnets to show up in your table.

To solve this problem, in this case, the easiest solution is to change the IP addresses used on the links between Pantera, Corvair, and GT40 to use the subnet of the 172.16.0.0 network, as shown in Figure 24-22.

Discontiguous networks aren't the only root cause of the never-getting-routes-to-remote-subnets problem, of course. (A bad cable does the trick, too.) Aside from basic connectivity problems, having a discontiguous network is probably the most common cause of this problem. Another possible cause is the use of VLSM. With IGRP and RIP 1, VLSM is not supported, so you cannot use different subnet masks for different subnets of the same major network. In other words, in Figure 24-22, you could not set the IP address and subnet mask for the link to the 172.16.1.0 network on Pantera to 172.16.1.64/26 and expect that subnet to be reachable by the other routers. Pantera

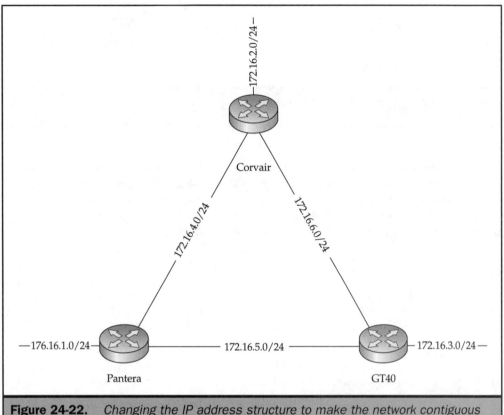

Figure 24-22.   *Changing the IP address structure to make the network contiguous*

would immediately realize that he is using two different masks for subnets of the same network, and would not even advertise his /26 subnet. Again, the router does this to reduce headache, not cause it—believe it or not.

Problem 2: Timer Misconfigurations

The second problem listed at the beginning of this section is routes that seem to appear and disappear at random on one or more of your routers. Typically, the cause of this problem is timer misconfigurations. If you change the timers on any of your routers in a given AS, you really *should* change the timers on all routers in that AS. Otherwise, this issue will be common.

To help you understand how to spot this problem, take a look at the network shown in Figure 24-23.

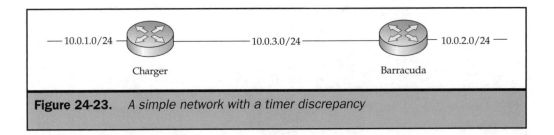

Figure 24-23. *A simple network with a timer discrepancy*

In this network, a problem is occurring on Barracuda where it seems to spontaneously lose routes to the 10.0.1.0 subnet on Charger. (This problem is known as *route flap*.) For instance, if you perform a *show ip route,* the route may show up, but another *show ip route* just seconds later may not show the route. For instance, imagine that at 12:00 P.M., you perform a *show ip route* and get the following output:

```
BARRACUDA#show ip route
Codes: C - connected, S - static, I - IGRP, R - RIP, M - mobile, B - BGP
       D - EIGRP, EX - EIGRP external, O - OSPF, IA - OSPF inter area
       N1 - OSPF NSSA external type 1, N2 - OSPF NSSA external type 2
       E1 - OSPF external type 1, E2 - OSPF external type 2, E - EGP
       i - IS-IS, L1 - IS-IS level-1, L2 - IS-IS level-2, ia - IS-IS
inter area
       * - candidate default, U - per-user static route, o - ODR
       P - periodic downloaded static route

Gateway of last resort is not set

     10.0.0.0/24 is subnetted, 3 subnets
I        10.0.2.0 [100/1600] via 10.0.3.1, 00:01:00, Ethernet0/0
C        10.0.3.0 is directly connected, Ethernet0/0
C        10.0.1.0 is directly connected, Ethernet0/1
```

But then you do another *show ip route* at 12:01 P.M., and get the following result:

```
BARRACUDA#show ip route
Codes: C - connected, S - static, I - IGRP, R - RIP, M - mobile, B - BGP
       D - EIGRP, EX - EIGRP external, O - OSPF, IA - OSPF inter area
       N1 - OSPF NSSA external type 1, N2 - OSPF NSSA external type 2
       E1 - OSPF external type 1, E2 - OSPF external type 2, E - EGP
       i - IS-IS, L1 - IS-IS level-1, L2 - IS-IS level-2, ia - IS-IS
inter area
       * - candidate default, U - per-user static route, o - ODR
       P - periodic downloaded static route

Gateway of last resort is not set
```

```
        10.0.0.0/24 is subnetted, 2 subnets
C          10.0.3.0 is directly connected, Ethernet0/0
C          10.0.1.0 is directly connected, Ethernet0/1
```

Thirty seconds later, however, the route will show up again. In real life, you would most likely become aware of these problems because a user would inform you that connectivity to a particular network is failing, but only sometimes. One way or another, however, the problem in this case is timer misconfigurations. After checking Charger's timers with the *show ip protocol* command, you know that Charger's timers are set to the defaults. However, when you perform the same command on Barracuda, this is what you get:

```
Routing Protocol is "igrp 1"
  Sending updates every 5 seconds, next due in 3 seconds
Invalid after 15 seconds, hold down 15, flushed after 35
Outgoing update filter list for all interfaces is
  Incoming update filter list for all interfaces is
  Default networks flagged in outgoing updates
  Default networks accepted from incoming updates
  IGRP metric weight K1=1, K2=0, K3=1, K4=0, K5=0
  IGRP maximum hopcount 100
  IGRP maximum metric variance 1
  Redistributing: igrp 1
  Routing for Networks:
    10.0.0.0
  Routing Information Sources:
    Gateway          Distance       Last Update
    10.0.3.1              100       00:00:15
Distance: (default is 100)
```

Because of this timer discrepancy, Barracuda is timing out the updates from Charger *very* quickly (every 40 seconds). The following *show* command outputs, however, really tell the story well:

```
BARRACUDA#show ip protocol
Routing Protocol is "igrp 1"
  Sending updates every 5 seconds, next due in 2 seconds
  Invalid after 15 seconds, hold down 15, flushed after 35
  Outgoing update filter list for all interfaces is
  Incoming update filter list for all interfaces is
  Default networks flagged in outgoing updates
  Default networks accepted from incoming updates
  IGRP metric weight K1=1, K2=0, K3=1, K4=0, K5=0
  IGRP maximum hopcount 100
```

CISCO ROUTING

```
    IGRP maximum metric variance 1
    Redistributing: igrp 1
    Routing for Networks:
      10.0.0.0
    Routing Information Sources:
      Gateway         Distance      Last Update
      10.0.3.1            100       00:00:02
    Distance: (default is 100)


BARRACUDA#show ip route
Codes: C - connected, S - static, I - IGRP, R - RIP, M - mobile, B - BGP
       D - EIGRP, EX - EIGRP external, O - OSPF, IA - OSPF inter area
       N1 - OSPF NSSA external type 1, N2 - OSPF NSSA external type 2
       E1 - OSPF external type 1, E2 - OSPF external type 2, E - EGP
       i - IS-IS, L1 - IS-IS level-1, L2 - IS-IS level-2, ia - IS-IS
inter area
       * - candidate default, U - per-user static route, o - ODR
       P - periodic downloaded static route

Gateway of last resort is not set


     10.0.0.0/24 is subnetted, 3 subnets
I        10.0.2.0 [100/1600] via 10.0.3.1, 00:00:02, Ethernet0/0
C        10.0.3.0 is directly connected, Ethernet0/0
C        10.0.1.0 is directly connected, Ethernet0/1
```

Notice that, two seconds after the last update from Charger, the route is in the routing table, just like it is supposed to be:

```
BARRACUDA#show ip protocol
Routing Protocol is "igrp 1"
  Sending updates every 5 seconds, next due in 3 seconds
  Invalid after 15 seconds, hold down 15, flushed after 35
  Outgoing update filter list for all interfaces is
  Incoming update filter list for all interfaces is
  Default networks flagged in outgoing updates
  Default networks accepted from incoming updates
  IGRP metric weight K1=1, K2=0, K3=1, K4=0, K5=0
  IGRP maximum hopcount 100
  IGRP maximum metric variance 1
  Redistributing: igrp 1
  Routing for Networks:
    10.0.0.0
  Routing Information Sources:
    Gateway         Distance      Last Update
    10.0.3.1            100       00:00:15
  Distance: (default is 100)
```

```
BARRACUDA#show ip route
Codes: C - connected, S - static, I - IGRP, R - RIP, M - mobile, B - BGP
       D - EIGRP, EX - EIGRP external, O - OSPF, IA - OSPF inter area
       N1 - OSPF NSSA external type 1, N2 - OSPF NSSA external type 2
       E1 - OSPF external type 1, E2 - OSPF external type 2, E - EGP
       i - IS-IS, L1 - IS-IS level-1, L2 - IS-IS level-2, ia - IS-IS
inter area
       * - candidate default, U - per-user static route, o - ODR
       P - periodic downloaded static route

Gateway of last resort is not set

     10.0.0.0/24 is subnetted, 3 subnets
I       10.0.2.0/24 is possibly down,
           routing via 10.0.3.1, Ethernet0/0
C       10.0.3.0 is directly connected, Ethernet0/0
C       10.0.1.0 is directly connected, Ethernet0/1
```

However, at 15 seconds, it is now marked in the table as *possibly down,* telling you that an update was expected and not heard within the configured invalid time. The route will still not be removed from the table until the flush timer expires, as shown here:

```
BARRACUDA#show ip protocol
Routing Protocol is "igrp 1"
  Sending updates every 5 seconds, next due in 0 seconds
  Invalid after 15 seconds, hold down 15, flushed after 35
  Outgoing update filter list for all interfaces is
  Incoming update filter list for all interfaces is
  Default networks flagged in outgoing updates
  Default networks accepted from incoming updates
  IGRP metric weight K1=1, K2=0, K3=1, K4=0, K5=0
  IGRP maximum hopcount 100
  IGRP maximum metric variance 1
  Redistributing: igrp 1
  Routing for Networks:
    10.0.0.0
  Routing Information Sources:
    Gateway         Distance      Last Update
    10.0.3.1             100      00:00:46
  Distance: (default is 100)

BARRACUDA#show ip route
Codes: C - connected, S - static, I - IGRP, R - RIP, M - mobile, B - BGP
       D - EIGRP, EX - EIGRP external, O - OSPF, IA - OSPF inter area
       N1 - OSPF NSSA external type 1, N2 - OSPF NSSA external type 2
       E1 - OSPF external type 1, E2 - OSPF external type 2, E - EGP
```

```
         i - IS-IS, L1 - IS-IS level-1, L2 - IS-IS level-2, ia - IS-IS
   inter area
         * - candidate default, U - per-user static route, o - ODR
         P - periodic downloaded static route

Gateway of last resort is not set

     10.0.0.0/24 is subnetted, 2 subnets
C       10.0.3.0 is directly connected, Ethernet0/0
C       10.0.1.0 is directly connected, Ethernet0/1
```

If you have this problem, of course the easiest solution is to set the timers on all of the routers in the AS to equal values. Obviously, other problems can cause route flap (like a link going up and down a lot); but in fairly stable networks, if routes are being removed from the table sporadically, make sure your timers are all set correctly.

Problem 3: Variance, Metric Components, and Weights

In some cases, you may have redundant links that are not load balancing like you would expect. Before looking any further, you need to make sure you know IGRP's rules regarding load balancing. These were explained in the "Load Balancing" section at the beginning of the chapter; but just to refresh your memory, I will briefly restate them here:

- The maximum paths (as in the *maximum-paths* command discussed in Chapter 23) defines how many paths IGRP will load balance over. The default is 4.

- All routes to a given destination must be less than the lowest-cost route multiplied by the variance. For example, if the lowest-cost route is 25,000 and the variance is 4, all routes to the same destination with a metric of 99,999 or less are considered acceptable for load balancing.

- The next hop router listed in the route entry must have a lower metric to the destination network than the lowest metric route to the destination network on the local router.

For starters, use the commands discussed in the "Problem 1: Discontiguous Networks and VLSM" section to make sure that you are actually *getting* a route to the network over the redundant connection. Once you have determined that redundant routes are being received for the network in question, you can figure out which of these rules are potentially causing you grief.

First, make sure your maximum paths are set to greater than 1. You can do this with the *show running-config* command. If you do not see a *maximum-paths* statement, then you are using the default of 4.

Once you have ensured that the router is allowing redundant paths, check to make sure that traffic sharing is enabled for load balancing. Again, use the *show running-config* command. If you see an entry for *traffic-share*, make sure it is not set to *traffic-share min*. If you do not see a *traffic-share* statement, again, it means everything is fine because you are using the default setting of *traffic-share balanced*, which enables load balancing.

Once you have ensured that load balancing is enabled, check your variance setting against the lowest and highest metrics that are advertised for your redundant paths. You can see the variance setting using the *show ip protocol* command, and you can see the metric for routes that are being advertised to you using the *debug ip igrp transactions* command.

If the variance is within an acceptable range and you are still not getting load balancing to function, then the final step is to ensure that the redundant router's local metric for the path in question is lower than the *best* local metric. Again, you can check this by using the *debug ip igrp transactions* command. If this turns out to be your problem, then you can improve one or more of the metric components (delay, typically) on the remote router to reduce the remote router's metric; however, remember that there are consequences associated with changing the metric components. (Again, refer to the earlier "Load Balancing" and "Metric and Maximum Hops" sections for a more thorough explanation.)

If load balancing is working some of the time but is not *always* functioning, then you may want to check to see whether the advertised metric (or local metric) is changing at all. If you are using the reliability or load weights on any routers along the path, it is possible for the metric to change enough to cause the route to be dropped from the table. If so, either change the weight on the reliability and load to enable small rather than large metric changes, or increase the variance on the local router.

Problem 4: Large Timers and Large Networks

In many cases, the primary complaint with IGRP is its slow convergence time. This is to be expected because IGRP uses much longer timers than RIP to reduce bandwidth use for routing updates. However, longer timers also mean slower convergence. To give you an example of how long convergence can take in IGRP, imagine that you have an IGRP router that fails. All directly connected neighbors will wait for an update. When they do not hear one, they will wait for the invalid timer to expire, and then will place the route in holddown. Once the route enters holddown, the route will be poisoned, with infinite metric updates from the directly connected routers being sent to all other routers. All routers will then begin propagating this poisoned route to immediately time the route out of the table and reduce convergence time.

At this point, between 180 and 270 seconds have passed since the route failed. Once the holddown timer expires on the directly connected routers (280 seconds later), they can learn a new, alternative route to the subnet. Assuming one is available, they will learn this route within the next 90 seconds and will immediately advertise the route to neighboring routers. Assuming that the addition of the route and the subsequent triggered update process takes two seconds for each hop, the entire network convergence process for a *single* failed route takes around 15 minutes, assuming that the default timers are used and the network spans a maximum default network diameter of 100 hops.

The primary method of reducing network convergence time is to reduce the holddown and invalid timers, but you must ensure that the timers allow adequate time to prevent routing loops. To make sure you allow for adequate time, remember that the invalid and holddown timers should be at least three times larger than the update timer. So, if you reduce the invalid or holddown timers, you must reduce the update timer as well. Also, note that in most cases, Cisco does not recommend setting the update timer to any lower than 15 seconds. Finally, ensure that you change the timers on all routers in the AS to equal values. If you do not, you will be reading the "Problem 2: Timer Misconfigurations" section again shortly after changing the timers.

You can view the timers on the router using the *show ip protocol* command (detailed in the "Problem 2: Timer Misconfigurations" section, as well as in Chapter 23), and you can change the timers using the *timers basic* command (also detailed in Chapter 23).

Problem 5: Holddown and Split Horizon

Assuming redistribution is not being used, the primary cause of routing loops in IGRP is the disabling of holddown and/or split horizon. Holddown can be disabled using the router configuration command *no metric holddown*. Cisco calls disabling holddown "Fast IGRP," but it is not recommended unless you are absolutely sure that no loops exist in the network. If you have no loops in the network, it basically means that no redundant paths exist; and if that is the case, why not just use static routing?

Split horizon is disabled by default on some interfaces, and this fact can lead to many routing loop problems with both IGRP and RIP. As mentioned in Chapter 23, the best way to ensure that routing loops don't form in Frame Relay is to apply your Frame Relay VCs to subinterfaces. In situations when this is not possible, you should use a *distribute-list* and/or *route-map* statements to ensure that routing loops do not form.

In any event, if you have routing loops, they will normally come to your attention because some or all of the packets to a particular destination fail to route correctly. In some cases, this problem will work itself out, only to occur again later. The best way to determine whether you have a routing loop is to examine the routing tables on all routers along the problem route's path. If you see routes looping, examine the router's configurations for the *no metric holddown* or *no ip split-horizon* commands. If you do not see these commands entered on the router, examine the timers and ensure that the holddown is set to at least three times the update. If you do not find a problem there, then make sure that none of the connecting interfaces have split horizon disabled by default. (If the interface is a Frame Relay interface, split horizon is disabled by default *unless* you are using point-to-point subinterfaces.)

If none of these solutions solves your problem, wait for about 15 minutes to see whether the problem sorts itself out. Although it isn't very common, occasionally, all of the stars will align and even with split horizon and holddowns enabled, a routing loop will occur. If the problem sorts itself out, then it was most likely a normal occurrence, and should rarely happen. (In other words, don't worry about it.)

Problem 6: Passive Interfaces and Unicast Updates

If you have a router that is not receiving routes from neighboring routers, there are a few ways that you can troubleshoot the problem.

First, ensure that you actually aren't receiving the route. In many cases, you may be receiving the route, but the route is not entered into the table because VLSM is being used or because of a discontiguous network problem. (See the "Problem 1: Discontiguous Networks and VLSM" section for more details in this case.) If the route is a redundant route (secondary route) that isn't being entered into the table for load-balancing purposes (problem 3), it may also appear as if you are not receiving the route. To determine whether the route is reaching the problem router at all, use the *debug ip igrp transactions* command, as detailed under problem 2, on both the sending and receiving routers. By watching the updates that are sent and received, you can correctly determine the cause of the problem.

If you are receiving the routes, but they are not being entered into the routing table, then you can take four major steps to determine why:

1. Ensure that no distribute lists are preventing the route from being entered into the table.

2. Ensure that VLSM is not causing the route to be viewed as invalid.

3. Ensure that you do not have a discontiguous network.

4. Ensure that a network statement has been entered on the destination router for the network on which the updates are received.

If it turns out that the updates are not being received at all, then you will need to dig a bit deeper.

First, you will want to ensure that connectivity between the two routers exists. (You should be able to ping the other router.) If you have connectivity, and the *debug ip igrp transactions* statement on the sending router shows that the routes are being sent but the destination router is still not receiving them, determine whether the routes are being broadcast or unicast. If they are being broadcast, ensure that the media you are using supports broadcasts (or has the ability to "simulate" a broadcast, like Frame Relay).

For a quick check to determine whether that is the problem, add a *neighbor* statement for the destination router. The *neighbor* statement causes the sending router to unicast an update to the receiving router. If this solves the problem, then you know the media either does not support broadcasts at all or has not been configured to simulate broadcasts. If updates are being unicast but are still failing, ensure that the IP address for the destination router is entered correctly. (This problem is more common than anyone would like to admit.)

If it turns out that updates are not being sent at all, then you need to ensure that a *network* statement has been entered for the network to which the sending router should be advertising. If a *network* statement has been entered but updates are still not being sent, ensure that the interface is not configured as a passive interface.

Problem 7: Default Networks

Because IGRP differs from RIP in that it doesn't understand the 0.0.0.0 route, it can sometimes be difficult to understand how to configure a default route in IGRP. However, there is an easy way to set up your default routes; you just need to remember a few simple rules.

First, IGRP specifies a particular network to be a default network. Because of this functionality, to define a default network, you must already have a route to that network.

Second, once you define a default network on any one router, that router automatically propagates the route to that network as an external route, which defines the route as a default network candidate on all routers in the AS. If a router hears of multiple default network candidates, it will choose the one with the lowest administrative distance and metric as its default route. Remember, IGRP does not load balance across default routes.

Third, you will typically set the default *network* statement on the router that *originates* the route to the network (usually, the router that is directly connected to that network). This ensures that the default network is correctly propagated as an external route to all routers in the AS. However, this causes a problem for the originating router because it is unable to use a directly connected network as the default route, which leads to the last rule.

The router that is directly connected to the network specified in the default network statement either must know of all other routes in the enterprise (usually, through redistribution), or it must have a different default route configured in order to forward traffic. Usually, the easiest way to configure a different default route on this router is to add a static route to the 0.0.0.0 network. Because a static route uses a lower administrative distance than a dynamic route, the static route will always be used by the router for the default route.

Summary

This chapter examined IGRP in detail, including metrics, timers, and autonomous systems, and explained how to configure and troubleshoot IGRP in both simple and complex environments. In addition, you learned the basics of route redistribution using IGRP in both one-way and two-way scenarios. The topics covered in this chapter and Chapter 23 are the building blocks for the more advanced topics covered in the next several chapters, so be sure that you understand the topics covered here before moving on.

For more information on IGRP, visit this URL: http://www.cisco.com/univercd/ cc/td/doc/product/software/ios122/122cgcr/fipr_c/ipcprt2/1cfigrp.htm.

The Complete Reference

Cisco

Chapter 25

Enhanced Interior Gateway Routing Protocol

This chapter examines routing with Enhanced Interior Gateway Routing Protocol (EIGRP). EIGRP is a balanced-hybrid routing protocol that improves on many of IGRP's shortcomings in enterprise networks, including the following:

- Reduced overhead. EIGRP reduces the network and router overhead needed for routing updates by multicasting routing updates, sending a routing update only when a change is detected (rather than sending the entire table periodically), updating only the routers that need to be aware of a topology change, and sending changes only to the routing table (rather than the entire table) when an update is necessary.

- Support for VLSM and CIDR. EIGRP includes the subnet mask in routing updates, allowing EIGRP to be used in VLSM scenarios.

- Support for discontiguous networks.

- Support for manual route summarization.

- Extremely short convergence times. Due to EIGRP's update algorithm, when a change occurs, EIGRP can achieve convergence in a matter of seconds in most reliable, well-designed networks.

- Loop-free topology generation.

- Support for multiple network protocols including IP, IPX, and AppleTalk. (Due to the need for brevity, however, only the IP version of EIGRP is discussed in this book.)

Like IGRP, EIGRP is a Cisco proprietary protocol. EIGRP is one of the most common interior gateway protocols (IGPs) in use today in large, rapidly changing networks. In fact, the benefits of EIGRP are so great in these environments that many companies have changed from some other vendor's routers to Cisco routers just to be able to use EIGRP. Although EIGRP is necessarily more complex than RIP or IGRP, it is less complex than OSPF, which (normally) makes EIGRP easier to implement than OSPF.

How EIGRP Works

Contrary to its name, EIGRP is *very* different from IGRP. In fact, about the only aspects EIGRP and IGRP have in common are their metric calculation (and even that is a bit different) and their use of autonomous systems (ASs).

Whereas IGRP uses a distance vector algorithm for updates, EIGRP uses a balanced-hybrid algorithm known as the Diffusing Update Algorithm (DUAL). DUAL is the defining trait of EIGRP; and although it is not dedicated a specific section in this chapter, DUAL operation is explained throughout the entire discussion of EIGRP.

The following discussion breaks EIGRP operation down into these components:

- Operational overview
- Terminology reference

- Operation
- Summarization
- Load balancing

Operational Overview

This section discusses EIGRP's basic operational procedures in a fairly nontechnical (and nonspecific) manner to ease you into its unique method of finding, selecting, and maintaining routes. Although DUAL takes a little getting used to, it isn't extremely complicated.

First, routers maintain three tables to hold information: the Neighbor table, the Topology table, and the Routing table. Upon bootup, the routers actively seek out neighbors (directly connected routers) and add them to the neighbor table. The route also transfers its entire routing table to its new neighbors. Rather than just picking a path or two and entering them into its routing table, however, at this point, EIGRP adds to the topology table all downstream routes to the *destination* (routes through routers that are closer to the destination than the local router).

From the topology table, it then chooses one or more routes as the "best" routes to the destination (based on the route metrics). If a topology change occurs, the routers closest to the topology change look in their topology table for an alternate route to the destination. If one is found, they begin using the route and update their upstream neighbors on the change to the route. If an alternate route is *not* found in the topology table, the router queries its neighbors for a new route. The neighbors look in their topology tables, attempt to find an alternate route, and reply with any information they have available. If no alternate route is listed in their topology tables, they query *their* neighbors, and so on.

Basically, as mentioned in Chapter 22, although EIGRP does keep some information on the network topology, it doesn't have to keep an entire map of the topology because it has enough sense to ask for directions.

Terminology Reference

This section provides a quick reference to some of the common terms used throughout the rest of the chapter. It is not meant to provide a detailed explanation of each of these topics. All of these topics are discussed in more detail in later sections. However, I feel it is important to have a section with simple definitions of the commonly used terms in EIGRP so that you don't have to hunt through the index to find the definition for a term. If you are reading this chapter for the first time (and this is your first exposure to the inner workings of EIGRP), you might want to move on to the next section and use this section as a reference as you encounter new terms. For ease of organization, I have broken this section up into three parts: basic terms, packet types, and tables.

CISCO ROUTING

Basic Terms

The following terms are general terms related to EIGRP/DUAL operation that you will see throughout the chapter:

- **Active state** A route is said to be in an active state when an input event causes a route recomputation. While in the active state, the router cannot change the route's successor, the distance it is advertising for the route, or the FD for the route. Once all routers have replied to the queries, the router can transition from an active to a passive state, either removing or recalculating the route in the process.

- **Distributed Update Algorithm (DUAL)** DUAL is the routing update, selection, and maintenance algorithm in EIGRP. DUAL performs all of EIGRP's routing calculations, and it is responsible for successor and feasible successor selection, hellos, updates, queries, replies, and acknowledgments.

- **Feasibility condition (FC)** This is the condition a route must meet to be listed in the topology table. The FC basically states that if the next-hop router's best local metric to the route in question is lower than the FD, the router in question is downstream in respect to the destination, and the route is allowed to be listed in the topology table as a successor or feasible successor. If a route does not meet the FC, the route is dropped. The FC is used to eliminate the possibility of routing loops.

- **Feasible distance (FD)** This is the metric of the lowest-cost route on the local router. The FD is used to determine whether an advertised route is allowed to be used as a successor or feasible successor.

- **Feasible successor** This is a next-hop router that does *not* have the best path to the destination, but is still closer to the destination than the local router. You can think of feasible successors as alternates in case the path through the current successor fails. Feasible successors are listed only in the topology table.

- **Input event** An input event is any event that could potentially affect the reachability of a route. An input event occurs when a metric on a directly connected link changes, a directly connected link changes state (goes up or comes down), or whenever a query, update, or reply packet is received.

- **Passive state** This is the normal state for routes in the DUAL algorithm. While in a passive state, the route can be updated and route evaluations can be performed.

- **Route evaluation (sometimes called a local computation)** A route evaluation occurs whenever a topology change (input event) occurs. The router reevaluates all routes affected by the topology change to determine the best successors and feasible successors. If a successor can be found in the topology table for the affected routes, the successor is used, and route recomputation does not occur.

If a valid successor is not found, the route must perform a route recomputation. Route evaluation does *not* require the router to enter an active state for the route.

■ **Route recomputation (sometimes called a diffusing computation)**   A route recomputation occurs when a successor for a route is lost and no feasible successors are listed in the topology table. When a router performs a route recomputation, it changes the FD to infinite, enters an active state for the route, and issues queries to its neighbors in an attempt to locate an alternate route. Once *all* neighbors have replied to these queries, the router either adds a new successor and feasible successors to the topology table, or the router removes the route from the topology table. A router may perform only one route recomputation at a time.

■ **Successor**   A successor is the next-hop router with the *best* path to a destination. Each route listed in the routing table has at least one successor. If multiple successors are chosen, EIGRP load balances between the successors. Successors are listed in both the routing table and the topology table.

Packet Types

EIGRP uses several different packet types to transmit different types of information: ACKs, hellos, queries, replies, requests, and updates. All of these except requests are discussed next. Requests are unused in EIGRP's current version (version 1).

■ **Acknowledgment (ACK)**   ACKs are sent in response to reliably sent EIGRP packets. They are always unreliably unicast.

■ **Hello**   Hellos are sent periodically by all EIGRP routers to establish adjacencies. (They are sent every 60 seconds over X.25, ATM, or Frame Relay links at 1.544 Mbps or lower, and every 5 seconds over all others, by default.) A router adds the neighbor to its table and then transmits its own hello to the neighbor candidate. If the AS number and K values are the same on both routers, an adjacency is formed, and the routers exchange routing information. Hellos are always unreliably multicast using the 224.0.0.10 address (an address processed only by EIGRP routers).

■ **Queries**   Queries are "question" packets sent to other routers in an attempt to find a valid path to a destination. Queries are sent when a router enters the active state for a route and begins the diffusing computation. Queries are either multicast to the 224.0.0.10 address or unicast directly to the neighbor. Queries always use RTP for reliable delivery.

■ **Replies**   Replies are sent in response to a query. Replies include information on the replying router's best path to a destination, or they include a "destination unreachable" response (a route listed with a 4,294,967,295 metric). Replies are always reliably unicast using RTP.

■ **Update**   An update contains information about routes. Updates are sent whenever an input event causes a change in a route. Updates only contain information about the routes that are affected by the input event (with the exception of initial updates). Updates are either reliably multicast to the 224.0.0.10 address or unicast to a specific router using RTP.

Note that, in many cases, EIGRP uses Reliable Transport Protocol (RTP) to transmit packets reliably. RTP is like EIGRP's own version of TCP. RTP is used only for EIGRP, and it ensures that packets are received in order and that any lost packets are retransmitted. RTP attempts to retransmit lost EIGRP packets up to 16 times (after which, the packet is dropped).

Tables

EIGRP maintains three tables for use by DUAL: neighbor, topology, and routing tables.

The Neighbor Table   The *neighbor table* contains every neighbor for which an adjacency has been established. EIGRP uses the neighbor table to maintain a record of all neighboring routers and RTP information for those routers. The neighbor table contains the following:

■ The neighbor's IP address.

■ The interface that is used to reach the neighbor.

■ The holdtime advertised by the neighbor (in other words, the neighbor's configured holdtime), minus how many seconds have elapsed since the last hello was sent. The *holdtime* is the amount of time the router will wait for a hello from a neighbor before declaring the neighbor dead and removing the neighbor from the neighbor table. (This action causes routes through the neighbor to be removed from the topology table, which may cause routes to be removed from the routing table). The default holdtime is three times the hello (15 or 180 seconds, depending on the interface).

■ The uptime (the amount of time the adjacency has been established with the neighbor).

■ The round trip time (listed as smooth round trip time, or SRTT). The SRTT is how long in milliseconds it takes, on average, for an ACK in response to a hello to be received from this neighbor.

■ The RTP sequence number for the last reliably sent packet from this neighbor. RTP uses sequence numbers to help ensure that packets are in the correct order and no packets are lost in transmission. This number increments by one every time a reliable packet is received from the neighbor.

■ The retransmission timeout (RTO) for the neighbor. The RTO is the amount of time, in milliseconds, that the router will wait for an acknowledgment to be received from this neighbor for reliably sent traffic. If the RTO expires before a reply is received, the router resends the RTP packet.

■ EIGRP queue length for this neighbor (how many EIGRP packets are waiting in the queue for this neighbor).

 If a router has multiple direct connections to a single remote router, the remote router actually appears as multiple neighbors in the neighbor table. This functionality occurs because adjacencies are based on IP address.

The Topology Table   The *topology table* contains information about every currently active route the router has heard that meets the FC. The topology table is used to pick successors and feasible successors and, therefore, to build the routing table. The topology table includes the following information about all valid routes:

■ The state of the route (active or passive)

■ The destination network address and mask

■ The local FD for the destination network

■ All feasible successors for the destination

■ The advertised distance for each feasible successor (the feasible successor's local metric for the route)

■ The locally calculated distance through each feasible successor (the final metric of the route through each feasible successor)

The Routing Table   The *routing table* for EIGRP is the same as the routing table for any other routing protocol. The routing table is used to make forwarding decisions, and contains the following:

■ The current default route

■ All usable routes (for EIGRP, either directly connected or routes with one or more valid successors)

■ The administrative distance for each route

■ The local metric for each route

■ The next-hop router (if not directly connected) for each route

■ The forwarding interface for each route

Operation

EIGRP operation is primarily based on the *DUAL finite state machine* (a fancy way of saying operational procedures). The DUAL process ensures that routes are always loop-free, without the need for additional functions to ensure that loops do not occur.

With DUAL, even short-lived loops should never occur. The basic operation of DUAL is as follows:

- Neighbors are discovered through hello messages (neighbor discovery).
- Routes are learned through updates and replies to queries (route discovery).
- Successors and feasible successors are chosen based on their metric (route selection).
- Routes are maintained and removed by the DUAL finite state machine (route maintenance).

Let's deal with each of these pieces in order, starting with neighbor discovery.

Neighbor Discovery

EIGRP does not send regular routing updates, like RIP or IGRP. Therefore, EIGRP must have some way of discovering routers that already exist on the network, as well as routers that have just come online. For instance, if you boot a router running EIGRP and that router does not have a way to locate its directly connected neighbors, the router will just have to wait until an update is sent to learn of new routes. Because EIGRP does not send periodic updates, it could take a very long time indeed for the new router to learn all routes. Therefore, in EIGRP, hellos are used to facilitate router detection.

In addition, hellos are needed for EIGRP to detect the failure of a router. Hellos are multicast from all EIGRP routers (unless the router is configured with a passive interface on the network in question) at regular intervals (by default, every 60 seconds for low-speed WAN connections and every 5 seconds for all others). Each router has a holdtime value configured; and if a hello is not heard from a neighbor before the holdtime for that neighbor expires, EIGRP assumes the neighbor has failed.

 In EIGRP, all metric weights (K values) must be the same for every router in the entire AS. If two routers are configured with different K values, they will never establish an adjacency.

An EIGRP hello message is unreliably multicast to the 224.0.0.10 address (a reserved address used only by EIGRP), and is processed only by directly connected EIGRP routers. The hello message contains the IP address the router is using on the interface the multicast was sent on, the AS number for the EIGRP process on the router, and the K values for the router's metric calculation. When an EIGRP router hears a directly connected router's hello message, the router examines the AS number and K values to see if they match the locally configured AS number and K values. If they do, the local router adds the remote router to its neighbor table. At the next hello interval, the local router then multicasts a hello message to the remote router. The remote router examines the AS and K values, and adds the local router to its neighbor table. Once both routers have each other listed in their respective neighbor tables, an adjacency is formed.

For an adjacency to be established, both routers must, of course, be using the same network protocol (IP, IPX, or AppleTalk) and have direct connectivity (be part of the same logical network). Because we are discussing only IP EIGRP, however, these settings are assumed.

Once an adjacency is formed, each router lists the other router in its neighbor table, including the neighbor's IP address and holdtime values, among other information. (See the section "The Neighbor Table" earlier in the chapter for a complete list of information included in this table.)

Unlike some other routing protocols (OSPF, in particular), EIGRP holdtime values do not have *to match. EIGRP establishes an adjacency with another router even if the holdtimes do not match. However, the difference in holdtime values can cause neighbors to be continually removed and re-added to the neighbor table, so equal holdtime values are strongly suggested.*

Route Discovery

Once an adjacency has been formed with an EIGRP router, the routers exchange routing information using EIGRP update packets. Update packets are always reliably multicast or unicast using Reliable Transport Protocol (RTP).

RTP is a proprietary protocol that EIGRP uses to transport both reliable and unreliable messages between EIGRP routers. Reliable messages differ from unreliable messages primarily in that reliable messages must be acknowledged with an ACK packet. If a reliably sent message is not responded to with an ACK within a certain time (known as the *retransmission timeout,* or RTO), EIGRP resends the reliably sent message (up to a maximum of 16 times). RTP also uses sequence numbers to ensure that packets are delivered in order. (Although failed routing updates are not usually a problem on LAN links, on WAN links, it can sometimes become an issue.) RTP keeps track of every neighbor that a reliable packet has been sent to, and expects an ACK from each neighbor. If all neighbors respond except one, RTP unicasts the repeat message only to that single neighbor. This RTP feature is useful to help reduce bandwidth use and CPU waste.

Each update packet can contain one or more routes, and each route contains different information depending on the route type. EIGRP defines two route types: internal and external.

An *internal route* is a route that originated from within the EIGRP AS to which it is now being propagated. Each internal route in an EIGRP update contains the following:

- **Next Hop**   The next-hop router address (which could be a different router from the advertising router)
- **Delay**   The cumulative EIGRP delay
- **Bandwidth**   The lowest EIGRP bandwidth along the path

- **MTU**   The lowest MTU along the path (Like IGRP, this field is included but not used.)
- **Hop Count**   The current hop count for the path
- **Reliability**   The reliability for the path, as measured on the exit interface for the route path on the next-hop router
- **Load**   The load for the path, as measured on the exit interface for the route path on the next-hop router
- **Prefix Length**   The length of the subnet mask for the route in question
- **Destination**   The network/subnet address for the route in question

Most of these fields are used for exactly the same purposes in IGRP and EIGRP, with the only completely new field being the Prefix Length because IGRP does not include any subnet mask information in updates. For example, the metric components (bandwidth, delay, reliability, and load) are all used to calculate the metric in exactly the same way in both protocols, with a minor exception: EIGRP applies a scaling factor to the bandwidth and delay components in the metric to make each of these parameters more important to the final metric calculation.

Note   *Because the metric calculation used in EIGRP is essentially the same as the calculation used in IGRP, please refer to Chapter 24 for a full discussion about how to calculate EIGRP's metric.*

To create a larger average metric, EIGRP multiplies by 256 both the bandwidth and delay parameters that were calculated using the formulas in Chapter 24. This scaling factor that EIGRP adds into the equation is the only difference between EIGRP's metric and IGRP's metric calculations. These scaling factors make the final metric calculation as follows (with the new scaling factors bolded): Metric = K1 × (Be × **256**) + (K2 × (Be × **256**))/(256 − load) + K3 × (Dc × **256**)) × (K5/(reliability + K4)

Note   *Some texts state that the scaling factor in EIGRP simply multiplies the total IGRP metric by 256. This is not the case; only the bandwidth and delay fields are affected by the scaling factor. The reason most texts state the scaling factor in this manner is that with the default K values, multiplying the IGRP metric by 256 yields the correct EIGRP metric.*

An *external route* in EIGRP is a route that has been redistributed into the EIGRP AS to which the update is being propagated. An external route contains all of the fields used in an internal route but also contains additional fields, described next:

- **Originating Router**   The router that is redistributing the route into EIGRP.
- **Originating AS Number**   The number of the AS the route originated from.

- **External Protocol Metric** The metric used in the redistributed protocol. For an IGRP route that has been redistributed into EIGRP, this metric will be the same as the original IGRP metric.

- **External Protocol ID** A code designating the protocol that has been redistributed into EIGRP. Valid external protocols are IGRP, EIGRP, Static Routes, RIP, OSPF, IS-IS (not covered in this book), EGP (not covered in this book), BGP (not covered in this book), IRDP (not covered in this book), and directly connected.

External routes are discussed in more detail in the "Redistributing EIGRP" section, later in the chapter.

During normal operation, EIGRP sends updates to neighbors only when required by DUAL, and only sends updates about routes that have changed to the neighbors that require the update. Note that EIGRP only sends changes in updates and does not send the entire routing table, with one major exception: when an adjacency is initially established. When an adjacency is established, the neighbors involved in the adjacency establishment send their entire routing tables in one or more update packets, via unicast RTP, to the other neighbor. This feature allows neighbors to become "current" on routes in the network very quickly, without causing undue network traffic or CPU use.

In addition, EIGRP uses no more than 50 percent of the configured bandwidth for updates, by default. (This parameter is modifiable, however, as you will see in the "EIGRP Configuration and Troubleshooting" section, later in the chapter.) So, on a 768-Kbps link, you can ensure that EIGRP uses no more than 384 Kbps for update traffic. All traffic exceeding the bandwidth limit is queued until EIGRP traffic is using less than 384 Kbps.

> **Note** *EIGRP advertises only the best routes listed in the routing table, not all routes listed in the topology table. In other words, EIGRP does not advertise multiple routes for the same subnet. This distinction may seem silly, but it isn't. If EIGRP has a single successor and five feasible successors, for instance, only a single route through the advertising router will be sent, not all six routes. The idea behind this concept is that neighbors don't need to know about every way to get to the remote network through a particular router, just that the router has a path to the remote network.*

In addition to updates, a router may learn of routes through the EIGRP query and reply process. This topic is discussed in more detail in the "Route Maintenance (The DUAL Finite State Machine)" section, later in the chapter.

Route Selection

To provide for more rapid convergence, EIGRP's route selection process is a bit different from distance vector protocols. First, when EIGRP initially hears of a route, it examines the metric for the route and sets the feasible distance (FD) to equal that metric. The feasible distance is used by DUAL to calculate the feasibility condition (FC). The FC basically states that for a route to be acceptable, the metric advertised for the route from the remote neighbor (the local metric on the remote neighbor) must be less than the lowest local metric (the FD) for the destination in question. If a route does not meet the FC, the route is not added to the topology table.

Using the FC, DUAL does not need to worry about routing loops because any route that meets the FC *cannot* (just based on the math) be through this router. If a new route to the destination is heard, and the route meets the FC, the route is added to the topology table. If the new route has a metric that is lower than the current best metric, that route is chosen as the *successor,* and the FD is updated. If the other routes in the table meet the new FC, those routes are configured in the topology table as *feasible successors.* If the new route instead has a metric that is greater than the current best metric but still meets the FC, that route is added to the topology table as a feasible successor. Only the successor (the best path for a destination) for a route is entered into the routing table.

Feasible successors (valid routes that are *not* the best path to the destination) are maintained in the topology table in case the successor fails or the metric to the successor increases. Also, if a router does not have any knowledge of a given route, the FD for the route is effectively infinite, and any advertised route is accepted into the topology table. In other words, when you first boot a router, it accepts all new routes. Once a route has been accepted and a successor chosen, the FD is updated and all new routes have to meet the FC to be accepted.

Admittedly, this explanation was a bit rushed, but I find that when explaining EIGRP route selection, it is best to quickly summarize the key points and then use an example to explain the process in detail. For this example, examine the network shown in Figure 25-1.

True EIGRP Metric Calculation

In most of the examples in this chapter, metric calculations have been simplified to place the focus on route selection, not metric calculation. Although this example simply takes the advertised metric and adds the cost of the exit interface to that to calculate the final metric, in reality, EIGRP takes the advertised values for the metric components, adds the delay of the link to the advertised delay, determines reliability and load based on local calculations as well as the advertised reliability and load, and reduces the minimum bandwidth (if necessary). Finally, EIGRP performs the full metric calculation using these parameters. The end result is essentially the same (because in EIGRP, all K values for the entire AS must be the same), but the math behind metric calculations is a bit more complicated than what is shown in this example. For a full discussion of IGRP/EIGRP metric calculation, please refer to Chapter 24.

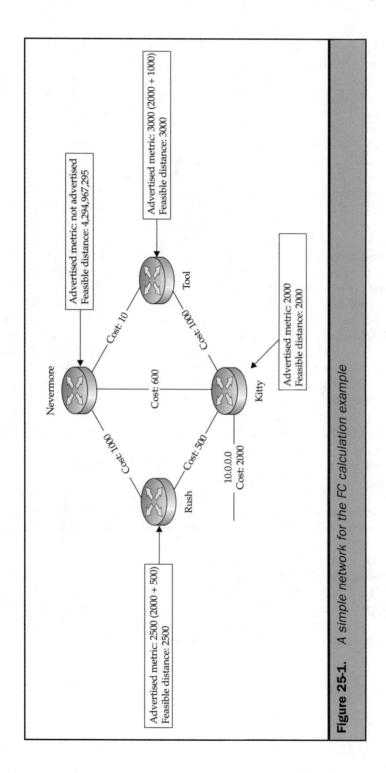

Figure 25-1.   *A simple network for the FC calculation example*

In this example, a new router (Nevermore) has just been added to the existing topology. The advertised metric listed by each router is the best metric on each router to network 10.0.0.0 at the moment Nevermore is added. The listed metric is the local metric that each router will advertise to all other directly connected routers in the topology. Each router, upon receiving the update, is responsible for adding to the local metric the "cost" associated with the interface through which the route was learned. In this way, the local metric equals the advertised metric plus the cost associated with the route's exit interface for the router in question.

For example, Rush's best path to the 10.0.0.0 network (and the only path at the time Nevermore is added) is through Kitty. Kitty is advertising a metric of 2,000 to Rush. The cost of Rush's link to Kitty is 500, so 2,000 + 500 = Rush's best metric for the route to the 10.0.0.0 network. So, once an adjacency has been formed between Rush and Nevermore, Rush advertises a metric of 2,500 to Nevermore for the route to network 10.0.0.0.

When you initially boot Nevermore, it does not know of any routes to the 10.0.0.0 network. Because of this, Nevermore will not advertise the 10.0.0.0 network in routing updates, and Nevermore's FD is set to infinite (4,294,967,295). When Nevermore boots, it almost immediately sends out hellos on all of its direct connections (Rush, Kitty, and Tool). Because this is the beginning of an adjacency, Rush, Kitty, and Tool send an entire update with all routes currently in their tables. These updates, however, will not all be received by Nevermore at the same time. Because the link to Tool is Nevermore's fastest link, let's assume the update from Tool arrives at Nevermore first.

Nevermore examines the update from Tool to see if it meets the FC. Again, the FC states that the advertised metric for a route must be less than the FD in order for the route to be entered into the topology table. Nevermore's current FD is 4 billion, and the advertised metric for the route through Tool is 3,000. 3,000 < 4,294,967,295, so the route meets the FC and is inserted into the topology table. Nevermore adds the cost of the link to Tool to the advertised metric to arrive at his local metric through Tool (3,000 + 10 = 3,010). At this instant, Nevermore's successor (the path with the best metric, and the only path at the moment) for the route to the 10.0.0.0 network is the route through Tool, and Nevermore's FD changes to 3,010 (his best local metric for the 10.0.0.0 network).

Next, Nevermore hears the update from Kitty with an advertised metric of 2,000. Nevermore checks to see if the route through Kitty meets the FC. 2,000 (Kitty's advertised metric) < 3,010 (Nevermore's current FD), so the route through Kitty is valid. Nevermore adds the cost of the link to the advertised metric to arrive at a local metric of 2,600 (2,000 + 600) and enters the route through Kitty in the topology table. At this point, Nevermore performs a local computation (discussed in detail in the next section) and examines his topology table to determine which paths to the 10.0.0.0 network should be chosen as successors and feasible successors.

The path through Kitty is determined to have the lowest cost (2,600 < 3,010) and is chosen as the successor. Because a new successor was chosen, the lowest-cost local path to the 10.0.0.0 network has changed, and the FD for the network must be updated to equal the new lowest-cost path (2,600). Nevermore reexamines its topology table to ensure that all listed routes meet the FC and finds that the path through Tool no longer meets the FC. (Tool's advertised metric is 3,000, while Nevermore's FD is 2,600. 3,000 > 2,600, so the route through Tool does not meet the FC.) Tool's route is removed from the topology table because it no longer meets the FC.

Finally, the update from Rush arrives at Nevermore with an advertised metric of 2,500. Nevermore checks the route from Rush to see if it meets the FC, finds that it does (2,500 < 2,600), and applies the link cost to the advertised metric to arrive at a local metric of 3,500 (2,500 + 1,000). Nevermore adds the path to the topology table, performs a local computation, and realizes that the path through Kitty is still the best path (2,600 < 3,500).

At this point, Nevermore sends an update about its known route to the 10.0.0.0 network advertising its best local metric (2,600). Now you get to see that EIGRP's route selection mechanism will not allow a loop to occur. On Kitty, the best path to the 10.0.0.0 network has a cost of 2,000, making Kitty's FD 2,000. When Kitty hears the update from Nevermore, Kitty checks the route to see if it meets the FC before adding it to the topology table. Because the advertised metric from Nevermore (2,600) is greater than Kitty's FD, Kitty simply ignores the path through Nevermore (and consequently back through Kitty). Because of the requirement that a route meet the FC before it is entered into the topology table, a route that leads back through your router will never be entered into the table.

Note *In this example, split horizon was not taken into account; but, by default, split horizon with poison reverse is still used for EIGRP (although it isn't really needed). If split horizon with poison reverse were enabled in this example, Nevermore would advertise the path to the 10.0.0.0 network back to his successor (Kitty) with a cost of 4,294,967,295.*

CISCO ROUTING

Just for Kicks

Using the example from Figure 25-1, try to determine which path to the 10.0.0.0 network Tool and Rush will choose as their successor after they hear the update from Nevermore. The answer and explanation are provided in a sidebar on the next page.

Answer

Rush still goes directly through Kitty because the cost of the route advertised from Nevermore is 2,600, which does not meet the FC on Rush. Tool, however, uses the path through Nevermore as his successor because the path through Nevermore has a cost of 2,610, while the path directly through Kitty has a cost of 3,000. Tool lists the path through Kitty in the topology table as a feasible successor to the 10.0.0.0 network because the advertised cost from Kitty of 2,000 meets the FC. If this explanation makes perfect sense, pat yourself on the back. Otherwise, you may want to go back and read through the example a few more times.

Route Maintenance (The DUAL Finite State Machine)

EIGRP routes are maintained using DUAL. The DUAL algorithm may seem complicated at first blush, but it really isn't all that difficult to understand once you break it down. First, let's break DUAL down into its components and then examine DUAL operation in a few examples.

For the explanation, I'll break DUAL into six major components (some of which have already been discussed):

- Neighbor discovery (hello messages)
- Loop prevention (FC)
- Route selection (successors and feasible successors)
- Route states (active and passive)
- Route computations (local and diffusing)
- DUAL operation (the DUAL finite state machine)

The first three components have already been discussed, but the last three are new, so they are covered in the following sections before walking through some examples.

Route States and Computations   DUAL routes can be in one of two states: active or passive. Note that a route may be in only one state at a time, and the state a single route is in is independent of all other routes. In other words, if one route is active, the other routes can be (and actually must be) passive. The reason the other routes must be passive is because DUAL can perform only a single diffusing computation at a time. Until the current active route returns to passive, another route will never enter the active state. (This functionality is covered in more detail in a moment.)

The passive state is the normal state for all DUAL routes. Passive routes can be reevaluated at any time, successors and feasible successors chosen, and the FD updated.

Passive routes are also used to route packets and respond to queries by remote routers. Passive routes are advertised to all neighbors. Basically, a *passive route* is a route that is not performing a diffusing computation.

Active routes are routes that are undergoing a diffusing computation (sometimes called a route recomputation). During a diffusing computation, the router queries all neighbors for a path to a failed network. *Queries* are a special type of EIGRP packet that must be responded to with a reply. Until all replies from all neighbors have been received, DUAL may not return to a passive state. While in the active state, the router cannot change the route's successor, the distance it is advertising for the route, or the FD for the route.

Basically, the primary difference between the active and passive states is that while in a passive state, a diffusing computation cannot be performed; and while in an active state, the route cannot be modified.

At this point, you are probably wondering what a diffusing computation is. Routes have two types of computations they can undergo, depending on whether a feasible successor exists for the route: a local computation and a diffusing computation. A *local computation* is the normal type, and it is performed any time an input event occurs. An *input event* is any event that could potentially affect the availability or cost of a route. Input events are any of the following:

- Change in state (up or down) of a directly connected link
- Change in cost of a directly connected link
- Reception of an update, query, or reply packet

Note   *In reality, the first two types of input events (change in state and change in cost of a directly connected link) are the same because when you change the cost parameters (bandwidth or delay) for a link, EIGRP resets the link. In other words, changing the cost parameters on the link actually shuts EIGRP down on the link and then brings EGRP back up on the link with the new cost.*

During a local computation, DUAL examines the topology table to determine the best successors and feasible successors for the route, and may update the FD if a new successor is chosen. If DUAL cannot find any successors or feasible successors for the route, DUAL *may* decide to perform a diffusing computation. Notice that I said "may" instead of "will." In some cases, DUAL may not need to perform a diffusing computation, but you will learn about this concept in more detail in a moment. In most cases, however, DUAL performs a diffusing computation if a successor or feasible successor for a previously entered route cannot be found.

A *diffusing computation* is when DUAL queries neighbors to attempt to determine if a network can be reached through an alternate route. Typically, a diffusing computation occurs when the primary route (the route through the successor) fails and no feasible successors exist in the topology table. Diffusing computations are needed because the FC may prevent a router from learning of all paths to a network while the successor for that route is still available.

For instance, in the previous example (Figure 25-1), once convergence has occurred, the only route to the 10.0.0.0 network that would be entered in Rush's topology table is the route through Kitty because the route through Nevermore does not meet the FC on Rush (2,600 > 2,500). If Rush's link to Kitty were to fail, it would need some way to learn of the route through Nevermore. Because Nevermore is not going to advertise the route at regular intervals (EIGRP advertises routes only when absolutely necessary), Rush needs some way of informing Nevermore of the failure of its primary route and asking for a new route.

Rush accomplishes this task by entering an active state and performing a diffusing computation. Rush sends a query to all of his remaining neighbors (only Nevermore, in this case) that basically says, "I have no idea how to reach the 10.0.0.0 network. Do you have a route to that network?" Nevermore responds with a reply containing the metric for his route to the 10.0.0.0 network. Rush then returns to passive and enters the route into the table, choosing Nevermore as his successor.

There are a couple of points to keep in mind while DUAL is performing this diffusing computation. First, if a particular route is currently active, DUAL may not change the successor or advertised metric for the route until the route returns to passive. Consequently, if a query about the route is received while DUAL is active, DUAL replies with the metric it had when it entered the active state. If the metric changes after the route returns to passive, DUAL sends an update to all neighbors (except the successor for the route) with the new metric.

Second, DUAL will not place more than one route in an active state at a time. If multiple routes require a diffusing computation, they are processed one at a time. Finally, DUAL queries from one router may cause other routers to transition to an active state and send queries to *their* neighbors (except for neighbors who share the link with the router that originated the query). In this manner, the query can be propagated across the entire AS until either an alternate route is found or the network is determined to be unreachable. All of these little details are examined further in the next section; but for now, just keep them in mind.

Now that the basic definitions are out of the way, let's take a look at the DUAL finite state machine.

DUAL Operation (The DUAL Finite State Machine)   The DUAL Finite State Machine is the algorithm that goes on behind the scenes to generate and maintain the DUAL routing topology. Basically, what happens is this:

- Neighbor adjacencies are established using hello packets.
- Neighbor failures are detected by lack of hello packets.
- When a neighbor relationship is established, the entire routing table is sent as an update to that neighbor.

■ Whenever an input event occurs, DUAL performs local computations to select the best routes. If the local computation cannot find any valid successors, DUAL may choose to perform a diffusing computation.

■ Whenever a change in the routing table occurs, all neighbors are updated with the change.

The DUAL algorithm actually defines multiple active states and multiple input event types to deal with the myriad of different events that can occur in the network. These events are detailed in J.J. Garcia-Luna-Aceves' papers "A Unified Approach for Loop-Free Routing Using Link States or Distance Vectors" (*ACM SIGCOMM Computer Communications Review*, Vol. 19, No. 4, September 1989) and "Loop-Free Routing Using Diffusing Computations" (*IEEE/ACM Transactions on Networking*, Vol. 1, No. 1, February 1993). Both papers are available on *www.acm.org* for a nominal fee. A slightly watered down (and EIGRP-specific) version of the algorithm is also provided in Jeff Doyle's book *Routing TCP/IP, Volume I* (Cisco Press, 1998).

The problem with using the diagrams and depictions provided by these core references is that, although they are highly accurate and take into account all possible events in an EIGRP network, attempting to understand them tends to cause the hair on your head to end up in your palms. To decrease the difficulty level, the flowcharts depicted in Figures 25-2 through 25-9 are a slightly simplified version of the algorithm. For those of you who would like to see the flowchart as one complete image, please visit *http://www.alfageek.com*. Although these charts do not take into account every possible event in EIGRP, they cover well over 90 percent of EIGRP events.

The charts function as follows. The chart begins at the top with Passive (Stable) in Figure 25-2, assuming that adjacencies have already been established and the network is in a converged state. Upon reception of an input event, follow the arrows down to the specific type of input event. Note that references to routers X, Y, and Z should be viewed as follows:

■ **Router X**   Any neighbor that is not the successor for the route in question

■ **Router Y**   The successor for the route in question

■ **Router Z**   The local router

The following terms are used to describe the input event types:

■ **Query from router (Router X) other than current successor (Router Y)**
A query was received for a route, but not from the current successor for that route. The importance of this event is that because the query was *not* received from the successor, the route through the successor is still valid.

■ **Update from router (Router X) other than current successor (Router Y)**
An update was received for a route, but not from the current successor for that route.

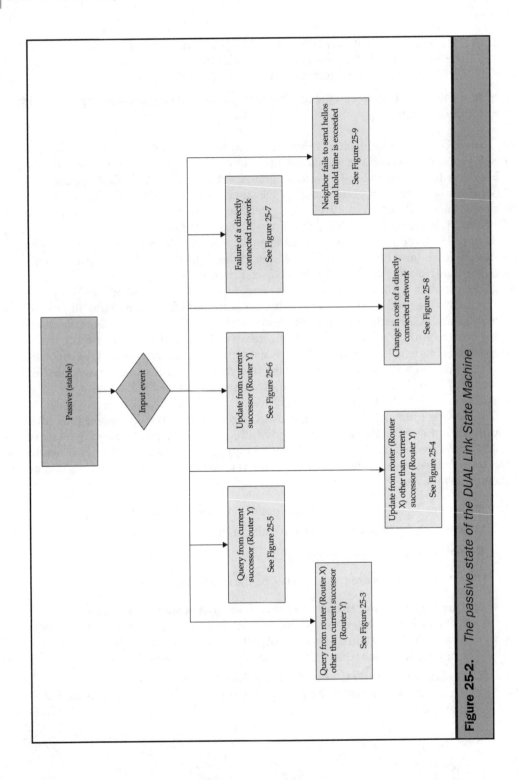

Figure 25-2. The passive state of the DUAL Link State Machine

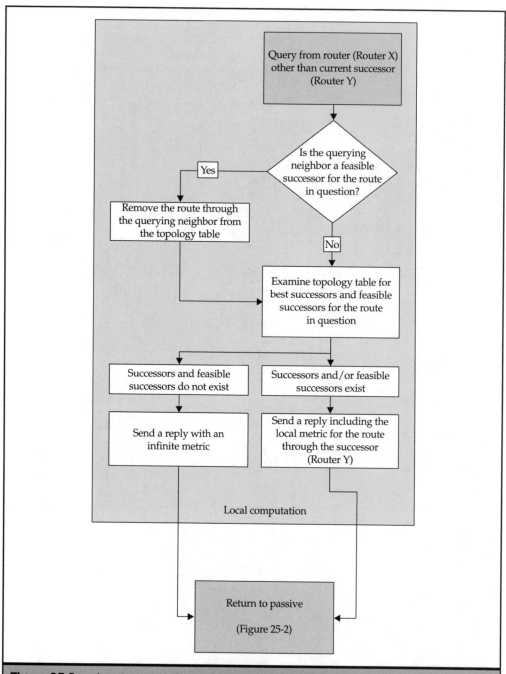

Figure 25-3. *Input event—Query from router (Router X) other than current successor (Router Y)*

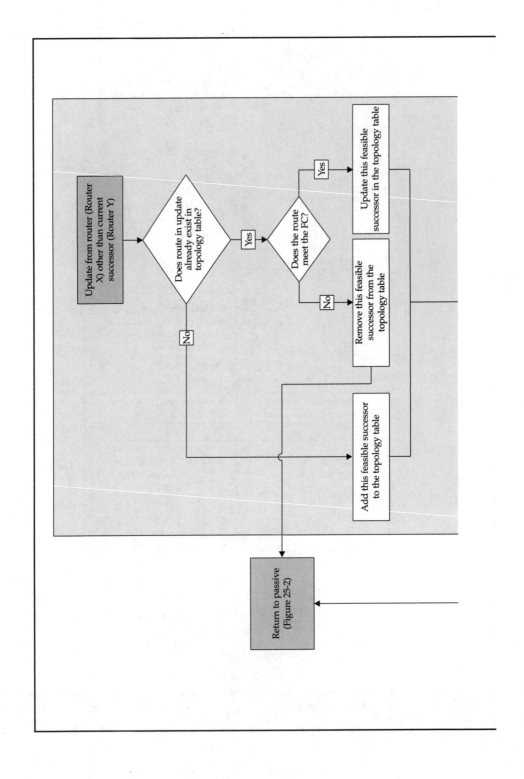

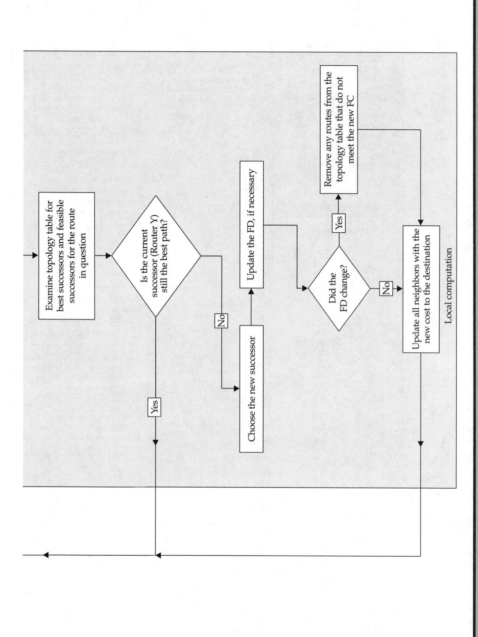

Figure 25-4.   *Input event—Update from router (Router X) other than current successor (Router Y)*

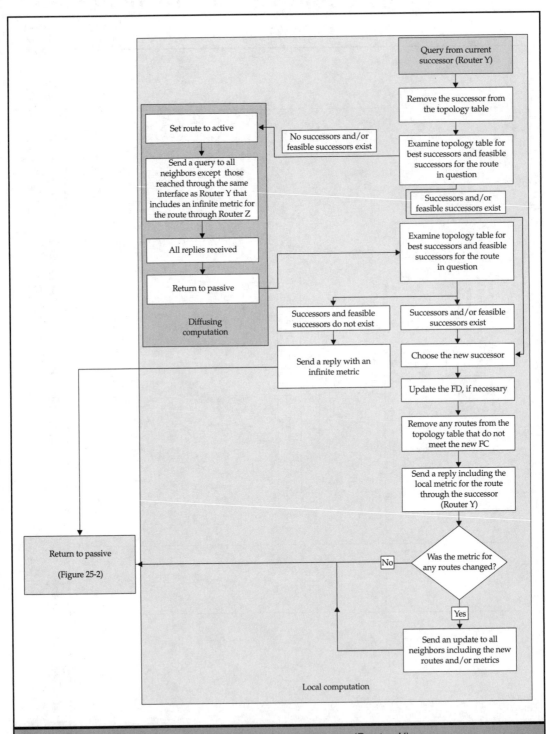

Figure 25-5. *Input event—Query from current successor (Router Y)*

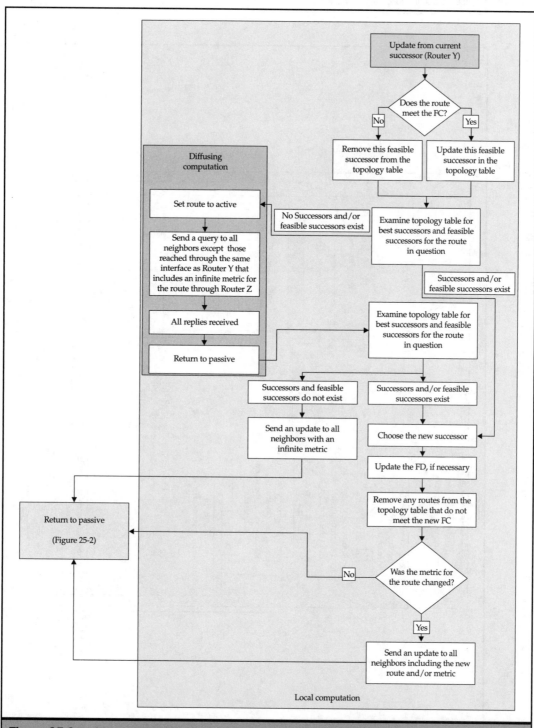

Figure 25-6. *Input event—Update from current successor (Router Y)*

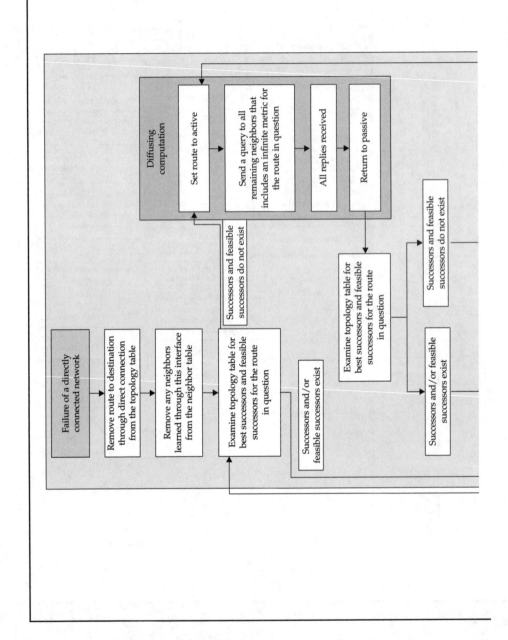

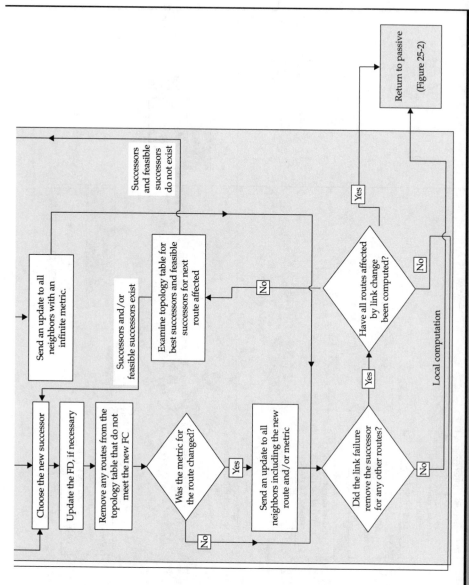

Figure 25-7.   *Input event—Failure of a directly connected network*

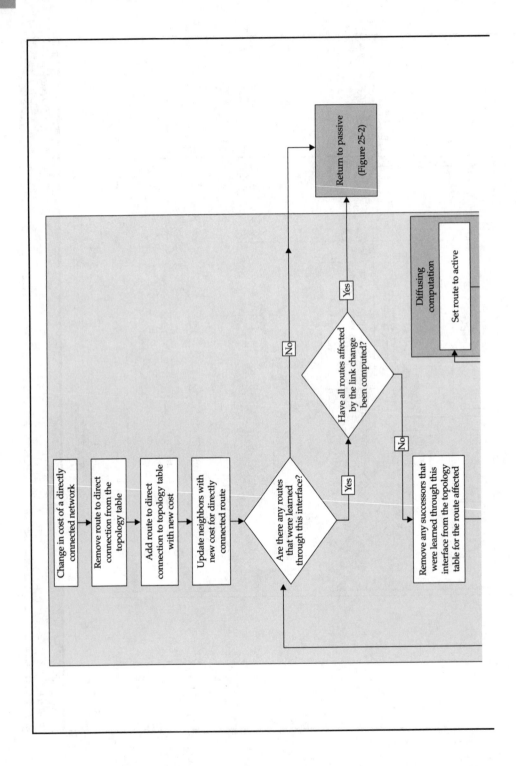

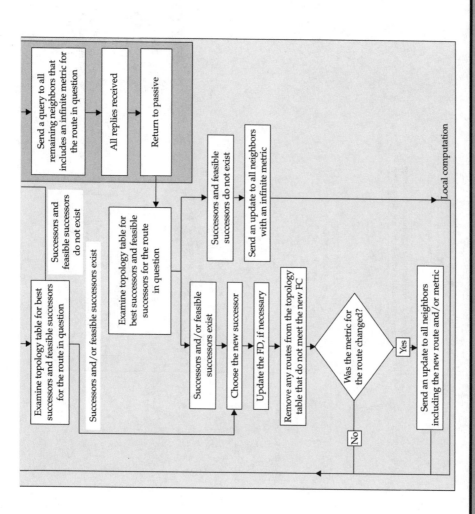

Figure 25-8.   *Input event—Change in cost of a directly connected network*

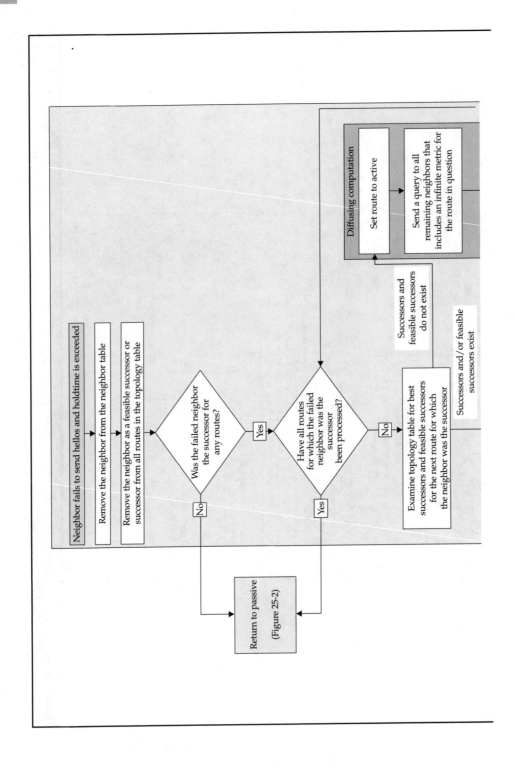

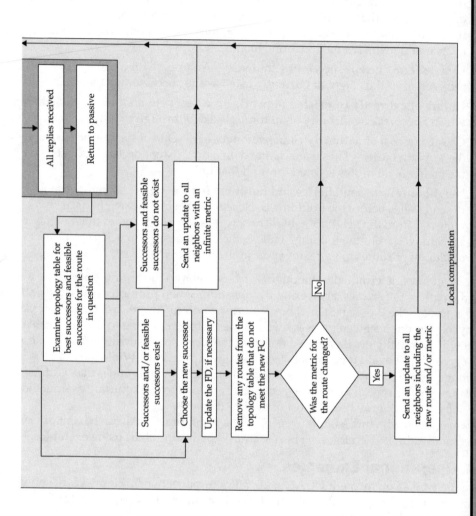

Figure 25-9. *Input event—Neighbor fails to send hellos and holdtime is exceeded.*

- **Query from current successor (Router Y)** A query for a route was received from the neighbor currently listed as the successor for that route. The importance of this event is that because the query *was* received from the successor, the route through the successor has failed. (Queries are sent only when a path to a particular network cannot be found at all.)

- **Update from current successor (Router Y)** An update for a route was received from the neighbor currently listed as the successor for that route.

- **Failure of a directly connected network** A direct network link has failed, which affects the availability of all routes learned through that link.

- **Change in cost of a directly connected network** One of the metric parameters (bandwidth or delay) has been changed on a directly connected link, affecting the metric of all routes learned through that link.

- **Neighbor fails to send hellos and holdtime is exceeded** A neighbor fails to send hellos before the holdtime is exceeded, causing the neighbor to be removed from the neighbor table. Thus, routes learned through that neighbor are removed from the topology table, which may cause a new successor to be chosen or routes to be removed from the routing table.

Once you have determined the input event type, refer to the appropriate figure and follow the path of the arrows, answering the questions when prompted, until the path leads you back to the Passive (Stable) box (returning you to Figure 25-2). In some cases, you may loop through a given process several times before returning to Passive (Stable), but you will eventually get there. Also note that the large boxes surrounding the paths under a given input event type are labeled "Local computation." While the router is performing these processes, the route is still passive. If DUAL determines that a diffusing computation needs to occur, it enters an active state and performs a diffusing computation, as is seen by the dark-gray box in most of the input events.

However, even with this "simple" version of the algorithm, DUAL can be a bit difficult to understand at first. Therefore, the next section explains the chart using examples.

DUAL Operational Examples

This section walks you through DUAL using an example network. Figure 25-10 shows the basic network, along with the costs for links. The current topology table for the 192.168.1.0 network for all routers is shown in Table 25-1.

The only routes you need to be concerned about in this case are the routes to the 192.168.1.0 network. Once again, the metrics have been simplified to place the focus on DUAL rather than metric calculation. The Local Cost and Advertised Cost columns are the metric calculated locally and the metric calculated at the neighbor in question, respectively. The S/FS column shows whether this neighbor is the successor or feasible successor. Finally, in this figure, convergence has already occurred. To show you how I arrived at these topology tables, however, I will show you the convergence process, beginning with Figure 25-11 and its associated topology tables, shown in Table 25-2.

In Figure 25-11, Bela boots, enters his direct connection to the 192.168.1.0 network as his successor for that network, establishes adjacencies with Miles and Primus, and

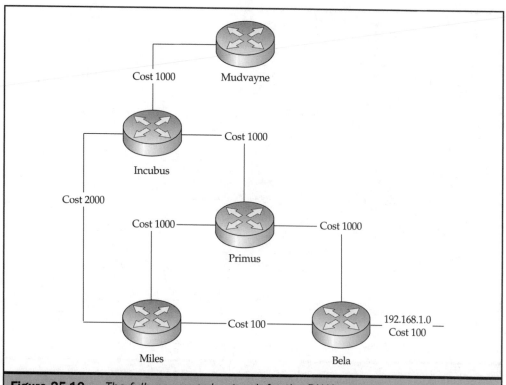

Figure 25-10.   *The fully converged network for the DUAL operation examples*

Router	Neighbor	Local Cost	Advertised Cost	S/FS	FD
Mudvayne	Incubus	3,100	2,100	S	3,100
Incubus	Primus	2,100	1,100	S	2,100
Incubus	Miles	2,200	200	FS	2,100
Primus	Bela	1,100	100	S	1,100
Primus	Miles	1,200	200	FS	1,100
Miles	Bela	200	100	S	200
Bela	Direct connection	100	100	S	100

Table 25-1.   *Topology tables for Figure 25-10*

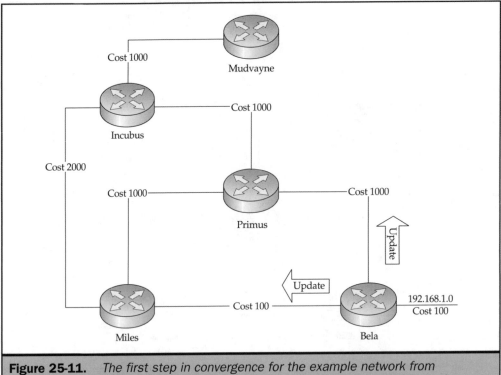

Figure 25-11. *The first step in convergence for the example network from Figure 25-10*

advertises the 192.168.1.0 network to his neighbors. On Miles and Primus (computations shown in Figure 25-12; the computation is the same for both routers), DUAL receives

Router	Neighbor	Local Cost	Advertised Cost	S/FS	FD
Mudvayne	—	—	—	—	4,294,967, 295
Incubus	—	—	—	—	4,294,967, 295
Primus	Bela	1,100	100	S	1,100
Miles	Bela	200	100	S	200
Bela	Direct connection	100	100	S	100

Table 25-2. *Topology tables for Figure 25-11*

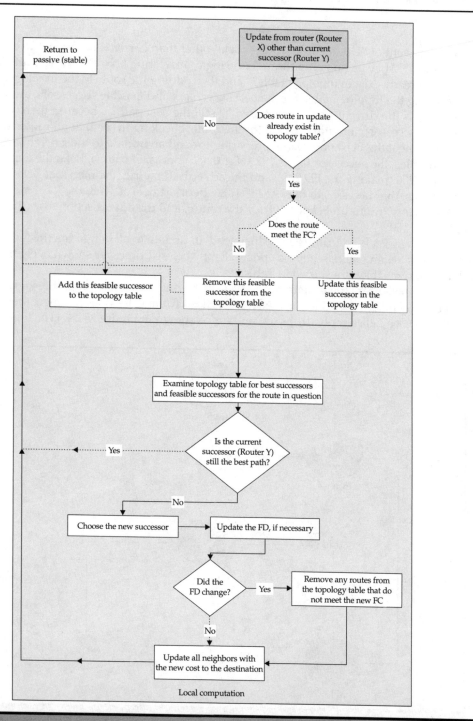

Figure 25-12. *Miles' and Primus' computations for Figure 25-11 (the process is identical for both routers)*

an input event of "Update from router (Bela) other than current successor (No current successor)." They begin going through the logic, answering "No" to "Does route in update already exist in topology table?" and then adding the route to the topology table; examining the topology table for the best successors and feasible successors; answering "No" to "Is the current successor (Router Y) still the best path?" (because there was no current successor); updating the FD for the route; checking to see that all routes to the network meet the FD; and, finally, preparing to send an update to their neighbors.

On Miles, the successor to the 192.168.1.0 network is, of course, Bela; the local metric is 200 (making the FD 200); and the advertised metric (the metric at Bela) is 100. On Primus, the successor to the 192.168.1.0 network is Bela as well, the local metric is 1,100 (giving Primus an FD of 1,100 for the route), and the advertised metric (the metric at Bela) is 100.

At this point, both Primus and Miles send an update to all neighbors (including Bela). Let's assume Miles sends his update slightly before Primus, as shown in Figure 25-13 (the topology tables are shown in Table 25-3).

When Bela hears the update from Miles, he begins walking through the logic and quickly determines that the metric advertised from Miles does not meet the FC (200 > 100), so Bela simply drops the update, as shown in Figure 25-14. (The logic says

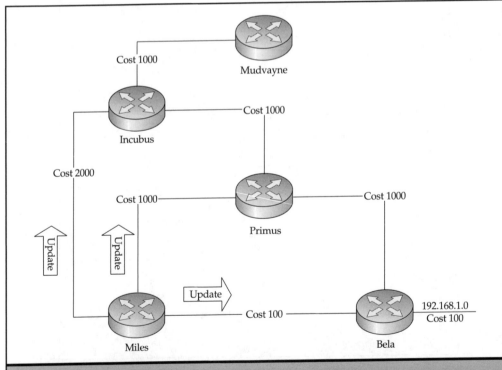

Figure 25-13. *Step two in convergence*

Router	Neighbor	Local Cost	Advertised Cost	S/FS	FD
Mudvayne	—	—	—	—	4,294,967,295
Incubus	Miles	2,200	200	FS	2,200
Primus	Bela	1,100	100	S	1,100
Primus	Miles	1,200	200	FS	1,100
Miles	Bela	200	100	S	200
Bela	Direct connection	100	100	S	100

Table 25-3.   *Topology Tables for Figure 25-13*

"Remove this feasible successor from the topology table"; but because Miles was not entered into the table as a feasible successor for the route, Bela just drops the update.)

When Primus hears the update from Miles, he begins walking through the logic, realizes that the metric advertised from Miles *does* meet the FC (200 < 1,100), and adds Miles into his topology table. Primus then examines the topology table for the best path, realizes that the path through Bela is still the best path, marks Miles as a feasible successor, and finishes the local computation (as shown in Figure 25-15).

Finally, Incubus hears the advertisement from Miles and, after examining the logic, enters Miles in his topology table as the successor to the 192.168.1.0 network with a local metric of 2,200 and a FD of 2,200. See Figure 25-16.

In Figure 25-17, Primus now sends his update to all neighbors (the topology tables are shown in Table 25-4).

Note   *In this example, a few steps have been omitted to keep the example as brief as possible while still showing all of the steps in the logic. For example, in reality, Miles and Primus would have sent their updates at approximately the same time. When Incubus hears the update from Miles (assuming Miles' update arrives first), he modifies the topology table and immediately sends out an update. Then he hears of the update from Primus; modifies the topology and routing tables (making Primus the new successor); and, because a new successor was chosen, sends out a new update. To reduce the complexity of the explanation, I consolidated these two updates from Incubus into a single update.*

At Miles, the update from Primus does not meet the FC (1,100 > 200), so it is dropped, as shown in Figure 25-18. At Bela, the update from Primus obviously does not meet the FC (1,100 > 100), so Bela drops the update as well (see Figure 25-18; the computations are the same for Miles and Bela).

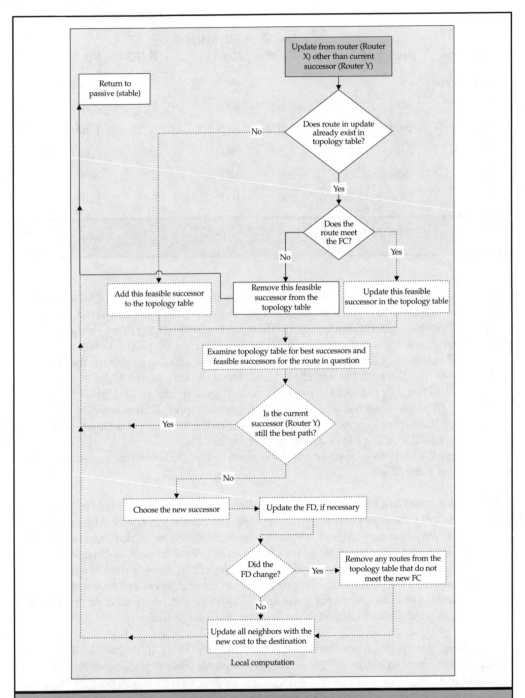

Figure 25-14.   *Bela's computation for Figure 25-13*

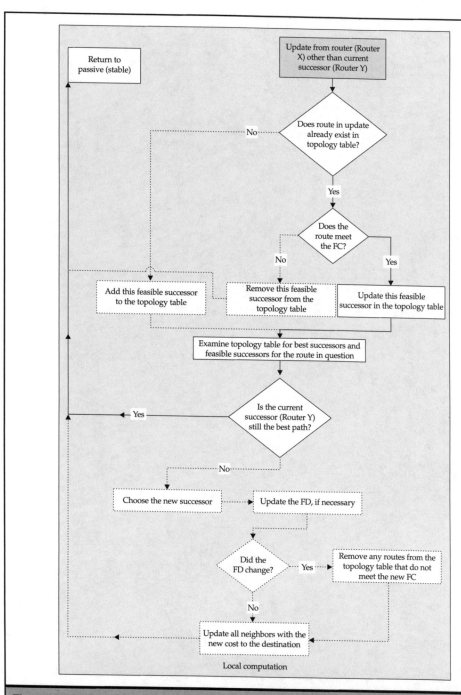

Figure 25-15. *Primus' computation for Figure 25-13*

CISCO ROUTING

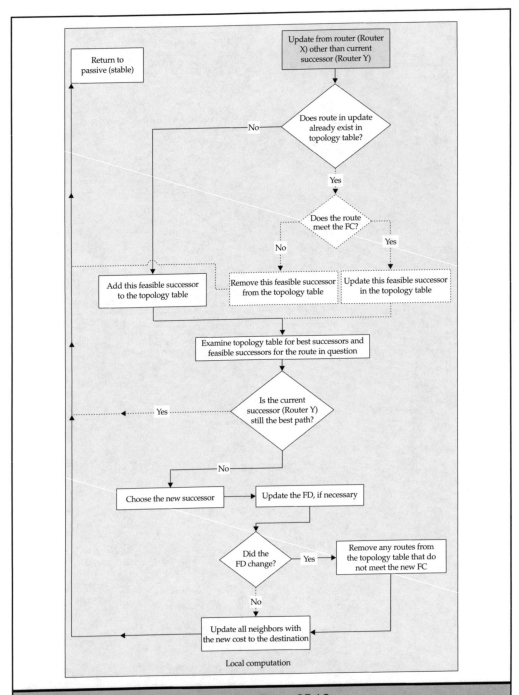

Figure 25-16.   *Incubus' computation for Figure 25-13*

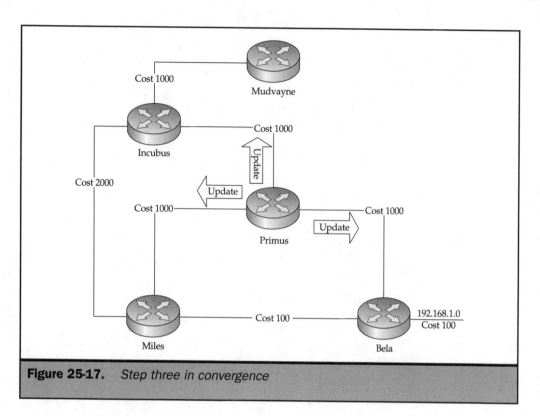

Figure 25-17.   *Step three in convergence*

At Incubus, however, the update not only meets the FC, but it turns out that the local metric for the path through Primus is better than the local metric for the path through Miles. See Figure 25-19.

Router	Neighbor	Local Cost	Advertised Cost	S/FS	FD
Mudvayne	—	—	—	—	4,294,967,295
Incubus	Primus	2,100	1,100	S	2,100
Incubus	Miles	2,200	200	FS	2,100
Primus	Bela	1,100	100	S	1,100
Primus	Miles	1,200	200	FS	1,100
Miles	Bela	200	100	S	200
Bela	Direct connection	100	100	S	100

Table 25-4.   *Topology Tables for Figure 25-17*

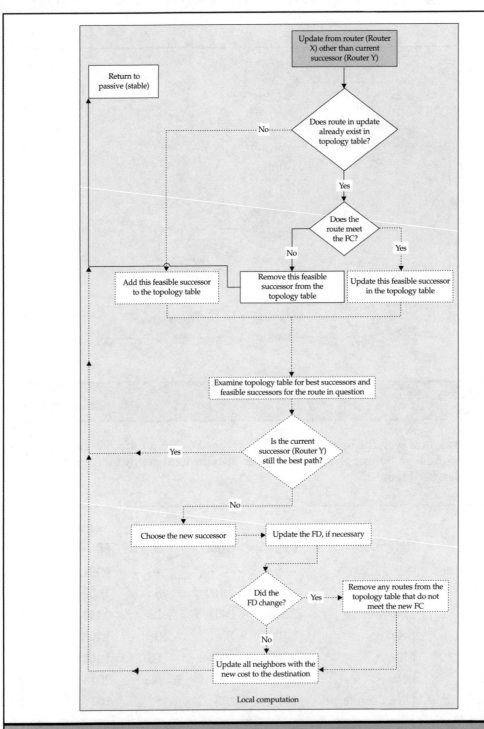

Figure 25-18. *Miles' and Bela's computation for Figure 25-17*

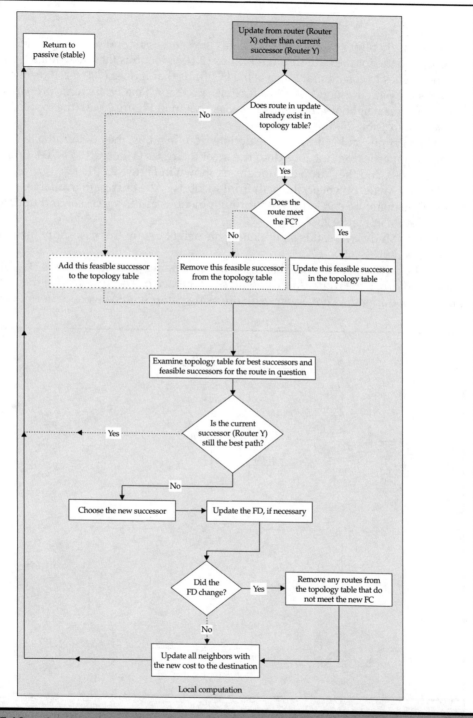

Figure 25-19. *Incubus' computation for Figure 25-17*

Incubus therefore chooses Primus as his new successor and changes the FD to equal the local metric for the path through Primus (2,100). Incubus then reevaluates the path through Miles to ensure that it meets the FC, finds that it does (200 < 2,100), and changes the path through Miles to a feasible successor. Finally, Incubus prepares the update for his neighbors and sends them, as shown in Figure 25-20 (the topology tables are shown in Table 25-5).

When Primus and Miles hear the update from Incubus, they determine that the advertised metric from Incubus does not meet their FCs (Miles: 2,100 > 200; Primus: 2,100 > 1,100), and they drop the update, as shown in Figure 25-21.

At Mudvayne, however, the path from Incubus is the only path available. Mudvayne therefore enters the path through Incubus as the successor and updates his FD, as shown in Figure 25-22.

Finally, Mudvayne sends an update to his neighbors (in this case, only Incubus), as shown in Figure 25-23.

Incubus will, of course, see that Mudvayne's update does not meet his FC (3,100 > 2,100), and drops the route. At this point, convergence has occurred. Note that throughout this entire convergence, no diffusing computations were performed; in fact,

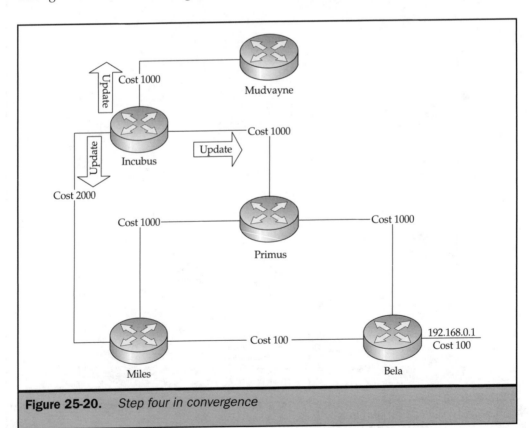

Figure 25-20.   *Step four in convergence*

Router	Neighbor	Local Cost	Advertised Cost	S/FS	FD
Mudvayne	Incubus	3,100	2,100	S	3,100
Incubus	Primus	2,100	1,100	S	2,100
Incubus	Miles	2,200	200	FS	2,100
Primus	Bela	1,100	100	S	1,100
Primus	Miles	1,200	200	FS	1,100
Miles	Bela	200	100	S	200
Bela	Direct connection	100	100	S	100

Table 25-5.   *Topology Tables for Figure 25-20*

there was only a single input event type. At this point, if nothing ever changed (no routes go down, no routers fail, and no costs change), no further updates will be sent! The routers would simply continue to send hellos at regular intervals, and the network would remain converged.

Also, realize that in real life, this entire process would most likely take place in under a second! The following are the two most powerful benefits of EIGRP: ultra-rapid convergence and minimal overhead. However, eventually, something in the network is going to change, and EIGRP has to be able to respond to the change rapidly and efficiently, find an alternate path, and prevent routing loops—all at the same time. To see how EIGRP accomplishes all of this, suppose the connection from Miles to Bela fails. As soon as Miles detects the failure, he begins the search for a new route, following the logic outlined in Figure 25-24 (the topology tables are shown in Table 25-6).

Basically, after removing the route through the failed connection, Miles looks for another successor candidate in his topology table. Finding none, Miles begins the process of switching the route to an active state and beginning a diffusing computation by sending a query to his remaining neighbors. Note that Miles also changed the FD for the route to infinite. Because he removed his only successor for the route, his best metric for the route changes to infinite, which changes his FD to infinite as well. This change occurs before Miles begins the diffusing computation (right after Miles removes the route through the failed network). See Figure 25-25.

Note   *Because, Miles cannot advertise a different metric for the route than the metric he was using when he began the diffusing computation (until the route transitions back to a passive state), if Miles were to receive a query about this network while he was still in an active state, he would advertise an infinite metric. In this example, Miles did not receive a query while in an active state, but it is still an important point.*

CISCO ROUTING

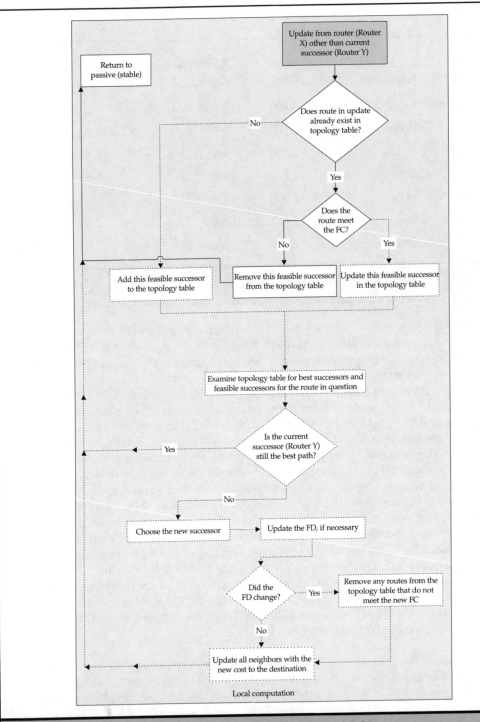

Figure 25-21. *Primus' and Miles' computation for Figure 25-20*

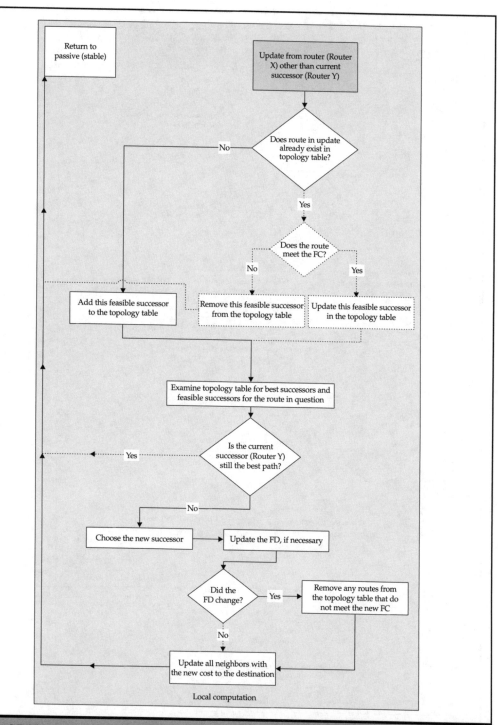

Figure 25-22. *Mudvayne's computation for Figure 25-20*

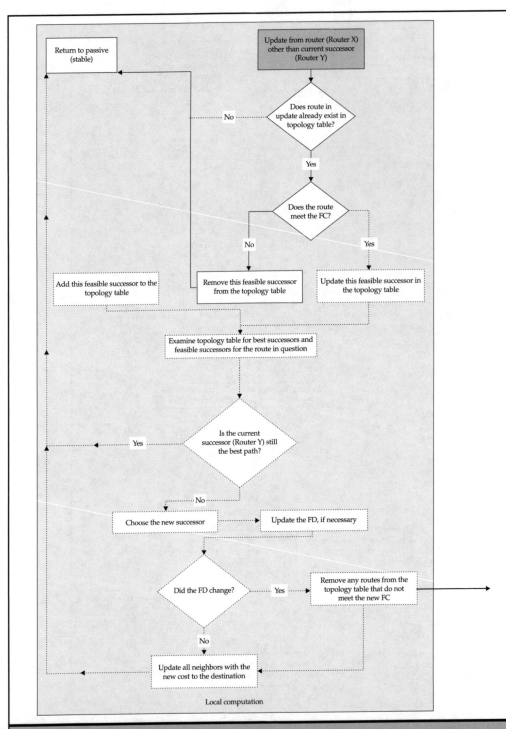

Figure 25-23. *The final step in convergence*

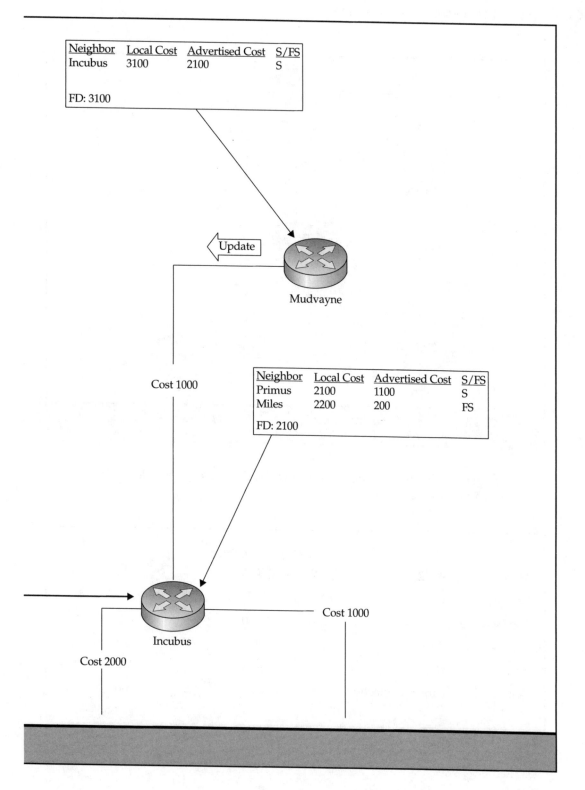

Neighbor	Local Cost	Advertised Cost	S/FS
Incubus	3100	2100	S

FD: 3100

Update

Mudvayne

Cost 1000

Neighbor	Local Cost	Advertised Cost	S/FS
Primus	2100	1100	S
Miles	2200	200	FS

FD: 2100

Incubus

Cost 1000

Cost 2000

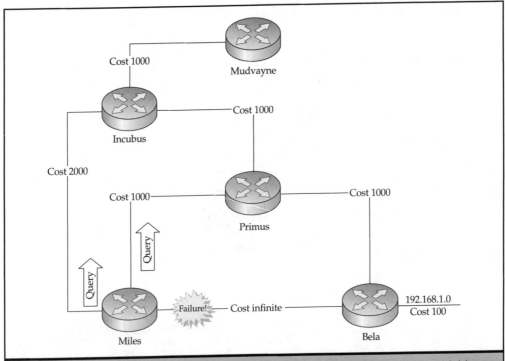

Figure 25-24. *Miles searching for a new route by sending queries to his neighbors*

At Primus and Incubus, the query is received and processed (see Figure 25-26 for the computations for both).

Router	Neighbor	Local Cost	Advertised Cost	S/FS	FD
Mudvayne	Incubus	3,100	2,100	S	3,100
Incubus	Primus	2,100	1,100	S	2,100
Primus	Bela	1,100	100	S	1,100
Miles	—	—	—	—	4,294,967,295
Bela	Direct connection	100	100	S	100

Table 25-6. *Topology Tables for Figure 25-24*

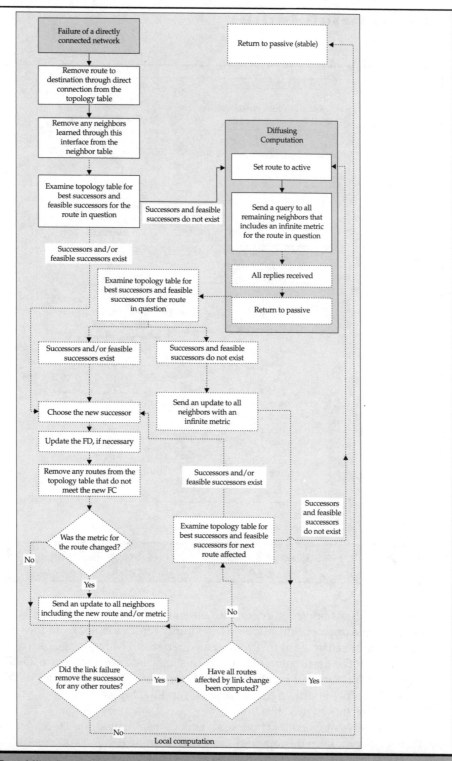

Figure 25-25. *Miles' computation for Figure 25-24*

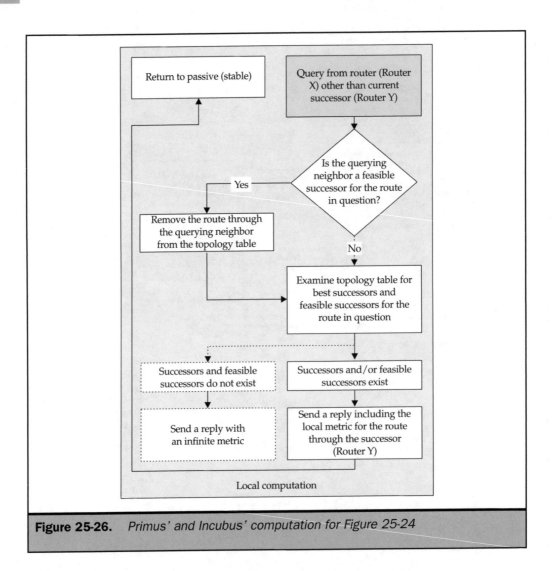

Figure 25-26. *Primus' and Incubus' computation for Figure 25-24*

Because Miles is not the successor for either of these routers, they will not need to perform any major reconfigurations. They simply remove Miles from the list of feasible successors (because, by design, if they get a query from Miles, it means that Miles can no longer reach the destination network). This logic is shown in Figure 25-27 (the logic is identical for Primus and Incubus). Then they reply with their metrics to the remote network. This reply is shown in Figure 25-28, and the topology tables for the network at this point are shown in Table 25-7.

Miles initially enters both Primus and Incubus as potential successors for the route, but then Miles examines the topology table to determine the best successor. Primus is picked as the best successor because the local metric at Miles for the path through Primus is 2,100, while the local metric at Miles for the path through Incubus is 4,100. Because

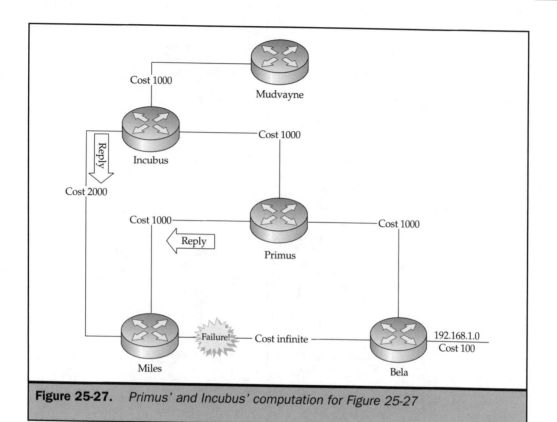

Figure 25-27. *Primus' and Incubus' computation for Figure 25-27*

Primus is the successor, the FD is set to 2,100. Consequently, the route through Incubus is removed from the table entirely because Incubus' advertised metric does not meet the FC. (The advertised metric for the path through Incubus is 2,100, which is *not* less than the FD, so Incubus does not meet the FC.) The logic for this step is shown in Figure 25-29.

Once Miles has chosen a successor, he sends out an update about the route (again, following the logic in the chart) to all of his neighbors. Primus and Incubus examine the advertised path to see if it meets the FC, realize that it does not, and throw the update away. This process is shown in Figure 25-30 (the topology tables are shown in Table 25-8), with the logic for Miles, Incubus, and Primus shown in Figures 25-31 and 25-32. In reality, Miles would have had to perform this process for each route affected by the downed link; but because you are concerned only with the route to the 192.168.1.0 network for this example, you would answer "No" to the "Did the link failure remove the successor for any other routes?" question.

One important point to note about this entire process is that Mudvayne was completely unaffected by this change. (In fact, Mudvayne doesn't even know a change occurred.) The ability to only affect a subset of the routers in a topology is another of EIGRP's major benefits: only those routers that need to know about a change will be affected by that change.

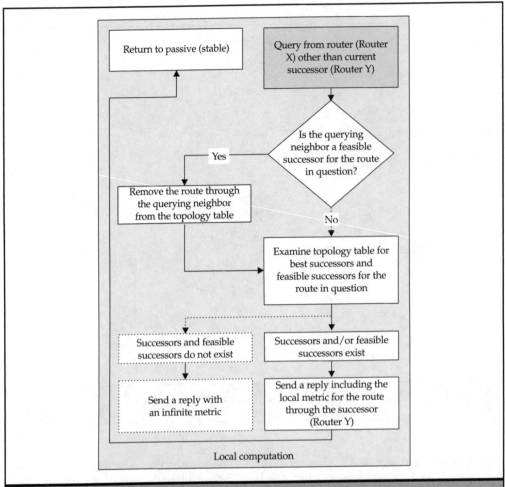

Figure 25-28. *Incubus and Primus reply to Miles' query*

Router	Neighbor	Local Cost	Advertised Cost	S/FS	FD
Mudvayne	Incubus	3,100	2,100	S	3,100
Incubus	Primus	2,100	1,100	S	2,100
Primus	Bela	1,100	100	S	1,100
Miles	Primus	2,100	1,100	S	2,100
Bela	Direct connection	100	100	S	100

Table 25-7. *Topology Tables for Figure 25-28*

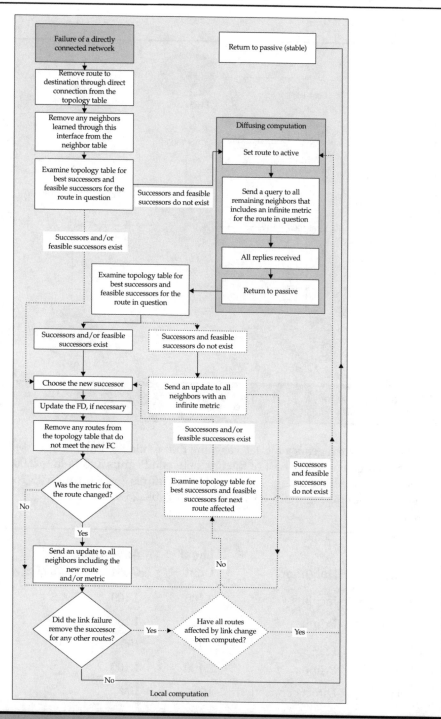

Figure 25-29. *Miles' computation for Figure 25-27*

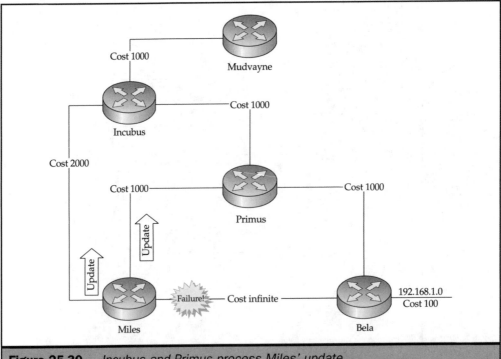

Figure 25-30.   *Incubus and Primus process Miles' update*

Now, to switch gears a bit, let's assume that while the link between Miles and Bela is down, you increase the cost of the link between Primus and Bela to 2,000. The first step and the resulting changes are detailed in Figures 25-33 through 25-36, and the topology tables for Figure 25-33 are shown in Table 25-9.

Router	Neighbor	Local Cost	Advertised Cost	S/FS	FD
Mudvayne	Incubus	3,100	2,100	S	3,100
Incubus	Primus	2,100	1,100	S	2,100
Primus	Bela	1,100	100	S	1,100
Miles	Primus	2,100	1,100	S	2,100
Bela	Direct connection	100	100	S	100

Table 25-8.   *Topology Tables for Figure 25-30*

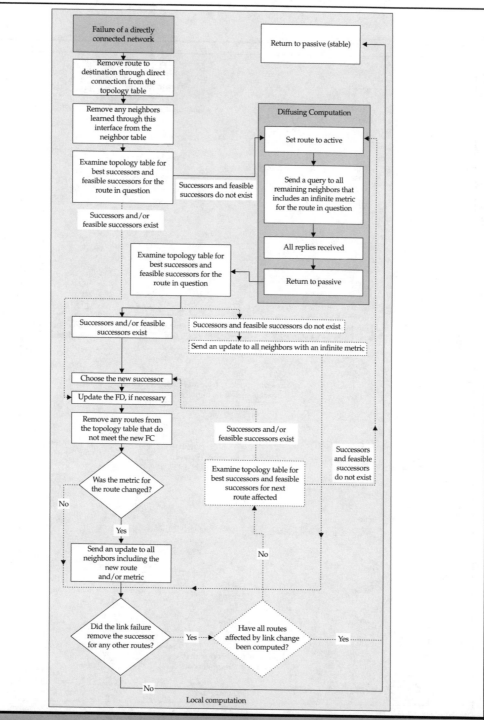

Figure 25-31.   *Miles' computation for Figure 25-30*

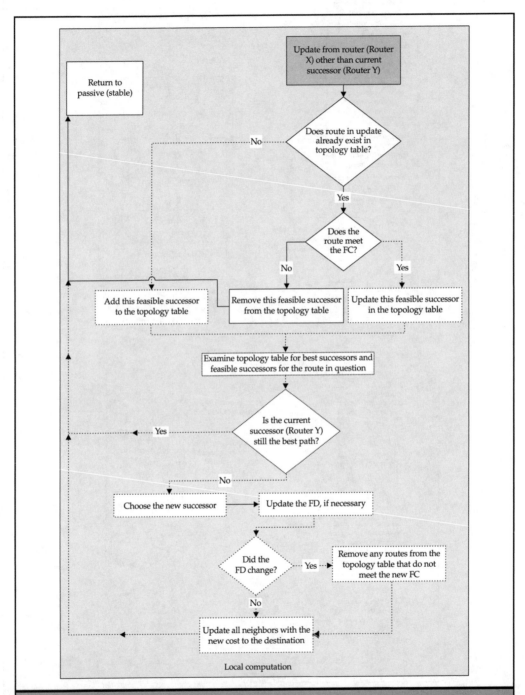

Figure 25-32. *Incubus' and Primus' computation for Figure 25-30*

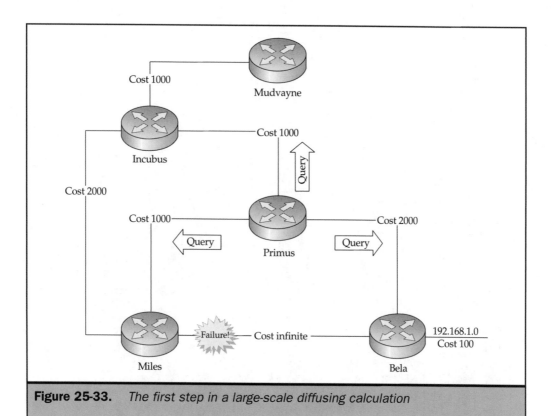

Figure 25-33. *The first step in a large-scale diffusing calculation*

Notice that a little change in the cost of a link caused a wide variety of events to happen. First, Primus removes and re-adds the route to the network that connects him to Bela. Thus, Primus also removes Bela from the topology table, which means Primus

Router	Neighbor	Local Cost	Advertised Cost	S/FS	FD
Mudvayne	Incubus	3,100	2,100	S	3,100
Incubus	—	—	—	—	4,294,967,295
Primus	—	—	—	—	4,294,967,295
Miles	—	—	—	—	4,294,967,295
Bela	Direct connection	100	100	S	100

Table 25-9. *Topology Tables for Figure 25-33*

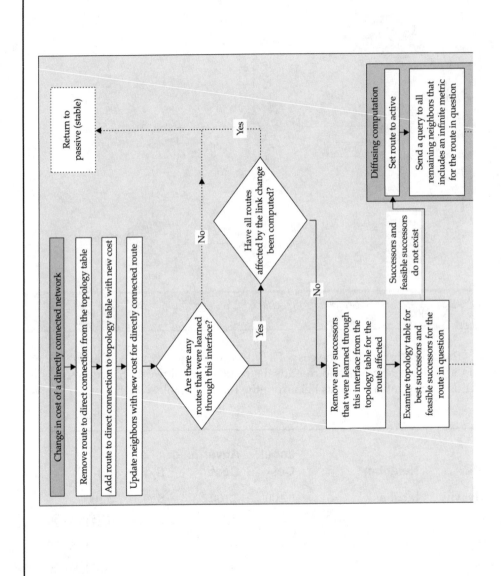

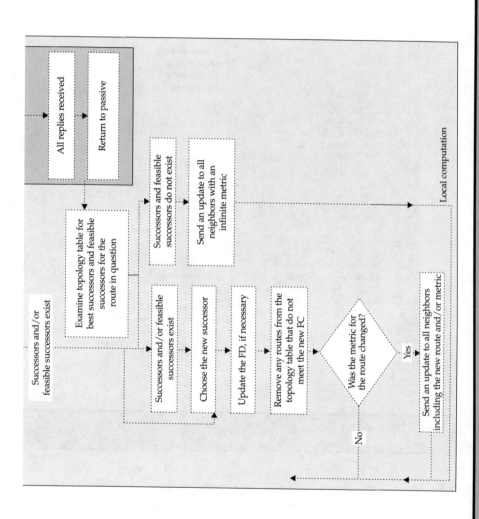

Figure 25-34. *Primus' computation for Figure 25-33*

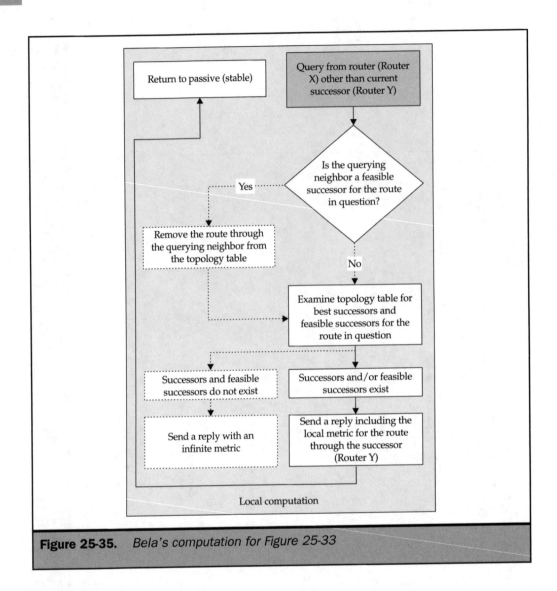

Figure 25-35. *Bela's computation for Figure 25-33*

no longer has a route to the 192.168.1.0 network (which is bad, because Primus is the only successor for that route for all of the other routers). Because Primus can no longer find a valid route in the topology table for the 192.168.1.0 network, Primus enters an active state for that network and queries all neighbors.

Although the query from Primus doesn't cause any real issues at Bela (he isn't using Primus to get to the 192.168.1.0 network anyway), it causes all kinds of interesting things to happen at Miles and Incubus. When they hear the query from Primus, they think to themselves, "Oh no, my successor to that network just died!" They look at the logic telling them what to do in such a situation and realize that they need to remove the path through

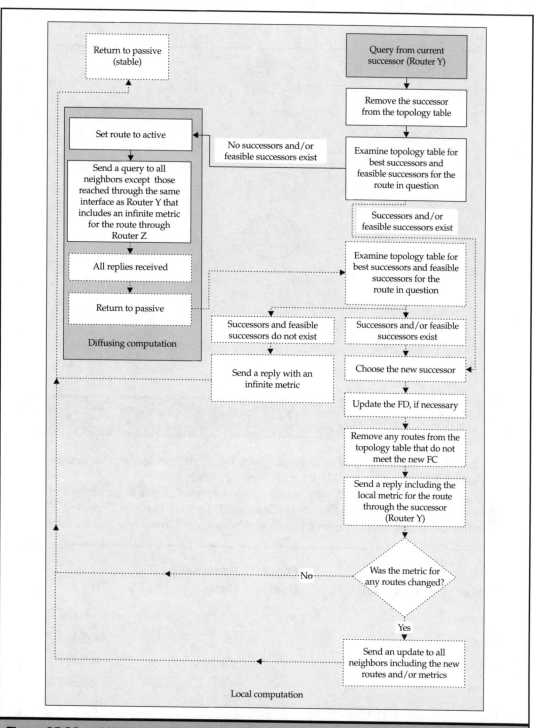

Figure 25-36. *Miles' and Incubus' computation for Figure 25-33*

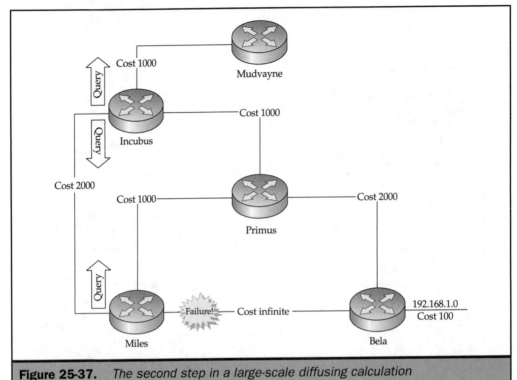

Figure 25-37.   *The second step in a large-scale diffusing calculation*

Primus from their tables and find an alternate path. This logic flow leads to the events shown in Figure 25-37 (see Table 25-10 for the topology tables).

Because the logic for the diffusing computation (shown for Miles, Incubus, and Mudvayne in Figure 25-38, and shown for Bela and Primus in Figures 25-39 and 25-40) in this situation states, "Send a query to all neighbors except those reached through the

Router	Neighbor	Local Cost	Advertised Cost	S/FS	FD
Mudvayne	—	—	—	—	4,294,967,295
Incubus	—	—	—	—	4,294,967,295
Primus	—	—	—	—	4,294,967,295
Miles	—	—	—	—	4,294,967,295
Bela	Direct connection	100	100	S	100

Table 25-10.   *Topology Tables for Figure 25-37*

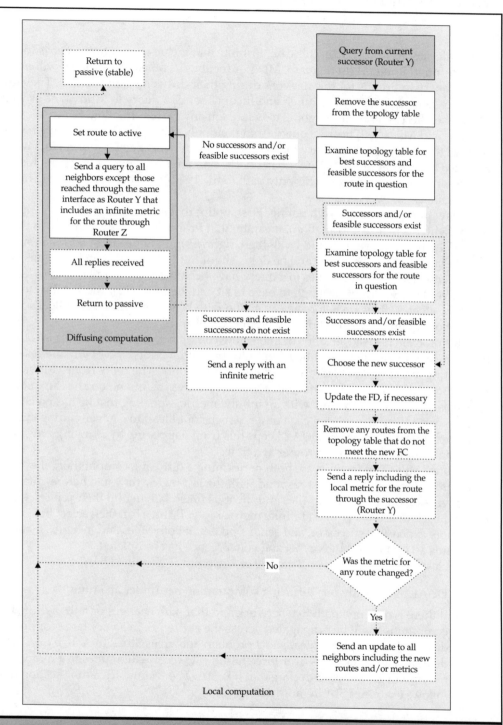

Figure 25-38. *Miles', Incubus', and Mudvayne's computation for Figure 25-37*

same interface as Router Y (in this case, Primus) that includes an infinite metric for the route through Router Z (in this case, Miles or Incubus)," Miles and Incubus will not send a query to Primus for this network during their diffusing computation. They do, however, send queries to each other, and Incubus sends a query to Mudvayne as well. Of course, these queries aren't going to help them any in this situation. Because Incubus and Miles both heard a query from their only successor (meaning they had to remove their only topology entry for the 192.168.1.0 network), neither of them have any idea how to get to the 192.168.1.0 network. And after hearing a query from his successor, Mudvayne removes his only path to the 192.168.1.0 network, making him just as clueless as everyone else.

A few points here are worth noting. First, notice that Bela sent a reply to Primus' query (with his metric information), meaning that Primus now knows a route to the 192.168.1.0 network. That should be all then, right? Primus should just add the route from Bela to the topology table and be done, shouldn't he? Unfortunately, the answer is no. Remember, until all replies to a query have been received, a router cannot leave the active state. So Primus is going to have to sit and wait on Miles and Incubus to tell him how clueless they are before he can get on with business. Primus' and Bela's calculations at this stage are shown in Figures 25-39 and 25-40.

This problem leads to all kinds of interesting complications. While Primus is waiting for an answer to his query, Incubus and Miles have both issued their own queries and are waiting for all replies before answering Primus' query. The reply from Mudvayne to Incubus will be almost instantaneous because Mudvayne has no one he can query, so he has no choice but to reply with an infinite metric, signaling that he has no clue. Incubus and Miles also queried each other, which could lead to a never-ending cycle of queries. Luckily, however, DUAL keeps this from happening by allowing only one diffusing computation on each router at a time.

Because Miles and Incubus are both performing a diffusing computation, they can't enter into another one and must respond with the last metric they had before entering the active state (infinite). Once Miles and Incubus finish tallying up their replies, they respond back to Primus with an infinite metric reply. Primus can then select Bela as his (now more expensive) successor, and sends updates to both Miles and Incubus. Incubus then sends an update to Mudvayne, and convergence has completed.

This last example shows you a couple of key points regarding DUAL:

- DUAL allows only one diffusing computation per router at a time.
- If there is only one path to a network and that path changes or fails, a diffusing computation will be propagated across the entire AS. In a very large (or poorly designed) network, this leads to a problem known as Stuck In Active (SIA). By default, if a router waits over three minutes to receive all replies to a query, the router will give up and declare that it is SIA. The "Summarization" section that follows examines SIA in more detail.

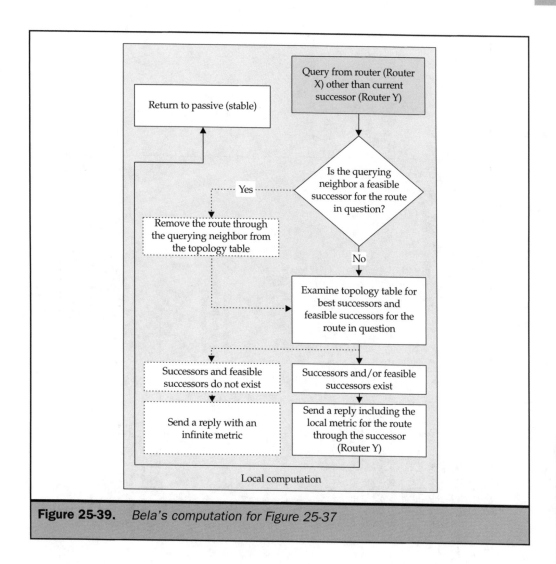

Figure 25-39.   *Bela's computation for Figure 25-37*

■ If a router must send an update or respond to a query while in an active state, the router responds with the last metric it had before entering the active state.

As these examples illustrate, in a well-designed network (and even in most poorly designed networks), EIGRP offers both fast convergence and efficiency, all without causing routing loops.

Now that you've learned about DUAL, let's take a look at some of the other benefits of EIGRP, beginning with summarization.

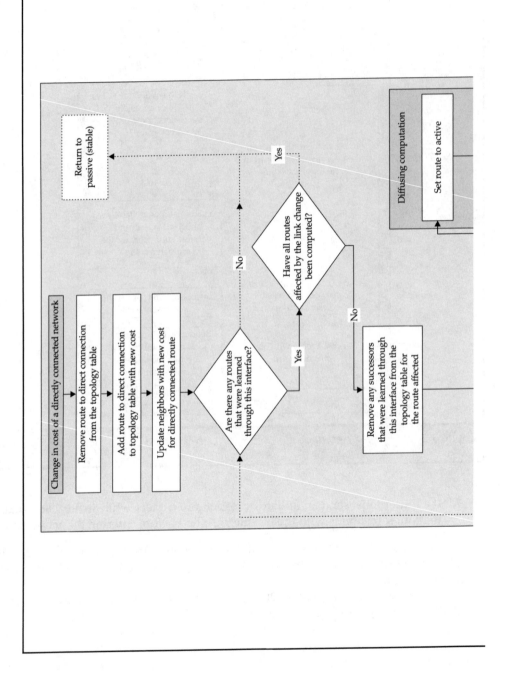

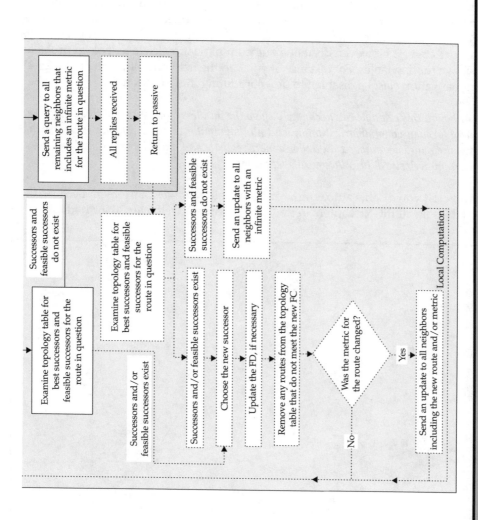

Figure 25-40. *Primus' computation for Figure 25-37*

Summarization

Unlike RIP 1 and IGRP, EIGRP includes masks with updates, which allows EIGRP to support discontiguous networks, VLSM, and CIDR. This functionality also allows EIGRP to perform manual summarization.

Normally, EIGRP will autosummarize, similar to IGRP or RIP 1, advertising an entire class-based network instead of subnets at network boundaries. To enable discontiguous networks, you may disable this function, causing EIGRP to advertise individual subnets of the major network. However, disabling autosummarization leads to its own problems with massive routing tables. For instance, if you had the network shown in Figure 25-41, autosummarization *must* be disabled to prevent routing into a black hole.

Note   *Although this example can adequately show the pitfalls of autosummarization and the need for manual summarization, it isn't all that similar to many complex networks that require manual summarization. For a more realistic and much larger example, please visit my web site at http://www.alfageek.com/.*

However, if you disable autosummarization, you now have a problem because the routing tables on all EIGRP routers in the AS will contain well over 30 individual

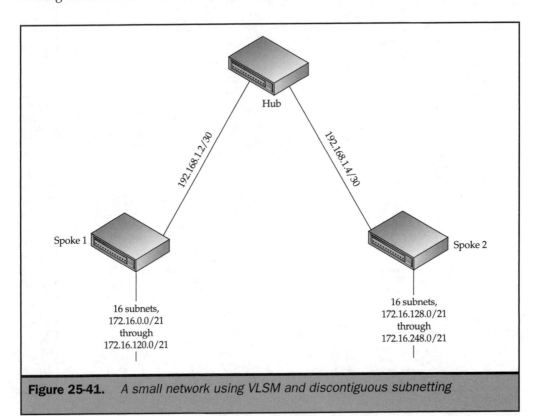

Figure 25-41.   *A small network using VLSM and discontiguous subnetting*

routes. Although this is problematic from a network use perspective (because of the bandwidth required for updates, and so on), it is even worse from a memory and CPU use perspective.

For instance, imagine that Spoke 1, connected to the 16 subnets of the 172.16.0.0 network, were to fail. Because it is directly connected to 16 subnets that have no other router connections, the Hub router will (most likely) be using it as the successor to all of those subnets. Because all 16 subnets will be listed in Hub's topology table, when the hellos from the failed router are not heard, it will remove the router from the table and enter the active state for each of the 16 routes, one at a time. Queries and replies will have to propagate across the entire enterprise, causing router CPU use on every router in the company for each of the 16 subnets.

However, if you manually summarize the routes, as shown in Figure 25-42, you will have the diffusing computation performed for only a single route (reducing the CPU and network use for the computation by 15/16, or 93 percent—instead of 16 computations, only one computation must be performed).

By preventing diffusing computations from involving large numbers of routers, summarization is very useful for resolving SIA errors. In a well-designed and

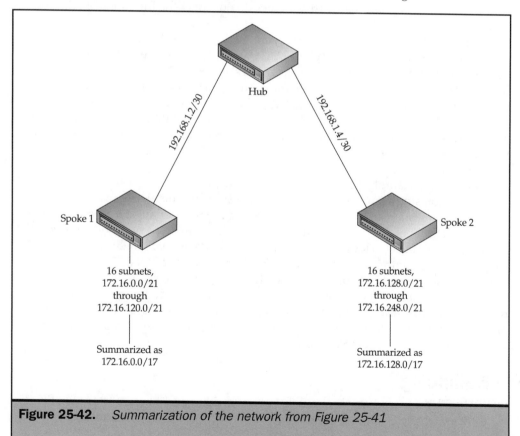

Figure 25-42.   *Summarization of the network from Figure 25-41*

configured network, a route will typically become SIA only because of the sheer number of routers a query must cross. (In a poorly designed network, other factors—such as saturated links, abnormally high resource use, and improper bandwidth settings—can also be a problem.)

If a query involves several hundred routers, for instance, it is very possible that one of the routers may be unable to respond to the query (or may not have received the query at all). In this case, an SIA timer starts, and once it expires (around three minutes later), the route is declared SIA. When a route becomes SIA, the neighbor(s) who did not reply are removed from the neighbor table; any paths through that neighbor are removed from the topology table, and, consequently, any routes through that neighbor are removed from the routing table (unless a feasible successor for the affected routes exists).

Thus, continual SIAs can cause serious instability and use problems for your network. Luckily, summarization can go a long way toward reducing SIA problems.

Manual summarization is extremely useful, but you need to be careful when manually summarizing to ensure that you do not produce conflicting routing entries. For instance, the network shown in Figure 25-43 cannot be easily summarized.

If you summarize this network, as shown in Figure 25-44, you will have some serious problems because the address ranges do not cover the full range of addresses connected to each router.

At Boston, I have tried to summarize subnets 172.31.1.0/24, 172.31.2.0/24, 172.31.3.0/24, and 172.31.4.0/24 with 172.31.0.0/22. The problem is that 172.31.0.0/22 actually matches the address range from 172.31.0.0 to 172.31.3.255, which doesn't include the 172.31.4.0/24 subnet at all. In addition, on Europe, I have attempted to summarize subnets 172.31.5.0/24, 172.31.6.0/24, 172.31.7.0/24, and 172.31.8.0/24 with 172.31.4.0/22. Again, this causes problems because 172.31.4.0/22 actually matches the address range from 172.31.4.0 to 172.31.7.255, which *does* include the 172.31.4.0 network (which does *not* connect to Europe) and does *not* include the 172.31.8.0 subnet (which *does* connect to Europe).

Now, in this example, EIGRP actually sorts the problem out because manual summarization suppresses routes for individual subnets from being advertised only if the routes in question are within the summarized range. In other words, on Boston, the 172.31.4.0/24 subnet does not fall within the summary range, so it will still be advertised as an individual subnet to Kansas. When Kansas hears the summary route for 172.31.4.0/22, it enters the route into its routing table, in addition to the 172.31.4.0/24 network, and uses the most specific route that matches any given packet to route the packet.

Keep this problem in mind when attempting to summarize routes manually. If you work out the numbering in binary before you attempt to summarize (making sure the proposed summary address matches the address range you need to summarize), you will save yourself considerable headaches.

Load Balancing

EIGRP load balances extremely similarly to IGRP, with the addition of DUAL. In EIGRP, if a route meets the FC, it will be entered into the topology table. Once all routes to a

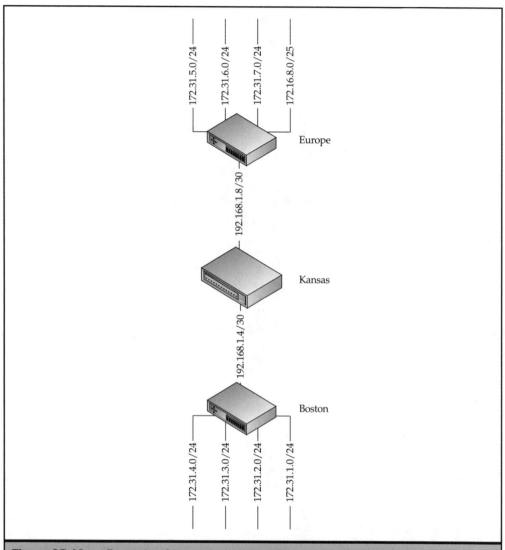

Figure 25-43.   *Example of a poorly designed IP scheme incompatible with summarization*

given destination are in the topology table, the route with the best metric becomes the successor, and the FD is based on its metric. Any routes that also meet the metric of the successor times the variance are also entered as successors to the destination, and unequal-cost load balancing is performed across all successors. All other routes meeting the FC but not within the variance range are marked as feasible successors.

To bring this concept home, examine the network shown in Figure 25-45.

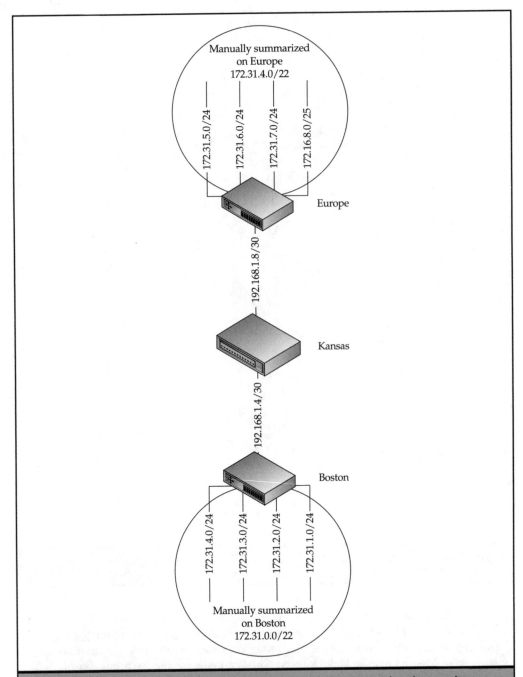

Figure 25-44. *Example of an unsuccessful attempt to summarize the previous network*

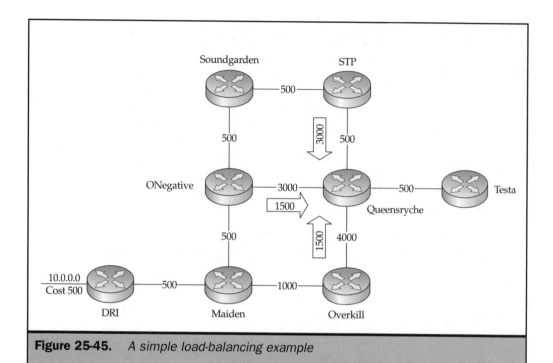

Figure 25-45.   *A simple load-balancing example*

In this network, Queensryche initially hears advertisements for the 10.0.0.0 network from ONegative, Overkill, and STP. The advertised metrics for each of these are shown in the figure, and the local distance is computed as shown here:

- **Through ONegative**   4,500
- **Through Overkill**   6,000
- **Through STP**   3,000

Assuming the variance on Queensryche was configured to 2, Queensryche chooses STP as its successor and sets the FD to 3,000. Because the advertised metric from both Overkill and ONegative meet the FC, they are added to the table. ONegative's metric is also less than the variance, so ONegative is entered as a successor as well, and unequal-cost load balancing is performed between both successors. Overkill is marked as a feasible successor. After building his routing table, Queensryche sends out an update to all neighbors (including Tesla) with Queensryche's new best local metric (3,000). Tesla, in this case, routes to the 10.0.0.0 network to Queensryche with an advertised metric of 3,500 (assuming, of course, that split horizon is disabled). Queensryche realizes that the path through Tesla does not meet the FC (3,500 > 3,000), and does not add Tesla to the topology table.

EIGRP Configuration and Troubleshooting

In contrast to the underlying complexity of DUAL, EIGRP configuration is actually fairly simple. (In some ways, it is even simpler than IGRP because there are fewer ways to "blow it up.") This section walks through EIGRP configuration and troubleshooting tasks, including the commands for your reference, and then runs through a few EIGRP case studies to show you how to avoid EIGRP problems. This section is broken down as follows:

- EIGRP configuration tasks
    - Enabling EIGRP and using network statements
    - Changing metric weights
    - Setting hello and holdtime intervals
    - Configuring the percentage of link bandwidth that EIGRP uses
    - Adding a metric offset
    - Configuring manual summarization
    - Configuring EIGRP stub routers
    - Configuring passive interfaces
    - Configuring EIGRP authentication
- Redistributing EIGRP
    - Redistributing EIGRP with RIP, and vice versa
    - Redistributing EIGRP with IGRP, and vice versa
    - Redistributing EIGRP into EIGRP
- EIGRP monitoring and troubleshooting commands
- EIGRP case studies
    - Basic EIGRP configuration
    - Configuring stub routers
    - Configuring manual summarization
    - Configuring load balancing

Configuration Tasks

This section briefly describes each common configuration task and the commands that are used in the task. You will notice that most of these are very similar to IGRP commands, so little explanation is needed. Unlike in other chapters, I do not provide a lengthy example for each command. Rather, I use the case studies at the end of the chapter to consolidate configuration and troubleshooting commands into a complete picture of EIGRP operation.

Enabling EIGRP and Using Network Statements

Enabling EIGRP is very similar to enabling IGRP. All you need to do is enter the *router eigrp [as number]* command on each router in the AS, and then use *network [network address]* statements to enable routing updates to be sent and received and the network(s) in question to be advertised. Finally, you may use the *default-network* statement to specify that a route be used as the default route (just like IGRP).

Changing Metric Weights

Except that the math is slightly different, metric weights in EIGRP are almost identical to IGRP. Like IGRP, by default, K1 and K3 are set to 1, and all other weights are set to 0 (disabled). To change the weights in EIGRP, use the same *metric weights [tos] [k1] [k2] [k3] [k4] [k5]* command described in Chapter 24. One point to remember when changing EIGRP weights, however, is that in EIGRP, routers must be using the same metric weights (and, of course, be configured with the same AS number) to establish adjacencies.

Setting Hello and Holdtime Intervals

Remembering from the "Operation" section earlier in the chapter, in EIGRP, the hello interval is how often the router sends hello messages (used to establish and maintain adjacencies) to other routers. The holdtime interval is how long the router waits to hear a hello from a neighbor before declaring the neighbor dead (causing paths from that neighbor to be removed from the topology table). Although EIGRP does not insist that these intervals be the same on all routers, configuring identical intervals is generally a good idea because misconfigured holdtimes and hello intervals can cause neighbors to be reset at regular intervals (causing routing failures and high CPU use).

To set the hello interval on an EIGRP router, use the *ip hello-interval eigrp [as number] [interval in seconds]* interface config mode command. The following example sets a hello interval on the FastEthernet 0/0 interface for AS 10 to be three seconds:

```
Router(config)# interface fastethernet 0/0
Router(config-if)# ip hello-interval eigrp 10 3
```

Note   *Unlike the regular routing updates sent by RIP and IGRP, EIGRP hello messages use very little bandwidth. (By default, EIGRP is not allowed to use over 50 percent of the link bandwidth.) So the default settings of 60 seconds for T1 or slower NBMA links and 5 seconds for all other links should be sufficient in most cases.*

If you change the hello intervals from the default settings, you may need to manually change the holdtime settings as well because EIGRP will not automatically change them for you. If you reduce the hello interval, holdtime modification is not required; but if you increase the hello interval beyond the default holdtime interval, you will need to change the holdtime. The recommended holdtime value is three times the hello interval. The default settings are 180 seconds for low-speed NBMA links and 15 seconds

CISCO ROUTING

for all other links. To change the holdtime values, use the *ip hold-time eigrp [as number]* *[interval in seconds]* interface config mode command. The following example sets the holdtime for the FastEthernet 0/0 interface to nine seconds for EIGRP AS 10:

```
Router(config)# interface fastethernet 0/0
Router(config-if)# ip hold-time eigrp 10 9
```

Configuring the Percentage of Link Bandwidth That EIGRP Uses

EIGRP version 1 has a parameter called *pacing* that determines how much bandwidth (as a percentage of total bandwidth) EIGRP will use on a link, at maximum. This parameter was implemented to keep EIGRP from using all available bandwidth when a significant input event occurs (like failure of a successor). Because of pacing, it is imperative that you set the bandwidth parameters correctly on all interfaces of an EIGRP router (using the *bandwidth* command discussed in Chapter 24). Failure to do so can cause EIGRP to use either too much or too little bandwidth on a given interface.

Therefore, it is also wise not to influence EIGRP route selection (read: reduce or increase EIGRP's metrics) by changing the bandwidth on an interface. If you must reduce a metric in order to influence route selection, use the *delay* parameter (set with the *delay* command discussed in Chapter 24). If you must increase metrics, use either *delay* or a metric offset (discussed in the upcoming "Adding a Metric Offset" section).

 EIGRP version 0 is used in older versions of the IOS—any versions earlier than 10.3(11), 11.0(8), and 11.1(3). Because it does not support pacing, it is therefore not recommended.

EIGRP pacing works by preventing EIGRP from transmitting over an interface for a certain percentage of a second, each second. By default, EIGRP uses a maximum of 50 percent of the available bandwidth on any link. To allow for delay-sensitive traffic, EIGRP allows a certain number of data bytes to be transmitted each fraction of a second, and processes data and EIGRP traffic in a round-robin fashion under times of congestion. For instance, with the default settings, EIGRP allows 1,500 bytes of data for every 1,500 bytes of EIGRP traffic.

To change the percentage of bandwidth EIGRP uses, use the *ip bandwidth-percent eigrp [as number] [interval in seconds]* interface config mode command. You can set the percentage to a value above 100 percent, but you should do so only when the bandwidth configured on an interface with the *bandwidth* command is lower than the actual bandwidth of the network media.

Adding a Metric Offset

Just like RIP and IGRP, you may add a metric offset to apply to routing updates sent or received that match certain criteria. Unlike IGRP, this method is preferred for increasing metrics entering or leaving a given interface (as opposed to adjusting the *bandwidth* parameter on the interface). To add an offset to one or more routes, use the *offset-list* command discussed in Chapter 23.

Configuring Manual Summarization

Configuring manual summarization is fairly easy, once you figure out the math associated with your addressing. The first step in configuring manual summarization is to disable autosummarization. You perform this task by using the *no autosummary* router config mode command. Once you have disabled autosummarization, use the *ip summary-address eigrp [as number] [summary address] [subnet mask]* interface config mode command to manually summarize routes advertised out each interface. All route advertisements about subnets that fall within the range of addresses specified by the summary address will be suppressed on that specific interface, and the summary address will be advertised instead. For instance, the following example summarizes the IP address range from 10.0.0.0 through 10.63.255.255 into one route for the 10.0.0.0/10 network on AS 4 on interface serial 0/0:

```
Router(config)# interface serial 0/0
Router(config-if)# ip summary-address eigrp 4 10.0.0.0 255.192.0.0
```

You may also configure a summary route to the 0.0.0.0 network with a 0.0.0.0 mask, which creates a default route on routers hearing the advertisement. Creating a default route in this manner on hub routers is useful in a true hub-and-spoke network because the summary route configures a default route on all spoke routers, but is not used on internal hub routers.

Tip   *Once you disable autosummarization, it is disabled for all routes advertised due to network statements on the local router for the entire AS. Make sure you manually configure summarization for all applicable routes after disabling autosummary, or your router may advertise a large number of individual subnets.*

Configuring EIGRP Stub Routers

Stub routers, as you learned in Chapter 24, are routers that do not require complete routes to all locations in the entire network—only routes to local segments and a default route to the core of the network. In EIGRP, rather than splitting these routers off into their own AS or using distribute lists to reduce the number of advertisements sent and received, you may designate these routers as *stub* routers, and EIGRP will do most of the work for you.

Note   *EIGRP stub designations are supported only with IOS 12.0(7)T and 12.0(15)S or later.*

When you configure a router as a stub router, EIGRP informs its neighbors of its stub designation and advertises only certain routes from the stub router. In addition,

CISCO ROUTING

other routers in the core will know that because the remote router is a stub router, it should not be queried when performing a diffusing update. The combination of reduced advertisements and query removal makes stub routers very helpful in controlling the EIGRP topology in a hub-and-spoke network.

> **Note**   *Before configuring a router as a stub router, ensure that the router has no neighbor routers except hub routers.*

When you configure a router as a stub router, you may specify the type of routes the router will advertise or, optionally, that no routes are advertised from the stub router. Stub routers can be configured to advertise summary routes, directly connected routes, static routes, or any combination of these (as well as no routes). To configure a router as a stub router, use the *eigrp stub [receive-only | connected | static | summary]* router config mode command. For instance, the following example configures a stub router to advertise connected and static routes only for EIGRP AS 1:

```
Router(config)# router eigrp 1
Router(config-router)# eigrp stub connected static
```

> **Tip**   *You must still redistribute static routes into EIGRP with the* redistribute static *command before the static option from the* eigrp stub *command will function. Similarly, network statements are still required for directly connected networks for the connected option to work. The point to remember is that configuring a router as a stub cannot increase the number of routes the router sends; it can only decrease the number of routes the router sends.*

Configuring Passive Interfaces

Passive interfaces in EIGRP work a bit differently than in most other routing protocols. In EIGRP, configuring an interface as passive does not prevent the router from sending updates out of that interface. Rather, configuring a passive interface prevents EIGRP from sending hellos out of that interface, which effectively forces EIGRP to fail to establish adjacencies to other EIGRP neighbors on that interface (which means EIGRP neither sends nor receives updates on that interface). To configure an interface on an EIGRP router to perform similarly to a passive interface in other routing protocols, you must use distribute lists. For example, the following configuration prevents routing updates from being sent out of the Ethernet 0/0 interface for EIGRP AS 1 (similar to the passive interface configuration for other routing protocols):

```
Router(config)# access-list 1 deny any
Router(config)# router eigrp 1
Router(config-router)# distribute-list 1 out Ethernet 0/0
```

> **Note**   *For more information on how passive interfaces work in EIGRP, please visit* http://www.cisco.com/warp/public/103/16.html.

Configuring EIGRP Authentication

EIGRP can use authentication with MD5 and keychains, similar to RIP 2 on a Cisco router (and covered in detail in Chapter 23). Like RIP, once authentication is configured, all routers in the AS should use the same keys to authenticate with each other. If two routers cannot agree on a key, routing updates will not be accepted between the routers.

Redistributing EIGRP

Redistributing in EIGRP uses the same commands as IGRP (discussed in detail in Chapter 24), but a few new features make EIGRP redistribution a bit different. First, realize that you must use the same caution when redistributing EIGRP as you do when redistributing any other protocol. Although EIGRP almost ensures that routing loops do not form within a single AS, when you combine multiple routing protocols and/or multiple ASs, routing loops may form due to differences in administrative distance.

 When EIGRP advertises a redistributed route, it is marked as an EIGRP external route, and the administrative distance is changed to 170 (as opposed to 90 for an internal route).

Redistributing RIP with EIGRP, and Vice Versa

When redistributing to and from RIP with EIGRP, the commands and process are identical to IGRP (covered in Chapter 24), with one major exception: EIGRP supports sending subnet masks in updates, but RIP 1 does not. This consideration can be important if you have discontiguous subnets or VLSM in use. With this problem, the only real practical solutions are often to change from RIP 1 to RIP 2, OSPF, or EIGRP, or to redesign your IP structure.

Redistributing EIGRP with IGRP, and Vice Versa

Redistributing from IGRP to EIGRP, or vice versa, is incredibly easy and very useful when transitioning from IGRP to EIGRP. Because the metrics used in IGRP and EIGRP are basically the same, IGRP can be redistributed into EIGRP with the correct metric applied. The only difference is that the scaling factor used in EIGRP is automatically applied to IGRP metrics that are redistributed into EIGRP, making the metric correct for EIGRP. One additional noteworthy point is that if the router running both EIGRP and IGRP has the same AS number configured in both protocols, the router automatically redistributes between the two protocols. Finally, IGRP, like RIP 1, does not support masks in updates, so keep this fact in mind when using VLSM.

Redistributing EIGRP into EIGRP

EIGRP also makes it very easy to redistribute one EIGRP AS into another. For instance, when redistributing one EIGRP AS into another, all metric parameters are retained, so the correct metrics are propagated among the different EIGRP ASs. Unfortunately, redistributing EIGRP among multiple EIGRP ASs is not useful to increase the maximum network size because the hop count is also propagated among ASs (meaning the maximum-hops rule still applies).

Redistributing EIGRP among multiple ASs is also not useful to decrease the query range because the query across one AS stops at the border routers, but the border routers simply issue a new query into the other AS. So, why would you redistribute EIGRP into EIGRP? Well, about the only major reason would be to reduce update traffic across a WAN link in a hub-and-spoke topology, but this isn't the suggested method. Because EIGRP supports stub routers, the suggestion is to configure the spokes as stubs and use summary routes to the 0.0.0.0 network from the hub routers to configure the stubs' routing tables. Configuring summary routes is shown in the case studies at the end of the chapter.

Although this chapter does not provide detailed information about redistribution, Chapter 24 does. The commands are the same, only the targets are slightly different. For additional information on EIGRP redistribution, visit http://www.cisco.com/univercd/cc/td/doc/cisintwk/ics/cs004.htm.

Monitoring and Troubleshooting Commands

This section covers some of the basics of common EIGRP troubleshooting commands, but it does not go through a full examination of troubleshooting EIGRP. The case studies later in the chapter require you to troubleshoot most common EIGRP problems.

The commands discussed in this section can be extremely useful, as long as you understand the fundamental operation of DUAL. Because EIGRP is such a fundamentally different protocol from the distance vector protocols that you have seen up to this point, many of the commands that were useful in RIP and IGRP (such as *show ip route*), although still useful, take a back seat to some of the other EIGRP monitoring commands. The commands covered in this section include the following:

- *clear ip eigrp neighbors [(optional) ip address | (optional) interface type and number]*
- *eigrp log-neighbor-changes*
- *show ip eigrp interfaces [(optional) interface type and number] [(optional) AS number]*
- *show ip eigrp neighbors [(optional) interface type and number]*
- *show ip eigrp topology [(optional) AS number] [(optional) ip address and subnet mask]*
- *show ip eigrp traffic [(optional) AS number]*
- *debug ip eigrp*
- *debug eigrp*

The clear ip eigrp neighbors Command

The *clear ip eigrp neighbors* command removes all neighbors from the neighbor table, which also removes all routes learned from the routing and topology tables. If you remove all neighbors from the neighbor tables, diffusing computations are not performed because the router has no neighbors to query. However, using the optional IP *address* or *interface* parameters only causes neighbors matching the IP addressees or learned through the interface specified to be removed, which has the potential to cause massive query traffic because DUAL enters a diffusing computation for each route removed. This command is useful when you want to examine the initial neighbor adjacency establishment in detail (by using this command in conjunction with the *debug ip eigrp* and *eigrp log-neighbor-changes* commands), or when you want to force an adjacency to be removed.

The eigrp log-neighbor-changes Command

The *eigrp log-neighbor-changes* router config mode command enables the logging of neighbor adjacency changes. By default, adjacency changes are not logged. This command can be especially useful in determining why neighbors are being removed (as you will see in the "Case Studies" section).

The show ip eigrp interfaces Command

This command is useful for displaying information about statistics related to each physical interface that is enabled for EIGRP. To help understand the command's output, take a look at the following output:

```
Router#show ip eigrp interfaces
IP-EIGRP interfaces for process 1

                  Xmit Queue    Mean   Pacing Time   Multicast    Pending
Interface   Peers Un/Reliable   SRTT   Un/Reliable   Flow Timer   Routes
Et0/0         1      0/0          4        0/10          50          0
Et0/1         1      0/0          6        0/10          50          0
Se0/0         1      0/0         56        0/15          50          0
Router#
```

The *Peers* section of this output shows how many adjacencies have been established over a given interface. The *Xmit Queue* section shows how many reliable and unreliable EIGRP packets are waiting in the transmit queue for a given interface. (Consistently high numbers here are typically a good indication that not enough of the line's bandwidth is allocated to EIGRP or that the *bandwidth* statement is incorrectly set.) The *Mean SRTT* section tells the smooth round trip time for this interface in milliseconds. The *Pacing Time* section shows the pacing times configured for reliable and unreliable traffic for a given interface. (Pacing time was described in the "Configuring the Percentage of Link Bandwidth That EIGRP Uses" section.) Finally, the *Pending Routes* section shows how many individual routes are waiting in the transmit queue for the particular interface.

By default, information is displayed for all interfaces and AS numbers. To view information for a single interface or AS, use the optional AS number and interface parameters.

The show ip eigrp neighbors Command

The *show ip eigrp neighbors* command displays the EIGRP neighbor table on the router. An example of the output of this command is shown here:

```
Router# show ip eigrp neighbors
IP-EIGRP neighbors for process 1
H   Address              Interface    Hold Uptime      SRTT   RTO  Q   Seq
                                      (sec)            (ms)        Cnt Num
1   169.254.15.1         Se0/0          12 00:19:16      56   336  0   32
0   10.0.3.2             Et0/0         143 4d21h          5   200  0   90
Router#
```

The H column in this output shows the host number for this neighbor, or, basically, the order in which the neighbors were added to the table. The Hold (sec) column indicates how much longer the local router will wait to hear a hello from this neighbor before declaring the neighbor down. The Uptime column tells how long this neighbor has been in the table. The SRTT (ms) column tells the smooth round-trip time for each individual neighbor. The RTO column tells the retransmission timeout for each neighbor. The Q Cnt column tells the current number of packets queued for each neighbor. Finally, the Seq Num column lists the last RTP sequence number heard from this neighbor. Each time a new reliable RTP packet is sent, the sequence number is updated. Massive jumps in this field can indicate a stability problem, whereas little or no change in this counter is a sign of either a very stable topology or a failure to receive RTP packets from the neighbor.

Again, by default, information is displayed for all neighbors. To view information for all adjacencies on a single interface, use the optional *interface* parameter.

The show ip eigrp topology Command

The *show ip eigrp topology* command shows the EIGRP topology table. The default output of this command is displayed here:

```
Router# show ip eigrp topology
IP-EIGRP Topology Table for AS(1)/ID(10.0.2.1)

Codes: P - Passive, A - Active, U - Update, Q - Query, R - Reply,
       r - Reply status

P 10.0.2.0/24, 1 successors, FD is 128256
        via Connected, Loopback1
P 10.0.3.0/24, 1 successors, FD is 281600
        via Connected, Ethernet0/0
```

```
P 10.0.0.0/8, 1 successors, FD is 128256
        via Summary (128256/0), Null0
P 10.0.1.0/24, 1 successors, FD is 61105920
        via 10.0.3.2 (61105920/61080320), Ethernet0/0
P 169.254.16.0/24, 1 successors, FD is 257024000
        via 169.254.15.1 (257024000/256512000), Serial0/0
>OUTPUT TRUNCATED DUE TO SIZE<
```

First, note the codes, listed at the top and for each route to the left of that route. The code indicators are described in the following list:

- **P—Passive**   The normal state.
- **A—Active**   The route is performing a diffusing computation.
- **U—Update**   An update is currently being sent to this network.
- **Q—Query**   A query is currently being sent to this network.
- **R—Reply**   A reply is being sent to this network.
- **r—Reply Status**   A reply is being awaited for this entry.

After the codes, for each route, you can see the FD, the successors for the route, and the local and advertised metrics (in the format local/remote).

To enable troubleshooting, there are a few additional modifiers to this command, described next:

- *show ip eigrp topology active*   Shows only routes in an active state. This command is very useful for determining which routes are in an SIA condition.
- *show ip eigrp topology all-links*   Shows all routes, including those with no successors.
- *show ip eigrp topology zero-successors*   Shows only routes with no valid successors.

You may also modify the *show ip eigrp topology* command with an AS number or IP address and mask combination to display specific entries.

The show ip eigrp traffic Command

You use the *show ip eigrp traffic* command to view EIGRP traffic statistics, as shown in the following output:

```
CCNA2600O# show ip eigrp traffic
IP-EIGRP Traffic Statistics for process 1
  Hellos sent/received: 221750/222809
```

```
Updates sent/received: 42/66
Queries sent/received: 3/5
Replies sent/received: 5/3
Acks sent/received: 45/32
Input queue high water mark 2, 0 drops

CCNA2600O#
```

The debug ip eigrp Command

This command has many permutations, all of which are extremely useful in understanding the inner workings of DUAL. These are described in the following list:

- *debug ip eigrp*   Displays all IP-related EIGRP debugging information available

- *debug ip eigrp [AS number]*   Displays all IP-related EIGRP debugging information available for a specific AS

- *debug ip eigrp neighbor [AS number] [ip address]*   Filters other EIGRP debugs to include messages only from a specific neighbor

- *debug ip eigrp notifications*   Enables debugging of EIGRP input event notifications

- *debug ip eigrp summary*   Enables debugging of EIGRP summary routes

The "Case Studies" section examines the output of these commands.

The debug eigrp Command

Like the *debug ip eigrp* command, this command has a number of options, most of which can be extremely useful in troubleshooting EIGRP problems. These are described here:

- *debug eigrp fsm*   Enables debugging of EIGRP finite state machine events. Very useful in examining DUAL's operation.

- *debug eigrp neighbors*   Enables debugging of neighbor changes (neighbor adjacency establishment and termination).

- *debug eigrp packets [ack | hello | query | reply | retry | stub | terse | update | verbose]*   Enables debugging of individual EIGRP packet types. Simply entering *debug eigrp packets* enables all possible packet debugging. Be very careful when enabling this debug because EIGRP can generate massive numbers of packets in a very short time. This command has the following optional modifiers:

- *ack*   EIGRP acknowledgment packets
- *hello*   EIGRP hello packets
- *query*   EIGRP query packets
- *reply*   Displays EIGRP reply packets
- *retry*   Displays EIGRP retransmissions
- *stub*   Displays EIGRP stub packets
- *terse*   Displays all EIGRP packets except hellos
- *update*   Displays EIGRP update packets
- *verbose*   Displays all EIGRP packets

The output of these commands is examined in the following section.

Case Studies

This section takes what is initially a fairly simple hub-and-spoke topology and configures a functional EIGRP AS while troubleshooting problems that arise along the way. The network used throughout the case studies is shown in Figure 25-46.

Four case studies are presented on this fairly simple network, listed here:

- Performing basic EIGRP configuration
- Configuring stub routers
- Configuring manual summarization
- Configuring load balancing

In each case study, you will run across one or more problems and tackle them using the troubleshooting tools you learned in the chapter. In addition, once you have finished each case study, you will verify your configurations. Without further ado, let's begin with the first case study.

> **Note**   *If you have a few routers to work with, to duplicate this configuration, you will need the following equipment: two routers (2610s were used in this example) with dual Ethernet and dual serial ports (these are Walsh and Frey), one router (another 2600 in the example) with dual Ethernet ports (Schmit), and one router (a 3640 in this example) with four FastEthernet ports and four serial ports (Henley). In a pinch, you could substitute loopback interfaces for the "client" networks and any port for the serial ports.*

CISCO ROUTING

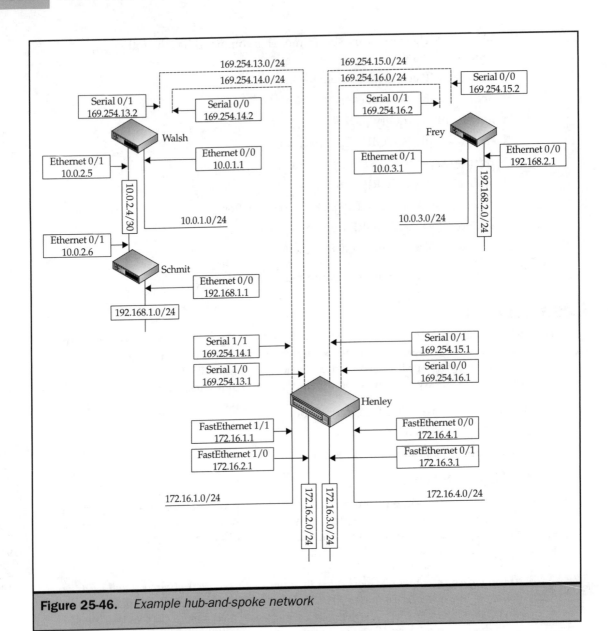

Figure 25-46. *Example hub-and-spoke network*

Basic EIGRP Configuration

In this case study, the goal is to initially configure all routers to use EIGRP in AS 1, and to have all networks advertised across the AS. All you need to do to meet this goal is to enter the correct *router eigrp* and *network* statements. To accomplish this task, perform the following actions on your routers:

Henley:

```
Henley(config)# router eigrp 1
Henley(config-router)# network 172.16.0.0
Henley(config-router)# network 169.254.0.0
```

Frey:

```
Frey(config)# router eigrp 1
Frey(config-router)# network 192.168.2.0
Frey(config-router)# network 169.254.0.0
Frey(config-router)# network 10.0.0.0
```

Walsh:

```
Walsh(config)# router eigrp 1
Walsh(config-router)# network 169.254.0.0
Walsh(config-router)# network 10.0.0.0
```

Schmit:

```
Walsh(config)# router eigrp 1
Walsh(config-router)# network 192.168.1.0
Walsh(config-router)# network 10.0.0.0
```

Once you perform these actions, your network should be configured (assuming all other settings are at the defaults), right? Well, let's see. From Henley, see if you can ping some remote networks:

```
Henley#ping 10.0.3.1

Type escape sequence to abort.
Sending 5, 100-byte ICMP Echos to 10.0.3.1, timeout is 2 seconds:
!U!U!
Success rate is 60 percent (3/5), round-trip min/avg/max = 32/34/36 ms
Henley#ping 10.0.1.1

Type escape sequence to abort.
Sending 5, 100-byte ICMP Echos to 10.0.1.1, timeout is 2 seconds:
!U!U!
Success rate is 60 percent (3/5), round-trip min/avg/max = 32/32/32 ms
Henley#
```

CISCO ROUTING

Okay, so something is obviously not working. Take a look at the topology table on Henley to see if you can get to the bottom of the problem:

```
Henley#show ip eigrp topology
IP-EIGRP Topology Table for AS(1)/ID(172.16.4.1)

Codes: P - Passive, A - Active, U - Update, Q - Query, R - Reply,
       r - Reply status

P 10.0.0.0/8, 4 successors, FD is 1787392
via 169.254.13.2 (1889792/384000), Serial1/0
via 169.254.14.2 (1889792/384000), Serial1/1
via 169.254.15.2 (1889792/384000), Serial0/1
via 169.254.16.2 (1889792/384000), Serial0/0
P 169.254.16.0/24, 1 successors, FD is 1761792
        via Connected, Serial0/0
P 169.254.0.0/16, 1 successors, FD is 1761792
        via Summary (1761792/0), Null0
P 169.254.15.0/24, 1 successors, FD is 1761792
        via Connected, Serial0/1
P 169.254.14.0/24, 1 successors, FD is 1761792
        via Connected, Serial1/1
P 169.254.13.0/24, 1 successors, FD is 1761792
        via Connected, Serial1/0
P 172.16.1.0/24, 1 successors, FD is 28160
        via Connected, FastEthernet1/1
P 172.16.2.0/24, 1 successors, FD is 28160
        via Connected, FastEthernet1/0
P 172.16.3.0/24, 1 successors, FD is 28160
        via Connected, FastEthernet0/1
P 172.16.4.0/24, 1 successors, FD is 28160
        via Connected, FastEthernet0/0
P 172.31.0.0/16, 1 successors, FD is 28160
        via Summary (28160/0), Null0
P 192.168.2.0/24, 2 successors, FD is 1787392
        via 169.254.15.2 (1889792/384000), Serial0/1
        via 169.254.16.2 (1889792/384000), Serial0/0
```

Notice the problem with the bolded section? The topology table has four equal-cost successors to the 10.0.0.0/8 network (both serial connections to Frey and Walsh). The problem is that Frey and Walsh are advertising summary routes for the 10.0.0.0/8 network because autosummary is enabled by default. On Henley, when you try to ping 10.0.3.1, it load balances packets across all four equal-cost successors in a round-robin fashion. For the packets that arrive at Frey, the pings are successful; but for the packets that arrive at Walsh, the packets are discarded and an Unreachable is sent.

What's wrong with this picture?   Looking at the topology table, can you see any other problems that may need to be addressed?

To solve this problem, disable autosummary on Walsh and Frey:

Frey:

```
Frey(config)# router eigrp 1
Frey(config-router)# no auto-summary
```

Walsh:

```
Walsh(config)# router eigrp 1
Walsh(config-router)# no auto-summary
```

Now your topology table on Henley (a few seconds later) should look like this:

```
Henley#show ip eigrp topology
IP-EIGRP Topology Table for AS(1)/ID(172.16.4.1)

Codes: P - Passive, A - Active, U - Update, Q - Query, R - Reply,
       r - Reply status

P 10.0.1.0/24, 2 successors, FD is 1787392
        via 169.254.13.2 (1889792/384000), Serial1/0
        via 169.254.14.2 (1889792/384000), Serial1/1
P 10.0.2.4/30, 2 successors, FD is 1787392
        via 169.254.13.2 (1889792/384000), Serial1/0
        via 169.254.14.2 (1889792/384000), Serial1/1
P 10.0.3.0/24, 2 successors, FD is 1787392
        via 169.254.15.2 (1889792/384000), Serial0/1
        via 169.254.16.2 (1889792/384000), Serial0/0
>OUTPUT TRUNCATED<
```

That's better. Now, try to ping:

```
Henley#ping 10.0.3.1

Type escape sequence to abort.
Sending 5, 100-byte ICMP Echos to 10.0.3.1, timeout is 2 seconds:
!!!!!
```

```
Success rate is 100 percent (5/5), round-trip min/avg/max = 32/34/36 ms
Henley#ping 10.0.1.1

Type escape sequence to abort.
Sending 5, 100-byte ICMP Echos to 10.0.1.1, timeout is 2 seconds:
!!!!!
Success rate is 100 percent (5/5), round-trip min/avg/max = 32/32/32 ms
Henley#
```

Great, so everything is working, right? Well, not quite. Did you catch the other problem with Henley's topology table? Just in case you didn't see the problem, I'll give you a hint: What would happen if you attempted to ping 192.168.1.1 from Henley? There is no entry for the 192.168.1.0 network in Henley's topology table, so packets to the 192.168.1.0 network from Henley (and Frey, for that matter) will fail. Well, let's get to the bottom of this problem. You know connectivity and a neighbor adjacency has been established between Henley and Walsh, so the problem has to be at Walsh. On Walsh, the *show ip eigrp topology* command gives the following output:

```
Walsh# show ip eigrp topology
IP-EIGRP Topology Table for AS(1)/ID(10.0.1.1)

Codes: P - Passive, A - Active, U - Update, Q - Query, R - Reply,
       r - Reply status

P 10.0.1.0/24, 1 successors, FD is 384000
        via Connected, Ethernet0/0
P 10.0.2.4/30, 1 successors, FD is 384000
        via Connected, Ethernet0/1
P 10.0.3.0/24, 2 successors, FD is 2809856
        via 169.254.13.1 (2809856/1889792), Serial0/1
        via 169.254.14.1 (2809856/1889792), Serial0/0
P 169.254.16.0/24, 2 successors, FD is 2681856
        via 169.254.13.1 (2809856/1889792), Serial0/1
        via 169.254.14.1 (2809856/1889792), Serial0/0
P 169.254.15.0/24, 2 successors, FD is 2169856
        via 169.254.13.1 (2809856/1889792), Serial0/1
        via 169.254.14.1 (2809856/1889792), Serial0/0
P 169.254.14.0/24, 1 successors, FD is 2169856
        via Connected, Serial0/0
P 169.254.13.0/24, 1 successors, FD is 2169856
        via Connected, Serial0/1
P 172.16.0.0/16, 2 successors, FD is 2172416
        via 169.254.13.1 (2809856/1889792), Serial0/1
```

```
         via 169.254.14.1 (2809856/1889792), Serial0/0
P 192.168.2.0/24, 2 successors, FD is 2172416
         via 169.254.13.1 (2809856/1889792), Serial0/1
         via 169.254.14.1 (2809856/1889792), Serial0/0
```

Okay, so it turns out that Walsh does not have a path to the 192.168.1.0 network either. To figure out why, run *show ip eigrp neighbors* and get the following:

```
Walsh# show ip eigrp neighbors
IP-EIGRP neighbors for process 1
H   Address            Interface    Hold Uptime   SRTT   RTO   Q  Seq
                                    (sec)         (ms)         Cnt Num
1   169.254.13.2       Se0/1         12 00:4:16    56    336   0  9
0   169.254.14.2       Se0/0         12 00:4:16    56    336   0  9
Walsh#
```

Hmm, it doesn't seem that Schmit is listed. To figure out why, you can use *debug eigrp packets hello* to see if you are receiving hellos from Schmit, as shown here:

```
2w4d: EIGRP: Received HELLO on Ethernet0/0 nbr 10.0.2.6
2w4d:    AS 1, Flags 0x0, Seq 0/0 idbQ 0/0
2w4d:          K-value mismatch
```

Aha, different K values are configured. You could also see this problem by using the *eigrp log-neighbor-changes* command, which would show the problem:

```
2w4d: %DUAL-5-NBRCHANGE: IP-EIGRP 1:
Neighbor 10.0.2.6 (Ethernet0/0) is down: K-value mismatch
```

To see what you have the K values configured to, you can use the *show ip protocol* command on both Walsh and Schmit, like so:

Walsh:

```
Walsh# show ip protocol
Routing Protocol is "eigrp 1"
  Outgoing update filter list for all interfaces is
  Incoming update filter list for all interfaces is
  Default networks flagged in outgoing updates
  Default networks accepted from incoming updates
  EIGRP metric weight K1=1, K2=0, K3=1, K4=0, K5=0
```

```
EIGRP maximum hopcount 100
EIGRP maximum metric variance 1
Redistributing: eigrp 1
Automatic network summarization is not in effect
Routing for Networks:
   10.0.0.0
   169.254.0.0
Routing Information Sources:
   Gateway          Distance       Last Update
   169.254.13.1           90       00:26:08
   169.254.14.1           90       00:26:08
Distance: internal 90 external 170
```

Schmit:

```
Schmit# show ip protocol
Routing Protocol is "eigrp 1"
   Outgoing update filter list for all interfaces is
   Incoming update filter list for all interfaces is
   Default networks flagged in outgoing updates
   Default networks accepted from incoming updates
   EIGRP metric weight K1=1, K2=1, K3=1, K4=1, K5=1
   EIGRP maximum hopcount 100
   EIGRP maximum metric variance 1
   Redistributing: eigrp 1
   Automatic network summarization is not in effect
   Routing for Networks:
      10.0.0.0
      192.168.1.0
   Routing Information Sources:
      Gateway          Distance       Last Update

Distance: internal 90 external 170
```

In this case, it appears that someone configured Schmit's weights to nondefault values, and Schmit and Walsh will not establish an adjacency with mismatched K values. To resolve this problem, you need to set Schmit to use the default K values, which you can do with the *no metric weights* command and verify with *show ip protocol*, as shown here:

```
Schmit(config)# router eigrp 1
Schmit(config-router)# no metric weights
```

```
Schmit(config-router)# ^Z
Schmit# show ip protocol
Routing Protocol is "eigrp 1"
  Outgoing update filter list for all interfaces is
  Incoming update filter list for all interfaces is
  Default networks flagged in outgoing updates
  Default networks accepted from incoming updates
  EIGRP metric weight K1=1, K2=0, K3=1, K4=0, K5=0
  EIGRP maximum hopcount 100
  EIGRP maximum metric variance 1
  Redistributing: eigrp 1
  Automatic network summarization is not in effect
  Routing for Networks:
    10.0.0.0
    192.168.1.0
  Routing Information Sources:
    Gateway          Distance      Last Update
    10.0.2.5               90      00:04:19
  Distance: internal 90 external 170
```

Now if you perform the *show ip eigrp neighbors* command from Walsh, Schmit will be listed in the table, and a route to the 192.168.1.0 network can be seen on Walsh with the *show ip eigrp topology* command, as shown next:

```
Walsh# show ip eigrp neighbors
IP-EIGRP neighbors for process 1
H   Address              Interface   Hold Uptime   SRTT   RTO  Q   Seq
                                     (sec)         (ms)        Cnt Num
2   10.0.2.6             Et0/1       12 00:1:16     4     200  0   17
1   169.254.13.2         Se0/1       12 00:42:21   56     336  0   9
0   169.254.14.2         Se0/0       12 00:42:21   56     336  0   9
Walsh# show ip eigrp topology
IP-EIGRP Topology Table for AS(1)/ID(10.0.1.1)

Codes: P - Passive, A - Active, U - Update, Q - Query, R - Reply,
       r - Reply status

>OUTPUT SKIPPED<
P 192.168.1.0/24, 1 successors, FD is 409600
        via 10.0.2.6 (409600/384000), Ethernet0/1
```

So, the basic configuration is complete, and all hosts can now communicate with all other hosts.

Configuring Stub Routers

In this case study, you want to configure a few routers for stub operation to reduce bandwidth use across your serial lines. In this particular network, you may decide that Walsh, Schmit, and Frey should all be stubs, and they should send information about directly connected routes only to the hub router, Henley. If you configure these routers using the *eigrp stub connected* command, however, you will have a few problems with the route to the 192.168.1.0 network again. To see the effects of this, set Walsh, Schmit, and Frey to be stubs and examine the results:

Frey:

```
Frey(config)# router eigrp 1
Frey(config-router)# eigrp stub connected
```

Walsh:

```
Walsh(config)# router eigrp 1
Walsh(config-router)# eigrp stub connected
```

Schmit:

```
Walsh(config)# router eigrp 1
Walsh(config-router)# eigrp stub connected
```

Now, examine the results of this on Henley:

```
Henley# show ip eigrp neighbors detail
IP-EIGRP neighbors for process 1
H   Address            Interface     Hold Uptime     SRTT   RTO  Q   Seq
                                     (sec)           (ms)        Cnt Num
3   169.254.13.1       Se1/0           12 01:42:06   56     336  0   26
      Version 12.0/1.1, Retrans: 0, Retries: 0
        Stub Peer Advertising ( CONNECTED ) Routes
2   169.254.14.1       Se1/1           12 01:42:06   56     336  0   26
      Version 12.0/1.1, Retrans: 0, Retries: 0
        Stub Peer Advertising ( CONNECTED ) Routes
1   169.254.15.1       Se0/1           11 01:41:24   52     336  0   18
      Version 12.0/1.1, Retrans: 0, Retries: 0
        Stub Peer Advertising ( CONNECTED ) Routes
0   169.254.16.1       Se0/0           11 01:41:24   52     336  0   18
      Version 12.0/1.1, Retrans: 0, Retries: 0
        Stub Peer Advertising ( CONNECTED ) Routes
Henley# show ip eigrp topology
IP-EIGRP Topology Table for AS(1)/ID(172.16.4.1)
```

```
Codes: P - Passive, A - Active, U - Update, Q - Query, R - Reply,
       r - Reply status

P 10.0.1.0/24, 2 successors, FD is 1787392
        via 169.254.13.2 (1889792/384000), Serial1/0
        via 169.254.14.2 (1889792/384000), Serial1/1
P 10.0.2.4/30, 2 successors, FD is 1787392
        via 169.254.13.2 (1889792/384000), Serial1/0
        via 169.254.14.2 (1889792/384000), Serial1/1
P 10.0.3.0/24, 2 successors, FD is 1787392
        via 169.254.15.2 (1889792/384000), Serial0/1
        via 169.254.16.2 (1889792/384000), Serial0/0
P 169.254.16.0/24, 1 successors, FD is 1761792
        via Connected, Serial0/0
P 169.254.0.0/16, 1 successors, FD is 1761792
        via Summary (1761792/0), Null0
P 169.254.15.0/24, 1 successors, FD is 1761792
        via Connected, Serial0/1
P 169.254.14.0/24, 1 successors, FD is 1761792
        via Connected, Serial1/1
P 169.254.13.0/24, 1 successors, FD is 1761792
        via Connected, Serial1/0
P 172.16.1.0/24, 1 successors, FD is 28160
        via Connected, FastEthernet1/1
P 172.16.2.0/24, 1 successors, FD is 28160
        via Connected, FastEthernet1/0
P 172.16.3.0/24, 1 successors, FD is 28160
        via Connected, FastEthernet0/1
P 172.16.4.0/24, 1 successors, FD is 28160
        via Connected, FastEthernet0/0
P 172.31.0.0/16, 1 successors, FD is 28160
        via Summary (28160/0), Null0
P 192.168.2.0/24, 2 successors, FD is 1787392
        via 169.254.15.2 (1889792/384000), Serial0/1
        via 169.254.16.2 (1889792/384000), Serial0/0
```

Oops, looks like you lost the route to the 192.168.1.0 network on Henley. Even worse, the topology table for Schmit would look like this:

```
Schmit# show ip eigrp topology
IP-EIGRP Topology Table for AS(1)/ID(192.168.1.1)

Codes: P - Passive, A - Active, U - Update, Q - Query, R - Reply,
       r - Reply status

P 10.0.2.4/30, 1 successors, FD is 384000
```

```
            via Connected, Ethernet0/1
P 10.0.1.0/24, 1 successors, FD is 409600
            via 10.0.2.5 (409600/384000), Ethernet0/1
P 169.254.14.0/24, 1 successors, FD is 2707456
            via 10.0.2.5 (2707456/2681856), Ethernet0/1
P 169.254.13.0/24, 1 successors, FD is 2707456
            via 10.0.2.5 (2707456/2681856), Ethernet0/1
P 192.168.1.0/24, 1 successors, FD is 384000
            via Connected, Ethernet0/0
```

Consequently, Schmit can't get to anyone who isn't directly connected to Walsh. These problems occur any time you configure a hub router as a stub. Technically, Walsh should be considered a hub router. For this reason, if you remove the stub designation from Walsh using the *no eigrp stub* command, all of your routes will function again. However, you still have the problem whereby too many updates are crossing the serial links to Walsh. To reduce this problem (as well as cut down the query range a bit), you need to configure manual summarization in the next case study.

Configuring Manual Summarization

With this particular implementation, configuring summary addresses is actually fairly easy. You can't summarize routes from Walsh or Frey because they contain portions of a discontiguous network that will not fit into a summary address, so you cannot reduce the number of routes sent to Henley. However, you can very easily reduce the query range and the number of routes sent from Henley to Walsh and Frey by using a single summary address to the 0.0.0.0 network, like so:

```
Henley(config)# interface serial 0/0
Henley(config-if)# ip summary-address eigrp 1 0.0.0.0 0.0.0.0
Henley(config-if)# interface serial 0/1
Henley(config-if)# ip summary-address eigrp 1 0.0.0.0 0.0.0.0
Henley(config-if)# interface serial 1/0
Henley(config-if)# ip summary-address eigrp 1 0.0.0.0 0.0.0.0
Henley(config-if)# interface serial 1/1
Henley(config-if)# ip summary-address eigrp 1 0.0.0.0 0.0.0.0
```

Now, if you look at the topology table on Walsh, you can almost immediately see the effects of this summary address:

```
Walsh# show ip eigrp topology
IP-EIGRP Topology Table for AS(1)/ID(10.0.1.1)
```

```
Codes: P - Passive, A - Active, U - Update, Q - Query, R - Reply,
       r - Reply status

P 0.0.0.0/0, 2 successors, FD is 2172416
        via 169.254.13.1 (2172416/28160), Serial0/1
        via 169.254.14.1 (2172416/28160), Serial0/0
P 10.0.1.0/24, 1 successors, FD is 384000
        via Connected, Ethernet0/0
P 10.0.2.4/30, 1 successors, FD is 384000
        via Connected, Ethernet0/1
P 169.254.14.0/24, 1 successors, FD is 2169856
        via Connected, Serial0/0
P 169.254.13.0/24, 1 successors, FD is 2169856
        via Connected, Serial0/1
P 192.168.1.0/24, 1 successors, FD is 409600
        via 10.0.2.6 (409600/384000), Ethernet0/1
```

Notice that our summary address reduced the number of routes on Walsh significantly. Yet, Walsh will still be able to route packets to all destinations. If Walsh does not have a more specific path to a given destination, he will simply send packets to Henley, who has routes to all destinations.

Configuring Load Balancing

In the current network, routes on Walsh and Frey that are routed through Henley are being load balanced by default because the dual serial links are configured for the same speeds (bandwidth 1544) and delay (20,000 microseconds) settings. But what if the serial 0/1 link on both Walsh and Frey were configured for a bandwidth of 512? The topology table on Walsh would then change to the following:

```
Walsh# show ip eigrp topology 0.0.0.0 0.0.0.0
IP-EIGRP topology entry for 0.0.0.0/0
  State is Passive, Query origin flag is 1, 1 Successor(s), FD is 2172416
  Routing Descriptor Blocks:
  169.254.14.1 (Serial0/0), from 169.254.14.1, Send flag is 0x0
      Composite metric is (2172416/28160), Route is Internal
      Vector metric:
        Minimum bandwidth is 1544 Kbit
        Total delay is 20100 microseconds
        Reliability is 255/255
        Load is 1/255
        Minimum MTU is 1500
        Hop count is 1
  169.254.13.1 (Serial0/1), from 169.254.13.1, Send flag is 0x0
```

```
        Composite metric is (5514496/28160), Route is Internal
        Vector metric:
          Minimum bandwidth is 512 Kbit
          Total delay is 20100 microseconds
          Reliability is 255/255
          Load is 1/255
          Minimum MTU is 1500
          Hop count is 1
Walsh#
```

The routing table on Walsh before the bandwidth change would have been as follows:

```
SerialRouter2#show ip route 0.0.0.0
Routing entry for 0.0.0.0/0
  Known via "eigrp 1", distance 90, metric 2172416,
candidate default path, type internal
  Redistributing via eigrp 1
  Last update from 169.254.13.1 on Serial0/1, 00:00:19 ago
  Routing Descriptor Blocks:
  * 169.254.13.1, from 169.254.13.1, 00:00:19 ago, via Serial0/1
      Route metric is 2172416, traffic share count is 1
      Total delay is 21000 microseconds, minimum bandwidth is 1544 Kbit
      Reliability 255/255, minimum MTU 1500 bytes
      Loading 1/255, Hops 1
    169.254.14.1, from 169.254.14.1, 00:00:19 ago, via Serial0/0
      Route metric is 2172416, traffic share count is 1
      Total delay is 21000 microseconds, minimum bandwidth is 1544 Kbit
      Reliability 255/255, minimum MTU 1500 bytes
      Loading 1/255, Hops 1
```

However, after the bandwidth change, the routing table changes to this:

```
SerialRouter2#show ip route 0.0.0.0
Routing entry for 0.0.0.0/0
  Known via "eigrp 1", distance 90, metric 2172416,
candidate default path, type internal
  Redistributing via eigrp 1
  Last update from 169.254.13.1 on Serial0/1, 00:00:19 ago
  Routing Descriptor Blocks:
  * 169.254.14.1, from 169.254.14.1, 00:00:27 ago, via Serial0/0
      Route metric is 2172416, traffic share count is 1
      Total delay is 21000 microseconds, minimum bandwidth is 1544 Kbit
      Reliability 255/255, minimum MTU 1500 bytes
      Loading 1/255, Hops 1
```

The route through the serial 0/1 interface would be removed from the table because it is no longer a successor and is relegated to a feasible successor. (It meets the FC, but does not meet the metric times variance successor rule.) To allow EIGRP to perform unequal-cost load balancing across these links, you should set the variance high enough to allow the path through the serial 0/1 interface to be used. Because the lowest metric is 2,172,416 and the metric for the serial 0/1 link is now 5,514,496, the variance should equal 5,514,496/2,172,416 = 2.5 or 3. Once you correctly set the variance, you can see the effect by performing another *show ip route* command and a *debug ip packets* command while pinging a remote network:

```
SerialRouter2(config)#router eigrp 1
SerialRouter2(config-router)#variance 3
SerialRouter2(config-router)#^Z
SerialRouter2#show ip route 0.0.0.0
Routing entry for 0.0.0.0/0
  Known via "eigrp 1", distance 90, metric 2172416,
candidate default path, type internal
  Redistributing via eigrp 1
  Last update from 169.254.13.1 on Serial0/1, 00:00:19 ago
  Routing Descriptor Blocks:
  * 169.254.13.1, from 169.254.13.1, 00:00:19 ago, via Serial0/1
      Route metric is 2172416, traffic share count is 1
      Total delay is 21000 microseconds, minimum bandwidth is 1544 Kbit
      Reliability 255/255, minimum MTU 1500 bytes
      Loading 1/255, Hops 1
    169.254.14.1, from 169.254.14.1, 00:00:19 ago, via Serial0/0
      Route metric is 2172416, traffic share count is 1
      Total delay is 21000 microseconds, minimum bandwidth is 1544 Kbit
      Reliability 255/255, minimum MTU 1500 bytes
      Loading 1/255, Hops 1
SerialRouter2#deb ip packet detail
IP packet debugging is on (detailed)
SerialRouter2#ping 172.16.1.1

Type escape sequence to abort.
Sending 5, 100-byte ICMP Echos to 172.16.1.1, timeout is 2 seconds:
2w2d: IP: s=169.254.14.2 (local), d=172.16.1.1 (Serial0/0), len 100,
sending
2w2d:     ICMP type=8, code=0
!
2w2d: IP: s=172.16.1.1 (Serial0/0), d=169.254.14.2 (Serial0/0), len 100,
  rcvd3
2w2d:     ICMP type=0, code=0
2w2d: IP: s=169.254.14.2 (local), d=172.16.1.1 (Serial0/0), len 100,
sending
2w2d:     ICMP type=8, code=0
```

```
!
2w2d: IP: s=172.16.1.1 (Serial0/0), d=169.254.14.2 (Serial0/0), len 100,
rcvd3
2w2d:      ICMP type=0, code=0
2w2d: IP: s=169.254.13.2 (local), d=172.16.1.1 (Serial0/1), len 100,
sending
2w2d:      ICMP type=8, code=0
!
2w2d: IP: s=172.16.1.1 (Serial0/0), d=169.254.13.2 (Serial0/1), len 100,
rcvd3
2w2d:      ICMP type=0, code=0
2w2d: IP: s=169.254.14.2 (local), d=172.16.1.1 (Serial0/0), len 100,
sending
2w2d:      ICMP type=8, code=0
!
2w2d: IP: s=172.16.1.1 (Serial0/0), d=169.254.14.2 (Serial0/0), len 100,
rcvd3
2w2d:      ICMP type=0, code=0
2w2d: IP: s=169.254.14.2 (local), d=172.16.1.1 (Serial0/0), len 100,
sending
2w2d:      ICMP type=8, code=0
!
2w2d: IP: s=172.16.1.1 (Serial0/0), d=169.254.14.2 (Serial0/0), len 100,
rcvd3
2w2d:      ICMP type=0, code=0
Success rate is 100 percent (5/5),
round-trip min/avg/max = 36/37/40 ms
```

Notice that three packets were sent over the serial 0/0 interface for every one packet sent over the serial 0/1 interface, just as you wanted. After all of these steps, the example network is configured as efficiently as possible without changing IP addressing structures.

Summary

This chapter examined EIGRP and DUAL in detail, and, through case studies, demonstrated some of the intricacies of EIGRP configuration and troubleshooting. Because EIGRP is such a widely used protocol in Cisco environments, the examination of EIGRP should help you support larger-scale internal routing needs. In the next chapter, we examine OSPF—the other choice for highly scalable internal routing.

The Complete Reference

Cisco

Chapter 26

Open Shortest Path First

This chapter examines routing with Open Shortest Path First (OSPF) version 2. OSPF is an open specification link-state routing protocol that was originally defined in RFC 1131 and was updated several times to reach its current specification defined in RFC 2328. OSPF is a very robust and scalable routing protocol; consequently, it is extremely complex. In fact, OSPF is so complex that I could spend 500 or more pages just attempting to describe all of the intricacies of its operation! (Even RFC 2328 is nearly 250 pages.) Therefore, rather than attempting to describe all aspects of OSPF in minute detail, this chapter focuses on practical concepts. Also, because OSPF is such a huge topic, this chapter is organized into four primary sections:

- OSPF overview
- How OSPF works
- Configuring OSPF
- Troubleshooting OSPF

OSPF Overview

This section briefly covers important concepts and terms related to OSPF, which will help you gain a basic understanding of OSPF operation before delving into the details of how it works. This section is organized into two major subsections for ease of reference and understanding:

- OSPF operational overview
- OSPF terms and concepts

OSPF Operational Overview

OSPF's operation is not all that complex when viewed from a high level. OSPF basically draws a complete map of an internetwork and then chooses the least-cost path based on that map. In OSPF, every router has a complete map of the entire network (with a few exceptions, covered in "How OSPF Works," later in the chapter). If a link fails, OSPF can quickly find and resolve an alternate path to the destination based on the map, without the possibility of routing loop formation. Because OSPF knows all paths in the network, OSPF can easily determine if a route would cause a loop.

OSPF is a link-state protocol: it bases its operation on the states of network connections, or links. In OSPF, the fundamentally important component in calculating the topology is the state of each link on each router. By learning what each link is connected to, OSPF can build a database including all links in the network, and then use the Shortest Path First (SPF) algorithm to determine the shortest paths to all destinations. Because each router contains the same exact map of the topology, OSPF does not require that updates be sent at regular intervals. Unless a change occurs, OSPF, like EIGRP, transmits practically no information.

Because of OSPF's operational characteristics, it has several distinct advantages over distance vector protocols, including the following:

- **Reduced overhead.** Like EIGRP, OSPF reduces the network overhead needed for routing updates by multicasting routing updates, sending a routing update only when a change is detected (rather than sending the entire table periodically) and sending changes to the routing table (rather than the entire table) only when an update is necessary.

- **Support for VLSM and CIDR.** Like EIGRP, OSPF also includes the subnet mask in routing updates.

- **Support for discontiguous networks.**

- **Support for manual route summarization.**

- **Short convergence times.** In a well-designed OSPF network, convergence after a link failure is very rapid because OSPF maintains a complete topological database of all paths in the OSPF area.

- **Loop-free topology generation.**

- **Hop counts limited only by router resource use and the IP TTL.**

Because OSPF is a completely open specification, it also allows for interoperability between different router manufacturers (something neither EIGRP nor IGRP can claim). Because of these factors, OSPF is the recommended Interior Gateway Protocol (IGP), according to the Internet Engineering Task Force (IETF).

| Note | *The IETF does not consider EIGRP a valid choice for an IGP recommendation because EIGRP is proprietary.* |

Despite all of these advantages, OSPF has some distinct disadvantages, covered later in this chapter in the section "How OSPF Works," in which OSPF is compared to RIP, IGRP, and EIGRP.

OSPF Terms and Concepts

This section defines some basic OSPF terms and concepts. Like in Chapter 25, the goal here is to give you a place to quickly find a term if you don't know its meaning. Once again, although I encourage you to read this section first before moving on to the rest of the chapter, you may find it easier to skip this section and return once the concept is formally introduced later in the chapter.

For ease of organization, these terms are grouped into ten sections:

- Fundamental terms
- Packet types
- LSA types

- Network types
- Databases and tables
- Router types
- Neighborship states
- Area types
- Destination types
- Path types

Fundamental Terms

This section examines basic operational terms related to OSPF. These terms are prerequisites to understanding OSPF operation (and the rest of this chapter):

- **Link** The direct connection to a network (the interface for a given network).
- **Link state** The status of a link (up, down, disabled, and so on).
- **Cost** The metric associated with a link. OSPF costs are based on the bandwidth of the link (10^8/bandwidth, by default).
- **Area** A boundary for the link-state database computation. Routers within the same area contain the same topology database. Areas are defined by *area IDs*. An area is not the same as an AS. An *area* is a subdivision of an OSPF AS. Each link may be assigned to a different area. If a router contains a link in a given area, it is considered an internal router (IR) in that area. If a router contains links to multiple areas within the same AS, it is considered an area border router (ABR). If a router contains links to different ASs, it is considered an autonomous system border router (ASBR).
- **Hello** A packet that is used to establish neighborships with directly connected routers. Hello packets are sent at a periodic interval known as the *hello interval* to maintain neighborships. Hellos are also used to verify two-way communication, advertise neighborship requirements, and elect designated routers (DR) and backup designated routers (BDR).
- **Dead interval** An interval used to determine when a neighbor should be considered down (similar to the holdtime in EIGRP).
- **Neighbor** A directly connected router that agrees on hello parameters and can establish two-way communication. For two routers to become neighbors, they must have the same hello and dead intervals, area ID, password, network mask for the link on which the hello was heard, and stub area flag. Not all neighbors will become adjacent.
- **Adjacency** A virtual connection to a neighbor over which link state advertisements (LSAs) may be transferred. Neighbors become adjacent

to the BDR and DR of broadcast-based networks, and with the remote endpoint
on NBMA point-to-point networks.

- **Link state advertisement (LSA)**   A topology link advertisement. LSAs are
included in link-state update packets (Type 4 packets); partial LSAs are included
in database description packets (Type 2 packets), link-state request packets
(Type 3 packets), and link-state acknowledgment packets (Type 5 packets).

- **Link-state request list**   The list that is used to track LSAs that need to be
requested. When a router recognizes that it does not have the most current
version of an LSA advertised (or does not have the LSA at all) in a database
description packet (Type 2 packet) or link-state request packet (Type 3 packet),
it adds the LSA to this list.

- **Link-state retransmission list**   The list that contains LSAs that have not been
acknowledged. When a router sends one or more LSAs to a neighbor in a
link-state update packet (Type 4), it waits for an acknowledgment of that packet.
If an acknowledgment is not heard before the retransmission timeout expires,
the router retransmits the LSA. Once an implicit or explicit acknowledgment is
heard, the router removes the LSA from the list.

- **Implicit acknowledgment**   An acknowledgment that occurs if a router detects
a link-state update packet (Type 4) from an adjacent neighbor that includes an
LSA listed in the link-state retransmission list for that adjacent neighbor.

- **Explicit acknowledgment**   An acknowledgment that occurs if a router receives
a link-state acknowledgment packet (Type 5) from an adjacent neighbor that
includes one or more LSAs listed in the link-state retransmission list for that
adjacent neighbor.

- **Flooding**   The process of forwarding LSAs out all applicable interfaces.
(Applicable interfaces depend on the LSA type.)

- **Router ID**   The router's identifier. This may be an interface IP address
(the default) or a statically defined number. Each router in an AS must have
a unique ID.

- **LSA sequence number**   The number that each LSA is assigned to identify
its version. In general, sequence numbers are incremented by 1 for every
change to an LSA. Higher numbers are therefore (usually) associated with a
more up-to-date LSA.

- **Priority**   The capacity of a router to become the DR or BDR in the election
process. In general, the router with the highest priority on a segment becomes
the DR. Priority ranges from 0 to 255, and the default value (for Cisco routers)
is 1. Routers with a priority of 0 are ineligible to become DR or BDR.

Packet Types

This section examines the OSPF packet types and their components. These types will become more important as we examine OSPF operation in detail:

- **OSPF packet headers (included in all types)**   Include basic information related to the router, such as the OSPF version, packet type, router ID, and area ID.

- **Type 1 packets (hello)**   Establish and maintain adjacencies. Hello packets include all information required to establish a neighborship, including hello and dead intervals, the password, the network mask for the link on which the hello was sent, the stub area flag, any elected DRs or BDRs, and any known neighbors.

- **Type 2 packets (database description)**   Build the link-state database on the router when an adjacency is initialized. Database description packets include LSA headers (not the entire LSA) in order for the receiving router to confirm that it has all the required LSAs.

- **Type 3 packets (link-state request)**   Request specific LSAs from neighbors. Link-state request packets are sent based on entries in the link-state request list.

- **Type 4 packets (link-state update)**   Supply LSAs to remote routers. They are flooded when an LSA changes or a link-state request is received. Type 4 packets must be acknowledged.

- **Type 5 packets (link state acknowledgement)**   Sent to explicitly acknowledge one or more LSAs.

LSA Types

This section examines the most common LSA types. Other types exist, but are not generally used.

- **Type 1 LSA (router link entry)**   Generated by every router for each area of which the router is a member. These LSAs contain the status of all of the router's links to a given area and are flooded to all links within the same area.

- **Type 2 LSA (network entry)**   Generated by DRs on every non–point-to-point (multi-access) network. Type 2 LSAs include all routers attached to the network for which the router is the DR.

- **Type 3 LSA (summary network link-state entry)**   Generated by ABRs and advertise internal networks from a specific area to other ABRs. The other ABRs then choose the best path to the network(s) based on all Type 3 LSAs received, and flood the best path into nonbackbone areas using Type 3 LSAs. Note that Type 3 LSAs may or may not be a summarized network entry. To summarize networks in Type 3 LSAs, you need to configure the ABRs to summarize the entries. Type 3 LSAs are not distributed to totally stubby areas (except for a single Type 3 for the default route).

- **Type 4 LSA (summary ASBR link-state entry)**   Used by ABRs to advertise the best paths to ASBRs. Type 4 LSAs are not flooded to stubby, totally stubby, and not-so-stubby areas (NSSAs).

- **Type 5 LSA (AS external entry, also called *external entries*)** Sent by ASBRs and advertise destinations external to the AS (destinations redistributed from another OSPF AS or another routing protocol). Type 5 entries are flooded throughout the entire AS, except for stubby, totally stubby, and not-so-stubby areas. Type 5 entries are split into two separate subtypes, depending on the metric calculation used:

  - **External Type 1 (E1)** E1 entries have their metric calculated as a sum of the redistributed route's cost and the cost of the links to the sending router. E1 entries are typically used when more than one ASBR is advertising a given external destination.

  - **External Type 2 (E2)** E2 entries have their metric calculated simply as the cost of the redistributed route. (Cost of internal links to the advertising ASBR are *not* taken into account.) E2 entries are subsequently "cheaper," and routers usually prefer them to E1 entries.

- **Type 7 (NSSA external link entry)** Generated only by ASBRs in NSSAs. Type 7 LSAs are flooded only throughout the NSSA. ABRs convert Type 7 LSAs into Type 5 LSAs for distribution to the rest of the AS. Type 7 LSAs also have two subtypes:

  - **NSSA External Type 1 (N1)** N1 entries have their metric calculated as a sum of the redistributed route's cost and the cost of the links to the sending router. N1 entries are typically used when more than one ASBR is advertising a given external destination.

  - **NSSA External Type 2 (N2)** N2 entries have their metric calculated simply as the cost of the redistributed route. (Cost of internal links to the advertising ASBR are *not* taken into account.) N2 entries are subsequently "cheaper," and routers usually prefer them to N1 entries.

 In addition to the LSA types listed here, there are also Type 6, Type 8, Type 9, Type 10, and Type 11 LSAs. However, because these LSAs are either unused at present or unsupported by Cisco routers, this chapter does not cover them.

Network Types

This section examines the OSPF network types, as defined by either the RFCs or Cisco (depending on the type in question). These concepts will be central to understanding the OSPF update process over various media:

- **Broadcast (multi-access)** A network that follows basic Ethernet conventions, where any host that is part of the same logical network can communicate with any other host. In this configuration, DRs and BDRs are elected, and neighbor and adjacency establishment is automatic. The default hello time in this network is 10 seconds, and the default dead interval is 40 seconds.

- **Point-to-point** A type of network in which a single WAN link (usually a Frame Relay PVC) connects two routers. More than two routers may be interconnected with multiple links. Each link needs to have its own logical network address. In this environment, a DR and BDR are not elected (or required), and neighbor adjacencies are automatically configured. The default hello time in this network is 10 seconds, and the default dead interval is 40 seconds.

- **Point-to-multipoint (NBMA with broadcast emulation, full mesh)** A type of network in which each router has a point-to-multipoint connection to every other router. Note that this is not a single multipoint connection from a central router to all routers (that would be a star topology), but a multipoint connection from each router to every other router (making the topology a full mesh). In this environment, you can set all routers to use the same logical network for the mesh, and enable broadcast forwarding on the multipoint connections. In this environment, a DR and BDR are elected, and neighbor adjacencies are automatically configured. The default hello time in this network is 10 seconds, and the default dead interval is 40 seconds.

- **Point-to-multipoint (NBMA, full mesh)** A type of network in which each router has a point-to-multipoint connection to every other router. Note that this is not a single multipoint connection from a central router to all routers (that would be a star topology), but rather a multipoint connection from each router to every other router (making the topology a full mesh). All routers use the same logical network address, but broadcast emulation is disabled. In this environment, a DR and BDR are elected, and neighbor adjacencies must be manually configured. The default hello time in this network is 30 seconds, and the default dead interval is 120 seconds.

- **Point-to-multipoint (NBMA, star, or partial mesh)** A type of network with one or more routers in a partial mesh or hub-and-spoke–style topology with point-to-multipoint links. In these topologies, all routers use the same logical network address. In this environment, a DR and BDR are not elected (or required), and neighbor adjacencies must be manually configured. The default hello time in this network is 30 seconds, and the default dead interval is 120 seconds.

- **Transit (all available topology types)** A type of network with OSPF routers attached. Transit networks may be used by OSPF to forward packets to other OSPF routers. In other words, a transit network may receive data packets that are destined for other networks and are just passing through.

- **Stub** A network with only one OSPF router attached. Stub networks receive data packets destined only for the stub network itself.

- **Virtual link** A link that is used to connect a remote area to the backbone area through another AS. Virtual links have no IP address, and are similar to tunnels through a nonbackbone area designed to forward packets to the backbone. In a well-designed OSPF network, permanent virtual links should not exist.

Note *Do not confuse transit networks with transit areas or stub networks with stub areas. The names are similar, but the concepts are very different.*

Databases and Tables

This section examines the different tables (also known as databases) used by OSPF. These concepts are central to understanding OSPF:

- **Neighborship database (neighbor table)**   Contains a list of all routers with which a two-way neighborship has been established, along with each neighbor's router ID, priority, neighborship state, router designation, remaining dead time, and interface IP address. If a router is directly connected to another router by more than one logical network, that router will have multiple neighborships established.

- **Link-state database (topology database)**   Contains all of the LSA entries for a given area, and is used to generate the routing table. All links in the entire area (plus external routes) should be listed in this database for every router in the area. Subsequently, every router in a given area should have an identical copy of the link-state database for that area.

- **Routing table (forwarding database)**   Contains the best valid paths to all known destinations. (The routing table in OSPF is the same as in any other routing protocol.) It is built by running the SPF algorithm on the link-state database.

Router Types

Understanding router types is central to understanding the OSPF update process. The different OSPF router designations are as follows:

- **Designated router (DR)**   The router chosen through the election process to be the primary "advertising router" for the individual logical network in question and all routers attached to that network. The DR is usually the router with the highest priority. DRs exist only on multi-access networks with more than one OSPF-speaking router. The DR establishes an adjacency with all routers on the network.

- **Backup designated router (BDR)**   The router chosen to be the secondary "advertising router" for the individual logical network in question and all routers attached to that network. The BDR is usually the router with the second-highest priority. BDRs exist only on multi-access networks with more than one OSPF-speaking router. The BDR also establishes an adjacency with all routers on the network.

- **DROther**   An OSPF router that is not the DR or BDR on a multi-access network with more than one OSPF router. DROthers establish adjacencies only with the DR and BDR on the logical network.

- **Internal router**   Any router that has all interfaces in the same area. All internal routers for a given area have a single, identical topology database.

- **Backbone router**   Any router that has at least one interface in Area 0 (the backbone area).

- **Area border router (ABR)**   Any router that has one or more interfaces in different areas. ABRs are used to summarize and forward packets along paths between different areas.

- **Autonomous system border router (ASBR)**   Any router that is redistributing routes from a different routing protocol or OSPF AS into the target OSPF AS.

Neighborship States

This section examines the different states of neighborship between OSPF routers. These concepts are central to the OSPF update mechanism:

- **Down**   The initial state of a neighbor. Neighbors in the down state do not exist in the neighbor table.

- **Attempt**   On NBMA networks, this is the state whereby neighborship establishment is attempted. On NBMA networks, neighbors must be manually configured. For these networks, hellos will be sent to all configured neighbors as a unicast to attempt to establish a neighborship.

- **Init**   The state that indicates that the router has seen a hello from this neighbor recently (the dead interval has not expired since the last hello was seen), but has yet to see its own router ID listed in the neighbor's hello packet (meaning that the neighbor has yet to see a hello from this router). Once a router reaches this state with a neighbor, it will begin including the neighbor's router ID in its hello packets.

- **2-way**   The state in which the router has seen its own router ID in the neighbor's hello packet, meaning that the neighbor is receiving his hello, and bidirectional communication has been established.

- **Exstart**   The state in which the neighbors have already chosen to form an adjacency and are in the process of determining how to transfer database description packets (OSPF Type 2 packets) to each other.

- **Exchange**   The state in which the routers exchange database description packets (OSPF Type 2 packets).

- **Loading**   The state in which routers send link-state request packets (OSPF Type 3 packets) for all LSAs listed in the link-state request list and receive link-state update packets (OSPF Type 4 packets) in response.

- **Full**   The state in which neighbors are fully adjacent and should have identical copies of the link-state database for the area.

Area Types

This section examines the OSPF area types. These concepts are central to understanding multi-area OSPF functionality:

- **Standard area** The most common type of area. Any area that is not the backbone area or some form of stub area is a standard area. Standard areas support Type 1 through 5 LSAs.

- **Backbone area (Area 0)** The hub in your OSPF AS. The backbone area is responsible for forwarding traffic between areas. All areas in a multi-area OSPF solution should have a connection to the backbone area.

- **Transit area** An area where traffic from other areas can travel through en route to their final destination. The backbone area is considered a transit area.

- **Stub area** An area where there is only one way to reach external (other AS) destinations. For this reason, a stub area does not require Type 4 or Type 5 LSAs. Instead, a single Type 3 LSA for a default route is inserted into the stub area to provide the path to external destinations. Stub areas require less network, CPU, and memory resources because they do not have to keep external LSAs in the topology table. Stub areas allow only Type 1 through 3 LSAs.

- **Totally stubby area** An area where there is only one way to reach external (other AS) and interarea destinations. In other words, in a totally stubby area, there is only one way to reach destinations external to the area. Therefore, a totally stubby area does not require Type 3, Type 4, or Type 5 LSAs. A single default route is inserted into the stub area to provide the path to all destinations that are not a member of the totally stubby area. Totally stubby areas therefore require even less resources than stub areas. Totally stubby areas allow only Type 1 and 2 LSAs (except for a single Type 3 LSA for a default route). Totally stubby areas are a non-RFC-defined (Cisco-defined) enhancement.

- **Not-so-stubby area (NSSA)** An area that requires the transmission of external LSAs from an ASBR within the area, but only has a single path to ASBRs in other areas. Because the NSSA area has only one path to external destinations reached by ASBRs in other areas, the NSSA area is normally able to be a stub area. However, because stub areas do not allow Type 5 LSAs and the NSSA contains an ASBR (which generates Type 5 LSAs), the NSSA is required to be a standard transit area (increasing resources required to support the area). In this case, the area can be configured as an NSSA. ASBRs in an NSSA generate Type 7 LSAs (instead of Type 5 LSAs) to advertise external destinations, and the ABR for the NSSA converts the Type 7 LSAs in Type 5 LSAs for the rest of the AS. NSSAs still do not allow Type 4 or 5 LSAs, and they use a single default route to reach external destinations advertised by ASBRs in other areas. NSSAs are RFC defined and described in RFC 1578.

Destination Types

These are the standard destination types discussed throughout the chapter:

- **Network**   A standard entry into the routing table. It defines networks to which packets can be routed (just like normal routing table entries).

- **Router**   A path to ASBRs and ABRs. Because OSPF needs to know where ABRs are located to correctly route interarea packets, an entry for each ABR is listed in an internal routing table. Similarly, an entry for each ASBR is listed in the internal routing table.

Path Types

Some standard path types used by OSPF are as follows:

- **Intra-area**   Routing table entries to an area of which the router is a member. If the router is an internal router, it contains only intra-area routes for a single area. If the router is an ABR, it contains intra-area routes for all attached areas. Intra-area routes are constructed from Type 1 and 2 LSAs.

- **Interarea**   Routes to destinations within another area in the same OSPF AS. Interarea routes are constructed from Type 3 LSAs.

- **E1 (External Type 1)**   Entries that are constructed from Type 5 E1 LSAs propagated by ASBRs.

- **E2 (External Type 2)**   Entries that are constructed from Type 5 E2 LSAs propagated by ASBRs.

Whew! Now that you've examined the very extensive list of definitions involved with OSPF, you're ready to learn how OSPF works.

How OSPF Works

If you read through the last section, at this point, you are probably a bit intimidated by OSPF's apparent complexity. Although I won't tell you that OSPF isn't very complicated, OSPF isn't quite as complex as it might initially appear. This section examines how OSPF works in both simple, single-area configurations and more complex, multiple-area ASs. Next, this section examines the advantages and disadvantages of OSPF as compared to other routing protocols. Finally, it examines redistribution using OSPF. Before you learn about how OSPF works in any of these situations, however, you need to understand the concepts of area and AS as they relate to OSPF.

In OSPF, an AS is a process domain. To route between process domains, an OSPF ASBR is required to perform route redistribution. (ASBRs are examined in more detail in the "Redistribution with OSPF" section of the chapter.) In Cisco routers, process domains are defined by *process ID*: an arbitrary number to define the AS on a single router. Note that in OSPF, the process ID has no significance beyond the router on which it is configured. In other words, the process IDs on two routers do not have to match in order for the routers to establish an adjacency. The real purpose of a process ID is to allow for multiple OSPF processes on a single router (with redistribution used to inject routes from one process into the other).

Note   *I highly discourage running multiple OSPF processes on a single router due to processing requirements and the possibility of database inconsistencies.*

OSPF also splits subsections of an AS into *areas*: sections of OSFP where all OSPF routers contain the same topology database and perform the same OSPF computations. Areas are needed to extend the scalability of OSPF beyond a few hundred routers. (The "OSPF in Multiple Areas" section, later in this chapter, examines this concept in more detail.) To route between areas, an ABR is required (also examined in more detail in the "OSPF in Multiple Areas" section of the chapter). To help you understand these differences, see the diagram in Figure 26-1.

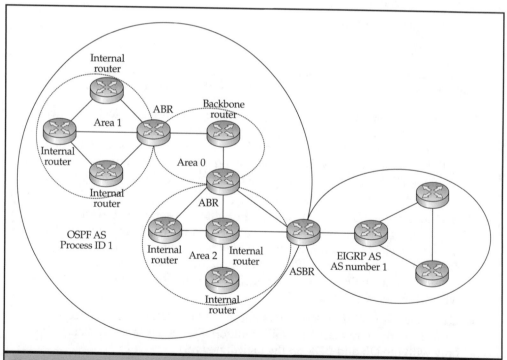

Figure 26-1.   *The difference between ASs and areas*

This example shows two separate ASs, with three areas in the OSPF AS. Routes between ASs are redistributed by the ASBR, which is a member of both ASs and runs both OSPF and EIGRP. Routes between OSPF areas are summarized by ABRs and injected into the backbone (Area 0), where the other ABR injects the best route into each respective area. These concepts (including the various router types) are expanded in later sections of this chapter (primarily in the "OSPF in Multiple Areas" section); but for now, just keep the differences in mind between areas and ASs.

OSPF in a Single Area

In a single area, OSPF operation is fairly straightforward. The tasks associated with OSPF in a single area can be divided into seven components:

- Establishing neighborships
- Forming adjacencies
- Building the link-state database
- Running the SPF algorithm
- OSPF's core logic
- Dealing with changes
- OSPF network types

Before discussing how neighborships are established, take a look at the router designations from the "Router Types" section, reprinted here for convenience:

- **Designated router (DR)**   The router chosen through the election process to be the primary "advertising router" for the individual logical network in question and all routers attached to that network. The DR is usually the router with the highest priority. DRs exist only on multi-access networks with more than one OSPF-speaking router. The DR establishes an adjacency with all routers on the network.

- **Backup designated router (BDR)**   The router chosen to be the secondary "advertising router" for the individual logical network in question and all routers attached to that network. The BDR is usually the router with the second-highest priority. BDRs exist only on multi-access networks with more than one OSPF-speaking router. The BDR also establishes an adjacency with all routers on the network.

- **DROther**   An OSPF router that is not the DR or BDR on a multi-access network with more than one OSPF router. DROthers establish adjacencies only with the DR and BDR on the logical network.

- **Internal router**   Any router that has all interfaces in the same area. All internal routers for a given area have a single, identical topology database.

- **Backbone router** Any router that has at least one interface in Area 0 (the backbone area).

- **Area border router (ABR)** Any router that has one or more interfaces in different areas. ABRs are used to summarize and forward packets along paths between different areas.

- **Autonomous system border router (ASBR)** Any router that is redistributing routes from a different routing protocol or OSPF AS into the target OSPF AS.

The next section deals with the first four steps of the neighborship establishment process.

Establishing Neighborships

Before examining the neighbor establishment process, you need to understand that OSPF is not a router-centric protocol. Rather, OSPF is a connection-centric protocol. In other words, OSPF defines different responsibilities to a router based on each connection to that router. Keep this fact in mind as you progress through the explanations that follow.

First, as defined at the beginning of the chapter, a *neighbor* is a directly connected router that agrees on hello parameters and can establish two-way communication. In OSPF, like EIGRP, two routers must become neighbors to have any chance of exchanging routing information. Unlike EIGRP, however, neighbors must become adjacent in order to transfer routing information, and not all neighbors will become adjacent (examined further in the upcoming "Forming Adjacencies" section). For now, just keep in mind that the primary reason a router forms a neighborship is to identify with whom it can establish two-way communication and with whom it needs to form an adjacency.

The neighbor establishment process differs slightly depending on the link type. Again, remember that OSPF is connection-centric—so on a single router, the neighbor establishment can vary between different interfaces.

For links to multi-access networks (like Ethernet), the OSPF router begins the neighbor establishment process by multicasting a hello packet to the reserved multicast address for AllSPFRouters, 224.0.0.5. All OSPF routers listen to this address and process packets destined for this address, while non-OSPF routers and end user devices simply ignore the packet. This hello packet contains the following information:

- The sending router's router ID

- The area ID for the interface on which the hello was sent

- The subnet mask for the interface on which the hello was sent

- The authentication type (none, cleartext, or MD5) and the associated password/key for the area

- The hello interval for the interface on which the hello was sent

- The dead interval for the interface on which the hello was sent

- The router's priority

- The current DR and BDR for the network on which the hello was sent
- Options bits (used to signify stub or NSSA status)
- Router IDs of known neighbors

The router ID uniquely defines the router in the OSPF AS. The router ID will be one of the following addresses, chosen in this order:

- If the router ID is statically configured (using the *router-id* command on Cisco routers), that address is used.

- If any software loopback interfaces (pretend interfaces) are defined, the highest IP address defined on any loopback interfaces is used.

- If either of the preceding is true, the router uses the highest IP address defined on any other interfaces.

Note   *It is generally better to statically define the router ID using either the manual* router-id *command or a loopback interface instead of letting the router pick an IP because changing a router ID leads to SPF computations on all routers in the area (which is undesirable). The "Advanced OSPF Configuration" section examines this concept in more detail.*

The rest of the fields are either self-explanatory or already described in the "OSPF Terms and Concepts" section of the chapter, except the options bits and the router IDs of known neighbors. The options bits in a hello are used only to signify whether the router is a member of a stub area or NSSA. The router IDs of the router's known neighbors are used to determine if two-way communication is possible with this neighbor.

Every ten seconds on multi-access networks (an interval that is configurable by changing the hello interval on Cisco routers), the OSPF router repeats the hello. If another OSPF router hears the hello, it checks to see if the required values in the hello (the hello interval, dead interval, subnet mask, area ID, authentication parameters, and options) match its configured values. If these values do not match between the two routers, a neighborship is not established. (The sending router's hellos are simply ignored.) If they do match, on the receiving router's next hello packet, it includes the router ID of the neighbor that sent the hello in the "Router IDs of known neighbors" section of the packet. When the original router sees its own ID in the neighbor's hello, it realizes that the neighbor is seeing its hellos, and two-way communication has been verified.

At this point, the routers have the opposite router entered into their neighbor tables, and expect to see a packet from the neighbor every hello interval. If a hello is not heard before the dead interval expires (40 seconds, or four times the hello interval for LAN interfaces, by default), the router assumes that the neighbor is down and removes the neighbor's listing in its hello packets and neighbor table. Removal of a neighbor can lead to negative consequences because, when the neighbor is removed from the neighbor table, all paths through that neighbor are removed from the topology table—which causes an SPF computation (examined in the "Dealing with Changes" section, later in the chapter).

This process is very similar no matter what the network type, but some of the specific steps vary slightly. For instance, in an NBMA network with broadcast emulation disabled, rather than multicasting to find neighbors, the router must be statically configured with neighbor addresses, and it will unicast hellos directly to those neighbors. The basic rules for establishing neighborships are as follows:

- Multi-access network, point-to-point network, or point-to-multipoint NBMA network (full mesh) with broadcast emulation enabled:

  - Send hellos every hello interval (ten seconds, by default) to the multicast address AllSPFRouters.
    State: down – init

  - If a hello from a neighbor candidate is heard, examine the required parameters for a match.
    State: init

  - If a match occurs, add the neighbor's router ID to the neighbors section of outgoing hello packets.
    State: init

  - If a hello is heard from a neighbor that contains your router ID in the known neighbors section, add that neighbor to the neighbor table.
    State: 2-way

  - If a neighbor is not heard from before the dead interval expires (40 seconds, by default), remove that neighbor from the neighbor table.
    State: init

  - If a hello from a neighbor is heard and the required parameters do not match, the following applies:
    If the neighbor is not in the neighbor table, do not add the neighbor.
    If the neighbor is in the neighbor table, remove the neighbor from the neighbor table and remove the neighbor from the list of known neighbors in outgoing hellos.
    State: init

- Point-to-multipoint NBMA network (any topology), broadcast emulation disabled:

  - Send hellos every *poll interval* (once every 60 seconds, by default) directly to the configured neighbor using a unicast.
    State: attempt

  - If a hello from the neighbor candidate is heard, examine the required parameters for a match.
    State: init

■ If a match occurs, add the neighbor's router ID to the neighbors section of outgoing hello packets.
State: init

■ If a hello is heard from a neighbor that contains your router ID in the known neighbors section, add that neighbor to the neighbor table.
State: 2-way

■ Continue sending hellos every hello interval (30 seconds, by default).
State: 2-way

■ If a neighbor is not heard from before the dead interval expires (120 seconds, by default), remove that neighbor from the neighbor table and restart the neighbor establishment process.
State: init

■ If a hello from a neighbor is heard and the required parameters do not match, the following applies:
If the neighbor is not in the neighbor table, do not add the neighbor.
If the neighbor is in the neighbor table, remove the neighbor from the neighbor table, remove the neighbor from the list of known neighbors in outgoing hellos, and restart the neighbor establishment process.
State: init

The only new term in these rules is the poll interval for NBMA networks without broadcast emulation. The *poll interval* specifies how long a router waits before sending hellos to neighbors that have yet to be heard from. By default, this interval is set to 60 seconds. Because graphical depictions are sometimes more helpful, the flowcharts in Figures 26-2 and 26-3 describe this process.

Forming Adjacencies

Once a set of neighbors has transitioned to the 2-way state, the neighbors must determine whether to establish adjacencies with each other. This determination is fairly simple if the network type does not require a DR or BDR (point-to-point or point-to-multipoint in a star or partial mesh). In a point-to-point or point-to-multipoint in a star or partial mesh topology, all routers on the logical network become adjacent. However, if a DR and BDR need to be elected, all routers on a network establish an adjacency only to the DR and BDR, and the process gets slightly more complex. Before you learn about the DR/BDR election process, you need to understand why DRs and BDRs are even needed.

In a multi-access network with more than one router, establishing an adjacency with all routers can be extremely resource intensive because, without DRs, when an OSPF router detects a change to a link, it advertises the new link state to all routers with which it has established an adjacency. These routers, in turn, advertise the change to the link state on the originating router to all routers with which they have established

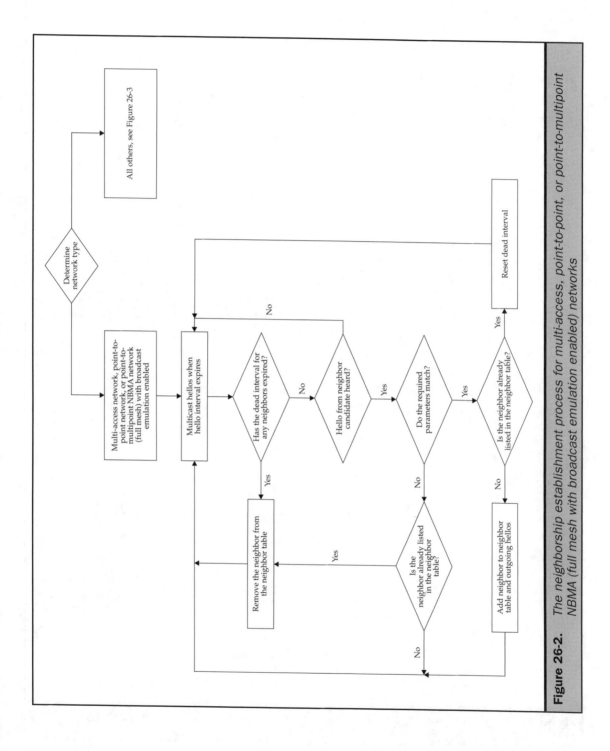

Figure 26-2. The neighborship establishment process for multi-access, point-to-point, or point-to-multipoint NBMA (full mesh with broadcast emulation enabled) networks

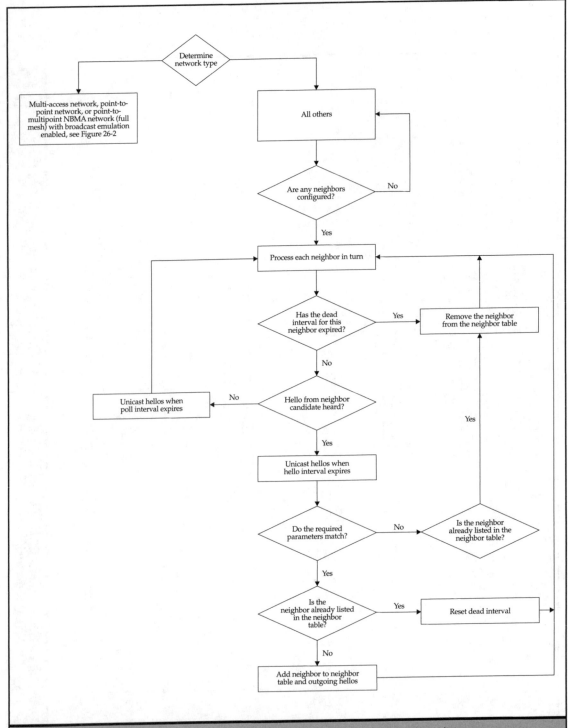

Figure 26-3. *The neighborship establishment process for all other network types*

an adjacency (including all routers that already received the update from the originating router)! With DRs, however, all routers on the network establish an adjacency on that network only with the DR and BDR of that network (which, in turn, have an adjacency established with all other routers on the network). The DR then advertises the change to all other routers on the network. For example, the network shown here is not using DRs.

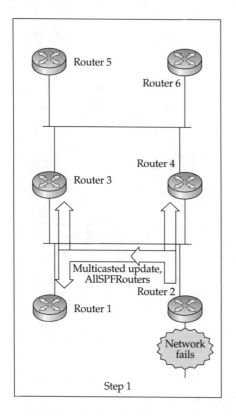
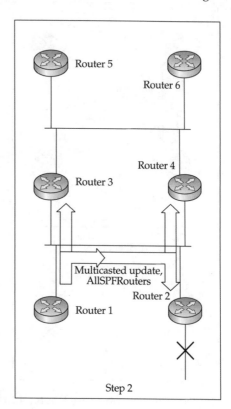

In this example, all routers are forced to establish an adjacency with all other routers. When the change occurs at Router 2, it will multicast an update to all OSPF routers, all of which will, at approximately the same time, remulticast the update back to all OSFP routers. In this example, six packets were used to transmit one change!

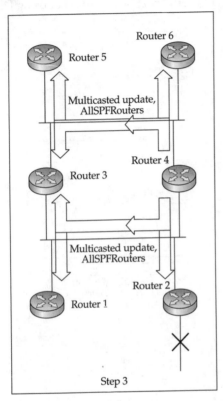

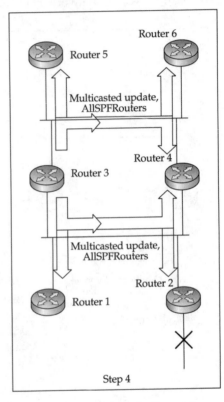

However, if you configure Router 4 to be the DR, you get the result shown in Figure 26-4. In this case, Router 2 establishes an adjacency only to the DR and BDR. When the change occurs, Router 2 multicasts an update to the AllDRouters address (224.0.0.6), which is monitored only by the DR and BDR. The DR then multicasts the update to both connected networks using the AllSPFRouters address. By using a DR, you reduce the number of transmissions to three update packets, reducing the traffic by 50 percent. Now, multiply the number of routers in the example by 10, and you can see how the DR helps tremendously in reducing network traffic.

Although the DR is the primary advertising router for a network, the BDR is there in case the DR fails. If there were no BDR, and adjacencies were not established with both the BDR and DR of each network, then if the DR were to fail, the entire election, adjacency building, and link-state database synchronization process would have to be repeated. With a BDR, however, the BDR has all of the same information that the DR has, and already has adjacencies with other routers (known as DROthers if they are not the DR or BDR), so it can just take over where the DR left off.

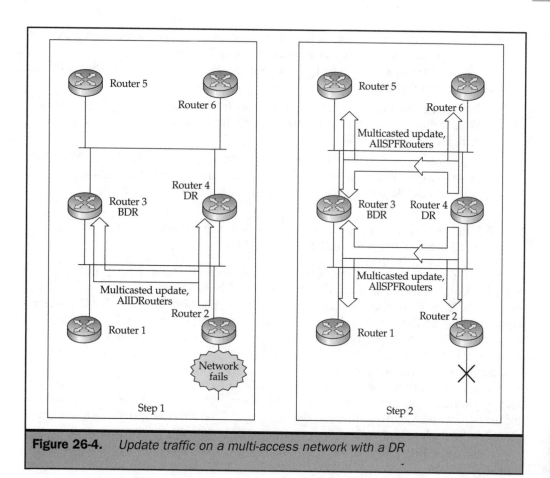

Figure 26-4. *Update traffic on a multi-access network with a DR*

Note *The DR and BDR are chosen for each individual multi-access interface, or link. In other words, it is possible for a router to be the DR on one interface and be a DROther on a completely different interface.*

So how does OSPF determine which routers should be the DR and BDR? The short answer is, it holds an election. The election process works as follows:

1. At initial interface activation, send all hellos to the network with the DR and BDR fields set to 0.0.0.0.

2. After the 2-way state has been reached for neighbors on a given network, each router examines the priority, BR, and BDR fields of the received hellos for that network.

3. Each router then builds a list of all routers claiming to be DR and BDR.

4. If only one router is claiming to be BDR, each router chooses that router as BDR. If more than one router is claiming to be BDR, then the router with the highest priority is chosen as BDR. If there is a tie in priority, the router with the highest router ID is chosen as BDR.

5. If no router claims to be BDR, the router with the highest priority is chosen as BDR. If there is a tie in priority, the router with the highest router ID is chosen as BDR.

6. If only one router claims to be DR, each router chooses that router as DR. If more than one router claims to be DR, then the router with the highest priority is chosen as DR. If there is a tie in priority, the router with the highest router ID is chosen as DR.

7. If no router claims to be DR, the new BDR is appointed DR. In this case, the election process repeats to choose a new BDR.

So, in the election process, the BDR is actually chosen first. If it turns out that there is no DR after the BDR is elected, then the new BDR becomes the DR, and an election is held for the BDR. One other aspect to notice about this election process is that if a router comes online and there is already a DR or BDR, the router will not hold an election, even if the new router's priority is higher than the current DR and BDR. In OSPF, the router that comes online first with the highest priority is elected DR, and the router that comes online first with the second-highest priority becomes BDR.

Once DRs and BDRs have been elected, all DROthers on the multi-access network form adjacencies only with the DR and BDR. Although this limitation saves on resources for the DROthers, it also requires that the DR and BDR be able to support the additional demands (including higher CPU and memory use) of their appointed positions.

Finally, because of the way the election process works, it is possible for the routers you want to be the DR and BDR to actually *not* be selected, especially if those routers have failed at one time or another. For instance, examine the network in Figure 26-5.

In this network, Router 4 should be elected as DR and Router 5 should be elected as BDR for both Net 1 and 2. However, what if Routers 4 and 5 are booted a few seconds later than all other routers? The election would have already occurred, and because all other routers have the same priority, the routers with the highest router IDs will be chosen as DR for the respective networks. Assuming that the router IDs in this case match the router numbers, Router 9 (with router ID 9) will be chosen as the DR for Net 2, and Router 6 (with router ID 6) will be chosen as the DR for Net 1. Because of the way the election process functions, Routers 4 and 5 would have no choice but to accept the current DRs.

So, to solve this problem, say you restart all routers and boot Routers 4 and 5 first. Now you have the correct routers configured as DR and BDR, but what happens if one of them fails? If Router 4 were to fail, Router 5 would become the new DR, Router 6 would become the new BDR for Net 1, and Router 9 would become the new BDR for Net 2. Then, if Router 4 were inserted back into the network, it would simply become a

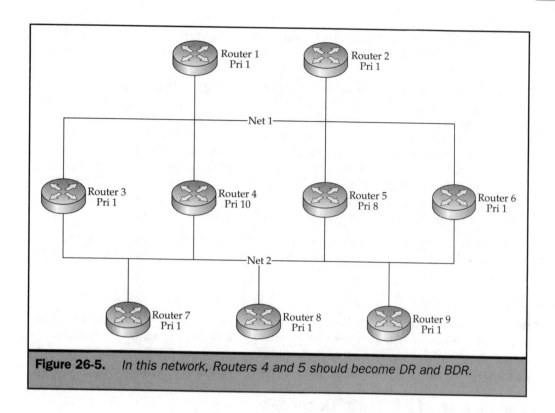

Figure 26-5.   *In this network, Routers 4 and 5 should become DR and BDR.*

DROther. Now, if Router 5 were to fail, the BDRs for each network would become the DRs, making Router 6 the DR for Net 1 and Router 9 the DR for Net 2. An election would be held for BDR, and Router 4 would get the job for both networks. At this point, if Router 5 were inserted back into the network, it would become a DROther. Finally, if Routers 9 and 6 were to fail after Router 5 were reinserted, Router 4 would become the DR and Router 5 would be elected the new BDR for both networks.

As you can see from this example, the election process can lead to a sort of cyclical situation where routers that shouldn't be the DR are chosen for that position. To resolve this issue, you have a few options:

- Reboot the routers or reset OSPF each time the wrong DR is chosen. This approach is particularly poor due to the administrative effort required, not to mention the amount of traffic caused by adjacency reestablishment and database synchronization in a large network.

- Add two more routers, and cluster Routers 4 and 5 with these new routers using Hot Standby Router Protocol (HSRP). This strategy works, but it is a particularly complicated and expensive way to resolve this simple issue. On the other hand, HSRP allows the routers to achieve a much higher level of availability, which can be useful if you require extremely high reliability. Unfortunately, HSRP configuration is beyond the scope of this book, so this solution is not an option in this context.

■ Increase the priority of Routers 3 and 6 to a value less than 8 (say, 2), and reduce the priorities of Routers 1, 2, 7, 8, and 9 to 0. This solution ensures that if Routers 4 and 5 are not available during the election, the next best candidates, Routers 3 and 6, are chosen. Because the priority for all other routers are set to 0, they will be considered ineligible to become either DR or BDR, so you will never have to worry about those routers usurping the throne. The flipside to this configuration is that if none of the routers with a positive priority (Routers 3–6) is available, no DR or BDR will be chosen, and no adjacencies on Nets 1 or 2 will be established until at least one of these routers is available.

Once the routers determine with whom they should establish adjacencies on each network, they will transition to the exstart state on each adjacency, covered in the next section.

> **Note** *Neighbors that choose not to become adjacent do not enter the exstart state and do not transfer link-state information to each other. They instead rely on their adjacency to the DR to transfer applicable link-state information to and from nonadjacent neighbors. Neighbors that reach the 2-way state with each other but choose not to form an adjacency remain in the 2-way state as long as valid hello packets remain to be heard.*

Building the Link-State Database

Once an adjacency has been established, adjacent neighbors can begin negotiating how to transfer their link-state databases to each other. This negotiation is known as the exstart state.

While in the exstart state, routers communicate with each other using database description packets (DD)—OSPF Type 2 packets—to negotiate DD sequence numbers. DD sequence numbers are used to ensure that DD packets are received in the correct order. Because both routers involved in the exchange will likely have different values for their sequence numbers, the exstart state is used to synchronize the DD sequence numbers used between the two routers. To perform this negotiation, the routers choose a master and slave for the DD packets. The only significance of the master/ slave relationship is that the master is the router that chooses the beginning sequence number. The slave simply resets its sequence number to equal whatever the master has configured.

> **Note** *Although there is no real practical significance in the choice of master or slave routers, the master is always the router with the highest router ID.*

Once the routers have negotiated the master/slave relationship, they enter the exchange state, where the following are transferred to build the link-state database:

DD, link-state request (LSR) (OSPF Type 3 packets), and link-state update (LSU) (OSPF Type 4 packets). The DD packets are transferred between the routers first and include the following:

- The router ID of the sending router.
- The area ID for the outgoing interface of the sending router.
- Authentication parameters.
- Options bits (used to signify stub or NSSA status).
- An I (Initial) bit, used to signify the initial DD packet. A value of 1 indicates that this is the first DD packet in a series.
- An M (More) bit, used to signify the last DD packet. A value of 0 indicates that this is the last packet in the series.
- An MS (Master/Slave) bit. A value of 1 signifies that the sending router is master.
- The DD sequence number.
- One or more link state advertisement (LSA) headers, each containing the following:
  - **Age**   The amount of time (in seconds) since the LSA was created on the originating router (the router that is directly connected to the network advertised by the LSA and that initially advertised the LSA).
  - **Options bits**   The bits used to signify stub or NSSA status for the area advertised by this LSA.
  - **Type**   The type code for the LSA.
  - **Link-state ID**   The use varies depending on the type of LSA.
  - **Advertising router**   The router ID of the originating router.
  - **Sequence number**   The sequence number for this specific LSA. This number is different from the DD sequence number, and is used to determine the "version" of the LSA.

The LSA headers sent in these DD packets are the most important components. These LSA headers do not contain the entire LSA—only enough of the LSA for the receiving router to examine its link-state database and determine if the LSA is listed. Basically, the receiving router uses the advertising router, link-state ID, and type fields to determine if the LSA is already in the database. If the receiving router notices a new LSA, it places that LSA in its link-state request list. If the receiving router has the LSA but the sequence number advertised in the sending router's LSA is newer (usually indicated by a higher sequence number for the LSA), the receiving router also places that LSA in its link-state request list.

For every DD packet sent by the master, the receiving router sends a DD packet with the same DD sequence number back to acknowledge the DD packet. If the master does not hear a DD packet in acknowledgment from the slave, it resends the DD packet.

Once all DD packets have been transferred, the state transitions to loading, and both routers send LSR packets to each other that include the LSA type, advertising router, and sequence numbers for each of the LSAs listed in the link-state request list. As each router receives these LSRs, it builds one or more LSUs containing the entire LSA for the requested entries. As LSAs contained in these LSUs are received, they must be either implicitly or explicitly acknowledged.

Explicit acknowledgment is when a router sends a *link state acknowledgment* (*LSAck*) (OSPF Type 5 packet) back to the other router to acknowledge one or more LSAs. Occasionally, however, the router may instead send an LSU back out to the network to advertise these new LSAs to other neighbors. In this case, the LSA is said to be implicitly acknowledged.

Once all LSUs have been transferred between these routers, the routers transition to the full state. If a change was made to the link-state database, the routers begin computing the routing table by running the SPF algorithm. Before examining the SPF algorithm, however, let's examine the details regarding the two primary LSA types that are covered in this section.

The most basic type of LSA is the Type 1 LSA, or router entry LSA. The router LSA describes all of the router's links to a given area, along with network types and metrics. It also informs the recipient of the LSA about the originating router's status as an ABR or ASBR. The router LSA propagated throughout the entire area contains the following information:

- **An LSA header (described earlier in this section under DD packet format)**   The Link ID field in the LSA header contains the IP address of the originating router.

- **A V (virtual link) bit**   This tells the receiving router that this router contains one or more virtual links to the backbone. Virtual links are described in more detail in the "OSPF in Multiple Areas" section, later in the chapter.

- **An E (external) bit**   This tells the receiving router that this router is an ASBR. ASBRs are described in more detail in the "OSPF in Multiple Areas" section, later in the chapter.

- **A B (border) bit**   This tells the receiving router that this router is an ABR. ABRs are described in more detail in the "OSPF in Multiple Areas" section, later in the chapter.

- **The total number of links**   This is the total number of links described by this LSA.

- **Link information for each link**   This includes the following:

  - The link type (point-to-point, transit, stub, or virtual link)
  - The link ID (value varies depending on link type; see Table 26-1)
  - Link data (value varies depending on link type; see Table 26-1)
  - Metric for the link (OSPF metrics are based on bandwidth in Cisco routers.)
  - TOS (Type of Service) fields, unused in Cisco routers

Link Type	Link ID	Link Data
Point-to-point	Neighboring router's router ID	Originating router's link IP address
Transit	IP address of DR's link to network	Originating router's link IP address
Stub	Network address	Network's subnet mask
Virtual	Neighboring router's router ID	MIB-II interface index value

Table 26-1.   *Relationship Between Link Type and Link Information Supplied in the LSA*

Note   *The TOS section is designed to allow OSPF to support Type of Service routing, varying the metric for specific routes based on the type of traffic that needs to be sent across that route. TOS routing using TOS fields in OSPF LSAs is unsupported in Cisco routers, so it is not discussed further.*

Each router generates one of these LSAs for each area of which it is a member. Although the link information fields of these LSAs (Link Type, Link ID, Link Data, and Metric) contain differing data depending on the link type, the basic information is the same. The link information fields describe the connection in enough detail for the SPF algorithm to identify and compute the shortest path to each of the remote router's connected links. The metric computed for each of the links is taken from the bandwidth of the links. By default, the metric is computed as 100,000,000/bandwidth in bps (with the result rounded down to the nearest whole number).

For instance, a T1 (1.544 Mbps) would have a metric of 64, whereas a Fast Ethernet link would have a metric of 1. Remember, a router LSA describes only the directly connected links on the router, so this metric is always the local metric on the originating router, no matter where the LSA is coming from. This concept is examined in more detail later in this section of the chapter.

Tip   *Although the default metric calculation does not take links over 100 Mbps into account, you can change this with the* auto-cost reference-bandwidth [speed in mbps] *router config mode command. This command is discussed further in the "Configuring OSPF" section, later in the chapter.*

The other primary LSA type is the Type 2 LSA, the network entry LSA (or simply *network LSA*). The network LSA is generated only by DRs, and it lists the network for which the router is the DR and all routers attached to that network. The network

CISCO ROUTING

LSA is also propagated throughout the entire area, and it contains the following information:

- **An LSA header (described earlier in this section under DD packet format)** The link-state ID field in the LSA header contains the IP address of the originating router (the DR).
- **A network mask** This is the subnet mask used on the advertised network.
- **A list of all routers to which the DR has an adjacency established** This should be all active routers on the network.

Running the SPF Algorithm

Now, to see how the SPF algorithm works, examine the simple single-area network shown in Figure 26-6.

First, you need to learn about how the various types of LSAs are propagated around the network. To understand the LSA propagation, you need to consider who will establish adjacencies with whom. For instance, on Net 4, Williams and Shaw establish adjacencies only with Peart and Lombardo, not with each other. (Like on a multi-access network, DROthers establishes adjacencies only with the DR and BDR.) Table 26-2 includes adjacency information.

In this network, Type 1 LSAs would be sent from all routers to all adjacencies describing the local links. For instance, at initial boot, Shaw (a DROther) would originate Type 1 LSAs to Ulrich, Peart, and Lombardo describing Shaw's links to Nets 2 and 4. Type 2 LSAs would be sent from all DRs to all adjacent neighbors describing the routers available on multi-access networks. For instance, Peart would originate a Type 2 LSA for Net 4 and send it to Shaw, Williams, Lombardo, and Sandoval.

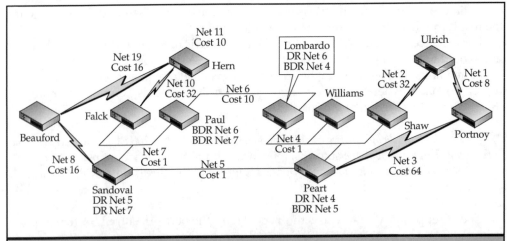

Figure 26-6. *A simple single-area network with multi-access and point-to-point links*

	Portnoy	Ulrich	Shaw	Peart	Williams	Lombardo	Paul	Sandoval	Falck	Beauford	Hern
Portnoy		Net 1		Net 3							
Ulrich	Net 1		Net 2								
Shaw		Net 2		Net 4		Net 4					
Peart	Net 3		Net 4		Net 4	Net 4		Net 5			
Williams				Net 4		Net 4					
Lombardo			Net 4	Net 4	Net 4		Net 6				
Paul						Net 6		Net 7	Net 7		
Sandoval				Net 5			Net 7		Net 7	Net 8	
Falck							Net 7	Net 7		Net 8	Net 10
Beauford								Net 8	Net 8		Net 9
Hern									Net 10	Net 9	

Table 26-2.   Adjacencies for the Network Depicted in Figure 26-6

To show you how LSAs propagate around the network to build the topology table on each router, let's examine the path of Type 1 LSAs from Hern to Portnoy. Hern boots and sends hellos every ten seconds to both Falck and Beauford. Falck and Beauford examine Hern's hello, add Hern to the neighbor table and the list of known neighbors in their hello packets, and forward a hello back to Hern. Hern receives the hellos from Falck and Beauford, adds them to his neighbor table and list of known neighbors, notices his router ID in the packet, and transitions the state of Falck and Beauford to 2-way. Hern then sends another hello packet back to Falck and Beauford. Falck and Beauford recognize their router IDs in the hello, and transition Hern's state to 2-way.

After this, Hern, Falck, and Beauford realize that they are connected over a point-to-point link, and they transition the state to exstart. (Adjacencies are always established over point-to-point links.) A master is chosen for each adjacency (Falck to Hern and Beauford to Hern), the state transitions to exchange, and DD packets are exchanged describing all known LSAs. Falck and Beauford realize that they do not have Hern's router LSA, and they request it (using an LSR packet) from Hern. Then Hern sends both Falck and Beauford his Type 1 LSA describing his links in an LSU packet. This LSA appears as follows (fields deemed insignificant for this example are not shown):

- **Type** 1 (router)
- **Advertising router** Hern (the originating router)
- **Link-state ID** Hern (the originating router)
- **Sequence number** 1 (first revision)
- **Number of links** 3
- Link to Net 9
  - **Link type** Point-to-Point
  - **Link ID** Beauford's serial IP
  - **Link data** Hern's serial IP
  - **Metric** 16
- Link to Net 10
  - **Link type** Point-to-Point
  - **Link ID** Falck's serial IP
  - **Link data** Hern's serial IP
  - **Metric** 32
- Link to Net 11
  - **Link type** Stub
  - **Link ID** Net 11's IP network/subnet address
  - **Link data** Net 11's subnet mask
  - **Metric** 10

While the Type 1 LSA is being sent from Hern, Hern realizes that he doesn't have most of Falck and Beauford's LSAs, and requests the appropriate LSAs using his own LSR. When Falck and Beauford receive Hern's LSUs, they send LSAcks back in response and transition to the full state. Hern receives the LSAcks and clears the link-state retransmission list. Hern receives his previously requested LSUs, sends LSAcks back to Falck and Beauford in response, and transitions Falck and Beauford to the full state.

Because Falck and Beauford have received a new LSA, they send an LSU to their configured adjacencies describing their new LSA. For Falck, an LSU is sent to the AllDRouters multicast address containing the new Router LSA for Hern, which is received by both Sandoval and Paul. For Beauford, an LSU is sent directly to Sandoval containing the new Router LSA for Hern. This LSA appears as follows:

- **Type** 1 (Router)
- **Advertising router** Hern (the originating router)
- **Link-state ID** Hern (the originating router)
- **Sequence number** 1 (first revision)
- **Number of links** 3
- Link to Net 9
  - **Link type** Point-to-point
  - **Link ID** Beauford's serial IP
  - **Link data** Hern's serial IP
  - **Metric** 16
- Link to Net 10
  - **Link type** Point-to-point
  - **Link ID** Falck's serial IP
  - **Link data** Hern's serial IP
  - **Metric** 32
- Link to Net 11
  - **Link type** Stub
  - **Link ID** Net 11's IP network/subnet address
  - **Link data** Net 11's subnet mask
  - **Metric** 10

Looking back, you may notice that the LSA did not change. This is because, unlike distance vector protocols, link-state protocols will not add anything to these LSAs. With link-state protocols, the LSA does not change as it is transmitted through the network. Each router just needs to know which links are connected to all other routers; they can compute the final metric themselves (because they can "see" the entire path back to the originator). When Sandoval receives the LSU from Falck, he immediately sends an LSU to the AllSPFRouters address on both Net 7 and Net 5 (he is the DR for both networks), as well as sending an LSU directly to Beauford. Falck and Beauford see the LSU from Sandoval as an implicit acknowledgment, and they remove the LSU from the link-state retransmission list.

However, because Falck and Beauford just received an LSU from Sandoval, they need to explicitly acknowledge the LSU (even though it didn't contain any new information) so that Sandoval can remove them from the link-state retransmission list. Using this same process, all routers in the network would propagate the new router LSA to all other routers, ensuring that all routers have an identical copy of the link-state database. After convergence of this LSA, a very simplified version of the link-state database would appear on all routers in the network, similar to Table 26-3.

You're probably wondering how to interpret this table. Well, the far-left column lists all routers in the network. At the top is a list of all networks in the area. Where the row for a given router and network meet is the cost to the direct link from that router to that network. If nothing is listed for a given router and a given network, that router does not have a direct link and must compute the shortest path to the destination with the SPF algorithm.

The SPF algorithm basically works like this: First, the router enters itself as root. Then it uses the fields in its LSAs and neighbor table to determine which routers it is connected to (through transit, point-to-point, or virtual links). Then, again using the information in the LSAs, it determines who *those* routers are connected to, and so on,

	Net 1	Net 2	Net 3	Net 4	Net 5	Net 6	Net 7	Net 8	Net 9	Net 10	Net 11
Portnoy	8		64								
Ulrich	8	32									
Shaw		32		1							
Peart			64	1	1						
Williams				1							
Lombardo				1		10					
Paul						10	1				
Sandoval					1		1	16			
Falck							1			32	
Beauford								16	16		
Hern									16	32	10

Table 26-3. *The Link-State Database for the Network Shown in Figure 26-6*

until all routers are accounted for and in the tree. Then the router adds any stub networks to the appropriate routers, and the calculation of the SPF tree is complete. From here, to compute a route to any given network is a piece of cake. SPF just has to compute the sum of the costs of the individual links along all paths to the destination (ensuring that no loops occur), and pick the best path. If two or more paths have an equal metric, SPF adds both to the routing table and load balances equally.

Note   *OSPF does not support unequal-cost load balancing.*

For instance, let's say Peart needs to add the route for Net 11 to his routing table. First, he computes an SPF tree with himself as the root. He performs this task by adding his transit links to himself, then adding the transit links from his neighbors, then the transit links from their neighbors, and so on. In Step 1, shown here, Peart adds his direct links to his neighbors.

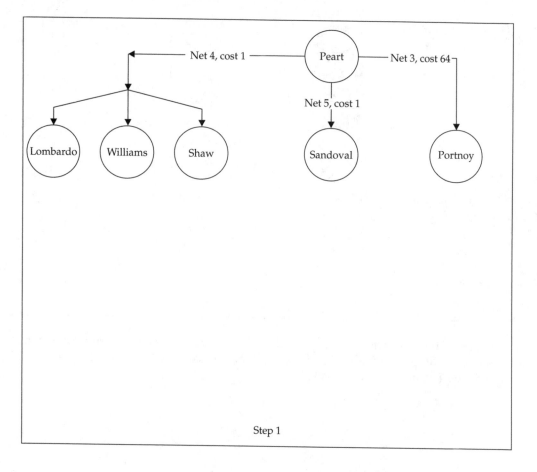

Step 1

In Step 2, he adds their links to their neighbors. Notice that in Step 2, the link from Ulrich to Portnoy is added, even though it looks like something that shouldn't happen until Step 3. The link is added in Step 2 because the link is one of Portnoy's links to his neighbor. Also notice that when this link is added, the direct link from Peart to Portnoy becomes grayed out. This occurs because SPF must choose a single best path to each router when building the tree, and the Peart>Shaw>Ulrich>Portnoy link is cheaper (1 + 32 + 8 = 41) than the Peart>Portnoy (64) link.

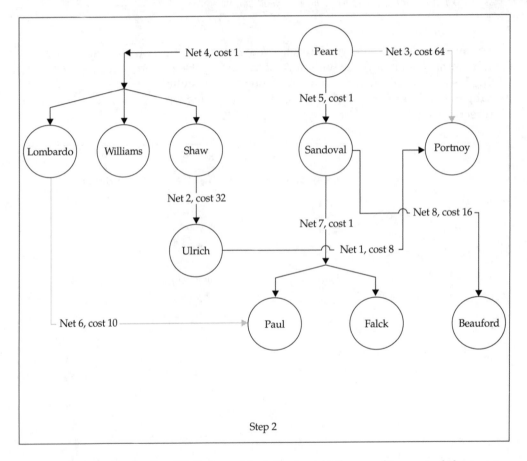

In Step 3, the final sets of links are added, the best paths are chosen, and the tree to all routers is complete. Finally, Peart adds stub networks (only Net 11 to Hern) and picks the best routes to each network.

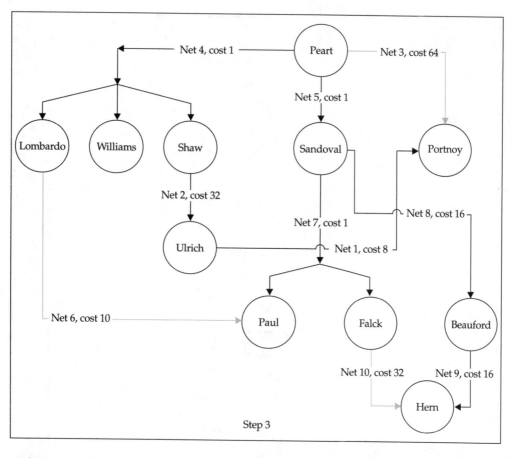

Step 3

So Peart's final path to Net 11 is Peart>Sandoval>Beauford>Hern, and Peart's routing table entry for Net 11 looks something like this:

Destination	Next Hop	Cost
Net 11	Sandoval	33

OSPF Core Logic

Now that you've seen most of the components of OSPF, let's spend some time examining OSPF's logic structure, shown in Figures 26-7 through 26-18.

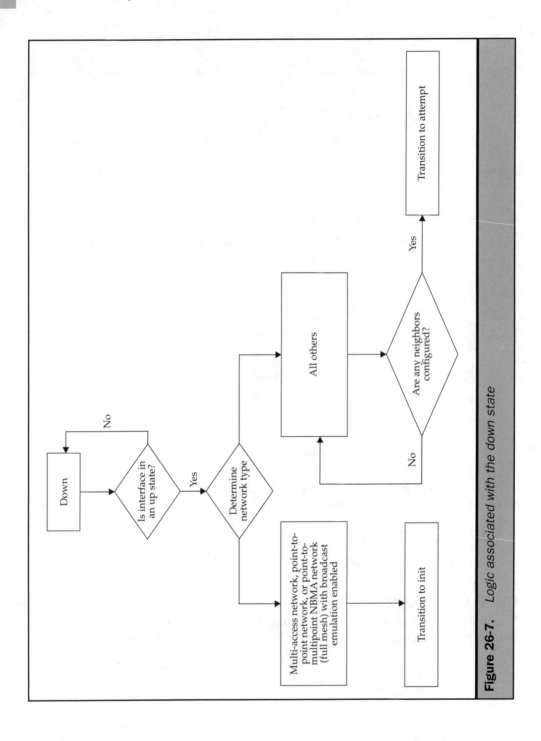

Figure 26-7.   Logic associated with the down state

> **Note**   *The logic used in the figures is slightly simplified to reduce confusion, but it is technically accurate (even if the real process isn't quite this organized).*

In the down state, not much happens. The interface is checked to ensure that it is actually available, the network type is determined, and then the process transitions to init or attempt (depending on the interface type).

> **Note**   *All of these logic steps are computed for each neighbor on each interface that is enabled for OSPF.*

If the link in question is an NBMA link in a star or partial mesh, or an NBMA link in a full mesh but with broadcast emulation disabled (manually configured neighbors), the router transitions to the attempt state, as shown in Figure 26-8.

> **Note**   *In some cases, a star or partial mesh NBMA topology may not make use of the poll interval or attempt state. You will learn more about this subject when you examine the link types in detail, later in this section of the chapter.*

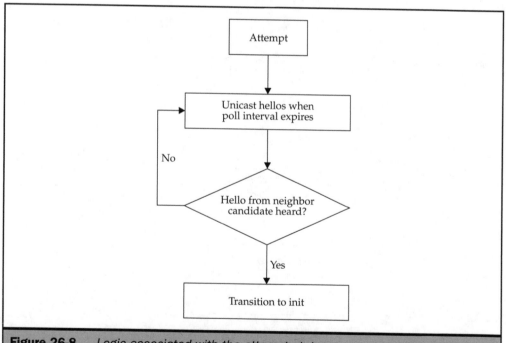

Figure 26-8.   *Logic associated with the attempt state*

Otherwise, the router transitions to the init state, shown in Figure 26-9.

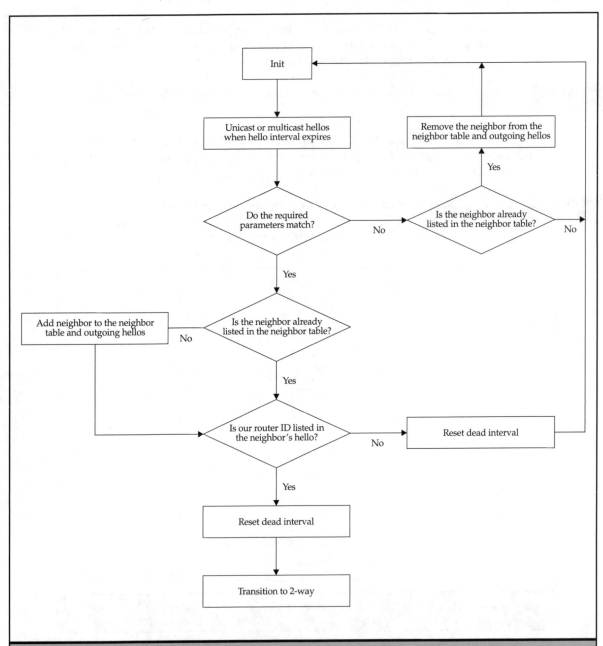

Figure 26-9.   *Logic associated with the init state*

The attempt state is extremely simple. If a hello is heard from a neighbor candidate, the router transitions to init. Otherwise, the router simply continues to send hellos every time the poll interval expires.

The init state is a bit more complex. First, the router sends hellos (as either a unicast or a multicast, depending on the link type) every time the hello interval expires. Then it checks the already-received hellos from the neighbor to determine if the required parameters match (the hello interval, dead interval, subnet mask, area ID, authentication parameters, and options). If so, the router examines the hello to determine if its router ID is listed. If a hello is heard from this neighbor where the required parameters do not match, the neighbor is removed from the neighbor table and outgoing hellos, and the state returns to init.

If the router examines the hello and finds that its router ID is listed, the router resets the dead interval and transitions the state to 2-way. If a hello was heard from this neighbor and the required parameters match, but the router ID is not listed in the hello, the router simply resets the dead interval and returns to init.

In the 2-way state (shown in Figure 26-10), the router first checks to ensure that the dead interval for the neighbor has not expired, that all of the required parameters in the hello match, and that the router's router ID is listed in the hello. If any of these parameters fail the checks, the router transitions this neighbor's state back to init (removing the neighbor from the neighbor table and outgoing hellos in the process). If the hellos pass these tests, the router next determines if a DR/BDR election needs to take place. The DR/BDR election needs to be held if the following are true:

- The network is an NBMA full mesh or broadcast-based multi-access network *and*

- There is no current DR or BDR

Otherwise, the router just uses the current DR and BDR, or (in the case of point-to-point or certain types of point-to-multipoint connections) establishes an adjacency with all routers. Assuming that DR/BDRs have either been elected or are not required, the router then has to determine whether to establish an adjacency with the neighbor. The rules associated with this are as follows:

- If the network does not require DR/BDRs, establish an adjacency (transition state to exstart).

- If your router is the DR or BDR for this network, establish an adjacency (transition state to exstart).

- If the neighbor in question is the DR or BDR for this network, establish an adjacency (transition state to *exstart*).

- In all other cases, return to 2-way.

Assuming that the router does decide to establish an adjacency, it will transition to the exstart state, the logic of which is shown in Figure 26-11.

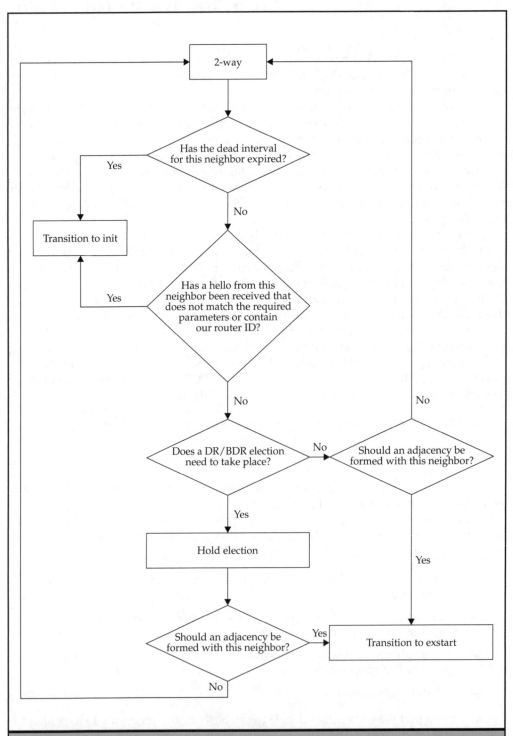

Figure 26-10.   *Logic associated with the 2-way state*

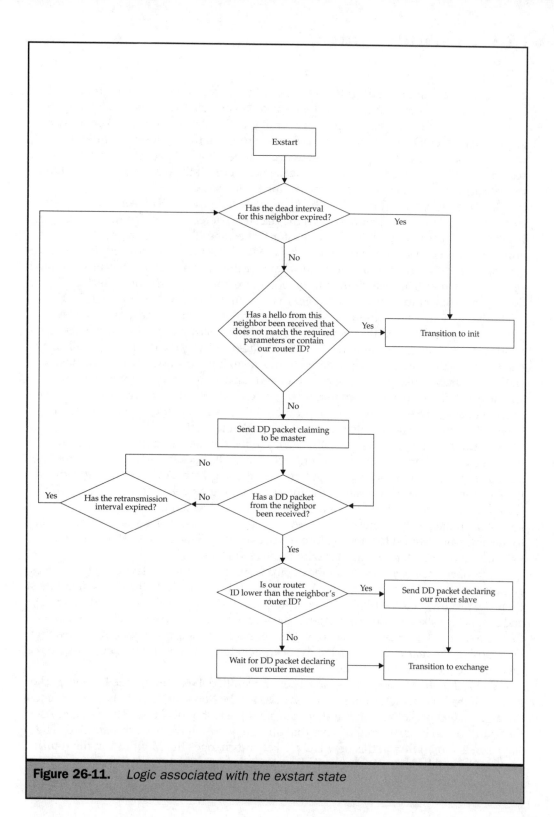

Figure 26-11.   *Logic associated with the exstart state*

In the exstart state, the neighbor's hellos are again examined to ensure that the dead interval has not expired and that the hellos meet all criteria to advance past the 2-way state, and the router sends a DD packet claiming to be master. It claims to be master by sending a blank DD packet with the M/S bit set to 1 and the beginning sequence number equal to its own beginning sequence number. If a DD packet is not sent in response, the router checks to ensure that the neighbor is still available and, if so (and the retransmit timer has expired), sends another DD packet.

Once the DD packet (most likely blank as well, with the M/S bit set to 1) is received from the neighbor, the router examines the packet to determine which router's router ID is lower. The router with the lower router ID becomes the slave, and sends an empty DD packet back to the master with the M/S bit set to 0 (acknowledging its slave status), and the sequence number set to whatever the master claimed was its beginning sequence number. The routers then transition to exchange, and the flowchart transitions from simple to really complex, as shown in Figures 26-12 through 26-17.

In the exchange state, you must first choose whether you are the master or slave. If you are the master, you first populate your database summary list with the headers of all LSAs in your link-state database. The *database summary list* is a listing of all LSA headers that still need to be transmitted to this neighbor for a transition to the next state to occur. Because you just populated the database summary list, you obviously still have entries, so you send an initial DD packet with the first sequence number and the first batch of LSAs. You do not mark these LSAs as transmitted yet because you have not received an acknowledgment of the packet from the slave.

Returning to the top, there are still entries in the database summary list (because you have yet to remove any), so you continue down. DD packets have been transmitted, so you check to see if an acknowledgment has been received. An acknowledgment in the DD transfer process consists of the master receiving a packet from the slave with the same sequence number as it sent to the slave. The slave, upon receiving a DD packet from the master, removes the LSAs in the master's packet, adds any new LSAs to the link-state request list, inserts into the packet any LSAs in its database summary list, and sends it back to the master with these new LSAs.

If the slave has no entries left in the database summary list, the slave simply removes all of the master's LSAs from the packet (adding any new LSAs to the link-state request list in the process), sets the M bit to 0 (indicating that the slave has finished sending LSAs), and sends the packet back to the master. If the master never receives the acknowledgment from the slave, it takes steps to resend the data or restart the adjacency process (depending on hellos). If the master does receive the acknowledgment, it adds any new LSAs to the link-state request list and returns to the start.

When all of the master's LSAs have been transmitted, he examines the last DD packet sent from the slave to determine if the slave has set the M bit to 0. If not, the master sends out a blank DD packet so that the slave can send the rest of his LSAs. The master sets the M bit to 0 to indicate to the slave that all of his LSAs have been transmitted. The slave sends more LSAs in this new DD packet and changes the M bit to 1 in the return packet if he still has more LSAs to send.

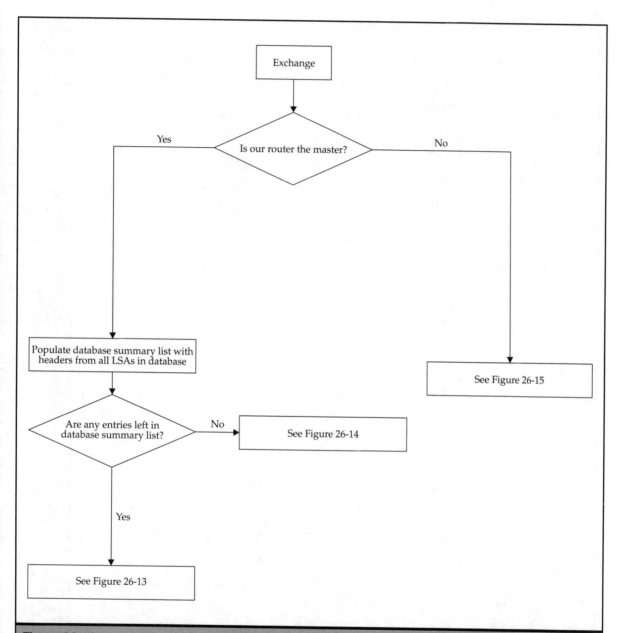

Figure 26-12. *Logic associated with the exchange state*

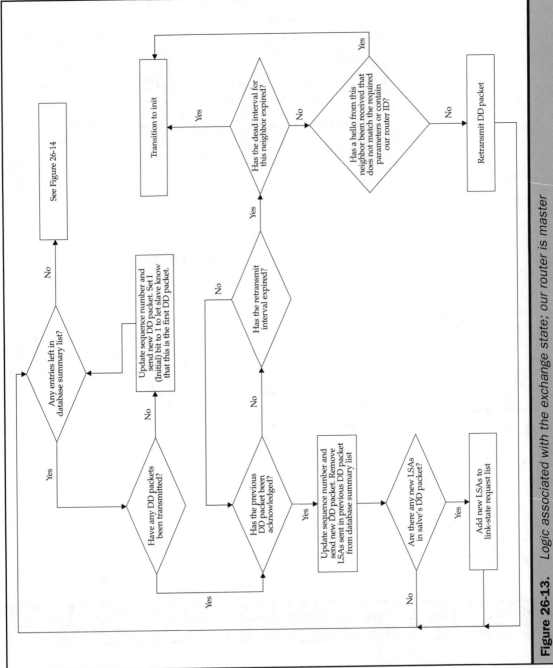

Figure 26-13. *Logic associated with the exchange state; our router is master*

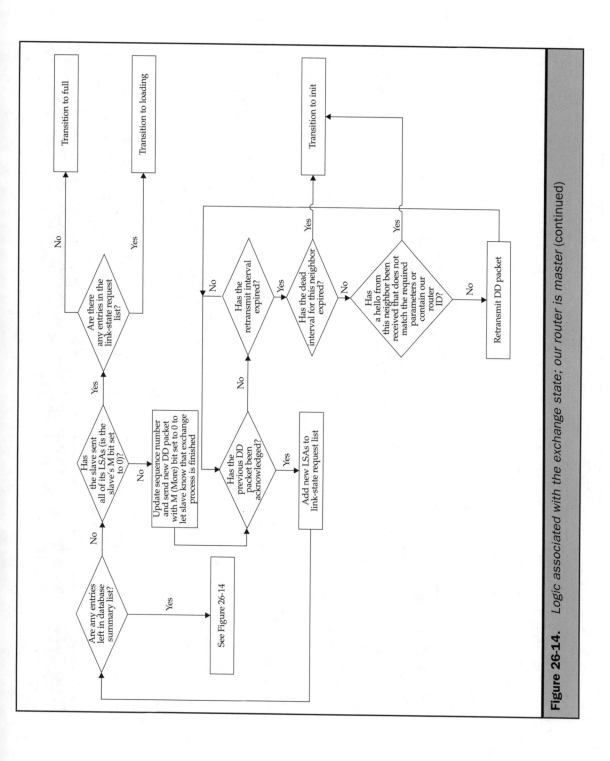

Figure 26-14. Logic associated with the exchange state; our router is master (continued)

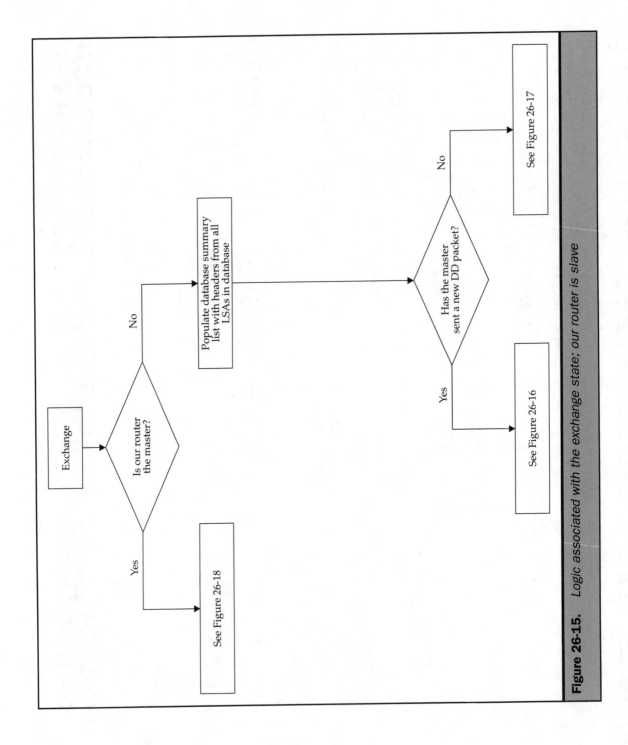

Figure 26-15. *Logic associated with the exchange state; our router is slave*

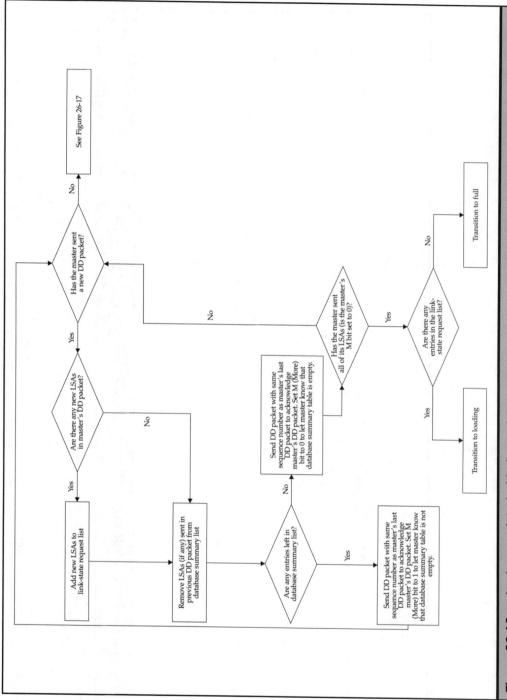

Figure 26-16. Logic associated with the exchange state, our router is slave (continued)

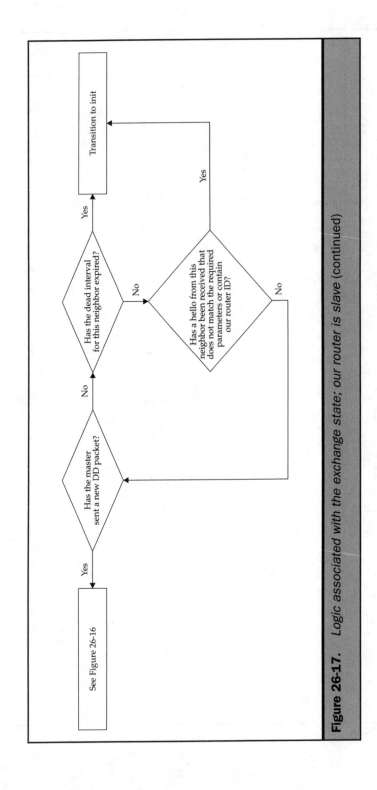

Figure 26-17. *Logic associated with the exchange state; our router is slave (continued)*

The master continues sending DD packets to the slave until the slave has depleted his database summary list. When the slave finishes sending LSAs and sets his M bit to 0, the master checks to see if there are any outstanding entries in the link-state request list. If there are, the master transitions to the loading state. If not, the master transitions to the full state. The slave performs the same process of checking the link-state request list and transition states appropriately.

Note
Even though it is not shown in the previous figures, the master and slave transmit LSRs and LSUs while still in the exchange state. I chose to simplify this area of the logic somewhat to reduce complexity. In actuality, if there are entries in the link-state request list, the router sends an LSR with these LSA requests every time the SPF delay timer expires (every five seconds, by default). This means that it is possible for the router to transition directly to the full state from the exchange state.

After the exchange state, one or both routers may transition to the loading state. The logic for the loading state is shown in Figure 26-18.

Although there are a lot of different paths in the logic diagram in Figure 26-18, the core functions of the loading state can be summarized fairly quickly:

- Ensure that failed neighbors are detected and removed.
- If there are entries in the link-state request list, do the following:
    - Send out LSRs for those specific LSAs.
    - If LSUs including the requested LSAs are not received before the retransmit interval expires, resend the LSRs.
    - After an update to the link-state database due to the new LSAs, send an LSU to all adjacencies with the new LSAs.
    - If an LSU was not sent to the router that you sent your LSR to, send an LSAck to acknowledge his LSU.
    - Respond to neighbors that do not acknowledge your LSUs by unicasting the LSUs to those neighbors directly.
- If there are no entries in the link-state request list, transition to full.

Note
The master and slave routers may transition to different states after this process. For example, when bringing a new router online, it is common for the new router to remain in the loading state for a while after the existing router has already transitioned to the full state.

CISCO ROUTING

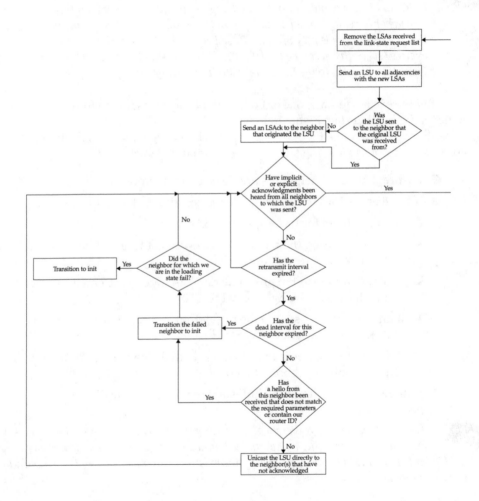

Figure 26-18. *Logic associated with the loading state*

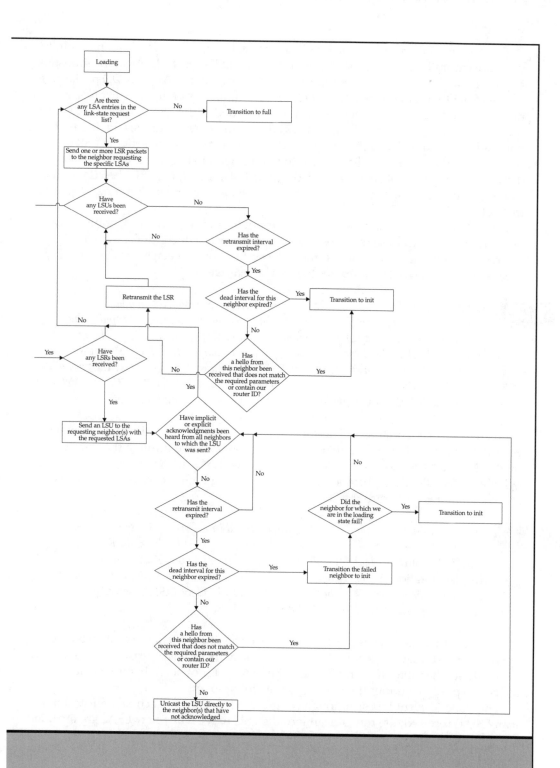

To sum up, if there are no entries in the link-state request list, transition to full. If there are entries in the link-state request list, create LSRs for the LSAs in the link-state request list and send them to the router that advertised the LSAs in its DD packets. If an LSU from that router is not heard before the retransmission interval expires, resend the LSR. If an LSU is heard, remove the LSAs included in the LSU from the link-state request list; send an LSU to all adjacencies advertising the new LSAs; and, if you did not implicitly acknowledge the LSU from the router to which you sent the LSR (by sending an LSU to him), send an LSAck to that router. Finally, if one or more routers do not implicitly or explicitly acknowledge *your* LSU, resend the LSU directly to those routers via unicast. The logic process for the full state is shown in Figures 26-19 through 26-22.

Finally, with the full state, as long as an input event is not received, nothing really needs to be done except send hellos every hello interval (which isn't shown in the figures to reduce the complexity). If the SPF tree has not yet been computed since the last change, the router computes the tree here as well.

Note   *Computing the SPF tree can take significant processor resources in a large area. What if several events were received in succession (such as with a link that rapidly changes state due to problems, a condition known as* route-flap*), each causing a change to the link-state database (and therefore causing an SPF computation)? Then processor use could rise to levels high enough to cause errors or packet loss. To reduce the possibility of this happening, Cisco routers employ a feature called SPF holdtime that (by default) does not allow more than one SPF computation in a ten-second interval.*

If an input event occurs, SPF performs different actions depending on the event type. If an interface fails, OSPF removes all neighbors known through that interface, modifies LSAs as necessary, and floods the changed LSAs in an LSU to all remaining neighbors. (The flooding process also takes into account LSU acknowledgment, as shown in Figures 26-19 through 26-22.)

Note   *A few types of events can occur that are not listed in Figure 26-15, but the most important events are covered. For a full discussion of OSPF, check out* Routing TCP/IP: Volume 1 *by Jeff Doyle (Cisco Press, 1998), and* OSPF: Complete Implementation *by John Moy (Addison-Wesley, 2000)—the man many consider to be the father of OSPF; or check out the primary RFC for OSPF, located at* ftp://ftp.isi.edu/in-notes/rfc2328.txt.

The next two input types are fairly simple. If the router ID of the router changes, the router must reset OSPF entirely (which is one of the primary reasons you should manually set the router ID); and if a neighbor's hello is not heard before the dead interval expires, the router transitions that neighbor to init.

The last two event types, although complicated looking, are actually fairly simple. If an LSU is received, the router examines the LSU to see if any new LSAs are included.

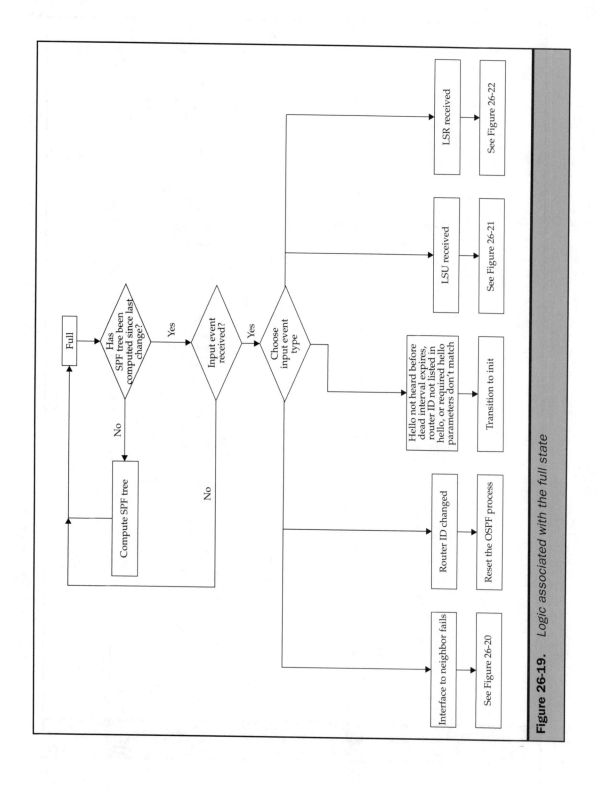

Figure 26-19. Logic associated with the full state

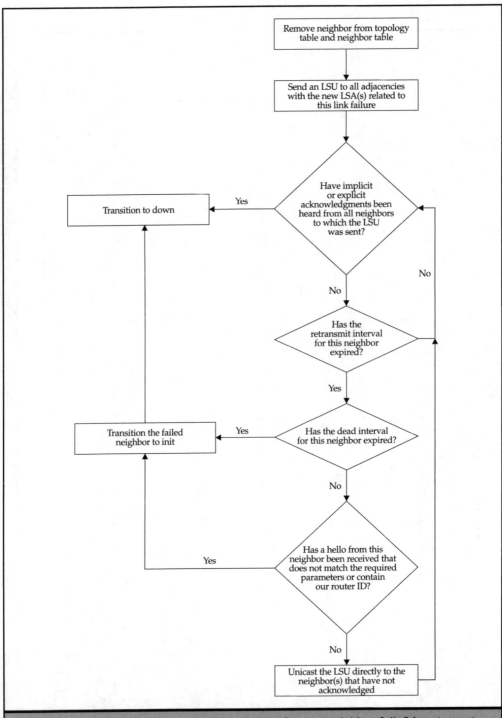

Figure 26-20. *Logic associated with the "Interface to neighbor fails" input event*

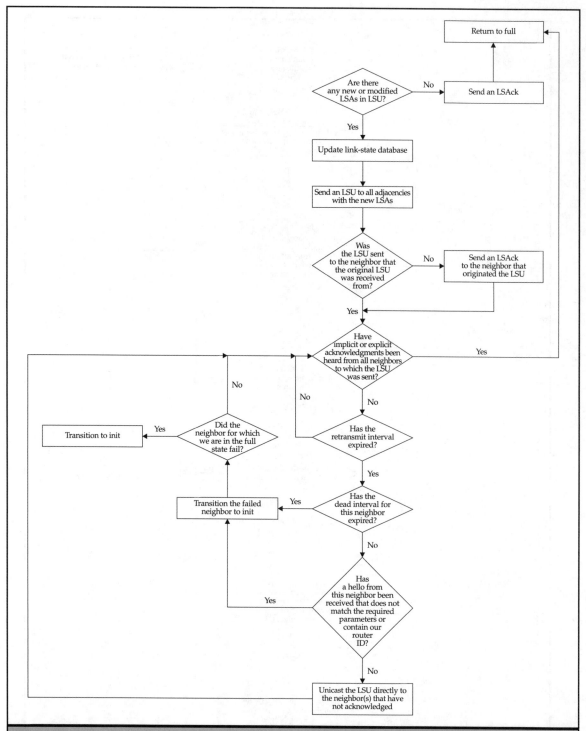

Figure 26-21. *Logic associated with the "LSU Received" input event*

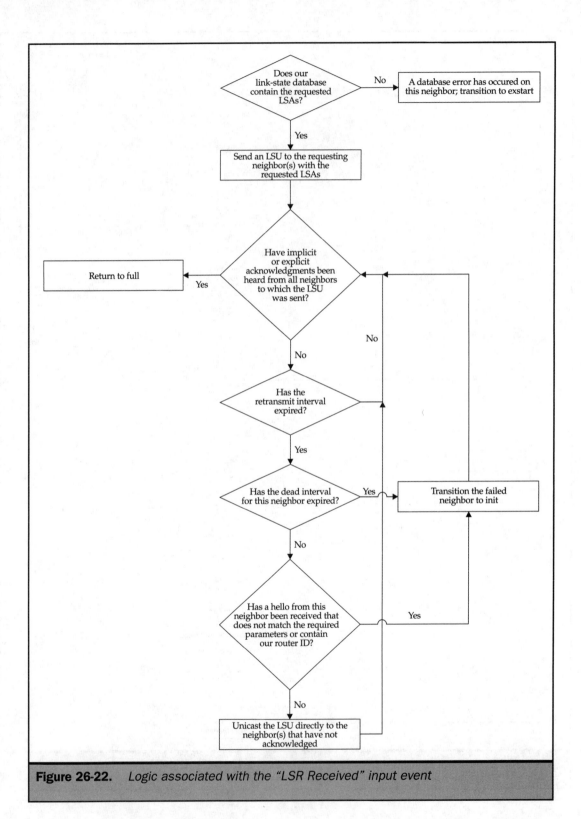

Figure 26-22. *Logic associated with the "LSR Received" input event*

If not, the router simply sends an LSAck back to the LSU originator. If there are new LSAs in the LSU, the router updates all adjacent neighbors; sends an LSAck to the originating router, if necessary; and waits for LSAcks from all routers to which it sent an LSU. Notice that the SPF computation process happens *after* the router updates its neighbors. This is a useful feature because it means that delays caused by the SPF calculation have minimal effect on the transmission of LSUs.

If an LSR is received, the router responds in almost exactly the same manner as it did in the loading state, sending LSUs as necessary and waiting for acknowledgments in response. One new logic step has been added to the diagram in Figure 26-22, however, that wasn't present in Figure 26-18. If the router receives an LSR for an LSA that is not listed in his link-state database, it means an error has occurred, and he transitions back to exstart. This problem occurs in the loading state as well—but to reduce complexity, it was left out of the diagrams.

Now that you understand the core logic behind OSPF, let's examine how OSPF responds to changes.

Dealing with Changes

One of the obstacles OSPF's designers had to contend with is how to deal with changes to LSAs. If a router receives an LSA, and that LSA is already present in the link-state database, how does the router know which LSA to trust? The primary method OSPF uses is sequence numbers. Each time the LSA is modified, its sequence number is incremented by 1. Therefore, the basic idea is that LSAs with higher sequence numbers are more recent. There are several problems with using sequence numbers as the method of determining which LSA is the most recent.

First, the sequence number space cannot be infinite. If it were, in an older OSPF network, the entire sequence number for a single LSA has the potential to completely fill a 1,500-byte packet's payload! What good is that? So, if sequence numbers can't be infinite, you have a choice of creating a circular sequence number space or a flat sequence number space.

With a flat sequence number space, the sequence number eventually reaches the maximum allowable number. In this case, the LSA is invalid forever (well, until the *maxage* expires, but I'll get to that in a moment). So, what if you make the sequence number space circular? With a circular space, once the sequence number reaches the maximum configured number, it "flips" back over to 0. This, however, leads to a fault in logic. I mean, if the number flips at the maximum number, then 45 is greater than 0, but 0 is also greater than 45!

The solution to this dilemma is to use what is known as lollipop sequence number spaces. OSPF version 2 uses a modified lollipop to represent sequence numbers. Although the whole theory of lollipop sequence numbers is a bit beyond the scope of this book, the basic idea is that when a router boots, it begins numbering LSAs it originates with a negative number. Once other routers hear this negative sequence number, they examine their databases to find the last sequence number issued by that router. If they find a "pre-reboot" sequence number, they inform the router of this (positive) number, and the router resumes numbering LSAs where it left off. If a previous LSA from this router cannot be found, the router simply keeps incrementing the negative numbers (eventually turning into a positive number) until it reaches the maximum sequence number. Once the maximum sequence number for an LSA is reached, all routers must flush the LSA from their databases so that a new LSA with the lowest possible positive number can be generated.

This whole process is well and good, but what happens when a router doesn't flush the maximum numbered LSA? If that router uses sequence numbers strictly as its guide, and the originating router issues a new sequence number with a lower value, that router simply throws out the new LSA because, from its perspective, the "new" LSA is actually an old LSA. This is the second problem with using sequence numbers as the only method of measuring the trustworthiness of an LSA.

To combat this problem, OSPF also tracks the age of the LSAs. There are three age values associated with LSAs: *maxage, age,* and *maxagediff*. The *age* is, obviously, how old the LSA is. The age of an LSA begins in the originating router at 0, and increments for each second the LSA spends inside the router. When the LSA is sent to another router, the current age is sent, and a small amount of time (one second, by default) is added for each outgoing interface. This added time for each interface is known as the *InfTransDelay*. The *maxage* is used to define the maximum amount of time an LSA can remain in the database without an update. By default, this is 3,600 seconds (60 minutes).

To prevent valid but stable LSAs from being timed out of the database, an interval, known as *LSRefreshTime,* is used to tell the originating router how long to wait between updates, even if no changes have occurred to the LSA. When the LSRefreshTime expires, the originating router sends out a new LSA with a value of 0 plus the *InfTransDelay* for the age. (The default value for *LSRefreshTime* is 30 minutes.)

However, if the router considered older LSAs more recent, the *LSRefreshTime* wouldn't help keep the *maxage* from being reached. So, routers employ an additional timer, known as *MaxAgeDiff,* to tell them when an LSA with a lower age should be considered more recent. If a new copy of an existing LSA is received, and the difference in the age of the existing LSA and the new LSA is greater than *MaxAgeDiff,* the LSA with the lower age is considered more recent. By default, the *MaxAgeDiff* is set to 15 minutes.

So how do all of these timers help in determining which packet is better? Well, when a router needs to remove an LSA from the database of all routers in the network (like when it reaches the last sequence number), it prematurely ages the LSA to equal *maxage* and sends it out in an update. Every router receiving this LSA removes the LSA from its databases and forwards the *maxage* LSA.

Now that you have examined how OSPF defines newer LSAs, let's look at OSPF network types in detail.

OSPF Network Types

As you have already seen, OSPF defines several different types of networks to optimize OSPF operation over various link types. Although point-to-point networks and broadcast multi-access networks are fairly easy to understand, point-to-multipoint networks running OSPF can be rather confusing. Point-to-multipoint networks on Cisco routers running OSPF come in three basic types:

- Point-to-multipoint (NBMA with broadcast emulation, full mesh)
- Point-to-multipoint (NBMA, full mesh)
- Point-to-multipoint (NBMA, star, or partial mesh)

Understanding the first type is fairly simple. The point-to-multipoint mode with broadcast emulation is one of the RFC-defined modes of operation, and it basically emulates a broadcast multi-access network. Because all routers are in a full mesh, the operation of OSPF can emulate a broadcast multi-access segment (like Ethernet), where all routers can communicate with each other. A DR and BDR will be elected, and all routers should use the same IP network/subnet address for the multipoint network. Figure 26-23 shows an example of this network.

The point-to-multipoint network in a full mesh without broadcast emulation is a Cisco-defined addition to OSPF. In this mode, a DR and BDR are still elected, and a single IP network/subnet number is used for the entire multipoint network, but neighbors must be statically defined because broadcast emulation is not enabled. This type of topology is useful in networks that do not support broadcast emulation, like some SVC-based networks.

Finally, the point-to-multipoint network in a star or partial mesh is the most complicated type of OSPF topology. In this type of network, OSPF operates differently depending on how the network is configured. In a true star, like the one shown in Figure 26-24, OSPF treats each link like a point-to-point link.

The problem with this network type (with most protocols) is that the multipoint link is using a single IP network/subnet address, but all routers cannot communicate with each other. If a "spoke" router learns of a path through another "spoke" router, it will not be able to send packets to that spoke because they cannot communicate, even though they are on the same logical network. For a star network, one solution is to use point-to-point subinterfaces and assign each subinterface a different IP network/subnet address. (You could also use static *frame-relay map* statements, but point-to-point subinterfaces are easiest.) Figure 26-25 shows this configuration.

A second solution is to use the RFC-defined multipoint mode for OSPF. This mode of operation actually creates host routes (/32 routes) for each spoke router, pointing back to the hub router. The hub router then has host routes pointing to each spoke router. Using this setup, each spoke can communicate with other spokes by virtue of the hub router's connections.

For partially meshed topologies, the process gets even more complicated. In this situation, you have the same problem as with a star topology, but only for some routers. For instance, take the network shown in Figure 26-26.

This network has two routers (Routers 2 and 3) that qualify as a full mesh with the hub router, but the other router (Router 4) is in a star-type arrangement with the hub

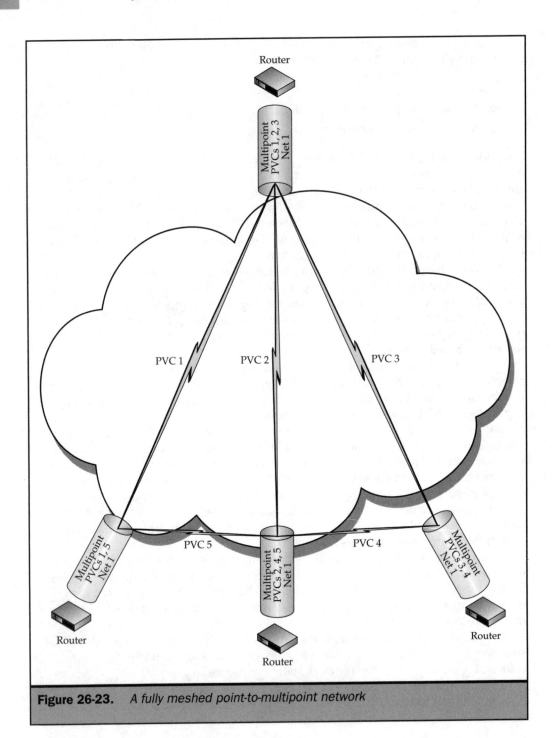

Figure 26-23. *A fully meshed point-to-multipoint network*

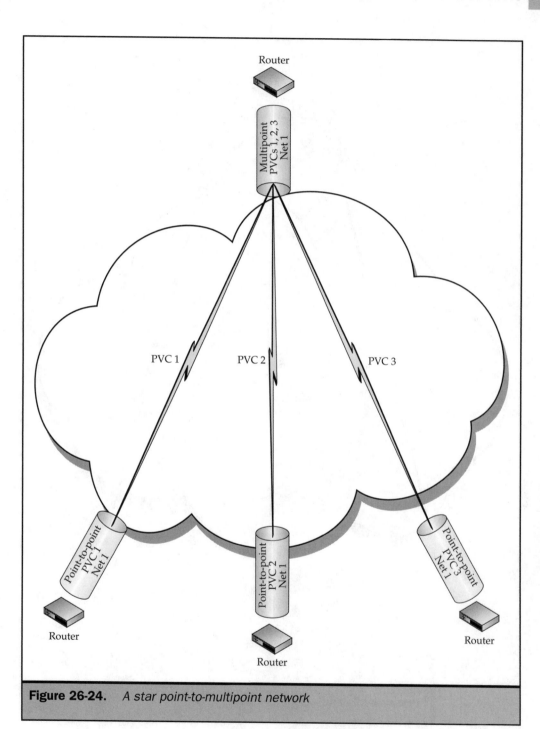

Figure 26-24. *A star point-to-multipoint network*

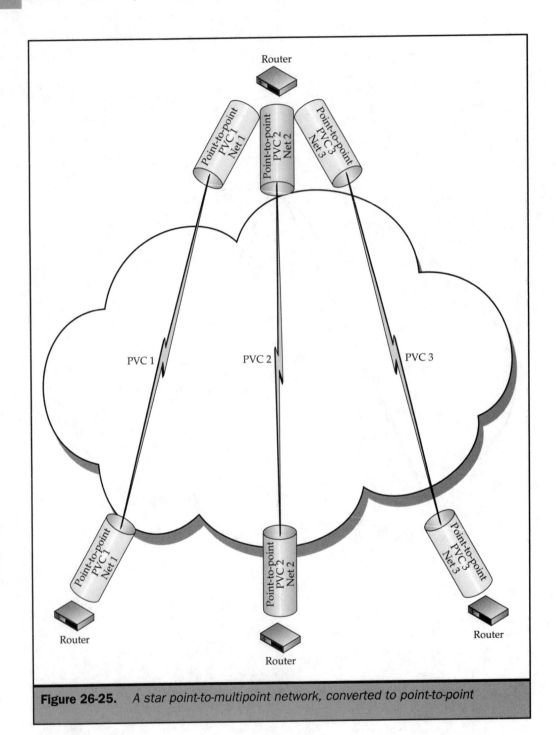

Figure 26-25. *A star point-to-multipoint network, converted to point-to-point*

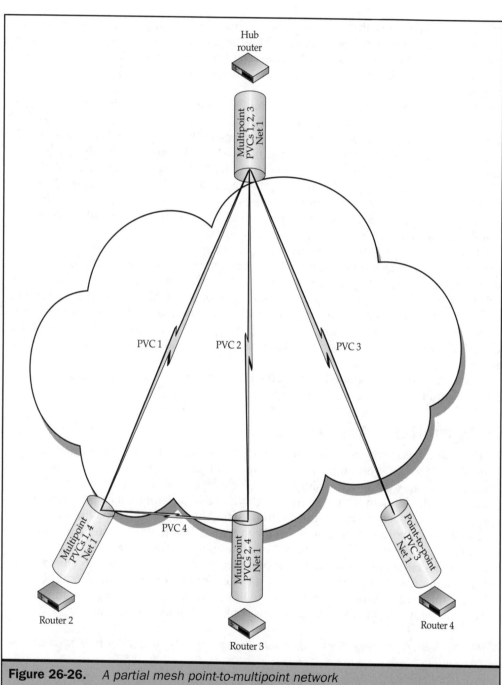

Figure 26-26.   *A partial mesh point-to-multipoint network*

router. The problem with this arrangement is that Router 4 will never be able to send packets through Routers 2 or 3, because it has no direct connection with those routers. Two possible solutions to this problem are to either add another PVC to make the network a full mesh, or split the network into two separate networks, as shown in Figure 26-27.

Another solution is to use the multipoint OSPF enhancement, allowing host routes to be used on the routers to reach the other routers.

One of the points to keep in mind when choosing your topology is the number of adjacencies that will be formed. In a large network, a large number of adjacencies can cause performance problems for all routers. With a full mesh topology, DR/BDR election occurs normally, so the number of adjacencies is kept to a minimum. In a star topology, the hub router simply treats each PVC to each router as a point-to-point network—meaning that the hub router will establish adjacencies with all spoke routers, but the spokes will not establish adjacencies with each other.

However, with a partial mesh, the number of adjacencies can vary widely. If you redefine the point-to-multipoint network as one or more full mesh multipoint networks and one or more point-to-point networks (as shown in Figure 26-27), DR/BDR elections will be held on the multipoint segment(s), and each router on the point-to-point segments establishes an adjacency with the opposite end. If you use the OSPF point-to-multipoint enhancement, each router establishes an adjacency with each other router at the end of each PVC. This situation can lead to an extreme number of adjacencies in a large, complex, partially meshed network.

Now that you've examined most of the functions of OSPF in a single area, let's take a look at multi-area OSPF configurations.

OSPF in Multiple Areas

OSPF areas are an enhancement that allows OSPF to support virtually unlimited numbers of routers. You might wonder why OSPF can't support unlimited routers with a single area. Well, with a single area, OSPF on each router must keep track of every individual link in the entire network. When a link changes, it causes an SPF computation on every router in the environment. Although OSPF consumes very few resources when the network is stable and converged, due to the requirement that every OSPF router retain full reachability and topology information for every link in the area, in a large environment, OSPF changes can cause all routers to consume extreme amounts of CPU and memory resources. For this reason, you can split OSPF into multiple areas, reducing the number of links any one OSPF router is required to keep up with (except ABRs) to only those links within its specific area.

  Remember that, regardless of the total network size, TTL limitations in IP prevent routing across more than 255 hops.

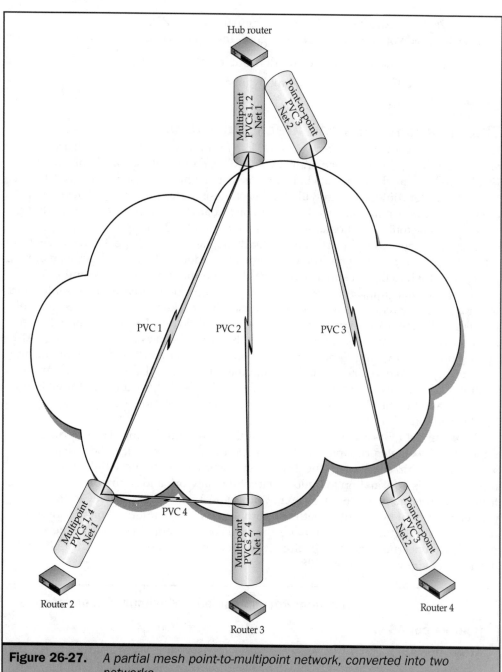

Figure 26-27. *A partial mesh point-to-multipoint network, converted into two networks*

Table 26-4 provides best practices for OSPF network sizes.

The discussion of multi-area OSPF designs is broken down into the following sections:

- The backbone area and area border routers
- Stub, totally stub, and not-so-stubby areas
- Virtual links

The Backbone Area and Area Border Routers

OSPF Area 0 is known as the *backbone area.* In a multi-area OSPF design, all areas should connect to the backbone. In this way, you can think of the backbone area as the OSPF core. In a nutshell, the backbone area is responsible for propagating all inter-area routes to border routers, which will then inject these routes into the nonbackbone areas. This responsibility means that all interarea paths transit through the backbone, making the backbone the most important area in your OSPF design.

The backbone area is the only area in OSPF that advertises Type 3 and 4 LSAs from other areas. Therefore, in a multi-area design, all border routers *must* have a link (either direct or virtual) to the backbone area to be able to route throughout the entire AS.

Area border routers are routers that live at the "edge" of an area, and have interfaces in at least two different areas (one of which must be the backbone). An ABR is responsible for taking all networks advertised in an area, creating Type 3 LSAs (summary LSAs) for those networks, and injecting the Type 3 LSAs into the backbone. In addition, ABRs are responsible for taking Type 3 LSAs learned from other ABRs and injecting the best route to any given network learned through these LSAs into the nonbackbone areas. Finally, ABRs are responsible for injecting Type 4 LSAs (ASBR summary LSAs) into backbone and nonbackbone areas. (ASBRs are discussed in more detail in the "Redistribution with OSPF" section.)

Type 3 and 4 LSAs are interesting in that they work on distance vector principles, rather than link-state principles. With the Type 1 LSAs, for instance, the router is advertising its direct links and the direct costs associated with those links. Other routers use this information to build a map of the network and figure out what the cost from point A to point Z is on their own. With Type 3 and 4 LSAs, however, the ABR just sends the cost of the entire path to internal routers. The internal routers add their cost to the ABR to this path, and that is the total cost to them. As far as the internal routers are concerned, the ABR is directly connected to the network. For instance, take the network shown in Figure 26-28.

	Minimum Recommended	Maximum Recommended
Routers per AS	20	1,000
Areas per AS	1	60
Routers per area	20	350

Table 26-4.   *Best Practices for OSPF Area and AS Sizes*

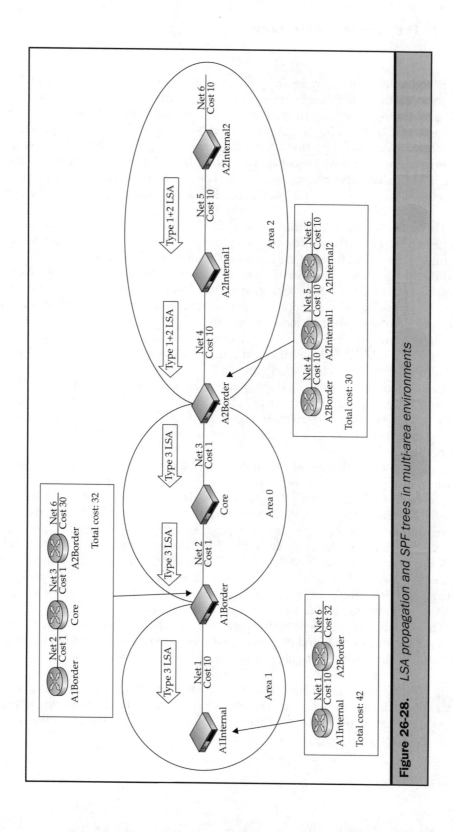

Figure 26-28. *LSA propagation and SPF trees in multi-area environments*

In the network in Figure 26-28, A2Border knows of the entire topology of Area 2 (because A2Border is a member of Area 2), so A2Border picks the best path to Net 6 using the standard SFP algorithm. A2Border then advertises Net 6 in a Type 3 LSA to Area 0 with a cost of 30 (A2Border's cost to Net 6). Core floods the Type 3 LSA to A1Border, which sees the path to Net 6 as going through Core and then A2Border. As far as Core and A1Border are concerned, A2Border is directly connected to Net 6 at a cost of 30. Similarly, A1Internal receives a Type 3 LSA from A2Border with a cost of 32 to Net 6. Again, as far as A2Internal is concerned, A2Border might as well be directly connected to Net 6.

This is the "distance vectorish" side of OSPF. When routing to other areas, internal routers do not know the entire path—only the path to the ABR. The ABR takes care of routing the packet through the backbone to the other ABR, which then takes care of routing the packet to the destination. This distance vector behavior is why all areas must be connected to a single backbone area in order to route inter-area traffic. In any other case, routing loops could occur; but because the network is forced into a hub-and-spoke topology, routing loops are avoided.

The obvious problem with this ABR scenario is that if the ABR for your area fails, you are left in a bit of a mess. Adding multiple ABRs easily solves this problem. For instance, consider the network in Figure 26-29.

In this network, there is no single point of failure to the stub networks (Nets 1 through 4) except, of course, for the router directly connected to the networks or the stub networks themselves. In Figure 26-30 you can see LSA propagation throughout the network. (To reduce confusion, some of the flooded LSAs in Area 1 are not shown.)

Again, note that the cost of each link along the path is not added to the LSA, except when crossing ABRs, for a very simple reason: every router can compute the SPF tree to its own ABRs. Therefore, all the router needs is the cost of the path external to the area to be provided by the ABR; it figures out the best path to the ABR on its own. This configuration allows for a great amount of redundancy. In this scenario, there are well over 20 different ways to reach Net 4 from Net 1. You can lose a single core router, an ABR from each area, and up to three WAN links before you experience any loss of connectivity! And OSPF works out the best path every time, without any possibility of looping. For example, in Figure 26-31, the best path is shown, along with two other possible paths should one or more critical WAN links fail.

The secondary path would be the path chosen if the link from the Core to the right ABR in Area 2 fails, and the tertiary path is the path that would be chosen if, in addition, the link from the right ABR in Area 1 to the Core were to fail.

Note   *To reduce confusion, I have not shown several other paths that the routers would use for equal-cost load balancing. For instance, for the tertiary path, there are four paths to Net 4 with a cost of 108. If the lowest-cost route were 108, OSPF would normally load balance equally along these four paths.*

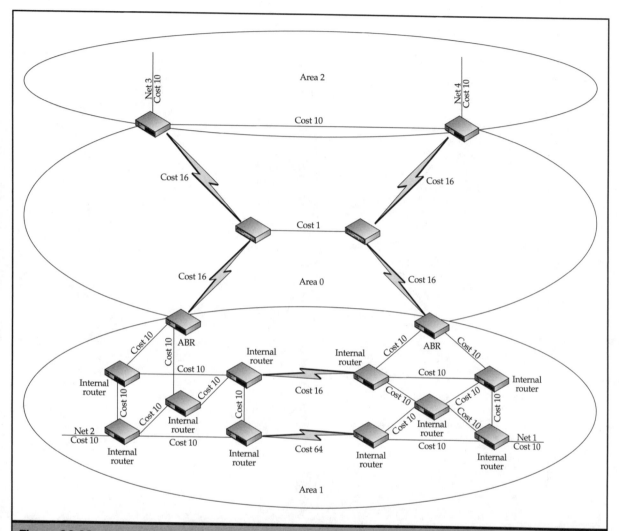

Figure 26-29. *An example of a highly redundant multi-area network*

Now let's take a look at what information is carried in these critical Type 3 LSAs:

- **An LSA header (described earlier in the "Building the Link-State Database" section, under DD packet format)**   The Link ID field in the LSA header contains the IP address of the network being advertised.

- **A network mask field**   This contains (who would've guessed) the network mask of the advertised network.

- **The metric for the advertised network**

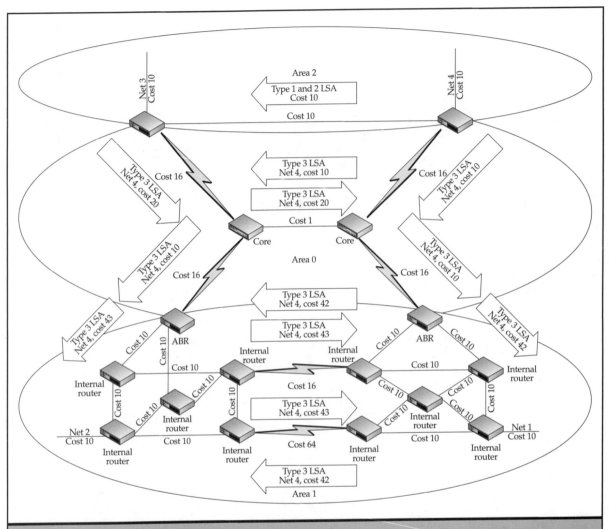

Figure 26-30. LSA propagation in the example network

Well that's pretty easy, huh? One point to realize, however, is that just because it is called a summary LSA doesn't mean that it will contain a summary route. In fact, OSPF does not perform autosummarization, so OSPF normally advertises each individual subnet with a Type 3 LSA to the backbone. (Ouch!) As a result, if you do not perform manual summarization, your topology tables could end up full of Type 3 LSAs. In addition, if you have a link that "flaps" and you are not summarizing, that link will

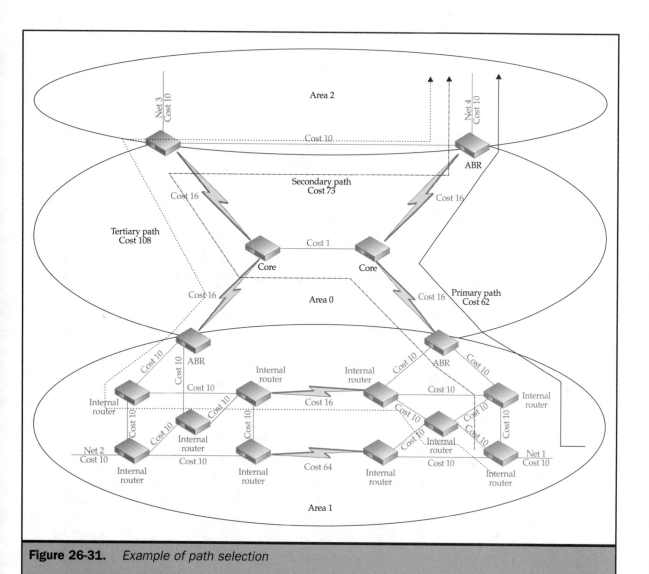

Figure 26-31. *Example of path selection*

cause lots of Type 3 LSAs for the specific subnet associated with that link to be sent by your ABR to the backbone, and then from the backbone to all other areas. (*Big* ouch!)

Therefore, it is suggested (well, almost required, really) that you design your areas hierarchically to support OSPF and perform manual summarization on your ABRs for each area. For instance, Figure 26-32 shows a poorly designed IP structure in an otherwise decent (if small) OSPF network.

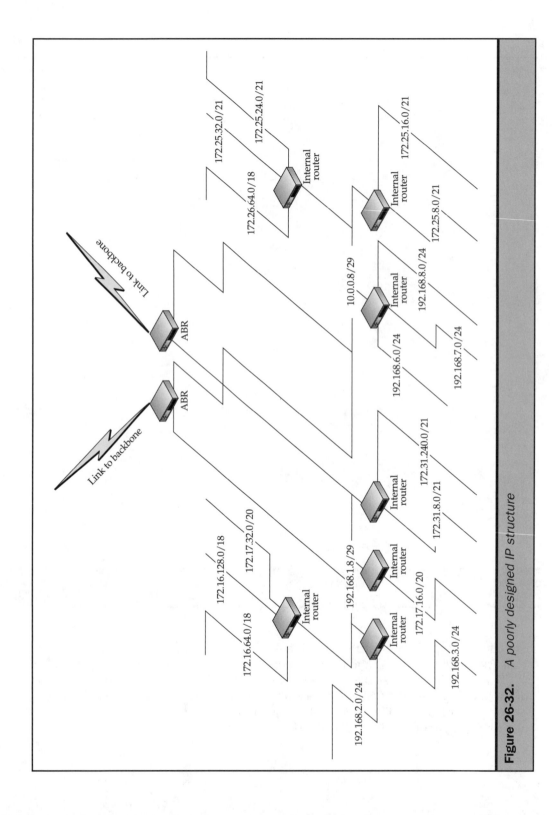

Figure 26-32. *A poorly designed IP structure*

If this section gives you trouble, be sure to reexamine Chapter 7 and the web sites suggested there.

Now, I think anyone can look at the IP structure in Figure 26-32 and recognize the mess. Without summarization, the ABRs would be forced to send a grand total of 18 Type 3 LSAs into the backbone to describe this network. Even worse, summarization cannot be easily performed on this network due to the massive gaps between some of the IP address spaces (assuming that other areas are using the addresses in these gaps).

For instance, you might want to summarize all of the 192.168 networks into one address. In the current design, you have six total 192.168 networks: 192.168.1.0, 192.168.2.0, 192.168.3.0, 192.168.6.0, 192.168.7.0, and 192.168.8.0. If you transfer these addresses to binary to find the matching bits (to determine the correct summarization mask), you get the following:

```
192.168.1.0: 11000000.10101000.00000001.00000000
192.168.2.0: 11000000.10101000.00000010.00000000
192.168.3.0: 11000000.10101000.00000011.00000000
192.168.6.0: 11000000.10101000.00000110.00000000
192.168.7.0: 11000000.10101000.00000111.00000000
192.168.8.0: 11000000.10101000.00001000.00000000
```

The part of these addresses that does not differ is the first 20 bits (11000000.10101000. 0000). Unfortunately, using these first 20 bits with a 20-bit mask not only summarizes the addresses you want, but also quite a few that you don't:

```
192.168.1.0:  11000000.10101000.00000001.00000000
192.168.2.0:  11000000.10101000.00000010.00000000
192.168.3.0:  11000000.10101000.00000011.00000000
192.168.4.0:  11000000.10101000.00000100.00000000
192.168.5.0:  11000000.10101000.00000101.00000000
192.168.6.0:  11000000.10101000.00000110.00000000
192.168.7.0:  11000000.10101000.00000111.00000000
192.168.8.0:  11000000.10101000.00001000.00000000
192.168.9.0:  11000000.10101000.00001001.00000000
192.168.10.0: 11000000.10101000.00001010.00000000
192.168.11.0: 11000000.10101000.00001011.00000000
192.168.12.0: 11000000.10101000.00001100.00000000
192.168.13.0: 11000000.10101000.00001101.00000000
192.168.14.0: 11000000.10101000.00001110.00000000
192.168.15.0: 11000000.10101000.00001111.00000000
```

So, if another area were using, say, 192.168.12.0, your summary would inadvertently include that address, and packets may not be routed correctly. To solve this problem, you would want to implement a hierarchical addressing scheme, like the one shown in Figure 26-33.

With this address structure, you can summarize the entire area into a single Type 3 LSA with minimal waste, all without changing the subnet masks in use (and the corresponding number of hosts supported). To see the address and mask you need to summarize the address into, use the same process as before. Take all of the network addresses in use and determine which portions of the addresses match.

```
172.16.64.0/18:    10101100.00010000.01000000.00000000
172.16.128.0/18:   10101100.00010000.01000000.00000000
172.17.16.0/20:    10101100.00010001.00010000.00000000
172.17.32.0/20:    10101100.00010001.00100000.00000000
172.17.48.0/24:    10101100.00010001.00110000.00000000
172.17.49.0/24:    10101100.00010001.00110001.00000000
172.17.50.0/24:    10101100.00010001.00110010.00000000
172.17.51.0/24:    10101100.00010001.00110011.00000000
172.17.52.0/24:    10101100.00010001.00110100.00000000
172.17.53.8/29:    10101100.00010001.00110101.00001000
172.17.53.16/29:   10101100.00010001.00110101.00010000
172.17.64.0/21:    10101100.00010001.01000000.00000000
172.17.72.0/21:    10101100.00010001.01001000.00000000
172.17.80.0/21:    10101100.00010001.01010000.00000000
172.17.88.0/21:    10101100.00010001.01011000.00000000
172.17.96.0/21:    10101100.00010001.01100000.00000000
172.17.104.0/21:   10101100.00010001.01101000.00000000
172.17.128.0/18:   10101100.00010001.10000000.00000000
```

Tip   *Actually, if you have a contiguous address structure like this one, you can save yourself some trouble by performing this process just for the first and last network address in the structure.*

In this case, you can actually summarize all of these addresses into a single advertisement for the 172.16.0.0/15 network, saving resources on all routers in the enterprise. Now, in truth, the 172.16.0.0/15 address does match a few more addresses than you are actually using. (It matches 172.16.0.0–172.17.255.255, and you are using only 172.16.64.0–172.17.191.255, with the space from 172.17.53.24–172.17.63.255 also vacant.) You have accomplished the number of hosts per subnet goals that you needed to and can view the wasted addresses as "room for growth."

As for actually summarizing these addresses, you should perform this task on both ABRs for your area. They inject the already-summarized routes into the backbone, and the backbone injects the summarized routes into other areas.

Now let's examine the other special OSPF areas.

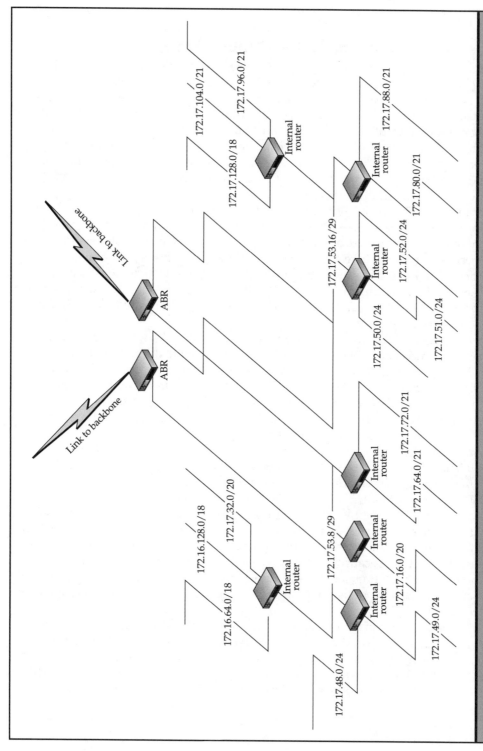

Figure 26-33. *A more efficient IP address structure for the network from Figure 26-32*

Stub, Totally Stub, and Not-So-Stubby Areas

These three area types are designed to allow fewer of a certain types of LSAs to be propagated into areas that don't require them. With a *stub area*, all routers must send packets to the ABR to reach routes from an external AS (which means no ASBRs can exist inside the stub area). Because routers do not require knowledge of routes that have been redistributed into the AS from an ASBR, Type 5 LSAs are blocked from entering the area by the ABR(s) for the area. The ABR simply sends a Type 3 LSA into the area advertising the 0.0.0.0 network (default route), and the internal routers send anything that isn't matched to an area within the AS to the ABR. The ABR then figures out how to reach the ASBR that is needed.

Because Type 5 LSAs are not needed, Type 4 LSAs are also superfluous, and are also not allowed. Type 4 LSAs advertise ASBRs to routers within an area. Because the internal routers in a stub area are just going to send packets to the ABR regardless, they don't need to know how to reach an ASBR. Stub areas are extremely useful in situations when more than one ABR exists to forward packets to the backbone, but no ASBRs exist within the area. The bottom line: if you can't make the area totally stubby (due to the need to load balance across multiple ABRs), make it stub if no ASBRs or virtual links exist in the area.

Totally stub areas take the stub concept one step further. What if you have an area that has only one way of reaching *any* destination external to the area, including areas internal to the AS? This situation is the case when you have only one ABR and no ASBRs. Routers have no choice but to send *all* packets not internal to the area to the ABR. In this case, you can not only block Type 4 and 5 LSAs, but you can also block all Type 3 LSAs (except for a single Type 3 for the default route). Totally stub areas allow you to maximize the efficiency for these areas.

Finally, *not-so-stubby areas* (NSSAs) are used when you have an ASBR internal to an area, but still have only one path to reach other external areas. In addition, the AS to which the ASBR is redistributing must not need to know about other paths external to the OSPF AS. For instance, in Figure 26-34, Area 1 qualifies as an NSSA.

Area 1 cannot be a stubby area because it has an ASBR in the area, but it has only one path to reach other external destinations (RIP), through the ABR. In addition, IGRP doesn't need to know any of the networks associated with OSPF; a default route to Area 1 will work fine (because it has no other links to any other routing domain). So you don't need the full functionality of a transit network, in which all ASBR-generated routes are advertised to out internal routers; you just need to be able to redistribute the IGRP routes into OSPF.

If you make Area 1 an NSSA, routes from IGRP will be distributed to routers in Area 1 as a Type 7 LSA. You can then choose to have the ABR either drop all Type 7 LSAs (not allowing other areas to learn of the IGRP routes), or convert the Type 7 LSAs into Type 5 LSAs for distribution to other areas (because only an NSSA can accept Type 7 LSAs) by telling the ASBR to set or clear the P bit in the LSA. If the P bit is set to 1, the

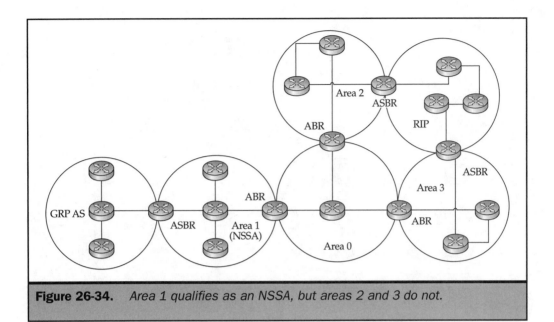

Figure 26-34.   *Area 1 qualifies as an NSSA, but areas 2 and 3 do not.*

ABR automatically translates the Type 7 LSA into a Type 5 LSA for distribution to Area 0. Otherwise, the Type 7 LSA is dropped, and no other areas will learn of the routes through IGRP.

On the other hand, Areas 2 and 3 should really *not* be NSSAs. Both areas have an ASBR for the RIP domain, meaning that RIP has more than one path to get to OSPF networks. Now, RIP is pretty horrible with multiple paths in the first place—but if you told RIP it had two default routes, it would make some pretty poor choices about which one to use at times. (Remember, load balancing is not performed over default routes.) So, you should really redistribute all OSPF routes back into RIP (except those that would cause RIP to loop), meaning that Type 5 LSAs from Area 1 should really be available to Areas 2 and 3.

Virtual Links

Virtual links are the bane of OSPF. Don't ever use them. See, wasn't that easy? Okay, so that was a bit harsh. Sometimes use of a *temporary* virtual link is acceptable; but in general, you should avoid them.

A *virtual link* is a pretend connection to the backbone. Virtual links are used when there is absolutely no way for an area to be directly attached to the backbone, or when the backbone has been split due to a failure. Figures 26-35 and 26-36 show both of these cases.

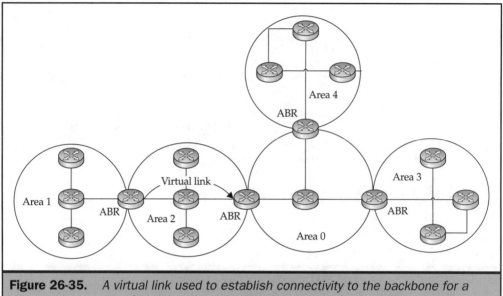

Figure 26-35. *A virtual link used to establish connectivity to the backbone for a remote area*

In Figure 26-35, there's an area (Area 1) that cannot have a direct connection to the backbone. To temporarily resolve this problem, you can add a virtual link between the ABR of Area 1 and the ABR of Area 2. Your virtual link acts like an *unnumbered* (a link without an IP address) point-to-point connection to Area 0, with the cost of the link equaling the cost of the path between the two ABRs. Once you add the virtual link, the two ABRs establish an adjacency with each other and transfer Types 3, 4, and 5 LSAs just as if the ABR of Area 1 were connected to the backbone.

Figure 26-36 shows the other situation that may require a virtual link. In this example, a link within the backbone fails, which would partition the backbone into two Area 0's that are unable to communicate with each other. In this case, once the backbone was partitioned, Area 4 would not be able to communicate with any other areas besides Area 1. To resolve this problem, you can create a virtual link between the two ABRs for Area 1, which patches Area 0 back together again.

One way or another, however, you should attempt to solve your problems without using virtual links. Use them only temporarily because they add complexity to any network. When trying to troubleshoot problems in an OSPF network that uses virtual links, you must troubleshoot each path between ABRs that host virtual links as well, which can become a nightmare. Therefore, designers typically view virtual links in an OSPF network as an indicator of a poor routing design.

If you need to establish a virtual link, keep in mind that you must do so through a transit area. That is, virtual links cannot be built through any type of stub area.

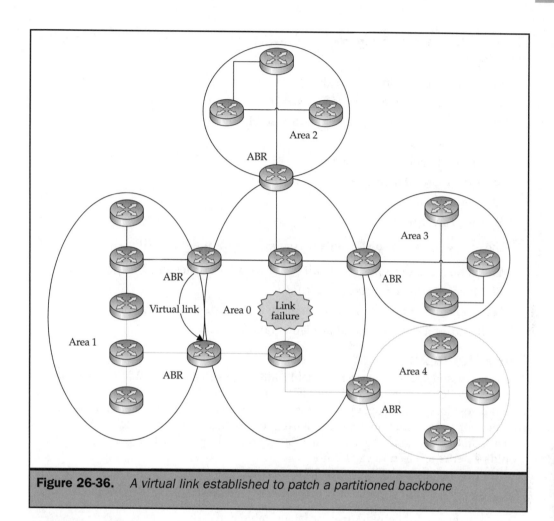

Figure 26-36. *A virtual link established to patch a partitioned backbone*

Comparing OSPF

Now it's time to examine the benefits and drawbacks of OSPF as compared to the other routing protocols already discussed.

OSPF Versus RIP

Comparing OSPF and RIP is almost unfair. The two protocols are designed for completely different environments. OSPF is designed for larger, more complex networks with good design principles, whereas RIP is designed for small networks in which a simple protocol can reduce configuration and design time. Basically, if you are small enough to use RIP, you probably shouldn't even consider OSPF. Rather, you would most likely be better off sticking with RIP, changing to EIGRP, or switching to static routing.

However, I've made the comparison anyway. The major advantages of OSPF over RIP are as follows:

- Much more scalable than RIP
- Supports VLSM (as opposed to RIP 1)
- Lower overall network use for fairly stable networks
- Better path selection
- Graceful avoidance of routing loops
- A more useful metric
- Hierarchical design
- Fast convergence

The following are the disadvantages of OSPF as compared to RIP:

- Hierarchical design doesn't work well with poorly designed IP structures.
- Much more complicated than RIP.
- Requires more processor and memory overhead.
- Requires more design and implementation time.

OSPF Versus IGRP

This comparison is a little more reasonable, although still a bit one-sided. The major advantage IGRP has over OSPF is that IGRP has a more robust metric and can perform unequal-cost load balancing. On the other hand, OSPF scales better and eliminates routing loops. Again, if you are using IGRP, you should probably switch to EIGRP rather than OSPF; and if you are using OSPF and thinking of switching to IGRP, you would be better advised to use EIGRP.

The advantages of OSPF over IGRP are as follows:

- Supports VLSM.
- Lower overall network use for fairly stable networks.
- Graceful avoidance of routing loops.
- Hierarchical design.
- Fast convergence.
- Metric is not as complicated as IGRP's composite metric.
- Vendor independent.

The OSPF disadvantages include these:

- Metric is not as flexible as IGRP's composite metric.
- Cannot perform unequal-cost load balancing.

- Hierarchical design doesn't work well with poorly designed IP structures.

- Much more complicated than IGRP.

- Requires more processor and memory overhead.

- Requires more design and implementation time.

OSPF Versus EIGRP

OSPF and EIGRP are actually fairly similar in a lot of ways. EIGRP, like OSPF, builds a topology table and maps out a path to destinations. Also like OSPF, EIGRP does not form routing loops under normal circumstances. There are, however, a few areas where OSPF stands out as the appropriate protocol (and vice versa). The following is the lowdown on the comparison of EIGRP and OSPF:

The OSPF advantages break down like this:

- Hierarchical design.

- Metric is not as complicated as EIGRP's composite metric.

- Does not succumb to SIA problems.

- Vendor independent.

The OSPF disadvantages are as follows:

- Metric is not as flexible as EIGRP's composite metric.

- Cannot perform unequal-cost load balancing.

- Hierarchical design doesn't work well with poorly designed IP structures.

- Requires more processor and memory overhead.

- Requires more design and implementation time.

Redistribution with OSPF

Unlike most other routing protocols, OSPF was designed from the ground up with redistribution taken into account. Routers that redistribute with an OSPF AS are known as *autonomous system border routers (ASBRs)*; and ASBRs generate a special type of LSA, a Type 5 LSA, to let the OSPF AS know about external routes.

The Type 5 LSA generated by an ASBR includes the following information:

- An LSA header, described earlier in the "Building the Link-State Database" section, under DD packet format. The Link ID field in the LSA header contains the IP address of the network being advertised.

- A network mask field, containing the network mask of the advertised network.

- An E bit, which specifies the type of external metric used for this route. If the E bit is set to 1 by the ASBR, the route type will be E2. Otherwise, the route type will be E1. The differences between E1 and E2 are as follows:

  - **External Type 1 (E1)**   E1 entries have their metric calculated as a sum of the redistributed routes cost and the cost of the links to the sending router. E1 entries are typically used when more than one ASBR is advertising a given external destination.

  - **External Type 2 (E2)**   E2 entries have their metric calculated simply as the cost of the redistributed route. (Cost of internal links to the advertising ASBR are *not* taken into account.) E2 entries are subsequently "cheaper," and are usually preferred by routers over E1 entries.

- The advertised metric for the network.

- A forwarding address, which is the address the router should use to send the packet to the ASBR. Usually, this is set to 0.0.0.0, which signifies that the originating ASBR should be used.

Note that, like the ABR and Type 3 LSAs, routes advertised by the ASBR with Type 5 LSAs take on distance vector properties. Routers in an OSPF AS will not know the entire path to the external network, just the path through the OSPF AS to reach the ASBR, which leads to the other LSA type used for redistribution, the Type 4 LSA.

The Type 4 LSA is very similar to the Type 3 LSA. It has the same basic format, and it is also generated by the ABRs for an area (*not* the ASBR). Type 4 LSAs tell routers how to reach ASBRs. Because the Type 5 LSA typically tells the routers which ASBR to send a packet destined for an external address to, but not how to get *to* the ASBR, the Type 4 LSA is used to advertise the path to the ASBRs located in other areas. Figure 26-37 illustrates this process.

In the example in Figure 26-37, the ASBR is redistributing the route learned through RIP to Net 4 as a Type 5 LSA into Area 3. The Type 5 LSA's link-state ID is equal to the network being advertised, the advertising router is the router ID of the originating ASBR (0.0.0.1, in this case), and the forwarding address is null (0.0.0.0), which means that routers should just send packets destined for Net 4 directly to the advertising ASBR. The ABR at Area 3 floods the Type 5 *unchanged* into Area 0; however, it also floods a Type 4 with the link-state ID equal to the ASBR's router ID and the advertising router equal to its router ID so that routers in Area 0 know where to send a packet destined for the ASBR's router ID. The ABR at the edge of Area 1 performs the same process, flooding the unchanged Type 5 into Area 1 but issuing a new Type 4 for the path to the advertising ASBR.

One final note about redistribution with OSPF: With an NSSA, Type 5 LSAs are not generated by the ASBR. Rather, the ASBR generates a Type 7 LSA for each redistributed network. The ABR for the NSSA then (depending on the setting of the P bit) either floods the information in the Type 7 LSA out to the backbone as a Type 5 LSA or drops the Type 7 LSA (meaning that only routers internal to the NSSA would be aware of the paths redistributed by the ASBR).

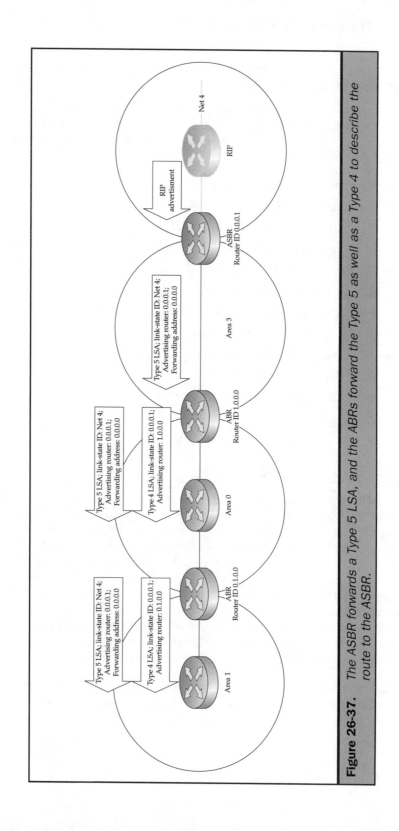

Figure 26-37. The ASBR forwards a Type 5 LSA, and the ABRs forward the Type 5 as well as a Type 4 to describe the route to the ASBR.

Configuring OSPF

OSPF, like most other routing protocols, requires very little work to get a basic network up and running. However, to get the most out of OSPF, you need to perform a little more than basic configuration. This section breaks down OSPF configuration tasks into the following sections:

- Enabling basic OSPF
- Configuring OSPF interfaces
  - Configuration tasks for all interfaces
  - Configuration tasks for NBMA interfaces
- Configuring areas
  - Configuring authentication
  - Configuring stub areas and NSSAs
  - Configuring ABRs and ASBRs
- Advanced OSPF configuration
  - Configuring virtual links
  - Modifying the reference bandwidth
  - Modifying the SPF timers
  - Changing administrative distances

Enabling Basic OSPF

Basic OSPF configuration consists of three primary tasks: starting the OSPF process with the *router ospf* command; enabling OSPF links with the *network* command; and setting the router ID with the *router-id* command or by setting a loopback interface IP address. Table 26-5 shows the specifics for each of these commands.

The *router ospf* command is fairly simple: just remember that the process ID has relevance only on the router on which it is configured. The *network* command, however, has a bit of a twist in OSPF. First, OSPF, unlike most other routing protocols, does not automatically choose all interfaces associated with the class-based network address when entering the *network* command. With OSPF, each subnet can be defined individually or all networks can be defined at once. Even better, you can use a combination of these two tactics. The *network* command basically works like this:

- The wildcard mask specifies the subnet or subnets that should be matched by the *network* statement. The wildcard mask works similarly to an *access-list* statement: any portions represented by a binary 0 must match exactly, and any portions that are represented by a binary 1 do not need to match.

■ *Network* statements are processed in the order entered. The first statement that matches a particular interface address is the one used.

■ The area portion of the *network* statement specifies the area in which interfaces matching the statement will be placed.

Command	Description	Mode
router ospf [process-id]	Enables OSPF on the router. Remember, OSPF process IDs are meaningful only to the router on which this command is run. Unlike EIGRP AS numbers, process IDs do not have to be the same on two routers for them to establish an adjacency.	Global config
network [IP address] [wildcard mask] area [area number]	Enables OSPF on specific networks. Unlike most other routing protocols, the *network* command in OSPF has many possibilities. (See the text that follows for a complete overview.)	Router config
router-id	Forces the router ID used by the OSPF process. Overrides all other router IDs (including loopback interfaces).	Router config
interface loopback [number]	Creates a software loopback interface (a virtual interface). A loopback interface's IP address will be chosen for the router ID in lieu of the *router-id* command. If multiple loopbacks are configured, the loopback with the highest IP address prevails.	Global config
ip address [ip address] [subnet mask]	Configures the IP address of the loopback interface.	Interface config

Table 26-5. *Basic OSPF Configuration Commands*

To expand on this, take a look at this configuration example:

```
Router(config)# interface Ethernet 0/0
Router(config-if)# ip address 192.168.1.1 255.255.255.0
Router(config-if)# interface Ethernet 0/1
Router(config-if)# ip address 192.168.2.1 255.255.255.0
Router(config-if)# interface Ethernet 1/0
Router(config-if)# ip address 192.168.3.1 255.255.255.0
Router(config-if)# interface Ethernet 1/1
Router(config-if)# ip address 192.168.4.1 255.255.255.0
Router(config-if)# interface Serial 0/0
Router(config-if)# ip address 172.16.1.1 255.255.255.0
Router(config-if)# interface Serial 0/1
Router(config-if)# ip address 172.16.2.1 255.255.255.0
Router(config-if)# interface Serial 1/0
Router(config-if)# ip address 10.0.0.1 255.255.0.0
Router(config-if)# router ospf 1
Router(config-router)# network 192.168.0.0 0.0.3.255 area 1
Router(config-router)# network 172.16.1.0 0.0.0.255 area 2
Router(config-router)# network 0.0.0.0 255.255.255.255 area 0
Router(config-router)# network 192.168.4.0 0.0.0.255 area 3
```

Note *You can enter the area number in either standard decimal or dotted decimal (like an IP address) notation. They both have the same meaning. For instance Area 3 is the same as Area 0.0.0.3, and Area 0.0.40.5 is the same as Area 10245.*

In this example, *Ethernet 0/0* (192.168.1.1), *Ethernet 0/1* (192.168.2.1), and *Ethernet 1/0* (192.168.3.1) will all be members of Area 1; *Serial 0/0* will be a member of Area 2; and all other interfaces will be members of Area 0. What about the *network 192.168.4.0 0.0.0.255 area 3* command? Wouldn't that command place 192.168.4.0 in Area 4? The short answer is no. Because the *network 0.0.0.0 255.255.255.255 area 0* command was entered before the *network 192.168.4.0 0.0.0.255 area 3* command, all interfaces that didn't match one of the first two commands would be placed in Area 0. Because everything matches the *network 0.0.0.0 255.255.255.255 area 0* statement, the router will never get to examine the last *network* statement; it is simply ignored. The moral of the story: be very careful with OSPF *network* statements, or you could end up with unintended results.

 Tip *If you are a whiz at binary, the actual operation performed on the wildcard mask/address pairs is a logical OR. The IP address entered in the network statement is ORed with the mask, and the IP address of each interface is ORed with the same mask. If the results are identical, the interface matches. If not, it doesn't.*

To set the router ID, you have several choices. First, you could leave the router ID alone, and the router will choose the highest IP address configured on any of the router's interfaces. This is problematic because, as you'll remember from the logic map for the full state (Figures 26-19 through 26-22), if the router ID changes, the entire OSPF process restarts. If the OSPF process restarts, packets to that router do not get routed, all of that router's LSAs must be removed from *all* routers in the area (and if it is an ASBR, all routers in the AS), and all of the neighbor adjacencies must be removed and reestablished!

Changing an IP address on an OSPF router that is using a live interface's IP address as the router ID can cause real havoc! And even though it is not as likely, the same results occur if the interface the router is using for its router ID fails (although just unplugging the cable doesn't cause this; the interface must physically fail). So, you have the ability to use a second method: setting a loopback interface's IP address as the router ID. Although it isn't likely that someone would change a loopback interface's IP, I still prefer the third method: statically configuring the router ID with the *router-id* command. The reason is simple: although it isn't likely that someone will inadvertently change the loopback's IP, it is still possible, and I don't like leaving things to Murphy and his laws.

Note   *Remember from the "OSPF in a Single Area" section, a router chooses its router ID as follows: if one is statically configured with the* router-id *command, it will be used; if not, the loopback interface with the highest IP address will be used; if there are no loopback interfaces, the highest interface IP address is used.*

Configuring OSPF Interfaces

Although basic interface configuration was covered in previous chapters and won't be discussed further here, OSPF does add a few layers of complexity to interface configuration that need to be addressed. First, let's take a look at general interface configuration tasks that apply to all interface types.

Configuration Tasks for All Interfaces

The configuration tasks that apply to all interface types include the following:

- Modifying the bandwidth for an interface
- Modifying the cost for an interface
- Setting the retransmit timer for an interface
- Modifying the transmission delay for an interface
- Modifying the router's priority on a network
- Modifying the hello and dead intervals for an interface

Table 26-6 describes the commands used in this section.

CISCO ROUTING

Command	Description	Mode
bandwidth [speed in Kbps]	Sets the bandwidth of an interface. Used to define the cost applied to that interface.	Interface config
ip ospf cost [cost]	Sets an arbitrary cost for an interface, overriding the default settings that are configured based on interface bandwidth.	Interface config
ip ospf retransmit-interval [seconds]	Sets the retransmit interval for the interface.	Interface config
ip ospf transmit-delay [seconds]	Sets the amount of time added to the age of all LSAs leaving an interface.	Interface config
ip ospf priority [priority]	Sets the priority of the router (the likelihood of the router being chosen as DR in an election) for the network(s) connected to the interface.	Interface config
ip ospf hello-interval [seconds]	Sets the hello interval for an interface.	Interface config
ip ospf dead-interval [seconds]	Sets the dead interval for an interface. (By default, the dead interval is four times the hello.)	Interface config

Table 26-6. *General OSPF Interface Configuration Commands*

Tip *All of these commands are applied in interface configuration mode for the interface in question, and for that reason, they apply only to the specific interface.*

The *bandwidth* command in OSPF is used primarily to configure the cost for your interfaces. Remember, by default, OSPF uses 100,000,000 (10^8) divided by the bandwidth in bps to determine the cost for an interface. Modifying the bandwidth, therefore, is the primary method of modifying the cost of an OSPF interface. Table 26-7 shows the corresponding OSPF costs for various interface speeds.

Note *For a complete discussion of the* bandwidth *command, please refer to Chapter 24.*

Link Speed	Cost
56 Kbps	1,785
64 Kbps (DS0)	1,562
128 Kbps	781
256 Kbps	390
512 Kbps	195
768 Kbps	130
T1 (1.544 Mbps)	64
E1 (2.048 Mbps)	48
4-Mbps Token Ring	25
10-Mbps Ethernet	10
16-Mbps Token Ring	6
T3 (44.736 Mbps)	2
100+ Mbps	1

Table 26-7.   *Default OSPF Cost Based on* Bandwidth

CISCO ROUTING

The *ip ospf cost* command takes a different tactic to set the cost of an OSPF interface. The *ip ospf cost* command sets the cost of an interface to a specific value, regardless of the bandwidth configured for the interface. The *ip ospf cost* command is extremely useful when you want to modify the cost OSPF applies to an interface, but you do not want to modify the metrics for other routing protocols (such as IGRP or EIGRP) running on the router that also calculate their cost based on the bandwidth of the interface.

The *ip ospf retransmit-interval* command sets the retransmission interval for reliably sent OSPF packets (including DD and Update packets). The default retransmission interval is 5 seconds, which should be sufficient in most cases. You may modify this value to any number between 1 second and 65,535 seconds. Like most timers, modifying the retransmission interval is a trade-off. If you have a really slow WAN link, you may want to increase the retransmission interval to prevent packets from being needlessly retransmitted, but this will also increase the latency involved when a packet actually requires retransmission. One way or another, setting extremely high values for the retransmission interval is not suggested.

The *ip ospf transmit-delay* command defines how much time, in seconds, is added to each LSA's *age* parameter when exiting an interface. The default setting is the minimum of 1 second (which typically covers transmission delays on any unsaturated link, with the possible exception of microwave links). You may increase this value as high as 65,535 seconds.

Note *Be very careful when increasing the transmission delay to any value above 10 seconds. Remember, the* maxage *for LSAs (by default) is only 3,600 seconds, so any value above 3,600 seconds will, obviously, cause the LSA to be dropped as soon as it is received.*

The *ip ospf priority* command sets the router's priority for the interface in question. Note that this command is useful only for network types that elect DR/BDRs. The router with the highest priority *normally* becomes the DR. (Refer back to the "OSPF in a Single Area" section for more details about DR/BDR elections.) The default priority for all interfaces is 1, and the maximum is 255.

Note *To make a router ineligible to become DR or BDR, set its priority to 0.*

Finally, the *ip ospf hello-interval* and *ip ospf dead-interval* commands configure the hello and dead intervals, respectively, for the given interface. The default setting for the hello interval is either 10 seconds or 30 seconds, depending on the network type. Similarly, the default dead interval is either 40 seconds or 120 seconds. If you modify these intervals, it is suggested that the dead interval be four times the hello interval. Also, you must ensure that all routers on a given network have the same hello and dead intervals defined (or they will not become adjacent).

Configuration Tasks for NBMA Interfaces

For NBMA interfaces, you may need to perform a few additional configuration tasks:

- Configuring the OSPF network type
- Configuring broadcast emulation
- Configuring neighbors

Table 26-8 describes the commands used to perform these tasks.

First, let's examine the *ip ospf network* command. This command defines the method OSPF will use to establish adjacencies on a network. If you choose the broadcast mode, OSPF treats the network like a broadcast-based multi-access network (Ethernet), and elects DR/BDRs to establish adjacencies. For this network type to work properly, the NBMA network must be in a full mesh and have broadcast emulation enabled.

If you choose the nonbroadcast network type, OSPF expects neighbors to be manually configured, but it still expects a full mesh and elects DR/BDRs as necessary. Use this network type when broadcast emulation is not enabled but a full mesh NBMA network exists.

Command	Description	Mode
ip ospf network [broadcast | non-broadcast | point-to-multipoint (non-broadcast)]	Defines the network type so that OSPF can choose the appropriate neighbor establishment technique.	Interface config
frame-relay map ip [ip address] [dlci] [broadcast]	Maps IP addresses to DLCIs and enables broadcast emulation over Frame Relay links.	Interface config
neighbor [ip address] [priority (priority)] [poll-interval (seconds)] [cost (cost)]	Defines neighbors. Used when broadcast emulation is not available or possible	Router config

Table 26-8.   *OSPF NBMA Configuration Commands*

Tip

You can also configure the nonbroadcast network type on broadcast-based multi-access networks like Ethernet to reduce the number of multicasts traversing the network. When using the nonbroadcast network type, hellos will be unicast rather than multicast.

If you choose the point-to-multipoint network type, OSPF treats the entire multipoint network as a collection of point-to-point networks, and creates host routes as necessary to enable communication between all routers on the network. A DR/BDR will not be elected. By default, OSPF treats these networks as broadcast capable, and does not require manual neighbor configuration. However, if your network is not capable of emulating a broadcast network, you can use the optional *nonbroadcast* keyword to force manual neighbor establishment.

Note

Point-to-point network types are not configured in this manner. OSPF automatically detects point-to-point networks.

On Frame Relay links, you may also need to configure broadcast emulation to get the benefit of automatic neighbor establishment. The *frame-relay map* command is used for this task, and because it was covered in detail in Chapter 22, it's not discussed further here.

Note

The dxi map *command, for configuring broadcast emulation on ATM PVCs, is also covered in Chapter 22.*

CISCO ROUTING

The *neighbor* command is used in OSPF to manually define neighbors. With OSPF, you can also set the priority for the neighbor, polling interval, and cost associated with the specific neighbor using the *priority (priority from 0–255), poll-interval (polling interval in seconds),* or *cost (cost from 0 to 65,535)* optional command modifiers.

Due to the nature of point-to-multipoint network adjacency establishment, polling intervals are not used and DR/BDRs are not elected, making the cost *keyword the only option on these networks. Similarly, fully meshed broadcast and nonbroadcast NBMA networks cannot make use of the* cost *keyword because all neighbors are assumed to have the same cost.*

Configuring Areas

After configuring your OSPF interfaces, you need to configure your areas themselves. In a single-area implementation, none of these steps is required; but in a multi-area design, you will at the very least need to configure ABRs. The configuration tasks involved with configuring OSPF areas are as follows:

- Configuring authentication
- Configuring stub areas and NSSAs
- Configuring ABRs and ASBRs

Table 26-9 describes the commands used to perform the tasks detailed in this section.

Configuring Authentication

Configuring OSPF authentication is fairly simple compared to EIGRP authentication. First, to enable authentication (authentication is disabled by default), you need to define the password (or key) that will be used for each interface. You do this with either the *ip ospf authentication-key* command (for plaintext passwords) or the *ip ospf message-digest-key* (for MD5-encrypted passwords) on each interface that will be using authentication.

Remember, for routers to establish adjacencies, authentication parameters for the two routers must match.

If you are using plaintext passwords, simply enter the *ip ospf authentication-key* command, followed by the password (which must be eight characters or less), like so:

```
Router(config)# interface Ethernet 0/0
Router(config-if)# ip ospf authentication-key noguesin
```

Then, to enable password protection for the entire area, use the *area [area ID] authentication* command, like so:

```
Router(config-if)# router ospf 1
Router(config-router)# area 1 authentication
```

Alternatively, you can use the *ip ospf authentication* interface config mode command to apply authentication only to specific interfaces. You may also remove authentication for specific interfaces by appending the *null* keyword, or change the authentication type for a specific interface to MD5 by using the *message-digest* keyword.

Configuring MD5 authentication is a little more complicated, but still fairly simple. First, like plaintext, you must initially define the password (known as a key in MD5) to be used for each interface. For MD5, you use the *ip ospf message-digest-key [key ID] md5 [key]* command. The *key ID* portion of this command defines the newness of the key and is used to transition from one key to another without disrupting OSPF. If more than one key ID is entered, OSPF uses both keys until all neighbors transition to the last key ID entered, at which point, OSPF transitions fully to using the new key. For instance, in the following example, OSPF uses key 1 with the password of *oldpassword* until all neighbors have transitioned to key 2 with the password of *newpassword*:

```
Router(config)# interface Ethernet 0/0
Router(config-if)# ip ospf message-digest-key 1 md5 oldpassword
Router(config-if)# ip ospf message-digest-key 2 md5 newpassword
```

MD5 keys are encrypted during transmission, and may contain up to 16 characters.

Once you have configured the key used for each interface, you may enable MD5 authentication for the entire area using the *area [area ID] authentication message-digest* command. Once again, you may also override the type of authentication used on a given interface with the *ip ospf authentication* command.

Configuring Stub Areas and NSSAs

The basic operation of configuring a stub area is fairly simple. Simply use the *area [area ID] stub* router config mode command on all routers in the area. Remember, for routers to establish an adjacency, they must both agree on how stubby an area is. To configure a totally stubby area, simply append the *no-summary* keyword.

NSSAs, on the other hand, are a little more complex. To configure an NSSA, you need to use the *area [area ID] nssa* command on all routers in the area. By doing this, you immediately block all Type 5 LSAs, but allow Type 3 and Type 7 LSAs. If you wish to remove Type 3 LSAs, use the *no-summary* keyword. This makes the NSA sort of a "totally not-so-stubby" area, allowing only Type 1, Type 2, and Type 7 LSAs (except a single Type 3 LSA for the default route.) Finally, if you want the ASBR to generate a

Command	Description	Mode
area [area ID] authentication [message-digest]	Enables authentication for the area	Router config
ip ospf authentication [message-digest | null]	Enables or disables authentication for a specific interface, overriding the authentication configured for the area as a whole	Interface config
ip ospf authentication-key [password]	Specifies the plaintext password to be used for each interface	Interface config
ip ospf message-digest-key [key ID] md5 [key]	Specifies the MD5 password to be used for each interface	Interface config
area [area ID] stub [no-summary]	Enables an area as a stub or totally stubby area	Router config
area [area ID] nssa [no-summary] [default-information-originate]	Configures an area as an NSSA and controls distribution of Type 7 LSAs	Router config
area [area ID] range [ip address] [mask] [advertise | not-advertise]	Summarizes or suppresses a range of network advertisements at an area boundary	Router config
summary-address [ip address] [mask] [not-advertise]	Summarizes or suppresses a range of network advertisements sent by an ASBR	Router config
default-information-originate [always] metric [metric] metric-type [1 | 2]	Causes an ASBR to redistribute a default route into the OSPF area	Router config
default-metric [metric]	Sets the default metric used for redistributed routes	Router config

Table 26-9.   *OSPF Area Configuration Commands*

Command	Description	Mode
redistribute [protocol] [process-ID] metric [metric] metric-type [1 | 2] [match (internal | external1 | external2)] [subnets]	Redistributes routes from or to OSPF	Router config
distribute-list [access list] out [interface type and number | routing protocol | AS or process-ID]	Controls the redistribution of routes on an ASBR	Router config

Table 26-9. *OSPF Area Configuration Commands* (continued)

Type 7 default route for the area, you can use the *default-information-originate* keyword. This strategy is useful when you want all routers in the NSSA to send all packets to unknown destinations to the ABSR rather than the ABR.

Configuring ABRs and ASBRs

Configuring ABRs in OSPF is simpler than it may seem at first. To use the default behavior of an ABR, which simply injects Type 3 LSAs for all internal networks (subnets included) at one network per LSA, all you need to do is configure the ABR to be a member of more than one area (one of which must be the backbone, of course). However, to configure the more efficient practice of summarizing networks at area boundaries, you must do a little more work.

To summarize networks at the ABR, use the *area [area ID] range [ip address] [mask] [advertise \| not-advertise]* command, specifying the summary IP network address in the *ip address* section and the summary mask in the *mask* section. The area portion specifies the area that will be summarized. For instance, to summarize 172.16.4.0–172.16.7.255 in Area 1 as one network, you issue the following command:

```
Router(config-router)# area 1 range 172.16.4.0 255.255.252.0 advertise
```

The advertise *keyword is not strictly required because the default behavior is to advertise the route. It is shown simply for illustrative purposes.*

To ensure that a specific network or range of networks are not advertised to the backbone by the ABR, you may add the *no-advertise* keyword to the command. For instance, the following example disables advertisement of a Type 3 LSA for networks 192.168.16.0–192.168.23.0 in Area 2:

```
Router(config-router)# area 2 range 192.168.16.0 255.255.248.0 no-advertise
```

Configuring ASBRs is a little more complex. First, to redistribute routes into OSPF, you need to enter router config mode for the OSPF process that you wish to redistribute routes into (the "target" process) and use the *redistribute [protocol] [process-ID] metric [metric] metric-type [1 | 2] [subnets]* command. The *protocol* section specifies the protocol you are redistributing into OSPF (which may include the *connected* or *static* keywords for directly connected or static routes). The *process-ID* section specifies either the process ID or AS number (for protocols, like EIGRP, that use those distinctions).

The *metric* section specifies the OSPF metric for the redistributed routes. (The default metric is 20, except for routes redistributed from BGP, in which case, the metric is 1.) The *metric-type* section determines which type of metric (E1 or E2) is used for the redistributed routes. (See the "OSPF Terms and Concepts" section, early in the chapter, for more information on the distinction between E1 and E2 routes.) Finally, the *subnets* optional keyword specifies whether individual subnets should be advertised. If the *subnets* keyword is not used, only major networks will be redistributed into OSPF.

To redistribute routes from OSPF into another routing protocol, the *redistribute* command is similar, but includes an addition or two. First, ensure that you are in router config mode for the routing protocol into which you wish to redistribute OSPF routes, and issue the *redistribute [protocol] [process-ID] metric [metric] [match (internal | external1 | external2)]* command. The *protocol* and *process-ID* sections are obviously the OSPF protocol and process ID that you wish to redistribute. The *metric* section is the metric (for the other routing protocol) that you wish to be assigned to the redistributed OSPF routes. You can use the *match* optional keyword when redistributing OSPF to determine which type of OSPF routes are redistributed into the target protocol/ process. By default, all routes will be redistributed; but you may specify that only routes internal to the area are redistributed, or that only E1 or E2 routes are redistributed.

Tip   *If you are redistributing from and to OSPF on a single ASBR, there's an easy way (without using distribute lists) to ensure that routes from another routing protocol that are redistributed into OSPF are not redistributed back out to the originating protocol. Use the* match internal *keyword in your* redistribute ospf *statement.*

Once you have enabled redistribution, you may find that, if using the *subnets* keyword, you have a large number of routes entered into OSPF for each ASBR. To reduce the number of routes the ASBR injects into OSPF, you may use the *summary-address [ip address] [mask] [not-advertise]* command to summarize the ASBR's

Type 5 (or 7) advertisements. Similar to the *area range* command discussed earlier, the *summary-address* command summarizes several subnet advertisements into a single advertisement about a block of subnets. If you use the optional *no-advertise* modifier, you can also specify (without using distribute lists) that specific networks/subnets are not redistributed into OSPF.

Once you have configured summarization, you may determine that you need to redistribute default routes from the other routing protocol into OSPF. To do this, use the *default-information originate* command on the ASBR. This command tells the router to advertise its best default route from other routing protocols back into OSPF. The *always* keyword informs the ASBR that it should always originate a default route back into the area, even when no default route from other protocols exist. This is useful when you want the OSPF routers to send packets to the ASBR for unknown destinations, even though a default route is not used in the other routing protocol. Finally, you can specify the metric and metric type using the *metric* and *metric-type* keywords, respectively. If these keywords are not used, the metric advertised into OSPF will be 10 and the type will be E2.

Finally, you may determine that you don't like the default metric advertised by OSPF for redistributed routes. By default, OSPF uses a metric of 20 for all redistributed routes except default routes (which get a metric of 10) and redistributed BGP routes (which get a metric of 1). To change this functionality, use the *default-metric [metric]* router config mode command. The metric value can be any number from 1 to 4 billion.

Advanced OSPF Configuration

Advanced OSPF configuration tasks do not need to be performed in every OSPF network. In most networks, performing actions like creating virtual links or changing the SPF timers should not be performed unless absolutely necessary. However, just in case you run into a situation in which you need to perform these functions, they are included in this section of the chapter.

Some of the advanced tasks that may be required in your OSPF network include the following:

- Configuring virtual links
- Modifying the reference bandwidth
- Modifying the SPF timers
- Changing administrative distances

Table 26-10 describes the commands used in this section.

Configuring Virtual Links

As mentioned in the "Virtual Links" section, virtual links are used to repair a partitioned backbone and to allow areas with no direct connection to the backbone access to the backbone through a transit area. In most cases, you shouldn't use a virtual link, but

Command	Description	Mode
area [area ID] virtual-link [router ID]	Creates a virtual link	Router config
ospf auto-cost reference-bandwidth [bandwidth in Mbps]	Changes the base cost for OSPF metric calculation	Router config
timers spf [delay] [holdtime]	Changes the default SPF route calculation timers	Router config
distance ospf [intra-area] [inter-area] [external]	Changes the default values for OSPF administrative distances	Router config

Table 26-10.   *Advanced OSPF Configuration Commands*

you might sometimes find it necessary to create a *temporary* virtual link. For instance, you may have a freak failure that causes a partitioning of your Area 0. To restore connectivity until a replacement router and/or link is available, you may be forced to resort to using a virtual link.

I have yet to see a case in which using a permanent virtual link was a good idea. In most cases, you are advised to perform a partial redesign of the network rather than create permanent virtual links.

In these cases, you will need to use the *area [area ID] virtual-link [router ID]* command on both routers that will participate in the virtual link. The area ID is the number of the area on the current router that will be participating in the virtual link, and the router ID is the remote router that will participate in the virtual link. An example of a virtual link is shown in Figure 26-38.

For instance, to create a virtual link between Copeland and Bruford in Figure 26-38, you would enter the following commands:

```
Copeland(config-router)# area 1 virtual-link 172.16.1.1
Bruford(config-router)# area 1 virtual-link 10.0.0.1
```

Modifying the Reference Bandwidth

In most modern networks, you will likely have one or more links running at speeds in excess of 100 Mbps. One problem with OSPF is that, by default, it recognizes all links at 100 Mbps or faster as having a cost of 1, which can lead OSPF to incorrectly route packets down a lower bandwidth link (a 100-Mbps link instead of a 1-Gbps link, for

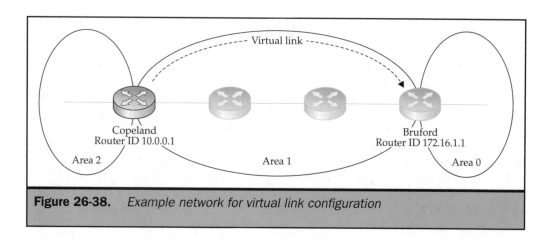

Figure 26-38. *Example network for virtual link configuration*

example). To resolve this problem, Cisco routers allow you to set the reference bandwidth used by OSPF on link calculation. The *reference bandwidth* is the amount of bandwidth that the SFP algorithm uses as the baseline for computing the cost of links. By default, the reference bandwidth is 10^8, or 100 Mbps.

You may change the reference bandwidth to any figure between 1 Mbps (10^6) and 4,294,967 Mbps (4.2 Tbps or 10^{12}) by using the *ospf auto-cost reference-bandwidth [bandwidth in Mbps]* command. One point to keep in mind, however, is that if you change the reference bandwidth on one router, you should change it on all routers in the AS to ensure proper metric calculation.

Modifying the SPF Timers

In some cases, the default timers used to control the SPF algorithm may not be acceptable. Cisco routers use two primary SPF timers: the holdtime and the delay.

As mentioned in the "OSPF Core Logic" section earlier in the chapter, the *holdtime* controls how often an SPF calculation can occur. By default, Cisco routers are configured to allow one SPF calculation every ten seconds. Reducing this value reduces convergence time for consecutive link changes, but it increases processor resources.

The *SPF delay* is the amount of time the SFP algorithm waits before computing the shortest path tree after receiving a topology change. By default, the delay is set to five seconds. Reducing the delay enormously increases processor resources used because SPF must recompute the tree more often when multiple changes occur. However, reducing the delay also slightly reduces convergence time for a single topology change.

You can change either of these values by using the command *timers spf [delay in seconds] [holdtime in seconds]*. The values do not have to be identical for all routers.

Note *Changing the default SPF timers is usually a case of "Yes, you can, no, you shouldn't." Although reducing the timers can lead to quicker convergence, the hit you take on processor use is usually not worth it.*

Changing Administrative Distances

By default, the administrative distance used for all OSPF routes is 110. Sometimes you might prefer OSPF routes to be considered better than other routes, such as IGRP routes (which have a default administrative distance of 100). However, OSPF classifies routes as one of three types, allowing you more flexibility: intra-area (within a single area), interarea (between areas within the same AS), and external (routes redistributed from other ASs).

For instance, you may decide that an OSPF intra-area route should be preferred over IGRP, and an interarea route should be preferred over RIP (distance of 120) while being less preferred than IS-IS (distance of 115), and external routes should be used only if no other routes exist to the destination. To configure this setup, you use the *distance ospf [intra-area] [inter-area] [external]* command. You set the intra-area distance to 99 (lower than IGRP), the interarea distance to 119 (less than RIP but greater than IS-IS), and the external distance to 254 (less than the maximum of 255 but more than all other protocols), like so:

```
Router(config-router)# distance ospf 99 119 254
```

Troubleshooting OSPF

Now that you have learned how to configure OSPF, let's examine some troubleshooting tools for OSPF. Like most protocols, the first step in troubleshooting OSPF is to verify the OSPF configuration. You can perform this task by using several *show* and *debug* commands, the most useful of which are shown in Table 26-11, with examples and more detailed descriptions of selected commands in the following paragraphs.

The *show ip ospf* command is useful for verifying basic OSPF functionality. The command output is shown here:

```
Router# show ip ospf
Routing Process "ospf 1" with ID 192.168.1.1
Supports only single TOS(TOS0) route
It is an area border and autonomous system boundary router
Redistributing External Routes from,
     rip with metric mapped to 5
     eigrp 1 with metric mapped to 1000
Number of areas in this router is 2
Area 0
     Number of interfaces in this area is 1
     Area has simple password authentication
     SPF algorithm executed 10 times
Area 1
     Number of interfaces in this area is 1
     Area has simple password authentication
     SPF algorithm executed 4 times
```

Command	Description	Mode
show ip route ospf [process-id]	Displays OSPF routes	User exec
show ip ospf [process-id]	Displays basic OSPF information for the selected process ID	User exec
show ip ospf border-routers	Displays information about OSPF BRs	Enable
show ip ospf [(optional) process-id] [(optional) area-id] database [optional modifiers]	Displays information related to the SPF database	User exec
show ip ospf interface [interface type and number]	Displays OSPF configuration for a specific interface	User exec
show ip ospf neighbor [interface type and number | neighbor router-id] [detail]	Displays information on OSPF neighbors	User exec
show ip ospf request-list [(optional) interface type and number] [(optional) neighbor router ID]	Displays the OSPF link-state request list	User exec
show ip ospf retransmission-list [(optional) interface type and number] [(optional) neighbor router ID]	Displays the OSPF link-state retransmission list	User exec
show ip ospf [process-id] summary-address	Displays configured summary addresses	User exec
show ip ospf virtual-links	Displays information on configured virtual links	User exec
debug ip ospf events	Debugs SPF events (adjacencies, calculations, and so on)	Enable
debug ip ospf packets	Debugs OSPF packet reception	Enable
debug ip ospf adj	Debugs the OSPF adjacency establishment process	Enable
ospf log-adj-changes	Provides basic adjacency change information	Router config

Table 26-11. *Useful OSPF* show *and debug* Commands

As you can see, with the preceding command, you can verify the areas for which OSPF is configured, the router's role in the network (as an ABR and/or ASBR), the router's ID, and the number of executions of the SPF algorithm on the router.

The *show ip ospf border-routers* command shows you information about all known border routers, like so:

```
Router# show ip ospf border-routers
OSPF Process 1 internal Routing Table
Destination Next Hop Cost Type Rte Type Area SPF No
172.16.10.10 192.168.1.10 10 ABR INTRA 0.0.0.1 14
10.189.221.14 192.168.1.11 20 ASBR INTER 0.0.0.1 14
```

For the most part, this output is self-explanatory, except for the *SPF No* section. The SPF number simply shows which SPF calculation was responsible for entering the particular route into the routing table.

The *show ip ospf database* command shows you the entire OSPF link-state database, like so:

```
Router# show ip ospf database

OSPF Router with id(192.168.1.1) (Process ID 1)

 Displaying Router Link States(Area 0.0.0.0)

 Link ID ADV Router Age Seq# Checksum Link count
10.0.0.1 192.168.1.6 1046 0x80002111 0x3ABD 1
>OUTPUT TRUNCATED<
 Displaying Net Link States(Area 0.0.0.0)

 Link ID ADV Router Age Seq# Checksum
10.0.0.1 192.168.1.6 1051 0x80000011 0x31A
>OUTPUT TRUNCATED<
 Displaying Summary Net Link States(Area 0.0.0.0)

 Link ID ADV Router Age Seq# Checksum
10.0.0.0 192.168.1.6 1052 0x80000012 0xAA01
>OUTPUT TRUNCATED<
```

Because most OSPF link-state databases are large, the *show ip ospf database* command has several permutations to help target the output. Table 26-12 describes these alternative commands.

Command	Description
show ip ospf [process-id area-id] database	Shows the OSPF database for a specific process and area.
show ip ospf [process-id area-id] database router [link-state-id]	Shows router (Type 1) LSAs. Append the link-state ID to show a specific LSA.
show ip ospf [process-id area-id] database network [link-state-id]	Shows network (Type 2) LSAs. Append the link-state ID to show a specific LSA.
show ip ospf [process-id area-id] database summary [link-state-id]	Shows network summary (Type 3) LSAs. Append the link-state ID to show a specific LSA.
show ip ospf [process-id area-id] database asb-summary [link-state-id]	Shows ASBR summary (Type 4) LSAs. Append the link-state ID to show a specific LSA.
show ip ospf [process-id area-id] database nssa-external [link-state-id]	Shows NSSA external (Type 7) LSAs. Append the link-state ID to show a specific LSA.
show ip ospf [process-id] database external [link-state-id]	Shows AS external (Type 5) LSAs. Append the link-state ID to show a specific LSA.

Table 26-12.   *The* show ip ospf database *Command Permutations*

The *show ip ospf interface* command shows OSPF information specific to each router interface (such as area membership, interface type, state, and so on), as shown here:

```
Router# show ip ospf interface Ethernet 0/0

Ethernet 0/0 is up, line protocol is up
Internet Address 192.168.1.1, Mask 255.255.255.0, Area 0.0.0.0
AS 1, Router ID 192.168.1.1, Network Type BROADCAST, Cost: 10
Transmit Delay is 1 sec, State OTHER, Priority 1
Designated Router id 192.168.1.10, Interface address 192.168.1.10
Backup Designated router id 192.168.1.20, Interface addr 192.168.1.20
Timer intervals configured, Hello 10, Dead 60, Wait 40, Retransmit 5
Hello due in 0:00:02
Neighbor Count is 14, Adjacent neighbor count is 2
 Adjacent with neighbor 192.168.1.20 (Backup Designated Router)
 Adjacent with neighbor 192.168.1.10 (Designated Router)
```

The *show ip ospf neighbor* command is another command with lots of interesting options. The base version of the command shows a summary of all neighbors, like so:

```
Router# show ip ospf neighbor

 ID Pri State Dead Time Address Interface
 192.168.1.10 20 FULL/DR 0:00:31 192.168.1.10 Ethernet0/0
 192.168.1.20 10 FULL/BDR 0:00:33 192.168.1.20 Ethernet0/0
```

If you want to see information on a specific neighbor, append the neighbor's router ID to the end of the command, like so:

```
Router# show ip ospf neighbor 192.168.1.10

Neighbor 192.168.1.10, interface address 192.168.1.10
 In the area 0.0.0.0 via interface Ethernet0/0
 Neighbor priority is 20, State is FULL
 Options 2
 Dead timer due in 0:00:34
 Link State retransmission due in 0:00:04
```

You can also have the command show all neighbors attached to a specific interface by appending the interface name and number to the end of the command. Finally, you can show detailed information on all neighbors (the same information shown with the *show ip ospf neighbor [neighbor ID]* version of the command) by appending the *detail* modifier to the command.

The *show ip ospf request-list* and *show ip ospf retransmission-list* commands show you the SPF request and retransmission lists, respectively:

```
router# show ip ospf request-list

 OSPF Router with ID (192.168.1.1) (Process ID 1)

 Neighbor 192.168.1.10, interface Ehternet0/0 address 192.168.1.1

 Type LS ID ADV RTR Seq NO Age Checksum
 1 192.168.1.10 192.168.1.10 0x80000111 8 0x4123
Router# show ip ospf retransmission-list

 OSPF Router with ID (192.168.1.1) (Process ID 1)
```

```
Neighbor 192.168.1.10, interface Ehternet0/0 address 192.168.1.1
Link State retransmission due in 3472 msec, Queue length 1

Type LS ID ADV RTR Seq NO Age Checksum
1 192.168.1.10 192.168.1.10 0x80000A1A 0 0xAFEF
```

The *show ip ospf summary-address* command shows all configured summary addresses on the ASBR, like so:

```
Router# show ip ospf summary-address
 OSPF Process 1, Summary-address
 10.0.0.0/255.0.0.0 Metric 1000, Type 0, Tag 0
```

The *show ip ospf virtual-links* command, as the name might suggest, shows configured virtual links on the local router, like so:

```
Router# show ip ospf virtual-links

Virtual Link to router 192.168.1.1 is up
Transit area 0.0.0.4, via interface Ethernet0/0, Cost of using 10
Transmit Delay is 1 sec, State POINT_TO_POINT
Timer intervals configured, Hello 10, Dead 40, Wait 40, Retransmit 5
Hello due in 0:00:03
Adjacency State FULL
```

The *debug* commands, as is normal with debugs, show extremely detailed information on OSPF events and packets. As always, debugs are best used only when absolutely necessary because of the high processor use they can cause. For adjacency monitoring, instead of debugging adjacencies, the *ospf log-adj-changes* command can give you basic adjacency change information without extremely high processor use. Once enabled, this command logs all adjacency changes to the configured syslog location(s).

Once you've verified OSPF operation, if you are still having problems, use the following tips:

■ Ensure that the problem is not related to basic connectivity issues (link problems, IP addressing problems, and so on).

■ If adjacencies aren't forming correctly, ensure the following:

    ■ OSPF is enabled on the interface (*show ip ospf interface*) and is not a passive interface.

CISCO ROUTING

- OSPF hello parameters (subnet mask, area ID, stub flags, hello intervals, and authentication keys) are identical on both routers. (The *debug ip ospf adj* command is useful for this purpose.)
- If the router's state is stuck in an attempt for one or more neighbors, ensure that there are no transient connectivity issues. (Perform a *show interface* and look for packet loss or perform a continuous ping.)
- If networks are listed in the OSPF database but do not show up in the routing table, check for mismatched network types (i.e., one router set to broadcast and the other set to point-to-point) with *show ip ospf interface*.
- If a default route is not being redistributed properly, ensure that the *default-information originate* command is entered on the ASBR.
- If SPF is taking up too much processor use, ensure that the timers are not set too low and make sure routes are summarized.
- If routes are not being advertised between areas properly (Type 5 LSAs), ensure that all areas have a connection to Area 0.
- If an inefficient path is being chosen to an external destination, ensure that all ASBRs redistributing the external route are advertising E1 routes (or N1 routes, in the case of an NSSA).

Although these tips will not solve every problem, they will help you identify some of the more common OSPF issues. Remember, most routing problems actually stem from misconfigurations. Use the troubleshooting commands discussed in this chapter to narrow down the problem, and then zero in on the cause.

Summary

This chapter examined OSPF at a detailed level. You should have acquired a solid understanding of OSPF operation, configuration, and basic troubleshooting. For more information on OSPF, be sure to check out RFC 2328 at *ftp://ftp.isi.edu/in-notes/rfc2328.txt*. For Cisco-specific OSPF information, be sure to check out *http://www.cisco.com/univercd/cc/td/doc/product/software/ios122/122cgcr/fipr_c/ipcprt2/1cfospf.htm*.

The final chapter examines access lists and network security.

The Complete Reference

Chapter 27

Access Lists

ccess lists, also known as access control lists (ACLs), are the primary packet-filtering mechanism on most Cisco routers. ACLs are also used for a number of other functions, including route filtering and targeting certain *show* commands. As a result, you need to have a solid understanding of ACLs to help you troubleshoot and configure Cisco equipment in most networks. ACLs give you control over which type of match the router makes for a given function, such as forwarding packets.

For instance, without some way to conditionally stipulate which packets are allowed into your network from another network (such as the Internet), you would be forced to either allow all packets (making your network an easy target for security violations) or deny all packets (making connectivity impossible). Security control through packet filtering is an ACL's primary function. However, you can use ACLs for most complex commands (such as route maps and distribute lists) to give you finer-grained control over the router's functionality. In short, whenever you need some type of conditional match to be performed, ACLs are usually your tool of choice.

This chapter covers ACLs, with a focus on their primary purpose: packet filtering. Although all of the concepts covered here apply to other commands that make use of ACLs, this chapter does not specifically target those commands. (See Chapters 22 through 26 on routing protocols for examples of distribute list ACLs.) Also, because to fully understand ACLs, you need to understand ACL configuration, troubleshooting and configuring ACLs are examined throughout this chapter rather than being discussed in separate sections. Finally, because TCP/IP is the primary focus of this book, IPX packet filtering is not discussed here.

Understanding Packet Filtering

The basic goals involved with packet filtering are fairly simple. You want to allow a remote network (usually a public network such as the Internet) access to only specific resources in your private network, while allowing access from your private network to the remote network. Simple, right? Unfortunately, this premise is easier envisioned than implemented. To see why, take a look at the example scenario in Figure 27-1, where you want packet filtering to be performed.

The network shown in the figure is not using NAT, and it is simply handing public IP addresses to the two workstations to provide Internet access. The problem with this scenario is that without packet filtering, every host on the private network can communicate with the Internet unhindered, and vice versa. Now imagine that your online banking information was on a shared drive on host 64.1.1.69. The only protection that host receives from prying eyes is the security of the operating system, which is usually highly suspect.

Before you learn how to solve this problem, let's examine some new terms associated with NAT and packet filtering. The *internal network* is the network you wish to protect from outside access. The *external network* is the remote network. The *internal interface* is the router interface attached to the internal network. The *external interface* is the interface attached to the external network. Finally, when dealing with filters, you will need to understand *direction*: the type of traffic to which the filter applies.

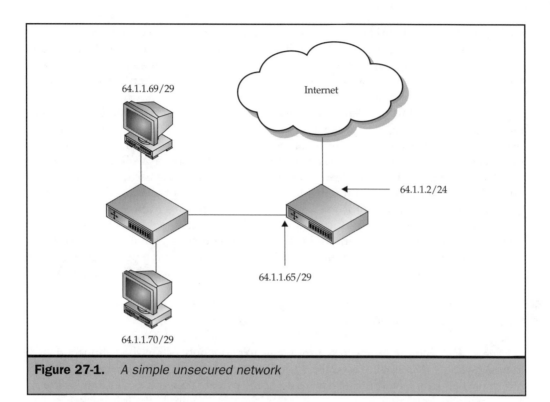

Figure 27-1.   *A simple unsecured network*

If the direction is *incoming,* then the filter applies to traffic entering the interface from the attached network. For an internal interface, incoming traffic is traffic entering the interface from the internal network. This traffic, if it were routed to the external network, would also be considered outgoing traffic on the external interface. If the direction is *outgoing,* then the filter applies to traffic leaving the interface on the attached network. For an internal interface, this is traffic that is forwarded out the internal interface to the internal network. This traffic, if it were routed from the external network, would also be considered incoming traffic on the external interface.

Now let's deal with solving the issues in the previous example. To help alleviate the problem of unhindered access to internal resources, you can use packet filtering on the router to deny access to the internal network from the Internet. For instance, you may want to eliminate all access to your internal network from the Internet, so you may choose to use a packet filter that denies every packet destined for the internal network. For this, you would use a packet filter that states "Deny all" or "Deny packets with a destination address of 64.1.1.64 – 71." You would then apply the packet filter to either incoming packets on the outside interface (64.1.1.2/24) or outgoing packets on the inside interface (64.1.1.65/29).

Problem solved, right? Well, not quite, unfortunately. This packet filter would not only (correctly) deny unauthorized access from the outside world, but also deny return packets from internal hosts to the Internet. In other words, when you send a packet to www.cisco.com from host 64.1.1.69, the TCP connection-establishment packets would be allowed to leave the private network unhindered en route to the Internet; but when the web server at Cisco responds, your filter would discard the packets! This is obviously not the appropriate outcome, so back to the drawing board.

One solution to this problem is to allow packets using client port numbers to enter the private network. Normally, clients should use ports listed only as "dynamic and private" by the IANA, which includes all ports between 49,192 and 65,535. However, many clients will use ports in the range of 1,024 through 49,191 as well (defined by the IANA as "registered"). This leads to problems in packet filtering because if you simply deny every packet with a port number lower than 49,192, you will most likely end up denying legitimate return traffic to clients as well as suspicious traffic.

So why not just allow all traffic using a destination port above 1023? Because several common server applications, many of which you do not want to be accessible from external hosts, use ports in this range (such as the Microsoft global catalog sever in Active Directory on ports 3268/3269, and Microsoft's SNA server on 1477 and 1478).

Note   *For a full listing of registered/well-known TCP and UDP port numbers, please visit http://www.iana.org/assignments/port-numbers.htm.*

The easiest solution to this little problem is to allow only established connections to pass through. By allowing established connections, you can tell the router to allow return packets from any external host to enter the private network. So, as long as the TCP session to the remote host remains established, the router will allow packets entry into the private network.

Established packet filtering works like so:

1. You set an established filter on the external interface for incoming packets that basically states, "Allow all established connections."

2. You send a packet to a remote host. Because the packet filter is set to check incoming packets only on the external interface, all packets from the internal network to the Internet are automatically allowed.

3. The remote host responds. The remote host should have the ACK or RST bits set in the response packet. The established filter allows all packets from the external network with these bits set to enter.

The functionality in step 3 obviously leaves a bit of a hole in the network security. For instance, an enterprising hacker could manually set the ACK bit in a TCP packet and enter your internal network. For this reason, Cisco devices also support a special type of packet filter known as a *reflexive ACL*. A reflexive ACL dynamically creates packet-filter entries for you to allow access to and from only the host and destination for which the session was established. Reflexive ACLs work like this:

1. You set the reflexive filter on the external interface for incoming packets that states, "Allow all established connections."

2. You send a packet to the remote host using a specific source IP (for instance, 64.1.1.69), a specific source port (for instance, 50,000), a specific destination IP (like 71.100.1.1), and a specific destination port (such as 80 for HTTP). The reflexive ACL builds a dynamic packet filter to allow all mirrored responses from the external network to enter the internal network until either the timeout has expired or the FIN bit in the session is seen. In this case, the reflexive ACL would create an entry that states, "Allow all packets from source address 71.100.1.1 with a source port of 80 and a destination of 64.1.1.69 with a destination port of 50,000 to enter the internal network."

3. The remote host responds, and the packet is forwarded.

4. When the session ends (either the FIN bit is seen or the timeout expires), the dynamic packet filter is removed.

Note   *On many proxy/NAT servers, including Microsoft's ISA server, mirrored and established connections are allowed automatically, whereas all other packets are denied as a side effect of the NAT/PAT table-creation process. Although this is not technically packet filtering, the functionality is the same.*

Reflexive filtering simplifies your life by letting you effectively allow internal traffic to access the external network. However, reflexive filtering has two small issues that you need to be aware of:

■ Reflexive filtering does not work for all applications. Specifically, applications that change port numbers during the session (FTP and RPC, for example) need to be statically allowed.

■ You can also use reflexive filters to allow only specific types of traffic to be used to access the external network from the internal network.

Let's examine the first point. Certain applications dynamically change ports during a session. These applications must be allowed to pass (if required) statically. For instance, FTP uses two ports, 20 and 21, during communications. When you establish an FTP connection, port 21 is used to transfer the Telnet-like control session. When you begin downloading a file, however, port 20 is used for the transfer. Reflexive ACLs don't know about the file transfer port; all they understand is the port initially used, so the ACL would block the FTP transfer. Remote Procedure Call (RPC), used by most Microsoft server applications, works similarly, but adds an additional complexity. After the initial connection, RPC chooses a random port over 1024 to communicate with.

To solve issues involved with these types of applications, you have two real choices:

■ Switch to a more robust firewall solution that understands application-level details (effective, but can be expensive).

■ Catalog these types of applications, the ports used, and open ports as necessary.

Some proxy/firewall solutions already include most of the application-level filters you may require. FTP support, for instance, is commonly available by default in most commercial products.

Because the first solution is beyond the scope of this book, let's focus on the second solution. The easiest way to implement it is to enable reflexive ACLs and then test to see which applications don't work. After some time, you will learn which applications you should configure static ACLs for in almost all situations (such as FTP); but, initially, this is the quickest way to get started. Once you determine which applications are not supported by reflexive ACLs, you can research the ports used by that application and configure ACLs to pass that specific traffic. For instance, for all clients to be able to use FTP, you would need to configure a static ACL that states, "Permit all traffic with a source port of 20 and a destination address equal to all addresses in your internal network."

For applications that pick random ports, such as RPC, the solution is a little more complex. If the application uses only a small range of ports, you simply need to open up ports in that range. However, if the application uses a large range of ports (like RPC), you will need to do a bit more work. The first step is to force the application to use a smaller range of ports. For RPC, for instance, you can modify the Windows registry on the hosts (usually servers) that require RPC connectivity through the firewall. Once you have defined a small range of random ports to use, you simply add a static ACL statement that permits those ports for communication to those specific servers.

For example, if you had five servers in the IP address range 70.1.1.5 through 70.1.1.9 that needed to communicate using RPC with another server at the IP address 140.0.0.1, first you would force RPC on all servers to use ports within a certain range, such as from 50,000 to 60,000. Then, you would add an ACL statement on the external interface of the firewall for the 70.1.1.0 network that stated, "Allow incoming packets with a source address of 140.0.0.1 using any port number with a destination address in the range of 70.1.1.5 through 70.1.1.9 with a destination port in the range of 50,000 through 60,000." On the firewall for the 140.0.0.0 network, you would simply mirror this statement.

For details on forcing RPC to use a specific range of ports on current Microsoft server platforms, please see article Q154596: http://support.microsoft.com/ default.aspx?scid=kb;EN-US;q154596.

With packet filtering, however, you can also get a bit more complex and choose to allow only certain types of traffic from *internal* hosts to exit to the *external* network. For instance, suppose you want to provide Internet access only for web browsing; but you do not want to give users the ability to use other standard Internet applications such as FTP, Telnet, and SMTP; nor do you want internal clients to be able to use unsupported (and sometimes bandwidth-hungry) applications such as Quake or Napster. In this case, you would create a reflexive ACL that states, "Permit internal hosts to access all external hosts using port 80 (HTTP) and 443 (HTTPS/SSL) ." You would then apply the ACL to the *internal* router interface for incoming traffic.

As a rule of thumb, try applying packet filtering as close to the network edge as possible. Also, try to filter traffic as it enters an interface (incoming direction) rather than as it exits an interface (outgoing direction). This strategy helps conserve router resources by eliminating the need for the router to perform processing on the packet before throwing it into the "bit bucket."

Using NAT and PAT with Packet Filtering

NAT can make your life a bit simpler by allowing you to reduce your focus on unauthorized access to hosts on the private network. By design, NAT and PAT generate translation tables dynamically (although static entries can also be defined) to allow hosts from the outside world to enter the internal network. Also by design, these dynamic entries are created only when an internal host needs to access the outside world. If an external host wishes to communicate with an internal host, and the NAT device does not have an entry in the translation table for the destination host, the packet is rejected. Therefore, as just a byproduct of NAT, internal resources are already given security similar to a reflexive ACL.

Note   For an overview of NAT, please refer to Chapter 6.

If you have hosts on your internal network that need to be accessible from hosts on the external network, like a web server, you simply add a static entry in NAT similar to the entry you would normally make on a standard packet filter, except you need to specify the external address and port associated with the web server, as well as the internal address and port associated with the web server. For instance, Figure 27-2 shows the network from Figure 27-1 with the addition of private addressing for the internal network (NAT is being performed by the router) and a web server on the internal network. You wish to stipulate that all external traffic is allowed to access the web server, but only for HTTP. (No access to the web server using any other protocol is to be allowed.) You also wish to allow internal clients to access external resources, but only for FTP, HTTP, HTTPS, POP3, and SMTP. All other communications should be denied.

The internal interface filters for Figure 27-2 (at the location marked with a 1) are listed here.

1. Internal Interface Filters for Figure 27-2

Type	Action	Direction	Source IP	Source Port	Destination IP	Destination Port
Nonreflexive	Permit	Inbound	Any	Any	Any	20–21
Nonreflexive	Permit	Inbound	Any	Any	Any	25
Nonreflexive	Permit	Inbound	Any	Any	Any	80
Nonreflexive	Permit	Inbound	Any	Any	Any	110
Nonreflexive	Permit	Inbound	Any	Any	Any	443
Nonreflexive	Permit	Inbound	192.168.1.100	80	Any	Any
Nonreflexive	Deny	Inbound	Any	Any	Any	Any

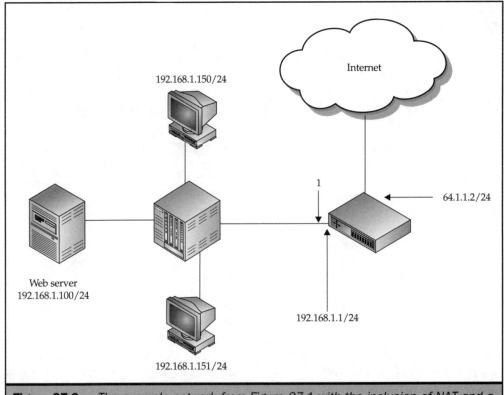

Figure 27-2.   *The example network from Figure 27-1 with the inclusion of NAT and a web server*

To support these requirements, you would create a NAT statement that allowed all public addresses to be used for dynamic translation. Then, you would add a static translation (known as a static mapping) that associates 64.1.1.2, port 80 with 192.168.1.100, port 80. Because of the nature of NAT, these two actions deny traffic entering the internal network from a remote host that is neither initiated by an internal host nor destined for HTTP on the web server.

Now, to keep internal hosts from using protocols other than FTP (20, 21), HTTP (80), HTTPS (443), POP3 (110), and SMTP (25), you simply apply a packet filter to *incoming* traffic on the *internal* interface that states the following:

- Permit traffic from any internal host using any source port that is destined for any external host using destination ports 20, 21, 80, 443, 110, or 25.

- Permit traffic from 192.168.1.100 using source port 80 with any destination address and port.

- Deny all other traffic.

Of course, once you take all of these steps to protect your network, you will probably realize that you are still left with one major vulnerability: the firewall itself. If a hacker were to compromise the firewall, all of your security settings would also be compromised. To help avoid this potentially disastrous occurrence, you should disable any unnecessary services on the router performing NAT/packet filtering. In a single firewall situation, if the firewall is performing PAT, you must be particularly careful of filtering packets on the external interface. By default, PAT on a Cisco router (also known as *NAT overload*) attempts to assign the same port number for translated packets. In other words, if an internal host is attempting to communicate with an external host using a source port of 12,000, NAT attempts to use the same port for the translated address.

However, because the source port may already be taken (another internal host may be using 12,000 as its source port), NAT may need to allocate a different port number. When NAT has to change the port of a session, it randomly assigns an unused port to the communication from one of three port ranges: 1–511, 512–1023, and 1024–65535. NAT always assigns a port to the translated packet that is in the same range as the pretranslation port.

For instance, if your client originally used port 443, NAT would choose a port between 1–511 for the packet translation, assuming, of course, that 443 was already in use. Because the port that NAT chooses for a particular client connection is unpredictable, packet-filtering ability on your external interface is limited. For this reason, careful consideration should be given to packet filtering for the router's external interface before implementing NAT.

Note *NAT configuration on a Cisco router is beyond the scope of this book. For more information, please visit http://www.cisco.com/warp/public/cc/pd/iosw/ioft/ionetn/prodlit/1195_pp.htm and http://www.cisco.com/warp/public/cc/pd/iosw/ioft/iofwft/prodlit/iosnt_qp.htm.*

Understanding DMZs

A *Demilitarized Zone*, or *DMZ* (also known as a *screened subnet*), is an additional security measure to help ensure that the availability of publicly accessible resources does not hinder the security of the internal network. A DMZ separates publicly accessible resources from completely private resources by placing the different resources on different subnets, allowing you to "lock down" your private resources and yet still provide security for the public resources.

Note *Although it will be considered blasphemy in some security circles, for the purposes of the rest of this chapter, the term "firewall" is used to represent any device that filters packets to control access, including both standard Cisco routers and specific firewall products like the PIX series of firewalls, unless specifically stated otherwise.*

DMZs can be arranged in several different ways, but this chapter discusses only three of the most common setups:

- **Single-firewall DMZ**   This type of DMZ consists of a single firewall with three or more interfaces, with at least one interface connected to the internal network, external network, and DMZ. These DMZs are the most cost efficient, but the security benefits (above and beyond a single firewall with no DMZ) are minimal.

- **Dual-firewall DMZ**   This type of DMZ is one of the most common, consisting of an external firewall, an internal firewall, and a DMZ between the two. Internal resources are protected by the internal firewall, whereas DMZ resources are protected by the external firewall. This type of DMZ structure can increase security substantially: the dual firewalls lend an extra layer of protection for internal resources. However, the drawback is that dual-firewall DMZs tend to be much more expensive than single firewall DMZs, and may require additional expertise (especially if you follow the recommendation that the firewalls be from different vendors).

- **Triple-firewall DMZ**   This type of arrangement uses three firewalls to protect the internal network, with both an internal and external DMZ. This type of firewall setup is used only in the most security-conscious organizations, and usually involves multitiered applications. Although extremely secure, this type of firewall arrangement is also extremely expensive to implement and maintain.

Let's examine the single-firewall solution first. Figure 27-3 shows a single firewall with three interfaces. The web servers and other publicly accessible servers live on the DMZ, while all internal resources are fully protected on the internal subnet. Also note that the SQL server used by the web server for order processing is on the internal network. The interface filters for the numbered locations in Figure 27-3 are listed here.

1. Internal Interface Filters for Figure 27-3

Type	Action	Direction	Source IP	Source Port	Destination IP	Destination Port
Nonreflexive	Permit	Inbound	SQL Server	SQL	Web server	Any
Nonreflexive	Deny	Inbound	SQL Server	Any	Any	Any
Nonreflexive	Permit	Inbound	192.168.1.X	80	Any	Any
Nonreflexive	Permit	Inbound	192.168.1.X	443	Any	Any
Nonreflexive	Permit	Inbound	192.168.1.X	20–21	Any	Any
Nonreflexive	Permit	Inbound	192.168.1.X	25	Any	Any
Nonreflexive	Deny	Inbound	Any	Any	Any	Any

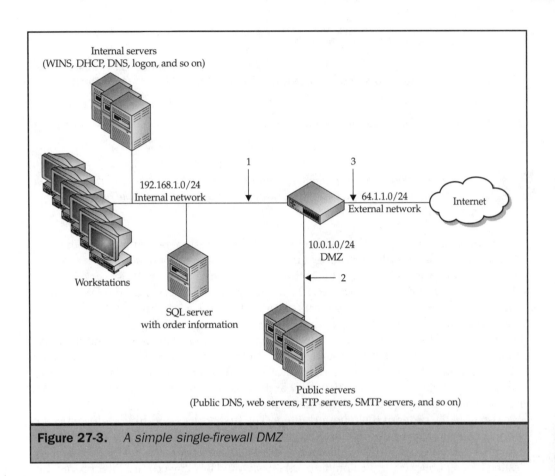

Figure 27-3. *A simple single-firewall DMZ*

2. DMZ Filters for Figure 27-3

Type	Action	Direction	Source IP	Source Port	Destination IP	Destination Port
Nonreflexive	Permit	Inbound	Web server	Any	SQL Server	SQL
Nonreflexive	Deny	Inbound	Any	Any	SQL Server	Any
Reflexive	Permit	Inbound	Any (est)	Any (est)	Any (est)	Any (est)
Nonreflexive	Deny	Inbound	Any	Any	Any	Any

3. External Interface Filters for Figure 27-3

Type	Action	Direction	Source IP	Source Port	Destination IP	Destination Port
Nonreflexive	Permit	Inbound	Any	Any	Public servers	Public ports
Reflexive	Permit	Inbound	Any (est)	Any (est)	Any (est)	Any (est)
Nonreflexive	Deny	Inbound	Any	Any	Any	Any

 Be aware that most firewall vendors, including Cisco, use a "first match" rule. All filters for a given direction on a given interface are examined in order, and the first matching filter is used.

Let's step through each of the packet filters for this design, beginning with the external filter. The first filter on the external interface allows any host to reach any public server on the DMZ using any of the ports for the services on those servers. In other words, if you had a web server, FTP server, and SMTP server on the DMZ, you would allow access to the web server using port 80 or 443, access to the FTP server using ports 20 and 21, and access to the SMTP server using port 25.

The second filter on the external interface is a reflexive filter to allow communication initially established by internal hosts to return to the internal hosts. The final filter on the external interface denies all other communications. (This is not required on Cisco ACLs because they always have an implicit deny at the end.) No outbound filters are applied to the external interface, so all outbound traffic is allowed.

Also note the order of the inbound filters. The first and second filters allow specific connections, while the third denies all connections. Because the deny filter is at the end, it functionally becomes a "Deny all others." If packets were to match the first filter (being destined for a public server on a public port), they would be allowed, and filter processing for this interface would end. If not, the next filter would be examined. If the packet matched the second filter (if the packet were a response packet to a session previously established by an internal or DMZ-based host), it would also be allowed, and filter processing for this interface would end. If the packet didn't match either permit filter, the packet would be denied. Again, just remember that filters are processed in order; and once a match is made, filter processing for that interface is complete.

For the DMZ, the first two inbound filters allow the web server (and only the web server) on the DMZ to communicate with the secured SQL sever on the internal network, using only SQL as the protocol. The stipulation that SQL be the only protocol used ensures that the other public servers, if hacked, cannot gain access to the secure database on the SQL server. The third outbound filter is a reflexive filter to allow all established communications from DMZ hosts to pass. The third filter is required because of the fourth outbound filter, which denies all other packets. Using this filtering technique, you help ensure that your secure SQL server is secure not only from external hosts, but also from inappropriate access from servers on the DMZ.

Finally, for the internal network, the first two inbound filters allow the SQL server to communicate only with the web server on the DMZ using the SQL protocol. This filter, although not strictly required, helps ensure the security of the SQL server by ensuring that the SQL server cannot communicate with any noninternal host other than the web server, and then using only SQL as the application. If you did not have this statement, an administrator physically seated at the SQL server could open a session to a web site using the SQL server.

Note *If the web server were actually communicating with the internal SQL server using SQL as the protocol, and the SQL server was a Microsoft SQL server, the port you would need to open would be 1433. However, because another type of application or another vendor's version of SQL (such as Oracle) may be used as the front end for the SQL server, I did not include the raw port number in this example. Ensure that you know the correct ports to open when using this technique.*

Although administrators surfing the web from a server, in and of itself, is really nothing to worry about, the remote chance exists that the session to the web site could be hijacked by a hacker; and because the session was previously established by the SQL server, the packets from the hacker to the SQL server would be allowed to enter the firewall. To help eliminate this possibility, you create two filters to shut off all SQL communications outside the internal network except for the required packets to the web server using the SQL ports.

Tip *You could apply this same philosophy and filtering technique to nearly all internal servers to eliminate the chance of unauthorized access. For instance, you could apply an additional filter before the third filter to deny the address block allocated to the internal servers from using external resources. Be aware, however, that using this filtering technique requires that your server administrators use standard workstations to download patches and other needed server files from the Internet, and then share the drives on the workstations (or use sneakernet) to transfer the files to the servers, causing additional administrative effort (which is potentially unnecessary, depending on your security policy).*

You use the next five filters on the internal interface to control access to the Internet for your internal hosts. These five filters allow any host on your internal subnet to access the Internet using ports 80 (HTTP), 443 (HTTPS/SSL), 20 or 21 (FTP), and 25 (SMTP), while denying all other access. Although the IP address block listed in the filters (192.168.1.X) is not strictly required (you could use the *any* statement for a single subnet), I included it to show you how to use the source address to allow users on one subnet to access the Internet while denying users on a different subnet access to the Internet. If you had another subnet (say, 192.168.2.0) with these filters, the second subnet would be denied connectivity to the DMZ and the Internet.

Now that you have examined filtering with a single firewall, let's examine dual-firewall DMZs. Figure 27-4 shows dual firewalls, internal and external, with a DMZ in between.

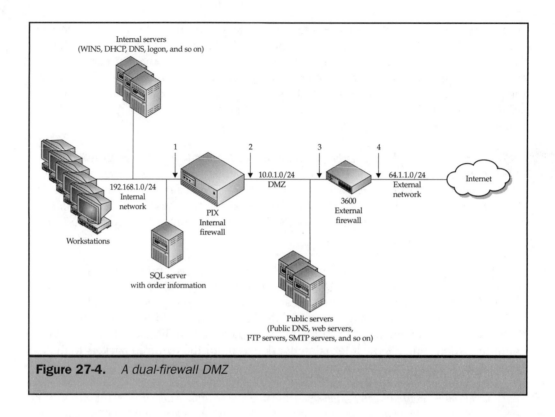

Figure 27-4. *A dual-firewall DMZ*

The interface filters for the numbered locations in Figure 27-4 are listed here.

1. Internal Interface Filters for Figure 27-4

Type	Action	Direction	Source IP	Source Port	Destination IP	Destination Port
Nonreflexive	Permit	Inbound	SQL Server	SQL	Web server	Any
Nonreflexive	Deny	Inbound	SQL Server	Any	Any	Any
Nonreflexive	Permit	Inbound	192.168.1.X	80	Any	Any
Nonreflexive	Permit	Inbound	192.168.1.X	443	Any	Any
Nonreflexive	Permit	Inbound	192.168.1.X	20–21	Any	Any
Nonreflexive	Permit	Inbound	192.168.1.X	25	Any	Any
Nonreflexive	Deny	Inbound	Any	Any	Any	Any

2. External Interface Filters for Figure 27-4

Type	Action	Direction	Source IP	Source Port	Destination IP	Destination Port
Nonreflexive	Permit	Inbound	Web server	Any	SQL server	SQL
Nonreflexive	Deny	Inbound	Any	Any	SQL server	Any
Reflexive	Permit	Inbound	Any (est)	Any (est)	Any (est)	Any (est)
Nonreflexive	Deny	Inbound	Any	Any	Any	Any

3. Internal Interface Filters for Figure 27-4

Type	Action	Direction	Source IP	Source Port	Destination IP	Destination Port
Nonreflexive	Permit	Inbound	Any	Any	Any	Any

4. External Interface Filters for Figure 27-4

Type	Action	Direction	Source IP	Source Port	Destination IP	Destination Port
Nonreflexive	Permit	Inbound	Any	Any	Public servers	Public ports
Reflexive	Permit	Inbound	Any (est)	Any (est)	Any (est)	Any (est)
Nonreflexive	Deny	Inbound	Any	Any	Any	Any

The external interface on the 3600 has basically the same ACLs as on the external interface of the firewall from Figure 27-3. A nonreflexive filter permits all communication to the public servers using proper port numbers, a reflexive ACL permits return traffic from internal hosts to pass, and a blanket *deny* statement takes care of unauthorized traffic. The internal interface of the 3600 allows all traffic to pass out of the network.

The external interface of the PIX is allowing communication only from the public web server to reach the SQL server with a nonreflexive access list. It then ensures that no communication from any device other than the web server can reach the SQL server with a nonreflexive access list that specifically denies all communication to the SQL server. Because the entry that permits the web server to access the SQL server is first in the list, as long as the correct ports are used, the web server will match the *permit* statement first and will be allowed to access the SQL server.

The next statement is a *reflexive* statement to allow all communications from the internal clients to access external resources. Finally, all other incoming external communications are denied with a blanket *deny* statement on the external PIX interface.

The last statements are applied to the internal interface of the PIX. These statements allow the internal clients to access external resources using HTTP, HTTPS/SSL, FTP, and SMTP. An additional statement allows the SQL server to communicate with the web server on the DMZ. Finally, a blanket *deny* rejects all other communications.

The advantage of a dual-firewall DMZ is an additional security of resources. When dual firewalls are used, it is more difficult to make unauthorized entry because two devices must be compromised to successfully access internal resources. If you use firewalls from different vendors, the security of the network improves even more substantially because the hacker must be proficient in compromising completely different platforms. The disadvantage is, of course, cost and possible configuration headaches. However, for medium-sized networks, dual firewalls are very popular.

Note   *The example in Figure 27-4 uses a dedicated firewall for the internal network in the form of a Cisco PIX firewall. Although basic firewall functionality is packet filtering (which can be performed by most routers), true firewalls have a host of features (including stateful packet filtering) that can improve overall network security. This chapter focuses only on packet filtering. For more information about dedicated firewalls, see http://secinf.net/ifwe.html.*

For triple-firewall scenarios, high security can be virtually ensured, even when complex filtering scenarios are required. For example, Figure 27-5 shows an example of a high-security, triple-firewall design allowing dual DMZs. The external DMZ is used to house resources that are completely public. The internal DMZ is used to house the SQL server, as well as extranet resources (semipublic) such as VPN servers and extranet web servers (like Outlook Web Access). The external firewall is used to protect Internet-accessible servers, while the midground firewall is used to protect extranet resources. Finally, the internal firewall protects purely internal resources, making this firewall configuration a formidable barrier against unauthorized access.

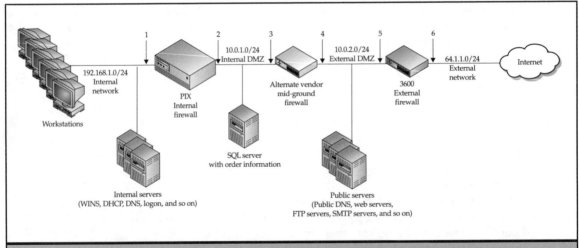

Figure 27-5.   *A triple-firewall DMZ*

The interface filters for the numbered locations in Figure 27-5 are listed here.

1. Internal Interface Filters for Figure 27-5

Type	Action	Direction	Source IP	Source Port	Destination IP	Destination Port
Nonreflexive	Permit	Inbound	192.168.1.X	80	Any	Any
Nonreflexive	Permit	Inbound	192.168.1.X	443	Any	Any
Nonreflexive	Permit	Inbound	192.168.1.X	20–21	Any	Any
Nonreflexive	Permit	Inbound	192.168.1.X	25	Any	Any
Nonreflexive	Deny	Inbound	Any	Any	Any	Any

2. External Interface Filters for Figure 27-5

Type	Action	Direction	Source IP	Source Port	Destination IP	Destination Port
Reflexive	Permit	Inbound	Any (est)	Any (est)	Any (est)	Any (est)
Nonreflexive	Deny	Inbound	Any	Any	Any	Any

3. Internal Interface Filters for Figure 27-5

Type	Action	Direction	Source IP	Source Port	Destination IP	Destination Port
Nonreflexive	Permit	Inbound	SQL server	SQL	Web server	Any
Nonreflexive	Deny	Inbound	SQL server	Any	Any	Any
Nonreflexive	Permit	Inbound	Any	Any	Any	Any

4. External Interface Filters for Figure 27-5

Type	Action	Direction	Source IP	Source Port	Destination IP	Destination Port
Nonreflexive	Permit	Inbound	Web server	Any	SQL server	SQL
Nonreflexive	Deny	Inbound	Any	Any	SQL server	Any
Reflexive	Permit	Inbound	Any (est)	Any (est)	Any (est)	Any (est)
Nonreflexive	Deny	Inbound	Any	Any	Any	Any

CISCO ROUTING

5. Internal Interface Filters for Figure 27-5

Type	Action	Direction	Source IP	Source Port	Destination IP	Destination Port
Nonreflexive	Permit	Inbound	Any	Any	Any	Any

6. External Interface Filters for Figure 27-5

Type	Action	Direction	Source IP	Source Port	Destination IP	Destination Port
Nonreflexive	Permit	Inbound	Any	Any	Public servers	Public ports
Reflexive	Permit	Inbound	Any (est)	Any (est)	Any (est)	Any (est)
Nonreflexive	Deny	Inbound	Any	Any	Any	Any

Although they are used only in the largest and most security-conscience networks, triple-firewall designs can be worth the investment in certain situations.

ACL Configuration

The first step in ACL configuration is to understand the syntax of access lists. For a standard access list, syntax is not extremely complex. For extended access lists, however, syntax can be quite a bit more confusing.

This section discusses access-list configuration using standard and extended ACLs in both the named and numbered styles.

Table 27-1 describes the commands used in this section.

Command	Description	Mode
access-list [access list number] [deny | permit] [source] [source wildcard mask] [log]	Creates a standard numbered IP access list	Global config
access-list [access list number] [deny | permit] [protocol] [source] [source wildcard mask] [(optional) port operators] [destination] [destination wildcard mask] [(optional) port operators] [established] [log]	Creates an extended numbered IP access list	Global config

Table 27-1.   *Access List Commands*

Command	Description	Mode
ip access-list standard [name]	Creates a standard named IP access list	Global config
[deny | permit] [source] [source-wildcard] [log]	Enters statements into a standard named IP access list	Access list config
ip access-list extended [name]	Creates an extended named IP access list	Global config
[deny | permit] [protocol] [source] [source wildcard mask] [(optional) port operators] [destination] [destination wildcard mask] [(optional) port operators] [established] [log]	Enters statements into an extended named IP access list	Access list config
ip access-group [access list number or name] [in | out]	Applies an access list to an interface	Interface config
show ip access-list [access list number or name]	Shows one or all IP access lists	User exec
show access-lists [access list number or name]	Shows one or all access lists	User exec

Table 27-1.   *Access List Commands* (continued)

The first command listed, *access-list*, is used to create both standard and extended numbered IP access lists. The deciding factor on whether the access list will be standard or extended is simply the number used to define the access list. If the number used is between 1 and 99, the access list will be a standard IP ACL; if the number is between 100 and 199, the access list will be an extended IP ACL.

Note   *On newer revisions of the IOS, numbered ACLs between 1300 and 1999 are also available for standard IP ACLs.*

Access list configuration in the numbered format follows a simple set of rules:

■ Access list *match* statements are entered using multiple *access-list* commands with the same number to create multistatement lists.

■ Access list *match* statements are processed in the order entered, with the first statement to match a packet being used.

■ An implicit *deny* is included at the end of every ACL, meaning that once applied to an interface, all packets that do not match any *permit* statements in an ACL will be automatically dropped.

■ Individual statements in numbered access lists cannot be modified. To remove an ACL statement, you must use the *no access-list [number]* command, causing all statements associated with that ACL to be removed.

You will examine these rules in detail as you explore the *access-list* commands. For starters, let's examine how to build a standard numbered IP access list.

Standard Access Lists

The syntax for the *access-list* command when building a standard numbered list is fairly simple: *access-list [access list number] [deny | permit] [source] [source wildcard mask] [log]*. The number must, of course, be in the range of 1–99. The *deny | permit* section specifies the action to take (drop or forward) for packets meeting this *match* statement. The *source* and *source wildcard mask* sections define which packets to match based on source address, and the optional *log* parameter instructs the IOS to log packets matching this access list to the syslog facility.

The only tricky parts to a standard IP access list are the *source* address and *wildcard mask* sections. The source address is the portion of the source address you wish to match. For instance, if you wanted to match all source addresses for the 172.16.0.0 network, you would enter *172.16.0.0* as the source address. Truthfully, you could put anything you wanted in the last two octets of the address (as long as you configure the mask correctly), but there is no point in entering a full address when you simply want to match the first two octets. The mask actually determines how much of the source address entered to match. The mask is written in *wildcard* format, which basically means that it is backward from what you would normally expect. In a wildcard mask, the portions of the address that are described by binary 0 in the mask must match exactly, while the portions that are binary 1 are ignored.

For instance, to match 172.16.0.0–172.16.255.255, you would enter the source address of 172.16.0.0 with a wildcard mask of 0.0.255.255. By the same token, to match every IP address between 192.168.1.128 and 192.168.1.255, you would enter a source address of 192.168.1.128 with a wildcard mask of 0.0.0.127. This combination matches the selected IPs because, in binary, all bits that are binary 0 must match the chosen source IP (192.168.1.128), while all binary 1's are allowed to be different. When converted to binary, this combination yields the following:

```
IP - 192.168.1.128:11000000.10101000.00000001.10000000
IP - 192.168.1.255:11000000.10101000.00000001.11111111
Mask - 0.0.0.127  :00000000.00000000.00000000.01111111
```

Based on this information, if you wanted to configure a packet filter using standard ACLs to block all transmissions from 192.168.1.1, allow all communications from the rest of the 192.168.1.0 network, and deny all others, you would create the packet filter with the following commands:

```
Router(config)#access-list 1 deny 192.168.1.1 0.0.0.0
Router(config)#access-list 1 permit 192.168.1.0 0.0.0.255
```

To specify a single host, you may also use the host *keyword in the ACL. For instance, in the previous command, rather than typing* access-list 1 deny 192.168.1.1 0.0.0.0, *you could have typed* access-list 1 deny host 192.168.1.1.

To ensure that this access list has been created, you can use the *show ip access-list* or *show access-lists* commands, like so:

```
Router# show access-lists
Standard IP access list 1
    deny   192.168.1.1
    permit 192.168.1.0, wildcard bits 0.0.0.255
Router# show ip access-list
Standard IP access list 1
    deny   192.168.1.1
    permit 192.168.1.0, wildcard bits 0.0.0.255
Router#
```

The order of the commands in an ACL is very important. For instance, if you reordered these statements, 192.168.1.1 would be allowed to communicate because that address matches 192.168.1.0 0.0.0.255. Remember, an ACL simply uses the first matching statement and ignores the rest.

Therefore, typically, you will want to put the most specific entries toward the top of the list. The trick is to examine the order of the list in your mind, one goal at a time, and ensure that all goals are satisfied by the list. This process necessarily becomes more difficult, however, the more complicated the list becomes.

For example, imagine that you wanted to configure a list that performs the following tasks:

- Allow all addresses in the range of 192.168.1.64 through 192.168.1.127.
- Allow 192.168.1.1 through 192.168.1.3.
- Allow all addresses in the range of 10.0.2.0 through 10.255.255.255.
- Deny all other addresses.

CISCO ROUTING

In this case, you might decide to try the following access list to meet these goals:

```
Router(config)#access-list 1 permit 10.0.0.0 0.255.255.255
Router(config)#access-list 1 permit 192.168.1.64 0.0.0.127
Router(config)#access-list 1 permit host 192.168.1.1
Router(config)#access-list 1 permit host 192.168.1.2
Router(config)#access-list 1 permit host 192.168.1.3
Router(config)#access-list 1 deny 10.0.1.0 0.0.0.255
```

Can you recognize the problems with this list? Let's examine each goal in order just to make sure you caught all of the issues.

First, you want to allow 192.168.1.64 through 127. Looking at the list, the second statement meets this goal. The statement of *access-list 1 permit 192.168.1.64 0.0.0.127* matches all IP addresses between 192.168.1.0 and 192.168.1.127, however.

The second goal was to allow 192.168.1.1 through 192.168.1.3 to communicate. Although the third, fourth, and fifth statements do this, they are not the most efficient way to accomplish this task—and besides, the second statement in the list would be matched before these statements are reached.

The third goal was to allow all addresses in the range of 10.0.2.0 through 10.255.255.255 to communicate. Although the first statement, *access-list 1 permit 10.0.0.0 0.255.255.255*, accomplishes this task, it matches all addresses from the 10.0.0.0 network.

The final goal was to deny all packets, which is not accomplished because of the following reasons:

- The second statement allows 192.168.1.0 through 192.168.1.63 to communicate.

- The first statement allows all addresses from the 10.0.0.0 network (10.0.0.0 through 10.255.255.255) to communicate, which is not the specified range.

To eliminate these problems, you would need to rebuild the list, like so:

```
Router(config)#access-list 1 deny 10.0.0.0 0.0.1.255
Router(config)#access-list 1 permit 10.0.0.0 0.255.255.255
Router(config)#access-list 1 permit 192.168.1.64 0.0.0.63
Router(config)#access-list 1 permit 192.168.1.0 0.0.0.3
```

Notice that, this time, you can accomplish the task with only four statements. The first statement denies the IP address range from 10.0.0.0 through 10.0.1.255, because the address and mask matches as follows:

```
IP - 10.0.0.0    :00001010.00000000.0000000.00000000
IP - 10.0.1.255 :00001010.00000000.00000001.11111111
Mask - 0.0.1.255:00000000.00000000.00000001.11111111
```

The second statement allows all addresses on the 10.0.0.0 network that do not match the first statement. The third statement permits all packets in the range of 192.168.1.64 through 192.168.1.127, as shown here:

```
IP - 192.168.1.64 :11000000.10101000.00000001.01000000
IP - 192.168.1.127:11000000.10101000.00000001.01111111
Mask - 0.0.0.63   :00000000.00000000.00000000.01111111
```

The fourth statement allows all packets from 192.168.1.1 through 192.168.1.3, as shown here:

```
IP - 192.168.1.1:11000000.10101000.00000001.00000001
IP - 192.168.1.3:11000000.10101000.00000001.00000011
Mask - 0.0.0.3  :00000000.00000000.00000000.00000011
```

Note *The last filter actually matches 192.168.1.0 as well; but because this address is unusable in a nonsupernetted class C network, there is no point in denying the address specifically.*

The implicit *deny* at the end of the access list takes care of the last goal of denying all other packets.

Once you have created your access list, you need to apply the list to an interface and choose a direction by using the *ip access-group [access list number or name] [in | out]* command. Remember that when applying an access list, you typically want to apply the list as close to the source of the packet as possible. So, if you wanted to use this list to match internal users entering the router, you would apply the list to the internal router interface with the direction of incoming, like so:

```
Router(config-if)# ip access-group 1 in
```

Note *Be aware that only a single access list can be applied to any given interface for inbound or outbound traffic (one ACL per direction). Consequently, you need to ensure that all required goals can be met with a single ACL.*

For a standard named access list, the process is almost identical, except for the process of modifying the ACL. When entering the ACL, you use the command *ip access-list standard [name]*. This command switches you to access list config mode for the named access list, like so:

```
Router(config)#ip access-list standard test
Router(config-std-nacl)#
```

Once in access list config mode, you enter the access list parameters using *permit* or *deny* statements, like so:

```
Router(config-std-nacl)# deny 10.0.0.0 0.0.255.255
Router(config-std-nacl)# permit 172.16.0 0.0.255.255
```

These statements are processed in the order entered, just like a numbered access list. The only difference is that, unlike a numbered access list, you may delete individual commands in a named access list by using specific *no deny* or *no permit* statements, like so:

```
Router#show access-lists
Standard IP access list test
    deny   10.0.0.0, wildcard bits 0.0.255.255
    permit 172.16.0.0, wildcard bits 0.0.255.255
Router#config t
Enter configuration commands, one per line.  End with CNTL/Z.
Router(config)#ip access-list standard test
Router(config-std-nacl)#no deny 10.0.0.0 0.0.255.255
Router(config-std-nacl)#^Z
Router#show access-lists
1w3d: %SYS-5-CONFIG_I: Configured from console by console
Standard IP access list test
    permit 172.16.0.0, wildcard bits 0.0.255.255
Router#config t
Enter configuration commands, one per line.  End with CNTL/Z.
Router(config)#ip access-list standard test
Router(config-std-nacl)#permit 192.168.1.0 0.0.0.255
Router(config-std-nacl)#permit host 192.168.2.1
Router(config-std-nacl)#deny 10.0.0.0 0.255.255.255
Router(config-std-nacl)#permit any
Router(config-std-nacl)#^Z

Router#show access-lists
Standard IP access list test
    permit 192.168.2.1
    permit 172.16.0.0, wildcard bits 0.0.255.255
    permit 192.168.1.0, wildcard bits 0.0.0.255
    deny   10.0.0.0, wildcard bits 0.255.255.255
    permit any
Router#
```

Tip
The primary difference between a named and a numbered access list lies in the ability to use descriptive names for named access lists and the ability to modify individual entries in named access lists. This second difference, however, is mostly nullified because you can simply copy your access list *statement into Notepad from a numbered access list (displayed in a* show run *or* show star *command) and rearrange the list as you please. Once you are done, simply delete the original access list with a* no access-list [number] *command and paste the modified access list into the terminal window.*

Notice how the *host* statement rose to the top of the access list even though it was entered after the other statements. This functionality is specific to access lists pertaining to a specific host because they are more specific and should generally be placed at the top of the order. Note also the *permit any* statement. The *any* keyword is a quick way to ensure that all packets match a filter. The *any* keyword is the same as specifying an IP of 0.0.0.0 (well, any IP, really) with a 255.255.255.255 mask.

Extended Access Lists

Now that you've learned about standard ACLs, let's take a look at the complexities involved with extended ACLs. In an extended ACL, you can specify many more parameters, most notably, protocols used (including IP, TCP, and UDP), destination addresses, and ports (for TCP and UDP).

Note
Extended numbered IP ACLs can use the 100 through 199 or 2000 through 2699 ranges.

With an extended ACL, you *must* specify a protocol to match. The protocol matched can be any IP protocol number from 1 to 255, or any of the following keywords:

- IP (to match all IP packets)
- ICMP
- TCP
- UDP
- EIGRP
- IGRP
- OSPF
- Other protocols beyond the scope of this book (PIM, IPINIP, GRE, NOS, or IGMP)

When matching against TCP or UDP, some other interesting possibilities open up: you may choose to match source or destination ports. When matching ports, you have the ability to match based on five operators:

- **lt**   Less than the listed number
- **gt**   Greater than the listed number

- **eq**   Equal to the listed number
- **neq**   Not equal to the listed number
- **range**   All ports in the range specified by two numbers

> **Note**   *You may also specify ports by using keywords rather than numbers. The keywords IOS recognizes for ports include* bgp, echo, finger, ftp, ftp-data, gopher, irc, nntp, pop2, pop3, smtp, syslog, telnet, whois, *and* www.

In addition, you may specify to allow packets (or deny them, although there is really no use in doing so) based on the packet being part of a previously established session using the *established* keyword. You can use the *established* keyword for your *reflexive* statements in ACLs in this manner.

To examine how all of these additional pieces fit together in an extended ACL, let's examine a set of filtering goals and walk through how to meet them. The required goals for filtering incoming packets on the external interface of your router are listed here:

- All packets destined for 192.168.1.1 using a destination TCP port of 80 should be allowed.

- All communications originating from within the private network to external web servers using HTTP and HTTPS should be allowed.

- All incoming communications from external host 10.1.1.1 using TCP and UDP source port numbers from 22,000 through 44,000 should be allowed.

- All incoming connections to host 192.168.1.200 should be allowed.

- All nonestablished sessions entering the external interface and using well-known destination ports to host 192.168.1.100 should be allowed.

- All other traffic should be denied.

Extended ACLs Versus Reflexive ACLs

Note that an extended ACL using the *established* keyword is not the same as a true reflexive ACL. A true reflexive ACL tracks sessions and builds temporary ACL entries based on those sessions—whereas the *established* keyword simply examines the ACK and RST bits in packets, and allows any packet that meets the filter criteria with those bits set. Simply using the *established* keyword, although it can reduce the complexity of the ACLs and thwart most entry attempts, will not stop a determined hacker who understands how to manipulate the ACK or RST bits in a packet, and should not be used on high-security networks. This book does not provide full coverage of reflexive ACLs; but for more information, check out *http://www.cisco.com/univercd/cc/td/doc/product/software/ios121/121cgcr/secur_c/scprt3/s cdreflx.htm.*

To accomplish the first goal, you enter a statement similar to the following:

```
Router(config)#access-list 100 permit tcp any host 192.168.1.1 eq 80
```

The *any* keyword tells the router to match any source IP. Because the port is not specified next, all source ports are matched. The *host 192.168.1.1* section tells the router to match the destination host 192.168.1.1. Finally, *eq 80* tells the router to match the destination port of 80. This filter does exactly what is required: it matches all communications from any host destined for 192.168.1.1 using TCP destination port 80 (HTTP).

To match the next goal, you use statements similar to the following:

```
Router(config)#access-list 100 permit tcp any eq 80 any established
Router(config)#access-list 100 permit tcp any eq 443 any established
```

Because you want to allow only packets using HTTP or HTTPS to return to hosts on the internal network, you need to match the source port for 80 and 443 rather than the destination. Also, because you want return packets only from sessions previously established by these hosts, you use the *established* keyword.

To allow all incoming communications from external host 10.1.1.1 using TCP and UDP source port numbers from 22,000 through 44,000, you need to enter two statements similar to the following:

```
Router(config)#access-list 100 permit tcp 10.1.1.1 range 22000 44000 any
Router(config)#access-list 100 permit udp 10.1.1.1 range 22000 44000 any
```

To allow all incoming connections to host 192.168.1.200, you use the following statement:

```
Router(config)#access-list 100 permit ip any host 192.168.1.200
```

Finally, to allow connections using well-known destination ports to host 192.168.1.100, you use the following statement:

```
Router(config)#access-list 100 permit ip any host 192.168.1.100 lt 1024
```

Once you finish building your ACL, you apply the ACL to the external interface of the router using the standard *ip access-group [access list number or name] [in | out]* command.

Named extended ACLs follow the same guidelines as numbered ACLs, so they are not discussed specifically. Simply apply the principles of extended ACLs to the commands listed in the standard named ACL section, and you will be well on your way to creating named extended ACLs.

Summary

This chapter has examined basic ACL and firewall configuration, with the goal of providing you with the tools necessary to configure packet filtering for IP on a Cisco router. Remember, you can use ACLs for other purposes as well, so a good grasp of the properties of ACLs will serve you well in many future configuration tasks.

The final section of this book contains an appendix that is an index of all commands used in the book. I sincerely hope you enjoyed reading this book as much as I enjoyed writing it, and hope it earns a spot in your technical library for many years to come.

The Complete Reference

Cisco

Appendix

Command Reference

This appendix contains a full alphabetical list of the commands covered in this book, from Chapters 14 through 27. Page numbers have been provided so you can quickly jump to the places in the book that discuss the command.

Command	Description	Mode	Pages
[deny | permit] [protocol] [source] [source wildcard mask] [(optional) port operators] [destination] [destination wildcard mask] [(optional) port operators] [established] [log]	Enters statements into an extended named IP access list.	Access list config	1037
[deny | permit] [source] [source-wildcard] [log]	Enters statements into a standard named IP access list.	Access list config	1037
accept-lifetime [start time] [end time]	Configures the times when the key will be accepted by the router.	Key chain config	739, 740
access-list [access list number] [deny | permit] [protocol] [source] [source wildcard mask] [(optional) port operators] [destination] [destination wildcard mask] [(optional) port operators] [established] [log]	Creates an extended numbered IP access list.	Global config	1036
access-list [access list number] [deny | permit] [source] [source wildcard mask] [log]	Creates a standard numbered IP access list.	Global config	1036, 1038
alias [mode] [alias] [command]	Creates a new command alias.	Global config	448–449
area [area ID] authentication [message-digest]	Enables authentication for the area.	Router config	1005, 1006
area [area ID] nssa [no-summary] [default-information-originate]	Configures an area as an NSSA and controls distribution of Type 7 LSAs.	Router config	1005, 1006

Table A-1.   *Command Index from* [deny \| permit] *to* area

Command	Description	Mode	Pages
area [area ID] range [ip address] [mask] [advertise | not-advertise]	Summarizes or suppresses a range of network advertisements at an area boundary.	Router config	1006, 1007–1008, 1009
area [area ID] stub [no-summary]	Enables an area as a stub or totally stubby area.	Router config	1005, 1006
area [area ID] virtual-link [router ID]	Creates a virtual link.	Router config	1010
arp [IP address] [MAC address] [ARP type] [interface name and number]	Adds a static ARP entry.	Global config	449
auto-summary	Enables autosumarization.	Router config	743, 744, 899
bandwidth [bandwidth in Kbps]	Defines the amount of bandwidth to reserve for a class (CBWFQ).	Policy map-class config	660
bandwidth [speed in Kbps]	Changes the bandwidth applied to routing metrics for a given interface.	Interface config	765, 766, 886, 891
banner [type] [delimiter]	Creates a banner message.	Global config	450–451
boot [bootstrap | buffersize | config | host | network | system] [source or modifiers]	Sets booting parameters.	Global config	384, 451–453, 460, 535
bridge [group number] protocol [STP protocol]	Configures the port to be a member of the specified bridge group.	Interface config	606
bridge [group number] route [protocol]	Enables routing for the bridge group.	Interface config	606, 609
bridge [irb | crb]	Configures the MLS mode.	Global config	606, 608
cdp advertise-v2	Enables CDP version 2 advertisements.	Global config	453
cdp enable	Enables CDP globally. (See also *cdp run*.)	Enable	394
cdp enable	Enables CDP on a specific interface.	Interface config	465

Table A-1. *Command Index from* area *to* cdp enable

Command	Description	Mode	Pages
cdp holdtime [time in seconds]	Sets the CDP holdtime.	Enable	398, 453
cdp run	Enables CDP globally. (See also *cdp enable.*)	Global config	453
cdp timer [time in seconds]	Sets the CDP update interval.	Enable	398, 453
channel-group [number]	Adds an interface to a Fast Etherchannel or Fast Gigachannel group.	Interface config	701
class [class name]	Inserts a class map into the policy map and enters policy map–class config mode (CBWFQ).	Policy map config	660, 661
class-map [name of class]	Defines a class map and enters class-map config mode (CBWFQ).	Global config	657
clear [target] [modifiers]	Erases information (usually statistical information).	Enable	432–433
*clear alias [alias name	all]*	Clears one or all command aliases.	Set-based	473
*clear arp [dynamic	static	permanent	(IP address)	all]*	Clears one or more entries from the ARP cache.	Set-based	473
clear banner motd	Removes the message of the day (MOTD) banner.	Set-based	474
clear boot auto-config [module]	Clears the list of configuration files for bootup.	Set-based	474
clear boot system all [module]	Clears the entire boot environment settings.	Set-based	474
clear boot system flash [boot device (bootflash | slot0 | slot1)] [module]	Clears the switch's boot image preference.	Set-based	474
*clear config [module	rmon	snmp	all]*	Clears configuration files on the switch.	Set-based	474–475, 580

Table A-1.   *Command Index from* cdp holdtime *to* clear config

Command	Description	Mode	Pages
clear counters	Clears statistical MAC address and port counters.	Set-based	475–476
*clear ip alias [alias name	all]*	Clears IP aliases on the switch.	Set-based	476
clear ip dns domain	Clears the domain name used for DNS queries.	Set-based	476
*clear ip dns server [server ip address	all]*	Clears the list of DNS servers.	Set-based	476
*clear ip eigrp neighbors [(optional) ip address	(optional) interface type and number]*	Clears the EIGRP neighbor table.	Enable	890–891
*clear ip permit [ip address	all] [subnet mask] [telnet	snmp	all]*	Clears the IP permit list.	Set-based	476
*clear ip route [destination	all] [gateway]*	Clears routing table entries.	Set-based	476
clear log [module]	Clears the system error log.	Set-based	476–477
clear log command [module]	Clears the system command log (command history).	Set-based	477
clear logging buffer	Clears the syslog buffer.	Set-based	477
clear logging server [server ip address]	Clears servers from the syslog server list.	Set-based	477
clear mls entry	Removes one or more entries from the flow cache.	Set-based	625
clear mls include	Removes one or more MLS-RP entries.	Set-based	625
clear timezone	Reverts to the default time zone (UTC).	Set-based	477
client [ip address] [mask]	Restricts access to the virtual server to specified IP addresses.	Virtual server config	654, 655

Table A-1.   *Command Index from* clear counters *to* client

Command	Description	Mode	Pages
clock [time in the format of hh:mm:ss military time] [month] [day] [year in yyyy format]	Sets the local time and date on a device.	Enable	433–434, 526
clock [timezone | summer-time] [value]	Sets clock parameters.	Global config	454
config-register [hex value]	Sets the configuration register.	Global config	390, 451, 454
configure [server name or ip address] [file] [optional RCP]	Configures the switch from a file located on a remote server.	Set-based	477–478
configure [terminal | memory | network]	Switches the CLI to global configuration mode from enable mode.	Enable	377–378
connect [remote host]	Opens a Telnet session to a remote host. (See also telnet.)	User	400–401, 427
copy [(file path) | tftp | RCP | flash | config | cfg1 | cfg2 | all] [(file path) | tftp | RCP | flash | config | cfg1 | cfg2]	Copies configuration files.	Set-based	478–479, 485, 537
copy [startup-config | running-config | tftp] [startup-config | running-config | tftp]	Copies a configuration file. (Syntax is from:to; see also write.)	Enable	377, 379, 380, 381, 536–537
custom-queue-list [number of queue list]	Defines a custom queue and applies the queue to the interface (CQ).	Interface config	663, 666
debug [target] [subtarget or modifiers]	Enables advanced diagnostics (debugging).	Enable	434–440
debug eigrp fsm	Enables debugging of EIGRP finite state machine events. Very useful in examining DUAL's operation.	Enable	894

Table A-1.   *Command Index from* clock *to* debug eigrp fsm

Command	Description	Mode	Pages
debug eigrp neighbors	Enables debugging of neighbor changes (neighbor adjacency establishment and termination).	Enable	894
*debug eigrp packets [ack	hello	query	reply	retry	stub	terse	update	verbose]*	Enables debugging of individual EIGRP packet types. Simply entering *debug eigrp packets* enables all possible packet debugging. Be very careful when enabling this debug, because EIGRP can generate massive numbers of packets in a very short period.	Enable	894–895, 901
debug ip eigrp	Displays all IP-related EIGRP debugging information available.	Enable	890, 891, 894
debug ip eigrp [AS number]	Displays all IP-related EIGRP debugging information available for a specific AS.	Enable	894
debug ip eigrp neighbor [AS number] [ip address]	Filters other EIGRP debugs to include messages only from a specific neighbor.	Enable	894
debug ip eigrp notifications	Enables debugging of EIGRP input event notifications.	Enable	894
debug ip eigrp summary	Enables debugging of EIGRP summary routes.	Enable	894
debug ip igrp events	Debugs IGRP updates.	Enable	797
debug ip igrp transactions	Debugs individual routes in IGRP updates.	Enable	797, 798, 805, 807
debug ip ospf adj	Debugs the OSPF adjacency establishment process.	Enable	1013, 1018

Table A-1. *Command Index from* debug eigrp neighbors *to* debug ip ospf adj

Command	Description	Mode	Pages
debug ip ospf events	Debugs SPF events (adjacencies, calculations, and so on).	Enable	1013
debug ip ospf packets	Debugs OSPF packet reception.	Enable	1013
debug ip rip	Debugs RIP routing updates.	Enable	744
debug mls rp [target]	Enables MLS debugging.	Enable	618, 621
debug spantree events	Debugs topology changes.	Enable	565
debug spantree tree	Debugs BPDUs.	Enable	565
debug sw-vlan vtp	Debugs VTP packets.	Enable	586
default [target] [subtarget]	Sets a command back to the defaults.	Global config	454–455
default-information-originate [always] metric [metric] metric-type [1 | 2]	Causes an ASBR to redistribute a default route into the OSPF area.	Router config	1006, 1007
default-metric [bandwidth in Kbps] [delay in tens of microseconds] [reliability] [load] [mtu in bytes]	Configures the default metric used for routes redistributed into this IGRP AS.	Router config	775
default-metric [metric]	Sets the default metric used for redistributed routes.	Router config	1006, 1009
delay [tens of microseconds]	Changes the delay applied to metrics for IGRP and EIGRP for a given interface.	Interface config	765, 766, 886
delete [file system]: [filename]	Deletes a file. May also require the *squeeze* command.	Enable	440–441, 580
delete [module number]/[device]:[filename]	Deletes configuration files.	Set-based	479–480
dir [file system]	Displays the list of files in a given file system.	Enable	441
disable	Exits enable mode.	Enable	441
disconnect [number of session]	Disconnects a Telnet session to a remote system.	User	401–402, 403

Table A-1. *Command Index from* debug ip ospf events *to* disconnect

Command	Description	Mode	Pages
distance ospf [intra-area] [interarea] [external]	Changes the default values for OSPF administrative distances.	Router config	1010, 1012
distribute-list [access list] [in | out] [(optional) interface type and number]	Applies a distribute list to filter routing updates.	Router config	738, 777–778, 780–781, 806, 1007
duplex [half | full | auto]	Sets the duplex of a link.	Interface config	699
dxi map [protocol] [address] [vpi] [vci] [broadcast]	Maps protocol addresses to ATM VCs.	Interface config	707
dxi pvc [vpi] [vci] [snap | nlpid | mux]	Configures PVC on serial ATM interfaces.	Interface config	707
eigrp log-neighbor-changes	Instructs EIGRP to log neighbor adjacency changes.	Router config	890, 891, 901
eigrp stub [receive-only | connected | static | summary]	Configures the router as a stub router for the EIGRP AS.	Router config	888, 904, 906
enable [0–15]	Switches the CLI to enable or privileged exec mode.	User	376, 377, 403
enable [password | secret] [password]	Sets the enable password (unencrypted) or secret (encrypted).	Global config	455–456, 531–532
encapsulation [isl | dot1q] [vlan number]	Sets the tagging method to use and the VLAN to associate with the subinterface.	Interface config	614, 700
encapsulation [type]	Sets the IPX encapsulation type.	Interface config	698, 699
encapsulation atm-dxi	Configures encapsulation on an ATM interface.	Interface config	707
encapsulation frame-relay [ietf]	Sets the Frame Relay encapsulation for a serial interface. Only add the IETF keyword to use IETF encapsulation.	Interface config	702

Table A-1.   *Command Index from* distance ospf *to* encapsulation frame-relay

Command	Description	Mode	Pages
end	Switches back to enable mode from any config mode.	Config	378, 456
erase [file system]	Erases all files in the destination file system of class B and C file systems.	Enable	440, 441–442, 580
exit	Switches back to the previous mode from any config mode.	Config	376, 378, 401, 403, 456
faildetect numconns [number] numclients [number]	Sets failure detection parameters for the server.	Real server config	653
fair-queue [congestive discard threshold] [dynamic queues] [reservable queues]	Configures and enables WFQ (WFQ).	Interface config	656–657
fair-queue [number of queues]	Specifies the maximum number of dynamic queues for the class-default class (CBWFQ).	Policy map config	661
frame-relay interface-dlci [dlci number]	Applies a DLCI to the interface or subinterface.	Interface config	702, 703
frame-relay map [layer 3 protocol] [layer 3 protocol address] [dlci] [broadcast] [ietf | cisco]	Maps a Frame Relay DLCI to an IP address.	Interface config	702, 703, 707, 971, 1003
full-duplex	Forces full-duplex operation on an interface (1900 series IOS).	Interface config	465–466
half-duplex	Forces half-duplex operation on an interface (1900 series IOS).	Interface config	465–466
history	Shows the command history.	Set-based	480
hostname [name]	Sets the router's host name. (See also *prompt*.)	Global config	456–457, 460, 524, 525

Table A-1. *Command Index from* end *to* hostname

APPENDIX

Command	Description	Mode	Pages
inservice	Enables the server to participate in the SLB server farm.	Real server config	651, 655
inservice	Enables the virtual server.	Virtual server config	652, 655
interface [interface name and number]	Switches the CLI from global config mode to interface config mode.	Config	378
interface bvi [bridge group number]	Enters interface config mode for a bridge virtual interface (BVI).	Global config	608
interface loopback [number]	Creates a software loopback interface (a virtual interface). A loopback interface's IP address is chosen for the router ID in lieu of the *router-id command.*	Global config	997
interface port-channel [number]	Creates a Fast Etherchannel or Fast Gigachannel link.	Global config	700
interface serial [number].[subinterface number] multipoint	Creates a Frame Relay multipoint subinterface.	Interface config	704
interface serial [number].[subinterface number] point-to-point	Creates a Frame Relay point-to-point subinterface.	Interface config	703
ip [target] [subtarget | modifier | or value]	Configures IP-specific settings.	Config	457–459, 466–467, 468
ip access-group [access list number or name] [in | out]	Applies an access list to an interface.	Interface config	1037, 1041, 1045
ip access-list extended [name]	Creates an extended named IP access list.	Global config	1037
ip access-list standard [name]	Creates a standard named IP access list.	Global config	1037, 1041

Table A-1.   *Command Index from* inservice *to* ip access-list standard

Command	Description	Mode	Pages
ip address [address] [mask]	Sets the IP address of the entire switch (switch IOS).	Global config	524, 525, 699
ip address [address] [mask] [secondary (optional)]	Sets the primary and/or secondary IP address on an interface.	Interface config	466, 606, 699
ip address [ip address] [subnet mask]	Configures the IP address of the loopback interface.	Interface config (loopback interface)	997, 998
ip bandwidth-percent eigrp [as number] [interval in seconds]	Configures the maximum bandwidth usable by EIGRP.	Interface config	886
ip classless	Configures classless route matching.	Global config	709
ip default-network [network number]	Sets a default network.	Global config	769, 773, 885
ip domain-list [domain]	Appends domain names to name queries from the router.	Global config	457–458
ip domain-lookup	Enables DNS name resolution (making the router a resolver or client to a DNS server).	Global config	458
ip hello-interval eigrp [as number] [interval in seconds]	Sets the hello interval used by EIGRP.	Router config	885
ip hold-time eigrp [as number] [interval in seconds]	Sets the holdtime used by EIGRP.	Router config	886
ip host [host name] [IP address]	Adds entries to the hosts table.	Global config	458
ip http [access-class | authentication | port | server] [modifiers or values]	Configures the web-based management interface on a device.	Global config	458–459
ip mask-reply	Enables IP subnet mask query replies.	Interface config	466
ip mtu [size in bytes]	Sets the MTU for IP on a specific interface. (See also *mtu*.)	Interface config	466, 468

Table A-1.   *Command Index from* ip address *to* ip mtu

Command	Description	Mode	Pages
ip name-server [IP address for preferred server] [IP address for secondary server] [IP address for tertiary server] [etc.]	Configures the DNS server or servers the router will use for name resolution.	Global config	458, 459
ip ospf authentication [message-digest | null]	Enables or disables authentication for a specific interface, overriding the authentication configured for the area as a whole.	Interface config	1005, 1006
ip ospf authentication-key [password]	Specifies the plaintext password to be used for each interface.	Interface config	1004, 1006
ip ospf cost [cost]	Sets an arbitrary cost for an interface, overriding the default settings that are configured based on interface bandwidth.	Interface config	1000, 1001
ip ospf dead-interval [seconds]	Sets the dead interval for an interface.	Interface config	1000, 1002
ip ospf hello-interval [seconds]	Sets the hello interval for an interface.	Interface config	1000, 1002
ip ospf message-digest-key [key ID] md5 [key]	Specifies the MD5 password to be used for each interface.	Interface config	1004, 1005, 1006
ip ospf network [broadcast | nonbroadcast | point-to-multipoint (nonbroadcast)]	Defines the network type so that OSPF can choose the appropriate neighbor establishment technique.	Interface config	1002–1003
ip ospf priority [priority]	Sets the priority of the router (the likelihood of the router being chosen as DR in an election) for the network(s) connected to the interface.	Interface config	1000, 1002
ip ospf retransmit-interval [seconds]	Sets the retransmit interval for the interface.	Interface config	1001

Table A-1. *Command Index from* ip name-server *to* ip ospf retransmit-interval

Command	Description	Mode	Pages
ip ospf transmit-delay [seconds]	Sets the amount of time added to the age of all LSAs leaving an interface.	Interface config	1002
ip redirects	Enables IP redirect messages.	Interface config	466
ip rip [send version | receive version] [1 | 2]	Modifies the version of RIP used for a specific interface.	Interface config	730
ip rip authentication key-chain [name of key chain]	Applies a key chain to a specific interface for RIP.	Interface config	740
ip rip authentication mode [md5 | text]	Configures encrypted or plaintext authentication for the RIP interface.	Interface config	740
ip route [destination network] [subnet mask] [next hop address | interface] [administrative distance] [permanent]	Configures a static route.	Global config	708, 709
ip slb natpool [pool name] [first IP address in pool] [last IP address in pool] netmask [subnet mask for pool]	Defines the address pool used for translations.	Global config	651
ip slb serverfarm [farm name]	Defines the server farm and enters server farm config mode.	Global config	651
ip slb vserver [virtual server name]	Defines the virtual server and enters virtual server config mode.	Global config	651
ip split-horizon	Enables split horizon.	Interface config	738, 767, 806
ip summary-address eigrp [as number] [summary address] [subnet mask]	Configures a manual summary address to be used for routes exiting a specific interface.	Interface config	887
ip summary-address rip [ip address] [mask]	Manually summarizes the routes advertised out the interface.	Interface config	743

Table A-1.   *Command Index from* ip ospf transmit-delay *to* ip summary-address rip

Command	Description	Mode	Pages
ip unreachables	Enables IP-unreachable messages.	Interface config	466
ip verify unicast reverse-path	Enables verification of IP source addresses (used to thwart address spoofing).	Interface config	467
ipx network [network number] encapsulation [type]	Sets the IPX encapsulation type for a specific network.	Interface config	698
key [key number]	Defines a specific key in the key chain.	Key chain config	739
key chain [name of chain]	Defines a key chain and enters key chain config mode.	Global config	739
key-string [password]	Configures the password for the key.	Key chain config	739
line [target] [line number(s)]	Enters line config mode of a specific line.	Global config	459, 532
lock	Locks a session.	Enable	442
*logging [(host name or IP address)	buffered	console	facility	history	monitor	on	source-interface	trap	synchronous]*	Sets syslog logging parameters.	Global config	382–384, 459, 529–531
logging dlci-status-change	Enables DLCI status change syslog messages for an interface.	Interface config	467
logging event link-status	Enables link status change syslog messages for an interface.	Interface config	467
logging subif-status-change	Enables syslog messages for subinterface status changes.	Interface config	467
login	Forces a line to use the configured login password.	Line config	532–533

Table A-1. *Command Index from* ip unreachables *to* login

Command	Description	Mode	Pages
loopback	Physically places a port in a loopback state.	Interface config	467
mac-address	Configures a custom MAC address.	Interface config	467–468
*match access-group [access list number	name (access list name)]*	Matches packets based on a named or numbered access list (CBWFQ).	Class map config	658
match input-interface [interface name and number]	Matches packets based on the interface the packet was received on (CBWFQ).	Class map config	658, 659
match ip-precedence [up to four precedence values, separated by spaces]	Matches packets based on the IP precedence of the packet (CBWFQ).	Class map config	658, 659
match protocol [protocol name]	Matches packets based on the protocol used (CBWFQ).	Class map config	658, 659
maxconns [number]	Sets the maximum number of connections to establish with this server.	Real server config	653
maximum-paths [number of paths]	Configures the maximum number of paths to load balance across.	Router config	738, 762, 767, 804
metric maximum-hops [number of hops]	Changes the maximum allowed hops for IGRP.	Router config	769
metric weights [tos weight] [K1 weight] [K2 weight] [K3 weight] [K4 weight] [K5 weight]	Sets the K values for IGRP and EIGRP.	Router config	768, 885, 902
mls aging fast threshold [number of packets] time [time in seconds]	Sets the MLS aging time for fast aging.	Global config	626
mls aging normal [time in seconds]	Sets the MLS aging time.	Global config	626

Table A-1.   *Command Index from* loopback *to* mls aging normal

Command	Description	Mode	Pages
*mls rp [ip	ipx]*	Enables the MLSP on the route processor (MLS-RP).	Config	599, 608–609, 611, 617
mls rp management-interface	Configures the interface used to propagate MLSP messages.	Interface config	611, 612
mls rp vlan-id [vlan number]	Configures the VLAN to associate with the port in question on the MLS-RP.	Interface config	611
mls rp vtp-domain [domain name]	Sets the VTP domain name for the MLS-RP.	Interface config	611, 614
mtu [size in bytes]	Configures the MTU for all protocols on an interface. (See also *ip mtu*.)	Interface config	468
name-connection	Names a suspended Telnet session.	User	403, 405
nat client [pool]	Enables NAT for the client connections.	Server farm config	651
nat server	Enables NAT for the server farm.	Server farm config	651, 655
neighbor [ip address]	Configures unicast updates to a specific neighbor (RIP and IGRP).	Router config	732, 767, 807
neighbor [ip address] [priority (priority)] [poll-interval (seconds)] [cost (cost)]	Defines neighbors in OSPF. Used when broadcast emulation is not available or possible.	Router config	1003, 1004
network [IP address] [wildcard mask] area [area number]	Enables OSPF on specific networks. Unlike most other routing protocols, the network command in OSPF has many possibilities.	Router config	996–998

Table A-1. *Command Index from* mls rp *to* network

Command	Description	Mode	Pages
network [network address]	Enables a routing protocol for a given network.	Router config	729, 731, 764, 765, 772–773, 788, 807, 808, 885, 896
no [command]	Negates a command.	Varies	376, 382, 383, 394, 398
ntp master [stratum]	Configures the device to become the NTP master.	Global config	526
ntp server [ip address of NTP server]	Configures the device to use an NTP server.	Global config	526
offset-list [(optional) access list] [in | out] [offset] [(optional) interface type and number]	Applies a metric offset to routes matching the statement.	Router config	733–735, 767
ospf auto-cost reference-bandwidth [bandwidth in Mbps]	Changes the base cost for OSPF metric calculation.	Router config	1010, 1011
ospf log-adj-changes	Provides basic adjacency change information.	Router config	1013, 1017
passive-interface [interface type and number]	Configures a specific interface to be passive.	Router config	731–732, 765, 767, 776
password [password]	Configures a login password for a line.	Line config	532–533
ping (with no modifiers)	Performs an extended ping.	Enable	442–443
ping [IP address | host name]	Performs a standard ping.	User	403–404
ping -s [ip address] [packet size] [packet count]	Performs an extended ping.	Set-based	480–481
policy-map [policy name]	Defines a policy map and enters policy map config mode (CBWFQ).	Global config	659

Table A-1. *Command Index from* network *to* policy-map

Command	Description	Mode	Pages
port group [number]	Configures the port to be a member of the specified port group.	Interface config	562
predictor [leastconns | roundrobin]	Sets the load-balancing method.	Server farm config	653
priority-group [group number]	Applies queuing to an interface (PQ).	Interface config	669
priority-list [list number] default [high | medium | normal | low]	Defines the default queue (PQ).	Global config	668–669
priority-list [list number] interface [interface name and number] [high | medium | normal | low]	Matches packets based on input interface (PQ).	Global config	668
priority-list [list number] protocol [protocol name] [high | medium | normal | low] queue-keyword [modifier] [value]	Matches packets based on protocol information (PQ).	Global config	668
priority-list [list number] queue-limit [high limit] [medium limit] [normal limit] [low limit]	Defines the packet size for each queue (PQ).	Global config	668
privilege [mode] [level]	Sets additional privilege modes and levels.	Global config	459
prompt [text for prompt]	Sets the system prompt and host name. (See also *hostname*.)	Global config	366, 456, 460
protocol [protocol] [address] [broadcast]	Configures an IP address for a PVC on an integrated ATM interface.	ATM VC config	708
pvc [name] [vpi/vci]	Configures a PVC on an integrated ATM interface.	Interface config	708
queue-limit [number of packets]	Specifies the maximum number of enqueued packets for the class (CBWFQ).	Policy map-class config	660, 661

Table A-1. *Command Index from* port group *to* queue-limit

Command	Description	Mode	Pages
queue-limit [number of packets]	Specifies the maximum number of enqueued packets for the class for each dynamic queue in the class-default class (CBWFQ).	Policy map config	660, 661
queue-list [list number] default-queue [queue number]	Defines the default queue (CQ).	Global config	665
queue-list [list number] interface [interface name and number] [queue number]	Matches packets based on input interface (CQ).	Global config	664
queue-list [list number] protocol [protocol name] [queue number] queue-keyword [modifier] [value]	Matches packets based on protocol information (CQ).	Global config	664
queue-list [queue list number] queue [queue number]	Defines the queues to be used in the custom queue list (CQ).	Global config	663
queue-list [queue list number] queue [queue number] byte-count [number of bytes]	Specifies the maximum number of bytes for the specific queue (CQ).	Global config	664
queue-list [queue list number] queue [queue number] limit [maximum packets]	Specifies the maximum number of enqueued packets for the specific queue (CQ).	Global config	663
quit	Ends your session.	Set-based	481
real [ip address]	Defines the servers that will participate in the SLB server farm and enters real server config mode.	Server farm config	651

Table A-1.   *Command Index from* queue-limit *to* real

Command	Description	Mode	Pages
reassign [number]	Defines the number of unanswered SYN packets to allow before assigning connections to other servers in the farm in case of failure or overload.	Real server config	653
redistribute [protocol and AS] [(optional) metric] [(optional) 1–15]	Configures redistribution into RIP.	Router config	773–775, 786–787, 788
redistribute [protocol] [process-ID] metric [metric] metric-type [1 | 2] [match (internal | external1 | external2)] [subnets]	Redistributes routes from or to OSPF.	Router config	1007, 1008
redistribute [routing protocol] [as number or process ID] metric [bandwidth in Kbps] [delay in tens of microseconds] [reliability] [load] [mtu in bytes]	Configures redistribution of another routing protocol into IGRP.	Router config	773–775, 786–787
reload [at | in] [reason]	Reboots the device.	Enable	443–444
reset [at or in] [hh:mm] [mm:dd] [reason]	Schedules a reload of the system.	Set-based	481–482
reset [module number or system]	Reloads the system or a specific module.	Set-based	481–482
reset cancel	Cancels a scheduled reset.	Set-based	481–482
resume [name | number]	Resumes a suspended Telnet session.	User	403, 405
retry [value in seconds]	Enables auto-retry and specifies the retry interval.	Real server config	653–654
rlogin [IP address | host name]	Begins an rlogin session.	User	405
router eigrp [as number]	Enables EIGRP and enters router config mode.	Global config	885, 896

Table A-1.   *Command Index from* reassign *to* router eigrp

Command	Description	Mode	Pages
router igrp [as number]	Enables IGRP routing for the AS number in question and enters router config mode.	Global config	764–765
router ospf [process-id]	Enables OSPF on the router. Remember, OSPF process IDs are meaningful only to the router on which this command is run. Unlike EIGRP AS numbers, process IDs do not have to be the same on two routers for them to establish an adjacency.	Global config	996, 997, 998, 1005
router rip	Enables RIP and enters router config mode for RIP.	Global config	729
router-id	Forces the router ID used by the OSPF process. Overrides all other router IDs (including loopback interfaces).	Router config	926, 996, 997, 999
send [line name and number]	Sends a message to one or all lines.	Enable	444–445
send-lifetime [start time] [end time]	Configures the times in which the key will be sent by the router.	Key chain config	739, 740
serverfarm [name]	Defines which real farm of servers should be associated with the virtual server.	Virtual server config	652
service [target] [modifier or value]	Configures various system services. (Individual targets are covered separately in this appendix.)	Global config	460–464
service compress-config	Performs compression of configuration files in NVRAM.	Global config	460

Table A-1. *Command Index from* router igrp *to* service compress-config

Command	Description	Mode	Pages
service config	Enables booting from the network. (See also *boot*.)	Global config	452, 460, 535
service password-encryption	Enables encryption of all passwords.	Global config	460
service tcp-small-servers	Enables simple TCP services (most notably, ping).	Global config	461
service timestamp [debug | log] [date | uptime] [options]	Configures timestamping.	Global config	461–464
service udp-small-servers	Enables simple UDP services (most notably, ping).	Global config	464
service-module t1 clock-source [internal | line]	Sets the clock source for an integrated CSU/DSU module.	Interface config	706
service-module t1 data-coding [inverted | normal]	Sets the data coding mode for an integrated CSU/DSU module.	Interface config	706
service-module t1 fdl [att | ansi]	Sets the FDL mode for an integrated CSU/DSU module.	Interface config	706
service-module t1 framing [esf | sf]	Sets the framing for an integrated CSU/DSU module.	Interface config	706
service-module t1 lbo [none | -7.5db |-15db]	Sets the line build out for an integrated CSU/DSU module.	Interface config	706
service-module t1 linecode [ami | b8zs]	Sets the line code for an integrated CSU/DSU module.	Interface config	706
service-module t1 timeslots [all | range] speed [56 | 64]	Sets the timeslots for an integrated CSU/DSU module.	Interface config	706
service-policy output [name of policy]	Applies a CBWFQ policy to an interface (CBWFQ).	Interface config	661

Table A-1.   *Command Index from* service config *to* service-policy output

Command	Description	Mode	Pages
session [module number]	Enters a specific module's CLI.	Set-based	482–483
set alias [alias] [command including parameters]	Sets a command alias.	Set-based	484
set arp [type] [MAC address] [IP address]	Adds a static ARP entry.	Set-based	484–485
set arp agingtime [time in seconds]	Sets the timeout for dynamic ARP entries.	Set-based	485
set banner motd [delimiting character]	Sets the message of the day (MOTD) banner.	Set-based	485
set boot auto-config [configuration file list]	Sets a configuration file search order.	Set-based	474, 485, 536
set boot auto-config [recurring | nonrecurring]	Configures clearing of the config register on reload.	Set-based	485–486
set boot config-register [boot option] [value]	Configures the configuration register using keywords.	Set-based	486
set boot config-register [hex value]	Configures the configuration register using a hex code.	Set-based	486
set boot sync-now	Synchronizes the config files of the primary and redundant supervisor modules.	Set-based	486
set boot system [device]:[filename] [prepend] [module number]	Sets the IOS image used for boot.	Set-based	486–487, 536
set cdp disable	Disables CDP.	Set-based	487
set cdp enable	Enables CDP.	Set-based	487
set cdp holdtime [value]	Configures the CDP holdtime.	Set-based	487
set cdp interval [value]	Configures the CDP interval.	Set-based	487
set cdp version [value]	Configures the CDP version.	Set-based	487

Table A-1. *Command Index from* session *to* set cdp version

Command	Description	Mode	Pages
set enablepass	Configures the enable password.	Set-based	487, 534
set interface sc0 [ip address] [subnet mask] [optional broadcast address]	Configures the in-band management interface.	Set-based	487–488, 527
set interface sc0 [up | down]	Enables or disables the in-band management interface.	Set-based	487–488
set interface sc0 dhcp [release | renew]	Enables DHCP IP address configuration for the in-band management interface.	Set-based	487–488
set interface sl0 [console ip address] [remote host ip address]	Configures the out-of-band management interface.	Set-based	487–488
set interface sl0 [up | down]	Enables or disables the out-of-band management interface.	Set-based	487–488
set ip alias [alias] [ip address]	Configures an IP alias (similar to a host table entry).	Set-based	476, 488
set ip dns [enable | disable]	Enables or disables DNS resolution.	Set-based	488
set ip dns domain [domain name]	Configures the DNS domain name.	Set-based	488
set ip dns server [ip address] [primary (optional)]	Configures the DNS server address.	Set-based	488
set ip http port [port number]	Sets the port used for the web-based management interface.	Set-based	488
set ip http server [enable | disable]	Enables or disables the web-based management interface.	Set-based	488
set ip redirect [enable | disable]	Enables or disables the sending of ICMP redirect messages.	Set-based	488

Table A-1.   *Command Index from* set enablepass *to* set ip redirect

Command	Description	Mode	Pages
set ip unreachable [enable | disable]	Enables or disables the sending of ICMP destination unreachable messages.	Set-based	488
set length [number of lines | default]	Sets the number of lines per screen.	Set-based	489
set logging buffer [number of messages]	Sets the number of commands retained in the command history buffer. (See also set logging history.)	Set-based	489
set logging history [number of messages]	Sets the number of commands retained in the command history buffer. (See also set logging buffer.)	Set-based	489
set logging level [facility] [level]	Sets the severity of items logged for each facility.	Set-based	489–492
set logging server [syslog server ip address]	Configures syslog servers to send messages to.	Set-based	492, 533–534
set logging server [enable | disable]	Enables or disables the sending of syslog messages to a remote server.	Set-based	492, 533–534
set logging server facility [facility type]	Configures the type of messages to send to the syslog server.	Set-based	492
set logging server level [level to log]	Configures the severity of items to send to the syslog server.	Set-based	492
set logout [time in minutes]	Configures the length of time a session can remain idle.	Set-based	492
set mls [enable | disable]	Enables or disables MLS on the MLS-SE.	Set-based	612, 625
set mls agingtime [value in seconds]	Sets the MLS aging time.	Set-based	626

Table A-1. *Command Index from* set ip unreachable *to* set mls agingtime

Command	Description	Mode	Pages
set mls agingtime fast [time in seconds] [number of packets]	Sets the MLS aging time for fast aging.	Set-based	626
set mls include [protocol] [address of MLS-RP]	Sets the MLS-RP the MLS-SE should use.	Set-based	612, 616, 617
set module [enable | disable] [module number(s)]	Enables or disables an entire module.	Set-based	492–493
set module name [module number] [name]	Configures the friendly name shown in the IOS for a given module.	Set-based	492–493
set ntp server [ip address | name]	Configures the switch to use an NTP server.	Set-based	528
set password [password]	Configures the console password.	Set-based	493
set port [enable | disable] [module number]/[port number]	Enables or disables ports on a given module.	Set-based	493–495
set port channel [module/port list] mode [mode]	Configures the port or ports to be a member of the specified port group.	Set-based	562–563
set port duplex [module number]/[port number] [full | half]	Configures one or more ports for full- or half-duplex operation.	Set-based	493, 494
set port flowcontrol [module number]/[port number] [receive | send] [on | off | desirable]	Configures flow control for one or more ports.	Set-based	494
set port name [module number]/[port number] [name]	Configures the friendly name shown in the IOS for a given port.	Set-based	494–495
set port negotiation [module number]/[port number] [enable | disable]	Enables or disables link negotiation protocol on Gigabit Ethernet ports.	Set-based	494, 495

Table A-1. *Command Index from* set mls agingtime fast *to* set port negotiation

Command	Description	Mode	Pages
set port speed [module number]/[port number] [4 | 10 | 16 | 100 | auto]	Configures port speed for a port or set of ports.	Set-based	494, 495
set prompt [prompt]	Configures the system prompt and host name. (See also set system name.)	Set-based	495, 496
set spantree [disable | enable] [vlan list | all]	Enables or disables STP.	Set-based	554
set spantree backbonefast	Enables backbonefast on the switch.	Set-based	559, 562
set spantree guard root [module/port]	Enables root guard on one or more ports.	Set-based	558–559
set spantree portfast [module/port] [enable | disable]	Enables or disables portfast on the specified port or ports.	Set-based	561, 562
set spantree portfast bpdu-guard	Enables or disables BPDU guard on the switch.	Set-based	561
set spantree priority [value]	Manually configures the STP priority.	Set-based	556
set spantree root [vlan list] dia [diameter value]	Sets the STP diameter for the root switch.	Set-based	555–556
set spantree root secondary	Configures the switch to be the STP secondary root.	Set-based	556
set spantree uplinkfast [enable | disable]	Enables or disables uplinkfast on the switch.	Set-based	560, 562
set summertime [enable | disable] [zone]	Enables or disables daylight saving time changes.	Set-based	495
set summertime date [date to begin] [month to begin] [hh:mm to begin] [date to end] [month to end] [hh:mm to end] [time offset]	Configures daylight saving time changes. (See also set summertime recurring.)	Set-based	495

Table A-1.   *Command Index from* set port speed *to* set summertime date

Command	Description	Mode	Pages
set summertime recurring [week of month to begin] [day of month to begin] [month to begin] [hh:mm to begin] [week of month to end] [day of month to end] [month to end] [hh:mm to end] [time offset]	Configures daylight saving time changes. (See also *set summertime date.*)	Set-based	495
set system baud [speed]	Sets the baud rate of the console port.	Set-based	495
set system contact [contact name]	Sets the system contact person.	Set-based	496, 528
set system countrycode [two digit ISO-3166 country code]	Sets the country code for the system.	Set-based	496
set system location [location]	Sets descriptive text regarding the system's physical location.	Set-based	496, 528
set system name [name]	Sets the system host name. (See also *set prompt.*)	Set-based	495, 496, 527
set time [day] [mm/dd/yy] [hh:mm:ss]	Sets the local time for the switch.	Set-based	496, 528
set timezone [name] [offset in hours] [offset in minutes]	Sets the time zone used by the switch.	Set-based	496
set trunk [module/port] [desirable | on | auto | nonegotiate]	Enables trunking on a port.	Set-based	584
set trunk [module/port] [vlan list]	Configures the VLANs to trunk across a port.	Set-based	584
set vlan [number] name [name]	Defines VLANs.	Set-based	582
set vlan [vlan number] [mod/port list]	Assigns ports to a specific VLAN.	Set-based	582
set vtp domain [domain name]	Sets the VTP domain name.	Set-based	580–581, 612

Table A-1.   *Command Index from* set summertime recurring *to* set vtp domain

Command	Description	Mode	Pages
set vtp mode [client | server | transparent]	Sets the VTP mode.	Set-based	581, 612
set vtp password [password]	Sets the VTP password.	Set-based	580
set vtp pruneeligible [vlan list]	Enables VTP pruning.	Set-based	582
set vtp v2 enable	Sets the VTP version.	Set-based	581
setup	Enters setup mode.	Enable	365, 445
show [changes | current | proposed]	Shows database changes and current config and proposed changes to database.	VLAN config	586
show [target] [modifiers]	Displays information. (Specific show targets are detailed elsewhere in this appendix.)	User	405–427, 432
show access-lists [access list number or name]	Shows one or all access lists.	User exec	1037, 1039
show alias [name]	Shows aliases for commands.	Set-based	496
show aliases [mode]	Displays configured command aliases.	Enable	405–406
show arp	Displays the full ARP table, regardless of upper-layer protocol.	Enable	406, 413
show arp [ip address] [host name]	Shows the ARP table.	Set-based	496
show async	Displays information related to asynchronous serial connections.	Enable	406
show boot [module number]	Shows booting environment variables.	Set-based	496, 498
show cdp [interface | neighbor | entry]	Displays CDP information.	User	391, 392, 393, 407

Table A-1.   *Command Index from* set vtp mode *to* show cdp

Command	Description	Mode	Pages
*show cdp [neighbors	port]*	Shows Cisco Discovery Protocol information.	Set-based	497
show clock	Displays the current system time and date.	User	407–408
*show config [all	system	module]*	Shows system configuration.	Set-based	497, 498–504
show debugging	Displays debugs configured on the router.	Enable	408
*show dhcp [lease	server]*	Displays configured DHCP information.	Enable	408
*show diag [slot number	detail	summary]*	Displays detailed diagnostic information.	Enable	408–409, 426
show environment [optional modifiers]	Displays environmental information. (Voltages, fan RPM, and so on. Modifiers available depend on the device.)	Enable	409–410
show file [device]:[filename]	Displays contents of file.	Set-based	497, 504–505
*show flash [flash device	devices	all	chips	filesys]*	Shows system flash information.	Set-based	497
*show flash: [all	detail	summary	err]*	Displays flash memory contents and information.	User	410–411
show history	Shows the command history.	User	411–412
show hosts [name]	Shows the hosts table for the device.	User	412
show interface	Shows sc0 and sl0 interface information.	Set-based	497
show interface [interface type and number] [modifiers]	Displays interface-specific information (not upper-layer protocol specific).	User	412–413

Table A-1.   *Command Index from* show cdp *to* show interface

Command	Description	Mode	Pages
show interface type [number]	Shows basic interface information.	Enable	586
show interface type [number] switchport	Shows interface VLAN membership and trunking.	Enable	586
*show ip [alias	dns	http]*	Shows IP information.	Set-based	497
show ip [modifiers]	Displays TCP/IP protocol suite–specific information. (Specific *show ip* targets are detailed elsewhere in this appendix.)	User	413–417
show ip access-list [access list number or name]	Shows one or all IP access lists.	User exec	1037, 1039
show ip arp	Displays the ARP table specific to IP. (See also *show arp.*)	User	406, 413
show ip eigrp interfaces [(optional) interface type and number] [(optional) AS number]	Displays EIGRP statistics specific to each interface.	Enable	890, 891
show ip eigrp neighbors [(optional) interface type and number]	Displays the EIGRP neighbor table.	Enable	890, 892, 901, 903
show ip eigrp topology [(optional) AS number] [(optional) ip address and subnet mask]	Displays the EIGRP topology table.	Enable	890, 892–893, 900, 903
show ip eigrp traffic [(optional) AS number]	Displays EIGRP traffic statistics.	Enable	890, 893–894
show ip interface [interface name and number] [modifiers]	Displays IP-specific information on an interface.	User	413–415
show ip ospf [(optional) process-id] [(optional) area-id] database [optional modifiers]	Displays information related to the SPF database.	User	1013, 1014–1015

Table A-1.   *Command Index from* show interface type *to* show ip ospf

Command	Description	Mode	Pages
show ip ospf [process-id area-id] database	Shows the OSPF database for a specific process and area.	User	1013, 1014–1015
show ip ospf [process-id area-id] database asb-summary [link-state-id]	Shows ASBR-summary (Type 4) LSAs. Appends the link-state ID to show a specific LSA.	User	1015
show ip ospf [process-id area-id] database network [link-state-id]	Shows network (Type 2) LSAs. Appends the link-state ID to show a specific LSA.	User	1015
show ip ospf [process-id area-id] database nssa-external [link-state-id]	Shows NSSA external (Type 7) LSAs. Appends the link-state ID to show a specific LSA.	User	1015
show ip ospf [process-id area-id] database router [link-state-id]	Shows router (Type 1) LSAs. Appends the link-state ID to show a specific LSA.	User	1015
show ip ospf [process-id area-id] database summary [link-state-id]	Shows network summary (Type 3) LSAs. Appends the link-state ID to show a specific LSA.	User	1015
show ip ospf [process-id]	Displays basic OSPF information for the selected process ID.	User	1013
show ip ospf [process-id] database external [link-state-id]	Shows AS external (Type 5) LSAs. Appends the link-state ID to show a specific LSA.	User	1015
show ip ospf [process-id] summary-address	Displays configured summary addresses.	User	1013, 1017
show ip ospf border-routers	Displays information about OSPF BRs.	Enable	1013, 1014
show ip ospf interface [interface type and number]	Displays OSPF configuration for a specific interface.	User	1013, 1015, 1017, 1018

Table A-1.   *Command Index from* show ip ospf *to* show ip ospf interface

Command	Description	Mode	Pages
show ip ospf neighbor [interface type and number | neighbor router-id] [detail]	Displays information on OSPF neighbors.	User	1013, 1016
show ip ospf request-list [(optional) interface type and number] [(optional) neighbor router ID]	Displays the OSPF link-state request list.	User	1013, 1016
show ip ospf retransmission-list [(optional) interface type and number] [(optional) neighbor router ID]	Displays the OSPF link-state retransmission list.	User	1013, 1016–1017
show ip ospf virtual-links	Displays information on configured virtual links.	User	1013, 1017
show ip protocols [protocol and AS or process ID]	Shows routing protocol details for the router.	Enable	743–744
show ip rip database	Displays all networks currently advertised by RIP.	Enable	744
show ip route [address] [mask] [longer-prefixes] [protocol]	Displays the IP routing table.	User	710, 711–712, 713
show ip route ospf [process-id]	Displays OSPF routes.	User	1013
show ip slb conns [vserver (virtual server name)] [client (ip-address)] [detail]	Displays connections handled by SLB (either all connections or only those handled by a specific virtual server or from a specific client).	Enable	656
show ip slb reals [vserver (virtual server name)] [detail]	Displays configured real servers, including connection statistics.	Enable	656
show ip slb serverfarms [name (serverfarm name)] [detail]	Displays information about server farms.	Enable	656

Table A-1.   *Command Index from* show ip ospf neighbor *to* show ip slb serverfarms

Command	Description	Mode	Pages
show ip slb stats	Displays SLB statistics.	Enable	656
show ip slb sticky [client (ip-address)]	Displays current sticky connections.	Enable	656
show ip slb vservers [name (virtual server name)] [detail]	Displays configured virtual servers, including connection statistics.	Enable	656
show ip sockets	Displays open sockets (IP address/transport protocol/port number triplets).	User	415
show ip traffic	Displays IP-specific traffic statistics.	User	415–417
show ipx [modifiers]	Displays IPX/SPX protocol suite–specific information. (Specific *show ipx* targets are detailed elsewhere in this appendix.)	User	417–418
show ipx arp	Displays the ARP table specific to IPX. (See also *show arp*.)	User	
show ipx interface [interface name and number] [modifiers]	Displays IPX-specific information on an interface.	User	417
show ipx servers	Displays IPX servers known via SAP broadcasts.	User	417–418
show ipx spx-spoof	Displays IPX keepalive spoofing information (usually used for DDR connections).	User	417
show ipx traffic	Displays IPX-specific traffic statistics.	User	417
show line [line name and number | line number | summary]	Displays information about asynchronous port (or line) use.	User	418–420, 426
show log [module number]	Shows log information.	Set-based	497, 505–506

Table A-1. *Command Index from* show ip slb stats *to* show log

APPENDIX

Command	Description	Mode	Pages
show logging [buffer]	Shows system logging information.	Set-based	497, 529, 530–531, 534
show logging [history]	Shows the logging configuration and buffer.	User	420–421, 529, 530–531, 534
show mac [module number]	Shows MAC information.	Set-based	475, 497, 506–507
show memory	Displays active processes in memory.	User	421
show mls aging	Shows MLS aging statistics.	Enable	626
show mls debug	Displays MLS debugging information (highly technical).	Set-based	621
show mls entry	Displays specific MLS switching table entries.	Set-based	621, 622–623
show mls flowmask	Shows the current flow mask used by the switch.	Enable	618
*show mls include [ip	ipx]*	Displays information on configured MLS-RPs.	Set-based	621, 623
show mls ip	Displays information about MLS for IP.	Set-based	621, 622
show mls ip [target] [subtarget]	Shows various IP MLS information.	Enable	618, 619
*show mls rp [ip	ipx] [address of MLS-RP]*	Displays information about a specific MLS-RP.	Set-based	621, 623
show mls rp [target] [subtarget]	Shows various MLS information (available only on switches with integrated routing, like 2948 and 4908).	Enable	618, 619, 621
show mls statistics	Displays MLS statistical information.	Set-based	621, 623–624

Table A-1.   *Command Index from* show logging *to* show mls statistics

Command	Description	Mode	Pages
show module [module number]	Shows module information.	Set-based	497, 507
show netstat [interface | icmp | ip | stats | tcp | udp]	Shows network statistics.	Set-based	497, 508–510
show ntp	Shows NTP information.	Set-based	530
show policy-map	Shows configurations for all CBWFQ policy maps, including all configured classes.	Enable	671
show policy-map [policy name]	Shows configurations for a single CBWFQ policy map, including all configured classes. Can also show information on a single class with the optional *class* keyword.	Enable	671
show policy-map interface [interface name and number]	Shows configuration for a single CBWFQ interface, including statistics. Can also show information on a single ATM or Frame Relay VC with the optional *vc* or *dlci* keywords.	Enable	671
show port [module/port]	Shows port information (including trunking status).	Set-based	586
show port [numerous optional parameters]	Shows port information.	Set-based	475, 497, 510–515
show port trunk	Shows which ports are trunking.	Set-based	586
show privilege	Displays the current privilege level.	User	421
show proc [cpu | mem]	Shows CPU and process use.	Set-based	497
show processes [cpu | memory]	Displays information about processes running on the device.	Enable	422

Table A-1. *Command Index from* show module *to* show processes

Command	Description	Mode	Pages
show protocols [interface name and number]	Displays protocol addresses for all or a specific interface.	Enable	422–423
show queue [interface name and number]	Shows queuing information for a single interface, including statistics. Can also show information on a single ATM VC with the optional *vc* keyword.	Enable	671
show queuing	Shows either all queuing statistics, or just queuing statistics for a particular queuing strategy for all interfaces.	Enable	671
show queuing interface [interface name and number]	Shows queuing statistics for a single interface. Can also show information on a single ATM VC with the optional *vc* keyword.	Enable	672
show reset	Shows schedule reset information.	Set-based	497
show running-config	Displays the configuration in RAM.	Enable	423–424, 425, 426
show sessions	Views current Telnet sessions established by the current user instance.	User	402, 425, 446
*show snmp [sessions	pending]*	Displays SNMP agent information.	User	425
show snmp contact	Displays the configured contact person for the device.	Enable	527
show snmp location	Displays the configured device location.	Enable	527
show spanning-tree	Shows detailed information on all interfaces and VLANs.	Enable	565

Table A-1. *Command Index from* show protocols *to* show spanning-tree

Command	Description	Mode	Pages
show spanning-tree interface [number]	Shows detailed information on a specific interface.	Enable	565
show spanning-tree vlan [number]	Shows detailed information on a specific VLAN.	Enable	565
show spantree	Shows summary information.	Set-based	565
show spantree [mod/port]	Shows information on a specific port or ports.	Set-based	565
show spantree [vlan]	Shows information on a specific VLAN number.	Set-based	565
show spantree backbonefast	Shows Spanning Tree backbonefast information.	Set-based	565
show spantree blockedports	Shows ports that are blocked.	Set-based	565
show spantree portstate	Shows spanning tree state of a Token Ring port.	Set-based	565
show spantree portvlancost	Shows spanning tree port VLAN cost.	Set-based	565
show spantree statistics	Shows spanning tree statistic information.	Set-based	565
show spantree summary	Shows spanning tree summary information.	Set-based	565
show spantree uplinkfast	Shows spanning tree Uplinkfast information.	Set-based	565
show startup-config	Displays the configuration in NVRAM.	Enable	425, 446
show summertime	Shows state of summertime information.	Set-based	497
show system	Shows system information (mainly environmental information, similar to the *show environment* command).	Set-based	497, 528–529
*show tcp [interface name and number	brief	statistics]*	Displays information on TCP sessions to the device.	User	425

Table A-1.   *Command Index from* show spanning-tree interface *to* show tcp

Command	Description	Mode	Pages
show tech-support	A combination of several commands, displays most commonly required technical data.	User	426
show tech-support [config | memory | module | port]	Shows system information for tech support.	Set-based	497
show terminal	Shows the console information. (See also *show line*.)	User	426
show test [diaglevel | packetbuffer | module number]	Shows results of diagnostic tests.	Set-based	498, 515–517
show time	Shows time of day.	Set-based	498
show timezone	Shows the current time zone offset.	Set-based	498
show traffic	Shows traffic information.	Set-based	498, 518
show users	Shows active admin sessions.	Set-based	498
show version	Shows the version of the IOS running on the device, as well as uptime and other statistics.	User	364, 426–427
show version [module number]	Shows version information.	Set-based	498
show vlan	Shows all VLANs.	Enable	586
show vlan	Shows a summary of all VLANs.	Set-based	586
show vlan [vlan]	Shows details on a specific VLAN.	Set-based	586
show vlan brief	Shows a summary of all VLANs.	Enable	586
show vlan id [vlan number]	Shows details on a specific VLAN.	Enable	586

Table A-1.   *Command Index from* show tech-support *to* show vlan id

Command	Description	Mode	Pages
show vlan name [vlan name]	Shows details on a specific VLAN.	Enable	586
show vlan trunk	Shows which VLANs are being trunked.	Set-based	586
show vtp counters	Shows statistical information on VTP.	Enable	586
show vtp domain	Shows information about your VTP domain, such as the name and revision number.	Set-based	580, 582, 586
show vtp statistics	Shows statistical information on VTP.	Set-based	586
show vtp status	Shows information about your VTP domain, such as the name and revision number.	Enable	580, 582, 586
shutdown	Disables (shuts down) an interface.	Interface config	448, 468
snmp-server contact	Configures the contact person displayed for the device.	Global config	526
snmp-server location	Configures the location displayed for the device.	Global config	527
spanning-tree [vlan list]	Enables STP (standard IOS).	Global config	554
spanning-tree [vlan list] root primary diameter [value]	Sets the STP diameter for the root switch.	Global config	555–556
spanning-tree [vlan list] root secondary	Configures the switch to be the STP secondary root.	Global config	556
spanning-tree backbonefast	Enables backbonefast on the switch.	Global config	559, 562
spanning-tree portfast	Enables portfast on the specified port.	Interface config	561, 562
spanning-tree portfast bpdu-guard	Enables BPDU guard on the switch.	Global config	561

Table A-1. *Command Index from* show vlan name *to* spanning-tree portfast bpdu-guard

Command	Description	Mode	Pages
spanning-tree priority [value]	Manually configures the STP priority.	Global config	556
spanning-tree uplinkfast	Enables uplinkfast on the switch.	Global config	560, 562
spantree [vlan list]	Enables STP (1900 series IOS).	Global config	554
speed [10 | 100 | auto]	Sets the speed of a 10/100 Ethernet link.	Interface config	699
squeeze [file system]	Deletes files marked for deletion in a given class A or B file system.	Enable	440, 441, 445
sticky [threshold in seconds]	Configures sticky connections.	Virtual server config	654
summary-address [ip address] [mask] [not-advertise]	Summarizes or suppresses a range of network advertisements sent by an ASBR.	Router config	1006, 1008–1009
switchport access [vlan number]	Assigns ports to a specific VLAN.	VLAN config	582
switchport allowed vlan [add | remove] [vlan list]	Configures the VLANs to trunk across a port (standard IOS).	Interface config	584
switchport mode trunk	Enables trunking on a port (standard IOS).	Interface config	584
synguard [number of invalid SYN packets] [interval in ms]	Configures Synguard.	Virtual server config	654
telnet [ip address | name] [optional modifiers]	Establishes a Telnet session. (See also *connect*.)	User	400, 405, 427, 428–429
terminal [options]	Sets terminal configuration for the current session.	User	427, 429–431
test [target]	Performs basic operation tests against the target.	Enable	445–446

Table A-1. *Command Index from* spanning-tree priority *to* test

Command	Description	Mode	Pages
timers basic [update in seconds] [invalid in seconds] [holddown in seconds] [flush in seconds]	Modifies RIP's routing timers.	Router config	735, 767, 806
timers spf [delay] [holdtime]	Changes the default SPF route calculation timers.	Router config	1010, 1011
traceroute [protocol] [address | name]	Performs a trace route.	User	431–432
traffic-share [balanced | min]	Enables or disables load balancing.	Router config	768, 805
trunk [desirable | on | auto | nonegotiate]	Enables trunking on a port (1900 series IOS).	Interface config	584
trunk-vlan [vlan list]	Configures the VLANs to trunk across a port (1900 series IOS).	Interface config	584
undebug [target] [subtarget or modifiers]	Disables advance diagnostics (debugging). Same as *no debug*.	Enable	440
undelete [index] [file system]	Undeletes a deleted file on class A and B file systems.	Enable	441, 446
variance [multiple]	Sets the variance for unequal-cost load balancing (IGRP and EIGRP).	Router config	767–768
version [1 | 2]	Globally sets the RIP version used.	Router config	730
virtual [virtual server IP address] [tcp | udp] [protocols to load balance] service (optional) [service name (optional)]	Configures the virtual server.	Virtual server config	651–652
vlan [number] name [name]	Defines VLANs.	VLAN config	582
vlan database	Enters VLAN config mode.	Global config	580, 581
vtp [client | server | transparent]	Sets the VTP mode.	VLAN config	581

Table A-1.   *Command Index from* timers basic *to* vtp

Command	Description	Mode	Pages
vtp domain [domain name]	Sets the VTP domain name.	VLAN config	580, 581
vtp password [password]	Sets the VTP password.	VLAN config	580
vtp pruning	Enables VTP pruning.	VLAN config	582
vtp v2-mode	Sets the VTP version.	VLAN config	581
weight [weight]	Sets the weight (or preference) of the real server.	Real server config	653
where	Shows current sessions. (See also *show sessions*.)	User	446
write [memory | network | terminal]	Copies or displays configuration files. (See also *copy*.)	Enable	446

Table A-1.   *Command Index from* vtp domain *to* write

Index

See also Appendix, "Command Reference"

Symbols and Numbers

D

E

INTERNATIONAL CONTACT INFORMATION

AUSTRALIA
McGraw-Hill Book Company Australia Pty. Ltd.
TEL +61-2-9417-9899
FAX +61-2-9417-5687
http://www.mcgraw-hill.com.au
books-it_sydney@mcgraw-hill.com

CANADA
McGraw-Hill Ryerson Ltd.
TEL +905-430-5000
FAX +905-430-5020
http://www.mcgrawhill.ca

GREECE, MIDDLE EAST, NORTHERN AFRICA
McGraw-Hill Hellas
TEL +30-1-656-0990-3-4
FAX +30-1-654-5525

MEXICO (Also serving Latin America)
McGraw-Hill Interamericana Editores S.A. de C.V.
TEL +525-117-1583
FAX +525-117-1589
http://www.mcgraw-hill.com.mx
fernando_castellanos@mcgraw-hill.com

SINGAPORE (Serving Asia)
McGraw-Hill Book Company
TEL +65-863-1580
FAX +65-862-3354
http://www.mcgraw-hill.com.sg
mghasia@mcgraw-hill.com

SOUTH AFRICA
McGraw-Hill South Africa
TEL +27-11-622-7512
FAX +27-11-622-9045
robyn_swanepoel@mcgraw-hill.com

UNITED KINGDOM & EUROPE (Excluding Southern Europe)
McGraw-Hill Education Europe
TEL +44-1-628-502500
FAX +44-1-628-770224
http://www.mcgraw-hill.co.uk
computing_neurope@mcgraw-hill.com

ALL OTHER INQUIRIES Contact:
Osborne/McGraw-Hill
TEL +1-510-549-6600
FAX +1-510-883-7600
http://www.osborne.com
omg_international@mcgraw-hill.com